Geoscience Instrumentation

Geoscience Instrumentation

Edited by
Edward A. Wolff
Enrico P. Mercanti

A Wiley-Interscience Publication
JOHN WILEY & SONS, INC.

New York
London
Sydney
Toronto

Library of Congress Cataloging in Publication Data:

Wolff, Edward A.
   Geoscience instrumentation.
   "A Wiley-Interscience publication.
   Includes bibliographies.
   1. Earth science instruments. I. Mercanti, Enrico P., joint author. II. Title.

QE49.5.W64   550'.28   73–18195
ISBN 0–471–95952–9

Printed in the United States of America

10 9 8 7 6 5 4 3 2 1

For the benefit of all mankind, now and for generations to come

Contributors

Elias Amdur
Honeywell Corporation
Minneapolis, Minnesota

A. R. Barringer
Barringer Research, Limited
Rexdale, Ontario, Canada

Michael Belton
Kitt Peak National Observatory
Tucson, Arizona

S. Ben-Yaakov
University of California at Los
   Angeles
Los Angeles, California

Ralph Bernstein
IBM Corporation
Gaithersburg, Maryland

Knute A. Berstis
National Oceanographic Instru-
   mentation Center
Washington, D.C.

Roscoe Braham, Jr.
University of Chicago
Chicago, Illinois

P. G. Brewer
Woods Hole Oceanographic
   Institution
Woods Hole, Massachusetts

Quentin Bristow
Geological Survey of Canada
Ottawa, Ontario, Canada

Neil Brown
Woods Hole Oceanographic
   Institution
Woods Hole, Massachusetts

George R. Carignan
University of Michigan
Ann Arbor, Michigan

Frank S. Castellana
Columbia University
New York, New York

Lawrence Chase
National Oceanographic Instru-
   mentation Center
Washington, D.C.

John C. Cook
Teledyne Geotech Company
Dallas, Texas

Vincent J. Cushing
Cushing Engineering Inc.
Silver Spring, Maryland

Robert H. Daines
Rutgers University
New Brunswick, New Jersey

Harold E. Dinger
Washington, D.C.

H. I. Ewen
Ewen-Knight Corporation
Weston, Massachusetts

Delvin S. Fanning
University of Maryland
College Park, Maryland

George R. Grainger
Planning Research Corporation
McLean, Virginia

Franklin S. Harris, Jr.
Old Dominion University
Norfolk, Virginia

J. C. Harrison
University of Colorado
Boulder, Colorado

Louis C. Haughney
NASA Ames Research Center
Moffet Field, California

A. A. J. Hoffman
Texas Christian University
Fort Worth, Texas

J. L. Hieatt
TRW Systems Group
Redondo Beach, California

R. A. Horne
A. D. Little, Inc.
Cambridge, Massachusetts

Michael C. Husich
NASA Goddard Space Flight
  Center
Greenbelt, Maryland

W. P. Johnson
Mark Products, Incorporated
Houston, Texas

I. R. Kaplan
University of California at Los
  Angeles
Los Angeles, California

W. N. Keller
TRW Systems Group
Redondo Beach, California

Wilfred K. Klemperer
NOAA Research Laboratories
Boulder, Colorado

Robert Lee
National Center for Atmospheric
  Research
Boulder, Colorado

C. Gordon Little
NOAA Research Laboratories
Boulder, Colorado

William Markowitz
Nova University
Dania, Florida

Mace Miyasaki
JHU Applied Physics Laboratory
Silver Spring, Maryland

Richard K. Moore
University of Kansas
Lawrence, Kansas

Forrest Mozer
University of California,
  Berkeley
Berkeley, California

Eugene A. Mueller
University of Illinios
Urbana, Illinois

William H. Myers
NOAA National Oceanographic
  Data Center
Rockville, Maryland

Joseph W. Noah
J. Watson Noah Associates
Alexandria, Virginia

Harry W. Otto
Deleware Division of Environmental
  Control
Dover, Deleware

H. Dean Parry
NOAA National Weather
  Service
Silver Spring, Maryland

Barbara Pijanowski
National Oceanographic Instru-
  mentation Center
Washington, D.C.

Kenneth A. Rayburn
University of Maryland
College Park, Maryland

Joseph B. Reagan
Lockheed Palo Alto Research
  Laboratory
Palo Alto, California

John C. Redmond
General Dynamics/Fort Worth
Fort Worth, Texas

Robert W. Rochelle
NASA Goddard Space Flight
  Center
Greenbelt, Maryland

R. K. Schisler
TRW Systems Group
Redondo Beach, California

Thomas Skillman
NASA Goddard Space Flight
  Center
Greenbelt, Maryland

Harold W. Smith
The University of Texas
Austin, Texas

R. Lawrence Swanson
NOAA Marine Sciences Research
  Center
Stonybrook, New York

W. James Trott
Naval Research Laboratory
Washington, D.C.

Allyn Vine
Woods Hole Oceanographic
  Institution
Woods Hole, Massachusetts

Donald R. J. White
Don White Consultants
Germantown, Maryland

R. S. Wilson
Woods Hole Oceanographic
  Institution
Woods Hole, Massachusetts

Joseph G. Wohl
The Mitre Corporation
Bedford, Massachusetts

Preface

In the beginning there was only Chaos until Geos (Mother Earth), the first goddess, appeared.*

This book is intended to be a reference work for engineers and scientists designing and using geoscience instrumentation. We have tried to compile a book useful to people working in the geoscience instrument field, people moving from one subdivision of the field to another, and people entering the field for the first time. The book provides a nomenclature and symbology which contribute to the unification of the field.

Geoscience instrumentation is used for the measurement of parameters associated with land, sea, air, and space including the increasingly more popular areas of space, oceanography, moon and planet, and pollution instrumentation.

Descriptive information and derivations are provided for the newcomer to the field. Design information is provided for the experienced designer. Because of space limitations in a single volume of this type, relatively long lists of references are provided at the ends of the sections for those readers who want to delve deeper.

An attempt has been made to include in each section a description of the underlying scientific and engineering theory to give the reader an understanding of the problems involved. Items included in such a description are the important factors and their meaning, the relationships between factors, and the requirements these factors create.

In the chapter on geoscience environment we have tried to review the parameters that are important in each domain and the interfaces with the other domains. This includes detailed information on each parameter including the underlying scientific theory relating to the parameter, the relationships between the parameter and other parameters, the variations of the parameter in space and time, and the information required from measurements to understand the physical processes.

---

* Greek mythology.

In the chapter on instrument platforms we have attempted to review the various types of platforms used and the interfaces with other parts of the system and the environment. This includes detailed information on each platform including external and internal environmental parameters which influence the design and performance of the platform. Attention is given to the problems imposed by the platform parameters in the design of instrument systems for use on the platform.

In the chapters on sensors the authors have tried to review the scientific phenomena being measured, review the various sensor types available for the parameter being measured and make general comparisons, and provide information on each sensor type. This includes the underlying scientific theory describing the operation of the sensor, the engineering problems involved in the design of the sensor, and the performance available from each sensor in terms of ranges of amplitude, frequency, resolution, and accuracy.

In the final chapter, a discussion of design problems, we have sought to review the relationship of each problem to the overall system and the required performance and to review the scientific and engineering fundamentals basic to each topic.

We both became aware of the need for a book of this type a dozen years ago when we were separately involved in the development of spacecraft geoscience instrumentation. This need was again apparent when we participated later in the work of the Institute of Electrical and Electronics Engineers Geoscience Electronics Group. The decision to proceed with a book was made at the urging of Mr. Richard F. Shea whose encouragement is gratefully acknowledged. Because of the wide diversity of knowledge required for the book we thought was needed and because of our own limited experience, we decided to design the book, seek contributors, and serve as editors rather than do the writing ourselves. We are pleased that so many competent leaders in their fields agreed the book was needed and accepted our invitation to contribute.

We have learned much in the process of assembling this book, and we hope the reader will also benefit.

Edward A. Wolff
Enrico P. Mercanti
January, 1973

Contents

The International System (SI) of Units

Parameter	Unit	Abbreviation	Equivalence
Current, electric	ampere	A	basic unit
Length	meter	m	basic unit
Luminous intensity	candela	cd	basic unit
Mass	kilogram	kg	basic unit
Matter, quantity	mole	mol	basic unit
Temperature	°Kelvin	°K	basic unit
Time	second	s	basic unit
Angle, plane	radian	rad	arc/radius (r)
Angle, solid	steradian	sr	area/r^2
Capacitance	farad	F	$A \cdot s/V$
Charge, electric	coulomb	C	$A \cdot s$
Conductance	mho	mho	$1/\Omega$
Energy	joule	J	$N \cdot m$
Force	newton	N	$kg \cdot m/s^2$
Frequency	hertz	Hz	$1/s$
Illumination	lux	lx	lm/m^2
Inductance	henry	H	$V \cdot s/A$
Luminous flux	lumen	lm	$cd \cdot sr$
Magnetic flux	weber	Wb	$V \cdot s$
Magnetic flux density	tesla	T	Wb/m^2
Period	second	s	—
Potential, electric (emf)	volt	V	W/A
Power	watt	W	J/s
Radioactivity	curie	Ci	$3.7(10)^{10}/s$
Resistance	ohm	Ω	V/A
Work	joule	J	$N \cdot m$

Parameter	Unit	Abbreviation	Equivalence
Acceleration, angular	—	—	rad/s^2
Acceleration, linear	—	—	m/s^2
Amount of substance	—	—	mol
Area	—	—	m^2
Conductivity	—	—	mho/m
Current density	—	—	A/m^2
Density	—	—	kg/m^3
Field strength, electric	—	—	V/m
Field strength, magnetic	—	—	A/m
Field intensity, gravitational	—	—	m/s^2
Field intensity, magnetic	—	—	A/m
Luminance	—	—	cd/m^2
Moment of inertia	—	—	$kg \cdot m^2$
Momentum, angular	—	—	$kg \cdot m^2/s$
Momentum, linear	—	—	$kg \cdot m/s$
Permeability	—	—	H/m
Permittivity	—	—	F/m
Potential, gravitational	—	—	J/kg
Pressure	—	—	N/m^2
Thermal conductivity	—	—	$W/m \cdot {}^\circ K$
Velocity, angular	—	—	rad/s
Velocity, linear	—	—	m/s
Volume	—	—	m^3

Prefix	Value	Abbreviation	Prefix	Value	Abbreviation
tera	10^{12}	T	centi	10^{-2}	c
giga	10^9	G	milli	10^{-3}	m
mega	10^6	M	micro	10^{-6}	μ
kilo	10^3	k	nano	10^{-9}	n
hecto	10^2	h	pico	10^{-12}	p
deka	10	da	femto	10^{-15}	f
deci	10^{-1}	d	atto	10^{-18}	a

Nonstandard Units Found in the Literature

Parameter	Unit	Abbreviation	Equivalence
Acceleration, gravity	galileo	Gal	cm/s^2
Angle, plane	degree	deg	0.017453292519943 rad
Energy	British Thermal Unit	Btu	$1.05587(10)^3 \cdot J$
Energy	calorie	cal	$4.184 \cdot J$
Energy	electron volt	eV	$1.6021(10)^{-19} \cdot J$
Energy	erg	erg	$10^{-7} \cdot J$
Energy per unit area	langley	Lg	$4.1855 \ J/cm^2$
Force	dyne	dyn	$10^{-5} \cdot N$
Frequency	revolutions per minute	rpm	$60 \cdot Hz$
Information rate	bits per second	b/s	—
Length	angstrom	Å	$10^{-10} m$
Length	foot	ft	0.3048 m
Length	micron	μ	$10^{-6} \cdot m$
Magnetic flux	maxwell	M	$10^{-8} \cdot Wb$
Magnetic flux density	gamma	γ	$10^{-9} \cdot T$
Magnetic flux density	gauss	G	$10^{-4} \cdot T$
Mass	pound	lb	0.45359237 kg
Mass	atomic mass unit	amu	$1.67(10)^{-24} g$
Power	horsepower	hp	$746 \cdot W$
Pressure	atmosphere	atm	$1.01325(10)^5 \cdot N/m$
Pressure	bar	bar	$10^5 \cdot N/m$
Pressure	millimeter of mercury	mmHg	$133.3224 \cdot N/m$
Pressure	torr	tr	$133.322 \cdot N/m$
Pulse rate	pulses per second	pps	—
Ratio (number)	percent	%	$100 \cdot n$
Ratio (power)	bel	B	$10 \log (P_1/P_2)$

Parameter	Unit	Abbreviation	Equivalence
Ratio (volume)	parts per billion	ppb	$10^{-9} v_1/v_2$
Ratio (volume)	parts per million	ppm	$10^{-6} v_1/v_2$
Ratio (volume)	parts per thousand	ppt, %	$10^{-3} v_1/v_2$
Temperature	degree celcius	°C	°K − 273.15
Temperature	degree Fahrenheit	°F	(9/5)(°K − 273.15) + 32
Temperature	degree Rankine	°R	(9/5)°K
Time	day (mean solar)	day	$86400 \cdot s$
Velocity	knot	kt	$0.51444 \cdot m/s$
Volume	gallon (liquid)	gal	$0.003785411784 \cdot m^3$
Volume	liter	1	$10^{-3} \cdot m^3$

List of Major Symbols

A	Absorption loss, amplitude, area, mass number, moment of inertia, availability
A_e	Effective area
A_i	Amplitude of ith component
A_R	Receiving antenna area
a	Absorption coefficient, activity, constant, diameter, length, axis
a_v	Shape factor
$\mathbf{B}$	Magnetic flux density vector
B	Bandwidth, base, constant, internal reflection losses, magnetic field intensity, number of bits
$\bar{B}$	Average flux density
B_s	Saturation flux
b	Constant, length, total scattering coefficient
b_t	Total relative damping force
C	Concentration, constant, damping coefficient, moment of inertia
C_D	Drag coefficient
C_k^n	Binomial coefficient
C_p	Specific heat at constant pressure
C_ρ	Heat capacity
C_T	Temperature fluctuation parameter
C_v	Velocity fluctuation parameter
c	Constant, propagation velocity
D	Declination, depth, direction, dispersion, distance, length, damping constant daughter material
$\bar{D}$	Average deviation
d_i	Deviation of ith measurement
d	Constant, declination, depth, diameter, length
$\mathbf{E}$	Electric field intensity vector
E	Elevation angle, energy, error, irradiance, potential, voltage, Young's modulus

$E(K)$	Spectral intensity
E_{as}	Asymmetry potential
E_b	Energy per bit
E_d	Dynamic error
E_h	Redox potential
E_j	Liquid junction potential
E_m	Membrane potential
E_n	Noise voltage
E_r	Reference voltage
E_s	Field strength, sensor voltage, static error
E_λ	Spectral emittance
e	Electron charge, naperian logarithm base, error
e_a	Ambient vapor pressure
e_n	Noise voltage
e_w	Water vapor pressure
F	Frictional force vector, magnetic field vector
F_D	Drag force
F	Faraday constant, noise factor, noise figure, total field intensity
F	Average noise factor
F_o	Operating noise factor
f	Constant, flattening, frequency, function
f_d	Doppler frequency, dynamical flattening
f_h	Hydrostatic flattening
f_u	Earth flattening
G	Gravity vector
G	Gain, gravitation constant, relative conductivity
G_R	Receiving antenna gain
G_T	Transmitting antenna gain
g	Coefficient, function, gravity acceleration
g_b	Bouguer gravity correction
g_e	Gravity at equator
g_f	Free air gravity correction
g_n^m	Gauss coefficient
g_r	Gravity at distance r
g_{STD}	Standard gravity
H	Magnetic field intensity vector
H	Ellipticity, gyro angular momentum, heat flow, height, horizontal field, hour angle
$^aH^+$	Hydrogen ion activity

cH^+	Hydrogen ion concentration
H_c	Saturating field
H_o	Height above reference
h	Elevation, height, Planck's constant, thickness
h_e	Effective height
h_n^m	Gauss coefficient
I	Current, inclination, radiation intensity
I_p	Particle current
i	Current
J	Current density, moment of inertia
j	Constant, $\sqrt{-1}$
K	Constant, deviation, moment of inertia, thermal conductivity
K_t	Temperature constant
k	Boltzmann's constant, bulk modulus, constant, spring constant, wave number
k_i	Ion selectivity ratio
L	Length, loss, radiance
L_c	Concentration length
l	Constant, length
M	Magnetic moment, magnetic polarization, magnification, magnitude, mass, mutual inductance, sensitivity
M_a	Modulation index
M_r	Mass within radius r
m	Average, constant, mass
m_e	Electron mass
m_i	Ion mass
m_n	Neutral atom mass
N	Noise power, number, spectral radiance
$N_{b\lambda}$	Blackbody spectral radiance
N_e	Electron density
N_o	Noise power density
$\mathbf{n}$	Unit vector
n	Constant-frequency plasma density
$\mathbf{P}$	Poynting vector
P	Compressional wave, pendulosity, photon number, parameter, power density, pressure, parent material
P_A	Altimeter pressure
P_b	Binomial distribution
P_c	Continuous micropulsations
P_I	Probability of interference
P_i	Irregular micropulsations

P_k	Poisson frequency function
P_{lm}	Legendre function
P_n	Normal distribution
P_R	Received power
P_T	Transmitted power
P_x	Chi-square distribution
P	Pressure, probability
$p(x)$	Probability density function
pE	Electron activity
Q	Dimensionless parameter, heat, sensor output
$\overline{Q}$	Average sensor output
Q_e	Sensor output with error
Q_i	Sensor output for ith measurement
Q_o	Initial sensor output, number of electrons
q	Heat flow
R	Distance, gas constant, radius, ratio, reflection loss, residual pressure, resistance, resolution, spectral response
R_c	Cell resistance
R_d	Reynolds number
R_E	Earth radius
R_e	Equatorial radius
R_m	Earth–moon distance, meter resistance
R_o	Resistance at zero temperature
R_p	Polar radius
R_s	System reliability
R_T	Ratio at temperature T
R_t	Target distance, thermistor resistance, total resistance
R_w	Wire resistance
r	Length, radius, range, reliability
r_a	Azimuth resolution
r_R	Range resolution
r_s	Resolution length
S	Area, salinity, Seebeck coefficient, sensitivity, shear wave, shielding efficiency, signal power, thickness
s	sun
T	Period, temperature, tension, transmission, volume transport, half-life
T_A	Antenna temperature, transit time
T_a	Dry bulb temperature
T_b	Body temperature
T_c	Cold temperature, constant load temperature

T_d	Time delay
T_e	Electron temperature, element temperature
T_h	Hot temperature
T_i	Temperature at point i
T_w	Wet bulb temperature
t	Thickness, time
U	Gravitational field, speed
$\mathbf{V}$	Velocity vector
V	Geomagnetic potential, speed, visual range, voltage, volume
V_F	Fall velocity
V_H	Horizontal velocity
V_r	Received voltage
$\mathbf{v}$	Velocity vector
v	Velocity
W	Flux, power
W_R	Received power
W_T	Transmitter power
X	Displacement
X_{FS}	Full-scale displacement
$\mathbf{x}$	Vector product
x	Abscissa
Y	Admittance
y	Ordinate
Z	Atomic number, depth, figure of merit, impedance, summation, vertical field
Z_c	Characteristic impedance
z	Vertical distance
α	Absorption, angle, attenuation, compressional wave velocity, constant, efficiency, error, temperature coefficient
β	Angle, beamwidth, constant, scatterance, shear wave velocity
Γ_{12}	Mutual-coherence function
γ	Ionic activity, magnetic field strength unit
γ_p	Proton gyromagnetic ratio
Δ	Angle
δ	Angle, constant
ε	Earth flattening, opacity, quantum efficiency
ε_0	Free space permittivity
ζ	Autocorrelation function, damping rate
η	Efficiency, refractive index, scattering cross section
θ	Angle, beamwidth
θ_g	Heat generated

λ	Decay constant, failure rate, Lamé constant, longitude, radius, wavelength
λ_0	Cutoff wavelength
μ	Absorption coefficient, angle, attenuance, gain, shear modulus, viscosity
υ	Collision frequency, constant, frequency, kinematic viscosity
υ_0	Threshold frequency
ξ	Tide height
π	Circumference/diameter ratio
ρ	Density, reflectance, resistivity
σ	Attenuation, electric conductivity, Poisson's ratio, radar cross-section, standard deviation
σ^2	Variance
σ_V	Volume scattering coefficient
τ	Period, time constant, transmittance
τ_b	Beam transmittance
Φ	Spectral intensity, radiant flux
ϕ	Beamwidth, latitude, phase, power per unit solid angle, power ratio, work function
χ	Nuclear susceptibility
ψ	Angle, geopotential
$\boldsymbol{\Omega}$	Earth rotation vector
Ω	Solid angle
ω	Angular frequency, angular velocity
ω_e	Earth rotation rate
ω_p	Natural frequency, plasma frequency
ω_r	Resonance frequency
$\boldsymbol{\nabla}$	Nabla or del operator
∂	Partial derivative
$*$	Complex conjugate
$\cdot$	Scalar product
$\ln$	Natural logarithm

Geoscience Instrumentation

Chapter I

Geoscience Instrumentation Systems

Edward A. Wolff

This chapter begins with a short history of the field followed by a definition of the geoscience instrument problem. The chapter includes a discussion of the external factors affecting the instrument and concludes with a discussion of the interrelationships between system components.

1.1 GEOSCIENCE INSTRUMENTATION HISTORY

"And it came to pass at the end of forty days, that Noah opened the window of the ark which he had made. And he sent forth a raven, and it went forth to and fro, until the waters were dried up from off the earth. And he sent forth a dove from him, to see if the waters were abated from off the face of the ground. But the dove found no rest for the sole of her foot, and she returned unto him to the ark, for the waters were on the face of the whole earth; and he put forth his hand, and took her, and brought her in unto him into the ark. And he stayed yet other seven days; and again he sent forth the dove out of the ark. And the dove came in to him at eventide; and lo in her mouth an olive-leaf freshly plucked; so Noah knew that the waters were abated from off the earth."

(Genesis VIII, 6-11)

Geoscience instrumentation may have begun in the stone age when early man cast a stick upon the water to measure the river current (or held up a wet finger to measure the air current).

Man's early efforts in instrumentation centered on the measurement of the land he occupied and the observation of the heavens he feared. The orientations of prehistorical monuments show that astronomical measurements were made very early.

Early evidence of the development of measurement systems which are prerequisites for modern instruments date to the third millennium before Christ. Babylon had adopted standards of length, weight, and volume by

1

2500 B.C. They had a 360-day year and used a gnomon sundial. They also studied the heavens and the movement of the sun, the moon, and the five known planets.

Equinoxes and solstices were determined by culminating stars by the Chinese in about 2300 B.C. The Chinese also used armillary spheres and quadrants.

The origin of the balance scale is thought to be 5000 to 2000 B.C. in Egypt. Egypt had an "L" shaped sundial in about 1500 B.C. The sundial is also mentioned in the Old Testament (*Isaiah* 38.8).

The Chaldean astronomer Berosus invented a hemispherical sundial in 300 B.C.

Much of the work which laid the foundation for geoscience instrumentation was done in ancient Greece. Geodesy and the measurement of the shape of the earth's surface had its beginning with the assumption by Pythagoras (sixth century B.C.) of a spherical earth and the measurement of its circumference by Eratosthenes. Eratosthenes (276–196 B.C.) invented or improved armillary spheres. Archimedes measured the density of matter in his bathtub about 250 B.C. Hipparchus, in about 130 B.C., made the first geoscience instrument. This was an astronomical instrument, the planispheric astrolabe, accurate enough to measure the precession of the equinoxes and the ratios of the distance to the moon and its diameter to the diameter of the earth with accuracy to 10 to 25 percent. By 200 B.C. Eratosthenes measured the circumference of the earth within three percent and the earth-sun distance within one percent. The astrolabe was used to measure the altitude of heavenly bodies. With this information and the knowledge of the orbits it was possible to measure the latitude and time.

During the middle ages the development of instrumentation was concentrated in Europe. Roger Bacon (about A.D. 1250) was an early advocate of experimentation and measurement. Optical devices with lenses appeared at this time. The origin of optical instruments is somewhat obscure. Water-filled glass spheres were used as burning glasses in ancient times. Convex lenses apparently were first used as spectacles in about 1280 and concave lenses appeared in about 1500. The camera was invented by Baptista Porta in 1553. The telescope was invented by Hans Lippershey in Holland in 1609, and a practical instrument for astronomy was developed by Galileo in 1610. The first reflecting telescope was built by Isaac Newton in 1668.

Magnetic instruments began early with the discovery of the lodestone. The compass came into existence as a floating magnet in about 1200. William Gilbert (1540–1603) studied magnetism and discovered magnetic inclination and its latitude dependency, and he deduced the magnetic field of the earth. John Mitchell (1724–1793) built the first magnetometer in England in about 1750, a torsional balance used to measure magnetic forces.

Perhaps the founder (if there is one) of geoscience instrumentation is Galileo Galilei (1564–1642). In addition to inventing and using the astronomical telescope, he invented the thermometer, a glass bulb containing air with the open end dipped in water. In 1643 Torricelli, a student of Galileo, invented the barometer, a sealed glass tube filled with mercury with the open end inverted in a jar of mercury. In 1781 Cavendish first analyzed the atmosphere and found it had 20.83% oxygen and 79.17% nitrogen.

In 1735 John Harrison made a temperature-compensated chronometer which permitted the measurement of longitude from astronomical and time measurements. This instrument was a great contribution to ocean navigation.

Most modern geoscience instruments use electricity for amplification and signal processing. The measurement of electricity began with the discovery by Oersted in 1820 that a current in a wire deflects a compass needle. This resulted in the construction of the first galvanometer by Schweigger in the same year. Electricity was finally used in geoscience instrumentation when L. Palmeri built an electromagnet-type seismograph in Italy in 1855 and made measurements on Vesuvius.

The progress in instrumentation in the twentieth century has been tremendous when compared with the progress which was painfully eked out prior to this century. The present state of development of the field should be apparent to the reader as he examines the pages that follow.

1.2 INSTRUMENTATION PROBLEM DEFINITION

A geoscience instrument is a device for measuring the value of a geoscience parameter. The geosciences whose parameters we are interested in measuring are those sciences pertaining to the domains of air, land, sea, and space. They include aeronomy, astronomy, astrophysics, geodesy, geology, geophysics, limnology, meteorology, and oceanology. The parameters themselves are numerous and vary from one domain to another. The important parameters for each domain are discussed in Chapter 2.

Often high measurement accuracy is required in spite of the complications posed by the operating invironment. The effects of the environmental parameters must therefore be considered before fabricating the hardware. These environmental parameters are discussed in Chapter 3.

This book considers both in situ sensors for measuring parameters at the sensor and remote sensors for measuring parameters at some distance from the sensor. The parameters whose measurement by in situ sensors is discussed in Chapter 4 are acoustic wave, atmosphere constituent, biology cloud, electric field, electromagnetic field, gravity field, heat, horizon, humidity, ion, magnetic field, motion, particle radiation, pH, position, precipitation, pressure, salinity, seismic wave, sun, telluric current, tide, time, turbidity, water constituent, water current, wave, and wind. The remote sensing techniques discussed in Chapter 5 are acoustic, radio, magnetic, gamma radiation, optical, radar, and radiometric.

In the early days of geoscience instrumentation it was difficult enough to consider only the measured parameter, and attention was focused on the development of sensors. Today the situation is often much different. An instrument mounted on an orbiting spacecraft can generate an enormous amount of data in a year. In order for a user to be able to interpret this data within his lifetime, it is necessary for the instrument designer to consider the data problem and the user before he builds the instrument. These data considerations include the utilization of outside information and the available techniques for signal processing and data processing. Signal processing is discussed in Chapter 6 and data processing is discussed in Chapter 7.

As the complexity of instruments increases, it becomes necessary to give more attention to the operation and maintenance of the instrument. Problems of electromagnetic interference, reliability and maintainability, human factors and cost become increasingly important. These auxiliary problems are discussed in Chapter 8.

1.3 EXTERNAL FACTORS

The modern geoscience instrument must be viewed as a complete system that includes many components between the parameter being measured and the operator of the instrument and the user of the measurement. This concept of the geoscience instrument is illustrated simply in Fig. 1-1. This figure shows that an instrument designer should consider many external factors early in the design effort. Some of these factors are shown in Table 1-1.

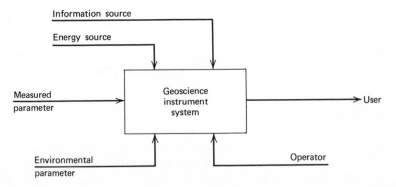

Figure 1-1 Simplified diagram of geoscience instrument system.

Table 1-1. External Geoscience Instrument Design Factors

Measured Parameter	*Environmental Parameters*
Range of values	Indentification of existing parameters
Spatial components	Range of values of each
Range of fluctuation	Spatial components
Rate of fluctuation	Range of fluctuation
Accuracy	Instrument response
Resolution	
Energy Source	*Information Source*
Types of energy available	Types of information required
Quantity of each	Form of each
Range of fluctuations	Quantity
Rate of fluctuations	Rate
Instrument requirement	Ability of system to handle the information
Operator	*User*
Functions required	Types of information required
Quantity of each	Form of each
Frequency	Quantity
Accuracy	Rate
Instrument capability	Ability of Instrument System to deliver
Other	
Life	
Cost	
Time	
Political	

The external factor which usually receives first consideration is the measured parameter. The first consideration for this parameter given in Table 1-1 is the range of values. This may seem like a trivial consideration. One does not normally think about using a household freezer thermometer to check the oven performance. However, consider the problem of the instrument to be dispatched to some inaccessible region of space to measure a heretofore unmeasured parameter whose amplitude is the subject of widely varying theoretical predictions. If increased complexity and cost are allowed, the instrument can be constructed to allow a change in scale by operator remote command, but this solution to the problem (like the solution to the other problems found throughout this book) involves the selection of engineering compromises.

The spatial component of the measured parameter must also be considered. Some parameters, like a magnetic field, are vector quantities, and it may be desirable to measure both the magnitude and direction of the vector. It is necessary to consider not only the value of the measured parameter at different points in space but also its direction which may place limitations on the sensor orientation.

It is also necessary to consider the fluctuation ranges of the measured parameter. If the amplitude is fluctuating greatly, an instrument with a large dynamic range may be required. Generally, a larger dynamic range also implies a lower accuracy, so that compromises may have to be made. The rate of the fluctuation of the measured parameter must also be considered. This fluctuation may be due to time variations in the parameter itself as well as variations induced by the motion of the instrument through a spatially varying field. The rate of fluctuation will determine the frequency bandwidth required in the instrument and hence the rate at which data is collected. This has many implications for the signal processing and data processing portions of the system.

The designer must also consider the accuracy with which the measurement must be made and the resolution required. Increased accuracy and resolution normally require a higher data content of the measurement and hence a more elaborate signal processing and data processing capability.

The nature of the environment in which the instrument must function must also be considered. There are a number of environmental parameters to be considered including temperature, pressure, humidity, shock, vibration, wind, ice, dust, salt water, and corrosive gases. In addition, consideration must be given to the range of values of each of the environmental parameters, their spatial components, their fluctuation amplitudes and rates, and the probable effects they will have on the instrument response. Some of the parameters such as shock and vibration primarily affect the mechanical design while other parameters such as temperature may affect both the mechanical and the electrical design because both the electrical and mechanical parts have thermal coefficients. Designing an instrument to operate over wide temperature ranges from $-40°C$ to $+90°C$, for instance, is very different from designing an instrument for operation in an air-conditioned laboratory.

Energy is required for the operation of most instruments. The designer should consider the type of energy that is available for his instrument, the quantity of each energy source available, the fluctuations in the energy source (if they exist), and the energy required by the instrument. When a continuous source of energy is unavailable, it may be necessary to consider energy-storage devices. The energy problem usually results in compromises between size, weight, cost, instrument capability and complexity, and operating lifetime.

The information available for use in the system is unfortunately one of the factors most frequently ignored until after the hardware is fabricated.

The designer should give early consideration to the type of information available for the system, the form of each information source, the quantity and rate of the information available, and the ability of the proposed instrument to handle this information. Examples of information available might include measurements of other environmental parameters made simultaneously in conjunction with the measurement of the desired parameter, measurements of sensor position, monitoring of various sensor characteristics, external calibration data, programs for data analysis and presentation, and annotations generated during the measurements. Generally, the greater the quantity of information generated by or required by the system the greater the need to give early consideration to this problem.

The operator and his problems should receive early attention. Consideration should be given to the functions the operator will be required to perform, the frequency and accuracy with which they must be performed, and the implications this has for the instrument and the necessary built-in functions. Operator requirements can vary widely from one operational situation to another. The requirements on the operator for using a hand-held mercury thermometer to measure the air temperature may be relatively minimal compared with the requirements on the operator required to analyze data from a spacecraft on a quick-look basis and transmit commands to the spacecraft while it is within the view of the ground station, normally a period of only a few minutes. In addition to simple instrument operation, the problems of interference, reliability, and maintainability must also be considered because they have a significant impact on the cost of the total system and its usefulness for the purposes intended.

Finally, the user must not be neglected. Careful thought must be given in the instrument design phase to the information the user requires, the form in which this information should be presented, the quantity of information, the rate of presentation, and the ability of the instrument to deliver what is required of it. This problem, like the others described above, involves an engineering compromise between what would be most desirable and the resulting cost, complexity, and time involved.

1.4 INSTRUMENT SYSTEM COMPONENT INTERRELATIONSHIPS

The geoscience instrument system illustrated in Figure 1-1 contains many components. An expanded diagram showing many of the identifiable functions and their interrelationships is shown in Figure 1-2.

The sensor is a transducer which responds to the sensed parameter and usually creates an electrical signal to be processed in the remainder of the instrument. Several important electrical characteristics should be considered. These are shown in Table 1-2. In some cases a choice is available between a sensor with an analog output signal or one with a digital output

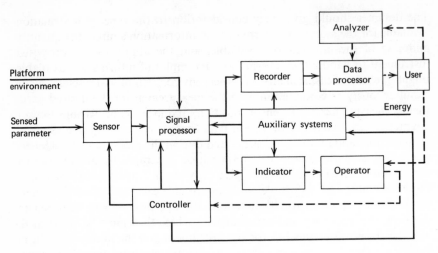

Figure 1-2 Geoscience instrument system.

signal. When this choice is available it is necessary to consider the signal processing and data processing portions of the system to determine which sensor output signal is preferred.

The electrical signal generated by the sensor can be either a voltage, a current, an impedance, or a frequency. Examples of these are a thermocouple used to measure temperature, a photomultiplier used to measure light, a thermistor used to measure temperature, and a rubidium vapor magnetometer used to measure magnetic field strength, respectively. For many parameters more than one type of output could be used, as in the thermocouple and the thermistor for temperature measurement illustrated above. The choice of the form and type of sensor output signal may depend upon the values of the various output signals available.

A knowledge of the characteristics of the fluctuations of the sensed parameter and the desired accuracy and resolution with which this parameter is to be measured can influence the type of transducer used for a sensor.

Several problems are associated with the design of sensors. One is the

Table 1-2. Electrical Signal Characteristics

Form:	Analog or digital
Type:	Voltage, current, impedance, frequency
Values:	Amplitude range
	Frequency range
	Accuracy
	Resolution

problem of accurately calibrating the sensor output in terms of the value of the sensed parameter. Inherent in this problem is the confidence that the sensor is truly responding to the sensed parameter and not to some other environmental factor. In other words, a pressure sensor designed to be operated in a rapidly changing thermal environment must be designed so that it does not become a better thermometer than pressure sensor.

A second problem involves the placement of the sensor so that it truly responds to the sensed parameter. For instance, a thermometer designed to measure the temperature of the atmosphere from an aircraft must be located so that it truly measures the external atmosphere rather than a boundary layer of air near the aircraft. It must also be designed so that it is not unduly heated by friction with the air through which it passes. These and similar problems encountered in the design of sensors are discussed in more detail for the particular sensors in Chapter 4.

The geoscience instrument system block diagram of Figure 1-2 shows blocks representing a number of system components interconnected with arrows. Those arrows represented by solid lines between blocks indicate electrical interconnections. For each of these interconnections the electrical signal characteristic can be described as shown in Table 1-2. These characteristics are considered in selecting the proper transducer for the sensor. They must also be considered when selecting the components for the other blocks of the instrument system of Figure 1-2. In other words, the characteristics of the signals and the interfaces between components must be thoroughly considered for all parts of the system.

The signal processor may contain devices for accomplishing several functions. Among these are amplification, detection, digitizing, filtering, modulating, synchronizing, and controlling. If the sensor is physically removed from the recorder, the signal processor may also include a telemetry link. Again, the instrument designer concerned with the signal processing portion of the instrument must consider the signal characteristics shown in Table 1-2. In addition, it is necessary to consider the automatic control functions of the controller which are required to be operated by the signal processor.

A controller within an instrument system can be used to control the signal processing, to control the sensor itself, or to control various auxiliary systems. Consideration must be given to the control functions required. Typical functions are changes in range, gain, information rate, or operating mode. The controller can also be used to operate the telemetry system; auxiliary systems such as power supplies, clocks, and platform controls; the recorder and the indicator.

Selection of a suitable indicator depends upon the parameters to be displayed, the operational problems involved, and the human factors associated with the display. Consideration must be given to the use to which the display information is put by the operator. For instance, if it is desired to control various parameters of the instrument system during the measurement in response to initial measurements, it will probably be necessary to give the operator a real time "quick-look" review of the instrument performance and to provide a control system with relatively rapid response.

The recorder shown in a separate block in Figure 1-2 is often associated with either the signal processor or the data processor. The considerations in the selection of the recorder are similar to those used in the selection of the indicator. Attention must be given to the characteristics of the electrical signals which must be recorded as well as the characteristics of the electrical signals which the recorder presents to the data processor.

The data processor shown in Figure 1-2 combines many functions in a single block. These functions include analysis, display, processing, and storage. Data analysis includes consideration of the mathematical techniques to be applied to the analysis of the sensed parameter and other operational and housekeeping data available from the auxiliary systems. Attention should be given to the method for including system calibration in the data processing operation and the manner in which the data is to be displayed to the user. This involves determination of the important parameters to be displayed and the form of this displayed data (analog or digital).

Data processing itself can be accomplished by a number of techniques including manual processing, desk calculators, general purpose digital computers, special purpose digital computers, or analog computers. Each of these has its own advantages and disadvantages in terms of speed, accuracy, and cost of computation. The use of high-speed computers sometimes requires the use of storage devices at the computer input and computer output. Storage devices are also used when there is an appreciable time delay between conduct of the experiment measurement and an analysis of the experiment data. Details of the data processing problem are given in Chapter 7.

The block diagram of Figure 1-2 contains a block labeled "Auxiliary Systems." These systems are normally associated with the instrument platform and provide information, signals, and controls essential to the conduct of the experiment. Such systems might include communications, navigation, power supply, attitude control, and timing systems. Some of these systems are discussed in Chapter 3, and the others in Chapter 4.

1.5 DESIGN PROBLEMS

Discussion of the instrument system illustrated in Figures 1-1 and 1-2 included a discussion of many of the problems involved in the design of the various components of an instrument system and the design of the interfaces between these components. There are other over-all design problems that require consideration and that are discussed in this book in Chapter 8. These include the problems of electromagnetic compatability, reliability and mantainability, human factors, and economic factors.

Electromagnetic compatability is concerned with the problem of creating an instrument which can function effectively in an environment which includes interference sources without itself creating interference for other equipment. Common sources of interference from which an instrument may have to be protected include radiated signals from nearby high power transmitters, electric power lines, or other equipment and interfering signals conducted into the instrument on the signal or power wires. Similarly, the instrument must be constructed so that there are no electromagnetic signals emanating from the instrument that can impair the performance of other equipment. For instance, instruments designed to operate in close proximity to sensitive magnetometers may have to be designed so that they do not disturb their surrounding magnetic field. Another example is an instrument designed to operate in space in proximity to other instruments designed to measure ions and other particles. Such instruments must be designed so that they do not emit contaminants.

The problem of reliability and maintainability becomes increasingly important as the instrumentation becomes more expensive and more inaccessible. As the cost of labor increases it becomes increasingly more desirable to improve the reliability of equipment and reduce the need for maintenance. And while it may be desirable to design most equipment so that any necessary maintenance can be done with ease and efficiency, the usual concept of maintainability ceases to be applicable when the instrument achieves a position in a space orbit.

Human factors concepts have been applied for some time to situations where the interaction between man and machine has an impact on human life. In other words, these concepts have been applied to the design of controls for operating aircraft. They have not been universally applied to the man-machine interactions between the operator and the instrument or the user and the instrument. The application of these concepts becomes increasingly more important as the cost of the instrument, the cost of the measurement, and the cost of the time of the operator and user increase.

This chapter contains many allusions to the various engineering compromises involved in the design of an instrument system. Many of these compromises have been described in terms of a compromise between performance and cost. Because of their importance, the economic factors are discussed in detail in Section 8.5.

Chapter 2

Geoscience Environment

2.1 ENVIRONMENTAL FACTORS

The geoscience environment encompasses that which exists: matter and energy. All we can or cannot see, taste, feel, smell, and touch is made up of mixtures, compounds, one of the known elements, or subatomic particles or energy which stimulate the senses. The sensors mankind has created or will invent in the future, being merely extensions of the human senses, can do nothing more than assess the properties of matter or react to the various forms of energy which impinge on them.

Mankind long ago developed an awareness of the limitations of human senses for making measurements and began the evolutionary, sometimes revolutionary, but always brilliant, development of extra-human senses to entend ranges, sensitivities, accuracies, reliabilities, frequencies of observations, and locations of observational points of measurements of our environment. With the advent of each new sensor it was learned that the environment was of greater scope than had been previously envisioned and defined. Soon after it was established that energy comprised part of our environment, it was learned that energy was of various types such as chemical, electrical, mechanical, and thermal. Similarly, the discovery of the elements was soon followed by the identification of ions, protons, electrons, atoms, and molecules. The rather restrictive constraints of the human eye have been extended almost exponentially in time with the telescope, microscope, and other imaging devices that have provided a view of new worlds that in succession have become defined and measured parts of our environment.

The geoscience environment today is conveniently divided into five components: earth, ocean, atmosphere, space, and planets. Properties of matter and characteristics of various forms of energy are unknown in many parts of the universe. Sensors must be developed and properly located to make measurements of these parameters. It is necessary to cope with properties and energies mankind does not know how to measure as well as with those mankind does not even know exist or need to be measured. Measurement techniques to provide a knowledge of both need to be evolved on a continuing basis. It is equally important to continue to measure properties and energy characteristics as a function of time and location. Geoscience instrumentation to make these measurements will continue to provide data which will unlock the secrets of the mechanism by which the universe operates.

Table 2.1-1 lists the in situ and remote sensors which are discussed in detail in Chapters 4 and 5. It shows their application to each of the five geoscience environment regions (earth, ocean, atmosphere, space, and planets). In addition, the type of platform on which each sensor type is usually accommodated is indicated. Detailed listings of measurements made, instruments used in conjunction with each sensor type, and purpose of each measurement would also be appropriate in the table, but space limitations prohibit their inclusion. Instead, these parameters are listed separately in Chapters 4 and 5.

Geoscience environment studies are characterized by their distinction between the earth as a planet, other planets, and the space that lies between. The earth environment is further subdivided into earth, ocean, and atmosphere environments. Actually, there is no clear line of demarcation between any of these five regions because of interactions.

Table 2.1-1a. Sensor Applications—In Situ

Sensed Parameter	Domain					Measurements	Sensors and Sensing Techniques
	Earth	Air	Water	Space	Planetary		
Acoustic wave	x	x	x		x	Acoustic pressure; particle dynamics; pressure gradient	Hydrophone; microphone
Age	x				x	Radioactivity; radioisotope content	Spectrometer; radiation detector
Atmospheric constituents		x			x	Sulfur dioxide; ozone; nitrogen dioxide; carbon monoxide; aldehydes; particles; sulfates; chlorides; soluble gases; aerosols; nitric oxide; other gases	Colorimetric, chemical, gravimetric, turbidometric, nephelometric, conductometric, polarographic, potentiometric, optical and chemiluminescence spectrometers; flame ionizer; gas chromatography; spectrograph; x-ray diffraction microscope; interferometer
Biology	x	x	x		x	Atmospheric pollutants; humidity; photochemical reactions	Plant color and form
Cloud		x			x	Shape; concentration; size; distribution; water and ice properties; temperature; density	Heated wire; paper tape; spinning bowl; dewpoint hygrometer; slide sampler; replicator; photograph; holograph; optical array; metal foil impactor; impact spectrometer; light spectrometer; membrane filter
Electric field	x	x		x	x	Plasma motion; auroras; radiation belts; solar and galactic cosmic rays; VLF whistlers; field strength and direction; atmospheric resistance	Langmuir probe; barium cloud; radar; particle detector; conductors
Electromagnetic field	x	x		x	x	Radio waves; infrared waves; visible and ultraviolet light; x-rays; gamma rays;	Receiver-antenna

Table 2.1-1a. (continued)

Sensed Parameter	Domain					Measurements	Sensors and Sensing Techniques
	Earth	Air	Water	Space	Planetary		
						radio astronomy; navigation; time; storm detection; tracking; radiation hazards; ionosphere; noise	
Geoid	x				x	Astronomic position; time; range; navigation	Laser; camera; radio frequency; radar; radio and atomic interferometers
Gravity	x	x			x	Earth shape and density; gravity influences on tides, altitude influence on gravity; gravity gradients	Free-fall device; gravity meters; accelerometer
Ground constituents	x				x	Silicates; halides; sulfides; carbonates; oxides; hydroxides; metals; organic matter; composition; particle size and shape	X-ray diffraction; electron microprobe; spark source; mass spectrometer; electron microscope; magnetic separator; neutron activator; petrographic microscope; hydrometer
Horizon		x		x		Local vertical; horizon location; earth position	Radiometric radiance balance sensor; photosensitive detector; thermal detector; bolometer and thermopile
Humidity		x			x	Vapor density; relative humidity; humidity/gas ratios; vapor pressure; dewpoint	Infrared hygrometer; thermometer; lithium chloride hygrometer; psychrometer; mechanical hygrometer; cold mirror dewpoint hygrometer; electrical hygrometer; carbon film hygrometer
Ion and electron	x	x	x	x	x	Charged particles; ions; electrons; fluxes; densities; energy ranges	Langmuir probe; ion trap; impedance probe; resonance probe; ion mass spectrometer

Table 2.1-1a. (continued)

Sensed Parameter	Domain					Measurements	Sensors and Sensing Techniques
	Earth	Air	Water	Space	Planetary		
Magnetic	x	x	x	x	x	Total field; component fields; relative and absolute fields; scalar and vector measurement; variable fields	Theodolite magnetometer; quartz horizontal magnetometer; earth inductor; magnetic field balance; variometer; fluxgate magnetometer; induction magnetometer; proton magnetometer; alkali vapor magnetometer; helium magnetometer; narrow-line magnetometer; superconducting magnetometer
Particle radiation	x	x	x	x	x	Solar wind; solar events; cosmic radiation; auroras; geomagnetically trapped particles; protons; electrons; alpha particles; fluxes; energy	Scintillation detector; photomultiplier; semiconductor detectors; electron multiplier
pH			x			Hydrogen ion concentrations; acidity; basicity	Colorimetric detector; electrochemical detector
Position	x	x	x	x	x	Altitude; velocity; acceleration: jerk	Potentiometer; capacitor; strain gage; differential transformer; variable permanence sensor; variable inductance sensor; piezoelectric sensor; vibrating wire sensor; velocimeter; electromagnetic log; moving coil; pitot-static tube; savonius rotor current meter; pendulum velocity meter; accelerometer; tachometer; pulse generator; drag cup: strobo-

Table 2.1-1a. (continued)

Sensed Parameter	Domain					Measurements	Sensors and Sensing Techniques
	Earth	Air	Water	Space	Planetary		
							scopic lamp; flyball sensor; pendulum
Precipitation		x			x	Rainfall rate and quantity; snowfall rate and quantity	Climatic raingage; tipping bucket raingage; weighing raingage; float rainfall rate gage; capacitor gage; photographic sensor; chemical sensor; momentum sensor; snow stake; radioactive sensor
Pressure	x	x	x	x	x	Absolute pressure; cage pressure; pressure changes	Liquid barometer; aneroid barometer; hypsometer: Bourdon tube; piezoelectric transducer; electrokinetic transducer; Mössbauer effect detector; thermal gage; radioactive ionization gage; thermionic ionization gage; variometer
Salinity			x			Salinity; titration; density; sound velocity; refractive index; electrical conductivity; temperature; pressure	Conductivity sensor-electrode cell; inductively coupled conductivity sensor; salinometer
Seismic wave	x				x	Earthquakes	Seismograph; geophone; vibration sensor
Sun		x		x		Solar constant; solar flares; prominences; magnetic fields; solar temperatures; solar corona	Telescope; silicon cell detector; photomultiplier detector; vacuum photoemissive sensor; bolometer; thermocouple; pyrometer; thermal detector; photodiode

Table 2.1-1a. (continued)

Sensed Parameter	Domain					Measurements	Sensors and Sensing Techniques
	Earth	Air	Water	Space	Planetary		
Telluric current	x				x	Natural and man-made electric currents in earth; geomagnetic micropulsations; earth resistivity	Electrodes in ground
Temperature	x	x	x	x	x	Temperature, heat transfer	Thermocouple; resistance thermometer; metallic element sensor; thermistor: optical, photoelectric and two-color radiation pyrometers; liquid-filled sensor; gas-filled sensor; vapor pressure thermometer; pyrometric cone; chemical sensor
Tide			x			Tide heights; estuarine circulation: coastal circulation; seiches	Tidestaff gage; mechanical float gage; gas purging gage; depth recorder; pressure gage
Time	x	x	x	x	x	Epochs; time intervals	Cesium, rubidium and helium atomic clocks; radio receiver; oscillator; quartz crystal
Visibility		x	x			Radiant flux; radiant intensity; reflectance; transmittance; illumination; visual ranges; meteorological ranges; runway visual range; slant visual range	Transmissometer; absorption meter; image converter; integrating nephelometer; angular scattering meter; laser; scatterometer
Water constituents			x		x	Chemical constituents; pH; salinity	pE sensor; dissolved oxygen sensor; automatic analyzer; chemical sensor
Water current			x			Water circulation; ocean flow patterns; turbulent flows	Drag force current meter; tracer detector; drogue; heat transfer sensor; pulse time and Doppler shift acoustic

Table 2.1-1a. (continued)

Sensed Parameter	Domain					Measurements	Sensors and Sensing Techniques
	Earth	Air	Water	Space	Plane-tary		
Water wave			x			Capillary, gravity and swell waves; wave heights; wave period; wavelength	sensors; impeller; Savonius rotor; magnetic field sensor; vane sensor Float gage; pressure sensor; strain gage; vibrating wire; accelerometer; electrical resistance staff; electrical capacitance staff; radar profiler; radar scatterometer; radiometer; infrared wave profiler; laser profiler; camera (stereo photogrammetry and Fourier optical transforms); sonar; ultrasonic altimeter
Wind		x			x	Wind velocities, directions and dynamics	Wind vane; pressure plate anemometer; rotating cup anemometer; propeller anemometer; fluidics anemometer; hot wire anemometer; Pitot-tube anemometer; sonic anemometer; aerodynamic drag anemometer; wind component meter; rawinsonde; pilot balloon; navigation aid wind finder; grenades; radio direction finder

Table 2.1-1b. Sensor Applications—Remote

Sensor Type	Domain					Measurements	Sensors and Sensing Techniques
	Earth	Air	Water	Space	Planetary		
Acoustic	x	x	x		x	Echo soundings	Echo sounders
Radio and magnetic	x	x	x	x	x	Field strength; field direction ground impedance; field phase relationship	Magnetic dipole field strength detector; magnetic field tilt angle detector; fluxgate magnetometer; proton magnetometer; optically pumped magnetometer
Gamma radiation	x	x		x	x	Gamma radiation intensity; gamma-ray analysis	Gas-filled radiation detector; ionization chamber; proportional counter; Geiger-Muller counter; scintillation detector; lithium drifted germanium detector
Optical	x	x	x	x	x	Imagery (photographs); spectral emissions; luminescence	Optical line scanner; infrared radiometer; correlation spectrometer; interferometer; scanning radiometer; Fraunhoffer line luminescence sensor; camera
Radar		x		x	x	Radio wave detection; ranging; precipitation; clouds; sea; ground; radar images; scattering coefficients; signal amplitude, direction, range, and frequency	Transmitter-receiver
Radiometric			x	x		Microwaves, clouds, sea, ground	Microwave radiometer

John C. Redmond

2.2 EARTH SCIENCE PARAMETERS

The solid earth is divided into three main segments: the crust, mantle, and core. The crust is that portion above the Mohorovičić discontinuity and is 10 to 50 km thick. It consists essentially of silicate rocks. The mantle of the earth is the portion between the bottom of the crust and the core. The core starts about 2900 km below the surface. The mantle is solid and consists of ferromagnesium silicate minerals. The core consists of iron or an iron-nickle-silicate mixture. The outer portion of the core is liquid, but a solid inner core may exist.

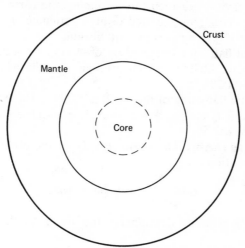

2.2.1 Size, Shape, Mass and Moment of Inertia of the Earth

Size and Shape. The early geodesists defined an oblate ellipsoid of revolution as the simple form that best fitted the earth and to which they could reference their measurements. Such a figure is specified by the semi-major axis (equatorial radius) and the flattening, f_0, given as

$$f_0 = \frac{R_e - R_p}{R_e},$$

(2.2–1)

where R_e and R_p are the equatorial and polar radii, respectively. In 1924 the International Union of Geodesy and Geophysics (IUGG) adopted as the "official" size and shape of the earth an ellipsoid of revolution specified by the semi-major axis, $R_e = 6,378,388$ m, and flattening, $f_0 = 1/297.0$. In 1967 the IUGG defined a more accurate ellipsoid as: $R_e = 6,378,160$ m; $f_0 = 1/298.25$. Most geodetic data in the world is referenced to the old formula.

Other non-ellipsoidal oblate surfaces of revolution ("spheroids") have been defined which are better fits to the earth. These normally deviate from the ellipsoids in the middle latitudes. For example, a formula for an earth spheroid is [9]

$$r = R_e[1 - K_1 \sin^2 \phi + K_2 \sin^4 \phi], \tag{2.2--2}$$

where r is the radius at latitude ϕ, and K_1 and K_2 are constants. The ellipsoid is much more commonly used.

In addition to the astrogeodetically measured flattening used in the 1924 international ellipsoid, there is the dynamical flattening, f_d, obtained from external gravity measurements assuming the earth's surface is an equipotential surface, and the hydrostatic flattening, f_h, calculated by assuming the earth is in hydrostatic equilibrium.

The dynamical flattening obtained by observing the secular changes in the orbits of earth satellites, which average over the whole earth [14], yield a value $f_d = 1/298.25$.

The hydrostatic flattening of the earth is $1/299.7$. There is thus a slight difference between f_d and f_h, which implies a significant strength in at least a part of the earth.

Mass and Mean Density. The mean density of the earth can be calculated from the relation [16]

$$\bar{\rho}G = \frac{3g_e}{4\pi R_e}[1 + \tfrac{3}{2}m + \tfrac{3}{7}mf], \tag{2.2--3}$$

where G is the gravitation constant, g_e the gravity at the equator, f the flattening, and

$$m = \frac{\omega^2 R_e}{g_e}, \tag{2.2--4}$$

where ω is the angular velocity of the earth.

Using values of g_e from the international gravity formula, Eq. 2.2–14, R_e and f from the international ellipsoid, and $G = 6.673 \times 10^{-8}$ (cgs units), the value of the mean density is calculated to be $\bar{\rho} = 5.514$ g/cm^3. More recent data for g_e, R_e, and m raises the value of the fourth figure slightly to 5.517 g/cm^3 [12].

Knowing the earth's mean density and shape, from which the volume can be computed, the mass of the earth is calculated to be $M = 5.977 \times 10^{27}$ g. The value of GM obtained from the tracking of the Ranger and

Mariner space probes (398,601.2 $\pm$ 1 in cgs units [25]) gives M and $\bar{\rho}$ as 5.973 $\times$ 10^{27} g and 5.514 g/cm^3, respectively.

Since the density of the crust material is considerably less than the mean density, there is an increase in density within the earth. The density change can be found from the measurement of the moment of inertia of the earth about its rotational axis. For a homogeneous ellipsoid this moment is

$$C = 0.4MR_e^2. \tag{2.2--5}$$

The value of C for the earth cannot be measured directly, but assuming the surface of the earth is an equipotential surface the value of C can be calculated from the relation

$$C = \frac{J_2}{H} MR_e^2, \tag{2.2--6}$$

where J_2 is the conventional notation for the second zonal harmonic of the gravitational field (see Section 2.2.2, p. 25) and H is the mechanical ellipticity given by

$$H = \frac{C - A}{C}, \tag{2.2--7}$$

where A is the moment of inertia about an axis in the equatorial plane. The value of H is derived from the precession of the earth's axis due to solar and lunar torques to give $H = 0.00327293$ [12]. Using this value of H with that of J_2 of 0.00108265 [13] yields

$$C = 0.33076\ MR_e^2. \tag{2.2--8}$$

The numerical coefficient of the right-hand side of this equation is lower than the value for a homogeneous ellipsoid, thus implying a density increase with depth. The density distribution can be calculated by iterative numerical methods that satisfy the boundary condition of total mass, axial moment of inertia, surface and average density, dimensions, and gravity. A more precise determination of the density variation with depth can be obtained from seismic data (Section 2.2.3).

2.2.2 Gravity

Description of the Earth's Gravitational Field. If the earth were a perfectly homogeneous sphere, its mass would have an external gravitational potential of the form

$$U = G\frac{M}{r}, \tag{2.2--9}$$

where G is the gravitational constant, M the mass of the earth, and r the radial distance to the center of mass. The gravity is the gradient of the potential. Since the earth is not a sphere, Eq. 2.2–9 gives only a crude approximation of its gravitational field.

A more accurate expression is obtained by assuming that the earth's surface is an ellipsoid of revolution and represents an equipotential surface. The potential for this gravity is termed the *geopotential* and is

$$\psi = U + \tfrac{1}{2}\omega^2 r^2 \cos^2 \phi', \qquad (2.2\text{--}10)$$

where ω is the angular velocity of the earth, r is a radius vector, and ϕ' is the geocentric latitude. The first term in this equation is the potential due to the gravitational attraction of the earth's mass and the second is the potential caused by the earth's rotation. The gravity for this geopotential is given by

$$g = \left(\left[\frac{\partial \psi}{\partial r} \right]^2 + \left[\frac{\partial \psi}{r\,\partial \phi'} \right]^2 \right)^{1/2}. \qquad (2.2\text{--}11)$$

Expanding Eq. 2.2–10 to second order, replacing r by the expanded expression for an ellipsoid of revolution, and differentiating gives [12]

$$g = g_e[1 + (\tfrac{5}{2}m - f + \tfrac{15}{4}m^2 - \tfrac{17}{14}mf)\sin^2 \phi' - \tfrac{1}{8}f(7f - 15m)\sin^2 2\phi'], \qquad (2.2\text{--}12)$$

where

$$m = \frac{\omega^2 R_e^3 (1 - f)}{GM}.$$

This equation can be written in the form

$$g = g_e(1 + \beta_1 \sin^2 \phi + \beta_2 \sin^2 2\phi), \qquad (2.2\text{--}13)$$

where ϕ is converted to geographic latitude. When taken only to the first order, this equation is known as Clairaut's equation.

In order to provide uniformity in world-wide gravity measurements, the IUGG in 1930 defined the international gravity formula [17]:

$$g = 978.0490(1 + 0.0052884 \sin^2 \phi - 0.0000059 \sin^2 2\phi)\,\text{cm/s}^2. \quad (2.2\text{--}14)$$

This equation shows that the change of gravity on the earth's surface varies approximately from 978 to 983 Gal (1 cm/s^2 = 1 Gal) with the high value measured at the poles.

A gravity formula utilizing more recent satellite data and an improved reference ellipsoid is given as [26]

$$g = 978.0404(1 + 0.0053015 \sin^2 \phi - 0.0000059 \sin^2 2\phi). \quad (2.2\text{--}15)$$

Mass variations within the earth cause departures of the actual gravity field from the simple symmetrical field given by the above gravity formulas.

Most gravity measurements are referenced to the world gravity base station at Potsdam, Germany [9], where the absolute value of gravity has been measured as 981.274 Gal and adopted as a standard (see Section 4.10).

The gravity at a point on the earth is measured with reference to the plumb-bob vertical, but neither the actual topographic surface nor the reference ellipsoid define a surface to which gravity is everywhere perpendicular. Thus, another surface, the geoid, is defined as the equipotential

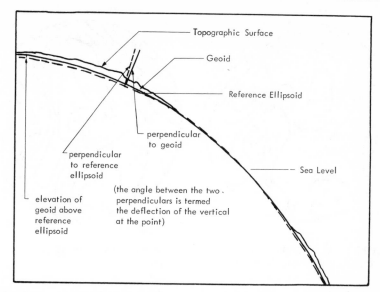

Figure 2.2-1 Arc segment of the earth showing the relationship between the topographic, reference ellipsoid, and geoid surfaces.

surface at mean sea level extended over the whole earth. Figure 2.2-1 shows the relationship between the three surfaces.

Being perpendicular to the sea-level gravity vector, the geoid's form is an indication of mass distributions within the earth. As shown on Figure 2.2-1 the geoid can be specified at a point by giving its elevation (above or below) the reference ellipsoid. These elevations of the geoid can be obtained from gravity measurements using Stokes' theorem [9]. This requires that gravity anomalies (difference between the measured gravity and the gravity calculated from the gravity formula, Eq. 2.2–14) be known over the whole earth. However, since the nearby anomalies have a greater effect, approximate values of the elevations can be obtained by considering the anomalies in a restricted local area [16]. Using a combination of satellite and terrestrial gravity data the geoid heights can be computed on a broad scale as shown in Figure 2.2-2 [14].

This figure shows that the geoid elevations (variations in the gravitational field) do not correlate well with the continents. This implies that the continents are in fairly good isostatic adjustment and there are other crustal or upper mantle mass inhomogenities in the earth.

Another method of describing the external gravity field of the mass of the earth which includes higher-order effects is by expressing the potential of the field as a series of spherical harmonics [13]:

$$U = \frac{GM}{r} \sum_{l=0}^{\infty} \sum_{m=0}^{1} \left(\frac{R}{r} \right)^l P_{lm}(\sin \phi) \left[C_{lm} \cos m\lambda + S_{lm} \sin m\lambda \right], \quad (2.2\text{–}16)$$

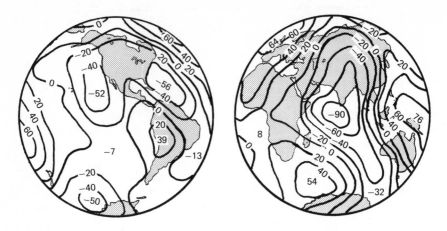

Figure 2.2-2 Geoid heights in meters referred to an ellipsoid of flattening 1/298.25, calculated by Stokes' formula from 300 nautical mile mean anomalies [14].

where r is the distance from the center of mass to the measuring point, R is the radius of a sphere with the same volume and mass as the earth, P_{lm} (sin ϕ) are Lengendre functions, C_{lm} and S_{lm} are constants, ϕ is the latitude, and λ is the longitude.

Satellite measurements of the harmonic coefficients have shown that some longitudinal variations in the earth's gravity field exist, but a definite

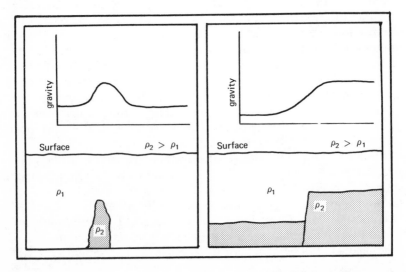

Figure 2.2-3 Effect of local differences in density on gravity (schematic representation).

triaxiality has not been established. It has also been shown that the gravity fields of the northern and southern hemispheres are not exactly symmetrical.

Local Gravity Field. Local variations in gravity are caused by differences in density of the near-surface earth materials. Figure 2.2-3 shows two structures in which a material of one density is imbedded in a material of a different density. The magnitude and shape of the gravity anomaly curve depends upon the density contrasts of the materials involved and their geometric arrangement. Table 2.2-1 gives some representative values of density for common earth materials.

Local gravity surveys are normally made with gravimeters that measure only a gravity difference, not an absolute gravity value (see Section 4.10). Local measurements are referenced to a single datum so that the individual measurements can be compared and anomalies identified. For local surveys, instrument readings are usually adjusted by three corrections: the free-air, Bouguer, and latitude corrections.

If the gravity measurements are not all made at the same reference datum (usually sea level), a correction must be applied to the measured gravity values to compensate for differences in elevations. The free-air correction is approximately

$$g_f = 0.3086h \quad \text{mGal}, \tag{2.2-17}$$

where h is the elevation above the reference datum in meters. The application of the free-air reduction gives the free-air gravity. Subtracting the gravity predicted by Eq. 2.2–14 from the free-air gravity gives the free-air gravity anomaly.

The free-air correction ignores the mass between the datum and the measurement elevation. The Bouguer correction, which considers the effect of this mass, depends on the density and geometry of the material

Table 2.2-1.   Density of Representative Crustal Materials [11]

Material	Density (g/cm^3)
Recent alluvium (dry)	1.54
(wet)	1.96
Sandstone	1.7–2.6
Shale	1.5–2.4
Limestone	1.6–2.2
Marble	2.7–2.9
Granite	2.5–2.8
Diorite	2.7–3.0
Gabbro	2.8–3.1
Dunite	3.3

between the measuring point and the datum. Assuming a cylinder with a height small compared to its radius and a density of 2.67 g/cm^3, the Bouguer correction is

$$g_b = -0.1108h \quad \text{mGal}, \qquad (2.2\text{–}18)$$

where h is in meters.

There is also a change in gravity with latitude due to the ellipticity of the earth that is given by Eq. 2.2–14. Several other corrections can be applied to the gravity measurements if necessary. When there is considerable topographic relief present, such as next to a mountain, corrections for the effect of the topographic masses should be applied [16].

Corrections can be made for regional gravity effects that are sometimes present. They are useful when examining the larger-scale earth features such as isostatic adjustments or departures of the earth from idealized shape. These corrections take into account the effect of regional variations of crustal or upper mantle mass differences due to gross geologic processes.

After a gravity survey is taken over an area and the latitude, free-air, Bouguer, and topographic corrections are applied to the measured values (in high-resolution work corrections would also have to be made for tidal effects), the resulting gravity values are close to those that would have been obtained had they all been made at the datum elevation. These values are called Bouguer gravities and any anomalies present (differences between the Bouguer gravity and the gravity value predicted by Eq. 2.2–14), are referred to as Bouguer anomalies. These anomalies are caused by regional geologic features such as unadjusted isostatic compensations within the earth.

2.2.3 Seismology

Earthquakes. When energy is released suddenly in the earth, either by a natural earthquake or an artificial explosion, some of it is converted into seismic waves. The ground motion caused by these seismic waves interacts with seismometers placed on the surface and a recording of the earth movements is made on a seismogram such as that shown on Figure 2.2-4. The frequency range of interest for seismic waves is shown on Figure 2.2-5. The amplitude of the measured seismic waves is a function of the energy of the source and its distance from the seismometer. Earthquake sources are normally referred to in terms of magnitude, which is related to the energy released by

$$\log_{10} E = 12 + 1.8M, \qquad (2.2\text{–}19)$$

where E is the energy and M the magnitude. Approximately one earthquake of magnitude 8 or greater occurs a year.

The focus of an earthquake is the location in the earth of the source,

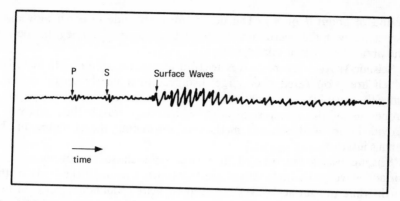

Figure 2.2-4 Recording of an earthquake on a seismogram. P is a compressional wave; S is a shear wave.

which is given by specifying the latitude and longitude of its epicenter (a point on the surface directly above the focus) and the depth to the focus. About 70 percent of all earthquakes occur in the upper 50 km of the earth and none have been observed deeper than about 700 km.

In addition to the disturbances caused by discrete sources, there is a low level of earth movement (microseism) that occurs continuously due to such things as meteorological effects, traffic, and ocean waves. The microseismic activity has a broad frequency range with a maximum at a

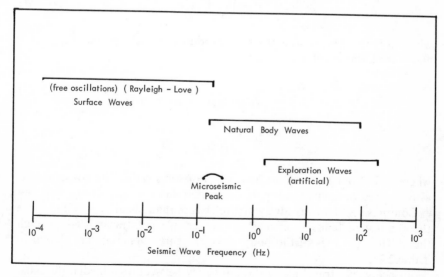

Figure 2.2-5 Frequency range of seismic waves.

period of about 7 or 8 s. The maximum amplitude of earth movement associated with the microseismic activity is about 10^{-3} cm near the oceans and about 10^{-4} cm at inland areas.

Seismic Waves. Seismic waves are transient strains in the earth materials which are propagated over large distances in a nearly elastic manner. Since the seismic waves penetrate the earth and are affected by the physical properties of the material through which they travel, their study has provided the most detailed method of examining the structure of the earth's interior.

Seismic waves are divided into two main classes: body waves and surface waves. The body waves are transmitted through the body of the earth while the surface waves travel at or near a boundary between two media.

In an infinite, isotropic, homogeneous medium two types of body waves are possible: compressional waves (P waves), in which the displacement of the material is in the direction of propagation; and shear waves (S waves) in which the displacement is transverse to the direction of propagation. Based on infinitesimal-strain theory, the velocities of propagation of the compressional (α) and shear (β) waves are derived as [6]

$$\alpha = \left(\frac{\lambda + 2\mu}{\rho} \right)^{1/2} = \left(\frac{k + \frac{3}{4}\mu}{\rho} \right)^{1/2}, \qquad (2.2\text{--}20)$$

$$\beta = \left(\frac{\mu}{\rho} \right)^{1/2}, \qquad (2.2\text{--}21)$$

where ρ is the density, k is the bulk modulus or incompressibility, and λ and μ are Lamé constants given by

$$\lambda = \frac{\sigma E}{(1 + 2\sigma)(1 + \sigma)}, \qquad (2.2\text{--}22)$$

$$\mu = \frac{E}{2(1 + \sigma)}, \qquad (2.2\text{--}23)$$

where E is Young's modulus and σ is Poisson's ratio. The constant μ is also referred to as the rigidity or shear modulus.

Compressional waves always travel faster than shear waves and since the rigidity of fluids is zero, shear waves are not transmitted through them. The velocities of the body waves in earth materials are shown in Table 2.2-2.

Surface waves travel along the boundary between two media, the most important of which is the external surface of the earth. There are several

Table 2.2-2. Body Wave Velocities for Representative Earth Materials [11]

	Compressional Wave Velocity (km/s)	Shear Wave Velocity (km/s)
Granite (Rockport, Mass.)	5.00	2.76
Rhyolite (Chaffee, Colo.)	4.10	2.46
Diorite (Jackson, Wyo.)	5.34	3.24
Anadesite (San Juan Co., Colo.)	5.66	3.22
Gabbro (Mellen, Wisc.)	6.80	3.37
Diabase (Fredick, Md.)	6.76	3.67
Basalt (USSR)	5.57	3.40
Dunite (Jackson Co., N.C.)	6.52	3.59
Periodotite (USSR)	7.40	4.02
Obsidian (Lake Co., Ore.)	5.82	3.57
Volcanic Tuff (Calif.)	1.41	0.83
Marble (Tate, Ga.)	5.34	2.83
Sandstone (baraker)	5.15	1.97
Sandstone (USSR)	3.22	1.92
Sandstone (India)	6.05	2.36
Limestone (Bavaria)	5.55	3.05

types of surface waves, such as Rayleigh and Love waves, that are possible depending upon the exact nature of the boundary [8].

Seismic waves are attenuated with passage through the earth. In the near-surface material this attenuation is often due to the gross heterogeneity and unconsolidated nature of the rock. In the deeper regions other mechanisms associated with the microstructure of the rock cause an attenuation but it is much smaller.

A measure of seismic wave attenuation can be specified by the dimensionless parameter Q defined as

$$Q^{-1} = \frac{1}{2\pi} \frac{\Delta E}{E}, \tag{2.2-24}$$

where E is the total amount of elastic energy stored per unit volume per cycle and ΔE is the energy dissipated per unit volume per cycle.

There is an interesting difference in attenuation between the upper and lower mantle. In the upper mantle a Q of about 80 is observed which starts to increase at about 400 km depth to a value of about 2000 in the lower mantle. This increase takes place over a vertical distance of several hundred kilometers [1] and indicates a change in the physical properties of the mantle material, the higher Q being generally associated with a more "rigid" material.

Seismic Data. A large earthquake produces a number of phases which make up the seismic wave train received at a seismometer. These phases are identified by the type of wave and the path taken. A P wave reflected from the earth's core and returned as a P wave is termed a PcP wave, if it returns as an S wave it is termed a PcS wave. If the P wave travels through the liquid outer core it acquires a K designation such as for a PKP wave —a K phase is a compressional wave. Passage through the solid inner core is designated as I for compressional waves and J for shear waves. A great number of these types of phases are possible as shown in Figure 2.2-6.

There is a general increase in velocity with depth which causes a bending of the ray paths of the body waves so that they are curved with the concave side up as shown in Figure 2.2-6. If there is a discrete discontinuity in physical properties between two media, a wave incident upon the discontinuity may be reflected, transmitted, directed along the boundary, transformed into other types of waves, or combinations of the above. For example, an S wave may originate from a source in the crust, travel to the mantle-core boundary and be reflected in two components, P and S (shown in Figure 2.2-6 as ScP and ScS). The angles of reflection and transmission are governed by Snell's law, which relates the angles of reflection and transmission to the velocities of the waves.

The curved boundaries in the earth complicate the effects of reflection

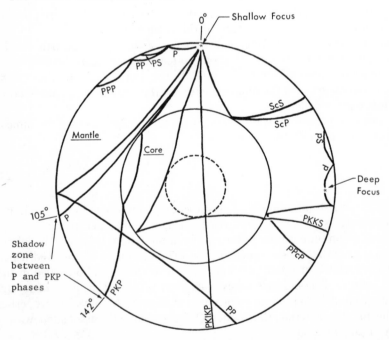

Figure 2.2-6 Examples of some of the different seismic phases that can travel in the earth.

and transmission. Thus, as shown in Figure 2.2-6, shadow zones exist at locations where phases of the seismic waves are not received directly. They are received at greatly diminished amplitudes as some of the energy is diffracted into the shadow zones.

Data from many stations are used to construct travel-time curves such as shown in Figure 2.2-7. The slopes of these curves give the apparent velocities of the phases. Curvature of a travel-time curve indicates a velocity change with increasing distance; this is related to passage of the waves at a greater depth in the earth. Thus, the velocities of the P and S waves as a function of depth in the earth can be derived as shown on Figure 2.2-8.

Earth Structure. There is a relatively sharp velocity discontinuity (from about 6.8 to 7.8 km/s) within a few tens of kilometers of the surface named the Mohorovičić discontinuity (the Moho) after its discoverer. Its location varies from about 10 km below the surface under the oceans to about 50 km below mountain roots. The P-wave velocity at the top of the mantle has been observed to vary from about 7.4 to 8.5 km/s with the lower velocities associated with high geologic activity.

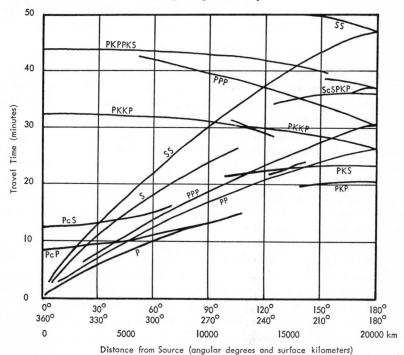

Figure 2.2-7 Travel-time curve for some of the possible seismic wave phases from a surface focus.

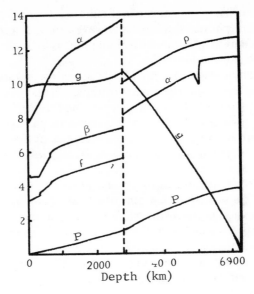

Figure 2.2-8 Distributions in a model of the earth of the P and S velocities α and β (km/s), pressure P (10^{12} dyn/cm^2), density ρ (g/cm^3), and gravity g (10^3 cm/s^2) [8].

Figure (2.2-8) shows a striking increase in the apparent velocity of the body waves at a depth of approximately 400 km. This discontinuity, which also correlates with the change in Q of the mantle, may represent a phase change to a more dense crystal form of the mantle material.

The effect of the outer core is easily perceived on seismic records. The P waves exhibit a definite decrease in amplitude from about 105° to 142°. This decrease in amplitude, a "shadow zone," is caused by diffraction of P waves by the core as shown on Figure 2.2-6. S waves are not observed to propagate through the core, which indicates that the core material is liquid or nearly so. There is some evidence to indicate that there is an inner core with a radius of about 1300 km. This inner core is believed to be solid, but the nature of the transition (gradual or sharp) between the outer and inner core is uncertain.

Since the velocities of the seismic waves depend upon the physical properties of the transmitting medium, it is possible to derive information about the material in the interior of the earth from a study of wave velocities. Several models of the earth have been constructed [4, 8]. Figure 2.2-8 shows the density distribution of the earth as a function of depth for one model.

2.2.4 Temperature

The effect of diurnal and seasonal temperature fluctuations, the main source of temperature changes at the surface of the earth, disappears rapidly with depth, becoming negligible below about 30 m. Neglecting these solar-radiation effects, the thermal state of the earth depends upon present and past heat sources in the earth and the thermal properties of the earth materials.

In constructing a model of the thermal state of the earth and its thermal history, the only parameters that can be measured directly are the temperature near the surface and the thermal properties and radioactive heat generation of crustal materials. The near-surface heat flow and thermal gradient are computed from these measurements. Additional information is derived from laboratory simulations of the deeper earth environment, other geological and geophysical data (e.g., seismic data that shows the presence of a liquid core), and theoretical analyses. Because of the uncertainty in the information relating to the thermal properties of the earth's interior, a rather wide range of thermal models of the earth may be postulated to satisfy the observed data.

Heat Sources. Natural radioactivity provides the main source of heat within the earth. In order for a radioactive isotope to produce a significant amount of heat over the history of the earth it must have a half-life comparable with the age of the earth and it must produce heat at a rate large enough to affect the earth's observed thermal state. The isotopes that satisfy these requirements include U^{235}, U^{238}, Th^{232}, and K^{40}. Table 2.2-3 gives an approximate average value of heat generation by these isotopes in several rock types. The data indicate that there is a correlation between the acidity of the rock and its radioactive isotope content and, consequently, there is a lower rate of radioactive heat generated with depth where the acidity of the rocks decreases.

This upward concentration of radioactive isotopes probably occurred during the period when the earth was differentiated into its major subdivisions of core, mantle, and crust. The radioactive isotopes, having large

Table 2.2-3. Approximate Average Present Rate of Heat Generation in Rocks (erg/g-year) [20]

Rock Type	By U^{235}	By U^{238}	By Th^{232}	By K^{40}	Total
Granite	7	110	84	34	235
Diorite	4	63	54	29	150
Basalt	1	24	41	6	72
Dunite	0.02	0.40	44	0.01	0.87
Chondritic meteorites	0.37	0.01	0.37	0.95	1.70

atomic radii, did not conveniently fit into the crystals which formed early in the solidification phase. The isotopes thus tended to remain in solution and be left to the later stages of solidification near the surface. The earth's mantle composition is assumed to be similar to that of chondritic meteorites, while the core is comparable to iron meteorites, which have a concentration of the radioisotopes several orders of magnitude less than the chondrites [21].

Other heat sources that may have been important in the thermal history of the earth include the original heat of formation, heat generated by the separation of the mantle and core from a once homogeneous earth, short-lived radioactive isotopes, and exothermic chemical reactions, but their significance to the earth's thermal history is uncertain.

Heat Flow. To obtain valid heat-flow data near the surface of the earth the measurement site must be in thermal equilibrium. On land it means having a deep hole that has reached and maintained thermal equilibrium; not being seriously affected by ground waters or convection. On the deeper parts of the ocean floor the overlying water acts as a thermal blanket effectively shielding the measurement site from unwanted thermal effects once the probe itself has come to thermal equilibrium.

Measurements of heat flow require that the temperature be measured at points (at least two) separated by some vertical distance to obtain a thermal gradient. The thermal conductivity of the material, K, at the measuring site is measured and the heat flow per unit area and time, H, is calculated from the one-dimensional equation

$$H = K \frac{\partial T}{\partial z},$$
(2.2–25)

where ∂T is the change in temperature with the change in vertical distance ∂z. The thermal gradient near the surface of the earth has been found to vary from about $10°$ to $100°C$ per kilometer. In the upper layers of the earth heat is transferred only by conduction, but at depths where the temperature is higher, radiative heat transport may become important.

The global mean of the surface heat flow is about 62 erg cm^{-2} s^{-1} (or 1.5 μ cal cm^{-2} s^{-1}). By comparison the sun supplies an average of about 3×10^5 erg cm^{-2} s^{-1}. Most of the heat flow from the earth occurs in normal regions with the anomalous regions such as hot springs and volcanoes contributing less than one-tenth of the total. The means of the continental and oceanic areas are 62.4 erg cm^{-2} s^{-1} and 61.6 erg cm^{-2} s^{-1}, respectively [18]. In both the continental and oceanic areas the mean heat flow has been found to decrease with the age of the province examined (see Table 2.2-4), but the time scales for the thermal decay in the two areas are different: the continents requiring 10^9 yr to come to an essentially constant value of about 50 erg cm^{-2} s^{-1} and the oceans requiring 10^8 yr to reach approximately the same value. In the continental areas the difference in heat flow with age can be correlated with radioactive isotope concentration. In the oceanic areas the heat flow can be correlated with distance from the oceanic ridges, which suggests that a significant portion of the heat flow in the oceanic areas may be associated with the cooling of hot material that is intruded and spread from the oceanic ridges.

Thermal Properties of Earth Materials. The thermal conductivities, specific heats, and melting points of crustal rocks are fairly well known for temperatures and pressures which can be achieved in the laboratory, but the values for these material properties become more uncertain with increasing depth in the earth, i.e., increasing pressure and temperature. Table 2.2-5 shows the thermal conductivities for earth materials at one atmosphere.

When the temperature rises to 2000°K or more in earth-type materials, a significant portion of the thermal energy may be transferred by radiation in addition to the normal conduction by lattice vibrations.

To include this radiative transport mechanism the thermal conductivity, K, can be written as [20]

$$K = K_0 + \frac{16\eta_1^2 k T^3}{3(\varepsilon + 120\pi\sigma/\eta)}, \qquad (2.2\text{-}26)$$

where η is the index of refraction, k the Stefan–Boltzmann constant, ε the opacity, and σ the electric conductivity.

Using absorption spectra data, Clark [5] calculated the component of heat transport due to radiation and found an increase for olivine from

Table 2.2-4. Heat Flow for Oceanic and Continental Provinces [24]

Province	Age $(10^6$ yr)	Mean Heat Flow (erg cm^{-2} s^{-1})
South Atlantic—oceanic		
Early cretaceous	140	47.7
Middle cretaceous	100	58.6
Late cretaceous	75	59.4
Outer ridge material	50	52.7
Ridge material	30	73.2
Near crest of mid-Atlantic ridge	10	90.0
North American—continental		
Precambrian shields	1000	39.8
Precambrian platforms	700	49.0
Calendonian orogenic belts	400	54.8
Mesozoic orogenic belts	160	64.9
Cenozoic basin and range	50	91.2

Table 2.2-5.   Thermal Conductivity of Rocks at One Atmosphere Pressure [2]

Rock Type	0°C	100°C	300°C
Quartz			
⊥ axis	11.2	7.9	5.0
⊥ axis	6.6	5.0	3.5
Granite	3.5	3.0	2.4
Anorthosite	1.7	1.7	1.8
Diabase	2.0	2.0	2.0
Gabbro	2.0	1.9	2.0
Dunite	12.4	9.4	- -
Glass			
Silica	3.3	3.5	4.1
Obsidian	3.2	3.5	4.0

2.97×10^6 at $1000°K$ to 20.2×10^6 erg/cm s°C at $2500°K$, and for diopside from 0.67×10^6 at $1000°K$ to 7.25×10^6 at $2500°K$.

The melting curves of the material in the mantle and core can be used to place limits on the temperature within the earth because the temperature in the solid part of the earth (mantle) is below its melting point while the temperature in the liquid part (outer core) is above its melting point. The melting curves with depth in the earth are obtained from a combination of theory and extrapolation of laboratory data (where data for pressures equivalent to about 500 km in the earth are available). Figure 2.2-9 shows the melting curves for diopside and iron.

The heat capacity of silicate rocks is approximately 10^7 erg/gm°C. This value is not as susceptible to change as is the thermal conductivity and is usually taken as being constant.

Temperature Distribution in the Earth.   The temperature distribution in the earth can be obtained by solving the generalized heat equation (in spherical coordinates)

$$\rho C_P \frac{\partial T}{\partial t} = \frac{1}{\rho^2} \frac{\partial}{\partial r}\left(r^2 K \frac{\partial T}{\partial r} \right) + \Theta_g(t, r), \qquad (2.2–27)$$

where t is time, T is temperature, r is radial distance from center of the earth, ρ is density, C_P is specific heat at constant pressure, K is thermal conductivity, and Θ_g is heat generated.

Equation 2.2–27 must be solved using preimposed initial conditions to satisfy the known and assumed boundary conditions. There are a large number of models that can satisfy the known conditions.

A representative temperature distribution for the earth is shown in Figure 2.2-9 ([20], Model 19). The melting curves for iron and diopside

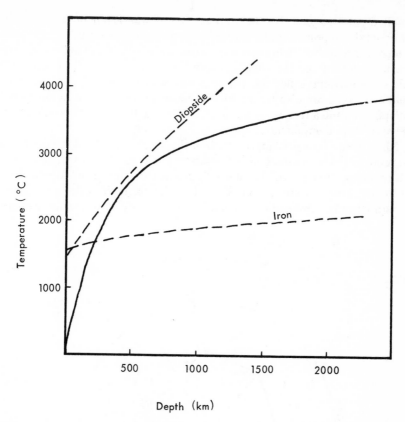

Figure 2.2-9 Temperature distribution in a model earth. The diopside and iron melting point curves are shown as dashed lines [20].

drawn on the figure show that the iron in the core is in the liquid phase and the silicates in the mantle may be approaching their melting point at a depth of about 600 km.

2.2.5 Magnetic Field

The magnetic field at the surface of the earth is a combination of a main field generated internally, an external field generated by the movement of charged particles well above the surface, and the field resulting from local concentrations of magnetic materials in the crustal rocks.

The maximum field strength at the earth's surface is about 0.6 G (or 60,000 gammas, γ). The external field is less than one percent of the total. The highly localized anomalies can have strengths comparable to the internally generated field.

Historically, the magnetic field of the earth was measured by means of a small dipole magnet. The field so measured was given in terms of the horizontal component of the field (*H*); the geographic azimuth of this component, with positive being measured from north to east, called the magnetic declination (*D*); and the angle between the total magnetic field vector (**F**) and the horizontal called the magnetic dip or inclination (*I*). The magnetic field at a point can also be specified in terms of north and east horizontal components and a vertical component. Figure 2.2-10 shows the relationships between these components.

Main Field. To a first approximation the magnetic field of the earth can be represented by a dipole near the center of the earth and inclined about 11° to the rotational axis. The internally generated magnetic field of the earth probably arises from a complicated hydromagnetic dynamo process resulting from a combination of a liquid, conductive core in a rotating planet [3].

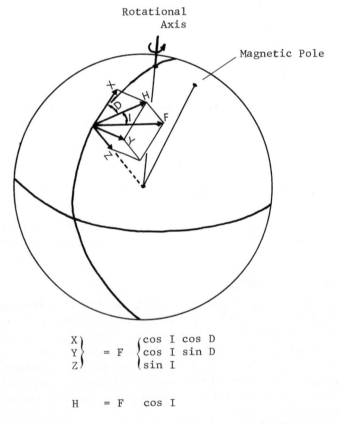

$$\left.\begin{matrix} X \\ Y \\ Z \end{matrix}\right\} = F \left\{\begin{matrix} \cos I \cos D \\ \cos I \sin D \\ \sin I \end{matrix}\right.$$

$$H = F \cos I$$

Figure 2.2-10 Relationship between components of the earth's magnetic field.

The geomagnetic potential, V, can be written in spherical harmonics as [5]

$$V = R_e \sum_{n=1}^{\infty} \left(\frac{r}{R_e} \right)^{n+1} \sum_{m=0}^{n} P_n^m(\theta) \left[g_n^m \cos(m\lambda) + h_n^m \sin(m\lambda) \right], \quad (2.2\text{-}28)$$

where R_e is the radius of the earth, r is the distance to the measuring point $P_n^m(\theta)$ are the normalized associated Legendre functions g_n^m and h_n^m are Gauss coefficients, θ is the colatitude, and λ is the longitude. The total vector field, $\mathbf{F}$, at a point is the gradient of this scalar potential.

The Gauss coefficients give the relative strengths of the components of the field, but unlike the gravity field the magnetic field undergoes secular changes that alter the values of these coefficients. Thus, when their values are quoted they must be dated as shown in Table 2.2-6.

Evaluation of the Gauss coefficients over the past 130 years has given an insight into the characteristics of the main magnetic field. These are [13, 10]:

1. To a first approximation the main field of the earth is presently represented by a dipole whose axis makes an angle of 11.5° to the rotational axis of the earth and passes through the meridian plane at a longitude of about 70° west.

2. The nondipole part of the field ($n \geq 2$) has a total intensity about one-tenth of the dipole field.

3. The dipole field has decreased nearly linearly at the rate of approximately 0.015 G per century (if this rate persists the dipole moment will disappear in 3991 A.D.) and moved westward at a rate of about 7° per century.

4. Concomitant with the decrease in the dipole field there has been an increase in the nondipole field at a rate of approximately 0.05 G per century and a westward movement of about 10° per century; however, these changes do not appear to be linear, are not the same everywhere on the earth, and are variable depending upon the particular component considered.

5. During the past 20 million years there have been at least 60 polarity reversals of the dipole field occuring at irregular intervals. The axis of the dipole has also wobbled with a maximum amplitude of about 20° from the earth's rotation axis.

Table 2.2-6. Gauss Coefficients of the Earth's Magnetic Potential (in Units of Gammas)

Author	Epoch	g_1^0	g_1^1	h_1^1	g_2^0	g_2^1	h_2^1	g_2^2	h_2^2
Gauss	1835	−32350	−3110	6250	510	2920	120	−20	1570
Adams	1845	−32190	−2780	5780	90	2840	100	40	1350
Fritsche	1885	−31640	−2410	5910	−350	2860	−750	680	1420
Dyson and Furner	1922	−30950	−2260	5920	−890	2990	−1240	1440	840
Cain et al.	1960	−30426	−2174	5761	−1548	3000	−1949	1574	201

Local Anomalies. The residual magnetism in rock is essentially due to the magnetism of its magnetite (iron oxide) content although minerals such as pyrrhotite, ilmenite, and several others can have high residual magnetism. Some ferromagnetic earth materials may be concentrated enough locally to significantly affect the earth's field in the vicinity. The strongest of these is probably the iron deposit near Moscow, U.S.S.R., where the local field from the iron results in a maximum vertical field of nearly 1.4 G.

Magnetic surveys conducted over the surface of the earth are used to detect iron ore bodies or to map the basement structure (igneous rock under the sedimentary layers) for oil exploration. Mapping of the basement structure is possible because of the generally higher magnetism of igneous rocks. Table 2.2-7 gives average magnetic susceptibilities for several rock types.

The permanent magnetism found in rocks may be aligned in a direction quite different from the present earth's field, which indicates that when the rocks acquired their magnetism (say an igneous rock cooling through the Curie point) the direction of the earth's field was in a direction different from that at present. Thus, observing the orientation of the magnetic field in rocks of different ages shows the movement of the magnetic pole or rock mass (continent) over geologic time.

2.2.6 Other Parameters

Electrical Properties. The electrical properties of crustal rocks are obtained by direct measurements both in the laboratory and in situ. Variations in these electrical characteristics are used as the basis for several types of geophysical explorations including resistivity, induced polarization, and induction surveys. Table 2.2-8 gives the electrical conductivities of some earth materials. The values of the conductivities were obtained by ac measurements and are, in general, valid over a wide range of frequencies.

Near the surface of the earth the electrical conductivity of rocks is strongly

Table 2.2-7. Average Magnetic Susceptibilities of Some Earth Materials [6]

Rock Type	Average Magnetic Susceptibility $\times 10^6$ (cgs)	Range of Susceptibilities $\times 10^6$ (cgs)
Basic igneous (basalt)	2596	44–9711
Acid igneous (granite)	647	3–6527
Metamorphic	349	0–5827
Shale	52	5–1478
Sandstone	32	0–1665
Limestone	23	2–280
Dolomite	8	0–75

Table 2.2-8. Electrical Conductivities of Some Earth Materials [27]

Material	Electrical Conductivity (mho/m)
Soils	10^{-4} to 10^{-1}
Sea water	4 to 5
Fresh water	10^{-4} to 10^{-3}
Sandstone (wet)	10^{-4} to 10^{-2}
Limestone	10^{-8} to 10^{-4}
Clay	10^{-2} to 10^{-1}
Granite	10^{-9} to 10^{-3}
Basalt	10^{-9} to 10^{-5}
Gneiss	10^{-7} to 10^{-3}
Galena (lead sulfide)	20 to 200
Hematite (iron oxide)	10^{-7} to 10^{-5}

influenced by their degree of water saturation. Variances of the water table or recent rainfall can have a decided effect on the electrical conductivity of the near-surface rocks.

The electrical conductivity of the mantle can be inferred indirectly from observing low-frequency electromagnetic waves originating above the surface of the earth and incident upon it. These external waves induce waves within the earth which when compared with the incident waves allow determinations of the electrical conductivity in the earth to be made. Since the depth of penetration of the waves is a function of their frequency, a broad range of frequencies allows the conductivity to be measured over a broad depth range. Analyzing waves with periods from a few days to six months, electrical conductivity profiles for the mantle were obtained as shown in Figure 2.2-11 [7].

For induced polarization-type surveys the dielectric constant of the earth material is important. The dielectric constant for rock changes somewhat with frequency; for example, the dielectric constant of granite changes from about 75 pF/m (picofarads/meter) at 100 Hz to 60 pF/m at 10 MHz. The dielectric constants for some earth materials are shown in Table 2.2-9.

Table 2.2-9. Dielectric Constants of Dry Rocks at 100 Hz [15]

Rock Type	Dielectric Constant (pF/m)
Limestone	92
Sandstone	50
Granite	75
Diorite	64
Serpentine	90

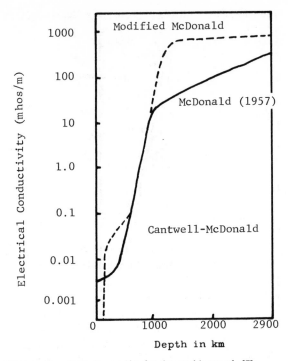

Figure 2.2-11 Electrical conductivity profiles for the earth's mantle [7].

Tidal Forces. The mutual gravitational attraction between the moon and the earth, in total, is exactly balanced by the centrifugal force due to rotation about their common center of mass, the barycenter. However, these two force vectors do not exactly balance at all points in the earth. The gravitational attraction of the moon is greater than the centrifugal force on the part of the earth nearest the moon while on the part farthest from the moon the opposite is true. These residual forces cause the tides in the earth as shown in Figure 2.2-12. In addition to the tides induced by the moon there are also tides due to the sun that act on the earth in a similar manner. Tide sensors are discussed in Section 4.26.

The tidal potential on the earth due to the moon can be written as

$$V = \frac{Gm}{R_m} \sum_{n=2}^{\infty} \left(\frac{r}{R_m} \right)^n P_n(\theta), \qquad (2.2\text{–}29)$$

where m is the mass of the moon, R_m the distance between the earth and the moon, r the distance from the center of the earth to the point of measurement, and $P_n(\theta)$ are Legendre polynomials (θ is the angle between R_m and r). The sun causes a similar tidal potential. It is usually sufficient to consider only

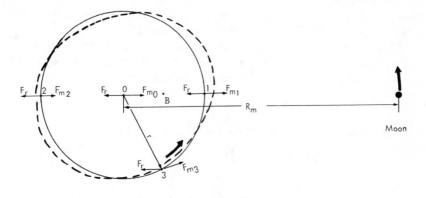

Earth

F_r - Centrifugal force due to rotation of 0 around B.

F_m - Gravitational force of moon on earth mass.

At Point   0, $F_r = F_{m0}$

          1, $F_r < F_{m1}$

          2, $F_r > F_{m2}$

          3, $F_r \neq F_{m3}$

Figure 2.2-12 Forces at several points on the earth due to gravitational attraction by the moon and the rotation about the barycenter, B.

the first term of this equation which is

$$V_2 = \tfrac{1}{2}Gm \frac{r^2}{R_m^3}(3\cos^2\theta - 1). \tag{2.2-30}$$

When the earth is deformed by the tidal disturbance its gravitational field will also be deformed. This deformation of the external gravitational potential of the earth can be written as an additional potential given by [25]

$$V^* = V \sum_{n=2}^{\infty} k_n \left(\frac{R_E}{r}\right)^{2n+1}, \tag{2.2-31}$$

where k_n are constants that reflect the rigidity of the earth (the k_n are actually the values, at the earth's surface, of more general constants $k_n(r)$) and R_E is the radius of the earth. Considering only the $n = 2$ term the height of the equilibrium ocean tide is given by

$$\xi^* = \frac{V_2 + V_2^*}{g} = (1 + k_2)\frac{V_2}{g}. \tag{2.2-32}$$

The theoretical maximum amplitude of this displacement is about one meter, but because of irregular shoreline effects, especially in bays, the observed oceanic tide can be many times this value.

Because the earth has some rigidity, the displacement of the solid surface will not be as great as the oceanic equilibrium tide which follows an equipotential surface. In treating this effect another constant h_2 is introduced and [25]

$$V_2^{**} = h_2 V_2, \tag{2.2-33}$$

so that the displacement of the solid surface is given by

$$\zeta^{**} = \frac{V_2^{**}}{g} = \frac{h_2 V_2}{g}. \tag{2.2-34}$$

The value of h_2 is about 0.6 so that the maximum displacement of the solid surface of the earth is approximately a half meter. The constants k and h are called Love numbers.

The tidal disturbances will affect gravity measurements made on the earth. Considering only the $n = 2$ terms the gravity change due to the tide is given by

$$\Delta g \approx \frac{\partial (V_2 + V_2^* + V_2^{**})}{\partial r}, \tag{2.2-35}$$

$$\approx (1 - \tfrac{3}{2}k_2 + h_2)\frac{2V_2}{R_E}. \tag{2.2-36}$$

In some texts the tidal disturbance on the measured gravity is given only by the first term of this equation. The change in gravity given by Eq. 2.2–36 can be measured directly with highly sensitive gravimeters. The maximum variation of gravity measured at the earth's surface due to the lunar tides is almost 0.2 mGal and due to the solar tide is about 0.08 mGal.

Another combination of the Love numbers can be obtained from tidal gages. This is because the tide observed with gages rigidly attached to the earth is actually the difference between the displacement of the equipotential or ocean surface and the displacement of the solid earth.

$$\zeta = \zeta^* - \zeta^{**} = (1 + k_2 - h_2)\frac{V_2}{g}. \tag{2.2-37}$$

The observed values of Δg and ζ yield values of $k_2 = 0.3$ and $h_2 = 0.6$.

Observations have shown that the tidal bulge on the earth is not exactly aligned with the moon, as shown on Figure 2.2-12. This displacement of the tidal bulge, called the luni-tidal interval, is due to the inelastic behavior of the solid body of the earth, to viscous damping in the ocean, or to a combination of both. Because it does not point directly at the moon, but is displaced slightly more than $2°$, the tidal bulge causes a couple which transfers energy from the earth to the moon. This couple slows the earth's rotation about its axis and increases the separation between the moon and the earth (which decreases the length of the lunar month).

Astronomical observations show that this energy transfer results in a deceleration of the earth's rotation by 4.81×10^{-22} rad/s^2 and a deceleration in the lunar orbit of 1.08×10^{-23} rad/s^2. From these values [23] the energy lost by earth is calculated to be 2.8×10^{19} erg/s and the energy gained by the moon is 0.1×10^{19} erg/s. The difference between these numbers, 2.7×10^{19} erg/s, is the energy dissipated in the earth. One-half to two-thirds of this energy is probably dissipated in shallow seas (the Bering Sea having by far the largest effect) [12]. Other possible dissipation mechanisms are the interaction of the oceanic tidal wave with irregular shorelines and the relative movement of crustal blocks. Evidently very little of the dissipation occurs in the solid body of the earth which behaves essentially elastically.

Geochemistry. The distinctions between and within the core, mantle, and crust are based on compositional differences caused by the fractionation of a once homogeneous earth and phase changes caused by high temperatures and pressures. The volumes and masses of the main components of the earth are shown in Table 2.2-10.

Determinations of the chemical composition of the earth are based on (1) measured chemical composition of the crust; (2) an assumed fractionation of a once homogeneous body; (3) observed physical properties of the earth; (4) laboratory simulations of the high pressures and temperatures found in the earth, and (5) solar and meteoritic compositions. The relative abundances of chemical elements in the sun, meteorites and the earth are shown in Table 2.2-11.

These tables show that a fractionation, in its simplest form, has moved iron to the core, ferromagnesium silicates to the mantle, and the more acidic rocks to the crust.

In addition to purely compositional changes there may also be phase changes, that is, changes in the crystal arrangement of a material resulting

Table 2.2-10. Volumes and Masses of Earth Shells [22]

Earth Shell	Thickness (km)	Volume ($\times 10^{27}$ cm^3)	Mean Density (g/cm^3)	Mass ($\times 10^{27}$ g)	Mass (%)
Atmosphere	- -	- -		0.000005	0.00009
Hydrosphere	3.80	0.00137	1.03	0.00141	0.024
Crust	30	0.015	2.8	0.043	0.7
Mantle	2870	0.892	4.5	4.056	67.8
Core	3471	0.175	10.7	1.876	31.5
Whole earth	6371	1.083	5.52	5.976	100.0

Table 2.2-11. **Relative Abundances of Chemical Elements in the Earth's Crust, Sun, Meteorites, and Earth [22]**

1	2	3	4	5
Sun	Meteorites	Earth's Crust	Whole Earth	Relative Abundance in Whole Earth (elements in column 4) (%)
H	O	O	Fe	35
He	Fe	Si	O	28
O	Si	Al	Mg	17
Fe	Mg	Fe	Si	13
Mg	S	Ca	Ni	2.7
N	Ni	Na	S	2.7
Si	Al	K	Ca	0.61
S	Ca	Mg	Al	0.44
C	Na	Ti	Co	0.20
Ca	Cr	H	Na	0.14
Al	Mn	P	Mn	0.09
Ni	K	Mn	K	0.07
Na	Ti	S	Ti	0.04
Cr	Co	C	P	0.03
			Cr	0.01

in different physical properties (e.g., density) with no change in composition. A phase change in the earth not associated with a compositional change would be expected to be manifest by a more gradual change in physical properties, although this is not always strictly true. Thus, the change at the Mohorovičić discontinuity more than likely reflects a compositional change because of its sharpness. Likewise the core-mantle boundary appears to be a compositional change associated with a phase change from a solid to a liquid. Within the mantle the phase change from an olivine- to a spinel-type structure, which results in an increase in density of about 10 percent, probably starts at about 400 km depth within the earth, and the transition is spread over a considerable vertical extent. The phase transition is the probable cause of the velocity discontinuity observed for the body waves and the change in Q between the upper and lower mantle.

References

1. Anderson, D. L., and C. B. Archambeau, "The Anelasticity of the Earth," *J. Geophys. Res.* **69**, 2071–2084, May 1964.

2. Birch, F., and H. Clark, "The Thermal Conductivity of Rocks and Its Dependence upon Temperature and Composition," *Am. J. Sci.* **238**, 529–558, Aug. 1940, and 613–635, Sept. 1940.
3. Bullard, E., and H. Gellman, "Homogeneous Dynamos and Terrestrial Magnetism," *Phil. Trans. Roy. Soc. London*, Series A, **247**, 213–278, Nov. 1954.
4. Bullen, K. E., *An Introduction to the Theory of Seismology*, 3rd Ed., Cambridge University Press, Cambridge, 1963.
5. Clark, S. P., Jr., "Radiative Transfer in the Earth's Mantle," *Trans. Am. Geophys. Union* **38**, 931–938, Dec. 1957.
6. Dobrin, M. D., *Introduction to Geophysical Prospecting*, McGraw-Hill Book Company, New York, 1952.
7. Eckhardt, D., K. Larner, and T. Madden, "Long-Period Magnetic Fluctuations and Mantle Electrical Conductivity Estimates," *J. Geophys. Res.* **68**, 6279–6286, Dec. 1963.
8. Haddon, R. A. W., and K. E. Bullen, "An Earth Model Incorporating Free Oscillation Data," *Phys. Earth Planet. Interiors* **2**, 34–49, 1969.
9. Heiskanen, W. A., and F. A. Vening Meinesz, *The Earth and Its Gravity Field*, McGraw-Hill Book Company, New York, 1958.
10. Hide, R., "Planetary Magnetic Field," *Planet. Space Sci.* **14**, 579–586, July 1966.
11. Jakosky, J. J., *Exploration Geophysics*, Trija Publishing Company, Newport Beach, California, 1957.
12. Jeffreys, H., *The Earth: Its Origin, History, and Physical Constitution*, Cambridge University Press, Cambridge, 1962.
13. Kaula, W. M., *An Introduction to Planetary Physics*, John Wiley and Sons, New York, 1968.
14. Kaula, W. M., "Tests and Combination of Satellite Determinations of the Gravity Field With Gravimetry," *J. Geophys. Res.* **71**, 5303–5314, Nov. 1966.
15. Keller, G. V., and F. C. Firschknecht, *Electrical Methods in Geophysical Prospecting*, Pergamon Press, Oxford, 1966.
16. Lambert, W. D., and F. W. Darling, "Density, Gravity, Pressure and Ellipticity in the Interior of the Earth," in *Internal Constitution of the Earth*, B. Gutenberg, Ed., pp. 340–363, Dover Publications, New York, 1951.
17. Lambert, W. D., "The International Gravity Formula," *Am. J. Sci.* **243A**, 360–392, 1945.
18. Lee, W. H. K., and S. Uyeda, "Review of Heat Flow Data," in *Terrestrial Heat Flow*, W. H. K. Lee, Ed., American Geophysical Union Monograph No. 8, pp. 87–190, 1965.
19. Love, A. E. H., *Some Problems of Geodynamics*, Cambridge University Press, Cambridge, 1911.
20. MacDonald, G. J. F., "Calculations on the Thermal History of the Earth," *J. Geophys. Res.* **64**, 1967–2000, Nov. 1959.
21. MacDonald, G. J. F., "Chondrites and the Chemical Composition of the Earth," in *Res. Geochem.*, P. H. Abelson, Ed., pp. 476–494, John Wiley and Sons, New York, 1959.
22. Mason, B., *Principles of Geochemistry*, Chapter 3, "The Structure and Composition of the Earth," pp. 26–58, John Wiley and Sons, London, 1952.

23. Munk, W. H., and G. J. F. MacDonald, *The Rotation of the Earth,* Cambridge University Press, Cambridge, 1960.
24. Sclater, J. G., and J. Francheteau, "The Duplications of Terrestrial Heat Flow Observations on Current Tectonic and Geochemical Models of the Crust and Upper Mantle of the Earth," *Geophys. J. Roy. Astron. Soc.* **20,** 509–542, Sept. 1970.
25. Sjogren, W. L., D. W. Trask, C. J. Vegos, and W. R. Wollenhaupt, "Physical Constants as Determined from Radio Tracking of the Ranger Lunar Probes," Jet Propulsion Laboratory Technical Report 32-1057, Dec. 30, 1966.
26. Uotila, U. A., "Harmonic Analysis of World-Wide Gravity Material," *Ann. Acad. Sci. Fennical* **67,** Oct. 1962.
27. Watt, A. D., F. S. Mathews, and E. L. Maxwell, "Some Electrical Characteristics of the Earth's Crust," *Proc. IEEE,* **51,** pp. 897–910, June 1963.

H. Dean Parry

2.3 METEOROLOGICAL PARAMETERS

Meteorology is the discipline of geoscience that treats the earth's atmosphere. Meteorology has three aims: (1) to forecast the weather; (2) to adapt human activities to normal or long-term average conditions through the use of climatological records, and (3) to understand the processes and phenomena that occur in the atmosphere.

2.3.1 Radiation

Weather and climate are manifestations of a giant heat engine, the earth's atmosphere. Every 24 hours radiant solar energy equivalent to the energy derived by burning about $5(10)^{14}$ kg of coal reaches the outer limits of the earth's atmosphere. Only 30 percent of this is reflected back into space and the remaining 70 percent "drives" the atmospheric processes.

Two laws that govern radiative processes from solids or liquids are used to understand the earth as a heat engine.

Planck's Law states

$$E_\lambda = \frac{c_1 \lambda^{-5}}{\exp(c_2/\lambda T) - 1},$$ (2.3–1)

where c_1 and c_2 are constants, T is the absolute temperature of the radiating body, λ is the wavelength, and E_λ is the spectral emittance which is the energy emitted per unit area, per unit time at a particular wavelength.

Figure 2.3-1 shows the theoretical emission curve for the approximate temperature of the sun and of the earth's surface (6000 and 300°K, respectively). Note that there are two separate abscissa scales and that the scale of the ordinate, E_λ/T^5, changes drastically for the two temperatures. From this curve it is clear that most of the sun's radiation is at a wavelength near 0.5 μm while most of the earth's radiation is at a wavelength near 10 μm. In order to be able to plot both the 6000°K curve and the 300°K curve on the same graph it is necessary to use two ordinate scales. The peak of the earth's radiation at 10 μm is below the intensity of solar radiation at the same wavelength [2].

The Stefan-Bolzmann Law, which quantifies the change of amount of radiation with temperature, states

$$E = \sigma T^4,$$ (2.3–2)

where E is the total radiation emitted by the body and σ is the Boltzmann constant which is determined from the radiative properties of the radiating

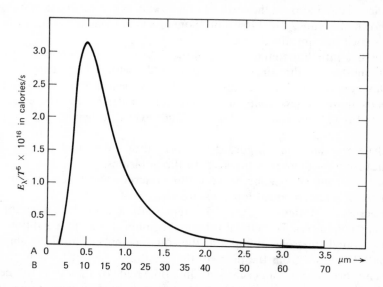

Figure 2.3-1 Theoretical blackbody radiation. Scale A corresponds to $T = 6000°$K, B to $T = 300°$K.

body. The fourth power of the absolute temperature of the earth's surface does not vary greatly between pole and equator; hence the total amount of long-wave radiation emitted by the earth is almost independent of latitude. In contrast to this the input of solar energy is concentrated at the equator.

2.3.2 Atmospheric Motions

Heat balance over the entire earth's surface is maintained by atmospheric motions that are largely responsible for changes in all other weather parameters.

Atmospheric motions occur in a continuous spectrum of sizes from global circulations encompassing the entire earth to turbulent effects which have dimensions of inches or less. Meteorologists have arbitrarily classified them as general circulation, macroscale motion, mesoscale motion, and microscale motion [7].

General Circulation. The largest atmospheric motions are global circulations which make up the so-called general circulation. If the earth were a perfect nonrotating sphere, huge hemispherical convection cells would probably exist. Air at the surface of the earth would flow meridianally from each pole toward the equator, rise there and divide into two currents each flowing toward a pole. Factors principally responsible for a much more complicated general circulation are the earth's rotation; orographic effects; water-land distribution with differing absorption, reflection, specific heat and roughness properties; polar ice caps and ocean currents. The actual circulation at the earth's surface can be represented schematically as shown in Figure 2.3-2. Flow aloft more or less mirrors that at the surface but is itself dynamically driven.

There is little horizontal motion in the doldrums, but there is a net upward motion of the air ocurring principally within huge cumuloform clouds concentrated in the intertropical convergence zone (ITC) which extends more or less continuously around the earth near the equator. Within the ITC massive cumuloform clouds extend to heights near 30 km in extreme cases.

The horse latitudes (Figure 2.3-2) have little horizontal motion but vertical motion there is antithetical to the doldrums.

In the strong shear zone between the westerlies and polar easterlies a number of semipermanent low-pressure areas exist.

Macroscale Motions. Figure 2.3-2 represents a smoothed diagram of average conditions. To maintain thermal equilibrium under conditions of concentrated solar energy input at the equator, movement of the surface layers of cold air from the poles to the equator is required. Figure 2.3-2 does not provide such an exchange. On a time scale of a few days to weeks huge volumes of air (air masses), having generally homogeneous properties in horizontal directions over millions of square miles, move out of their

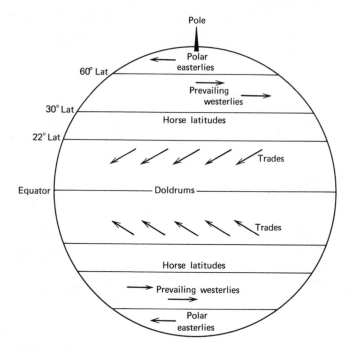

Figure 2.3-2 Atmospheric circulation.

source regions. Cold masses form over arctic regions while most warm air masses originate over tropical oceans. When a cold air outbreak occurs, a cold air mass moves southward or southeastward invading the temperature latitudes. The edge of an air mass normally is a sharp discontinuity called a front. Temperature changes of 20°C occasionally occur at a location as a strong front moves by. If cold air is advancing, the boundary is called a cold front. If cold air is receding, the boundary is called a warm front. Fronts and the weather changes associated with them are surprisingly long-lived. Cold fronts are by no means unknown in the Hawaiian Islands or in the Yucatan Penninsula of Mexico. Both of these locations are roughly at 20°N latitude and the cold front may have started at 60 or 70°N latitude.

The storms of temperate latitudes originate as a breaking wave on the frontal surface [1]. Figure 2.3-3 is a schematic representation of such an extratropical cyclone as it forms in the northern latitudes during a period of 36 to 48 hr.

This circulation is the "low" (pressure area) on the weather map. The circulation is counterclockwise (to an observer looking down from above) in the northern hemisphere and has a typical diameter of 800 to 2500 km. Cloudiness and precipitation are normally associated with the low, especially

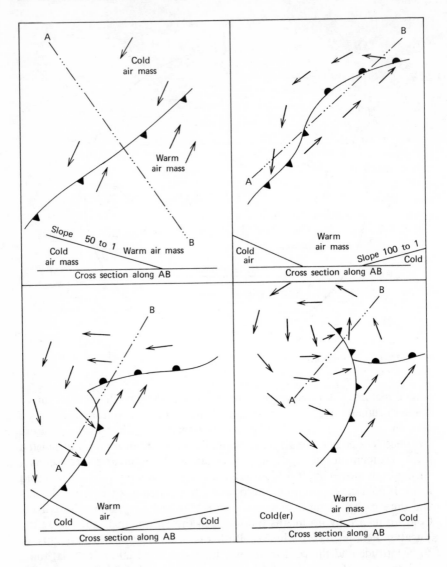

Figure 2.3-3 Wind flow.

in the southeast quadrant (in the northern hemisphere). The counterpart of the low is the "high" (pressure) circulation. This encompasses the air mass and often is concentric with it. A typical diameter is 2500 to 5000 km and circulation in the northern hemisphere is clockwise. Fair weather usually prevails in "highs." The low- and high-pressure systems are called *macro-*

scale circulations. They are transients superimposed on the general curculation.

Another macroscale circulation which originates over the equatorial oceans (typical latitude of 5°) is the tropical cyclone. In the Western Atlantic and Eastern Pacific tropical cyclones are called hurricanes, in the Western Pacific they are called typhoons, in waters near Australia these storms are called willy willies, and in the Indian Ocean they are known as cyclones. Tropical cyclones have no frontal systems associated with them, are 500 to 800 km in diameter, have winds within them of 100 to 300 km per hour, and produce torential rains, mountainous seas, and storm tides. A unique feature of the tropical cyclone is its eye, a central core some 30 km in diameter, in which the winds are calm and skies are usually clear. Separating the eye from the active storm is the wall of the eye. This cylindrical boundary, ordinarily a solid sheet of very black clouds, is a region of very sharp wind shear and extreme turbulence.

Mesoscale Motions. Spawned in many a hurricane is a small but extremely violent disturbance called a tornado. Typical diameter is 1.5 to 5 km, life span is less than 8 hr, pressure difference between the center of the funnel and the nearby atmosphere can be more than 10 cm of mercury and winds can be in excess of 500 km per hour [3]. Tornadoes are not limited to hurricane areas, but occur in much of the world. A "sea-going" tornado is called a waterspout. The tornado is an example of a *mesoscale* circulation. Other meteorological circulations in the mesoscale class are thunderstorms, land and sea breezes, isolated showers of rain or snow, local wind storms, and large areas of fog.

Microscale Motions. *Microscale* phenomena are typically less than 1.5 km in the greatest dimension. Examples of weather variations at this scale are dust devils, patches of fog, air temperature differences that exist within a few meters of horizontal distance [8], and turbulence of all wavelengths. Turbulence occurs in a wide spectrum of sizes. A wavelength of hundreds of meters can affect large airliners. Wavelengths of a few centimeters can affect only astronomic observations by generating strong stellar scintillations.

2.3.3 The Equation of Motion and the Pressure Parameter

Atmospheric motions can be related to other atmospheric parameters by writing the equations of motion for the atmosphere. Applying Newton's second law for a unit mass of the atmosphere and using vector notation:

$$\dot{V} = 2V \times \mathbf{\Omega} + G - \frac{1}{\rho} \nabla p + F. \qquad (2.3\text{--}3)$$

In Eq. 2.3–3 all boldface letters are vectors and the dot indicates the total time derivative. V is the velocity, $\mathbf{\Omega}$ is the earth's rotation, G is the apparent gravity which is the gravitational force corrected for the centripetal force produced by the earth's rotation, F is the total frictional force or forces and/ or terrain constraints acting on the unit mass, p is the pressure, and ρ is the density [9].

The first term on the right is the coriolis force, an apparent acceleration due to the fact that the motion is referred to an accelerating (i.e., rotating) set of coordinates. Equation 2.3–3 contains all significant forces and apparent forces acting on the atmosphere. Negligible at the macro- and mesoscales are sound waves and motion induced by man and his devices. The continuity equation for an incompressible fluid (a satisfactory assumption for large-scale atmospheric motions),

$$\mathbf{V} \cdot \mathbf{V} = 0, \tag{2.3–4}$$

and the equation of motion form the basis for numerical weather forecasting for the macroscale and contribute much to mesoscale studies.

For no acceleration and no friction the equation of motion for only horizontal forces is

$$2V_H \times \mathbf{\Omega} = \frac{1}{\rho} \mathbf{V}_H p. \tag{2.3–5}$$

This is the geostrophic approximation much used in practical forecasting [10]. It shows that winds blow perpendicular to the pressure-gradient vector and parallel to the lines of equal pressure. An observer looking in the direction toward which the wind is blowing will have low pressure to his left in the northern hemisphere (Buys Ballot's Rule).

2.3.4 Atmospheric Stability and the Roles of Temperature and Moisture

Temperature. Perhaps the most important element governing the vertical motions of the atmosphere is temperature. The earth's atmosphere consists of a series of shells, more or less concentric with the center of the earth and with each shell having a specific temperature distribution. From the ground up to 8 to 20 km, depending on the latitude, the temperature generally decreases with increasing height. There are minor reversals of this which are called inversions. The rate of decrease with altitude is also variable. The region in which this general decrease occurs is called the troposphere. Its upper boundary, the tropopause, is higher at the equator than at the poles where it is diffuse and indistinct. It may not exist over the south pole. In middle latitudes there are frequently multiple tropopauses that overlap one another to produce a shingle-like structure. The stratosphere is the layer from the tropopause upward to about 25 km in which the temperature remains constant or increases slightly with increasing height.

The next layer above the stratosphere, the mesosphere, extends to about 80 km. The layer has a temperature maximum with kinetic temperatures reaching about 10° C near the middle of the layer. Temperature decreases from the mesosphere maximum to the mesopause and then increases rapidly through the thermosphere which is the next higher atmospheric layer. The outer shell of the atmosphere is the exosphere where atmospheric gasses are not firmly held by the earth. In the exosphere the kinetic temperature and mean free paths are such that molecules of gas escape from the earth and move into free space. Figure 2.3-4 is a highly schematic representa-

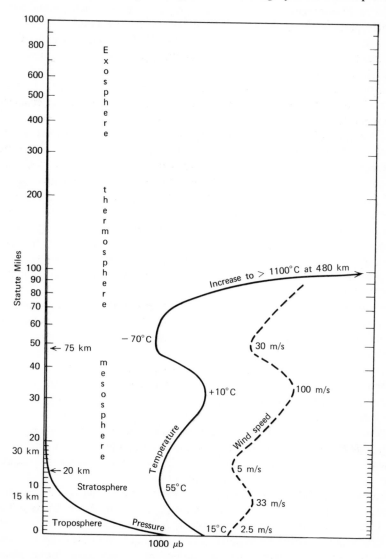

Figure 2.3-4 Typical profiles of temperature, wind, and pressure.

tion of average vertical profiles of temperature wind speed and pressure in the middle latitudes.

This complex vertical temperature distribution is largely a result of the differential absorption of radiation by earth and the atmosphere. The existence of the higher layers of warm temperature is primarily due to the fact that atmospheric gasses absorb certain bands of the solar radiation spectrum. This absorption is a function of atmospheric pressure and the characteristics of the molecules which make up the gas. Gases do not obey Planck's Law or the Stefan-Bolzmann Law. Radiation coming from the outer limits of the atmosphere passes through regions of increasing pressure. The first time a combination of atmospheric gas and pressure favorable for absorption of a band of radiation is reached, the band is absorbed. All other bands of the spectrum pass through to either be filtered out at lower levels or to reach the earth's surface.

Most solar radiation reaches the surface and warms it. Air next to the surface is warmed either by contact with the surface or by absorbing long wave radiation emitted by the surface. The heated air is forced aloft by denser unheated air and the atmosphere "boils" with vertical convections.

As the air is forced upward it moves into progressively lower pressure regions, expands and cools adiabatically (i.e., without addition or removal of heat since there is no heat source or sink available to it). Air rising through the atmosphere cools at a nearly constant rate of 10°C per km (5.5°F per 1000 ft) by adiabatic expansion. The average actual lapse rate of, or decrease in, temperature with height in the troposphere is 6.5°C per km (3.7°F per 1000 ft) [4].

Air moving downward (upward) in the average atmosphere would be heated (cooled) at the adiabatic lapse rate of 10°C per km and would be warmer (colder) than its surroundings and hence would return to its original height and remain stable.

Air rising through an inversion becomes progressively colder than its surroundings and hence sinks back to its original level. An inversion in this manner puts a "lid" on the atmosphere by damping out vertical motions from below.

If the lapse rate in the dry air is at the adiabatic rate, air that moves up or down is at the temperature of the surrounding air. There is no damping of vertical motions and any heating from below can produce convective currents which penetrate to considerable heights.

Moisture and Stability. The amount of water vapor which can exist in the air is a function of temperature. Normally the air in the lower layers of the atmosphere contains more water than air in upper layers because the air near the surface is warmer and has a greater capacity, and water vapor enters the atmosphere at the earth's surface from bodies of water, snow, wet ground, and transpiration from plants. When heated surface air rises, it cools at the adiabatic lapse rate, finally becoming saturated with some of the water vapor condensing and releasing its latent heat of condensation. This makes the air even warmer than its surroundings so that its upward motion is accelerated. In this manner water vapor accelerates vertical motions which in turn produce precipitation. After saturation, rising air is said to cool pseudoadiabatically. The pseudoadiabatic lapse rate is not constant as is the adiabatic, but is a function of temperature and pressure, varying from 3.5°C per km with high temperatures near sea level to the adiabatic rate above 15 km where the amount of water vapor in the air is too small to release significant amounts of heat.

2.3.5 Atmospheric Dynamics

The Mechanism of Precipitation and Cloud Formation. Precipitation is produced by proper combinations of flow (wind) heat (temperature) and moisture. Principal mechanisms are:

1. Convection initiated by surface heating.
2. Orographic effects. This effect occurs when wind blows against sloping terrain and is thereby forced to rise.
3. Frontal action. At both warm and cold fronts, the warm air is lifted, typically producing cloudiness and frequently precipitation.

Sources of Energy Which Intensify Storms. Cold air underrunning warm air in an occluding cyclone produces a decrease in potential energy by lowering the center of gravity of the colder, heavier air. This decrease in potential energy is converted to an increase in kinetic energy that increases wind circulation around the storm. This kind of temperature distribution, baroclinicity, is one of three principal energy sources for an extra-tropical cyclone (the storms or lows of temperature latitudes). The other two are the drawing of energy by the storm from the general circulation and energy released by phase change of atmospheric water vapor. In a tropical cyclone, temperature differences contribute, in all probability, to the storm's intensification. Tropical cyclones disintegrate as they move onto land, which may imply that much of the energy comes from water vapor picked up over the sea or that increased friction over land destroys the storm. Temperature distribution is a vital but not sole factor governing intensification of these storms. A unique temperature distribution is also a necessary but probably not a sufficient condition for the formation and intensification of tornadoes. The energy of storms derives from a combination of conditions of temperature, humidity, and wind.

2.3.6 Meteorological Parameters

Meteorological parameters are classified and selected on the basis of three criteria: the degree of usefulness in weather forecasting, the usefulness in defining climate, and the contribution made to man's understanding of physical phenomena occurring in the atmosphere.

Meteorology is in a state of rapid transition from a descriptive and inexact science to an exact physical science based on mathematical physics and using the capabilities of modern electronic computers. Useful meteorological parameters are those which quantitatively define physical characteristics and properties of the atmosphere. Atmospheric movement, temperature, and amount of atmospheric water vapor are the most important and primary meteorological parameters. Two additional primary parameters of only slightly less importance are atmospheric pressure (which relates to motion through Newton's second law and is useful in depicting and understanding the field of motion) and radiation which is a prime mover of the atmospheric heat engine.

Table 2.3-1 lists the primary parameters and their characteristics. In particular, characteristics which are an unchangeable part of nature are listed. Figures in the "accuracy" column represent a best guess and are not intended to define the accuracy requirement.

The large number of remaining meteorological parameters are called secondary because they are more often "effect" than "cause" in the weather production chain. The details included in Table 2.3-2 are the name of the parameter, its definition, name of measuring instrument, description of the physical principle employed in measuring it, the range of the measurement in the units normally used and the principal field where the parameter is used. Parameters pertaining to the ozonosphere and the ionosphere have been omitted since the operational meteorologist finds little use for them. Reference 5 contains more information on these parameters.

2.3.7 Sampling Meteorological Parameters

The complexity and wide distribution of all meteorological parameters disturbed by various scales of motion make it impossible to completely measure the atmosphere. It is necessary to resort to sampling techniques.

Meteorologists refer to data which correctly samples phenomenon as representative [9]. In evaluating a typical weather map an attempt is made to interpret the data and explain inconsistencies. There will always be a few pieces of data inconsistent with remaining data. For example, a coastal station may report a southeast sea breeze of 4 m/s while other stations in the area under the influence of a macroscale high report northwest winds of 2.5 to 5 m/s. None of these wind measurements is incorrect. The one which is inconsistent reflects a mesoscale motion, whereas the bulk of the data relates to a macroscale motion.

The sampling process must be adapted to the scale of the phenomenon by employing a suitable "filter." This filter can be a simple subjective judgement for charts analyzed "by hand," as in the example given previously. Filtering

Table 2.3-1. Primary Meteorological Parameters and Their Characteristics

Parameter Name (Scale)	Nature of Phenomenon to be Measured	Working Range (Extremes)	Accuracy Maximum Permissable		Averaging Time Conventionally Used or Recommended
			RMS Random error	Bias	
A. SURFACE MEASUREMENTS					
Short wave hemispherical radiation (global and macro)	Total electromagnetic radiation received from the sun on a unit horizontal area.	96 langlys/hr 104 langlys/hr	± 5 %	2 %	Integrated over a full day
Short wave radiation (meso and micro)	Not required operationally for these scales.				
Net radiation (all scales) long and short wave direct and diffuse	Total electromagnetic radiation long and short wave, both diffuse and direct passing up and down through the atmosphere	Not in general operational use in the United States	8 %	5 %	Integrated for a full day or between sunset and sunrise or sunrise and sunset
Wind speed and direction (global)	Horizontal component of motion of air at a standard height. Height standardized at 20 feet in U.S.A.	0–50 m/s 360° (0–150 m/s 360°)	5 m/s 20°	0.5 m/s 2°	> 1 day
(macro)	World Meteorological Organization (WMO) recommends 10 meters. Motion is not steady state or	0–50 m/s 360° (0–150 m/s 360°)	2.5 m/s 10°	0.5 m/s 2°	WMO recommends 10 min
(meso)	constant. It may be represented by a power spectrum having frequencies from microhertz to	0–50 m/s 360° (0–150 m/s 360°)	1.5 m/s 5°	0.5 m/s 2°	1 minute
(micro)	kilohertz.	0–25 m/s 360° (0–150 m/s 360°)	0.5 m/s 2°	0.5 m/s 2°	15 seconds
Atmospheric pressure (global)	Hydrostatic pressure of the air. Pressure variations due to acceleration forces are not included.	700–1060 mb For a given station (i.e., at a fixed height)	1 mb	2 mb	> 1 day #

\# averaging done as data processing and not as part of measurement process

Table 2.3-1. (continued)

Parameter Name (Scale)	Nature of Phenomenon to be Measured	Working Range (Extremes)	Accuracy Maximum Permissable		Averaging Time Conventionly Used or Recommended
			RMS Random error	Bias	
Atmospheric pressure (macro)	Station pressure is referred to the published height of the reporting station. Sea level pressure is the pressure which would be measured if the station were at sea level. For stations not at sea level this is a computed value	Required range is considerably less, circa 150 mb	0.5 mb	2 mb	1 minute #
(meso)			0.3 mb	2 mb	1 minute #
(micro)			0.1 mb	2 mb	1 minute #
Air temperature (global)	Temperature of the air at a location or the average temperature along a sample line or throughout a sample volume. The sensing process must exclude all effects other than the temperature of the air. WMO specifies that "surface" temperature be measured at a height of 2 m above the ground	$-45°$ to $50°C$	$1°C$	$0.5°C$	> 1 day #
(macro)		$(-55$ to $40°C$) Required range will be smaller for a given location	$0.5°C$	$0.5°C$	1 hour #
(meso)			$0.25°C$	$0.5°C$	5 minutes #
(micro)			$0.5°C$	$0.5°C$	1 minute #
Atmospheric moisture (global) (macro) (meso) (micro)	Amount of water vapor in the atmosphere. Height of measurement is same as for temperature. Moisture parameter is usually given in terms of dew point or relative humidity	Dew point -30 to $25°C$ $(-55$ to $40°C)$	Bias and scale are not as important as degree of saturation and amount of water in the air. The dewpoint error RMS and bias should be less		> 1 day # 1 hour # 5 min. #

Table 2.3-1. (continued)

Parameter Name (Scale)	Nature of Phenomenon to be Measured	Working Range (Extremes)	Accuracy Maximum Permissable		Averaging Time Conventionally Used or Recommended
			RMS Random error	Bias	
Atmospheric Moisture			than the greater of 0.25°C and $\dfrac{(T-T_d)}{10}\dfrac{300}{T}$ Where T is the temperature and T_d the correct dew point		

B. UPPER AIR MEASUREMENT

Parameter Name (Scale)	Nature of Phenomenon to be Measured	Working Range (Extremes)	RMS Random error	Bias	Averaging Time Conventionally Used or Recommended
Radiation, short and longwave	Radiation is not operationally measured by upper air equipment. Satellites measure radiation incident on the outer limit of the earth's atmosphere and infrared emmitted by the earth's surface and the atmosphere.				
Wind speed and direction (global and macro)	Primary requirement for these scales is the wind pattern along the following constant pressure surfaces: 850, 700, 500, 300, 250, 200, 100 and perhaps 50 millibars. In each case the two dimensional field is needed.	0–100 m/s 360° 0–175 m/s	2.5 m/s vector error	0.5 m/s vector error	6 hours # 1 to 2 min averaging time during data reduction.
Wind speed and direction (meso) (micro)	Required for these scales is a height profile of the horizontal wind speed and direction. The Mesoscale data	0–100 m/s360° (0–150 m/s/360°) (0–50 m/s 360°)	1 m/s vector error 0.5 m/s vector	0.5 m/s 0.5 m/s	meso 5 min # micro 1 min #

* Further averaging during data analysis.

Table 2.3-1. (continued)

Parameter Name (Scale)	Nature of Phenomenon to be Measured	Working Range (Extremes)	Accuracy Maximum Permissable		Averaging Time Conventionly Used or Recommended
			RMS Random error	Bias	
	should be to 8 km and the microscale to about 3 km.	(0–100 m/s 360°)	error		
Air temperature (global) and (macro)	Patterns of air temperature at 6 to 8 constant pressure or constant height surfaces to 30 km, and mean temperature of the layers. Between surfaces.	−90 to +35°C	±1°C	±1°C	No averaging in the normal data processing time lag of thermistor 4 s (63%) at sea level and 300 m per min rise.
Air temperature (meso)	Temperature height profile to 8 km	−50 to +35°C (−60 to +40°C)	±½°C ±0.2°C	±1°C ±0.1°C	
Air temperature (meso)	Temperature height profile to 3 km	−50 to +35°C	±0.2°C	±0.1°C	
Atmospheric moisture (global and macro)	Patterns of moisture at 5 constant pressure or height surfaces between sea level and 15 km.	−25 to 85°C Dew point	±2°C	2°C	Averaged only by instrument lag. Time constant increases as temperature decreases. Typical value at sea level 300 m/min and 10°C is 2 s.
Atmospheric moisture (meso)	Dew point – Height profile to 8 km for meso and 3 km for micro.	−25 to 85°C less range required	±1°C	±1°C	
(micro)			±5°C	±1°C	

Table 2.3-2. Secondary Meteorological Parameters

Parameter Name and Definition	Name of Measuring Instrument and Principle Employed	Range in Units Normally Used	Use Code
CEILING—Height of base of cloud which summed with cover of all clouds having lower bases, covers at least half the sky.	Rotating beam ceilometer (RBC)—Right triangle calculation of height using length of baseline and angle from end of baseline to a spot of light on cloud. Coverage estimated by human observer.	RBC 0–1.2 km. Observer estimates to 8 km. All pilot reports relayed.	Av
CLOUD COVER AMOUNT —Number of eighths of the celestial hemisphere above the observer which is occupied by clouds.	Estimated by observer.	8th's of the hemispherical dome.	Av G
CLOUD BASE HEIGHT— Height of bottom of cloud body.	Rotating beam ceilometer (RBC) see ceiling. Also a LIDAR or light radar device. The latter is not yet in common use.	See ceiling. Lidar has range to at least 7.5 km.	Av
CONDENSATION NUCLEI COUNT—Number of condensation nuclei present in a given volume.	Aitken nuclei counter—Chamber in which moist air is cooled by expansion. Water condenses on nuclei to form droplets which fall to bottom of chamber where they can be counted. A technique using an electron microscope indicates the classic Aitken technique gives values one third too low.	Number per cm^3.	P
DROP SIZE SPECTRUM (for rain, drizzle and fog)— Number of drops of each size over the entire range of sizes occurring in rain, drizzle or fog.	Inertial collector—Air stream containing drops is deflected. Inertia of drops causes them to follow an undeflected path and strike a collector. Microphotography also used.	Number per size interval normalized by volume.	P
ELECTRIC POTENTIAL GRADIENT—Change in electrical potential per unit distance as measured normal to the equipotential surfaces.	Electrometer (modified)—For example, a rotating semicircular disc to collect any charge on the foil of the electrometer which is not in equilibrium with the air charge at the level being measured.	0 to 350,000 volts per meter. Normal range in fair weather (no lightning or convective clouds 50 to 400 volts/meter.	P
PRECIPITATION AMOUNT—Depth of equivalent liquid water which falls on a horizontal	Precipitation is caught in a cylindrical vessel which has a known area horizontally exposed at its top. Volume of	0 to 75 cm. 0 to 5 m for a storage gage.	G H

Table 2.3-2. (continued)

Parameter Name and Definition	Name of Measuring Instrument and Principle Employed	Range in Units Normally Used	Use Code
surface during a given time interval.	liquid is measured. Weight of either solid or liquid may be measured. A tipping bucket is balanced so that it will tip when 0.25 mm precipitation runs into it. An electrical contact is made by each tip. These contacts can be counted electronically.		
PRECIPITATION DETECTION—Determination of present occurrence of precipitation.	Precipitation detector—A device which undergoes a strong change in electrical impedance if and only if precipitation falls on it.	Binary (yes-no)	G
PRECIPITATION TYPE DETECTOR—Determination of whether precipitation is rain, snow, hail, freezing rain, drizzle, freezing drizzle.	Human observer is the only fully satisfactory detector.	Type of precipitation	
PRECIPITATION RATE—Amount of precipitation per unit time.	Precipitation rate device—Continuous records of a weighing or tipping bucket gage can be used. For liquid precipitation a funnel with a small outlet has been used. The greater the amount of water in the funnel, the higher the rate. Latter device used in England.	0–75 cm	G H
RIVER STAGE—Difference between height of some reference point and height of water surface in the river.	River gage—Measuring stick or height of a float on surface of river is detected mechanically or electrically or depth of water in river changes impedance of a staff standing vertically in the river.	0–20m	H
RUNWAY VISIBILITY—Maximum distance average (normalized) pilot can see runway lights.	Transmissometer—Measures transmisivity (transmission of light flux per unit length) by means of a photomultiplier which measures amount of light received from a known source at a known distance.	200–600 m	Av
RUNWAY VISUAL RANGE (RVR)—	Derived from the transmissometer measurement.		Av
SEA SURFACE TEMPERATURE—Temperature of the upper monomolecular layer of the sea surface.	Usually obtained by an observer dipping a bucket of water from the sea surface and measuring its temperature.	0–30°C	G

Table 2.3-2. (continued)

Parameter Name and Definition	Name of Measuring Instrument and Principle Employed	Range in Units Normally Used	Use Code
	Sometimes condenser water intake line temperature is used.		
SNOW DEPTH (on ground) —Depth of snow cover on the level ground. Average depth in undrifted areas is needed.	Human observer with a measuring stick, or a snow pillow. The snow pillow is a rubber bag containing some air. As the depth of the snow increases the weight of the snow on the bag increases air pressure within it. This pressure is interpreted as snow depth.	0–10 m	G H
SOIL MOISTURE CONTENT—Amount of liquid water in a given volume of soil.	Scale—Soil weighed then dried in oven then weighed again. Nuclear moisture detector— Hydrogen slows down neutrons more than other atoms. Fast neutrons are emitted by a nuclear source. Slow (thermal) neutrons are scattered back. Number returned indicates a mount of water present.	Grams/gram g/m^3	
SOIL TEMPERATURE— Temperature at various depths in the soil.	Soil thermometer—Any one of several kinds of thermometers placed at various depths in the soil.	0–65°C	G
PRECIPITATION TYPE DETECTOR—Determination of whether precipitation is rain snow, hail, freezing rain, drizzle, freezing drizzle.	Human observer is the only fully satisfactory detector thus far. Attempts have been made to develop automated devices.	Type of precipitation.	G Av
SUBLIMATION NUCLEI COUNT—Number per unit volume of nuclei on which water vapor will sublimate.	Nuclei counter—Similar to condensation nuclei counter but cooling across the freezing point of water.	Number/unit volume.	P
VISIBILITY—Greatest distance over more than half the horizon (in not necessarily contiguous sectors) which large objects (or at night low intensity, circa 25 watt, lights) can be seen.	Transmissometer—Derived from transmissivity. On-site observer can determine it also.	0–16 km fractions of km.	Av

Use Code for Secondary Parameters
Av = Aviation
P = Cloud physics
H = Hydrology
G = General public service

can also be effected by adjusting the distance between observation points. A rule of thumb is that for undulatory motion, data points should not be more than $\frac{1}{5}$ wavelength apart. Essentially the same rule holds for lows and highs if the average diameter of circulations is substituted for wavelength.

Averaging measurements over a time period also filters out phenomena of certain scales. In general only large scale phenomena remain after long time averages have been taken. Instrument lag has an effect similar but not identical to a time average. While the conventional "box car" time average gives all data the same weight, the instrument lag weights most recent data more heavily with the effect of older data fading away exponentially. Again, a rough rule of thumb indicates that instrument lag time is about $\frac{1}{3}$ the averaging time.

A critical aspect of a satisfactory sampling process is the avoidance of generating incorrect data by aliasing. The conventional method of computing upper winds from tracking radiosondes creates aliasing errors equal in magnitude to the total instrumental error of the tracking system [6]. The parameter to be measured, the measuring instrument, its exposure, network spacing, and the frequency of sampling must be regarded as a system to obtain valid and useful meteorological measurements.

References

1. Bjerknes, V., J. Bjerknes, H. Solberg, and T. Bergeron, *Physikalische Hydro-dynamik*, Springer, Berlin, 1933.
2. Brunt, David, *Physical and Dynamical Meteorology*, Cambridge University Press, Cambridge, England, 1939.
3. Dergarabedian, P., and F. Fend Fendell, "On Estimation of Maximum Wind Speeds in Tornadoes and Hurricanes," *J. Astron. Sci.* **17**, No. 4, Jan. 1970.
4. Holmboe, J., G. Forsythe, and W. Gustin, *Dynamic Meteorology*, John Wiley and Sons, New York, 1948.
5. Massey, H. S. W., and R. L. F. Boyd, *The Upper Atmosphere*, Hutchinson of London, 1958.
6. Morrissey, E. G., and F. B. Muller, "Spectral Aspects of Upper Wind Measurement Systems," Canadian Meteorological Memoirs, Meteorological Branch Department of Transport, Toronto, 1968.
7. Parry, H. D., and J. Giraytys, "Electronics and Meteorological Data Acquisition Systems—A Forecast for the 1970's" *IEEE Trans. Geosci. Electronics,* **GE-8**, No. 2, April 1970.
8. Parry, H. D., "The Problem of Atmospheric Temperature Measurement," in *Temperature, Its Measurement in Science and Industry,* Charles M. Werzfeld, Ed., Reinhold Publishing Corp., New York, 1962.
9. Petterssen, S., *Weather Analysis and Forecasting,* McGraw-Hill Book Company, New York, 1956.
10. Scherhag, R., *Neue Methoden der Wetteranalyse und Wetter Prognose,* Springer, Berlin, 1948.

William H. Myers

2.4 OCEANOLOGY PARAMETERS

2.4.1 The Hydrosphere

The hydrosphere is the aqueous envelope of the earth including oceans, lakes, streams, underground waters, and the aqueous vapor in the atmosphere. By far the greatest portion of this aqueous envelope consists of sea water. The oceans and seas cover about 71 percent of the earth's surface.

The land, oceans, and atmosphere comprise an extremely complex system of actions and interactions. Energy from the oceans is transferred to the atmosphere, resulting in large- and small-scale atmospheric circulation patterns. Conversely, the winds of the atmosphere provide the energy to produce waves and wind-driven currents. Precipitation that falls on land either as rain or snow eventually reaches the ocean, bringing with it the chemicals leached from the soil. Water evaporated from the ocean surface falls from the atmosphere as precipitation on oceans and the land. Land masses form natural barriers separating the oceans and their marginal seas, deflecting ocean currents, and preventing the migration of many animals from one ocean to another. The oceans, in turn, erode the land and sometimes produce devastating effects on coastal areas.

Geography. When the earth is viewed from space, focusing on Antarctica, the oceans appear continuous, being separated into "gulfs" by the continents. There are three "gulfs": the Atlantic Ocean, the Pacific Ocean, and the Indian Ocean. In addition to the land masses which partially define ocean boundaries, certain arbitrary limits have been established. The Atlantic Ocean, which lies between the continents of North and South America, Europe, and Africa, is generally conceded to include the Arctic Sea. To the south, the Atlantic Ocean is separated from the Pacific Ocean by an arbitrary boundary extending from Cape Horn (South America) along meridian 68°E to Antarctica. To the north, the limit between the Atlantic and Pacific is the Bering Strait. The boundary between the Atlantic and Indian Ocean is on meridian 20°E from the Cape of Good Hope (Africa) to Antarctica. The Indian Ocean is separated from the Pacific Ocean by an arbitrary line running from the Malay Peninsula through Sumatra, Java, Timor, Cape Londonderry (Australia), and Tasmania, thence following meridian 147°E to Antarctica.

The Pacific Ocean. Of the three oceans, including adjacent seas, the

Pacific Ocean is the largest. It is almost as large as the Indian and Atlantic Oceans combined (Pacific area = 180×10^6 km^2; Atlantic area = 107×10^6 km^2; Indian area = 74×10^6 km^2). The greatest depths are found in the Pacific Ocean. A depth of 11,512 m was recorded in 1962 in the Mindanao Trench near the Philippines. Other significant depths in trenches in the Pacific Ocean include: the Challenger Deep (10,859 m) in the Marianas Trench off Guam, a depth of 11,030 m near the Challenger Deep, and a depth of 10,628 m in the Tonga Trench, south of the Samoan Islands. Of the three oceans, the Pacific Ocean also has the greatest mean depth (4282 m). River drainage into the Pacific Ocean is considerably less than into the Atlantic and Indian Oceans, primarily because the large rivers drain into the Pacific Ocean's peripheral seas. The Pacific Ocean is consequently more oceanic in nature than the others. The land area drained by the Pacific Ocean is 18×10^6 km^2. A large number of islands is another characteristic of the Pacific Ocean.

The Atlantic Ocean. In contrast to the Pacific Ocean, the Atlantic Ocean is a rather narrow body of water contorted into somewhat of an S-shape by the continental margins surrounding it. The land area draining into the Atlantic Ocean (including the adjacent seas—Arctic, Mediterranean, and Black) has been estimated to be 67×10^6 km^2. This is about four times the area drained by the Indian and Pacific Oceans. Excluding continental-type islands (for example, the British Isles, the Canadian Arctic Archipelago, and the Falkland Islands), there are relatively few oceanic islands in the Atlantic Ocean. Of the three oceans, the Atlantic Ocean has the least mean depth (3926 m).

The Indian Ocean. Of the three oceans, the Indian Ocean is the smallest. Unlike the Pacific Ocean, the Indian Ocean has few islands and, unlike the Atlantic Ocean, has few large adjacent seas. The land area drained by the Indian Ocean is, however, about equal (17×10^6 km^2) to that of the Pacific Ocean. The mean depth of the Indian Ocean is 3963 m. Surface conditions in the Indian Ocean vary greatly during the year, primarily because of the effect of the monsoons.

2.4.2   The Oceanic Lithosphere

The Continental Shelf. The Continental Shelf is the shallow water platform adjacent to the continents (or some large islands), sloping gently to a point where a marked increase of slope to greater depths is noted. The shelf can vary from zero km, as in certain parts off the Pacific coast of the United States and the west coast of South America, to 1290 km, as on the northern coast of Siberia where the shelves are the widest in the world. The average width of all the shelves is about 71 km and the average depth of termination approximately 130 m. The average slope angle is 0°7′. Although the shelves slope gently seaward, they do not necessarily have an even-graded profile. There may be terraces, ridges, hills, depressions, and in some cases, deep depressions or troughs cutting across them. The shelves are important as fishing grounds and as a source of oil and the minerals found on and beneath their floors. The most common sediments found on the shelves are sand, mud, gravel, pebbles, cobbles, and boulders. In addition, there are seaweed and other marine plants and the sponges which grow on the shelf in tropical areas.

The Continental Slope. At a depth of about 130 m bottom slopes become much steeper. This boundary area, referred to as the Continental Shelf break (or shelf break) marks the beginning of the Continental Slope. The average gradient for the upper 2000 m of the slope is about 4.25° ; however, there is great regional variation. For example, slopes off major deltas average about 1.3° for the upper 2000 m, those off "fault coasts" with insignificant shelves average 5.6°, those off young mountain range coasts average 2.6°, and those off stable coasts lacking major rivers average about 3°. In the three major oceans, slopes are steepest in the Pacific Ocean and least steep in the Indian Ocean. Muds are the predominant sediments of Continental Slopes with lesser amounts of sand and gravel.

The Ocean Floor. Topographically, the deep ocean floor can be categorized by two regions: the oceanic ridges and rises and the abyssal floor comprised of the abyssal plains and abyssal hills. The most spectacular feature of the ocean floor is the mid-ocean ridge, a continuous feature extending through the Atlantic, Indian, Antarctic, and South Pacific Oceans; the Norwegian Sea; and the Arctic Basin for a distance of over 56,000 km. The elevation of the ridge is about 1 to 3 km above the adjacent ocean floor, and, in most places, the ridge is more than 1,500 km wide. The first abrupt gradient change between the abyssal hills and the ridge defines lateral boundaries. The sediments on the higher elevations of the ridge, at a depth of less than 4.5 km, are predominantly carbonate oozes, fine deposits composed primarily of the shells and debris of marine organisms. At depths greater than 4.5 km, red clay is the major sediment type. Sediment types less frequently encountered are siliceous ooze and residual sand.

Abyssal hills are small hills rising from the ocean floor. They vary in height between 90 and 365 m and width from 3 to 10 km. In the North Atlantic, the abyssal hills form two strips parallel to the Mid-Atlantic Ridge along almost its entire length.

Abutting the abyssal hills are the abyssal plains extending seaward from the base of the continental rise. The abyssal plains are extremely flat portions of the ocean bottom. The sediments of the abyssal plains vary according to the source area. They are generally derived from turbidity currents descending from the continental margin through submarine canyons. In the upper

15 m, silt, sand, and gravel comprise from 2 to 90 percent of the sediments. The remaining sediments are clay-sized material derived from turbidity currents and normal pelagic sedimentation.

Other topographic features of the deep ocean are: seamounts, which are isolated circular or elliptical elevations with rather steep slopes; guyots or tablemounts, which are seamounts with flat tops; and trenches, which are long and narrow depressions in the ocean floor that attain the greatest depths of the oceans.

2.4.3 Physical/Chemical Characteristics

Temperature. Temperature is the most widely, directly sensed variable of the oceans. Consequently, it is the one oceanographic parameter most widely understood and for which we have the greatest abundance of data. By far the greatest concentration of observations is for surface and near-surface waters where the major fluctuations in temperature occur. However, recent advances in electronic oceanographic instrumentation, such as the expendable bathythermograph (XBT) and the salinity-temperature-depth recorder (STD), are contributing greatly to our knowledge of temperature perturbations in the entire water column.

In the upper layers of the oceans, heat is gained primarily from solar and sky radiation and is lost by radiation from the sea surface. Fluctuations in the temperature of surface waters are cyclic, generally being influenced by seasonal and diurnal effects. Seasonal effects are most notable in temperate latitudes. In winter, there is a deep mixed (isothermal) layer overlying a weak thermal gradient. During spring, weak gradients develop at varying depths. By summer, a strong shallow gradient develops. When fall cooling begins, surface temperatures decrease, the mixed layer becomes progressively deeper, and the *thermocline* (layer of great temperature change with depth) becomes sharp. Diurnal temperature variations occur generally in the upper 10 m of the ocean, but they are small, usually less than 0.4°C, and, at the most, about 1°C in calm and fair weather. Diurnal heating is more pronounced at central latitudes during the summer. During the day, with solar heating, negative gradients appear and reach their maximum by about 3:00 p.m. As the surface layer starts to cool, a shallow isothermal layer overlies a weak negative gradient. With continued cooling the isothermal layer deepens, and by midnight the upper layer is again isothermal.

Average surface temperatures in the oceans vary annually from 30°C in tropical and subtropical areas to − 1.7°C near the poles. The ocean surface in the Northern Hemisphere is somewhat warmer than the surface in the Southern Hemisphere, possibly because of the greater abundance of land in the former area.

Chemistry. Of all the chemical properties of sea water, salinity is the most universally measured and widely used in studies of various oceanographic processes. Salinity is defined as the total amount of solid material in grams contained in 1 kg of sea water when all the carbonate has been converted to oxide, all bromine and iodine replaced by chlorine, and all organic matter completely oxidized. The major salt constituents (comprising 99.95% of the total salt) of sea water are given in Table 2.4-1.

Surface salinity is greater in the Atlantic, particularly in the North Atlantic, than in the Pacific. Based on 10° latitudinal bands, average surface salinity in the Atlantic Ocean varies from a minimum of 30.50‰ between 80 and 90°N to a maximum of 36.75‰ between 20 and 30°N. In the Pacific Ocean it varies from 31.00‰ between 60 and 70°N to 35.66‰ between 20 and 30°S. Surface salinity extremes occur in coastal regions and some partially enclosed seas. For example, in the Gulf of Bothnia salinity may be as low as 5.00‰; in the Red Sea it may exceed 40.00‰.

At a depth of approximately 700–800 m an intermediate salinity minimum exists in the three major oceans. This salinity minimum, known as the Sub-Antarctic Intermediate Water, originates from surface water in the region south of 45°S. In this region, water masses of low salinity and low temperature, and consequently of relatively high density, sink along the Antarctic Convergence. The salinity minimum due to this Sub-Antarctic Water extends over the entire breadth of the Atlantic Ocean between 45°S and 20°N and over the Pacific and Indian Oceans between 45°S and the Equator. In most ocean regions, salinity increases with depth below the intermediate salinity minimum and forms a salinity maximum between 1500 and 4000 m. In the Atlantic Ocean, subsurface outflow of relatively high saline waters from the Mediterranean strongly influences the upper portion of this water mass. Below 4000 m in the three major oceans, a water type of Antarctic origin with slightly lower salinity can be traced.

Table 2.4-1. Major Salt Constituents of Sea Water [4]

Values in Grams/Kilogram[a]	
Component	Concentration
Cl^-	18.980
Na^+	10.543
SO_4^{--}	2.465
Mg^{++}	1.272
Ca^{++}	0.400
K^+	0.380
HCO_3^-	0.140
Br^-	0.065
H_3BO_3	0.024

[a] Referred to a chlorinity of 19‰.

Practically all the elements in the periodic table are found in sea water. They may be arbitrarily divided into two categories: those of intermediate concentration and trace elements. The former is comprised of nine elements: strontium, oxygen, silicon, fluorine, nitrogen, argon, lithium, phosphorous, and iodine; of the latter, 43 have been determined to exist in sea water. The elements oxygen, nitrogen, phosphorous, and silicon in their various forms have been studied most because of their importance to living resources of the oceans.

Density. The density of sea water depends on three variables: temperature, salinity, and pressure. In surface waters, density is decreased by heating, precipitation, melting of ice, or runoff from land; it is increased by cooling, sea ice formation, and evaporation. When the density of surface water increases beyond that of the underlying layers, vertical convection currents penetrate to greater and greater depths until the density achieves uniformity from top to bottom. Continued increases in density of surface waters lead to an accumulation of densest water near the bottom. Continuation of this process in an area that is in free communication with other areas causes this great-density water to spread to other regions. If a layer of deep or bottom water of greater density is already present, the sinking water spreads at a level above this layer. The temperature of the surface water in lower and middle latitudes is so high that the density remains low, even in regions where excess evaporation causes high salinities. In these areas, convection currents are limited to a rather thin layer near the surface and do not contribute to the formation of deep or bottom water. The most dense water is usually formed in high latitudes. Because this water sinks and fills all ocean basins, the deep and bottom water of all oceans is cold. However, in a few isolated basins in middle latitudes relatively warm deep and bottom waters are encountered. For example, a temperature of 59°C was recently reported in the Persian Gulf at 38°E-21°N; however, this high temperature may be caused by a through-bottom heat flux. Ascending motions can exist anywhere in the ocean, but they are particularly conspicuous along the western coasts of continents where prevailing winds carry surface waters away from the coasts. In these areas, the upwelling of subsurface water brings water of greater density to the surface.

2.4.4 Ocean Circulation

Oceanic circulation involves systems of horizontal circulation gyres with intense flow at and near the sea surface coupled with slow circulations of deep water with opposing flows at different levels joined by vertical motions.

The Atlantic Ocean current regime consists of two major gyres: the North Equatorial Current and the South Equatorial Current. The North Equatorial Current, flowing westward between latitudes 10 and 20°N is joined by a branch of the South Equatorial Current, and part of the combined flow enters the Caribbean Sea. This flow passes into the Straits of Florida as the Florida Current. It continues as an intense jet giving rise to the Gulf Stream. The Gulf Stream flows northward along the coast of the United States, but is deflected eastward at about Cape Hatteras. East of Cape Hatteras the lateral boundaries of the Gulf Stream are difficult to define, since they are characterized by many meanders and eddies. It passes the Grand Banks at about latitude 40°N and becomes part of the complex set of currents known as the North Atlantic Current. The Gulf Stream is the greatest of the permanent ocean currents. Off New England, it probably is about 5 km deep and has a transport volume well in excess of 100×10^6 m^3/s.

Portions of the North Atlantic Current flow toward the Arctic along the coast of Norway as the Norwegian Current and along Iceland as the Irminger Current. Outflow from the Arctic Sea runs along Greenland as the East Greenland Current. The East Greenland Current, with waters from the Irminger Current and the North Atlantic Current, rounds Greenland, flowing northward into the Labrador Sea and Baffin Bay. The outward flow from the Labrador Sea constitutes the Labrador Current, which flows southward by the coast of Labrador as a countercurrent along the westward edge of the Gulf Stream.

The South Equatorial Current, flowing westward between 20°S and a few degrees north of the Equator, divides near South America. One branch joins the North Equatorial Current; the other continues as the Brazil Current flowing from Cape San Roque southward. Off the estuary of the La Plata, the eastern part of the Brazil Current turns southeastward, working in with the Falkland Current flowing from the southwest. Near the coast, the Falkland Current deflects the Brazil Current to the east. A sharp convergence line forms between the opposing currents somewhat in the manner of the relation of the Gulf Stream to the Labrador Current.

In the Pacific Ocean there also exists a North Equatorial Current and a South Equatorial Current, both occurring at approximately the same latitudes as their counterparts in the Atlantic Ocean. Because of the great width of the Pacific Ocean, the currents flow predominantly east-west. The North Equatorial Current, flowing from east to west, divides north of Mindanao. One branch flows northward to become the Kuroshio; the other turns sharply southward into the Equatorial Countercurrent. The Kuroshio Current is the most remarkable of the currents in the North Pacific Ocean. Like the Gulf Stream, it is very narrow, being on the average about 80 kilometers wide, and attains high surface velocities. Near Japan the Kuroshio volume transport is between 40 and 50×10^6 m^3/s. The Oyashio Current flowing southwest to south-southwest is the Pacific counterpart of the Atlantic Ocean's Labrador Current. The boundary between the Kuroshio and Oyashio is a convergence region consisting of numerous vortices. The Oyashio derives its cold waters from the Sea of

Okhotsk and, in part, from the Bering Sea. The Alaska Current is a well-developed current flowing along the Aleutians and extending into the southern Bering Sea. In the eastern part of the North Pacific Ocean, the California Current flows southward from the coast of California.

The South Equatorial Current of the Pacific Ocean, occupying about the same latitudes as the Atlantic's South Equatorial Current, is the source of the East Australian Current which flows southward adjacent to Australia. The major current regime in the eastern part of the South Pacific Ocean is the Peru, or Humboldt, Current flowing northward along the west coast of South America. The volume transport of the East Australian Current is estimated at 10–25×10^6 m^3/s, that of the Peru Current about 15–20×10^6 m^3/s.

The circulation of the North Indian Ocean varies with monsoon winds. During the northern summer, prevailing winds blowing from the ocean onto Asia create the strong, narrow Somali Current flowing northeastward along the African coast. When the wind blows off Asia onto the ocean, as in the northern winter, the Somali Current disappears and a fairly broad, weak southwestward flow develops. In the South Indian Ocean north of 20°S, a South Equatorial Current flows westward, bends south as it approaches Africa, and becomes the Agulhas Current. It flows between Madagascar and the continent with a volume transport of about 20×10^6 m^3/s.

The Antarctic Circumpolar Current flows around Antarctica in an almost unrestricted sense due to the lack of continental obstructions. However, because of some channels and island arcs, much of the current is deflected northward just after passing Cape Horn, becoming the Falkland Current.

The Atlantic, Pacific, and Indian Oceans have Equatorial Countercurrents embedded in the system of North and South Equatorial Currents. However, in the Indian Ocean, the countercurrent occurs only during the northern winter.

2.4.5 Waves and Tides

The surface of the oceans is seldom still. It rises and falls primarily in response to gravitational forces, earth tremors, and wind. These physical forces give rise to wave motions such as tides (by gravitational attraction), tsunamis (by submarine earthquakes), and sea and swell (by winds). The most commonly occurring wave motions are the tides and wind-induced waves.

Unlike many physical characteristics of the oceans, tides are remarkably predictable. They are controlled by the physics of the solar system, particularly the sun-earth-moon relationship. The moon has the most dominant effect on tides because of its proximity to the earth, but the sun, despite its great distance from the earth, has an appreciable effect because of its mass. At the times of new and full moon, the moon's gravitational pull is aligned with that of the sun, and the two bodies acting together produce the greatest vertical range of tide. These tides are referred to as spring tides. Near first- and third-quarter moon, neap tides, which have the smallest vertical range, occur. Generally, spring and neap tides occur at intervals of about 14 days.

Three types of tides are usually distinguished: diurnal tides, semidiurnal tides, and mixed tides. The diurnal tide, common in parts of the northern Gulf of Mexico and southeast Asia, has one high and one low water per tidal day. The semidiurnal tide, common on the Atlantic coasts of the United States and Europe, has two high waters and two low waters per tidal day. The mixed tide, most common along the Pacific coast of the United States, is characterized by appreciably differing successive high and low water stands. The mixed tides may be predominantly diurnal or semidiurnal.

The daily tide range along the shores of continents contiguous with major oceans is usually 1.2–3.0 m. There are, however, some notable tides. In the Bay of Fundy, between Nova Scotia and the Canadian mainland, the tide ranges up to 18.3 m. On the coast of France in the inner part of the Bay of St. Malo, spring tides range up to 12 m. On the south shoreline of the North Sea, along the Pacific continental coast of Alaska, and in northwestern Australia, 6.1 m ranges are common.

Wind-induced waves are the most familiar type of ocean wave. The winds blowing across the ocean surface impart energy, and the dimensions of the resulting waves are a function of velocity and duration of the winds combined with the fetch (the distance the wind has blown over the water). In the area where waves are being actively generated by the wind, a complex pattern of the ocean surface, commonly referred to as *sea*, is formed. The resulting surface is an irregular superposition of short waves with a broad range of frequencies, amplitudes, and propagation directions. The waves grow with increasing wind velocity to the maximum that winds of a particular velocity can generate. For example, in the North Atlantic Ocean, shipboard wave records have shown that a force-4 wind produces waves of 6-s period (time elapsed between successive crests or troughs) and 0.61 m height (vertical distance from crest to trough); wind force 6 produces 7-s, 1.52 m waves; and wind force 8 produces 11-s, nearly 6.1 m waves.

As waves travel from the generating area, they lose energy, decrease in height, and increase in period. The shorter waves become so reduced in height that they disappear, but the longer waves decrease slowly and travel over long distances as *swell*. Waves generated in the South Pacific have been recorded on the coast of California, about 11,000 km away.

The largest and most destructive waves are those caused by earthquakes or submarine slumping. These waves are called seismic sea waves or tsunamis. Seismic shock produces waves that, in the deep ocean, can have lengths of up to 200 km, periods of 10 to 20 min, and wave heights of up to 0.5 m. Generally, they are unnoticed by ships on the open sea, but they may form huge destructive breakers when they encounter certain bottom configurations in shallow water. The Japanese and Hawaiian Islands are especially prone to these waves which cause extensive damage to life and property.

Subsurface waves, known as internal waves, are found between layers of different density or within layers where vertical density gradients are present. Internal waves may be several hundred feet high, as in deep ocean waters, or about 6–15 m in height in the main thermocline. They are important factors in water mixing and transport.

2.4.6 Biology

The living resources of the oceans affect man both positively and negatively —positively by providing him with a source of food, drugs, and other useful items; negatively by thwarting some of his efforts as he thrusts seaward. Of primary concern to those who work in the marine environment are the fouling, boring, and predatory marine organisms. Although many protective measures have been developed to counteract or minimize their threats, they still exert a sizable influence on man's activities in the sea.

Fouling generally occurs in the shallow marine areas, and certain instruments that function well in deep ocean waters may have their sensitivities lessened or may be made completely inaccurate with the accumulation of marine growth. Fouling is known to have an acoustic dampening effect on underwater transducers, often reducing their sensitivity by as much as 10 dB. Electrical conductivity cells are sometimes desensitized by fouling. Protective coatings intended to reduce corrosion are often damaged by fouling. Some organisms increase the corrosion rate of unprotected metal by creating oxygen concentration cells at points of adhesion. In one case, a penetration of 1.6 mm occurred in stainless steel after 111 days of marine exposure.

Marine borers are not as much of a hazard, since they primarily attack wooden structures. However, some boring organisms attack concrete structures, and some have been known to attack lead cable sheath. Some of the larger fish, especially sharks, have attacked instruments, bitten and severed mooring lines, and either attacked or collided with submersible vehicles.

References

1. Defant, Albert, *Physical Oceanography*, Vol. I, Pergamon Press, New York, 1961.
2. DePalma, John R., "Marine Fouling Research, A State-of-the-Art Report," U. S. Naval Oceanographic Office Informal Manuscript Report No. 0-1-64 (unpublished).

3. Dubach, H. W. and R. W. Taber, *Questions About the Oceans,* National Oceanographic Data Center, Publication G-13, Washington, D. C.
4. Fairbridge, Rhodes W., Ed., *The Encyclopedia of Oceanography,* Rheinhold Publishing Corp., New York, 1966. Particularly articles by R. W. Fairbridge, K. Hidaka, D. W. Hood, T. Ichiye, E. C. LaFond, J. G. Pattullo, F. A. Richards, F. B. Shepard, W. G. Van Dorn, B. Warren, and W. S. Wooster.
5. Gross, M. Grant, *Oceanography,* Charles E. Merrill, Columbus, Ohio, 1967.
6. Hull, E. W. Seabrook, Ed., *Ocean Science News,* Volume 13, No. 21, Nautilus Press, May 21, 1971.
7. Moskovits, George, "Rock Borers and Their Relation to Concrete Structures in the Gulf of Mexico," *Offshore* **3,** No. 1, September 1955.
8. National Oceanographic Data Center, Report of the Joint NODC/NOIC Workshop on Subsurface Current Measurements, August 1970 (unpublished).
9. Pierson, Willard J., Jr., "Ocean Waves," in *Science and the Sea,* Vol. II, U. S. Naval Oceanographic Office, Washington, D. C., 1970.
10. Snoke, L. R. and A. P. Richards, "Marine Borer Attack on Lead Cable Sheath," *Science* **124,** September 7, 1956.
11. Sverdrup, H. V., M. W. Johnson, and R. H. Fleming, *The Oceans,* Prentice-Hall, Englewood Cliffs, New Jersey, 1965.
12. Van de Ree, Andries, "Geological Oceanography," in *Science and the Sea,* Vol. II, U. S. Naval Oceanographic Office, Washington, D. C., 1970.

Enrico P. Mercanti

2.5 SPACE SCIENCE PARAMETERS

2.5.1 The Atmosphere

Space environment studies distinguish between the earth as a planet, other parts of the solar system, and the space that lies between. There are no clear lines (surfaces) of demarcation even though forces, energy and mass are transported through the surfaces. The combined effects of the gravitational pull of the earth on the constituents of the atmosphere and a temperature inversion which occurs at an altitude of 80–100 km in the mesosphere provide a boundary between the atmosphere included in meteorological studies and the adjacent space environment.

The lower part of the atmosphere, called the troposphere, which extends to an altitude of about 12 km, is generally characterized by a decrease of temperature of 6 to 7° per kilometer. The principal mechanisms of energy transfer in the troposphere are radiation and convection. The tropopause is that altitude where the temperature decrease with altitude becomes significantly less.

With the tropopause the dividing line, the region above the troposphere is referred to as the stratosphere and extends to an altitude of approximately 50 km. The stratosphere is characterized by an abundance of ozone and absence of water vapor, and provides a temperature inversion which helps the gravitational attraction of the earth keep atmospheric constituents in the immediate vicinity of the surface of the earth.

The mesosphere region commences at an altitude of about 50 kilometers and extends to an altitude of 80 km. In this region an even sharper temperature drop occurs. Because it is adjacent to the higher-temperature, lower-density space environment and the atmosphere of earth, the mesosphere is generally considered thermally unstable.

Beyond the mesosphere, the thermosphere extends to an altitude of approximately 700 km. It is characterized by an increase in temperature to approximately 1500°K. At the higher altitudes (exosphere and magnetosphere) the temperature remains essentially constant, varying slightly because of solar influences.

In the exosphere the predominant constituent is helium at lower altitudes and hydrogen at higher altitudes.

The study of the region above the mesosphere is divided into several subdisciplines including micrometeroids and cosmic dust; magnetic fields; solar plasma; solar radiation in the visible, ultraviolet, and gamma regions; emission of solar particles including hydrogen, helium, and traces of metals such as beryllium and lithium; trapped radiation including high-speed protons and electrons in the inner and outer zones; man-made radiation caused by events on earth or in the atmosphere such as explosions of nuclear devices; cosmic rays; ionospheric properties; ring currents; and hydromagnetic waves.

2.5.2 The Ionosphere

The lower part of space referred to as the ionosphere begins at the upper boundary of the stratosphere at an altitude of approximately 50 km and extends to the boundary of the magnetosphere at an altitude of about 500 km. The ionosphere is divided into five regions; the D region from 50 to 90 km, the E region from 90 to 120–140 km, the F1 region and the F2 region, from 140 to about 1000 km, and the protonosphere from about 1000 km to a donut shaped outer boundary about 2000 km above the earth at the equator. The regions are defined in terms of the processes by which they are formed and the density and gradients of electrons or species of ions they contain. The D, E, and F1 regions are composed mainly of oxygen and NO^+ molecular ions. The F2 region is primarily comprised of the O^+ atomic ion. The protonosphere contains primarily positive hydrogen ions. The F1 region tends to appear in the daytime when the solar zenith angle is small. It is influenced by changes in solar activity, including solar eclipses, and by ionospheric storms. The F2 layer is usually the most dense ionosphere region with 10^4 to almost 10^7 electrons per cubic centimeter. Its maximum concentrations vary greatly from day to night and variations between summer and winter have also been noted. The protonosphere is that part of the earth's ionosphere in which ions of atomic hydrogen predominate. It extends from the top of the F-Region to a magnetic field boundary aligned within the magnetosphere.

Micrometeoroids are considerably smaller than the sun, the planets, the satellites, the comets, and the larger asteroids. Meteoroids are larger than micrometeoroids and smaller then these larger bodies. Measurements of this environment from numerous satellites, indicate that the number of micrometeoroids in space is low compared with prior measurements and theoretical analyses.

2.5.3 The Magnetosphere

The geomagnetic field has long been used as an aid to navigation and has served mankind by shielding it from some of the energetic charged particles or cosmic rays that bombard earth from the sun and from outer space. Except at the poles its deflecting force repels all but the most powerful of the charged particles. The field also acts as a trap holding protons and electrons circling about the field lines. The field, in a gross time sense, is constant, but sensitive measurements reveal that it is constantly changing in magnitude and direction. There are slow variations with a period of years, daily variations, and rapid variations lasting minutes or seconds. Most of these changes are quite small equal to a fraction of a percent of the surface field where the magnitude is about 30,000 gammas. Many speculative theories have been constructed to explain the existence of the main field. A number of scientists have attempted to attribute the existence of the field to some characteristic of the interior of the earth.

The magnetosphere is the region above about 500 km where magnetic field lines connect back to earth. The magnetotail is part of the magnetosphere. The field is distorted by the solar wind to a progressively greater degree at increasing vertical distances from the surface of the earth. The boundary of the magnetosphere, the magnetopause, is of the order of 100 km thick on the side of the magnetosphere toward the sun. Magnetic field lines are dragged out by the solar winds into a very long magnetotail known to extend well beyond the moon.

Beyond the magnetopause is a hydromagnetic bowshock characterized by sudden changes in the magnetic field, charged particle density, and particle temperature. The region between the shock and the magnetopause is the magnetosheath. Beyond the shock is the interplanetary medium primarily controlled by the sun rather than by earth phenomena.

Other planets and the sun have magnetic fields that slightly influence the space environment in the vicinity of earth. The exact mechanism of the sun's magnetic field is as little known as the origin of the earth's magnetic field.

The interplanetary magnetic field magnitude ranges between 1 and 10 gamma.

2.5.4 Interplanetary Space

Tremendous clouds of ionized hydrogen are emitted by the sun both on a steady-state basis and with increased intensity during solar storms. These clouds often contain high-speed particles having velocities of the order of hundreds of kilometers per second and are sometimes referred to as solar winds or solar plasma. The flow of plasma from the sun is detectable at all times. The plasma number density observed from one experiment was between 5 and 10 ions per cubic centimeter. The thermal energy of the ions was small compared to kinetic energies and the energy density always exceeds the locally measured magnetic field energy density.

The plasma contributes to an energy flux that is absorbed in the upper atmosphere causing a variation in the density of the upper atmosphere. Thus, the upper atmosphere is influenced by solar cycles and other solar phenomena.

Van Allen radiation is composed of naturally occurring energetic particles trapped by the earth's geomagnetic field. Energetic particles are called "magnetic" relative to the distribution of ambient particles. While electrons in the solar wind have not been measured, it is generally assumed that they are swept along through interplanetary space with the same velocity as protons. The dominant constituent in the inner zone of the Van Allen radiation belt is the energetic proton, although the sun also frequently generates fluxes of lower energy protons, alpha particles, and heavier nuclei. The production of these lower-energy solar cosmic rays is usually associated with a large flare on the visible side of the sun and usually accompanied by high radio noise emission. However, not all large flares result in a detectable increase in cosmic rays on the earth. The number of low-energy solar cosmic-ray events per year tends to follow the number of sun spots but there are departures from this tendency.

Galactic cosmic rays consist mainly of hydrogen nuclei protons, with helium nuclei alpha particles the second most abundant species. The elements lithium, beryllium, and boron, which occur in low concentrations on the sun and other stars, are by some reports also present in cosmic radiation. The galactic cosmic ray flux has not varied much over long periods of time. Investigations on micrometeorites give some evidence that the average flux has remained essentially constant over the last billion years. The short-term fluctuations generally attributable to solar activity have a negligible effect on the long-term average.

A Forbush decrease is a world-wide reduction of a few days duration in galactic cosmic-ray intensity. Decreases of intensity ranging from a few percent to more than 30 percent have been observed. Forbush decreases are generally associated with magnetic storms believed to be caused by an ionized solar gas cloud ejected from the sun which envelops the earth. The cloud carries a magnetic field with it which effectively shields the earth from primary cosmic rays. Irregularities in the magnetic field deflect these rays.

There is also an 11-yr cycle in the flux of primary particles incident on the atmosphere. In addition to the sun, stars, stellar explosions, extra-galactic radio noises, and other sources generate energetic particles.

2.5.4 Planetary Space

Most theories relate the composition of planets and their atmospheres to the sun. Very little is known about the interiors of planets except the earth, Jupiter, and Saturn. Based on mean density measurements, Jupiter and Saturn are believed to be principally composed of hydrogen. Seismic-wave measurements of the earth indicate an increase in density with depth with a sharp increase at about 3000km. Seismic waves do not propagate at lower depths than this, giving rise to the theory of a molten core, possibly of liquid iron. Mercury has approximately the same density as the earth; that of the moon is about 60 percent that of earth. More information on the planets is given in Section 2.6.

More, of course, is known about the atmospheres of planets, but much remains to be learned. The side of Mercury that always faces the sun exhibits a temperature of 600–700°K. At this temperature most lighter gases would escape; however, light polarization and other studies indicate the likelihood of an atmosphere principally composed of argon produced by a radioactive decay of potassium.

The existence of an atmosphere on Mercury is uncertain. The characteristics of the Venusian atmosphere have proven difficult to define precisely. Because the planet is covered with clouds, the only constituent definitely identified has been CO_2 and there is some evidence of the existence of CO and nitrogen. Little is known of the cloud cover itself and the characteristics of the atmosphere below the cloud layer.

Mars, on the other hand, has a clearer atmosphere and has been studied more fruitfully, especially in the field of radiation. CO_2 is present in the atmosphere. Infrared measurements of the polar caps confirm that they are composed of ice, and other measurements show the presence of slight amounts of water vapor.

Jupiter is believed to contain ammonia, hydrogen, and carbon with the last two also observed on Saturn, Uranus, and the sun. The other planets have exhibited methane and hydrogen via spectroscopic analysis.

References

1. Corliss, W. R., *Scientific Satellites*, National Aeronautics and Space Administration, Washington, D. C., 1967.
2. Corliss, W. R., *Space Probes and Planetary Exploration*, D. Van Nostrand, Princeton, New Jersey, 1965.
3. Johnson, F. S., *Satellite Environment Handbook*, Stanford University Press, Stanford, California, 1961.
4. Kendrick, I. B., Ed. *TRW Space Data*, TRW Systems Group, Redondo Beach, California, 1967.
5. LeGalley, D. P., and A. Rosen, *Space Physics*, John Wiley and Sons, New York, 1964.
6. Stecker, F. W., *Cosmic Gamma Rays*, National Aeronautics and Space Administration, Washington, D. C., 1971.
7. White, R. S., *Space Physics*, Gordon and Breach, Science Publishers, New York, 1970.
8. Wolff, E. A., *Spacecraft Technology*, Spartan Books, Washington, D. C., 1962.

Michael J. S. Belton

2.6 PLANETARY PARAMETERS

A careful analysis of the planetary data below, Tables 2.6-1 to 2.6-5, may reveal small numerical inconsistencies because the different sources of information often disagree as to the values of fundamental parameters and constants.

2.6.1 The Terrestrial Planets

Mercury. Mercury is similar to the earth's moon in many respects. This is particularly true for photometric and polarimetric properties and probably indicates that their surface microstructures are similar. This similarity also extends to the character of the observed surface markings of which a crude map has been prepared [16]. The basic composition of the planet must be different from that of the moon and the rest of the terrestrial planets because of its relatively high mean density, which is probably due to a higher proportion of iron in its composition. There is no direct evidence of an atmosphere on Mercury, nor of a magnetic field. Mercury has no known satellites.

Venus. Venus is surrounded by a massive CO_2 atmosphere which supports dense clouds (Table 2.6-6). The planetary rotation is retrograde and slow. Venus has no known satellites, nor has any intrinsic magnetic field been detected (Table 2.6-7).

Mars. Mars has the highest probability for indigenous life. The planet has variable polar caps and a thin CO_2 atmosphere which exhibits a highly varied range of meteorological phenomena (Table 2.6-8). The planet has two natural satellites.

2.6.2 The Giant Planets

Jupiter. Jupiter is the largest, most massive planet in the solar system and is the easiest of the outer, giant planets to observe. It is composed primarily of hydrogen and probably has a basic atomic composition similar to that of the sun (Table 2.6-10). It has a very strong dipole magnetic field and well-developed radiation belts (Table 2.6-11). The internal structure is probably fluid throughout. In some manner, probably conversion of gravitational potential energy, it is producing prodigious amounts of internal energy. The atmospheric structure is not well understood. Observations of the β Sco occultation by Jupiter have provided a first detailed look at the structure of the upper stratosphere.

Saturn. Because of its rings Saturn is one of the most intriguing objects in the solar system (Table 2.6-12). The presence of a substantial internal source of energy is strongly suspected although not firmly established. There is no evidence at present which indicates that the planet possesses a magnetic field or radiation belts. The former is anticipated on grounds of the planet's otherwise strong resemblance to Jupiter. The lack of nonthermal radio emission from well-developed radiation belts may be connected with the presence of the rings.

Uranus and Neptune. Uranus and Neptune are similar in size, composition, and general physical properties. They nevertheless are markedly different in terms of their rotational properties and also in their attendant satellite systems. Only hydrogen and methane lines have been observed on these planets

Table 2.6-1. Orbital Parameters of Planets [1, 3, 21, 55]

Planet		Mercury	Venus	Mars
Epoch		1900 Jan. 0.5 ET	1900 Jan. 0.5 ET	1900 Jan. 0.5 ET
Semi-major axis (AU)	(a)	0.3871	0.7233	1.5237
Eccentricity	(e)	0.2056	0.0068	0.09331
Inclination	(i)	$7°.0029$	$3°.3937$	$1°.8503$
Longitude of ascending node	(Ω)	$47°\ 08'\ 45''.4$	$75°.7806$	$48°.786$
Longitude of perihelion	($\tilde{\omega}$)	$75°\ 53'\ 58''.9$	$130°.1642$	$334°.2182$
Mean anomaly at epoch		$102°\ 0.279$	$324°.7671$	$319°.5294$
Mean longitude at epoch		—	—	—
Mean daily motion		$4°.0923$	$1°.6021$	$0°.5240$
Sidereal period		87.969 days	224.701 days	686.98 days
Synodic period		115.88 days	583.92 days	779.94 days
Mean orbital speed (km s^{-1})		47.9	35.1	24.1
Minimum distance to sun (AU)		0.3075	0.7184	1.3815
Maximum distance from sun (AU)		0.4667	0.7282	1.6660
de/dt (century^{-1})		2.046×10^{-3}	4.769×10^{-5}	9.4×10^{-5}
di/dt (century^{-1})		$6''.699$	$1°.000 \times 10^{-3}$	$2''3$
dΩ/dt (century^{-1})		$1°.1852$	$0°.9000$	$2''.786 \times 10^3$
d$\tilde{\omega}$/dt (century^{-1})		$1°.5555$	$1°.4064$	$6''.626 \times 10^3$

Jupiter	Saturn	Uranus	Neptune	Pluto
1960 Jan. 0.0 ET	1950 Jan. 0.5 ET	1950 Jan. 0.5 UT	1950 Jan. 0.5 UT	1950 Jan. 0.5 UT
5.203	9.539	19.1818	30.0579	39.44
0.0483	0.05589	0.04724	0.00858	0.2494
1° 18′ 31″.45	2° 29′ 33″.1	46′ 21″	1° 46′ 45″	17° 10′
99° 26′ 36″.19	112° 47′ 10″	73° 28′ 50″	130° 40′ 50″	109° 00′
12° 43′ 15″.34	91° 05′	169° 03′	43° 52′	224°
—	—	—	—	—
238° 2′ 57″.32	158° 18′ 13″	98° 18′ 31″	194° 57′ 08″	165° 36′ 09″
4″.988	2″.008	42″.234	21″.53	14″.29
11.862 yr.	29.45772 yr.	84.013 yr.	164.79 yr.	248.4 yr.
1.092 yr.	1.035 yr.	369.66 days	367.49 days	366.74 days
13.1	9.7	6.80	5.43	4.74
4.9508	9.006	18.275	29.800	49.28
5.4148	10.07	20.088	30.316	29.60
1.6418×10^{-4}	3.45×10^{-4}	—	—	—
− 20″.506	14″.0	3″	− 34″	—
1° .01	3140″	30′	1″.097	—
1° .61	7050″	1″.61	40′	—

Table 2.6-2.   Gross Physical and Gravitational Parameters of Planets

Planet	Mercury	Venus	Mars	Jupiter
Mass (g)	3.32×10^{26} [3]	4.90×10^{27} [3]	6.43×10^{26} [3]	1.90×10^{30} [59]
Mean radius (km)	2434	6050 [2]	3394	69663 [32]
Mean density (g cm^{-3})	5.5	5.3	3.94	1.357 [59]
Gravity (rotation not included) (cm s^{-2})	374	892	371	2575
Centrifugal term in gravity (at equator (cm s^{-2})	—	—	-1.7	-225
Escape velocity from surface (km s^{-1})	4.3	10.5	5.0	59.8
Rotational period	58.646 days [67]	242.98 days [17]	$24^h39^m22^s.6689$	$9^h50^m30^s.003$ $9^h55^m40^s.632$ $9^h55^m29^s.37$
Gravitational harmonics	—	—	J_2 1.96×10^{-3} [4] C_{21} $< 3 \times 10^{-6}$ S_{21} $< 3 \times 10^{-6}$ C_{22} -5×10^{-5} S_{22} 3×10^{-5}	J_2 0.02206 [82] J_4 0.00253
Observed oblateness	—	—	0.012	0.060 [33]
Dynamical oblateness	—	—	0.00525	0.96518
Polar radius (km)	—	—	—	67526
Equatorial radius (km)	—	—	—	71802
Inclination of pole of rotation to orbit pole	$< 10°$ [56]	$6°$	$24°.936$	$3°$ $4'$ [55]
Internal heat flux (ergs cm^{-2} s^{-1})	2×10^{-1} [77]	—	—	7×10^3 [59]
Internal structure	[44]	[44]	[44]	[31]

Table 2.6-3.   Surface Parameters of Planets

Planet	Mercury	Venus	Mars
Subsolar temperature (°K)	613 [61]	—	305
Darkside temperature (°K)	111 [58]	747 [4]	170
Polar cap temperature (°K)	—	—	148 [18]
Electrical/thermal skin depth (cm)	0.9 λ [56]	—	—
Inverse thermal inertia (erg^{-1} cm^2 s$^{\frac{1}{2}}$°K)	1.5×10^{-5}	—	4×10^{-6}
Electric loss tangent	0.009	—	—
Dielectric constant	1.6–4	~ 4.5	3.5
Conductivity (erg cm^{-1} s^{-1}K^{-1})	4×10^2	—	$\sim 8 \times 10^3$
Thermal skin depth (cm)	~ 10	—	—
Density of surface material (g cm^{-3})	~ 1.5	~ 2	~ 1 [73]

	Saturn	Uranus	Neptune	Pluto
	5.48×10^{19}	8.73×10^{28}	1.03×10^{29}	$\sim 1.0 \times 10^{27}$ [22]
	56,800	25,640	24,950	< 3400 [28]
	0.70	1.21	1.57	> 6.1
	1081	867	1088	100–1200
	− 176	− 67.6	− 28	—
	35.6	—	23.4	—
(Equatorial)	$10^h 14^m$ Equatorial	10.8 hr	15.8 hr	6.39 days [78]
(Mean cloud)	$10^h 36^m 39^s$ 57° latitude			
(Radius)				
	J_2 0.02571	—	J_2 0.0038	—
	J_7 0.0052			
	0.108 [59]	0.01	0.026	—
	0.105	—	0.02	—
	53,550	25,700	24,700	—
	60,000	25,380	25,200	—
	26°74	97° 53′	28° 48′	—
	4×10^3	—	—	—
	—	—	—	—

Table 2.6-3. (continued)

Planet	Mercury	Venus	Mars
Altitude range of large scale topography (km)	~ 3 [36]	4 [15]	15 [66]
Terrain	—	Features present [66]	Highly varied [54]
Mean surface slopes	—	$\sim 5°$	3° [62]
Composition (%)	—	—	K 4
			Th 0.00065
			U 0.002

Table 2.6-4. Atmospheric Parameters of Planets

Planet	Mercury	Venus	Mars
Composition: major constituent [fraction]	[35]	CO_2: 0.97% [35]	CO_2: 0.80% [35]
Observed: minor constituents [mixing ratio]		H_2O: 10^{-6} CO: 4.6×10^{-5} HCL: 6×10^{-7} HF: 5×10^{-9} O_2: 10^{-5}	CO: 8×10^{-4} H_2O: 10^{-3} variable O_2: 1.3×10^{-2} [72] O_3: ? [35]
Mean surface pressure	$< 10^{-6}$ mb [9]	100 bars	$\gtrsim 5.5$ mb (value for CO_2)
Mean surface temperature ($^\circ$K)	~ 600	747 [4]	230
Mean atmospheric temperature ($^\circ$K)	—	—	199
Tropospheric lapse rate ($^\circ$K km^{-1})	—	9 [24]	4.6
Mean tropopause altitude (km)	—	58	16–25
Mean tropopause pressure (mb)	—	380	0.5–1
Tropopause temperature ($^\circ$K)	—	275	—
Stratospheric lapse rate ($^\circ$K km^{-1})	—	4	—
Stratospheric scale height (km)	—	4.5	6.5
Stratospheric temperature ($^\circ$K)	—	215	130
Stratopause level (mb)	—	—	—
Stratopause temperature ($^\circ$K)	—	—	—
Mesopause altitude (km)	—	90–100 [51]	~ 90 [50]
Mesopause temperature ($^\circ$K)	—	190	~ 110
Exospheric scale height (km)	—	—	20 [6]
Exospheric temperature ($^\circ$K)	—	640 [79]	~ 350
Peak daytime ionospheric electron density (cm^{-3})	—	5×10^5 [64]	1.5×10^5 [40]
Altitude of daytime ionospheric peak (km)	—	140	145
Topside plasma scale height (km)	—	10	38

Jupiter	Saturn	Uranus	Neptune
H_2 (67 km-atm) [52] He ($<$ 34 km-atm)	H_2 (190 km-atm) [52]	H_2 (480 km-atm) [10]	H_2 (?) [59]
CH_4 (10^{-3}) [35] NH_3 (2×10^{-4}) CH_3D (not yet available [8]	$CH_4 (\leqslant 2 \times 10^{-3})$ NH_3 $(<2 \times 10^{-5})$	$CH_4 (\sim 3.5$ km-atm) [59] NH_3 (?)	$CH_4 (\sim 6$ km-atm)
2 atm [35]	~ 2 atm [59]	~ 4 atm	—
150–170	~ 90	118 [10]	—
—	—	—	—
2 [20]	0.8	0.73	0.8
300	~ 300	~ 300 [70]	~ 350 [70]
140	—	—	—
—	~ 80	~ 67	~ 52
0.2–0.3	—	—	—
—	~ 20	~ 30	~ 30
24	~ 55	~ 64	76 [59]
6.5 [20]	—	—	—
145	—	—	—
—	—	—	—
130	—	—	—
—	—	—	—
150 [34]	130–170 [53]	—	—
[34]	—	—	—
—	—	—	—
—	—	—	—
—	—	—	—

Table 2.6-5.   Photometric Parameters of Planets

Planet		Mercury	Venus	Mars
Solar constant at mean distance of planet (erg cm^{-2} s^{-1})		9.28×10^6	2.66×10^6	6.0×10^5
Calculated effective temperature (°K)		527	235	259
Observed effective temperature (°K)		557	260	250
Radiometric bond albedo		0.1 [56]	0.77 [37]	0.24 [39]
Bond albedo:	V	0.056	0.80	0.146
	B	—	0.71	0.077
	U	—	0.49	0.053
Visual photometric magnitudes (at a distance of 1 AU and zero phase)	U(0.353μ)	—	− 2.77 [37]	+ 0.20
	B(0.448μ)	+ 0.57	− 3.75	− 0.23
	V(0.554μ)	− 0.36	− 4.46	− 1.50
	R(0.690μ)	− 1.21	—	—
	I(0.820μ)	− 1.73	—	—
Disk brightness temperatures (°K)	1 mm	—	—	170 [73]
	3 mm	295 [56]	290 [26]	—
	1 cm	340	400	240
	10 cm	360	650	180
	20 cm	—	600	—
	40 cm	—	400	—
Geometric albedo	p(U)	—	—	—
	p(B)	0.076 [30]	—	—
	p(V)	0.100	—	—
	p(R)	0.145	—	—
	p(I)	0.179	—	—

Jupiter	Saturn	Uranus	Neptune	Pluto
5.1×10^4	1.53×10^4	3.77×10^3	1.54×10^3	8.94×10^2
105 [59]	71 [59]	55 [59]	44 [59]	42 [59]
134	97	55	—	—
0.45	0.61	0.42	0.42	~ 0.14
—	—	—	—	—
—	—	—	—	—
—	—	—	—	—
− 8.05 [37]	− 6.98 [38]	− 6.35 [59]	− 6.25 [59]	+ 0.06 [59]
− 8.54	− 7.75	− 6.63	− 6.46	− 0.21
− 9.39	− 8.80	− 7.19	− 6.87	− 1.01
—	—	—	—	—
—	—	—	—	—
155 [59]	140 [59]	130 [59]	~ 140	—
—	—	110 [59]	~ 88 [59]	—
140	140	180	—	—
260	200	280	—	—
—	50 cm: 385	—	—	—
—	73 cm: 1690	—	—	—
0.305	0.166 [38]	0.48	0.46	—
0.422	0.297	0.55	0.49	—
0.504	0.429	0.51	0.40	—
—	—	—	—	—
—	—	—	—	—

Table 2.6-6. Venus: Cloud Parameters

Pressure height of top clouds	7 mb [27]
Pressure height for single scattering	50 mb [29]
Pressure height for CO_2 line formation	100–200 mb [35]
Temperature for CO_2 line formation	240°K
Composition of clouds	Controversial [35]
Nature of upper cloud aerosols:	
Mean diameter	2 μ [29]
Refractive index	1.44
Shape	Spherical
Other cloud levels	Uncertain [63]

Table 2.6-7. Venus: Magnetic Field Parameters

Upper limit to magnetic dipole moment	$< 1.6 \times 10^{23}$ emu [13]
Upper limit to radiation belts	[75]

Table 2.6-8. Mars: Cloud Parameters

Upper haze layers altitude	$\sim$ 60 km [54]
Upper haze layers pressure height	0.01–0.1 mb
Optical depth of upper haze layers	$\ll$ 1 [43]
Distribution of high-altitude haze layers	Mid latitudes
Planetwide dust (yellow) clouds:	
Optical thickness	$\gg$ 1 [68]
Vertical extent	$>$ 10 km
Development	[54]
Sporadic yellow clouds:	
Statistics	[76]
White clouds:	
General properties	[54]
Statistics	[80]

Table 2.6-9. Mars: Magnetic Field Parameters

Surface field	60×10^{-5} G
Upper limit on trapped radiation	— [74]

Table 2.6-10. Jupiter: Cloud Parameters[a]

Visible clouds and upper haze layer	[5]
Visible clouds and lower layers	[35, 45]

[a] Little consensus exists in this subject; therefore only references are given.

Table 2.6-11. Jupiter: Magnetic Field and Radiation Belt Parameters

Dipole moment (assumed centered)	4×10^{30} G cm^3 [20]
Atmospheric field strength	
equator	12 G
poles	24 G
Magnetosphere dimensions	
solar direction	50 R$_J$
Radiation belts[a]	[20]

[a] Difficult to define; therefore only reference is given.

Table 2.6-12. Saturn: Cloud Parameters

Visible cloud composition	NH_3 crystals [59]

and it is thought that the estimates of CH_4 abundance are high. There is no evidence for a solid surface and even the presence of clouds in their visible atmospheres is in doubt. Early suspicions of an enhanced He content appear to be groundless. Some high-altitude haze may be present in the atmosphere of Uranus which is reported to exhibit limb brightening in the near IR [81], but another explanation in terms of pressure-induced hydrogen opacity may be possible. Both planets exhibit strong limb darkening in the visual, which makes it difficult to measure their radii. The occultation of BD17° 4388 by Neptune in 1968 provided the relatively precise measurement of the diameter of Neptune. The value quoted for Uranus is based on a scaling of visual measurement with the sense and magnitude indicated by the Neptune results. There have been no efforts to develop a theory of the upper atmospheres of these planets. Neither is an ionospheric theory available. There are no indications of an internal heat source or of intrinsic magnetic fields.

2.6.3 Pluto

The mass of Pluto is estimated from its perturbing effect on the motion of Neptune. Its brightness varies indicating rotation, but the orientation of the axis of rotation is unknown as is the reason for the brightness variations. The mean dayside subsolar surface temperature is estimated at 50°K rising to approximately 70°K when the planet is at perihelion. The overall dimensions of the planet are not known but an upper limit on the diameter has been set from observations of a "near miss" occultation of a star by the planet. No atmosphere has been detected although it is possible that the planet could retain atmospheres of H, He, and Ne. The planet has no known satellites. Its orbit has the greatest eccentricity and inclination in the family of planets, and crosses the orbit of Neptune. This circumstance has led to the suggestion that Pluto is an escaped satellite of Neptune [48]. There is no reason to suspect that Pluto has a magnetic field.

2.6.4 The Satellites of the Planets

Mars: Phobos and Deimos. These tiny satellites are solid rock, with an irregular shape [54]. They are heavily cratered and provide a yardstick for measuring the true cratering history of Mars. See Tables 2.6-13, 2.6-14, and 2.6-15.

Jupiter: The Galilean Satellites. In addition to the four satellites discovered by Galileo, J5 (Amalthea) is included in this group (Table 2.6-16).

Table 2.6-13. **Phobos and Deimos: Orbital Parameters**

Satellite	Phobos	Deimos
Semi-major axis	2.743 Mars radii	6.891 Mars radii
Eccentricity	0.021	0.0028
Inclination to planets equator	0°57′	1°18′
Sidereal period	0.31891 days	1.26244 days
Nodal regression rate	158° yr^{-1}	6°.374 yr^{-1}

Table 2.6-14. **Phobos and Deimos: Gross Physical Parameters**

Dimensions: Phobos	25×21 km^2 [54]
Deimos	13.5×12.0 km^2
Shape	Irregular

Table 2.6-15. **Phobos and Deimos: Photometric Parameters**

Satellite	Phobos	Deimos
Photometric visual magnitudes (at 1 AU and zero phase)	12.1	13.3
Visual geometric albedo	0.05[54]	0.05

Table 2.6-16. **Galilean Satellites: Orbital Parameters [59]**

	Semi-Major Axis (Jovian radii)	Eccentricity	Inclination	Sidereal Period
J1 (Io)	6.058	0.0	1°6	11^{h}57^{m}22^{s}70
J2 (Europa)	9.638	0.0003	28°1	1^{d}18^{h}27^{m}33^{s}51
J3 (Ganymede)	15.374	0.0015	11°0	3^{d}13^{h}13^{m}42^{s}05
J4 (Callisto)	27.044	0.0075	15°2	7^{d}3^{h}42^{m}33^{s}35
J5 (Amalthea)	2.605	0.0028	27°3	16^{d}16^{h}32^{m}11^{s}21

Little is known about the surface environments of these satellites. The mean densities are small compared to the earth's moon. Ganymede, for example, is larger than Mercury but is only half as massive. The diameter of Io is known with greatest precision due to a well-observed occultation of a star in the β Sco system. This fortunate occulation has also allowed a stringent upper limit to be placed on the surface pressure for Io's atmosphere. It is unlikely that the other Galilean satellites have atmospheres in excess of this (Table 2.6-17). Surface markings have been visually observed and, in the case of Io, photographed. Io has dark polar regions. Dark markings on the other satellites similar to lunar maria are reported. Callisto has bright polar regions. The light curves are generally variable, as is color. These variations presumably indicate lateral surface inhomogeneities. The periodicity of their light curves is in agreement with the requirements of synchronous rotation. Ellipse radiometry of Ganymede indicates that the thermophysical properties are not startlingly different from the moon. See Table 2.6-18.

The low densities have, however, given rise to physical models quite unlike those constructed for the earth's moon [46]. A picture emerges of an icy body with a near solar composition (but excluding H and He). They may have suffered extensive melting and differentiation due to radioactive heating and dissipation of tidal energy. They may therefore have light siliceous cores surrounded by a slushy mantle. A thin crust of H_2O ice or undifferentiated protomaterial probably comprises the external layer.

Jupiter: The Outer Satellites. The outer satellites comprise a total of seven bodies divided into two distinct groups by virtue of their orbital

Table 2.6-17. Galilean Satellites: Gross Physical and Atmospheric Properties [59]

	J1	J2	J3	J4	J5
Mass (g)	7.2×10^{25}	4.7×10^{25}	1.6×10^{26}	9.7×10^{25}	—
Mean radius (km)	1829 [33]	1550	2640	2500	90–250[a]
Mean density (g cm^{-3})	2.82	3.02	2.08	1.48	—
Gravity (cm s^{-2})	145	131	149	103	—
Escape velocity (km s^{-1}) from surface	2.3	2.01	2.80	2.27	—
Thermal inertia (erg cm^{-2} s$^{-1/2}$ K)	—	—	3×10^4 [5]	—	—
Rotational period	Orbital period	Orbital period	Orbital period	Orbital period	—
Atmospheric surface pressure	0.2 [7]	—	—	—	—

[a] assumes albedo in range 8–65%

Table 2.6-18. Galilean Satellites: Photometric Parameters

		J1	J2	J3	J4	J5
Photometric magni-						
tudes[a]	U	+ 0.57	− 0.14	− 0.83	+ 0.21	—
(at 1 AU and						
zero phase)	B	− 0.73	− 0.66	− 1.33	− 0.34	—
	V(Δv)	− 1.90 (.21)	− 1.53 (.34)	− 2.16 (.16)	− 1.20	+ 6.3
Geometric albedo:	U	0.17	0.41	0.24	0.11	—
	B	0.49	0.58	0.34	0.17	—
	V	0.80	0.72	0.40	0.21	—
Disk brightness						
temperature	10μ[57]	∼ 140°K	∼ 130°K	∼ 140°K	∼ 160°K	—
	20μ[57]	127°K	119°K	136°K	149°K	—

[a] Both magnitudes and colors are variable (Δv) [59].

characteristics. No information is available concerning their physical properties with the exception of visual magnitudes and the fact that they must all be extremely small, that is a few tens of kilometers in diameter at most. See Tables 2.6-19 and 2.6-20.

Table 2.6-19. Outer Jovian Satellites: Orbital Parameters [55]

Object	Semi-Major Axis (Jovian radii)	Eccentricity	Inclination	Sidereal Period (days)
J6 (Hestia)[a]	160.7	0.15798	28°.436	250.6
J7 (Hera)[a]	164.4	0.20719	27°.75	260.1
J8 (Poseidon)[b]	326	0.291–0.660	155°–146°	735
J9 (Hades)[b]	332	0.275	157°	758
J10 (Demeter)[a]	164	0.14051	28°.4	260
J11 (Pan)[b]	313	0.20678	163°.377	692
J12 (Andrastea)[b]	290	0.168702	146°.7338	617

[a] Members of inner group.
[b] Members of outer group.

Table 2.6-20. Outer Jovian Satellites: Photometric Properties

	J6	J7	J8	J9	J10	J11	J12
Photometric visual magnitudes: (at 1 AU and zero phase)	+ 7.0	+ 9.3	+ 12.1	+ 11.6	+ 11.9	+ 11.4	+ 12.1

Saturn's Satellites. There are ten known satellites although the reality of one, Janus, is still in some doubt (Table 2.6-22). Only Phoebe in an eccentric retrograde orbit, is irregular. It is most likely that all the satellites are in synchronous rotation about the parent body. Only for Titan, which is not variable, is this conclusion in doubt. The albedos and spectra of the satellites are very similar (again excepting Titan) and indicate an icy (either H_2O or NH_3 or both) surface (Table 2.6-23). Titan is much redder and is the only satellite which has been observed to possess an atmosphere. The composition of Titan's atmosphere is partly known: CH_4 has been observed and H_2 may have been detected (Table 2.6-24). If the latter observation is confirmed then the surface pressure should be quite high. There is visual evidence for cloud phenomena on Titan. Brightness temperatures for Titan have been measured which may indicate the presence of a "greenhouse" effect in the satellite's atmosphere. Iapetus is a most remarkable satellite in that it varies in brightness by a factor of 6 between one elongation and the other but without any appreciable change in color. It is conjectured that the "leading" surface has been modified by collision with orbital debris.

Saturn: The Rings. Saturn's rings are composed of a layer of solid fragments in differential Keplerian motion. The inclination of the ring plane is, as far as can be determined, identical to the equatorial plane of Saturn. Its geometrical thickness is about 1 km. The rings are divided into five regions which have azimuthal symmetry. These regions are designated D, C, B, A, and D' in order of increasing distance from the planet (Table 2.6-25). The existence of the D' ring, extending out to the orbit of Enceladus, is controversial. The radial distribution of brightness is complex with azimuthally symmetric "gaps" of various intensities being present. The most predominant are the Cassinni Division separating the A and B rings and the Enke Division in the A ring.

The optical thickness (Table 2.6-26) of the rings is similarly complex. There is evidence that the gaps are not entirely clear of material. Physical models of the rings are very uncertain and should be used with caution. The fragments probably approach kilometer size in the A and B rings[19]. The inner rings C and D could be composed of much smaller fragments originating from collisions in the outer rings [19]. The mean space density of the ring system is limited by stability arguments to less than 1.04 g/cm^3 [12]. Spectroscopic evidence indicates the H_2O ice is a constituent of the ring fragments.

Uranian System. Uranus has five satellites in a dynamical system which is noted for its regularity (Table 2.6-27). However, the three largest satellites, Titania, Oberon, and Ariel, are variable in brightness (Table 2.6-28) [69]. This is a remarkable fact when it is noted that they were observed from near the pole of their orbits. This presumably implies that the axes of rotation

Table 2.6-21.   Saturn Satellites: Orbital Parameters [59]

Name	Semi-Major Axis (Saturn radii)	Eccentricity	Inclination	Period (days)
Janus	2.817	0	0	0.749
Mimas	3.271	0.0201	1°31'.0	0.942422
Enceladus	4.195	0.00444	1'.4	1.370218
Tethys	5.192	0	1°05'.6	1.887802
Dione	6.653	0.00221	1'.4	2.736916
Rhea	9.289	0.00098	21'	4.517503
Titan	21.525	0.029	20'	15.945452
Hyperion	26.128	0.104	17'–56'	21.276665
Iapetus	62.727	0.02828	14°.72	79.33082
Phoebe	228.169	0.16326	150°.05	550.45

Table 2.6-22.   Saturn Satellites: Physical Parameters [59][a]

Name	Mass (g)	Radius (km)	Density (g cm^{-3})
Mimas	3.8×10^{22}	~ 240	~ 0.7
Enceladus	8.5×10^{22}	280	~ 1.0
Tethys	6.5×10^{23}	600	~ 0.7
Dione	1.05×10^{24}	410	~ 3.6
Rhea	1.7×10^{24}	725	~ 1.1
Titan	1.4×10^{26}	2425	~ 2.3
Iapetus	1.4×10^{24}	850	~ 0.9

[a] Data not available on omitted satellites.

Table 2.6-23.   Saturn Satellites: Photometric Parameters [30]

	Mimas	Enceladus	Tethys	Dione[b]	Rhea[c]	Titan[c]	Hyperion	Iapetus[b]
Photometric magnitudes[a]: (at 1 AU and zero phase)								
U	—	—	+ 1.79	+ 1.90	+ 1.32	+ 0.89	+ 5.72	+ 2.47
B	—	+ 2.84	+ 1.45	+ 1.60	+ 0.97	+ 0.14	+ 5.30	+ 2.19
V	+ 2.6	+ 2.22	+ 0.72	+ 0.89	+ 0.21	− 1.16	+ 4.61	+ 1.48
Geometric albedo:								
U	—	—	0.64	0.75	0.60	0.06	—	—
B	—	0.53	0.76	0.87	0.73	0.12	—	—
V	0.49	0.54	0.84	0.94	0.57	0.21	—	0.04–0.25
Disk brightness temperature:								
10 μ	—	—	—	—	—	132°K	—	—
20 μ	—	—	—	—	96°K	93°K	—	110–117°K

[a] Photometric distance modulus: $+9^{m}.55$.
[b] Variable.
[c] Variable [11].
[d] [2.6-57].

Table 2.6-24. Titan: Atmospheric Parameters

Composition (amount above hypothetical reflecting layer)	$CH_4(\geq 200\text{m-atm})$
	H_2 (?) [71]

Table 2.6-25. Saturn's Rings: Physical Parameters

Dimensions:	
Inner edge of C ring	7.2×10^4 km [59]
Inner edge of D ring	9.2×10^4 km
Outer edge of B ring	11.78×10^4 km
Inner edge of A ring	12.04×10^4 km
Outer edge of A ring	13.64×10^4 km
Estimated fragment size	$\leqq 1$ km [19]
Mean thickness of rings	~ 1 km
Mean space density	$\sim 1 \text{ g cm}^{-3}$ [12]
Total mass of rings	$\lesssim 10^{-2}$ (Saturn)
Composition	H_2O; NH_3(?) ices [59]
Inclination of ring plane	$0°$

Table 2.6-26. Saturn's Rings: Photometric Parameters [19]

	D	C	B_3	B_2	B_1	A_2	A_1
Photometric magnitudes: (mag. arc s^{-2})							
V	—	—	7.00	6.67	6.55	6.89	7.40
B	—	—	7.85	7.52	7.40	7.74	8.25
Geometric albedo:							
V	—	—	0.82	0.82	0.82	0.82	0.82
B	—	—	0.65	0.65	0.65	0.65	0.65
Vertical optical thickness	—	< 0.18	0.32	0.61	1.0	0.37	0.17

Table 2.6-27. Uranus Satellites: Orbital Parameters [59]

	Semi-major axis (Uranus radii)	Eccentricity	Inclination[a]	Period (days)
Titania	17.64	0.0024	$\leqq 1'$	8.706
Oberon	23.59	0.0007	$\leqq 1'$	13.463
Ariel	7.72	0.0028	$\leqq 1'$	2.520
Umbriel	10.75	0.0035	$\leqq 1'$	4.144
Miranda[b]	5.25	—	—	1.413

[a] To the mean orbital plane of the satellites.
[b] Present data indicate an inclined orbit.

Table 2.6-28. Uranus Satellites: Physical and Photometric Parameters [59]

	Titania	Oberon	Ariel	Umbriel	Miranda
Mass (g)[a]	$\sim 4.3 \times 10^{24}$	$\sim 2.6 \times 10^{24}$	$\sim 1.2 \times 10^{24}$	$\sim 5.1 \times 10^{23}$	$\sim 8.7 \times 10^{22}$
Radius (km) [1]	~ 500	~ 400	~ 300	~ 200	~ 100
Photometric magnitude: V [30] (at 1 AU zero phase)	+ 1.30	+ 1.49	+ 1.7	+ 2.6	+ 3.8
Color magnitude: B-V [30]	+ 0.62	+ 0.65	—	—	—
U-B	+ 0.25	+ 0.24	—	—	—

[a] Masses are very uncertain and with the following radii lead to unrealistic densities.

of, at least, the three largest satellites are not normal to their orbital planes. Physical properties are very uncertain with, for example, masses known only to an order of magnitude, and dimensions to a factor of about 3.

Neptunian System. Both satellites of Neptune have unusual orbital properties (Table 2.6-29). Their orbits have large inclinations to the equatorial plane of Neptune. Triton, the larger of the two, is in a relatively unstable [49] retrograde orbit while that of Neried, although direct, is highly eccentric. Triton should be capable of retaining an atmosphere but no observational evidence exists. See Table 2.6-30.

Table 2.6-29. Neptune Satellites; Orbital Parameters [59]

	Semi-Major Axis (Neptune radii)	Eccentricity	Inclination	Period (days)	Orbital Speed (km s^{-1})
Triton	14.21	0	159°57′	5.877	6
Neried	222.48	0.7493	27°48′	359.9	0.5–4

Table 2.6-30. Neptune Satellites: Physical and Photometric Parameters

		Triton	Neried
Mass (g) [59]		1.4×10^{25}	3.4×10^{22}
Radius (km)		1890	~ 150
Photometric magnitude:	V	$- 1.16$ [30][a]	+ 4.0
(at 1 AU and zero phase)			
Color magnitude:	B-V	+ 0.77	—
	U-B	+ 0.40	

[a] Variable.

References

1. Allen, C. W., *Astrophysical Quantities*, University of London, The Athlone Press, 1963.
2. Anderson, J. D., D. L. Cain, L. Efron, R. M. Goldstein, W. G. Melbourne, D. A. Handley, G. E. Pease, and R. C. Tausworthe, "The Radius of Venus as Determined by Planetary Radar and Mariner V Radar Tracking Data," *J. Atmos. Sci.* **25**, 1171, 1968.
3. Ash, M. E., I. I. Shapiro, and W. B. Smith, "Astronomical Constants and Planetary Ephemerides Deduced from Radar and Optical Observations," *Astron. J.* **72**, 338, 1967.
4. Avduevsky, V. S., M. Ya. Morov, M. K. Rozhdestvensky, N. J. Borodin, and V. V. Kerzhanovitch, "Soft Landing on Venera 7 on the Venus Surface and Preliminary Results of Investigations of the Venus Atmosphere," *J. Atmos. Sci.* **28**, 263, 1971.
5. Axel, L., "Inhomogeneous Models of the Atmosphere of Jupiter," *Astrophys. J.* **173**, 451, 1972.
6. Barth, C. A., G. W. Hord, and A. L. Lane, "Mariner 9 Ultraviolet Spectrometer Experiment: Initial Results," *Science* **175**, 309, 1971.
7. Bartholdi, P., and F. Owen, "The Occultation of Beta Scorpii by Jupiter and Io. II. Io," *Astron. J.* **77**, 60, 1972.
8. Beer, R., C. B. Farmer, R. H. Norton, J. V. Martonchik, and T. G. Barnes, "Jupiter: Observation of Deuterated Methane in the Atmosphere," *Science* **175**, 1360, 1972.
9. Belton, M. J. S., D. M. Hunten, and M. B. McElroy, "A Search for an Atmosphere on Mercury," *Astrophys. J.* **150**, 1111, 1967.
10. Belton, M. J. S., M. B. McElroy, and M. J. Price, "The Atmosphere of Uranus," *Astrophys. J.* **164**, 191, 1971.
11. Blanco, C., and S. Catalano, "Photoelectric Observations of Saturn Satellites Rhea and Titan," *Astron. Astrophys.* **14**, 43, 1971.
12. Bobrov, M. S., "Physical Properties of Saturn's Rings," in *Surfaces and Interiors of Planets and Satellites*, Ed. A. Dollfus, Academic Press, New York, 1970.
13. Bridge, H. S., A. J. Lazarus, C. W. Snyder, E. J. Smith, L. Davis, Jr., P. J. Coleman, and D. E. Jones, "Mariner V: Plasma and Magnetic Fields Observed Near Venus," *Science* **158**, 1670, 1967.
14. Cain, D. L., "The Shape of Mars," paper given at the 3rd Annual Meeting, American Astronomical Society—Div. of Planet. Sci., Kona, Hawaii, March, 1972.
15. Campbell, D. B., R. B. Dyce, R. P. Ingalls, G. H. Pettengill, and I. I. Shapiro, "Venus: Topography Revealed by Radar Data," *Science* **175**, 514, 1972.
16. Carmichel, H. and A. Dollfus, "La Rotation et la Cartographie de la Planète Mercure," *Icarus* **8**, 216, 1968.
17. Carpenter, R. L., "A Radar Determination of the Rotation of Venus," *Astron. J.* **75**, 61, 1970.
18. Chase, S. C., H. Hartzenkeler, H. H. Kieffer, E. Miner, G. Münch, and G.

Neugebauer, "Infrared Radiometry Experiment on Mariner 9", *Science* **175**, 308, 1971.

19. Cook, A. F., F. A. Franklin, and F. D. Palluconi, "Saturn's Rings—A Survey," JPL Tech. Memo. 33–488, 1971.
20. Devine, N. "The Planet Jupiter 1970," NASA SP–8069, 1972.
21. Duncombe, R. L., "The Motion of Venus 1750–1949 " *Astron. J.* **61**, 266, 1956.
22. Duncombe, R. L., W. J. Klepszynski, and P. K. Seidelman, "Note on the Mass of Pluto," *Pub. Astron. Soc. Pacific* **82**, 416, 1970.
23. Dyce, R. B., G. H. Pettengill, and I. I. Shapiro, "Radar Determinations of the Rotations of Venus and Mercury," *Astron. J.* **72**, 351, 1967.
24. Fjeldbo, G., A. J. Kliore, and V. R. Eshleman, "The Neutral Atmosphere of Venus as Studied with the Mariner V Radio Occultation Experiments," *Astron. J.* **76**, 123 1971.
25. Gill, J. R., and B. L. Gault, "A New Determination of the Orbit of Triton, Pole of Neptune's Equator and Mass of Neptune," *Astron. J.* **73**, 595, 1968.
26. Goldstein, R. M., "Radio and Radar Studies of Venus and Mercury," *Radio Sci.* **5**, 391, 1970.
27. Goody, R. M., "The Scale Height of the Venus Haze Layer," *Planet. Space Sci.* **15**, 1817, 1967.
28. Halliday, I., R. H. Hardie, O. G. Franz, and J. B. Priser, "An Upper Limit for the Diameter of Pluto," *Astron. J.* **70**, 676, 1965.
29. Hansen, J. E., and A. Arking "Clouds of Venus: Evidence for Their Nature," *Science* **171**, 669, 1971.
30. Harris, D. L., "Photometry and Colorimetry of Planets and Satellites," in *Planets and Satellites*, Eds. G. P. Kuiper, and B. M. Middlehurst, *Solar System III*, Chapter 8, p. 272, University of Chicago Press, Chicago, 1961.
31. Hubbard, W. B., "Structure of Jupiter: Chemical Composition, Contraction, and Rotation," *Astrophys. J.* **162**, 687, 1970.
32. Hubbard, W. B., R. E. Nather, D. S. Evan, R. G. Tull, D. C. Wells, V. G. W. Citters, B. Warner, and P. VandenBout, "The Occultation of Beta Scorpii by Jupiter and Io. I. Jupiter," *Astron. J.* **77**, 41, 1972.
33. Hubbard, W. B., and T. C. Van Flandern, "The Occulation of Beta Scorpii by Jupiter and Io. III. Astrometry," *Astron. J.* **77**, 65, 1972.
34. Hunten, D. M., "The Upper Atmosphere of Jupiter," *J. Atmos. Sci.* **26**, 826, 1969.
35. Hunten, D. M., "Composition and Structure of Planetary Atmospheres," *Space Sci. Rev.* **12**, 539, 1971.
36. Ingalls, R. P., and L. P. Rainville, "Radar Measurements of Mercury: Topography and Scattering Characteristics as 3.8 cm," *Astron. J.* **77**, 185, 1972.
37. Irvine, W. M., "Monochromatic Phase Curves and Albedos for Venus," *J. Atmos. Sci.* **25**, 610, 1968.
38. Irvine, W. M., and A. P. Lane, "Monochromatic Albedos for the Disk of Saturn," *Icarus, 15*, No. 1, 18–26, August, 1971.
39. Irvine, W. M., T. Simon, D. H. Menzel, C. Pikoos, and A. T. Young, "Multicolor Photoelectric Photometry of the Brighter Planets. III. Observations from Boyden Observatory," *Astron. J.* **73**, 807, 1968.

40. Kliore, A. J., D. L. Cain, G. Fjeldbo, B. L. Seidel, and S. I. Rasool, "Mariner 9 S-Band Martian Occultation Experiment. Initial Results on the Atmosphere and Topography of Mars," *Science* **175**, 313, 1971.
41. Kuiper, G. P., *The Atmospheres of the Earth and Planets*, University of Chicago Press, Chicago, 1952.
42. Kuiper, G. P., "The Planet Mercury: Summary of Present Knowledge," *Comm. Lunar and Planet. Lab.* **8**, 165, 1970.
43. Leovy, C. B., B. A. Smith, A. T. Young, and R. B. Leighton, "Mariner-Mars 1969: Atmospheric Results," *J. Geophys. Res.* **76**, 297, 1971.
44. Levin, B. J., Internal Constitution of Terrestrial Planets, in *Surfaces and Interiors of Planets and Satellites*, Ed. A. Dollfus, Chapter 8, p. 462, Academic Press, New York, 1970.
45. Lewis, J. S., "The Clouds of Jupiter and the NH_3-H_2O and NH_3-H_2S Systems," *Icarus* **10**, 365, 1969.
46. Lewis, J. S., "Satellites of the Outer Planets: Thermal Models," *Science* **172**, 1127, 1971.
47. Lorell, J., G. H. Bonn, E. J. Christensen, J. F. Jordan, P. A. Laing, W. L. Martin, W. L. Sjogren, I. I. Shapiro, R. D. Rosenberg, and G. L. Slater, "Mariner 9 Celestial Mechanics Experiment: Gravity Field and Pole Direction of Mars," *Science* **175**, 317, 1971.
48. Lyttleton, R. A., "On the Possible Results of an Encounter of Pluto with the Neptunian System," *Mon. Not. Roy. Astron. Soc.* **97**, 108, 1936.
49. McCord, T. B., "Dynamical Evolution of the Neptunian System," *Astron. J.* **71**, 585, 1966.
50. McElroy, M. B., "The Upper Atmosphere of Mars," *Astrophys. J.* **150**, 1125, 1967.
51. McElroy, M. B., "The Upper Atmosphere of Venus," *J. Geophys. Res.* **73**, 1513, 1968.
52. McElroy, M. B., "Atmospheric Composition of the Jovian Planets," *J. Atmos. Sci.* **26**, 798, 1969.
53. McGovern, W. E., "Exospheric Temperature of Jupiter and Saturn," *J. Geophys. Res.* **73**, 6361, 1968.
54. Masursky, H., R. M. Batson, J. F. McCauley, L. A. Soderblom, R. L. Wildey, M. H. Carr, D. J. Milton, D. E. Wilhelms, B. A. Smith, T. B. Kirby, J. C. Robinson, C. B. Leovy, G. A. Briggs, T. C. Duxbury, C. H. Acton, B. C. Murray, J. A. Cutts, R. P. Sharp, S. Smith, R. B. Leighton, C. Sagan, J. Veverka, M. Noland, J. Lederberg, E. Levinthal, J. B. Pollack, J. T. Moore, W. K. Hartmann, E. N. Shipley, G. de Vaucouleurs, and M. E. Davies, "Mariner 9 Television Reconnaissance of Mars and its Satellites: Preliminary Results," *Science* **175**, 294, 1971.
55. Michaux, C. M., "Handbook of the Physical Properties of the Planet Jupiter," NASA SP-3031, 1967.
56. Morrison, D., "Thermophysics of the Planet Mercury," *Space Sc. Rev.* **11**, 271, 1970.
57. Morrison, D., and D. P. Cruikshank, "Thermal Properties of Galilean Satellites" *Icarus*, **18**, 224, 1973.

58. Murdock, T. L., and E. P. Ney, "Mercury: The Darkside Temperature," *Science* **170**, 535, 1970.

59. Newburn, Jr., R. L., and S. Gulkis, "A Brief Survey of the Outer Planets Jupiter, Saturn, Uranus, Neptune, Pluto, and Their Satellites," NASA (JPL) TR 32–1529, 1971.

60. Parkinson, T. D., and D. M. Hunten "Martian Dust Storm: Its Depth on 25 November 1971," *Science* **175**, 323, 1971.

61. Petit, E., "Planetary Temperature Measurements," in *Planets and Satellites* Ed. G. P. Kuiper and B. M. Middlehurst, *Solar System III*, Chapter 10, p. 400, University of Chicago Press, Chicago, Illinois, 1961.

62. Pettengill, G. H., A. E. E. Rogers, and I. I. Shapiro, "Martian Craters and a Scarp as Seen by Radar," *Science* **174**, 1321, 1971.

63. Rasool, S. I., "The Structure of the Venus Clouds—Summary," *Radio Sci.* **5**, 367, 1970.

64. Rasool, S. I., and R. W. Stewart, "Results and Interpretation of the S-Band Occultation Experiments on Mars and Venus," *J. Atmos. Sci.* **28**, 869, 1971.

65. Rogers, A. E. E., M. E. Ash, C. C. Counselman, I. I. Shapiro, and G. H. Pettengill, "Radio Measurements of the Surface Topography and Roughness of Mars," *Radio Sci.* **5**, 465, 1970.

66. Rogers, A. E. E., and R. P. Ingalls, "Radar Mapping of Venus with Interferometric Resolution of the Range-Doppler Ambiguity," *Radio Sci.* **5**, 425, 1970.

67. Smith, B. A., and E. J. Reese, "Mercury's Rotation Period: Photographic Confirmation," *Science* **162**, 1275, 1968.

68. Smith, E. J., T. L. Davis, P. J. Coleman, Jr., and D. E. Jones, "Magnetic Field Measurements Near Mars," *Science* **149**, 1241, 1965.

69. Steavenson, W. H., "The Satellites of Uranus," *Brit. Astron. Assoc. J.* **74**, 54, 1964.

70. Trafton, L. M., "Model Atmospheres of the Major Planets," *Astrophys. J.* **147**, 765, 1967.

71. Trafton, L. M., "On The Possible Detection of H_2 in Titan's Atmosphere," *Astrophys. J.*, **175**, 285, 1972.

72. Traub, W. A., and N. P. Carleton, "Observations of O_2 on Mars and Venus," paper given at the 3rd Annual Meeting, Am. Astron. Soc., Div. of Planet. Sci., Kona, Hawaii, 1972.

73. Troitskii, V. S., "On the Possibility of Determining the Nature of the Surface Material of Mars from its Radio Emissions," *Radio Sci.* **5**, 481, 1970.

74. Van Allen, J. A., L. A. Frank, S. M. Krimigis, and H. K. Hills, "Absence of Martian Radiation Belts and Implications Thereof," *Science* **149**, 1228, 1965.

75. Van Allen, J. A., S. M. Krimigis, L. A. Frank, and T. P. Armstrong, "Venus: An Upper Limit on Intrinsic Magnetic Dipole Movement Based on Absence of a Radiation Belt," *Science* **158**, 1674, 1967.

76. Vaucouleurs, G. de, *Physics of the Planet Mars*, Faber and Faber Limited, London, 1954.

77. Walker, J. C. G., "The Thermal Budget of the Planet Mercury," *Astrophys. J.* **133**, 274, 1961.

78. Walker, M. F., and R. Hardie, "A Photometric Determination of the Rotational Period of Pluto," *Pub. Astron. Soc. Pacific* **67**, 224, 1955.
79. Wallace, L. W., "Analysis of the Lyman-Alpha Observations of Venus made from Mariner 5," *J. Geophys. Res.* **74**, 115, 1969.
80. Wells, R. A., "Some Comments on Martian White Clouds," *Astrophys. J.* **147**, 1181, 1967.
81. Westphal, J., Comment at the 3rd Annual Meeting, Am. Astron. Soc., Div. Of Planetary Sci., Kona, Hawaii, 1972.
82. Wildt, R., Planetary Interiors, in *Planets and Satellites*, Eds. G. P. Kuiper, and B. M. Middlehurst, *Solar System III*, Chapter 5, p. 159, University of Chicago Press, Chicago, 1961.

Chapter 3

Instrument Platforms

3.1 PLATFORM DESCRIPTIONS

The design of a geoscience instrument is greatly influenced by the nature of the platform upon which it is mounted. The instrument and platform must be designed to accommodate the environment of the domain in which they are placed. This means that consideration must be given to the range of values and fluctuations of a number of parameters including temperature, pressure, humidity, shock, vibration, wind, ice, dust, salt water, and corrosive gases. The problems involved in the utilization of instruments on platforms designed for land, air, sea, and space are described in Sections 3.2 through 3.5, respectively.

Instrument platforms normally carry a number of auxiliary systems with which the instrument must interface. These include communications and telemetry systems for remotely operated platforms; power supplies; and positioning systems including propulsion, navigation, guidance, and control. These auxiliary systems are described in Sections 3.6 through 3.8.

The instrument platform may contain other equipment with which the instrument must operate. This might include synchronization and timing equipment and other instrumentation systems designed for the concurrent measurement of other parameters. In addition, it is often desirable to gather housekeeping-type data to provide a measure of the over-all platform performance. Frequently these data are necessary to properly interpret the measurement of the geoscience instrument. Most of the sensors used for acquiring this housekeeping data are similar to some of the sensors described in Chapter 4.

The installation of an instrument on a platform often involves the solution of electromagnetic compatability problems. There is also an engineering compromise between the instrument performance, reliability, maintainability, and cost. These design problems are discussed in Chapter 8.

The reader will note that while the environments are different, the approach to the solution of the platform problem is identical for the domains of air, land, sea, and space.

John C. Cook

3.2 GROUND PLATFORMS

3.2.1 Seismological Observatories

For more than 75 years, continuous recording of seismic vibrations from natural events such as earthquakes has been performed in many countries.

The Worldwide System. The worldwide system consists of 125 seismic observatories throughout the world [7]. Figure 3.2-1 is a block diagram of the system. The basic geophysical sensors are velocity-sensing, short-period seismometers tuned to 1 second and long-period seismometers tuned to 30 seconds. Section 4.22 describes seismometers. The entire system of six seismographs covers a three-decade frequency range.

Recording is photographic. Three mirror galvanometers inscribe helical traces upon each drum-transport which carries three sheets of 30 cm by 92 cm photographic paper and can record three channels of data for 24 hr. Constant-frequency alternating current for the drum drives is provided by a crystal-controlled oscillator and frequency dividers, which also produce time-mark deflections on the photographic record by means of a program generator and beam-deflecting optics. A 2-second mark is made each minute, and a 5-second mark each hour except every sixth hour. A radio receiver and manual phasing adjustment are provided so that the operator can synchronize the timing system with national frequency standards daily or more often.

Means are provided for both electromagnetic and manual calibration of the six seismographs. Electromagnetic calibration is accomplished by applying an accurately known pulse of current to a small forcing coil attached to the inertial mass of each seismometer. The resulting deflection on the photographic record is a measure of seismograph sensitivity and also indicates motional sense.

Operating power required is about 100 W at 155 V to 130 V, 25 to 60 Hz. It is converted to 24 V dc, part of which is used to keep a bank of nickel-cadmium batteries charged. The remainder is used for system operation. If line power fails, the batteries automatically supply the needs of the station for up to 12 hr.

A typical worldwide station installation requires indoor shelter for the electronics, and a darkened room for the galvanometers, recorders, and photographic processing. The galvanometers and recorders are placed on

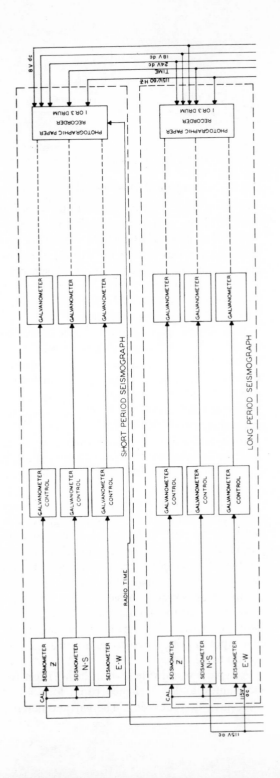

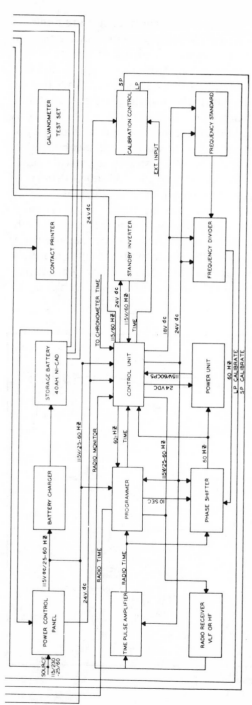

Figure 3.2-1 Block electrical diagram of a "world wide" seismological system.

firm masonry piers to minimize vibration. The seismometers are mounted in the standard orientations (E-W and N-S for the horizontal types) on one or more rigid masonry piers anchored to bedrock. The long period sensors are placed in an air-tight underground vault to prevent disturbance from sudden changes of temperature or barometric pressure. They may be at a distance of up to a few kilometers from the recording station, and should be as far as possible (tens of kilometers) from sources of seismic noise such as surf, traffic and machinery. Records must be changed and developed daily. Adjustment and calibration daily to weekly are recommended.

Intermediate Observatories. Figure 3.2-2 illustrates a typical, intermediate observatory layout [11, 12]. An array of 10 to 30 short-period (1 s) vertical-component seismometers is deployed over an area of 3 to 10 square kilometers in a seismically quiet area far from cultural activity. This array provides signal enhancement and information on source azimuth. Additional 3-component clusters of short-period, intermediate-band, broad-band, and long-period seismometers are also provided for the study of various seismic wave types at various levels of sensitivity. The seismometer output signals are carried on the surface via shielded cables up to

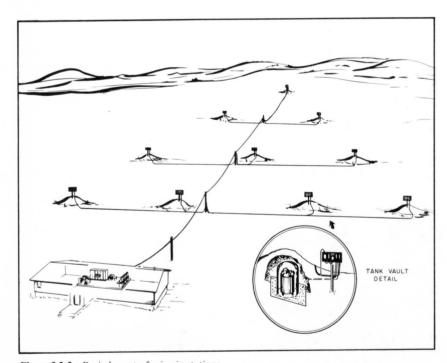

Figure 3.2-2 Basic layout of seismic stations.

several km to a central group of amplifiers. Both ends of each conductor are connected to the instrumentation through lightning-surge protectors such as standard telephone fuses or r.f. chokes, and are shunted to ground through carbon spark gaps or gas-discharge tubes, and zener diodes. Nevertheless, local electrical storms can cause surge damage.

The Large Aperture Seismological Array (LASA). The LASA is a major example of a fixed geophysical instrument platform [10, 14].

Figure 3.2-3 is a map of the LASA. It consists of 21 sub-arrays, arranged at the corners of concentric squares B, C, D, E, and F around the central

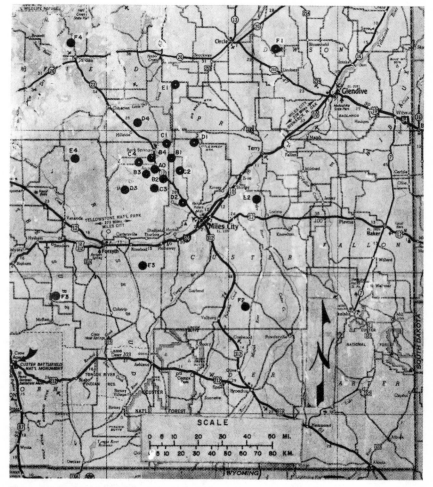

Figure 3.2-3 Map of LASA seismological observatory.

sub-array AO. The entire system is 200 km in diameter. Except for E3, which is larger, each sub-array is 7 km in diameter and includes about 16 short-period vertical seismometers. Each seismometer is at the bottom of a drilled, cased hole, either 60 or 150 m deep. A solid-state amplifier in each well-head vault connects via buried cables to a central buried walk-in vault (SEM) at the center of the sub-array. At the SEM the data are digitized for telemetry via overhead lines and microwave links to the LASA Data Center at Billings, Montana. Here the data are recorded and telemetered continuously to the Seismic Array Analysis Center in Alexandria, Virginia, for automatic filtering, compositing, and computer analysis.

In addition to the large short-period seismic array, the LASA includes an array of 21 three-component sets of long-period seismometers. One set occupies a buried walk-in vault (Figure 3.2-4) near the center of each sub-array. There are also two interlacing arrays of microbarograph sensors for

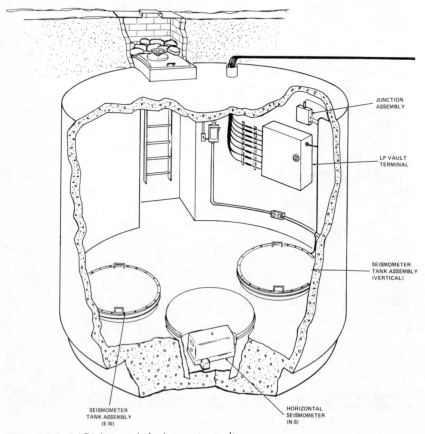

Figure 3.2-4 LASA long-period seismometer vault.

the study of infrasonic signals generated by volcanos, tornadoes, magnetic storms, and large explosions. Figure 3.2-5 indicates the equipment of these arrays, including supplementary wind sensors and the wind-noise attenuators (several hundred meters of galvanized pipe with intake orifices each 3 m diameter) used to average out short-wavelength atmospheric disturbances.

Movable Seismic Observatories. At least three types of movable, high-quality seismic stations have been developed. The Long-Range Seismological Measurements (LRSM) program uses large trailer-van stations[8]. Where commercial power is not available, a 15-kW diesel generator is operated continuously near the van. Standby power is provided by a storage battery bank. A crew of two visits the site daily for 4 to 10 hr for record-changing, calibration, maintenance, and preliminary data reduction. Records are shipped to a central laboratory weekly.

The sensors, consisting of one 3-component set of short-period seismometers and one set of long-period seismometers, are installed below ground on concrete pads, protected by steel tank vaults covered with bags of vermiculite insulation. Shielded, 4-conductor surface cables lead to an isolated amplifier housing and the van. Lightning protectors are provided at each cable termination.

Figure 3.2-6 illustrates the main components of an LRSM "portable"

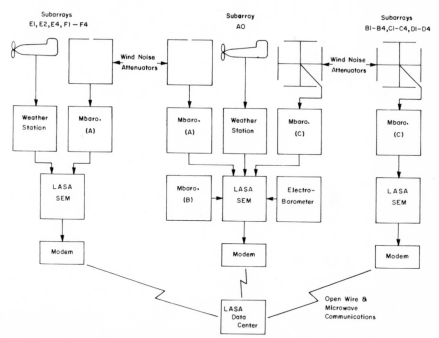

Figure 3.2-5   Block diagram of LASA microbarograph arrays.

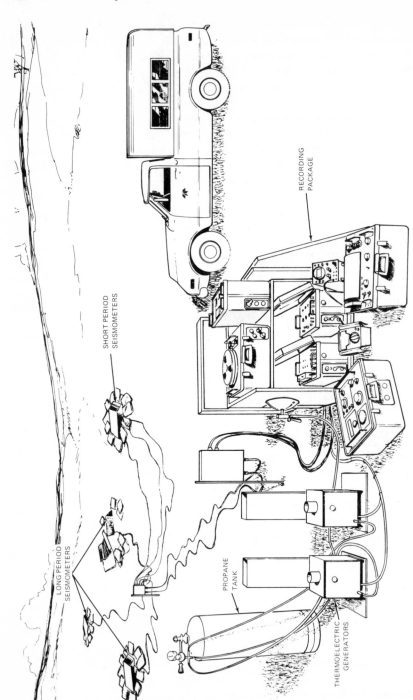

Figure 3.2-6 Typical field layout of the portable seismological observatory (opened for testing).

system [6]. Also shown are two thermoelectric (flame) power generators and a propane tank capable of heating and running the station for a week, where commercial power is not available. A standby storage battery capable of operating the station for a day is provided. The primary recording means is a 14-channel 0.75 mm/s magnetic tape recorder and the standard unattended recording period is a few days. All of the instrumentation is housed in luggage-type containers. The protective field box folds flat. All can be carried in a camper truck or as excess baggage for air travel. The compact short-period seismometers used are suitable for direct burial, but the long-period seismometers require an insulated box or tank vault. Solid-state amplifiers are used exclusively [19].

The third example is a seismic station housed in a small 2-wheel trailer. Eight of these stations have been used as an array [13]. A single buried short-period seismometer is employed as the sensor for each station. The supporting equipment includes a phototube amplifier and a magnetic tape recorder capable of recording 7 days on one reel of tape. A special 35-mm film recorder is also used which can operate for 50 days without attention. This is done by galvanometric recording of a helical line on a 45 cm length of 35-mm film stretched around a drum. Upon completion of 48 turns of the helix, the exposed film is automatically wound into a takeup cannister and is replaced with fresh film from a supply reel. A small photographic darkroom in the rear half of the trailer holds the film recorder; standing room is provided by a drop-floor connected to the trailer by a large, light-tight bellows. Commercial power is generally used, stabilized by a floating standby storage battery and charger. However, battery power can be provided for up to 7 days of unattended operation. The power demand is only 30 W per station.

Exploration Seismic Platforms. For seismic prospecting on land, the sensing geophones are connected in clusters for signal enhancement and the clusters are deployed in a one-dimensional or two-dimensional array of up to 96 channels to provide spatial filtering [1]. Lightweight, multiconductor cables carry the signals to a central recording facility, which may be a light truck or a set of back-packed instruments placed on the ground. Time on station is typically a few hours. Recording is generally done on magnetic tape. Wide-band digital recording is favored for its precision and great dynamic range. The separate units, comprising the tape deck, the digitizer/formatter, the multiplexer/amplifier, and power supply, can also be back-packed as separate, sealed cases.

Controllable seismic-wave generators have largely displaced dynamite in drilled holes. These are of many types, but generate either a seismic impulse or a swept-frequency train of seismic oscillations [2] at the surface of the ground. Such sources are synchronized with the recording equipment by radio. They weigh many tons and are mounted on heavy-duty rubber-tired vehicles.

3.2.2 Weather Stations

Ground-based weather platforms can be divided into major groups including regional forecast centers, first-order observatories, second-order observatories, and cooperative stations. The entire system is linked together by teletype and other circuits. In a sense it is one huge geophysical sensor platform of many essential parts, none of which can be effective alone.

The 400 first-order weather stations are the primary instrument platforms. Lesser stations have fewer sensors and much less communication equipment. Most second-order stations measure only 3 to 5 weather parameters to fill necessary gaps in the first-order network. Second-order stations are all located at small-town and rural airports. The cooperative stations generally report only unusual local rainfalls and river levels, frequently by telephone, and are run by volunteers.

First-order weather observatories are installed at major airports [18]. A typical station occupies a room with 30 square meters of floor area near the top of the building with a good view of the horizon and the sky for estimates of visibility and cloud cover. The sensors are mounted on masts or other structures remote from buildings and other obstruction. Supporting equipment includes repeater indicators and chart recorders for air temperature, wind speed, wind direction, dewpoint, precipitation, light transmissivity, and cloud cover (total solar intensity); facsimile recorder for the rotating-beam ceilometer; radar repeater scope connected by microwave link to the regional weather radar set; facsimile receiver for weather maps; four teletype receivers for incoming local reports; two typing transceivers for radar warnings, weather bulletins, and forecasts; and a facsimile transmitter to send local reports to the airlines and the control tower. Additional equipment provided at selected stations includes WSR-57 weather radar system; rawinsonde upper-air balloon-tracking and receiving equipment; and the Automatic Meteorological Observing System (AMOS), which collects, processes, records, and disseminates sensor data automatically.

3.2.3 Air Quality Stations

Two kinds of protective structures are used for typical air-quality sampling stations, a 12-sided geodesic dome hut 7.5 m in diameter, or a modular rectangular temporary building [4]. The structure is automatically heated and air-conditioned.

The air-quality sensors are analytical instruments arranged for continuous on-stream operation with both the sample and the reagent flowing continuously. They are supplied with outside air via a glass manifold. The instruments include a carbon monoxide sensor, measuring the absorption of infrared wavelengths characteristic of CO; nitrogen oxide analyzers, employing specific reactions of NO and NO_2 with colored reagents, and measurement by a ratio photometer; sulfur dioxide sensor, similar to the NO sensors; total oxidant analyzer, similar to the NO sensors, employing oxidation of KI solution to colored free iodine; total hydrocarbon analyzer, measuring the ionic conductivity of a hydrogen flame produced by carbon atoms; high-volume and tape-filter particulate samplers; and special weather sensors including a wind gust vane sensor to evaluate air mixing, a sun photometer to estimate total atmospheric transparency, and a covered precipitation sampler opened automatically by falling moisture to measure rain-collected pollutants. These sensor outputs are correlated with reports from local weather stations.

Strip-chart recordings are made continuously of all gaseous monitor outputs. The station is attended and charts are changed daily. A digital magnetic tape record is also made every five minutes of each sensor output averaged over a five-minute period. Weekly records are compiled monthly.

3.2.4 Ionosphere Observing Stations

Ground-based observations of ionospheric structure are performed with ionosonde equipment at geophysical stations throughout the world [3] for research and to provide data for radio propagation forecasts [15].

Ionosonde equipment consists of a $1 \times 1 \times 2$-m-high cabinet containing a 1-kW pulsed radio transmitter, a radio receiver, and a 35-mm film recorder employing an intensity-modulated cathode ray sweep. A broad-band, vertically-aimed half-rhombic antenna is mounted outdoors. Echo intensity is plotted across the film as a function of range (height) and successive sweeps are placed along the film as the system is slowly tuned from about 0.25 to 25 MHz. The entire operation is automatic and is repeated hourly (or more often as desired). The equipment need only be checked and the film changed once each week.

3.2.5 Magnetic Observatories

Phenomena in the ionosphere and in the magnetosphere beyond it produce magnetic variations at the earth's surface. These variations affect compass navigation and radio communications. They are recorded continuously at stations throughout the world [20]. At some stations observations are also made of higher-frequency phenomena having periods between 0.2 s and 100 s for research purposes [9].

A typical station is located in an area of uniform magnetic gradient several kilometers from major electrical transmission lines and machinery and is served by branch power and telephone lines. The sensors consist of N-S, vertical, and azimuthal magnetic-balance magnetographs recording on a drum of photographic paper; large sensing coils oriented N-S, E-W, and vertical and mounted on massive, rigid foundations; and pairs of lead-plate electrodes buried in the earth several hundred meters apart on N-S and E-W lines for electric-potential sensing. The coils and electrodes are connected to the central recording building by buried cables. Small, remote wooden houses shelter the coils and the magnetographs. They are maintained at summer temperatures by automatic ac electric heaters. No steel is used in construction, and nonmagnetic concrete aggregates and road material are used. Vehicles are parked at least 100 m from the sensors.

Since the magnetographs are self-contained, the principal supporting equipment is in the micropulsation system [9]. It consists of amplifiers, filters, analog recorders, digital recorders, and a digital computer to assist in the study of the data.

3.2.6 Combination Geophysical Observatories

Many research institutions maintain a variety of multipurpose geophysical sensor platforms. One is located near the Arctic circle, in a tectonically active area near glaciers and is equipped for work in several disciplines [5]. A unique aeronomy facility is the rocket launching station 50 km north of Fairbanks, Alaska.

Another combination geophysical observatory is the South Pole station maintained for Antarctic studies and exploration [16]. Figure 3.2-7 illustrates the main features of the original station. The chief function of the station is life support for 18 to 22 men over the 6-month polar night. Buildings are prefabricated of 10-cm-thick plywood-insulation-metal sandwich board. Diesel generators provide power and their heat melts snow for culinary water. Oil heaters warm each building. There are several radio circuits to the outside world. In the polar summer the snow surface outside is scraped smooth by tractor for the landing of ski-equipped supply aircraft. A physician-dentist is in residence and the crew is normally changed each year [17].

Geophysical facilities and studies emphasize meteorology, for which there are externally mounted sensors, a preparation room for rawinsonde instrumented balloons, and a rawin tracking antenna dish in an elevated radome. Aurora, airglow, and geodetic star measurements (for ice-drift estimates) are taken at an elevated observatory with transparent-dome and slit observing ports. There is ionosonde equipment, as well as small magnetic and seismological observatories reached by under-snow tunnels to their isolated locations. Glaciological studies (snow stratigraphy) have been made in a deep "snow mine" excavated for the culinary water supply.

The original station shown in Figure 3.2-7 has become deeply buried and endangered by the weight of many years' snows. In the replacement station all buildings stand free inside wide steel-roofed tunnels excavated by machine deep into the snow, or inside strong steel-arch protective enclosures [20].

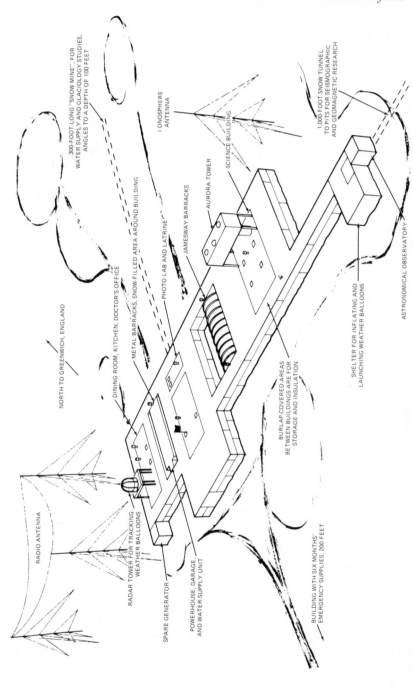

Figure 3.2-7 The original Amundsen-Scott South Pole station.

References

1. Clifford, O. C., "Presidential Address—The Revolution in Petroleum Exploration, 1955–," *Geophys.* **24**, 1–11, Feb. 1959.
2. Crawford, J. M., W. E. N. Doty, and M. R. Lee, "Continuous Signal Seismograph," *Geophys.* **25**, 95–105, Feb. 1960.
3. Davies, K., "Ionospheric Radio Propagation," National Bureau of Standards Monograph #80, U. S. Government Printing Office, Washington, D. C., April 1, 1965.
4. Environmental Health Service, U. S. Department of Health, Education and Welfare, National Air Pollution Control Administration, Cincinnati, *Continuous Air Monitoring Program* (booklet), GPO 810-314-2, 1967.
5. Geophysical Institute, University of Alaska, College, Alaska, Annual Report, 1968–69.
6. Geotech Teledyne Corporation, Garland, Texas, "Operation and Maintenance Manual, Portable Seismograph System, Model 19282," 29 April 1966.
7. Geotechnical Corporation, "Instrumentation of the World-Wide Seismograph System, Model 10700," Contract CGS-910, U. S. Coast and Geodetic Survey, (booklet), Washington, D. C., 1962.
8. Geotechnical Corporation, Garland, Texas, "Interim Report on Operating Procedures," Technical Report TR 62-22, Project VT/074, 1962.
9. Green, A. W., Jr., and A. A. J. Hoffman, "Micropulsation Instrumentation Systems at the Dallas Geomagnetic Center," *IEEEE Geosci. Electron.* **GE-5**, 3–17, March 1967.
10. Green, P. E., Jr., R. A. Frosch, and C. F. Romney, "Principles of an Experimental Large Aperture Seismic Array (LASA)" (and 3 following articles), *IEEE* **53**, 1821–1859, Dec. 1965.
11. Gudzin, M. G., and J. H. Hamilton, "Wichita Mountains Seismological Observatory," *Geophys.* **36**, 359–373, June 1961.
12. Gudzin, M. G., and E. G. Holle, "Seismological Observatories," *IRE* **50**, 2216–2224, Nov. 1962.
13. Lehner, F. E., and F. Press, "A Mobile Seismograph Array," *Bull. Seis. Soc. Am.* **56**, 889–897, Aug. 1966.
14. Philco Ford Corporation, "Montana Large Aperture Seismic Array," Reports MLM (9 volumes), Contract F19628-68-C-0401, 1969.
15. Rishbeth, H., and O. K. Garriott, *Introduction to Ionospheric Physics,* Academic Press, New York, 1969.
16. Siple, P. A., "Man's First Winter at the South Pole," *The National Geographic Magazine* **113**, 439–478, April 1958.
17. U. S. Naval Support Force, Antarctica, *Support for Science,* Washington, D. C., April 1968.
18. U. S. Weather Bureau, Department of Commerce, Washington, D. C., *The Instrumental Program of the U. S. Weather Bureau* (booklet), Dec. 1961.
19. Watkins, C. D., "Evaluation of the Portable Seismograph System," Technical Report TR 65-7, Geotech Teledyne Corporation, Garland, Texas, 1965.
20. Wienert, K. A., *Notes on Geomagnetic Observatory and Survey Practice,* UNESCO, Paris, 1970.

Louis C. Haughney

3.3 AIRBORNE PLATFORMS

3.3.1 Uniqueness of Airborne Platforms

The use of aircraft as research instrument platforms flourished with the introduction of high-speed, high-altitude, jet aircraft. Many airplanes, from the small, single-engine type to the fast, multi-engine jet, are now recognized as valuable research platforms with unique benefits. Airborne platforms fill the gap between ground stations and balloons, rockets, and spacecraft. They provide for research that is impracticable from the ground but that, on the other hand, does not require the instruments to be entirely out of the earth's atmosphere.

The mobility and the altitude capability of an airplane are its principal advantages over ground stations. Observations can be made in inaccessible regions [11, 21]. Latitude and longitude surveys over large areas can be accomplished in short periods; a jet airplane flying at 480 kt covers eight degrees of arc over the earth's surface in one hour. (1 knot = 1 nautical mile per hour = 1 arc minute per hour = 0.5148 m/s.) The aircraft can often climb above clouds that obscure ground observations of airglow, auroras and stars.

Infrared (IR) observations of the upper atmosphere and outer space become feasible at 10 to 12 km where the airplane is often above the tropopause and consequently above 95 % or more of the atmospheric water vapor that absorbs incoming infrared radiation. The height of the tropopause varies inversely with latitude, from about 15 km in the equatorial regions to 8 km in the polar areas; also it is lower in winter than in summer. The amount of water vapor at altitudes typical of ground and airborne platforms is given in Table 3.3-1.

High-speed jet aircraft can keep up with the earth's rotation in the higher latitudes. For example, an airplane with a ground speed of 900 km/hr can remain at local midnight on an east to west flight at 60°N. Universal and local time effects in auroral substorms can be separated by means of a constant local time flight [1].

In contrast to balloons, rockets, and unmanned satellites, airplanes allow the investigator to fly with his equipment. Thus, he may evaluate his measurements in real time, change the mode of observation during flight to fit the prevailing conditions, and even repair his equipment.

Table 3.3-1.   **Precipitable Water Vapor above Certain Altitudes**

Altitude (km)	Observation Platform	Precipitable H_2O Vapor (μm)	Latitude	Season
0	sea level	5,000[a]	30–50°N, S	Not given
4	mountain top	600[a]	30–50°N, S	Not given
12	aircraft	8–40[b]	37–43°N	January
12	aircraft	2–12[c]	55–70°N	Nov., Dec.

[a] From Table 1 [43].
[b] [41].
[c] Kuhn (unpublished data).

Coordinated measurements from ground stations, aircraft, and spacecraft give a quantitative altitude profile of geophysical phenomena and often reveal cause and effect relationships [16, 25, 38]. For effective aircraft-satellite coordination, the aircraft group must be supplied with the latest available satellite ephemeris. Precisely coordinated observations between aircraft and sounding rockets depend upon how closely the airplane must approach the launch base and the subtrajectory of the rocket.

The scope of airborne geophysical research is indicated in Table 3.3-2. The examples listed illustrate research problems adaptable to aircraft and the instruments used.

3.3.2   Aircraft Used for Scientific Research

The largest aircraft used regularly in the United States for scientific research are the four-engine CV 990 and KC 135 jet transports, which have altitude capabilities of 12.4 and 12 km and ranges of 6000 and 9800 km, respectively. Smaller "flying observatories" include the Sabreliner 265-60, Lear Jet 23, B57A, RB57F, DC 6, DC 4, C130, C121, Queen Air A80, and Aero Commanders 500B and 680E. Small-scale programs are often conducted on light aircraft leased from general aircraft operators. A complete list and description of airborne research facilities in the United States is available [51]. Detailed accounts of some individual facilities can be obtained [6, 26, 27, 48, 49].

3.3.3   Environment in a Research Aircraft

Commercial jet transports fly at altitudes up to 12.4 km; the equivalent altitude inside the pressurized cabin reaches about 2.5 km. Near 12 km, the cabin air is very dry, about 10% relative humidity at 20°C. Some aircraft can fly above 14 km. For these flights, personnel undergo low-pressure physiological training and use oxygen continuously even though the cabin may be pressurized. In unpressurized aircraft, personnel generally use oxygen above 3.5 km.

Table 3.3-2. Examples of Airborne Geophysical Research

Research field	Instrumentation	Ref.
Earth resources		
General bibliography		50
Terrain signatures		
Snow fields	K-band radar	65
Natural subjects	Thermal mapper, 3.7–5.5 μm	10
Surface albedo	Pyranometers	60
Oceanography		
Bathymetry	Neon laser	34
Sea waves	Laser profilometer	54
Chlorophyll concentration	Ebert spectrometer	
	IR radiometer	13
Sea surface temperature	Radiation thermometers	52, 57
Magnetic surveys		
General review	Magnetometers	35
Anomalies, sea floor	Fluxgate, vector	5
spreading	magnetometer	
Aerial photography		
General review	Aerial cameras	59, 63
Gravimetry		
Gravimetry	Air-sea gravity meter	44
Lower atmosphere		
Meteorology		
IR reflectance of clouds	Filter wedge spectrometer,	36
and contrails	0.68–2.4 μm	
Radon 222 collection,	Ram filters	
air mass movement	Geiger counters	56
Hurricanes	Radars	
	Vortex thermometers	33
Cloud seeding	Pyrotechnic seeding	
	systems	58
Lee wave flow	Reverse flow thermometer	
	Aircraft systems instruments	64
CO_2 content	IR gas analyzer	39
Cloud physics, microstructure	Laser nephelometer	9
Atmospheric structure	Lidar	14
Particle detection	Optical array detectors	40
	Precipitation particle	
	probes	
Atmospheric optics		
Sunset green flash	Questar telescope	
	35-mm camera	61
	Heliostat	
Upper atmosphere		
Constituents		
IR scattering	Interferometer, 60–400 μm	
	IR radiometer	23
H_2O vapor	Bolometer radiometer	41
H_2O vapor	Interferometer	46

Table 3.3-2. **(continued)**

Research field	Instrumentation	Ref.
CO_2	Nondispersive gas analyzer	29
Aurora and airglow		
Conjugate auroras	All-sky and narrow field cameras	7
Latitude survey	Tilting filter photometer	20
"Soft" electrons		
O_2 1.27 μm emission	Scanning spectrometer	53
OH (6,1) and (7,2) and	1-meter Ebert spectrometer	
OI 5577Å emissions	5-channel photometer	18
Conjugate photoelectron effect	6-channel scanning photometer	30
Cosmic rays		
Latitude survey	Neutron monitor	12
Altitude survey	Ionization chamber, thin-walled, air-filled	28
Astronomy, solar physics		
Solar constant and irradiance	Multichannel, filter radiometers	45
Solar constant and irradiance	Pyrheliometers Radiometers Monochromators Spectrometers	62
Solar constant and irradiance	Prism-grating spectrophotometer	2
Solar IR spectrum		
Far IR brightness temperature	Scanning Michelson interferometer	22
	Heliostat	
Atlas of solar IR spectrum, 0.85–3.3 μm	2 30 cm telescopes 2 heliostats 4-m spectrometer	8, 42
Solar eclipses		
IR coronal lines	$\frac{1}{2}$-m Ebert-Fastie spectrometer	
	Heliostat	47
Outer corona brightness and polarization	Camera, polarizers	55
Astronomy, planetary, stellar, etc.		
IR spectroscopy	30 cm Cassegrain telescope Block 20 cm^{-1} interferometer Heliostat	43
Far IR spectrum of galactic center	30 cm open port telescope Ge bolometer, LHe cooled, 1.5–350 μm	4

An airplane rolls, pitches, and yaws even in calm air. Roll is the motion about the longitudinal axis; pitch, that about the transverse axis; and yaw, that about the vertical axis. The amplitudes and the periods of these motions depend upon the autopilot's tuning and sensitivity, the air's calmness, and the aircraft's weight, moment of inerita, and speed. Roll has the largest amplitude of the three; pitch, the least. Typical amplitudes, for a four-engine jet transport are 0.5° to 1° in roll and yaw with periods of about 5 s and 40 to 100 s [22]. Optimum tuning of its autopilot reduces the amplitude to ± 0.2° for the 5-s roll and to ± 0.7° for the longer period [6]. These motions affect mostly optical experiments making long exposures on small, localized targets, such as the sun, the planets, and the stars. Platforms or gimballed mirrors stabilized by gyros reduce these motions to about ten arc seconds. (See Section 3.3.9)

Aircraft vibration during flight generally does not bother most installations since much of it is dampened by the stand holding the equipment. An exception occurs in the case of interferometers, which may be sensitive to microphonics of acoustic frequencies. Acoustic insulation or isolating mounts may be needed for interferometers or for any other optical instrument that requires a stability better than 10 arc seconds. Severe vibrations may occur during takeoffs and landings. Electrical connections, plug-in circuit boards, and the like tend to work loose unless tightly seated.

3.3.4 Limitations on Aircraft Operation

The airplane becomes an integral part of the experimental system, a complex component with many parts, any of which may malfunction and thus postpone or even cancel a flight. Preventive maintenance and uncompromising inspection by the aircraft's crew are needed to achieve the maximum possible flight time. Weather conditions also delay flights. The predicted weather must be favorable at an alternate landing site as well as at the intended terminal, and the chief pilot must reserve enough fuel to reach the alternate. This provision often limits the flight range in areas where suitable landing fields are scarce.

The frequency of flights depends on the availability of flight and ground crews, the stamina of the airborne scientists, and the funding. Pilots' flying time is controlled on a daily, a weekly, and a monthly basis for safety reasons. Even if extra flight crews were available, daily flights on an extended basis would put a severe physical strain on the participating experimenters unless they too have replacements. On programs lasting two weeks or more, a schedule averaging one flight every two days has proved most efficient. Flying three or more consecutive days or nights is avoided except in very special circumstances.

Operations at advance bases present logistics problems. Length and bearing capacity of runways, availability of landing aids, fuel, hangar, and ground support equipment, accessibility of communications and transportation links to the home base, as well as availability of experimenters' needs like liquid nitrogen, dry ice, and even lodging, are items that must be considered in selecting an advance base. Overflights of other countries and basing in other countries require the permission of those governments.

Airborne researchers cannot always fly whenever and wherever they wish. All flights are governed by Federal Air Regulations (FAR) in the United States and by equivalent rules in other countries. In most of the United States, Visual Flight Rules (VFR) apply when the airplane is flying under 7.3 km, is clear of clouds, and has a prescribed range of visibility. Under VFR, the flight plan can be changed at will to look at targets of opportunity, such as particular clouds or surface features. Above 7.3 km, however, an aircraft is always under Instrument Flight Rules (IFR), which require the chief pilot to file a flight plan and to maintain continuous coordination with Air Traffic Control (ATC). An en route change in the flight plan requires the approval of ATC. The procedure for a complex change takes up to 15 min, during which time the aircraft must continue on its original route. On and near regular air routes, it is difficult to obtain approval for a specific flight plan at a certain time [17].

3.3.5 Navigation and Air Data Systems

Reduction of airborne data requires a time history of many aircraft parameters; for example, position, altitude, heading, attitude, ambient pressure and temperature, and wind velocity. Position versus time, which is almost always needed, is given by the navigational systems. These are described in Section 3.8. Post-flight navigational data are recovered from the navigator's log and charts, from sequential photographs of the navigator's instrument panel, and, where available, from the inertial navigation system (INS) tapes. In many studies, aerial photography with a time data insert on each frame suffices to identify the area overflown, but that method obviously applies only when the ground is visible. Ground-based radars may also give a precise fix on the aircraft's position [37].

Ambient air pressure and temperature sensors are mounted flush with the fuselage or extended in pitot tubes. They provide the input for the air data system, which computes pressure altitude, true airspeed, Mach number, and ambient air temperature. The pressure sensors are aneroid types; the temperature sensors, resistance thermometers or thermocouples. The sensors' outputs must be corrected by a mechanical or an electronic computer for the effect of airflow over them and their position on the airplane. The pressure altimeter gives the output of the pressure sensors in terms of altitude according to a standard atmospheric model. Before takeoff and landing, it is set to the barometer reading at each terminal. Radio and radar altimeters determine the actual altitude above the underlying terrain by measuring the time of flight of radio waves and pulses, respectively, from the airplane to the ground and return.

Some outputs from the navigational and the air data systems do not yet have the accuracy and resolution desirable for scientific and detailed studies of lower atmospheric phenomena; particularly, computed wind velocity and true values of ambient pressure. The difficulty arises not only in cross-calibration among airplanes of a multi-plane operation but also in the calibration of one airplane's sensors.

3.3.6 Special Apertures and Windows

For precise optical observations, the regular aircraft windows are neither of good enough quality nor, generally, in the proper position. Apertures are desirable for upward and downward viewing. The size and the number of viewports that can be cut depends upon the framework of the aircraft. Typical observation windows in a high-altitude jet airplane are single panes, 30 by 35 cm clear aperture and 25 mm thick to withstand the pressure differential of 0.6 kg/cm^2 and the temperature differential of 50 to 80°K across them at 12 km. Fogging and icing of the windows' inside surfaces, which are likely to occur at high altitudes, can be prevented by directing warm air from the ship's air conditioning system across the window.

It is important to keep the windows clean. In most localities, they are covered whenever the airplane is on the ground, especially when it is parked outside a hangar. Condensation of nighttime dew may leave dirt on the glass. The windows can also become dirty during takeoff and climb to altitude. Bottom windows are obviously susceptible to soiling and even damage during takeoff. If structurally possible, the windows are covered with external, sliding shutters that can be cranked opened from inside the airplane after takeoff.

Optical observations in the visible and the near infrared region, to 2.5 µm, can be made with windows of borosilicate crown glass (Pyrex), quartz, or fused silica. At longer wavelengths, materials like arsenic trisulfide, sapphire, Irtran, polyethylene, and the like must be used. They are not as strong as the other materials and are more expensive. Consequently, such IR windows are often mosaics composed of smaller and thinner pieces of those materials. (See Section 3.3.11 for an alternate way of installing IR equipment.)

All optical windows must undergo environmental tests in pressure and temperature chambers before each series of flights. As soon as possible after takeoff on each flight, a crewman ought to inspect the windows for leakage around the frame or for damage. For safety reasons, extreme care must be taken to avoid hitting the window when one holds a small instrument like a camera close to it.

3.3.7   Electrical Power

The electrical power generated by aircraft is either 28 V dc or 220 V, 3ϕ, 400 Hz ac. To provide 60 Hz power, inverters or frequency converters must be added to the aircraft, either static, solid-state devices or rotary, motor-generator sets. In the power distribution circuit, the aircraft structure normally serves as the return wire. Ground loop interference may be a problem; pulsating loads like paper punches and camera shutter motors may produce electrical noise on neighboring experiments. Good internal grounding, careful shielding, and relocation of leads or even of equipment reduce the interference. At times, operation of the aircraft's radio trans-mitters is detected by sensitive instruments. If batteries are needed, either dry cells or special spill-proof wet batteries must be used; Ni-Cd cells are very suitable for airborne experiments. The problem of interference is discussed in Section 8.2.

When the airplane is on the ground electrical power is supplied by a ground power unit (GPU) or, for light loads, through extension cords from the hangar. After the engines are started, the airplane's circuit is switched from the GPU to the aircraft's generators. A 1- or 2-s interruption of power occurs during the transfer. More information on power supplies is given in Section 3.7.

3.3.8   Timing Systems

Most scientific research aircraft carry a time code generator (TCG) syn-chronized to a time signal radio station, such as WWV. Occasional monitor-ing of WWV can be made with the ship's HF radio. For continuous mon-itoring, however, the airplane should carry a separate WWV receiver and antenna. The TCG should have a stand-by battery since momentary inter-ruptions of aircraft power do occur. Distribution circuits for timing signals need buffering and isolating units that prevent the false triggering of the TCG by a feedback signal from an experiment.

3.3.9   Stabilized Platforms

When high pointing accuracy is required, stabilized platforms are recom-mended. One type stabilizes only the optical axis of the system. The optical instrument is mounted on a stand attached securely to the aircraft structure. The stabilizing element is a mirror mounted in a gyro-controlled, two-axis gimbal. The gyro on one axis, which is aligned parallel to the aircraft longitudinal axis, nullifies the roll motion. The gyro on the other axis, which is in the plane normal to the roll axis, nullifies only those components of the pitch and the yaw motions that are parallel to it. The line of sight goes from the object, the sun for example, through the window to the center point of the mirror and then along the optical axis of the instrument to the detecting element. Stability of ± 10 arc seconds can be maintained for minutes even in light turbulence. In a second type of stabilizing system, the instrument is mounted in a gyro-controlled, three-axis gimbal. All objects initially in the instrument's field of view are about equally stabilized. With this second system, however, the size and the weight of the equipment that can be supported is limited by the gimbals and the torquers available.

3.3.10 Data Handling, Recording, and Telemetry Systems

The handling of data collected on aircraft does not differ much from that gathered by a ground-based program of equal complexity and observing time. As mentioned in Section 3.3.5 the data must often be correlated with aircraft position. The experimenter on an airplane observes the environment in which he is taking data in real time, and his "quick-look" at the data often suggests which segments of the data should be analyzed first.

Because most airplanes carry recorders on board, telemetry systems are rarely used. If possible, an experiment should include both a strip chart recorder and a magnetic tape recorder. Complex data-acquisition systems made of standard components and computers are used on many research aircraft. Telemetry from small aircraft has been used for low-altitude surveys [37].

3.3.11 Equipment Location

Table 3.3-2 gives examples of the various scientific instruments that have been installed in airplanes. Most of them are regular laboratory types used in ground observations. Cryogenic and vacuum systems present no major problems. Among the largest installations in an aircraft have been the 900 kg ionospheric sounder carried permanently in a KC-135 [31] and an optical system consisting of a four-meter spectrograph, a 30 cm telescope, and a gyrostabilized mirror that was carried in a CV-990 [15].

Many experiments can be mounted in the passenger cabin where the investigators can operate them. Some detecting instruments are placed in external pods and booms or in the nose and the tail sections. Equipment placed in unpressurized areas can become very cold at the higher altitudes, − 55°C or lower at 12 km. As the airplane descends, moisture or even frost may condense on the cold surfaces. To prevent this, electrical heaters are placed on the exposed instruments [32].

Small probes, antennas, and detectors are often mounted directly on the outside of the aircraft. Their aerodynamic effects must be studied to determine if streamlining is required and if the structure is adequate.

Some instruments are flown in ports open to the outside atmosphere but closed to the inside of the airplane. This open-port method is often used with IR experiments where a large, thick window may absorb a significant part of the IR intensity incident upon it [3, 6, 45]. Care must be taken so that the cavity does not act as an organ pipe because of the airflow over it and excite resonant vibrations in the aircraft structure.

3.3.12 Equipment Installation

Aircraft safety regulations are one of the most stringent interface requirements of airborne instrumentation. Equipment must be installed so that it cannot break loose during a minor crash landing or a sudden stop on the runway. The equipment installation in the passenger sections must meet the same specifications as do the passenger seats and belts. The loads or forces that each item must sustain before its structure yields or breaks loose depend upon the type of aircraft. Typical factors for a large four-engine jet aircraft are 9 g forward, 7 g downward, 2 g upward, 1.5 g lateral, and 1.5 g aft. A stress analysis of each installation must be made to show that it meets loading constraints.

In passenger airplanes used for research, the framework containing the experimental equipment is attached to the wall and floor tracks that hold the seats. It cannot be indiscriminately bolted to the aircraft ribs or floor. Hardware items, such as screws, nuts, and bolts, must be certified to National Aerospace Standards (NAS), which stipulate very rigid quality control during manufacture. Aluminum is recommended for the bracketry in order to conserve weight. The installation should not be overdesigned since conservation of weight is essential when maximum range or altitude is important.

References

1. Akasofu, S. -I., "Auroral Observations by the Constant Local Time Flight," *Planet. Space Sci.* **16**, 1365–1370. Nov. 1968.
2. Arvesen, J. C., R. N. Griffin, Jr., and B. D. Pearson, Jr., "Determination of Extraterrestrial Solar Spectral Irradiance from a Research Aircraft," *Appl. Optics* **8**, 2215–2232, Nov. 1969.
3. Aumann, H. H., C. M. Gillespie, Jr., and F. J. Low, "The Internal Powers and Effective Temperatures of Jupiter and Saturn," *Astrophys. J.* **157**, 169–172, July, 1969.
4. Aumann, H. H., and F. J. Low, "Far-Infrared Observations of the Galactic Center," *Astrophys. J.* **159**, Pt. 2, L159–L164, March 1970.
5. Avery, O. E., G. D. Burton, and J. R. Heirtzler, "Aeromagnetic Survey of the Norwegian Sea," *J. Geophys. Res.* **73**, 4583–4600, July 15, 1968.
6. Bader, M., and C. B. Wagoner, "NASA Program of Airborne Optical Observations," *Appl. Optics,* **9**, 265–270, Feb. 1970.
7. Belon, A. E., J. E. Maggs, T. N. Davis, K. B. Mather, N. W. Glass, and G. F. Hughes, "Conjugacy of Visual Auroras during Magnetically Quiet periods," *J. Geophys. Res.* **74**, 1–28, Jan. 1, 1969.

8. Bijl, L. A., G. B. Kuiper, and D. P. Cruikshank, "Arizona-NASA Atlas of Infrared Solar Spectrum, Report V," Communications of the Lunar and Planetary Laboratory, No. 161, University of Arizona, Tucson, 1969.

9. Blau, H. H., Jr., M. L. Cohen, L. B. Lapson, P. von Thüna, R. T. Ryan, and D. Watson, "A Prototype Cloud Physics Laser Nephelometer," *Appl. Optics* **9**, 1798–1803, Aug. 1970.

10. Blythe, R., and E. Kurath, "Infrared Images of Natural Subjects," *Appl. Optics* **7**, 1769–1777, Sept. 1968.

11. Buchau, J., J. A. Whalen, and S. -I. Akasofu, "Airborne Observation of the Midday Aurora," *J. Atmos. Terrest. Phys.* **31**, 1021–1026, July 1969.

12. Carmichael, H., M. A. Shea, and R. W. Peterson, "III. Cosmic-Ray Latitude Survey in Western USA and Hawaii in Summer, 1966," *Can. J. Phys.* **47**, 2057–2065, Oct. 1, 1969.

13. Clarke, G. L., G. C. Ewing, and C. J. Lorenzen, "Spectra of Backscattered Light from the Sea Obtained from Aircraft as a Measure of Chlorophyll Concentration," *Science* **167**, 1119–1121, 20 Feb. 1970.

14. Collis, R. T. H., "Lidar," *Appl. Optics* **9**, 1782–1788, Aug. 1970.

15. Cruikshank, D. P., F. A. De Weiss, and G. P. Kuiper, "High-Resolution Solar Spectrometer for Airborne Infrared Measurements," Communications of the Lunar and Planetary Laboratory, No. 126, University of Arizona, Tucson, 1968.

16. Curtis, W. R., and P. K. Rao, "Gulf Stream Thermal Gradients from Satellite, Ship, and Aircraft Observations," *J. Geophys. Res.* **74**, 6984–6990, Dec. 20. 1969.

17. Danielson, E., R. Bleck, J. Shedlovsky, A. Wartburg, P. Haagenson, and W. Pollock, "Observed Distribution of Radioactivity, Ozone, and Potential Vorticity Associated with Tropopause Folding," *J. Geophys. Res.* **75**, 2353–2361, April 20, 1970.

18. Dick, K. A., G. G. Sivjee, and H. M. Crosswhite, "Aircraft Airglow Intensity Measurements: Variations in OH and OI (5577)," *Planet. Space Sci.* **18**, 887–894, June 1970.

19. Drummond, A. J., and J. R. Hickey, "The Eppley-JPL Solar Constant Measurement Program," *Solar Energy* **12**, 217–232, Dec. 1968.

20. Eather, R. H., "Latitudinal Distribution of Auroral and Airglow Emissions: The 'Soft' Auroral Zone," *J. Geophys. Res.* **74**, 153–158, Jan. 1, 1969.

21. Eather, R. H., and S. -I. Akasofu, "Characteristics of Polar Cap Auroras," *J. Geophys. Res.* **74**, 4794–4798, Sept. 1, 1969.

22. Eddy, J. A., R. H. Lee, P. J. Lena, and R. M. MacQueen, "Far Infrared Airborne Spectroscopy," *Appl. Optics* **9**, 439–446, Feb. 1970.

23. Eddy, J. A., and R. M. MacQueen, "Infrared Scattering Observations in the Upper Atmosphere," *J. Geophys. Res.* **74**, 3322–3330, June 20, 1969.

24. Estes, J. E., and B. Golomb, "Oil Spills: Method for Measuring Their Extent on the Sea Surface," *Science* **169**, 676–678, 14 Aug. 1970.

25. Evans, J. E., "Coordinated Measurements on Auroras," in *Auroral Phenomena*, Ed. M. Walt, pp. 130–149, Stanford University Press, Stanford, California, 1965.

26. Friedman, H. A., F. S. Cicirelli, and W. J. Freedman, "The ESSA Research

Flight Facility: Facilities for Airborne Atmospheric Research," Environmental Science Services Administration Technical Report ERL126–RFF1, Aug. 1969.

27. Gassman, G. J., and E. W. Pittenger, "Upper Atmosphere Research Using Aircraft," Air Force Cambridge Research Laboratories—66–292, April 1966.

28. George, M. J., "New Data on the Absolute Cosmic Ray Ionization in the Lower Atmosphere," *J. Geophys. Res.* **75**, 3693–3705, July 1, 1970.

29. Georgii, H. W., and D. Jost, "Concentration of CO_2 in the Upper Troposphere and Lower Stratosphere," *Nature* **221**, 1040, March 15, 1969.

30. Glass, N. W., J. H. Wolcott, R. L. Wakefield, and R. W. Peterson, "Airborne Observations of the Night Airglow," *Ann. Geophys.* **26**, 179–186, Jan.–Feb.–March 1970.

31. Gowell, R. W., and R. W. Whidden, "Ionospheric Sounders in Aircraft," Air Force Cambridge Research Laboratories Instrumentation Papers, No. 144, 1968.

32. Hariharan, T. A., "Airborne Polarimeter for Atmospheric Radiation Studies," *J. Sci. Instr., J. Phys. E Ser. 2*, **2**, 10–12, Jan. 1969.

33. Hawkins, H. F., and D. T. Rubsam, "Hurricane Hilda, 1964, Part 2, Structure and Budgets of the Hurricane on Oct. 1, 1964," *Monthly Weather Rev.* **96**, 617–636, Sept. 1968.

34. Hickman, G. D., and J. E. Hogg, "Application of an Airborne Pulsed Laser for Near Shore Bathythermic Measurements," *Remote Sensing of Environment* **1**, 47–58, March 1969.

35. Hood, P., and S. H. Ward, "Airborne Geophysical Methods," in *Advances in Geophysics*, Vol. 13, Ed. H. E. Landsberg and J. van Mieghem, pp. 1–112, Academic Press, New York, 1969.

36. Hovis, W. A., Jr., L. R. Blaine, and M. L. Forman, "Infrared Reflectance of High Altitude Clouds," *Appl. Optics* **9**, 561–563, March 1970.

37. Howell, R. L., "Equipment and Techniques for Low-Altitude Aerial Sensing of Water-Vapor Concentration and Movement," *Remote Sensing of Environment* **1**, 13–18, March 1969.

38. Johnson, R. G., R. E. Meyerott, and J. E. Evans, "Coordinated Satellite, Ground-Based, and Aircraft-Based Measurements on Auroras," in *Aurora and Airglow*, Ed. B. M. McCormac, pp. 169–189, Reinhold Publishing Company, New York, 1967.

39. Kelley, J. J., Jr., "Observations of Carbon Dioxide in the Atmosphere over the Western United States," *J. Geophys. Res.* **74**, 1688–1693, March 15, 1969.

40. Knollenberg, R. G., "The Optical Array: An Alternative to Scattering or Extinction for Airborne Particle Size Determination," *J. Appl. Meteorology* **9**, 86–103, Feb. 1970.

41. Kuhn, P. M., M. S. Lojko, and E. V. Petersen, "Infrared Measurements of Variations in Stratospheric Water Vapor," *Nature* **223**, 462–464, Aug. 2, 1969.

42. Kuiper, G. P., and D. P. Cruikshank, "Arizona—NASA Atlas of Infrared Solar Spectrum—A Preliminary Report," Communications of the Lunar and Planetary Laboratory, No. 128, University of Arizona, Tucson, 1968.

43. Kuiper, G. P., F. F. Forbes, and H. L. Johnson, "A Program of Astronomical Infrared Spectroscopy from Aircraft," Communications of the Lunar and Planetary Laboratory, No. 93, University of Arizona, Tucson, 1967.

44. La Coste, L. J. B., "Measurement of Gravity at Sea and in the Air," *Rev. Geophys.* **5**, 477–526, Nov. 1967.

45. Laue, E. G., "The Measurement of Solar Spectral Irradiance at Different Terrestrial Elevations," *Solar Energy* **13**, 43–57, April 1970.

46. Louis, J. F., and R. M. MacQueen, "Spectroscopic Observations of Water Vapor near the Tropopause," *J. Appl. Meteorology* **9**, 722–724, Aug. 1970.

47. Münch, G., G. Neugebauer, and D. McCammon, "Infrared Coronal Lines, II. Observation of [Si X] λ1.43 μ and [Mg VIII] λ3.03μ," *Astrophys. J.* **149**, Pt. 1, 681–686, Sept. 1967.

48. National Aeronautics and Space Administration, Manned Spacecraft Center, "Earth Resources Aircraft and Sensors," Houston, 1970.

49. National Aeronautics and Space Administration, Manned Spacecraft Center, "Earth Resources Program Synopsis of Activity," Houston, 1970.

50. National Aeronautics and Space Administration, Scientific and Technical Information Division, "Remote Sensing of Earth Resources, A Literature Survey with Indexes," NASA SP-7036, Washington, 1970.

51. National Center for Atmospheric Research, Research Flight Facility, "Aircraft and Instrumentation in Atmospheric Research," NCAR Technical Note TN-6, Boulder, Colorado, 1970.

52. Noble, V. E., and J. C. Wilkerson, "Sea Surface Temperature Mapping Flights—Norwegian Sea, Summer, 1968," *Remote Sensing of Environment* **1**, 187–193, Summer 1970.

53. Noxon, J. F., "Auroral Emission from O$_2$($^1\Delta_g$)," *J. Geophys. Res.* **75**, 1879–1891, April 1, 1970.

54. Olsen, W. S., and R. M. Adams, "A Laser Profilometer," *J. Geophys. Res.* **75**, 2185–2187, April 20, 1970.

55. Pepin, T. J., "Observations on the Brightness and Polarization of the Outer Corona During the 1966 November 12 Total Eclipse of the Sun," *Astrophys. J.* **159**, Pt. 1, 1067–1075, March 1970.

56. Prospero, J. M., and T. N. Carlson, "Radon-222 in the North Atlantic Trade Winds: Its Relationship to Dust Transport from Africa," *Science* **167**, 974–977, 13 Feb. 1970.

57. Saunders, P. M., "Corrections for Airborne Radiation Thermometry," *J. Geophys. Res.* **75**, 7596–7601, Dec. 20, 1970.

58. Simpson, J., W. L. Woodley, H. A. Friedman, T. W. Slusher, R. S. Scheffee, and R. L. Steele, "An Airborne Pyrotechnic Cloud Seeding System and Its Use," *J. Appl. Meteorology* **9**, 109–122, Feb. 1970.

59. Smith, J. T., Jr., Editor-in-Chief, *Manual of Color Aerial Photography*, 1st ed., American Society of Photogrammetry, Falls Church, Virginia, 1968.

60. Stanhill, G., "Some Results of Helicopter Measurements of the Albedo of Different Land Surfaces," *Solar Energy* **13**, 59–66, April 1970.

61. Taylor, J. H., and B. T. Matthias, "Green Flash from High Altitude," *Nature* **222**, 157, April 12, 1969.

62. Thekaekara, M. P., R. Kruger, and C. H. Duncan, "Solar Irradiance Measurements from a Research Aircraft," *Appl. Optics* **8**, 1713–1732, Aug. 1969.

63. Thompson, M. M., Editor-in-Chief, *Manual of Photogrammetry*, 3rd ed., 2 volumes, American Society of Photogrammetry, Falls Church, Virginia, 1966.

64. Vergeiner, I., and D. K. Lilly, "The Dynamic Structure of Lee Wave Flow as Obtained from Balloon and Airplane Observations," *Monthly Weather Rev.* **98,** 220–232, March 1970.

65. Waite, W. P., and H. C. MacDonald, "Snowfield Mapping with K-Band Radar," *Remote Sensing of Environment* **1,** 143–150, March 1970.

Allyn Vine

3.4 OCEAN PLATFORMS

The characteristics of ocean platforms frequently control the design and use of instruments used in the sea. Thus a particular platform behaves somewhat as a bandpass filter permitting only certain kinds of measurements or measurement programs to be carried out, and is thus a major component of a measurement system. In addition, the largest ocean platforms also serve as bases for personnel.

Ocean platforms can be categorized as general-purpose ships, special-purpose ships, submersibles, and buoys. Most ocean research has been done from conventional ships of small or medium size. Recently there has been a trend to develop and use more specialized ships to broaden the variety of kinds of measurements that can be done at sea. Certain features such as reliability, safety, comfort, and general compatibility with the ocean are essential to all good ocean platforms.

Maintaining or installing large equipment on seagoing craft often requires extensive work or shipyard availability. If equipment can be made modular and portable then installation time, cost and maintenance can be reduced. If measuring instruments can be towed rather than mounted below the hull, the time and cost of a dry dock or marine railway can be saved. For short-time use it is often possible to borrow or rent special equipment.

3.4.1 General-Purpose Ships

The general-purpose surface ship is the great workhorse platform for making measurements at sea. Surface ships are important for three reasons.

First, ships operate on the air-water interface with its great density discontinuity. Thus they can carry large loads and can simultaneously use acoustic devices in the water, and they use electromagnetic communications, navigation devices, and ordinary vision in the air. The frequent roughness of the wave surface is, of course, a negative and frequently severe factor but that does not negate the advantages of surface ships.

Second, in the hands of skilled operators, good surface ships can be reliable and versatile.

Third, they exist in large numbers, and are available with operating personnel. Numerous shore establishments for maintenance, insurance, and operations are prepared for them.

Pictures and important information on the principal ocean going research ships of the world have been compiled [12].

Ships are designed to obtain a balance between speed, range, maneuverability, payload, sea-keeping, layout convenience, low first cost, and low upkeep. In any particular situation these factors are strongly influenced by the technical, legal, and financial customs and laws of the organization operating the ship.

In several countries there has been a concentration of ocean-wide range research ships 55 to 75 m long with speeds of 5 to 7 m/s. About one-fourth of the internal volume or area of those ships is available for measurement programs and one-fourth to one-half of the 40 to 60 people aboard are responsible for the measurement program rather than for running the ship.

Some considerably larger research and development survey ships and many smaller ships are being operated.

3.4.2 Special-Purpose Ships

Occasionally it is essential to design a ship to be better for one type of measurement even it it is degraded for other uses or measurements. Such craft are referred to as special-purpose ships.

Drilling Ships. A typical drilling ship is the 130m-long Glomar Challenger [16]. Drilling ships with their versatile propulsion systems have the capability of micromaneuvering to locate and remain within a few tens of meters horizontally from a given beacon on the bottom of the ocean. In addition, with their capability of handling long pipe, they can be rigged to lower or retrieve objects weighing many tons. Such ships can retrieve samples of rock or sediments from hundreds to thousands of meters below the ocean bottom.

Stable Platforms. The desire to do more precise work at sea and to minimize angular and translational motion caused by wave motion has resulted in a variety of relatively stable platforms to meet scientific and commercial requirements. One of the earliest and most used of such craft is Flip (floating instrument platform) (Figure 3.4-1) [2].

The 110-m-long hull is a compromise shape between a ship and an

Figure 3.4.1 Flip, a 117-m long special-purpose platform to provide great stability when in the vertical mode. This craft is a mannable buoy without major propulsion. Marine Physical Laboratory of the University of California.

elongated cylindrical buoy; it is towed to position horizontally, then upended to provide a buoy that has a 90-m draft with a heave period of 29 s and a pitch period of about 47 s. As these buoy periods are much longer than commonly encountered (6-to 12-s wave periods), the buoy damps about 90% of the surface wave motion. Such buoys are enlargements of narrow-necked instruments and spar buoys that have been used for many years as mechanical filters to minimize wave action.

It is, of course, possible to build a variety of single or multileg semisubmersible buoys or ships that are more stable than conventional mono-hulls [2]. These have been built as stable platforms for drilling or radar.

Small Boats. A wide variety of boats or small ships are used particularly for inshore biological or environmental measurements. Almost any craft that floats is satisfactory for simple measurement programs in protected waters providing the equipment can be carried aboard. For more complex or extensive measurement programs the selection of an appropriate small ship is as important as the selection of a proper large ship.

Because small boats are highly weather dependent, the area and season are important in selecting an appropriate boat. Of particular importance is whether it is to be operated as a day boat or whether a crew is required to operate around the clock. In the past, most small work boats were 4 to 6 m/s heavy-displacement hulls. Now there are a variety of lightweight planing hulls that can operate at speeds of 10–12 m/s and can take maximum advantage of daylight hours and good weather. Still newer and faster small craft such as hydrofoil and air-cushion vehicles can be used.

3.4.3 Major Equipment or Software

Just as a large computer requires appropriate software to efficiently couple its unique capabilities to real problems, so a research or survey ship requires considerable special equipment or "major software" to efficiently couple its unique capabilities to measurement problems at sea. Examples of such major equipments are handling devices, winches for tension and electric cables, and electronics for navigation and acoustic survey. Once at sea on a rolling ship the mechanical characteristics frequently control what can be accomplished under the prevailing conditions of wind and sea.

Electrical and acoustical problems most frequently encountered are electrical power supplies with large transients and fluctuations in both frequency and voltage.

Winches on research and survey ships are commonly made to carry 6 to 10 km of steel cable, of about 12.5 mm wire and about 6.25 mm wire. About one-fifth of those winches have a slip ring assembly and carry cable with an internal electric conductor. Nearly all ships have one or more kinds of winches that can handle cables long enough to reach bottom on the continental shelves. Particular attention needs to be given to the handling of heavy or awkward equipment if the weather is apt to be rough.

3.4.4 Buoys

Buoys have been used for navigation for many decades. While most of these are very simple, some buoys used in shallow water have remarkable reliability. At the other end of the size and permanence scale are small sophisticated free-drifting sonobuoys with a design life of only a few hours. In the last decade a variety of anchored and drifting instrument buoys [18] has entered the mainstream of ocean-measurement programs. Emphasis on global weather and estuarine pollution will make buoys even more common.

Buoys are fundamental to measuring programs. They can be as small and as low priced as the problem and numbers permit. In some cases it has been shown that expendable buoys are more cost effective than recoverable buoys. The lowly drift bottle or card buoy still remains a potent device for measuring average currents.

For most problems buoys can be unmanned and hence free of overhead costs, legal requirements, and acoustic backgrounds normally associated with manned ships and propelled devices.

Although buoys usually float on the surface, they can rest on the bottom or float at mid-depth in the ocean. They can be free floating, anchored, or within reason be propelled by sail or engine. The free-falling and free-ascending drop-sonde buoys provide new types of measurements.

Buoys have the potential of providing time series at a variety of widely dispersed points. By using relative drift techniques with surface and subsurface buoys it is possible to deduce a great deal about vertical and horizontal shear motion.

A buoy designed to follow the motion of the surface of the water will have to endure large motions and accelerations during storms, including forces that tend to invert it. Small shallow buoys tend to follow the motion of the water surface. Vertically shaped or spar buoys tend to damp out the high-frequency waves of short wavelength. However, because it is sometimes difficult to design spar buoys with natural heave periods longer than ocean wave periods, some spar-shaped buoys can achieve vertical resonance.

Some common difficulties with making measurements from buoys are the pitch, heave, and yaw motion of the buoy and its anchoring system, if any. These motions can introduce apparent discrepancies in the frequency characteristics of observations [19]. The design or behavior of a buoy depends on several factors, principally size of payload and expected sea state. If the buoy is to be anchored (Figure 3.4-2), the mechanical design problem multiplies about as the product of the water depth times the square of the current speed. Incorporation of sensors along mooring cables usually complicates handling procedures.

Figure 3.4-2 Monster 12-m diameter saucer shaped general purpose buoy able to remain anchored in fast currents. CONVAIR-General Dynamics.

3.4.5 Large Submarines

Several navies have furnished considerable submarine time for research and survey. Submarines were used for gravity measurements during the 1930–1960 era before successful surface ship gravimeters were available. Submarines and particularly nuclear submarines are unique in their ability to traverse below polar ice. With high-frequency high-resolution upward-looking echo sounders and scanning sonar and closed-circuit optical equipment they have permitted studies of the underside of the ice, the marine life under the ice, and the depth to the bottom. The ability of a submarine to operate below the surface in a quiet fashion, either stopped or proceeding slowly, is of great advantage in acoustical experiments such as listening to marine life or making acoustic transmission measurements. Also, the vertical and azimuthal motions encountered under such conditions are slow enough to permit directional sonars to remain pointing in the same direction for modest time durations. The use of large submarines to examine continental slope bottoms has been limited.

3.4.6 Small Submersibles

The small submersible has become a recognized working vehicle for measurements [1]. Descriptions and operating concepts of typical submersibles [4, 6, 8, 14, 15] are available (Figures 3.4-3 and 3.4-4).

Common features that are available in varying degrees are ability to get below the surface and wave action; sufficient maneuverability to locate, approach, and circle an object of interest; windows through which the eye can observe the situation or subject in situ in a dynamic fashion rather than a snap-shot fashion; the ability to carry a mechanical arm or other samplers so that measurements and samples can be made in a purposeful fashion; and design depths that range from a hundred meters to the deepest part of the ocean at some 12,000 m.

3.4.7 Specialized Submersible Equipment

Much of the instrumentation for a deep submersible is similar to that for lowered instruments except for possible repackaging. There are several unique kinds of small submersible equipment.

Releases permit dropping heavy instruments or catchers or those that might become entangled and hinder surfacing. Life-support systems, although much less stringent than for space vehicles, still require a high degree of reliability and special packaging. External manipulators are used to select, sample, or do simple work. Precision depth gages, of both absolute and relative type, are used for both navigation and for some experiments.

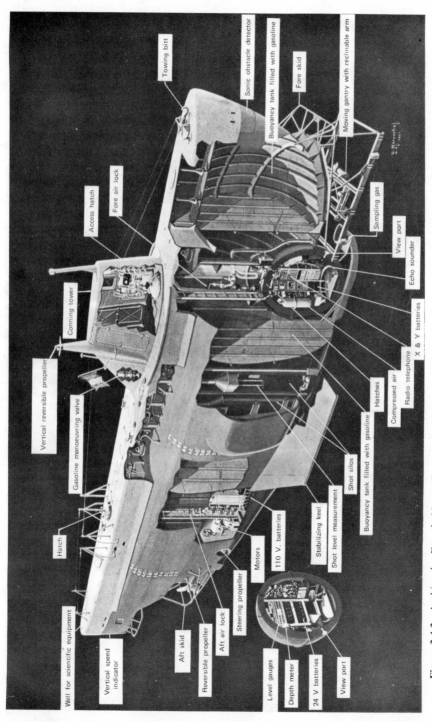

Figure 3.4-3 Archimede. French Navy (Groupe des Bathyscaphes) operated three-man bathyscaphe built to dive to any depth in the ocean. Archimede has periodically been making scientific dives in the great trenches of the world's oceans.

Figure 3.4-4 Alvin, a 2–3-man submersible with mid-depth range 2000 m capability. Speed reduced to obtain greater maneuverability and versatility for making a variety of scientific measurements. Woods Hole Oceanographic Institution.

References

1. Ballard, Robert, and K. O. Emery, Research Submersibles in Oceanography, Marine Technology Society, Contribution #2456, Woods Hole Oceanographic Institution, Woods Hole, Mass. 1970.
2. Brahtz, John, Ed., *Ocean Engineering-System Planning and Design*, John Wiley and Sons, New York, 1968.
3. D'Arcangelo, Amelio M., Ed., *Ship Design and Construction*, Society of Naval Architects & Marine Engineers, New York, 1969.
4. Eliot, Frederick, The Design and Construction of the Deepstar 2000, Marine Technology Society Symposium, Vol. "The New Thrust Seaward," 1967.
5. Knight, Austin M., *Modern Seamanship*, 11th ed. D. Van Nostrand, New York, 1945.
6. *Manned Submersibles and Underwater Surveying*, SP–153, U.S. Government Printing Office, Washington, D.C., 1970.
7. *Marine Technology Society Journal*, Marine Technology Society, Washington, D.C.

8. Mavor, J., H. Froehlich, Wm. Marquet and Wm. Rainnie, "Alvin, 6000 ft. Submergence Research Vehicle," *Trans. Soc. Naval Architects and Marine Engineers* **74,** 1966.
9. Meyers, Holm, and McAllister, *Handbook of Ocean and Underwater Engineering*, McGraw-Hill Book Company, New York, 1969.
10. Nelson, Stewart B., *Oceanographic Ships Fore and Aft*, Office of the Oceanographer of the Navy, Washington, D.C., 1971.
11. *Ocean Industry*, Gulf Publishing Company, Houston, Texas.
12. *Oceanographic Vessels of the World*, National Oceanographic Data Center & U.S. Navy Hydrographic Office, Washington, D.C., 1961 (with additions and deletions).
13. *Oceanology International*, Industrial Research Inc., Beverly Shores, Indiana.
14. Picard, Auguste, *Earth, Sky and Sea*, Oxford University Press, Oxford, England, 1956.
15. Shenton, Edward, *Exploring the Ocean Depths*, W. W. Norton and Co., New York, 1968.
16. Taylor, D. M., "Glomar Challenger Makes Exciting Discoveries," *Ocean Ind.* **3,** No. 10, 35–50, October 1968.
17. *Underwater Science and Technology Journal*, Iliffe Science and Technology Publications Ltd., Guildford, Surrey, England.
18. Walden, R. G., "Oceanographic and Meteorological Buoys," *Underwater Sci. Tech. J.* **2,** No. 3, 1970.
19. Webster, T. Ferris. "Some Perils of Measurement from Moored Ocean Buoys.," in *Transactions of the 1964 Buoy Technology Symposium*, Marine Technology Society, Washington, D.C., 1964.

Enrico P. Mercanti

3.5 SPACE PLATFORMS

Space platforms are labeled according to configuration and orbital path. Sounding rockets are used to place sensors in the near-earth space environment. Spacecraft are called satellites if they orbit the earth or other cosmic bodies and probes if their destination is some distant planet. Landers are probes designed to make ground contact with other celestial bodies; soft landers touch ground gently to minimize damage while hard landers make measurements until impact when the spacecraft is destroyed. Probes can carry provisions for returning the spacecraft to earth. A probe is a flyby if its course carries it near a planet but not so close as to be captured by its gravitational field into a satellite or collision trajectory. To accomplish its mission, a spacecraft may be used in a combination of capacities. For example, a lunar lander may assume successive roles of a satellite orbiting earth, a probe headed for the moon, a satellite orbiting the moon, and finally, a lander on the lunar surface.

Instruments carried on space platforms depend on the mission. Spacecraft have been instrumented to make observations of earth not feasible with ground based devices. These observations provide information on pollution, geological features of the surface, agriculture, forestry, oceans, population concentrations, transportation, and earth resources. Data are used to make ecological assessments of the earth's environment and its resources.

Some spacecraft carry climatological instrumentation exclusively. Balloons, aircraft, and rockets are used to provide local meteorological information while spacecraft provide the higher-altitude, wider-scope data needed for forecasting.

Spacecraft have been used to determine constituents and their properties in the space environment of earth and in interplanetary space. Space science measurements made by geophysical observatories over a seven-year period are given in Table 3.5-1. Much of the information obtained has been used to design spacecraft suitable for manned space flight and to determine the influence of the sun and the interplanetary medium on the earth.

Space platforms are useful for stellar astronomy because measurements can be made free of atmospheric absorption, atmospheric distortions, and man-made radiation. Only satellites can provide the long-range, undistorted surveys needed for astronomical studies. Spacecraft astronomical instruments include multichannel filter photometers, spectrophotometers, and scanning spectrographs.

Spacecraft for solar system studies have made possible in situ measurements and collection of specimens and have contributed significantly to understanding the nature of the sun and its planets.

Spacecraft have carried sensors to determine solar constituents and their properties. Measurements have included electromagnetic emissions, solar plasma, the sun's magnetic field, and solar cosmic rays.

The communication satellite is useful for transmitting signals reliably over long distances for lengthy time periods. Such space platforms serve

Table 3.5-1. Space Science Measurement Devices and Experiments

Aeronomy	Fluxgate magnetometers
Accelerometers	Proton precession magnetometers
Ram-pressure gages	Alkali-vapor magnetometers
Ionization gages	Helium magnetometers
Neutral mass spectrometers	
Radiometers and photometers	*Micrometeoroids*
Spectrometers (dispersive)	Piezoelectric detectors
Star trackers and stellar	Capacitor detectors
refractometers	Light-flash detectors
	Pressurized cells
Ionospheric physics	Wire-grid detectors
Satellite-to-satellite propagation experiments	Light-transmission-erosion detectors
Topside-sounders	Time-of-flight experiments
Passive radio receivers	
Electric-field meters	*Solar physics*
Standing-wave impedance probes	Filter photometers
rf impedance probes	Spectrophotometers (nondispersive)
Langmuir probes	Dispersive spectrometers and spectro-
Planar ion traps	photometers
Spherical ion traps	Faraday-cup plasma probes
Ion mass spectrometers	Spectroheliographs
	Coronagraphs
Radiation	Radiation counters
Geiger counters	Curved-surface electrostatic analyzers
Proportional counters	
Ionization chambers	*Astronomy*
Channel multipliers	Spectrometers and spectrophotometers
Scintillators	Radio-astronomy receivers
Cerenkov detectors	
Cadmium-sulfide cells	*Cosmic-ray*
Solid-state detectors	Basic detectors (e.g., ionization chambers)
Current collectors	Geiger-counter telescopes
Telescopes (various types)	Proportional-counter telescopes
Magnetic spectrometers	Cerenkov-scintillator telescopes
Faraday-cup probes	Solid-state telescopes
Electrostatic analyzers	Scintillator E-versus-dE/dx telescopes
Ionization chambers plus Geiger counters	Spark chambers
Emulsions	
Spark chambers	*Biology*
Scintillation chambers	Weightlessness and zero-g experiments
	Weightlessness-radiation experiments
Magnetometers	Biological-rhythm experiments
Search-coil magnetometers	

as useful aids in navigation.

Space platforms are useful tools for earth observations, space studies, interplanetary explorations, solar physics research, weather forecasting, astronomy, and communications.

3.5.1 Space Platform Requirements and Limitations

Environmental Exposures. A space platform must survive the series of environments to which it is exposed. Prior to launch, the space platform is subjected to tests in simulated environments which provide only an approximation of the true environment to be encountered. Frequently a flight-quality prototype space platform is subjected to wider temperature excursions, higher accelerations, greater shock loads, higher duty cycles, and other more severe environmental and operational exposures than are predicted for space flight. Environments to which a space platform can be exposed are ground shipment from the point of assembly or final test to launch site, launch, earth escape trajectory, interplanetary trajectory or orbital flight path, entry into the atmosphere of a planet, soft landing, secondary launch, re-entry, and final soft landing. Not all space environments can be duplicated on earth. An example is the near vacuum of interplanetary space.

Environmental parameters include atmosphere (composition, properties) temperature (storage, launch, flight, re-entry), acceleration (launch, space maneuvers, re-entry, landing), shock (launch, space maneuvers, landing), humidity (storage, outgassing in space), gravitational forces (launch, orbit or flight path, re-entry), radiation effects (solar, trapped radiation zone, man-made), magnetic field effects, electric field effects, space particles (micrometeoroid), contamination, pressure, life (operational, storage, degradation, reliability), preservation of internal atmosphere (manned flight), and communication barriers (shrouds, noise, life of electrical components, interference).

Some environmental parameters such as life, pressure, and internal atmosphere are defined in mission requirements, some can be calculated in advance and others can only be surmised or based on observations from previous space platforms. The last group of parameters is especially difficult to predict on first visits to unknown regions and when new launch vehicles are used.

Gravitational effects can be predicted for those bodies of the solar system with known masses. Table 3.5-2 lists masses of the other planets relative to that of earth. The masses of the known planetary satellites in the solar system plus the countless asteroids, comets, meteors, and meteorites have only been determined in isolated cases.

Atmospheric characteristics include the variation of density and pressure with altitude, temperatures due to solar and planetary radiation and atmospheric convection and composition. These affect aerodynamic drag, lift, and temperature of the space platform and govern design for re-entry, launch, heat shielding, and dielectric protection.

Radiation is caused by solar and cosmic electrons and protons and

Table 3.5-2. Solar System Properties [10]

Body	Semi-Major Axis to Sun (AU)	Period (Earth Years)	Mean Diameter[a]	Mass[b]	No. of Natural Satellites	Surface Excape Velocity[c]
Sun	—	—	109.0	3×10^5	—	55.0
Mercury	0.387	0.241	0.38	0.054	0	0.371
Venus	0.723	0.616	0.97	0.815	0	0.915
Earth	1.00	1.00	1.00	1.00	1	1.00
Mars	1.52	1.88	0.52	0.108	2	0.449
Jupiter	5.20	11.9	11.0	318.0	12	5.38
Saturn	9.54	29.5	9.03	95.1	9	3.26
Uranus	19.2	84.0	3.72	14.5	5	1.97
Neptune	30.1	165.0	3.38	17.0	2	2.24
Pluto	39.5	248.0	1.02	0.8	0	0.85
Earth's moon	—	0.075	0.27	0.012	0	0.212

[a] Expressed in earth diameters: 12.75632×10^3 km.
[b] Expressed in earth mass units: 6.04125×10^{27} g.
[c] Expressed in earth escape velocity: 11.17854 km/s.

requires space platform shielding to protect equipment and personnel from exposure. Radiation pressure must be compensated for by the attitude-control system to offset torques which, if uncorrected, can alter the flight course. Solar cells, used to power many long-life space platforms, are particularly susceptible to radiation damage. The radiations requiring consideration for manned flight are x-rays, gamma rays, beta particles (0.1 to 1.0 MeV), neutrons (to 10 MeV), protons (to 7100 MeV), and alpha particles (to 5 MeV). Individuals are normally subjected to a 0.001 rem per day background radiation dose but can tolerate a 25-rem dose on a single emergency basis.

Weight Limitations. Energy is required to propel a space platform from zero velocity at the launch site to orbital or escape velocity. The energy provided by the launch vehicle is calculable from the integral of a thrust-time curve and constitutes the limiting factor in space platform weight. While use of larger launch vehicles permits use of larger space platforms, the fuel system weight and volume (which causes drag resistance) also launched with the payload puts a practical limit on payload weight.

This weight restriction is further aggravated if the space platform carries propellant to launch the craft from the surface of another planet.

Power Limitations. Batteries such as nickel-cadmium, silver-cadmium, mercuric-oxide, and silver-zinc are used to power space platforms for missions of short time duration. For longer flight periods, solar energy is converted to electrical energy by solar cells mounted in arrays. The solar cell, a sandwich of *n*- and *p*-type semiconductors, creates an electromotive force to produce a current when electron-hole pairs are created by photons absorbed adjacent to the *p-n* or *n-p* junction. A typical cell converts less than 10 percent of the impinging solar energy (about 1300 W per square meter at 1 AU) to electrical energy. The practical size of the solar array limits the power available for the space platform. Weight is a limitation since solar arrays weigh almost 10 kg per square meter and the platforms require up to 1000 W.

Since solar cells derive their energy from sunlight, orbits with long eclipse periods require use of batteries until solar exposure is resumed. This further limits the power because some solar cell energy must be diverted to recharge batteries.

Orbital Limitations. Launch sites can restrict the inclination of the orbit for an earth orbiting satellite due to safety regulations which prohibit an ascent trajectory over populated land. In the United States, polar or near-polar orbiting space platforms must be launched in a southerly or south-westerly direction from the Western Test Range in California. Those destined for low inclination orbits are launched in a southeasterly direction from the Eastern Test Range in Florida. In both cases, the ascent trajectory is over an ocean.

After a space platform has been injected into an elliptical orbit, perturbing forces change orbital characteristics. These are taken into account in establishing orbital parameters to prevent a premature end of the satellite due to atmospheric drag (which reduces the velocity, forcing uncontrolled re-entry into the earth's atmosphere with the subsequent aerodynamic heating burning the vehicle). Forces that increase altitude are also controlled when orbital perigee tolerances are included in mission requirements.

Significant forces which cause perturbations are the earth's bulge (recession of nodes and advance of perigee), the earth's pear shape (slow changes in eccentricity), solar and lunar graviation (changes in perigee), and atmospheric drag (dependent on altitude, cross section of platform and atmospheric density).

Solar radiation pressure and magnetic drag induce second-order forces which are negligible for most satellites. Some of these forces can be counteracted by propulsion systems. Normally orbit and life are so interrelated that it is not possible to specify both in mission requirements.

Orbits and flight paths are further restricted by a combination of considerations including launch time, destination of space platform, eclipse periods desired, relationship of orbit or flight path to earth-sun line or other celestial body, and launch site.

These parameters are usually analyzed in advance with results presented in terms of launch window constraints.

Other Limitations. Some space platforms carry sensors with very precise pointing requirements difficult to achieve with conventional attitude control systems.

Frequently, large sensors exposed to the sun cannot tolerate the high temperature thereby induced and have no means for conducting or radiating the heat absorbed. Special protective devices are used in such cases or the platform is designed to spin about an axis perpendicular to the sun.

In a space environment, coated or layered sensors may degrade due to outgassing or other environmental factors. The solution is a compromise between sensitivity and life.

Many instruments are sensitive to interfering electrical or electromagnetic signals. Use of these is restricted to space platforms with extendable booms to provide the necessary isolation from interfering space platform subsystems.

Communications via space platform telemetry is restricted by location of ground stations and visible passes over them to permit data acquisition. If data must be transmitted during the ascent trajectory through the atmosphere, it is necessary to station a telemetry receiving ship in the ocean under the flight path. Communications from orbit depend on the relative locations of ground receiving stations. Space platforms in highly elliptical orbits ranging in altitude from 250 km (perigee) to apogees of 150,000 km can transmit to receiving stations almost continuously. However, low-altitude orbits (i.e., 250 km perigee and less than 1500 km apogee) permit short intermittent passes over ground stations. In such cases, data is stored on tapes and quickly played back during the short station passes. Tape recorders are limited in recording ability and cannot record data at rates as high as can be transmitted in real time. An alternative solution is a network of high altitude communication relay satellites to receive data from space platforms and transmit it to ground receiving stations.

3.5.2 Space Platform Design

Table 3.5-3 is a listing of the more common requirements specified for space missions. Each space platform is unique. Weights and shapes have varied and there is no "typical" space platform. A combination of requirements from Table 3.5-3 defines the configuration of a space platform to meet the objectives of a specific mission. Design compromises are made if two or more requirements cannot be simultaneously met.

Flight Path, Time, and Type. During launch, the primary forces on the vehicle are rocket thrust, aerodynamic drag, aerodynamic lift, gravity, and centrifugal force. In space, the primary forces are aerodynamic drag and the gravitational forces of the earth, sun, and moon. The horizontal velocity required to escape the earth's gravitational field is approximately 11,200 m/s and the rocket energy required is ideally 6.25×10^7 J/kg.

Table 3.5-3.   Space Platform Design Requirements

Flight path, time, and type requirements
Flight paths (probes, flybys, landers, rendezvous, re-launch, re-entry)
Orbits (satellites) (orbital parameters, eclipse periods, orbit relationship to earth-sun line)
Launch date
Life
Payload weight (sensors, instruments, personnel)
Space platform attitude (spinning, earth pointing, sun pointing, star-tracking precision)
Tracking

Launch requirements
Launch vehicle interfaces
Shroud interfaces (mechanical, thermal, electrical, separation)
Electrical (power, communications, interference)
Deployment of appendages (after shroud and launch vehicle separation)
Tracking

Communications requirements
Type of telemetry (PFM, PCM, FM, PDM, PAM, Special Purpose: frequency, amplitude or phase modulated)
Telemetry format and sensor channel requirements
Bit rate
Real time or tape recorder
Commands
Signals (timing, switch, mode, status)
Housekeeping data

Power requirements
Peak power load
Powet time profile
Power regulation
Type of power (voltage, dc, ac, frequency, phase)

Attitude control requirements
Space platform attitude (spinning, pointing, tracking)
Precision of stabilization
Changes in attitude (i.e., from stable platform to spinning)

Environmental control requirements
Thermal (sunlight, eclipse, re-entry)
Magnetic shielding
Electrical
Types of cabling (coaxial, shielded, twisted pairs)
Radiation protection
Seals
Outgassing vents

Structural requirements
Fields of view of sensors
Sensor view directions (steady-state, variable)
Sensor isolation requirements (interference, magnetic effects, radioactive sources)
Pyrotechnics (squibs, other explosive devices)
Rotating components
Moving components (shutters, doors, steppers)
Materials

Thermal control requirements
Passive
Active
Precision
Life
Solar cells ⎫
Batteries ⎪
Sensors   ⎬ may require severely limited
Seals   ⎪   temperature excursions
Bearings   ⎭

Data requirements
Analog
Digital
Other

At an altitude of 35,800 km, the satellite velocity is such that the space platform rotates with the same period as the earth and the "synchronous" spacecraft appears to remain stationary if it is orbiting over the equator. The motion of space platforms in orbital flight is derivable from the laws of conservation of energy and momentum. Thus at the lowest altitude (perigee) the satellite velocity is the greatest because of the requirement for an increased centrifugal force to counteract the higher gravitational force exerted by the earth.

The time of launch fixes the elliptical path (orbit) of the satellite in inertial space because the location of the launch site moves in inertial space, rotating 360° daily.

Changes in orbits or flight paths occur because of changes in aerodynamic drag and the sun-moon-earth resultant gravitational pull. These can be compensated for by on-board propulsion systems which usually utilize an inert gas such as argon, nitrogen, Freon or krypton. The same propulsion systems are used to deliberately change orbits or flight paths as desired.

A satellite orbit cannot remain constant because of perturbing forces. If long life is desired, the launch vehicle can impart an initial small radial velocity component to the spacecraft leading to a gradual rise in perigee altitude, an accompanying less rapid decrease in apogee altitude, and eventual circularization of the initially elliptical orbit.

Launch Requirements. The space platform interface with the launch vehicle is mechanical and electrical. Usually the platform includes the interstage structure with provisions for separation after burnout of the rocket. The electrical interface is used to supply power to the spacecraft for heating and testing and includes communication and command links with the ground.

The shroud, made up of two halves held together by bands cut by explosive devices, is smoothly shaped to minimize aerodynamic drag during ascent. The shroud is insulated to prevent heat damage to the spacecraft inside.

Spacecraft booms, solar arrays, and antennas are hinged or coiled for stowage inside the shroud. A typical arrangement uses spring loaded hinges with the appendages tied close to the spacecraft body by bands which are cut by remotely commanded explosive devices. In other cases, appendages are extended by electric motors or gas pressure or hydraulic devices.

Because of launch vehicle vibration, clearance is provided between the shroud and spacecraft.

During launch, it is desirable to continuously know spacecraft position and status. Ground telemetry receiving stations and tracking radars are used to satisfy these needs. Power for transmission is derived from batteries on both the launch vehicle and the space platform.

Communications Requirements. The space platform communication system transmits data from sensors and platform status information such as temperatures, pressures, circuit continuity, and data used to control space platform position. These data are subsequently used for sensor data evaluation and spacecraft assessment, position determination, and dynamic evaluations (velocity, acceleration).

Sensors and their associated instruments generate either analog or digital signals. Sometimes the analog signals are converted to digital signals and transmitted to ground serially.

Commands received by the spacecraft are usually either on–off power commands or impulse commands to control the operating mode.

Directional and omnidirectional antennas are used for communications. Typical frequencies are 136 and 2200 MHz. Tracking transponders, used to fix spacecraft position with accuracies in the order of kilometers, transmit and receive at 1700–2270 MHz.

The communication systems of large space platforms often include timing and control signals used for counting, switching, changing operations, and otherwise causing desired operational variations in sensors or spacecraft.

Power Requirements. Power requirements vary from a few watts to close to a kilowatt for missions of long duration. Solar cells are often mounted on moving arrays which are continuously pointed at the sun or on the spinning spacecraft body. Solar cells are covered with thin (0.15-cm) sheets of silicon glass or quartz to protect them from radiation and particle damage and their back sides are conductively connected to radiating panels for dissipating unconverted solar energy.

During eclipse portions of flight paths or for peak power periods, energy cells are used to provide power. Nickel-cadmium batteries are used if eclipses are frequent. Silver-cadmium batteries provide more energy per unit weight and are less magnetic but cannot tolerate high duty cycles; hence, they are used most on space platforms in highly elliptical orbits. Mercuric-oxide batteries have a slightly higher energy-to-weight ratio and can withstand higher temperatures. Silver-zinc batteries provide the most energy per unit weight but they have the disadvantage of generating gas except when discharging.

Other types of power plants have been tried for space platform applications. Some use heat derived from nuclear reactors, radioisotopes, or chemical processes. Energy conversion is by thermoelectric, thermionic, fuel cell, or other means. In each case, the unconverted heat generated is radiated to outer space.

Many scientific space platforms provide power at 28 V dc. Because of varying power input and drain, the voltage is allowed to vary about

± 15%. Sensors requiring more voltage regulation are supplied with individual power regulators at the expense of efficiency. Converters are used for other voltages. Fuses, circuit breakers, under voltage cutoffs, and other regulation devices are used to provide protection against power surges and decays.

Attitude Control Requirements. After separation from the shroud and the launch vehicle, the space platform is in an uncontrolled state, tumbling randomly. Often the spacecraft is deliberately spun prior to injection into its space flight path for stability. Gas jets or inertia wheels can be initially used to despin the craft. Next, the attitude control system locks on to one or more targets such as the local vertical, the sun, a star, the earth's magnetic field, or a horizon to orient the space platform in the desired direction. Finally, the attitude control system maintains pointing within prescribed tolerances.

Devices used to perform these maneuvers include pulse jets, rockets, inertia wheels, deployable appendages which change the spacecraft moment of inertia, electromagnets which react with the earth's field to create torques, weights distributed to utilize gravity gradients in a vertical profile, aerodynamic forces which produce a weathervane effect, and energy absorbers such as springs, fluids, and electromagnets. The pulse jets can be derived from high-pressure inert gases, boiling liquids, subliming solids, gases heated by nuclear or electrical means, magnetically or electrically induced plasmas, and explosive devices. Signals from sensors drive these devices to maintain attitude control. Angular accuracies better than $\pm 2°$ are typical. Sun sensors, silicon *p-n* junction cells (solar cells), are used for orienting an axis with respect to the sun. Infrared horizon detectors are used to position an axis perpendicular to the earth's radius vector. A single degree of freedom gyroscope can be used to orient an axis with respect to the orbital plane.

Some of the more powerful activating devices are used to provide velocity increments needed to change flight paths or orbits. Rockets are usually employed to decelerate space platforms for re-entry and soft landings.

Environmental Control Requirements. There is a need to protect equipment and personnel in flight from permanent, gradual, or temporary damage.

With operational temperature excursions for most spacecraft components limited to the 0–40°C range, it is necessary to compensate for the extremes of sustained solar and eclipse exposures. Cold conditions are usually remedied by means of electrical heaters actuated by temperature sensors. For higher temperatures, conductive paths, radiation to outer space, and refrigerants are used. Surfaces exposed to the sun are covered with low-absorptivity high-emissivity coatings or insulated with multiple-layer materials to minimize heat absorption or transmission. The absorbed heat is conducted to panels with high emissivities that radiate to space. If the spacecraft is positioned, the temperature sensitive component can be located to receive little solar exposure. Sensors can be provided with temperature actuated covers or doors to reduce high heat inputs. Temperature extremes can be reduced by controlling power to the experiment.

Components with high magnetic fields can be positioned so that their effects cancel and magnetic shielding material can be applied. Use of nonmagnetic materials such as titanium or stainless steel is effective in reducing magnetic fields. Sensitive or offending instruments can be mounted on long booms where magnetic effects are reduced by a factor proportional to the third power of the separating distance.

Radiation protection is obtained by shielding. Solar cells are usually covered with a thin sheet of glass or quartz to prevent degradation. Systems, including sensors, can be designed to degrade at a rate consistent with mission objectives or space platform life.

Materials or contaminants in materials can sublime in a vacuum environment generating gases that can damage sensors or cause erroneous readings. Electrical components can arc due to corona discharge. Space platforms are designed to contain venting paths directed away from sensitive instruments and components. Frequently, instruments are not turned on until many hours after injection into orbit to prevent corona arcing damage.

Spacecraft contain large numbers of cables. To prevent undesirable electrical and electromagnetic interference effects, cables are shielded or twisted and instruments are grounded. Even with those precautions it is still necessary at times to operate only one of two or three instruments at one time to prevent erroneous data from being generated by electrical interference.

Charged particles in space can collect on space platform surfaces causing currents, interference with operations, and a buildup of spacecraft potential. Sometimes the effect is transient, but on some space platforms it has been permanent. Protective grids and conducting shields or coatings can be applied to minimize the effects of charged particles.

Seals are used to prevent escape of lubricants and other gases or liquids. Rubber and certain plastics have been successfully used. Some space platforms simply provide enough of the sealed material to satisfy life requirements at a known leak rate.

Structural Requirements. Instrument and attitude control sensors must be mounted so that required fields of view are maintained. If requirements cannot be met on the spacecraft body, the sensors are mounted on appendages or booms.

Booms also provide isolation of components if required for magnetic, electrical, radioactivity, or outgassing compensation or protection.

Commandable explosive devices are used to extend booms, solar arrays, antennas, and other appendages and to uncover sensors or uncage rotating or scanning mechanisms.

Materials selection is based on strength, weight, magnetic, thermal, and outgassing properties. Aluminum, titanium, and stainless steel are commonly used for structural elements. Special rubber and plastic compounds have been successfully used for seals and electrical isolation. Gold is used as a plating for thermal control purposes. Many paints with varying absorptivities and emissivities are applied to exposed surfaces. Aluminized mylar sheets provide good insulation characteristics if used in multiple layers.

Rotating, moving, and deployed components are used of necessity rather than preference. Because of their motions, they are more susceptible to failures than passive components. If used, telemetry readouts are incorporated to indicate position. Booms usually actuate microswitches to signal achievement of proper position; potentiometer-type devices can be incorporated to indicate intermediate positions. Frequently, motor, pneumatic, or explosive actuated appendages are provided with an auxiliary total ejection system which separates the entire component from the space platform in the event the unit fails to achieve its desired position.

References

1. "Assessment and Control of Spacecraft Magnetic Fields," Publication SP-8037, National Aeronautics and Space Administration, Washington, D.C., 1970.
2. Corliss, W. R., *Scientific Satellites*, National Aeronautics and Space Administration, Washington, D.C., 1967.
3. Corliss, W. R., *Space Probes and Planetary Exploration*, D. Van Nostrand, Princeton, New Jersey, 1965.
4. Daniels, G. E., "Terrestrial Environment (Climatic) Criteria Guidelines for Use in Space Vehicle Development," Publication TM X-53872, National Aeronautics and Space Administration, Washington, D.C., 1970.
5. "Design Handbook Series 3-0, Space Vehicles," Publication USAF-AFSC D-11 3-2, Department of Defense, Washington, D.C., 1969.
6. "Earth Albedo and Emitted Radiation," Publication SP-8067, National Aeronautics and Space Administration, Washington, D.C., 1971.
7. "Earth's Ionosphere," Publication SP-8049, National Aeronautics and Space Administration, Washington, D.C., 1971.
8. "Entry Gasdynamic Heating," Publication SP-8062, National Aeronautics

and Space Administration, Washington, D.C., 1971.

9. Johnson, F. S., *Satellite Environment Handbook*, Stanford University Press, Stanford, California, 1961.

10. Kendrick, I. B., Ed., "TRW Space Data," TRW Systems Group, Redondo Beach, California, 1967.

11. LeGalley, D. P. and A. Rosen, *Space Physics*, John Wiley and Sons, New York, 1964.

12. "Lunar Surface Models," Publication SP-8023, National Aeronautics and Space Administration, Washington, D.C., 1969.

13. "Magnetic Fields—Earth and Extraterrestrial," Publication SP-8017, National Aeronautics and Space Administration, Washington, D.C., 1969.

14. "Mars Surface Models (1968)," Publication SP-8020, National Aeronautics and Space Administration, Washington, D.C., 1969.

15. "Meteoroid Environment Model—1970 (Interplanetary and Planetary)," Publication SP-8038, National Aeronautics and Space Administration, Washington, D.C., 1970.

16. "Meteoroid Environment Model—1969 (Near Earth to Lunar Surface)," Publication SP-8013, National Aeronautics and Space Administration, Washington, D.C., 1969.

17. "Models of Earth's Atmosphere (120 to 10000 km)," Publication SP-8021, National Aeronautics and Space Administration, Washington, D.C., 1969.

18. "Models of Mars Atmosphere," Publication SP-8010, National Aeronautics and Space Administration, Washington, D.C., 1968.

19. "Models of Venus Atmosphere," Publication SP-8011, National Aeronautics and Space Administration, Washington, D.C., 1968.

20. "Nuclear and Space Radiation Effects on Materials," Publication SP-8053, National Aeronautics and Space Administration, Washington, D.C., 1970.

21. "Solar Electromagnetic Radiation," Publication SP-8005, National Aeronautics and Space Administration, Washington, D.C., 1971.

22. "Spacecraft Aerodynamic Torques," Publication SP-8058, National Aeronautics and Space Administration, Washington, D.C., 1971.

23. "Spacecraft Gravitational Torques," Publication SP-8024, National Aeronautics and Space Administration, Washington, D.C., 1969.

24. "Spacecraft Magnetic Torques," Publication SP-8018, National Aeronautics and Space Administration, Washington, D.C., 1969.

25. "Space Radiation Protection," Publication SP-8054, National Aeronautics and Space Administration, Washington, D.C., 1970.

26. "Spacecraft Radiation Torques," Publication SP-8027, National Aeronautics and Space Administration, Washington, D.C., 1969.

27. Stecker, F. W., "Cosmic Gamma Rays," National Aeronautics and Space Administration, Washington, D.C., 1971.

28. Wildner, D. K., "Space Environment Criteria Guidelines for Use in Space Vehicle Development," Publication TM X-53957, National Aeronautics and Space Administration, Washington, D.C., 1969.

29. White, R. S., *Space Physics*, Gordon and Breach, Science Publishers, New York, 1970.
30. Wolff, E. A., *Spacecraft Technology*, Spartan Books, Washington, D.C., 1962.

Robert W. Rochelle

3.6 COMMUNICATIONS AND TELEMETRY

The form of the communications and telemetry link between the geoscience instrument platform and the data-collection facility depends on the type of platform. Some ground-based platforms use land lines connected through the commercial telephone service. Line-of-sight radio communication links are used where it is impractical to run a land line. Satellite communications links are available to transfer data from remote platforms. Airborne platforms using aircraft usually store the data aboard for later processing, but the data can be sent down to a data-collection facility over existing telemetry channels for real-time use. Ocean platforms can communicate over terrestrial radio links and also via satellites. Space platforms use the very-high frequency (VHF), ultra-high frequency (UHF), L, S, C, and X bands for their communications with ground terminals. As the quantity of data required and the number of platforms increase, there is a trend to the use of higher frequencies to support the ever-increasing communication needs.

3.6.1 Design Considerations

The communications between the geoscience instrument platform and the data-collection facility usually utilize a radio-frequency link. Since most platforms are unattended and in remote regions, the efficiency of the link design is very important. Efficient design reduces the power requirement and can be a large factor in extending platform life.

The primary items that influence the link are the platform transmitter, platform antenna, modulator, communication distance (range), receiver antenna, received noise power, and demodulator.

Received Signal Power. The space loss due to the distance separating the platform and the receiving station is the dominating factor in the calculation of received power. In wildlife telemetry the range between an instrumented animal and the data-collection facility can be only a few hundred meters while in space exploration the range between the instrumented spacecraft and the ground station can be as much as 10^{10} km. Since the received power falls off as the inverse square of the distance, the difference between these two requirements can be as great as 200 dB. This dictates extreme differences in the platform communication system design.

The power per unit area of the incident wave at the receiving site is given by

$$P = \frac{W_T G_T}{4\pi r^2}, \qquad (3.6-1)$$

where P is the power per unit area in watts per square meter, W_T is the power into the transmitting antenna terminals in watts, G_T is the power gain of the transmitting antenna, and r is the range between transmitter and receiver in meters.

The receiving antenna intercepts this incident wave and captures a portion of the energy. The received power available at the receiving antenna terminals is given by

$$W_R = A_R P, \qquad (3.6-2)$$

where W_R is the received power available at antenna terminals, and A_R is the effective area of receiving antenna.

The effective area of the receiving antenna is defined as

$$A_R = \frac{\lambda^2 G_R}{4\pi}, \qquad (3.6-3)$$

where G_R is the gain of receiving antenna, and λ is the wavelength of the incident radiation.

Combining Equations 3.6–1 through 3.6–3, the received power becomes

$$W_R = W_T G_T G_R \left(\frac{\lambda}{4\pi r} \right)^2. \tag{3.6-4}$$

This equation is useful for optimizing wavelength when either fixed-gain antennas, fixed-area antennas, or both, are used. If both the platform and the receiving station have fixed-gain antennas, the received power is increased when the wavelength is increased (decreased frequency). As an example, in a platform to low earth-orbiting satellite link, the platform antenna gain pattern is fixed to provide the proper coverage, and the spacecraft antenna gain is fixed to cover a certain area on the earth's surface. To minimize platform power, it is desirable to utilize as low a transmitting frequency as possible. Correspondingly, if both the platform and the receiving site have fixed-area antennas (such as in deep-space exploration), the received power is increased for decreasing wavelength or increasing frequency. Increasing the frequency is generally favorable until either the pointing requirements on the antennas become too severe or the signal is attenuated due to atmospheric effects [1]. If one of the antennas has a fixed area and the other has a fixed gain, the received power is independent of frequency. In this case the radio noise becomes the major factor in determining the best operating frequency.

Antenna Characteristics. The platform antenna design is strongly influenced by the gain requirements of the link and the pointing requirement on the platform. The antenna gain in a given direction is defined as the ratio of the power radiated from the antenna per unit solid angle in that direction to the power per unit solid angle radiated from an isotropic antenna with the same input power. The gain can be expressed as [3]

$$G(\text{dB}) = 10 \log \frac{\phi}{W/4\pi}, \tag{3.6-5}$$

where ϕ is the power radiated per unit solid angle, and W is the total power radiated. The isotropic antenna is an idealized antenna which radiates uniformly in all directions. In making actual antenna pattern measurements a plot is made of the electric-field strength as a function of angle with a small test transmitter driving the antenna. The gain is calibrated by substituting a reference antenna of known gain in place of the test antenna.

A useful expression for calculating the gain of an antenna knowing the half-power (3 dB) beamwidths is given by

$$G \simeq \frac{3 \times 10^4}{\theta_1 \theta_2}, \tag{3.6-6}$$

where θ_1 is the beamwidth of the major axis in degrees and θ_2 is the beamwidth of the minor axis in degrees.

The beamwidth in degrees of a parabolic reflector antenna at the half-power points is given approximately by

$$\theta \simeq \frac{70\lambda}{d}, \tag{3.6-7}$$

where the wavelength, λ, and the diameter, d, are in the same units.

Received Noise Power. Every receiver has a limitation in its sensitivity due to the noise background. This noise background can perturb the signal and cause errors and interference in the received data. At frequencies below 30 MHz atmospheric noise resulting primarily from lightning discharges is the major noise contributor. At higher frequencies galactic or radio noise is the prime noise source. Above 1000 MHz the radio stars act as thermal noise sources, and their power is inversely proportional to the square of the frequency. Below 1000 MHz the radiation is nonthermal and the power is proportional to frequency raised to the -2.8 power [2]. The noise from sources external to the receiver is termed antenna noise.

There is also a contribution to the received noise power caused by random motions of electrons in both the passive and active components of the receiver. These noise sources are lumped into an equivalent noise source at the input termination of the receiver for convenience in analysis. Figure 3.6-1 shows the equivalent circuit of a receiver with equivalent noise generators inserted in series with the antenna impedance, R_A, and the re-

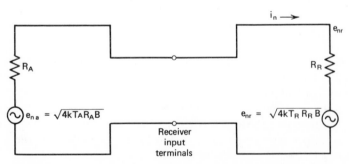

Figure 3.6-1 Equivalent noise circuit for antenna and receiver.

ceiver input termination, R_R. The ratio of the signal power developed across R_R to the noise power at that same point is the signal-to-noise power ratio of the receiver. The noise voltage developed across the input termination due to both antenna noise and receiver noise is

$$e_{nr} = i_r R_R, \qquad (3.6\text{–}8)$$

but,

$$i_r = \frac{[e_{na}^2 + e_{nr}^2]^{1/2}}{R_A + R_R}, \qquad (3.6\text{–}9)$$

where e_{na} is the equivalent noise voltage in the antenna, and e_{nr} is the equivalent noise voltage in the receiver. The noise voltages add as the square root of the sum of the squares since the two sources are uncorrelated.

Substituting,

$$e_{nr} = \frac{R_R}{R_A + R_R} [4k(T_A R_A + T_R R_R) B]^{1/2}, \qquad (3.6\text{–}10)$$

where k is Boltzmann's constant (1.38×10^{-23} $J/^\circ K$), T_A is the equivalent noise temperature of the antenna, T_R is the equivalent noise temperature of the receiver, and B is the intermediate-frequency bandwidth of the receiver.

For maximum power transfer of the signal into the receiver, R_A should equal R_R. When this relation is true, e_{nr} is

$$e_{nr} = [k(T_A + T_R) R_R B]^{1/2}. \qquad (3.6\text{–}11)$$

The noise power delivered to R_R is then

$$N_R = \frac{e_{nr}^2}{R_R} = k(T_A + T_R) B. \qquad (3.6\text{–}12)$$

Receiver Noise Factor. A significant parameter of any receiver is its average noise factor, $\bar{F}$. If an antenna at standard temperature, $T_0 (290^\circ K)$, is connected to the receiver, the average noise factor is the ratio of the total noise power delivered to R_R to the noise power available from the antenna, or

$$\bar{F} = \frac{k(T_0 + T_R) B}{k T_0 B} = 1 + \frac{T_R}{T_0}. \qquad (3.6\text{–}13)$$

Since the antenna is not usually at the standard temperature, an operating noise factor, $\bar{F}_0$, is needed. This factor is the ratio of the noise power actually delivered to R_R to the noise power of the antenna if it were at standard temperature. Thus,

$$\bar{F}_0 = \frac{k(T_A + T_R) B}{kT_0 B} = \frac{T_A}{T_0} + \frac{T_R}{T_0}, \tag{3.6-14}$$

but

$$\frac{T_R}{T_0} = \bar{F} - 1, \tag{3.6-15}$$

so

$$\bar{F}_0 = \bar{F} - 1 + \frac{T_A}{T_0}. \tag{3.6-16}$$

Therefore, to find the total noise power in the receiver, it is only necessary to multiply the operating noise factor, $\bar{F}_0$, by $kT_0 B$ or,

$$N = kT_0 B \bar{F}_0. \tag{3.6-17}$$

The noise power density (power per unit bandwidth) is

$$N_0 = kT_0 \bar{F}_0 \tag{3.6-18}$$

Input Signal-to-Noise Power Ratio. The signal-to-noise power ratio can be found by dividing the received signal power on the receiver input termination by the total noise power at that same point, or

$$\frac{S}{N} = \frac{W_R}{N_0 B} = \frac{W_T G_T G_R (\lambda/4\pi r)^2}{k(T_A + T_R) B}. \tag{3.6-19}$$

It is standard practice to express this power ratio in decibels for ease of manipulation.

For digital communications a more useful ratio than the signal-to-noise power ratio is the ratio of the energy per bit to the noise power density,

$$\frac{E_b}{N_0} = \frac{W_T G_T G_R (\lambda/4\pi r)^2}{k(T_A + T_R)}. \tag{3.6-20}$$

This parameter is discussed in Section 6.3.4 under PCM performance measurement.

3.6.2 Link Analysis

The usual practice is to construct a table listing the parameters in the link analysis. As an example, in analyzing the performance to be expected in communicating between a sensor platform and a low earth-orbiting spacecraft, Table 3.6-1 illustrates a typical format for listing the various parameters which affect the link. The upper portion of the table lists the significant

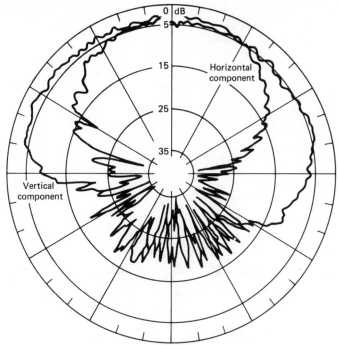

Figure 3.6-2 Circularly polarized pattern of spacecraft antenna.

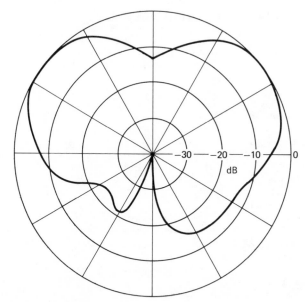

Figure 3.6-3 Pattern of platform antenna.

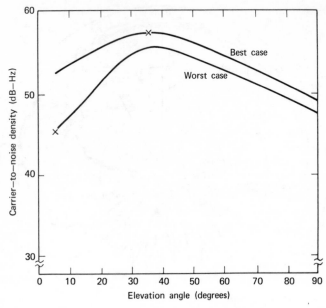

Figure 3.6-4 Carrier-to-noise density of platform-to-spacecraft link for a 1 W platform transmitter.

Table 3.6-1. Analysis of Platform-to-Spacecraft Link

	Worst Case	Best Case
Frequency (MHz)	400	400
Elevation angle	5°	35°
Range (km)	3456	1779
Transmitter Power (W)	1	1
Receiver noise figure (dB)	3	3
Transmitter power (dBW)	0.0	0.0
Transmitter line loss (dB)	− 0.5	− 0.5
Transmitter antenna gain (dB)	3.0	3.0
Free space loss (dB)	− 155.4	− 149.6
Polarization loss (dB)	− 2.7	− 0.2
Multipath loss (dB)	− 3.7	0.3
Receiving antenna gain (dB)	4.0	4.0
Receiving line loss (dB)	− 0.5	− 0.5
Received signal power (dBW)	− 155.8	− 143.5
Thermal noise density (dBW/Hz)	− 201.0	− 201.0
Signal-to-noise density (dB-Hz)	45.2	57.5

parameters used to calculate the gains and the losses. The worst case and the best case calculations are shown to indicate the expected variations in signal-to-noise power ratio. In this example it is assumed that the spacecraft receiver has a circularly polarized antenna with an elevation-plane pattern as shown in Figure 3.6-2 and that the platform has an antenna with an elevation-plane pattern as shown in Figure 3.6-3. For the worst case it is assumed that the multipath signal partially cancels the direct signal and that the polarization loss is maximum. For the best case it is assumed that the multipath signal reinforces the direct signal and that the polarization loss is minimum.

Table 3.6-1 gives the expected signal-to-noise power density for the worst case at an elevation of 5° and the best case at 35°. These two points are plotted in Figure 3.6-4 along with the best case and worst case curves as a function of the platform-to-satellite elevation angle. This type of curve is extremely valuable in the analysis of any link since it gives the bounds of the expected signal-to-noise power ratio.

References

1. Krassner, G. N., and J. V. Michaels, *Introduction to Space Communication Systems*, pp. 89–96, McGraw-Hill Book Company, New York, 1964.
2. Shklorsky, I. S., *Cosmic Radio Waves*, Harvard University Press, (translation) Cambridge, Mass., 1960.
3. Wolff, E. A., *Antenna Analysis*, John Wiley and Sons, New York, 1966.

Michael C. Husich

3.7 POWER SUPPLIES

3.7.1 Requirements

The design factors which exercise the greatest influence on platform power supplies are platform mission, environment, regime, equipment power requirements, platform design characteristics, and characteristics of system components.

Within these major categories, there are parameters which have major impacts on design. For the category of platform mission, platform life is of major significance. The environment regime determines the extent to which the platform and the power supply are stressed. Equipment requirements in terms of power, duty cycle, and types of voltages influence power supply element sizing. Platform design characteristics dictate the volume and weight allocation allowable to the electric power supply. Characteristics of system components available for use pace the design in terms of weight, size, power handling capabilities, and operating life.

The power supply system must be designed to efficiently utilize the energy source, to store energy during periods when the source is inadequate, and to convert the received and stored energy into electrical energy. The system must also be able to dissipate all of the excess energy it collects from the source.

3.7.2 Energy Sources

The three basic sources of energy used most for platform power supplies are chemical energy, nuclear energy, and solar energy.

Chemical energy can be obtained by the use of combustion, fuel cells or primary batteries. Chemical combustion can be used to operate thermoelectric, thermionic, or thermomechanical engines.

Nuclear energy sources use radioisotopes, nuclear fission, or nuclear fusion to obtain heat, ionization, or plasmas. Isotopes are suitable for heat generation for power supplies requiring relatively low power over long periods of time. Nuclear fission reactors are suitable for providing large power for long periods of time.

Solar energy can be obtained for power supplies by the use of solar energy collectors and the use of photons. With solar mirrors, it is possible to achieve temperatures of 2000°C or higher; such devices might prove practical for operating thermoelectric, thermionic, or thermomechanical engines.

The quantum energy of the photons in solar radiation is used in solar cells which are photovoltaic devices that produce a voltage when illuminated by the sun. Other photoelectric phenomena such as photoemissivity and photochemical processes have not had wide use.

3.7.3 Energy Storage

For platforms which do not have a continuous, reliable source of energy, the power supply must store energy for use when the source energy is insufficient. Energy can be stored in the form of thermal, mechanical, electrical or chemical energy.

Thermal energy can be stored by the heat from fusion of solids, by the use of materials with high heat capacities, and by the dissociation of chemicals. Mechanical energy can be stored by the use of springs and flywheels. Electrical energy can be stored by the use of electrical capacitance. Chemical energy can be stored by the use of storage batteries. Galvanic cell batteries are the most practical method of storing energy on platforms. Nickel-cadmium batteries have been used almost exclusively, but other types, such as silver-cadmium and silver-zinc, are possible [1].

The Ni-Cd cell has the widest application. Although the index of energy per unit weight and volume for the Ag-Cd and Ag-Zn cells is greater than that for the Ni-Cd, limited cycle and operating life has restricted their application. In some applications the need to restrict the presence of strong residual magnetic fields has influenced the selection of these two types, because of their nonferromagnetic composition, over the Ni-Cd cell with its better cycle and operating life characteristics.

In application, these three types of cells are hermetically sealed or are contained within hermetically sealed containers. Each type has its own unique voltage and current characteristics during discharge and charge. Major considerations in the application of these three cells are battery charge and overcharge regime, battery operating temperature, and depth of battery discharge.

The sealing requirement for space application imposes stringent requirements on battery charge control. Excessive overcharge can lead to the generation of excessive oxygen pressures. For these cells, the voltage limits for charge must be selected so that the rate of oxygen generation does not exceed the cells oxygen recombination rate. The sensitivity of Ag-Cd and Ag-Zn batteries to the charge control method is greater than that for Ni-Cd cells.

The charging voltages for these batteries is a function of temperature. In some charge regulator designs, when the battery is expected to operate over a wide temperature range, battery temperature feedback is utilized to vary the value of charge voltage.

3.7.4 Energy Conversion

When the energy source is not electrical, the power supply must convert the source energy into useful electrical energy. The three principal methods of converting energy are the photon, chemical, and thermal methods.

The photon method involves utilization of photovoltaic cells. These are silicon cells doped with boron and arsenic which produce an output voltage when illuminated by visible light.

Chemical devices for the conversion of energy include chemical batteries such as the nickel-cadmium battery, which transform chemical energy into electrical energy, and fuel cell batteries such as the hydrogen-oxygen cell.

Chemical batteries are used for short-term applications. They provide a high power-to-weight ratio at a relatively low cost.

The zinc-manganese dioxide dry battery has been widely used. It is reliable but has a relatively short shelf life if not refrigerated. The magnesium-manganese dioxide battery has a shelf life in excess of 4 yr and twice the capacity of the zinc-manganese dioxide battery.

Alkaline dry battery systems have been used for special applications. The zinc-mercuric oxide battery is one example. Although its low-temperature performance is poor and its cost is relatively high, it has found use in small devices such as hearing aids and air-sea rescue equipment. Another battery in this category is the zinc-silver oxide battery. These batteries sometimes contain built-in heaters to overcome the low-temperature problems. Water-activated batteries have been used where lower power drain is required. They have relatively high capacities of 45 watt hours per kg. The chemical reactions in these batteries are usually exothermic so that they perform well at low temperatures. The lead acid battery is the most widely used battery in this category. The nickel-cadmium battery has a high cycle life, a long shelf life, a rugged construction, and good reliability. Although it does not have a high capacity, it can withstand considerable overcharging. Another battery in this category is the cadmium–silver-oxide battery. It has twice the capacity of the nickel-cadmium battery but its cycle life and reliability are not as good. A fourth battery is the zinc–silver-oxide battery. It has three times the capacity of the nickel-cadmium battery but its cycle life is less than 100 cycles under controlled conditions.

Fuel cells operate somewhat like batteries. A fuel and an oxidant are fed into the cell when electrical energy is required. The fuel cell then converts the chemical energy directly into electrical energy without the use of moving parts. Theoretical fuel conversion efficiencies of better than 70 percent and capacities as high as 3500 watt hours per kg are possible with hydrogen-oxygen systems. The fuel cell consists of an anode at which oxidization takes place, a cathode at which the oxidizing agent is consumed, and an electrolyte in which the electrodes are immersed. In regenerative systems, the chemical energy which is converted into electrical energy is restored, using energy from an external source which is converted into chemical energy. Some fuel cells use electrolytes in the form of a solid sheet called an ion-exchange membrane.

One of the most common types of low-temperature fuel cells using an ion-exchange membrane uses hydrogen as a fuel, oxygen or air as the oxidizing agent, and water as the reaction product.

Three methods of thermal conversion are the thermomechanical, thermoelectric, and thermionic methods. Thermomechanical systems make use of heat to operate a heat engine (such as a steam turbine) which rotates an electric generator to produce electricity.

The efficiency of such a system is given by the Carnot equation

$$\eta = \frac{T_h - T_c}{T_h}, \tag{3.7-1}$$

where T_h is the temperature of the hot source and T_c is the temperature of the cold reservoir, both in degrees Kelvin. Conventional systems generally operate at temperatures below 1500°C with efficiencies up to 40%.

Thermoelectric generation of electric power is obtained when one of two junctions between two dissimilar conductors is heated. The efficiency of a thermoelectric converter depends upon the Seebeck coefficient which is a measure of the voltage developed for a given temperature difference between hot and cold junctions. The efficiency is inversely proportional to the thermal conductivity and the electrical resistivity of the junction materials. The former tends to reduce the difference in junction temperatures and the latter tends to impede the flow of electricity. Thermoelectric materials are often compared in terms of a figure of merit defined by

$$Z = \frac{S^2}{\rho K}, \tag{3.7-2}$$

where S is the Seebeck coefficient in volts per degree centigrade, ρ is the electrical resistivity in ohm-centimeters, and K is the thermal conductivity. The efficiency is given by

$$\eta = \frac{T_h - T_c}{T_h} \left(\frac{A - 1}{A + T_c/T_h} \right), \tag{3.7-3}$$

where

$$A = \left[1 + Z \left(\frac{T_c + T_h}{2} \right) \right]^{1/2}. \tag{3.7-4}$$

Thermionic devices are based on the electron emission from metals at high temperatures (called the Edison effect after its discoverer), an effect which is the basis of the operation of conventional radio vacuum tubes. Heat is supplied to an emitter electrode, generally fabricated from tungsten or tantalum. This heat causes the emitter to emit electrons. The electrons are collected at a cold collector electrode. The excess electrons are returned to the emitter through an external circuit in which they can be made to do

Table 3.7-1. Work Function and Melting Point

Metal	Work Function (V)	Melting Point (°C)
Aluminum	3.0	660
Carbon	4.8	3500
Cesium	1.7	28
Copper	4.3	1083
Gold	4.8	1063
Lithium	2.2	186
Nickel	5.0	1455
Platinum	6.3	1773
Potassium	1.9	62
Silver	4.6	960
Sodium	2.0	97
Tantalum	4.1	2850
Tungsten	4.6	3370

useful work. The maximum current density, J, emitted by an electrode at temperature T is given by

$$J = 120T^2 \exp - (\rho\phi/kT) \ \text{A/cm}^2 \qquad (3.7-5)$$

where e is the electron charge, k is Boltzmann's constant, and ϕ is the work function of the cathode, the amount of energy required to lift an electron from the highest occupied energy level in the cathode to a point just outside its surface. A desirable cathode is one with a low work function and a high melting point, so that it can be operated at high temperatures. Table 3.7-1 shows the work function and melting point of some common metals.

Magnetohydrodynamic (MHD) generators make use of the fact that a magnetic field can be made to interact with a flowing ionized fluid to produce an electric current. The ionized fluid is called a plasma. It is generally obtained by heating a gas to high temperatures. The current density which can be obtained from a plasma with a conductivity σ moving at a velocity $\mathbf{v}$ is given by

$$\mathbf{J} = (\mathbf{v}\mathbf{H} - \mathbf{E})\,\sigma, \qquad (3.7-6)$$

where $\mathbf{H}$ is the applied magnetic field and $\mathbf{E}$ is the retarding electric field due to the load. MHD systems have a high theoretical efficiency because they can operate at high temperatures.

3.7.5 Solar Cells

Solar cells are made of semiconductor materials which have impurities introduced into their crystal structure. This provides electrons which are free to move about in the material or holes which can accept these electrons. A material which contains an excess number of electrons is called an *N*-type material. A material which contains an excess number of holes is called *P*-type. These electrons and holes are called majority carriers. There are also minority carriers which occur in pairs in both *N*- and *P*-type materials. The minority carriers are very mobile; the majority carriers are not.

If a block of *N*-type and a block of *P*-type material are brought together, the majority carriers diffuse through the material and create a potential gradient across the boundary between these two materials. If a minority carrier gets into the region of this potential difference, it will immediately be swept to the oppositely charged area. In a steady-state condition the two currents, the majority carrier current and the minority carrier current, are equal. However, new hole pairs can be generated by bombarding this junction with light beams. Light energy which strikes crystals in the structure excites electrons and the additional energy causes some to break loose. Those electrons in the vicinity of the potential gradient are swept to the other side. The light upsets the balance between the diffusion current of the majority carriers and the current of the minority carriers, and a voltage difference can be measured across the crystal with a pair of electrodes. It is the minority current which does the work in a photocell.

Figure 3.7-1 illustrates the output current and voltage relationship for a representative solar cell. The factors primarily affecting these two electrical characteristics are the spectral signature of the illuminating source, the intensity of the illuminating source, the cell temperature, and the impingement angle of the illuminating source on the cell. The solar spectral characteristics are constant for platforms above the atmosphere.

The short-circuit current (T_{se}) of a typical cell varies directly with the illumination level whereas the open-circuit voltage (V_{oc}) of a solar cell varies logarithmically with illumination level. The open-circuit voltage output of a cell is affected more by the temperature of the solar cell than is the cell current.

The short-circuit current temperature coefficient is positive whereas the open-circuit voltage temperature coefficient is negative [2].

The angle of illumination incidence has a significant effect upon the electrical characteristics of the solar cell. The output power varies approximately as the cosine of the angle of incidence. As the angle of incidence increases, the output falls below that predicted by the cosine law because of increased reflection of useful spectral energy.

The spacecraft solar cell usually has a glass or quartz cover bonded to its light sensitive or active surface. Since the cell is sensitive to particle irradiation in the space environment (Van Allen belts), this covering reduces particle irradiation damage that reduces the power output of the cell. The cover glass can incorporate an optical filter to reject energy in the

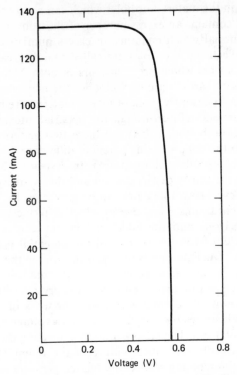

Figure 3.7-1 Voltage versus current characteristics, 2×2 cm N/P silicon solar cell, $30°$C, simulated sunlight.

solar spectrum to which the cell is insensitive to yield lower cell operating temperatures.

3.7.6 High-Energy Generation

Two methods have been used to obtain high voltages, the charge transport method and the variable capacitance method. The charge transport machines operate on the same principle as the Van de Graaff generator, in which a belt of insulating material is used to carry charges to a high-voltage electrode where they are collected to increase the high-voltage electrode potential.

Variable capacitance generators make use of the relationship between capacitance, charge, and voltage given by

$$Q = CV, \tag{3.7-7}$$

where V is the voltage across a capacitor of capacity C containing a charge Q. If the capacitor is initially charged with a charge Q and the capacitance is then decreased, the voltage across the capacitor will increase. Systems of this type show promise for the high-vacuum conditions of space where it is possible to have a high impedance across the capacitor to prevent the charge from leaking.

3.7.7 Systems

Each platform has a unique configuration. A typical spacecraft power supply configuration includes a solar-cell array as a power source, an energy-storage device consisting of a rechargeable battery, battery charge control or voltage limiting devices, and power conditioning devices such as converters and inverters as shown in Figure 3.7-2.

Since the output voltage of the solar array varies over a wide range as a function of array temperature and current loading, the shunt regulator serves as a controlled variable load on the solar array to maintain the output voltage within the desired limits.

Because the various rechargeable batteries require prescribed rates of current charge and maximum charge voltages, a charge regulator is provided.

During solar eclipse, the battery supplies power to the spacecraft via the discharge regulator that converts the widely variable battery voltage to a more highly regulated range consistent with the voltage requirements of various loads.

The dc/dc converter and dc/ac inverter provide the multiplicity of voltages required by platform equipments. The generation of these voltages within the power system eliminates the requirement that each equipment include the power conversion process.

The basic system is usually expanded to provide for equipment redundancy when it is desirable to enhance system reliability.

Attendant with the use of standby electronics is the need for detection and control schemes to identify equipment failures, to command faulty equipment out of service, and to bring the appropriate standby equipment into operation.

References

1. Bauer, P., "Batteries for Space Power Systems," National Aeronautics and Space Administration, SP-172, 1968.
2. Ralph, E. L., "Performance of Very Thin Silicon Solar Cells," 6th *IEEE Photovoltaics Conference Record*, 98–116, March 1967.

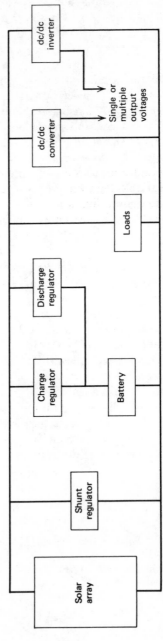

Figure 3.7-2 Spacecraft power supply configuration.

Ralph Bernstein

3.8 POSITIONING

Positioning involves determining the location of the instrument platform. Navigation techniques include piloting, electronic navigation, and celestial navigation. Piloting involves determining position by visual observations and relative measurements to known landmarks. Magnetic and gyro compasses and speed logs are commonly used. Electronic navigation determines platform position using electronic signals transmitted between the vehicle and fixed or mobile references. These systems include radio beacons, direction finders, radar, loran, and satellites. Celestial navigation involves position determination by reference to the sun, moon, planets, and stars. Manual and automatic sextants (star trackers) are used for this type of navigation.

Rho-theta position fixes are determined by a range (ρ) and bearing (θ) relative to a single reference (Figure 3.8-1a).

Theta-theta position fixes are determined by the intersection of lines with given angular relationship (θ_1 and θ_2) with respect to two references. The accuracy of this type of system is maximum when the lines of position intersect at right angles (Figure 3.8-1b).

Rho-rho position fixes are determined by the intersection of two or more circles of constant range. The two fixes resulting from two range measurements can be resolved by a third range measurement or by knowledge of approximate position. Maximum accuracy is obtained when the unknown position is within the triangle formed by the three stations (Figure 3.8-1c).

Hyperbolic position fixes are determined by the intersection of two lines of position that are defined by constant differences in distance from two or more stations (Figure 3.8-1d). The difference in distance is determined by the measurement of the difference in radio transmission reception times or phase from multiple fixed-position transmitters. Maximum accuracy is when the lines of position intersect at right angles.

Inertial position fixes are determined by sensing platform accelerations and known gravity and earth rotation forces. Position determination is self-contained, and accuracy degrades with operating time.

Typical electronic positioning and navigation systems and their performance characteristics are summarized on Table 3.8-1.

Ground-based electronic systems derive relative or absolute position from fixed-position ground systems.

Table 3.8-1. Summary of Common Electronic Positioning and Navigation Systems

	Category	Applications	Equipment Name	Range	World coverage (%)
Ground-based systems					
(a) Loran-A	Hyperbolic, passive	Medium-range ship and aircraft navigation	Loran-A	1,100 km (gnd wave) 2,200 km (sky wave)	10–15% (90 stations)
(b) Loran-C	Hyperbolic, passive	Medium/long-range ship and aircraft navigation	Loran-C	2,200 km (gnd wave) 5,500 km (sky wave)	10–15% (30 stations)
(c) Omega	Hyperbolic, passive	Long-range ship, submarine, and aircraft	Omega	11,000 km	100% (8 stations)
(d) VLF	Range-range passive	Long-range ship, submarine, and aircraft	VLF	11,000 km	50%
(e) Consolan/ Consol	Bearing, passive	Long range bearing to station ship and aircraft	Consolan (U.S.) Consol (Europe)	92–2600 km 37–1850 km (day) 37–2400 km (night)	2 stations (U.S.) Extensive in Western Europe (5 stations)
(f) Decca	Hyperbolic, passive	Short/medium-range ship and aircraft navigation	Decca	460 km	5%
(g) Autotape	Range-range, active	Short-range ship and aircraft precise positioning	Autotape	55 km	Special
(h) Hydrodist	Range-range, active	Short-range ship and aircraft precise positioning	Hydrodist (Telluro-meter)	0–40 km	Special
(i) RPS	Range-range, active	Short-range precise positioning	Range positioning system	160 km	Special
(j) Radar	Range-bearing, active	Short-range ship and aircraft positioning and collision avoidance	Radar	24 km	Wherever land or beacons occur
(k) Doppler Sonar	Range-bearing, active	Shallow water ship and submarine navigation, docking	Doppler sonar navigation	250 m (depth)	Shallow water or relative to water mass
Inertial navigation systems	Inertial, passive	Self-contained ship, aircraft, and spacecraft navigation	Ship's inertial navigation system (SINS)	Unlimited	Unlimited
Satellite systems					
(l) Navsat/ Transit, NNSS	Doppler freq. shift, passive	World wide positioning	Navy Navigation Satellite System	World wide	100%

Fix Time (min.)	Sensitivity	Accuracy	Corrections	Frequency	Mult. User Cap.	Auto/ Manual	Ref.
1–2	150 m	1.0–7.4 km	Propagation, terrain	1750–1950 kHz	No limit	M	5, 7
0.5	1.5–15 m	0.46–3.7 km	Propagation, terrain,	100 kHz	No limit	M or A	5, 7
Inst. − 0.5		0.93 km (day) 1.8 km (night)	Propagation	10.2 − 13.6 kHz	No limit	M or A	1, 5, 18
Inst. − 0.5		3.7 km (day) 7.4 km (night)	Propagation Freq. Std.	16–24 kHz	Nomlimit	M or A	15
1 min 1 min	0.03° (day)	0.3° day 0.7° night	Bearing conversion	192, 194 kHz 250–350 kHz	No limit	M	5, 7
1 min	15 m	0.4 km (day) 1.8 km (night)	Propagation terrain	70–130 kHz	No limit	A	7
Inst.	0.1 m	1.5 m		2900–3100 MHz	1 user per system	A	3
Inst.	<0.03 m	1.5 m		2800–3200 MHz	1 user per system	A	4
Inst.		3–15 m (at 80 km)		9300–9500 MHz	Multiple	A	7
0.5 min		30–91 m (range) 1° (bearing)		3000–9000 MHz 500–3000 pps	Virtually unlimited, some interference	M	5
Inst.	114 dB (1 μV) 1° bearing	0.5% of speed 1° bearing	Propagation	300 kHz	Virtually unlimited, some interference	A	13
Inst.	10^{-5}g	Errors increase with time	Gyro drift, accel. bias Periodic position		1 user per system		2, 5, 6
1–10 min	18 m	0.04–0.73 km	Ship velocity, antenna height	150 MHz 400 MHz	Unlimited	A	16, 17

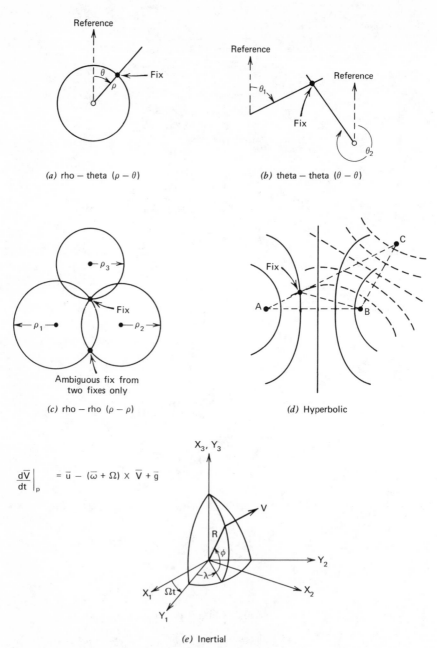

(a) rho − theta (ρ − θ)

(b) theta − theta (θ − θ)

(c) rho − rho (ρ − ρ)

(d) Hyperbolic

$$\left.\frac{d\overline{V}}{dt}\right|_p = \overline{u} - (\overline{\omega} + \Omega) \times \overline{V} + \overline{g}$$

(e) Inertial

Figure 3.8-1 Common position fixing/navigation systems.

3.8.1 Loran-A

Loran-A is an electronic, long-range hyperbolic navigation system utilizing manual time difference pulse matching to determine intersecting lines of position from three ground-based radio transmitting stations.

A Loran chain consists of three synchronized radio pulse transmitting shore stations. The center station is designated as the master and the end stations, slaves. Radio receivers on the platform receive the master and slave pulsed transmissions. The time difference in reception of a master/slave transmission defines a hyperbola of possible positions (see Figure 3.8-2). Position is determined by the intersection of two hyperbolic lines of position, derived from two time difference measurements between the master and two slave stations (see Figure 3.8-3).

3.8.2 Loran-C

Loran-C is an electric extra-long-range hyperbolic navigation system similar to Loran-A with performance and operational improvements. It uses eight closely spaced pulses and correlation techniques to detect weak signals with more reliability and accuracy. Automatic signal tracking can be implemented after initial signal locking. Both pulse and phase signal matching are performed for "coarse" and "fine" time difference matching, and sky wave contamination is eliminated. The use of automatic signal tracking with a digital computer provides continuous position.

3.8.3 Omega

Omega is a pulsed, phase-comparsion long-range hyperbolic navigation system using shore based transmitters operating at 10.2 kHz continuous

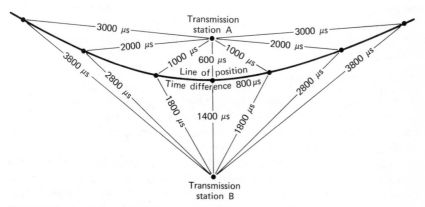

Figure 3.8-2 Example of hyperbolic line of position resulting from a constant transmission time difference (800 μs) from one transmission pair (station A and B).

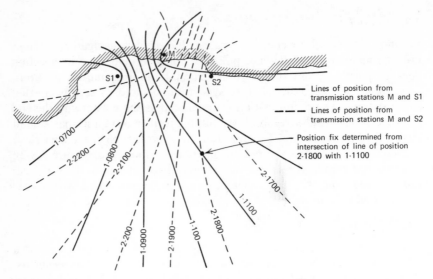

Figure 3.8-3 Loran fix obtained from intersection of two lines of position from two transmission paths.

wave (cw). Phase comparison of the periodically transmitted frequencies provides an isophase reference, which uniquely determines the location of the receiver.

Only three stations are needed for a position fix; each station transmits a signal for about 1 s every 10 s that is phase locked to a common time standard. Station transmitters for position determination are selected on the basis of signal reception and line of position intersection angles. Computerized systems compute the lines of position, and the geographic latitude and longitude at the intersection of the lines of position. Because of the low frequency, signals can be received down to 15 m below the water surface.

3.8.4 VLF Navigation

VLF Navigation is a range-range phase-comparison, long-range navigation system using shore-based VLF radio transmitters, precise platform frequency standards, and VLF tracking radio receivers. Only two transmitters are needed to determine a position fix. The phases of two received transmissions are measured at a known geographic location, and the changes in phase of the received transmissions are measured when the platform has moved to a new, unknown location. The phase determination is accomplished by precise phase comparison of the received transmissions with the platform frequency standard. Due to the low-frequency characteristics of the transmitted signals, long range (greater than 8,000 km) positioning can be accomplished.

3.8.5 Consolan/Consol

Coded transmissions from Consolan shore stations provide bearing information to a user. Consolan transmissions result in a pattern of alternating dot sectors and dash sectors, each about 12° in width. The bearing to the transmitting stations can be determined by counting the number of dots or dashes and using Consolan charts and tables.

Consolan is most accurate in an area that is perpencicular to the line of transmission towers, having an unused bearing sector in the direction of the tower line. A line of position is determined from one Consolan bearing. The intersection of a second line of position, obtained from a second consolan provides a fix.

Consol is similar to Consolan, and is used extensively in Western Europe.

3.8.6 Decca

Decca is a hyperbolic navigation or surveying system using phase comparision of radio transmissions to provide distance data to the transmitters. A master transmitter is located in the center of three equally spaced slave transmitters. The Decca operates either in a hyperbolic mode as in the Decca Navigator, or in a circular mode, as in the Two-Range Decca.

Each of the four stations transmits a continous wave at a different frequency in the band 70–130 kHz and with frequency ratios 5, 6, 8, and 9. Four receivers with suitable phase comparison circuits are used to determine the phase relationships between the received slave transmissions relative to the master transmissions. Two slave stations provide a unique position fix, the third provides a check or alternate fix in unfavorable areas.

3.8.7 Autotape

Autotape is an electronic line of sight range-range system used for automatic precision surveying. A master station on board the platform transmits a signal that initiates two slave transmitters (responders). By the use of sequential operation and phase comparison, the ranges to the responders can be precisely determined. Ranges can be determined every second and displayed or recorded with a 10-cm resolution.

3.8.8 Hydrodist

Hydrodist is an electronic line of sight range-range position fixing system used for precision surveying. Two master stations are carried on the survey platform and two remote instruments are located at previously coordinated control points. One master and one remote form a link from which precise range information can be derived.

3.8.9 Range Position System

The Range Position System (RPS) is similar to Autotape. The elapsed time between the transmitter interrogation and the transponder response provides range-range information to the transponders at known locations.

An RPS navigation system can be added to RPS, providing the capability for automatically determining the position of a platform.

3.8.10 Radar

Radar (from Radio Detection and Ranging) is a short-range azimuth (bearing) and range-position fixing system. A radar generates an electromagnetic signal which is reflected from a target and sensed by the platform radar receiver. Transmitting energy over 360° in azimuth permits a display of targets in the range of the radar.

The operating range of radar is limited by the earth curvature, and is thus dependent upon the height of the antenna and the target. A typical ship radar range is about 20 km.

Radar provides range and bearing information relative to detected land areas. Some radars have transponder sensor interrogation capability to activate navigational aid beacons which then transmit identification information, providing precise positioning relative to the accurately located beacons.

3.8.11 Doppler Sonar

Doppler sonar navigation systems determine seaborne platform speed relative to the ocean bottom or relative to ocean water mass reflections. This data combined with platform heading from a magnetic compass or gyrocompass is used to compute position.

A signal transmitted from a moving object, reflected from a stationary object and received by an observer has an apparent shift in frequency. The shift is proportional to the velocity of the moving object relative to the stationary object (Figure 3.8-4b).

A Doppler navigation system has both a transmitter and receiver on the platform (see Figure 3.8-4c) directed at an angle θ to the ocean bottom. A strong sonar energy return from the ocean bottom can be obtained to a depth of about 250 m to provide speed information relative to the earth. In deeper water, the Doppler sonar systems lock on to scattering layer returns about 10 m below the hull and provide speed relative to the water mass.

Fore and aft Doppler sonar systems are used to cancel out errors due to pitch and roll.

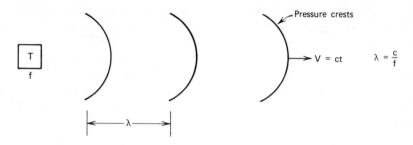

(*a*) Stationary transmitter

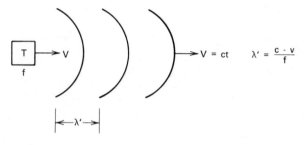

(*b*) Moving transmitter

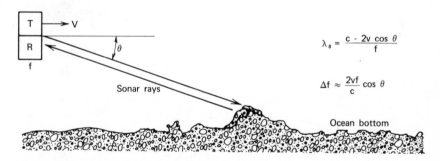

(*c*) Transmission directed at angle θ to ocean bottom

Figure 3.8-4 Doppler sonar speed sensing.

3.8.12 Navsat/Transit/NNSS

The Navigation Satellite System (Navsat) is a world-wide precise, navigation and position-fixing system employing earth orbital satellites, ground tracking stations, ground injection stations, and platform receiver and computer (see Figure 3.8-5). This system was called Transit and is also known as the Navy Navigation Satellite System (NNSS).

The platform position relative to the orbiting satellite is determined by measurement of the Doppler shift of the continuously transmitting satellite. Precise satellite position at the instant of transmission and several Doppler velocity measurements provide sufficient lines of position to compute a unique platform position.

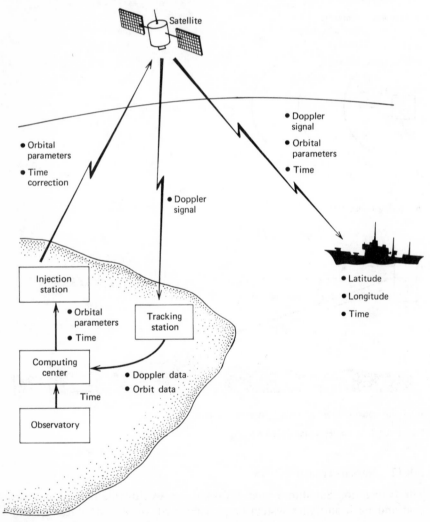

Figure 3.8-5 Simplified characterization of Navsat system.

The satellites orbit the earth at about 1110 km altitude with a period of 105 min and transmit two frequencies. Orbit (ephemeris information) is transmitted by the satellite and updated every 2 min, providing platform systems with the satellite position. Two frequencies are used to determine the propagation velocity of the transmitted radio waves. Doppler shift is measured at 400 MHz. The phase difference between the 150 and 400-MHz signals is indicative of the meteorological conditions of the atmosphere and is used to correct the propagation velocity. Predicted satellite position for the next 16 hr is computed for every 2 min and transmitted to the satellite by the ground injection station. The satellite stores this data and transmits its position and universal time every 2 min. In this manner, relative position to the satellite is determined by the platform system by Doppler measurements and absolute position by the combination of the relative position and the absolute satellite position provided by the satellite.

3.8.13 Dead Reckoning

Dead-reckoning navigation is the process of determining a platform's approximate position on the basis of its previous position and the true course steered and distance traveled. This type of navigation is affected by ocean and wind currents, equipment errors, and sea condition. Heading is provided by magnetic or gyroscopic compass, and speed by engine rotation, pitot tube, Doppler, electronic, or mechanical speed sensors. Two common electric heading and speed sensors are the gyrocompass and electromagnetic (EM) log.

A device that accepts the heading and speed inputs and computes the dead reckoning position is a Dead-Reckoning Analyzer (DRA). It computes the incremental latitude and longitude from the platform speed and heading, and updates the previous latitude and longitude. The Dead-Reckoning Tracer (DRT) provides a graphic record of the course. The accuracy of dead-reckoning navigation depends upon many variables, including weather and ocean conditions and types of equipment used. Errors increase progressively with time, and errors in the range 1–35 km are not unusual.

The gyrocompass is a north-seeking gyroscope. Suitable sensors and actuators keep the gyroscope spin axis fixed with respect to the meridian of the earth so it always points to true North.

An EM log is an electronic speed-sensing system used for seaborne platforms. The operation is based upon the principle that a conductor produces an electrical voltage when moved through a magnetic field. A sensor system (rodmeter) is mounted below the hull. A vertical magnetic field is produced in the water by coils in the sensor. Two conductors on the rodmeter pick up the voltage induced from the relative motion of the water and rodmeter. From Faraday's law, the voltage induced is directly proportional to the water flow velocity relative to the magnetic field (*B*). Since *B* is a constant, a linear relationship exists between the induced voltage and the platform speed. The device provides relative water speed information only. The operating range of the EM log is 0–20 m/s, with a resolution of 5 mm/s. Relative speed accuracies of 5 cm/s are realizable.

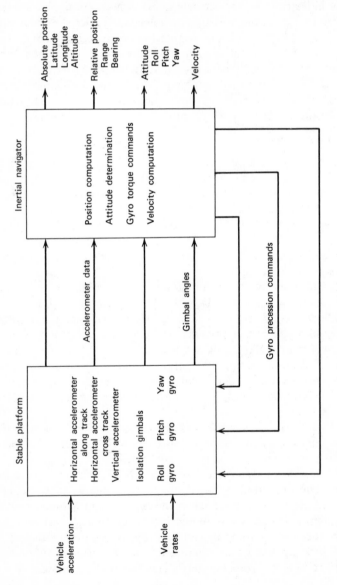

Figure 3.8-6 Functional diagram of an inertial navigation system.

3.8.14 Inertial Systems

Inertial navigation involves detection of platform acceleration and computation of velocity and position by integrating the sensed acceleration. An inertial navigation system consists of a stable platform and an inertial navigator (see Figure 3.8-6). The stable platform consists of a triad of gyro-stabilized accelerometers, the outputs of which are processed in the inertial navigator which computes instrument platform position and attitude.

Gimbals isolate the stabilized structure containing the accelerometers and gyros from vehicle rotations. Strap-down inertial navigation systems do not use gimbal isolation of the accelerometers and gyros, but maintain a computed orientation vector instead. The gyro outputs are used to maintain the platform gimbals in an attitude-stabilized state.

References

1. Brogden, J. W., "The Omega Navigation System," *Navigation: J. Inst. Navigation* **15**, 115–118, Summer 1968.
2. Broxmeyer, C., *Inertial Navigation Systems,* McGraw-Hill Book Company, New York, 1964.
3. Cubic Industrial Corporation, "Autotape Model DM-40 Electric Positioning System," Catalog CB 11-31/2-66.
4. Decca Survey Systems, Incorporated, "Hydrodist Type MRB-2," Catalog Sheet No. 58.
5. Dunlap, G.D., and H.H. Shufeldt, *Dutton's Navigation and Piloting,* 12th ed., United States Naval Institute, Annapolis, 1969.
6. Fernandez, M., and G.R. Macomber, *Inertial Guidance Engineering,* Prentice-Hall, Englewood Cliffs, New Jersey, 1962.
7. Ferrara, A.A., "Electronic Positioning Systems for Surveyors," Coast and Geodetic Service, Tech Memo C8GSTM-3, May 1967.
8. Griswold, L. W., "Underwater Logs," *Navigation: J. Inst. Navigation* **15**, 115–118, Summer 1968.
9. Hastings, C.E., and A.L. Comstock, "Pinpoint Positioning of Surface Vessels Beyond Line-of-Sight," presented at National Marine Navigation Meeting, Institute of Navigation, Nov. 21, 1969.
10. Iacona, M. A., "Navigation in Oceanography," *Navigation: J. Inst. Navigation* **15**, 244–256, Fall 1968.
11. Kayton, M., and W.R. Fried, *Avionics Navigation Systems,* John Wiley and Sons, New York, 1969.
12. Kuebler, W., "Marine Electronic Navigation Systems—A Review," *Navigation: J. Inst. Navigation* **15**, 268–273, Fall, 1968.
13. The Marquardt Company, "Precision Shipboard Navigator/Pulse Doppler Sonar Model MRQ-2015A," Catalog, July 1970.

14. Milne, C.R., "Navigational Instruments: Progress During the Last Fifty Years," *J. Sci. Instr.* Ser. 2, **1,** 1045–1052, Nov. 1968.
15. Ruppert, G. N. P. A., R. Bernstein, and C. O. Bowin, "Precise Positioning of a Ship at Sea Utilizing VLF Transmissions," *Navigation: J. Inst. Navigation* **16,** 110–128, Summer 1969.
16. Stansell, T.A., Jr., "The Navy Navigation Satellite System: Description and Status," *Navigation: J. Inst. Navigation* **15,** 229–243, Fall 1968.
17. Stansell, T. A., Jr., "Transit the Navy Navigation Satellite System," *Navigation: J. Inst. Navigation,* **18,** No. 1, Spring, 1970.
18. Swanson, E. R., "Omega," *Navigation: J. Inst. Navigation,* **18,** No. 2, October, 1970.

Chapter 4

In Situ and Laboratory Sensors

4.1 SENSOR PARAMETERS

In situ sensors are used to measure parameters of the environment in which the sensor is immersed. The problem of sensor design is thus a problem of ascertaining with certainty that the sensor truly responds to the desired parameter and is insensitive to other parameters.

Sensors for the measurement of 31 parameters are included in this chapter in alphabetical order. In those cases where parameters are measured in more than one domain (land, sea, air, or space) all of the pertinent domains are included in the same section. The relationships between the sensor and the other portions of the instrument system are described in Chapter 1. The important parameters of each of the various domains are described in Chapter 2, and the problems associated with the various instrument platforms are described in Chapter 3.

The discussion of the in situ sensors of this chapter includes a review of the scientific phenomena being measured and the detail with which the parameter is known. The sensor discussions also include a review of the various types of transducers available for each of the parameters and include comparisons of the advantages and disadvantages of each. An attempt has been made to present the underlying scientific theory describing the operation of each sensor, the engineering problems involved in the design of the sensor, and the performance available from each sensor. Sensor performance is usually described in terms of its transfer function, sensitivity, accuracy, or response time.

A sensor responding to an input parameter, P, and producing an output quantity, Q, has a transfer function which expresses the relationship between the input and output as

$$Q = f(P). \tag{4.1-1}$$

The resolution of a sensor is a measure of the ability to discriminate between two nearly equal values of P separated by an amount ΔP. The smaller ΔP, the greater the resolution. The precision of a sensor is a measure of the number of significant figures with which the parameter P is measured.

The sensitivity of a sensor is a measure of the change in the output quantity

191

produced by a change in the input parameter. It can be written as

$$S = \frac{dQ}{dP}. \qquad (4.1\text{–}2)$$

The accuracy of a sensor is a measure of the absence of errors. When errors are present the expression of Eq. 4.1–1 becomes

$$Q_e = f(P) + E, \qquad (4.1\text{–}3)$$

where E is the error given by

$$E = Q - Q_e. \qquad (4.1\text{–}4)$$

Several types of errors are possible. A static error can have the form

$$E_s = g(Q), \qquad (4.1\text{–}5)$$

where the function $g(Q)$ can be a constant, a linear function of Q, or something more complex. The error may also have a hysteresis component that depends on the previous parameter values.

A dynamic error has the form

$$E_d = g(Q, t) \qquad (4.1\text{–}6)$$

and depends upon time. Noise and drift are special cases of dynamic error and are examples of errors that depend upon other environmental parameters.

Errors can be classified according to their origin as systematic or random. Systematic errors can arise from instrumental errors, the environment, or the operator. Random errors are usually of unknown origin and are reduced by the use of statistical techniques.

The average, or arithmetic mean, of a finite number, N, of measurements is the most probable value. It is given by

$$\bar{Q} = \frac{1}{N} \sum_{i=1}^{N} Q_i. \qquad (4.1\text{–}7)$$

The deviation of any one measurement is given by

$$D_i = Q_i - \bar{Q}. \qquad (4.1\text{–}8)$$

The average deviation is

$$\bar{D} = \frac{1}{N} \sum_{i=1}^{N} |D_i|. \qquad (4.1\text{–}9)$$

The variance is given by

$$\sigma^2 = \frac{1}{N} \sum_{i=1}^{N} D_i^2, \qquad (4.1\text{–}10)$$

and the mean square deviation or standard deviation is

$$\sigma = \left(\frac{1}{N} \sum_{i=1}^{N} D_i^2 \right)^{1/2}. \tag{4.1-11}$$

When the errors are due to the combined effect of a large number of independent random causes, the result is a normal distribution or Gaussian disbribution. For any one measurement, the probability that the result is between x and $x + dx$ is

$$P(x, x + dx) = p(x) dx, \tag{4.1-12}$$

where $p(x)$ is the probability density function. For the normal distribution the normal probability density function for a zero mean and a standard deviation σ is given by

$$P_n(x) = \frac{1}{\sigma (2\pi)^{1/2}} \exp \left(\frac{-x^2}{-2\sigma^2} \right). \tag{4.1-13}$$

If two independent variables have the same variance,

$$P_n(x, y) = \frac{1}{2\pi\sigma^2} \exp \left(\frac{-x^2 + y^2}{2\sigma} \right). \tag{4.1-14}$$

The distance R from the origin,

$$R = (x^2 + y^2)^{1/2}, \tag{4.1-15}$$

has a Rayleigh distribution given by

$$P_n(R) = \frac{R}{2\sigma^2} \exp \left(-\frac{R^2}{2\sigma^2} \right) \tag{4.1-16}$$

If the result of a random measurement is one of two alternatives, one of which is correct, the result of many measurements is a binomial distribution. If the probability of a correct result is p, then the probability of an incorrect result is $1 - p$ and the probability of k correct results in N measurements is given by the binomial distribution

$$P_b = C_k^N p^k (1 - p)^{N-k}, \tag{4.1-17}$$

where the binomial coefficient is given by

$$C_k^N = \frac{N!}{k!(N-k)!}. \tag{4.1-18}$$

The average is

$$m = Np, \tag{4.1-19}$$

and the variance is

$$\sigma^2 = Np(1 - p). \tag{4.1-20}$$

If there is a constant probability vdt that one of many independent random events will occur in a short period of time, dt, the probability that k events will occur in a time interval T is given by the Poisson frequency function

$$P_k = \frac{m^k}{k!} \exp(-m), \qquad (4.1-21)$$

where the average number of events is

$$m = vT, \qquad (4.1-22)$$

and the variance is

$$\sigma^2 = m. \qquad (4.1-23)$$

The sum, s, of the squares of N independent random variables each having a zero mean and unity variance has the probability density function called a Chi-square distribution given by

$$P_x(s) = \frac{s^{n/2-1}}{2^{N/2}\Gamma(\frac{1}{2}N)} \exp(-\frac{1}{2}s). \qquad (4.1-24)$$

The mean of s is

$$\bar{s} = N, \qquad (4.1-25)$$

and the variance of s is

$$\sigma^2 = 2N. \qquad (4.1-26)$$

The response time of a sensor is another important characteristic. If there is a finite delay time, t_d, but no distortion, the output of Eq. 4.1–1 becomes

$$Q(t) = f[P(t - t_d)]. \qquad (4.1-27)$$

Normally, there is distortion that depends upon both the form of the parameter and the sensor response characteristics. A first-order sensor response is one that depends only on the input and the first derivative of the output. The response to a step change, ΔP, in the parameter is an exponential monotonic change to the new output value. This response has the form

$$Q(t) = Q_0 + \Delta Q(1 - e^{-t/\tau}), \qquad (4.1-28)$$

where Q_0 is the initial value, $\Delta Q = f(\Delta P)$ is given by Eq. 4.1-1, and τ is the time constant of the sensor, the time it takes to change by a factor $e = 2.718. \ldots$ In an elapsed time equal to one time constant the sensor output changes 63 percent of the amount it would change in an infinite length of time. Thus the error in the sensor output is found from Eq. 4.1–28 or

$$E(t) = Q_0 + \Delta Q - Q(t) = \Delta Q e^{-t/\tau}. \qquad (4.1-29)$$

If the input is changing linearly with time as αt the error is given by

$$E(t) = \alpha \tau (1 - e^{-t/\tau}). \qquad (4.1-30)$$

If the parameter has a periodic time variation with a corresponding sensor output $A \cos \omega t$, the error is

$$E(t) = A \cos \omega t + \frac{A - e^{-t/\tau}}{1 + (\omega \tau)^2} - \frac{A \cos\left[\omega t + \tan^{-1}(\omega \tau)\right]}{\left[1 + (\omega \tau)^2\right]^{1/2}}. \qquad (4.1-31)$$

Note that the error for the linear and periodic variations does not even disappear after an infinite time but that it is reduced for all three cases as the time constant τ is reduced.

A second-order sensor response is one that depends on the input and the first and second derivatives of the output. The response of this type of sensor to a step change at the input has an error

$$E(t) = \frac{1}{\left[1 - k^2\right]^{1/2}} e^{-k\omega t} \sin\left[\omega t(1 - k^2)^{1/2} + \tan^{-1}(1 - k^2)^{1/2}/k\right],$$

$$k < 1, \qquad (4.1-32a)$$

$$E(t) = \frac{2}{\omega} e^{-\omega t}, \qquad k = 1, \qquad (4.1-32b)$$

$$E(t) \simeq e^{-\omega t/2k}, \qquad k > 1. \qquad (4.1-32c)$$

Equation 4.1–32a represents an oscillation at a frequency ω that is damped in time by the constant k. The value $k = 1$ is the critical damping case with no oscillation. The case $k > 1$ is the overdamped case.

The transient response to a step change described above is only one of several important sensor characteristics. The important parameters discussed in each of the sections of this chapter include:

1. The nature of the environment. What is known and unknown?
2. The need for the sensor. What can it contribute?
3. Is it passive or active?
4. Type of output (voltage, current, impedance, frequency).
5. Form of output (analog or digital).
6. Dynamic range.
7. Frequency response.
8. Transient response.
9. Transfer function.
10. Effect of value on parameter being measured.
11. Accuracy (absolute and relative).
12. Sensitivity.
13. Resolution.

14. Stability.
15. Effect of overload, time to recover from overload, damage point.
16. Sensitivity to parameters other than that being measured.
17. Power consumption.
18. Environmental limitation (operating and storage).
19. Reliability.
20. Life (elapsed time or number of measurements).
21. Availability.

Comparative information for sensors for a given parameter for items 3 through 21 above are given in tabular form where practical.

W. James Trott

4.2 ACOUSTIC WAVE SENSORS

4.2.1 Nature of Acoustic Waves

Acoustic waves are mechanical or elastic waves that depend upon the inertia and compliance of the medium for propagation. The periodic strain is in the direction of propagation in longitudinal waves and perpindicular to the direction of propagation in shear or transverse waves. Longitudinal acoustic waves can propagate in solids, liquids, and gases. Shear waves propagate in solids and some highly viscous liquids; media that can support shear strain. Rayleigh waves are shear waves that propagate along the surface of a solid. They are the principal disturbance propagated along the surface of the earth from earthquakes. An acoustic wave can be characterized by certain detectable parameters which vary periodically in space and time. Generally a sensor measures the acoustic or excess pressure, the instantaneous pressure at the sensor in the presence of the acoustic wave minus the static pressure. Other sensors measure the particle velocity or particle acceleration.

The inertia of the medium is a function of its density ρ; the compliance is a function of the reciprocal of the elastic modulus. By analogy to a transmission line the density, ρ, is the unit volume equivalent of the inductance, L, per unit length of the transmission line and $1/k$ (where k is the bulk modulus) is the unit volume equivalent of capacitance, C, per unit length of the transmission line. The propagation velocity of a wave in a transmission line is $c = (LC)^{-1/2}$ and the characteristic wave impedance is $Z_c = (L/C)^{1/2}$. Thus for a plane, longitudinal wave the propagation velocity is $c = (k/\rho)^{1/2}$ and the characteristic wave impedance is $Z_c = (k\rho)^{1/2} = \rho c$. The characteristic wave impedance is the ratio of the acoustic pressure to the particle velocity. The particle acceleration is the time derivative of the particle velocity.

Neglecting second-order effects the density, ρ, is the mean density of the medium and the modulus is the adiabatic elastic modulus or (γk) in air where γ is the ratio of the specific heat at constant pressure to the specific heat at constant volume. Only in special cases such as low-frequency calibration of a microphone in a coupler will the acoustic pressure vary isothermally. In water the difference between the adiabatic and isothermal bulk modulus is less than 1 percent.

Underwater sound speed varies with temperature, salinity and depth as

$$c = 1449 + 4.6T - 0.055T^2 + 0.0003T^3 + (1.39 - 0.012T)(S - 35) + 0.017d$$

$$(4.2-1)$$

where T is the temperature in degrees centigrade, S is the salinity in parts per thousand, and d is the depth in meters. Underwater sound absorption is a function of temperature and frequency as shown in Figure 4.2-1 [18].

4.2.2 Hydrophone

In underwater acoustics the sensor is a hydrophone. Unlike air acoustics where a condenser is so widely used, the hydrophone is generally a piezoelectric type sensor with a built-in preamplifier. A maximum operating frequency range, a minimum equivalent noise pressure level, and stability with age, change in ambient temperature and operating depth are basic requirements. A piezoelectric ceramic such as lead metaniobate or lead zirconate-titanate yields a hydrophone having a lower equivalent noise pressure level for the same directional response and frequency bandwidth than would be obtained with a crystal such as lithium sulfate; the latter, however, is more stable with age, varying temperature and depth. Lead metaniobate changes sensitivity with depth, 0–1600 m, by 1.5 percent, lead zirconate-titanate by 2 or 3 percent.

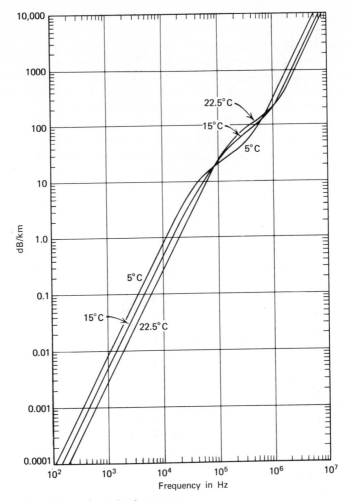

Figure 4.2-1 Absorption in sea water.

A piezoelectric sensor electrically is a capacitor when operating well below the first resonant frequency. This sensor is coupled into a high-impedance input preamplifier. The sensor is sometimes coupled to the sound field through castor oil and a butyl rubber boot. In this design the butyl rubber minimizes water vapor penetration and the castor oil absorbs some water, thus prolonging the life of the hydrophone. The instrument loses low-frequency sensitivity when water reduces the electrical leakage resistance across the sensor to the range of the capacitor impedance.

In seismic prospecting a string of hydrophones, 30 m to several km long, is towed behind a boat and signals from each hydrophone are recorded. The frequency range of interest is 10 to 1000 Hz [17].

In these two examples the sensor is omnidirectional in its sensitivity to the acoustic signal. The volume of piezoelectric material is segmented and electrically connected to obtain the desired impedance for preamplifier or cable input. For a particular element configuration of discs, cylindrical tube, or spherical shell the product of the sensitivity squared (M^2) (volts per microbar) times the electrical capacitance, C, is a constant independent of the number of segments but varying with the type of piezoelectric material. For a radially poled spherical shell of piezoelectric ceramic,

$$M^2C = 0.02\, g_{31}^2\, e_{33}^T\, (r/t)\, V, \qquad (4.2\text{--}2)$$

where g_{31} is the piezoelectric stress coefficient, e_{33}^T is the dielectric constant for constant stress, r is the outer radius of the shell, t is the wall thickness, and V is the volume of the sensor.

For a cylindrical shell 0.02 is replaced by 0.03. For a stack of discs 0.02 is replaced by 0.01 [13].

4.2.3 Underwater Applications of Acoustic Sensors

High resolution for depth sounding, bottom profiling, fish detection, or object locating requires an active system consisting of a directed sound source and a directional hydrophone; one transducer is often used as both source and hydrophone. Side-looking sonar in the form of a long line or rectangular array looks broadside from the ship and produces a bottom profile as a result of the forward motion of the ship. [2, 3, 7]. A typical sonar emits a pulse of a few miliseconds duration, 10 pulses per second with a beamwidth in the plane of the line array of approximately 2°. Resolution of a few meters is possible out to 1 km with a pulse peak power of about 500 W.

Scanning sonar sweeps a volume of the sea by phasing a group of sensors in a cylindrical, spherical, or planar array in the desired direction.

The beamwidth between half-power points is $30°(\lambda/d)$ for a 2-point source, $50°(\lambda/d)$ for a rectangular source, and $60°(\lambda/d)$ for a circular piston source, where λ is the wavelength of the sound in the medium and d is the dimension of the source.

As this indicates the 2° beamwidth for side-looking sonar requires a sensor of 25 wavelengths. A 2° beamwidth at 5 kHz requires a rectangular source length of 8 m.

Sound attenuation in sea water rises sharply with frequency as shown in Figure 4.2-1. Long-range capability, therefore, requires low frequencies. Deep penetration in bottom profiling also requires low frequencies.

One way to overcome the large source requirement for high-resolution low-frequency bottom profiling is to use the nonlinear characteristics of wave propagation at finite amplitudes [16, 22]. The wave equation is generally derived for infinitesimal amplitudes using the mean value of the density and the bulk modulus. For high-intensity sound $c = (k/\rho)^{1/2}$ and $Z_c = (k\rho)^{1/2}$ are a function of the sound intensity. Two high-intensity signals from a common source interact within the near sound field of the source and thus produce a difference frequency and a sum frequency. The resulting end-fire effective source within the interacting sound field produces a highly directional difference frequency sound beam. Primary frequencies in the range of 200 kHz with a source level greater than 100 dB re 1 μbar produce a 10- to 15-kHz difference frequency having a source level of 60 dB re 1μbar with a 2° beamwidth. Although the efficiency is less than 1 percent the smaller size of the sonar makes low-frequency high-resolution bottom profilers applicable to small boat installation.

Acoustic imaging over short distances at high frequencies can be achieved by acoustic holography and the image converter tube [12, 15]. In acoustic holography a source produces coherent sonic radiation. An array of hydrophones detects the sound scattered by the object. The pattern of amplitude and phase of the sonic field is converted into electrical signals from the hydrophones which in turn are combined with the original signal driving the source to produce a luminous oscilloscope display that is a hologram photograph. The photograph is then projected as in optical holography to produce a visual image of the object in three dimensions.

The image converter tube uses a 0.5 to 5-MHz source. The reflected or transmitted sound is detected by a piezoelectric ceramic plate backed by an electron beam tube. The electron beam is modulated by the sound-induced electric charges on the ceramic. An 800-line picture requires 800 pulses of 10 μs. duration each. The system can detect acoustic signals down to 10^{-9} to 10^{-11} W cm^{-2}.

4.2.4   Signal Processing

In all of these examples signal processing is required to enhance the information signal in the presence of noise, to identify the bearing or range of the echo source or to measure the relative speed of the echo source in relation to the sonar. In communication sonar the speech modulates the frequency of the sonar signal. Filtering enhances the signal-to-ambient-noise ratio, (S/N). The range of prevailing noise and some of the noise sources in the sea are shown in Figure 4.2-2 [18, 21]. More sophisticated techniques are used to detect the signal when the S/N ratio is less than 1 [10, 14]. A continuous-strip-chart record of successive echos in seismic prospecting, bottom profiling, or military sonar can resolve an image through optical correlation when one echo would be indeterminate in the noise. Side-looking sonar detects a sunken ship or a cable by showing an echo and a shadow or absence of return signal due to the slant angle of observing the bottom.

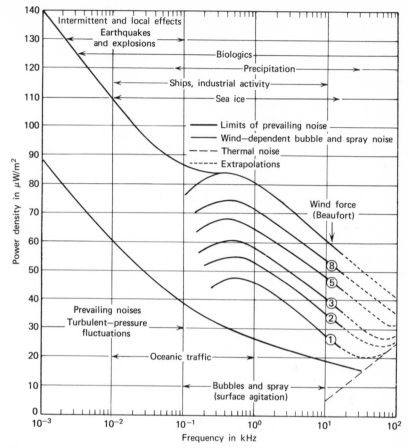

Figure 4.2-2 Noise power density in the sea.

Sonar detects relative movement of the target ship as a difference in frequency of the returned echo from the transmitted signal. The difference in frequency, known as the Doppler shift is [18]

$$\Delta f = (2f_0/c) \left[|V_s|\cos \theta_s - |V_t| \cos \theta_t \right], \qquad (4.2\text{-}3)$$

where f_0 is the transmitted signal frequency, c is the speed of sound, $|V_s|$ is the speed of the source ship, $|V_t|$ is the speed of the target ship, and θ_s and θ_t are the angles of the velocity vectors from the radius vector from the source to the target.

For fish and similar object locating sonar of high resolution, the sonar must be stabilized in pitch, roll, and yaw. The sonar may transmit a narrow beam, 30°, to increase the sound source level with the available electrical power. The receiving beam may be 1° or 2° and scan the region insonified [19]. Some sonar recorders delineate the bottom profile for improved detection of schools of fish feeding on the bottom [8]. Digital read-out of the depth sounders has been used to automate hydrographic surveys [4].

4.2.5 Microphone

In air acoustics the sensor is a microphone. The acoustic wave pressure usually actuates a diaphragm. If only one side of the diaphragm is exposed to the acoustic wave, the diaphragm approximates a circular piston with directional response described in Section 4.2.3. If both sides of the diaphragm are exposed to the acoustic wave, the pressure gradient of the acoustic wave is detected. The directional response is a function of the cosine of the angle of reception. The maximum sensitivity occurs for wave propagation normal to the diaphragm and the minimum for wave propagation parallel to the diaphragm. Since the particle velocity in a wave is a vector quantity in the same direction as the pressure gradient, the pressure-gradient microphone can also be called a velocity microphone. To convert the pressure or pressure gradient to an electrical signal, the diaphragm can be one plate of a capacitor, making an electrostatic or condenser microphone. The diaphragm can be coupled to a coil in the gap of a permanent-magnet structure, making an electrodynamic microphone. The diaphragm can be in the form of a ribbon (one-turn coil) in the gap of a permanent-magnet structure, making a ribbon microphone. The diaphragm can be coupled to a piezoelectric crystal that is actuated in compression, shear, or twist, making a crystal microphone. The sensitivity of a condenser microphone is constant from about 10 Hz to 15 or 20 kHz.

The lower frequency limit is generally a mechanical limitation due to a capillary vent hole from the outside to the back side of the diaphragm (pressure microphone) that provides for temperature and atmospheric pressure changes. The upper frequency limit is a mechanical limitation due to the resonance of the mass of the diaphragm and the combined compliance of the diaphragm and the air volume behind the diaphragm. The diaphragm of a condenser microphone can be a stretched metal foil or metalized mylar plastic. The back plate is massive in comparison to the diaphragm so that the acoustic wave induces movement only in the diaphragm. In the condenser microphone the acoustic pressure or pressure gradient modulates the value of the capacitance. This capacitance variation can be used to modulate a dc potential bias or it can be part of a tuned circuit that frequency modulates a carrier signal.

4.2.6 Applications of Sensors in Air Acoustics

The acoustic power radiated by a mosquito is the order of 10^{-12} W. The acoustic power radiated by the Saturn rocket during launch is 4×10^8 W.

The increase in the application of power, machines, and vehicles, with a concomitant increase in noise, has been at the rate of 15 percent per decade. Probably the largest effort in air acoustics is the application of sensors to measurement and analysis of noise pollution for determining the cause and the means for reducing the radiated noise [6, 11], for determining the characteristics of noise which cause annoyance [6], and for finding the relation of noise intensity and time of exposure in relation to industrial deafness [20].

Sensors are used in biological research to record and study the means of communication among insects, animals, and birds and to study the means of object localization by bats [1]. In the study of bats the frequency range of interest is 20 to 100 kHz. For studies in noise pollution the audible frequency range 20 Hz to 20 kHz is of primary interest. Sonic boom investigations require a frequency range 0.03 to 1000 Hz [9]. Infrasonic waves, 10^{-3} to 20 Hz, produced by natural catastrophies and nuclear explosions provide a means for their detection and localization. To achieve low self-noise these sensors are designed to detect over a narrow frequency band, 1 Hz or less. One sensor design consists of an insulated flask as a Helmholtz resonator with a heated thermistor in the throat of the flask measuring the movement of the air due to the infrasonic wave. A group of sensors spaced 500 to 1000 m apart yields directional information [5].

References

1. Busnell, R. G., Ed., *Acoustic Behavior of Animals*, Elsevier, New York, 1963.
2. Chesterman, W. D., P. R. Clynick, and A. H. Stride, "An Acoustic Aid to Sea Bed Survey," *Acustica* **8**, 285, 1958.
3. Clay, C. S., J. Ess, I. Weissman, "Lateral Echo Sounding of the Ocean Bottom on the Continental Rise," *J. Geophys. Res.* **69**, 3823, 1964.
4. Cooke, C. H., "Digital Read-Out Echo Sounder," in Proceedings of the Conference on Electronic Engineering in Oceanography, paper #6 (British), 1966.
5. Fehr, Uri, "Instrumentation for an Array of Infrasonic- and Hydromagnetic-Wave Sensors," *J. Acoust. Soc. Am.* **41**, 587, 1967.
6. Harris, C. M., Ed., *Handbook of Noise Control*, McGraw-Hill Book Co., New York, 1957.
7. Haslett, R. W. G., and D. Honnor, "Some Recent Developments in Sideways-Looking Sonars, in Proceedings of the Conference on Electronic Engineering in Oceanography, paper 5 (British), 1966.
8. Haslett, R. W. G., "A High-Speed Echo-Sounder Recorder Having Seabed Lock," *J. Brit. I. R. E.* **24**, 441, 1962.
9. Hilton, D. A. and J. W. Newman Jr., "Instrumentation Techniques for Measurement of Sonic Boom Signatures," *J. Acoust. Soc. Am.* **39**, s36, 1966.
10. Horton C. W., Sr., *Signal Processing of Underwater Acoustic Waves*, U. S. Government Printing Office, Washington, D. C., 1969.

11. King, A. J., *The Measurement and Suppression of Noise with Special Reference to Electrical Machines*, Chapman and Hall, London, 1965.
12. Korpel, A., "Acoustic Imaging and Holography," *IEEE Spectrum* **5**, 45, 1968.
13. Liddiard, K. C., Australian Defence Scientific Service Technical Note CPD 158.
14. Middleton, D., *An Introduction to Statistical Communication Theory*, McGraw Hill Book Co., New York, 1960.
15. Silverman, D., "Seismic Holography—Oil Finding Tool of the Future?" *Ocean Industry* **40**, Jan 1970.
16. Tucker, D. G., "The Exploitation of Non-Linearity in Underwater Acoustics," *Sound and Vibration* **2**, 429, 1965.
17. Tolstoy, I., and C. S. Clay, *Ocean Acoustics*, McGraw-Hill Book Co., New York, 1966.
18. Urick, R. J., *Principles of Underwater Sound for Engineers*, McGraw-Hill Book Co., New York, 1967.
19. Voglis, G. M., and J. C. Cook, "Underwater Applications of an Advanced Acoustic Scanning Equipment," *Ultrasonics* **4**, 1, 1966.
20. Ward, D., "The Concept of Susceptibility to Hearing Loss," *J. Occupational Med.* **7**, 595, 1965.
21. Wenz, G. M., "Acoustic Ambient Noise in the Ocean: Spectra and Sources," *J. Accoust. Soc. Am.* **34**, 1936, 1962.
22. Westervelt, P. J., "Parametric Acoustic Array," *J. Acoust. Soc. Am.* **35**, 535, 1963.

Edward A. Wolff

4.3 AGE SENSORS

4.3.1 Radioactivity

The age of natural materials and the date of geologic events can often be determined by the measurement of the decay of radioactive isotopes contained in the material. The radioactivity can be expressed in terms of the half-life, T, the time required for half of any given number of atoms to decay. Radioactivity can also be expressed in terms of the decay constant, λ, the fraction of the number of atoms that decay per unit of time. The relationship between T and λ is

$$T = \frac{\ln 2}{\lambda} = \frac{0.693}{\lambda}. \tag{4.3-1}$$

Elements are characterized by their atomic number, Z, and their mass number, A. The atomic number is the number of protons in the nucleus, and the mass number is the number of nucleons in the nucleus. Isotopes are different forms of an element having the same Z but different A. These isotopes can be changed by the emission of either an alpha or beta particle. Alpha particles are the nuclei of helium with a mass of 4 and a positive charge of 2. Therefore an isotope that decays by alpha emission experiences a decrease of Z by two and a decrease of A by four. A beta particle is an electron with a charge of -1 and a mass $1/2000$ that of a proton. Therefore, when an isotope decays by beta emission Z increases by 1 and A remains essentially unchanged.

The original radioisotope is called the parent and the isotope to which it decays is called the daughter. The relationship between the relative amounts of parent (P) and daughter (D) atoms, the decay constant, and the time of decay is given by

$$\frac{P}{P + D} = e^{-\lambda t}, \tag{4.3-2a}$$

or

$$D = P(e^{\lambda t} - 1), \tag{4.3-2b}$$

and the age or decay time is

$$t = \frac{1}{\lambda} \ln\left(\frac{D}{P} + 1\right). \tag{4.3-3}$$

The accurate determination of the age of materials therefore requires the precise knowledge of the decay constant. Any uncertainty in this decay constant produces a comparable uncertainty in the age measurement. The accurate measurement of age also requires an accurate measurement of the quantity of both the parent and daughter isotopes in the material. The measurement of these isotope quantities can involve complex procedures for the separation of these elements from the material sample and the necessity for accurately measuring very small amounts. If the daughter element is also present in the environment, care must be taken to ensure that it does not contaminate the sample.

An accurate determination of age also requires a knowledge of the amount of daughter material in the sample at time $t = 0$ when the radioactive decay process began. Age determination is convenient if the assumption of zero initial daughter material is valid.

An accurate measurement of age requires that the parent and daughter element content of the sample not be disturbed by chemical, mechanical, or other forces during the decay time. In other words, there should be no increase or decrease in the parent and daughter elements except by radioactive decay.

The various methods which have been developed for the measurement of age have different advantages and disadvantages with respect to the types of materials, the age of the materials, their environments, and the time spans over which they have been decaying. Accurate age measurements rarely exceed four or five half-lives. Information on the measurement of solids, liquids, and gasses is given in Sections 4.11, 4.29, and 4.4, respectively.

4.3.2   Uranium—Lead

One of the common methods of measuring age is the measurement of the decay of uranium and thorium into lead. This decay can be indicated symbolically as

$$^{238}U \xrightarrow{\quad T = 4.5 \times 10^9 \text{ yr} \quad} {}^{206}Pb + 8\,{}^{4}He + \text{energy}, \tag{4.3-4}$$

$$^{235}U \xrightarrow{\quad T = 7.14 \times 10^8 \text{ yr} \quad} {}^{207}Pb + 7\,{}^{4}He + \text{energy}, \tag{4.3-5}$$

$$^{232}Th \xrightarrow{\quad T = 1.39 \times 10^{10} \text{ yr} \quad} {}^{208}Pb + 6\,{}^{4}He + \text{energy}. \tag{4.3-6}$$

The emission of successive alpha particles during the decay of uranium to lead results in the creation of successive isotopes. The complete decay series for both uranium and thorium are shown in Figure 4.3-1. Also shown is the half-life of each decay process. In Figure 4.3-1 it can be seen that the decay of the intermediate elements is relatively rapid so that the total decay is governed by the decay of the parent isotope for long decay times.

For materials containing known proportions of uranium and thorium it is possible to determine the quantity of parent element by measuring the alpha radiation. The daughter lead content can be measured spectrochemically. This method is sometimes called the lead-alpha method.

Age can also be computed by measuring the ratio of the lead and uranium isotopes and solving the equation

$$\frac{^{207}Pb}{^{206}Pb} = \frac{^{235}U\,(e^{\lambda_1 t} - 1)}{^{238}U\,(e^{\lambda_2 t} - 1)}, \tag{4.3-7}$$

where λ_1 and λ_2 are the decay constants for ^{235}U and ^{238}U, respectively. The measurement of age using ^{238}U and ^{206}Pb in Eq. 4.3–2 has the advantage that these isotopes are most abundant so that the determination of age is less sensitive to errors in analysis. The disadvantages are the susceptibility of this uranium to being leeched from the sample and the fact that the gaseous radon (^{222}Rn) that forms during the decay has a half-life of 3.8 days and can escape from the sample. The use of lead ratios in Eq. 4.3–7 is also subject to error due to loss of radon.

The use of ^{235}U $-$ ^{207}Pb with Eq. 4.3–2 to determine age has the advantage that the radon formed in this decay process is ^{219}Rn which has a half-life of only 4 s and is therefore less likely to escape from the sample before it decays. Its disadvantage is its relative scarcity.

In principle, age can be determined by measuring the ratio of any one of the members of the decay series to the stable end product. The convenient properties of ^{210}Pb have lead to the use of the measurement of the ratio of ^{210}Pb to ^{206}Pb. This method is subject to the same errors as the ^{238}U method.

These methods have been used for ages of 3×10^5 to 3×10^9 yr.

4.3.3 Uranium—Helium

The helium that results from the alpha decay of radioactive uranium and thorium shown in Eqs. 4.3–4 through 4.3–6 can be used to measure age using Eq. 4.3–2. The amount of helium in the earth's crust has been increasing since the amount produced by radioactive decay exceeds the amount lost to the atmosphere. This method suffers from possible inaccuracy due to the diffusion of helium through the material into the atmosphere. This release of helium can depend on the temperature history of the sample, especially if there have been high temperatures. It is also possible that radiation damage due to the radioactive uranium can damage the sample enough to permit the release of additional helium. This method has been used for ages of 5×10^5 to 10^9 yr.

4.3.4 Uranium—Thorium

Uranium in sea water decays to ^{230}thorium which precipitates on the ocean bottom. This ^{230}thorium has a half-life of 80,000 yr as it decays to radium. The age can be determined by measuring the thorium and radium and using Eq. 4.3–2. The accurate measurement of age using this method requires a relatively constant rate of thorium deposition. It also requires that the thorium and resulting radium remain together in the sediment and do not migrate. This method can be used for ages up to 4×10^5 yr.

4.3.5 Rubidium—Strontium

The beta decay of rubidium (^{87}Rb) yields strontium (^{87}Sr). This process, illustrated in Figure 4.3-2, has a half-life of approximately 4.5×10^{10} yr.

Rubidium is a relatively rare element generally found in nature with

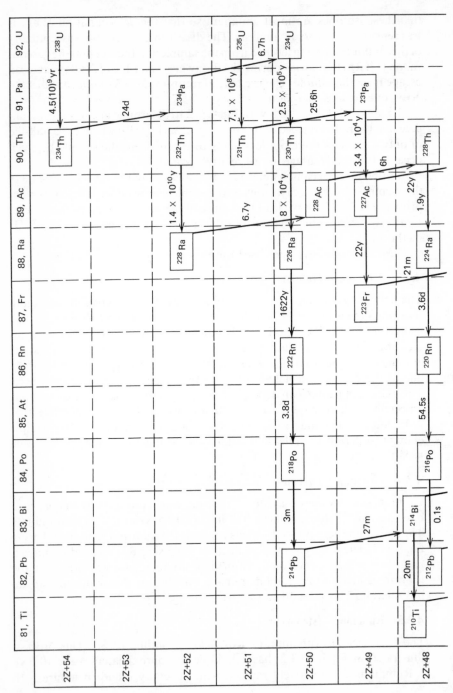

Figure 4.3-1 Uranium, thorium decay.

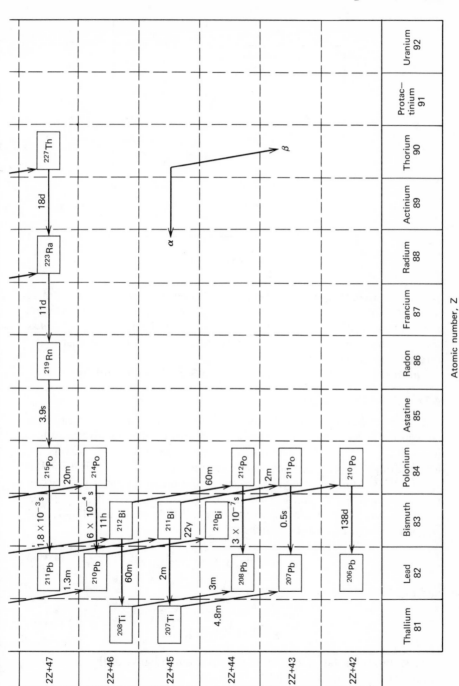

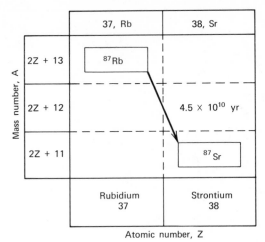

Figure 4.3-2   Rubidium-strontium decay.

potassium. Strontium is a rare element associated with calcium. Quantitative measurements of rubidium can be made by optical spectrography or flame photometry. Both rubidium and strontium can be measured by mass spectrometry. Quantitive measurements are difficult to make because of the rarity of the elements.

This method has been used for ages from 3×10^7 to 3×10^9 yr.

4.3.6   Potassium—Argon

The radioactive isotope of potassium (^{40}K) is unstable with respect to beta emission and electron capture. That is, a ^{40}K atom will either be transmuted to a calcium atom (^{40}Ca) by emission of a beta particle or will become an argon atom (^{40}Ar) by capture of an electron. This process is illustrated in Figure 4.3-3. Approximately 11 percent of the transmutations yield argon and the remainder calcium. The half-life of ^{40}K is approximately 1.3×10^9 yr.

If the decay constants for decay to ^{40}Ca and ^{40}Ar are denoted by λ_C and λ_A, respectively, Eq. 4.3–1 becomes

$$\lambda = \lambda_C + \lambda_A = \frac{0.693}{T}, \tag{4.3-8}$$

and Eq. 4.3-3 becomes

$$t = \frac{1}{\lambda} \ln \left(1 + \frac{\lambda\,^{40}\mathrm{Ar}}{\lambda_A\,^{40}\mathrm{K}} \right). \tag{4.3-9}$$

Potassium-argon dating has become important because of the ubiquitous nature of potassium minerals.

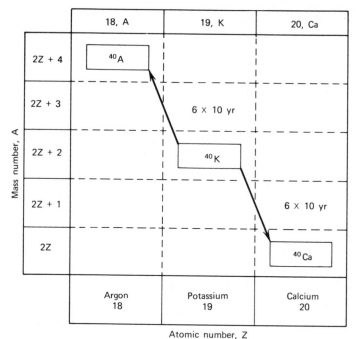

Figure 4.3-3 ^{40}Potassium decay.

The possible loss of argon gas by diffusion can be one source of error in this dating method. The possibility of contamination from atmospheric argon must also be considered.

Potassium can be measured using flame photometry and argon can be measured using mass spectrometry.

This method has been used for ages of 6×10^5 to 6×10^9 yr.

4.3.7 Radiocarbon

Radioactive carbon (^{14}C) decays to nitrogen (^{14}N) with a half-life of 5600 yr as illustrated in Figure 4.3-4.

Radiocarbon is produced in the upper atmosphere by the bombardment of cosmic rays which form neutrons that combine with the nitrogen according to the reaction

$$n + {}^{14}N = {}^{1}H + {}^{14}C. \tag{4.3-10}$$

The radioactive carbon combines with oxygen to form radioactive carbon dioxide. This mixes in the atmosphere and oceans and enters the earth life cycle to become part of living matter everywhere. This continual accretion of radiocarbon terminates on death so that the radiocarbon technique can be used for dating previously live material.

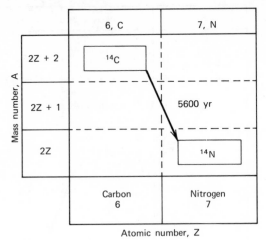

Figure 4.3-4 Radiocarbon decay.

In the analysis of radiocarbon, samples of organic matter are treated with diluted acid to remove carbonates and burned to create carbon dioxide. This carbon dioxide is purified and reduced to produce a coarse carbon whose radioactivity is measured with a radiation counter.

Errors in radiocarbon could result from variations in cosmic-ray flux over the years and variations in the thoroughness and speed with which radioactive carbon is mixed in the air and ocean. There is also the possibility that the combustion of fuel containing radioactive carbon can change the level of radioactive carbon in the atmosphere.

This method has been used for ages from 2000 to 20,000 yr.

4.3.8 Radiation Damage

Some of the energy dissipated by alpha, beta, and gamma rays passing through crystals is expended in the dislocation of atoms from their normal crystal lattice positions. These dislocations change the crystal properties. It is thus possible to measure the crystal properties, determine the rate of the radiation by some other means, and thereby determine the time at which the radiation began (generally the time of crystal formation). The accuracy of this method depends upon knowing the relationship between radiation dosage and radiation damage and knowing the rate of the radiation.

One crystal property affected by radiation is thermoluminescence, the emission of light at elevated temperatures. Errors can be caused by impurities and the thermal history of the sample.

References

1. Faul, Henry, Ed., *Nuclear Geology,* John Wiley and Sons, New York, 1954.
2. Hamilton, E. I., *Applied Geochronology,* Academic Press, New York, 1965.
3. Hamilton, E. I., and R. M. Facqukar, Eds., *Radiometric Dating for Geologists,* John Wiley and Sons, New York, 1968.
4. Libby, W. F., *Radiocarbon Dating,* University of Chicago Press, Chicago, 1952.
5. "Radioactive Dating and Methods of Low-Level Counting", International Atomic Energy Agency, Vienna, 1967.
6. Schaeffer, O. A., and Zahringer, J., Eds., *Potassium Argon Dating,* Springer-Verlag, New York, 1966.
7. Smales, A. A., and L. R. Wager, Eds., *Methods in Geochemistry,* Interscience, New York, 1960.

Harry W. Otto

4.4 ATMOSPHERIC CONSTITUENT SENSORS

Chemical analysis of the atmosphere is extremely complex and is frequently hampered by a lack of specificity and sensitivity to low concentrations of certain pollutants. Thus, the use of relatively sophisticated analytical procedures and equipment is frequently required.

4.4.1 The Atmosphere

The principal result of complete combustion of fossil fuels is the return of carbon dioxide and water vapor to the atmosphere. Most fuels contain impurities that are released as gases and particulates when burned. For example, sulfur present in fuels in converted to sulfur dioxide. In addition, combustion of fossil fuels is usually incomplete so that hydrocarbons, oxygenates including aldehydes, and carbon monoxide are emitted as only partly oxidized products. Other products of combustion are nitrogen oxides, which result from the reaction of atmospheric nitrogen and oxygen under the high temperature attained during combustion.

Particulate matter, generally characterized as organic soluble and insoluble material, in polluted atmospheres is higher than natural background by factors of 10 to 100. Table 4.4-1 gives typical examples of the actual elements present. The total particulate concentrations range from 100 $\mu g/m^3$ in areas of low pollution up to 4000 $\mu g/m^3$ during very severe pollution periods. Mean values for urban areas typically range from 200 to 800 $\mu g/m^3$. The natural particulate, including sea salt and dust, is usually of the order of 5 to 50 $\mu g/m^3$. The quantity of pollutants emitted in the U. S. air space yearly (Table 4.4–2) ranges to 142 billion kilograms [109]. Typical measured atmospheric concentrations are shown in Table 4.4-3 [48].

Pollutants undergo complex changes in the atmosphere as shown in Figure 4.4-1 in simplified form [1, 2, 49, 86, 87]. The major ingredients of smog are shown at the top of Figure 4.4-1. The curved lines in the chart represent changes in concentration with time that occur as the reactions proceed.

The reactions involving the oxides of nitrogen, oxygen, ozone, and sunlight are represented by the solid lines on the left of the diagram. These

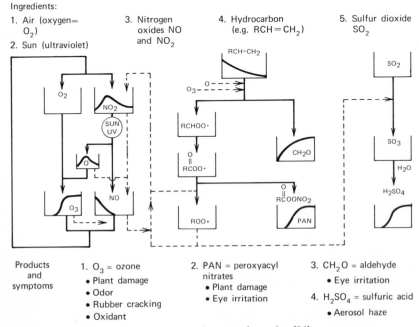

Figure 4.4-1   Schematic reaction sequence for smog formation [86].

reactions, energized by sunlight, quickly establish a photochemical equilibrium between NO, O_3, and NO_2. The addition of hydrocarbons introduces the reactions represented by the broken lines in Figure 4.4-1. Thus, the addition of hydrocarbon completely alters the cyclic character of the inorganic reaction sequence.

Hydrocarbons emitted into the atmosphere represent a multitude of individual hydrocarbons, each of which react at differing rates. Laboratory studies have shown olefins to be very reactive while paraffins by contrast are relatively unreactive. The reaction of olefins is initiated by the highly reactive oxygen atom (O) and ozone (O_3), resulting in the formation of two fragments (CH_2O and $RCHOO \cdot$).

Table 4.4-1. Important Elements in Particulate Pollutants

	Micrograms/Cubic Meter			
Element	Los Angeles[a]	Cincinnati[a]	Baltimore[a]	Portland[b]
Silicon	4	—	—	—
Calcium	2	16	—	—
Aluminum	4	4	4	—
Iron	14	12	15	4
Magnesium	2	7	—	—
Lead	3	3	1	0.4
Manganese	0.1	0.3	0.3	0.1
Copper	0.1	0.9	0.4	0.2
Zinc	—	2	—	—
Titanium	0.1	1	0.5	0.01

[a][57].
[b][17].

Table 4.4-2. The Major Sources of Air Pollution in the U.S.—1966 [109]

	Billions of Kilograms Per Year				
Pollutants	Total	Motor Vehicles	Power Generation and Other Industry	Space Heating	Refuse Incineration
Hydrocarbons	19	12	5	1	1
Nitrogen oxides	13	6	5	1	1
Carbon monoxide	72	66	3	2	1
Sulfur oxides	26	1	21	3	1
Particulates	12	1	9	1	1
Total	142	86	43	8	5

Table 4.4-3.   Concentration of Air Pollutants at CAMP Sites in Various Cities—
1964 [48]

	Maximum Daily Concentration-ppm				
Pollutants[a]	Chicago	Los Angeles	New York[b]	Philadelphia	Washington
Hydrocarbons	6	—	—	6	7
Nitrogen oxides	0.50	0.55	0.22[c]	0.28	0.38
Carbon monoxide	27	23	15	21	13
Sulfur dioxide	0.79	0.10	1.00[c]	0.43	0.22
Particulates (g/m³)	714	594	532	411	—
Oxidants	0.07	0.13	—	0.09	0.07

[a] Measured by the air monitoring stations of the Continuous Air Monitoring Program (CAMP) except for New York.
[b] Pollutant concentrations monitored by the New York City Department of Air Pollution Control at 170 East 21st Street.
[c] Concentration estimated from approximate interrelationships among data.

Peroxy radicals, the species ending in $-OO \cdot$ can give up oxygen atoms to nitric oxide (NO) to regenerate nitrogen dioxide (NO_2). They can also provide oxygen atoms for reaction with SO_2 to form sulfur trioxide (SO_3) which becomes sulfuric acid aerosol mist in the presence of moisture. The oxidation of nitric oxide by oxygen atoms from peroxy radicals upsets the delicately balanced oxides of nitrogen cycle. Thus, the presence of hydrocarbon is important to ozone formation in polluted air. Peroxyacyl radicals can react with nitrogen dioxide to form a peroxyacyl nitrate (PAN). This compound and its analogs are eye irritants and phytotoxicants. These chemical symptoms of smog can be reproduced in the laboratory. Figure 4.4-2 shows the typical profiles that are characteristic of photochemical smog produced by irradiation of dilute automotive emissions and nitrogen oxides.

4.4.2   Inorganic Gaseous Pollutants

Concentrations of these materials in the atmosphere are usually less than a part per million (by volume). The methods for detection must be sensitive and accurate at these low concentrations. Minimum interference, short sampling times, high degree of specificity and ease of calibration are important considerations in the selection of an analytical method.

Colorimetric Methods. A variety of colorimetric methods have been adapted to the analysis of atmospheric sulfur dioxide. The modified West-Gaeke [62, 81, 99] colorimetric method for sulfur dioxide is probably more specific than most other methods employed in the analysis of SO_2 in the atmosphere. SO_2 in an air sample is absorbed in a solution of sodium tetrachloromercurate to form nonvolatile dichlorosulfitomercurate ions.

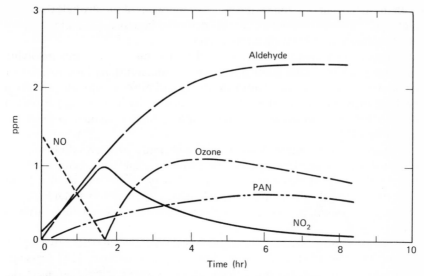

Figure 4.4-2 Laboratory irradiation of automotive exhaust emissions.

Addition of acid-bleached pararosaniline and formaldehyde complexes with dichlorosulfitomercurate ions to produce red-purple pararosaniline methysulfonic acid which is determined spectrophotometrically. However, nitrogen dioxide and ozone interfere if present in the atmosphere at concentrations greater than SO_2 [28]. Interference of nitrogen dioxide can be eliminated [53, 100, 108]. The method has a sensitivity of 0.005 ppm with an air sample of 7.5 liters.

The fuchsin-formaldehyde method [106] is based on a color reaction between sulfur dioxide and a bleached sulfuric acid-fuchsinformaldehyde chromogenic reagent. The method requires a large air sample (40 liters) in order to have a sensitivity of 0.01 ppm. Inorganic sulfides, thiosulfates, and thiols interfere with the analysis but can be removed [31] through precipitation.

The Stratman method [89] and the barium chloranilate method [13, 44] have been used for ambient air analysis. Both methods are relatively complex and require large air samples to maintain sensitivity. In addition, moisture interferes with the SO_2 absorption on silica gel in the Stratman procedure so that samples should be obtained over a relatively short time. The barium chloranilate method suffers also from serious interference from phosphates, fluorides, and chlorides and requires a preliminary separation in order to apply the analysis. In addition, the stability of the barium sulfate colloidal suspension, the barium ion concentration, and the pH of the solution are factors that limit the application of the barium chloranilate method.

In the atmosphere, ozone is the major component of the strongly oxidizing materials termed "oxidants." Ozone analysis, therefore, generally consists of determining the total oxidants present in the air [38].

Ozone is most frequently determined by iodometric methods involving the liberation of iodine from potassium iodide solutions [16, 73]. Other oxidants such as peroxyacyl nitrates (PAN) and nitrogen dioxide produce a positive interference but this can be avoided if the analysis is performed quickly before these slower reacting oxidants have a chance to react appreciably [74].

Reducing compounds such as sulfur dioxide and hydrogen sulfide provide the most serious interference with the iodometric methods, being approximately on a mole to mole equivalency with ozone. This interference, however, can be largely eliminated, even when SO_2 is present in a hundred-fold excess over oxidant, by incorporating an absorber upstream in the analytical train [75].

Other procedures for ozone analysis based on chemical oxidation employ phenolphthalein [34] and sodium diphenylaminesulfonate [14] reagents. Both of these procedures suffer from the same interferences that affect the KI methods, but in addition are sensitive to interference from halogens and acid gases.

The lack of a reliable primary standard is a problem common to all analytical methods for ozone. Ozone is highly reactive making it virtually impossible to maintain exact low concentrations for calibration.

The most sensitive procedure for the determination of nitrogen dioxide is based on the Griess-Ilosvay reaction by which a colored complex is formed among α-naphthylamine, nitrite ion, and sulfanilic acid. The procedure has been modified [54, 76] by the addition of a promoter to speed color development. No serious interferences exist from sulfur dioxide, nitrogen oxides and other gases that might be present in polluted atmosphere. A slight interference has been noted due to ozone [77].

Analysis for nitrogen dioxide can be accomplished with accuracy unless nitrogen is first. The efficiency of this conversion is 95–100 % but is subject to humidity variations [103].

A procedure for measuring nitrogen dioxide has been proposed [40] in which the nitrogen dioxide present in air is absorbed in an alkaline solution and determined colorimetrically as the azo dye. The absorbed nitrogen dioxide is used to diazotize sulfanilamide in phosphoric acid and it is then coupled with N-(1-naphthyl)-ethylenediamine dihydrochloride. The method [67, 78] tends to yield low values, apparently because of the conversion of nitrogen dioxide to nitrogen tetroxide in aqueous solution. For accurate results, corrections for this conversion must be applied.

Atmospheric carbon monoxide has been measured colorimetrically [83] based on the use of an indicating gel. The method employs silica gel which is impregnated with a sulfuric acid digest of palladium metal or palladium oxide. The gel will detect carbon monoxide at levels of 0.5 to 1.0 ppm (by volume). Another method [12] for carbon monoxide makes use of the blackening which results when an air sample containing carbon monoxide is passed over a paper strip coated with selenium sulfide. This technique has been adapted to an instrumental technique for continuous monitoring.

A method based on 3-methyl-2-benzothiazolone hydrozone reagent has been applied to the analysis of aldehydes in the atmosphere [3, 4, 37, 80]. This procedure suffers in that the derivatives of each individual aldehyde have characteristic molar absorptivity maxima. However, since formaldehyde often makes up more than 50% of the total aldehyde, the inherent uncertainty is not serious.

Procedures have been developed to measure atmospheric formaldehyde at concentrations as low as 0.01 ppm using the chromotropic acid method [4, 5]. Formaldehyde reacts with the acid to form a purple monocationic chromogen in concentrated sulfuric acid, a procedure that has very little interference from other aldehydes. Although olefinic and aromatic hydrocarbons interfere substantially, their influence can be substantially reduced by using an aqueous sodium bisulfite solution as the scrubbing medium in place of chromotropic-sulfuric acid [6].

Volumetric Methods. Volumetric methods of analysis can be quite specific in certain situations, but are usually employed to provide information for a type of pollutant rather than a specific one. Volumetric methods including alkalimetry, acidimetry, and iodometry have been employed in atmospheric analysis of pollutants.

The hydrogen peroxide method [41, 42] gives erroneous results when strong acidic gases other than SO_2 or alkaline gases are present in the air sample. The titrimetric method is preferred to the West-Gaeke method if SO_2 is the principal gaseous air pollutant and if long storage of samples prior to analysis is required.

In the iodometric method [32, 45], sulfuric acid mist or sulfates and unsaturated hydrocarbons do not interfere. However, appreciable errors are encountered from oxidizing and reducing species such as nitrogen dioxide, ozone, oxidants, and hydrogen sulfide.

Gravimetric Methods. The simplest and most rapid of the analytical methods employs the principle of weighing. A determination of atmospheric fall-out can be made by collecting the particulate in cylindrical jars over a 30-day period. The usual collecting agent is either water or a water/alcohol mixture. The coarse material is removed by a preliminary separation by passing the collection agent through 30-mesh screening followed by filtration to remove suspended matter and evaporation to a residue.

Dynamic collection methods employ inertial separation of particulates from atmospheric samples [110]. Using these methods particulates can be separated, measured gravimetrically and further examined as a function of size separation. Particles in the atmosphere generally range in size from 0.1 μm to 10–50 μm.

Another method [47, 101] employs impaction of particulates on the surface of vibrating cyrstals. The added mass of particulates alters the total mass of the crystal thereby changing the vibration frequency slightly. This change is related to mass and can be measured very accurately.

A system based on selective combustion of hydrocarbons and a water sorption detector has been used to measure hydrogen, methane, reactive hydrocarbons, and nonreactive hydrocarbons [47]. Atmospheric air samples are first dried and then passed through a combustor. The hydrocarbons present are selectively converted to carbon dioxide and water vapor. The water vapor is then detected and measured with a sorption detector. This gravimetric method has afforded a method of detection for hydrocarbons at concentrations of 0.01 ppm with a response of 5 min.

Nephelometry or Turbidimetry. Precipitates can be examined by nephelometry [46] and turbidimetry [66]. Atmospheric samples are collected in a suitable medium and precipitated. These procedures have been applied to sulfate and chloride determination. The sensitivities of the turbidometric method and the nephelometric method are 50 μg and 2 μg, respectively, in the determination of atmospheric sulfate. These methods are also dependent upon the amount, size, stability and suspension of the precipitate, parameters which can be difficult to control reproducibly.

Interferences from soluble sulfates, hydrogen sulfide, and sulfuric acid aerosols are encountered in using the turbidimetric barium sulfate method [96], [105]. In addition, the stability of the colloidal barium sulfate suspension, sulfate concentration, barium ion concentration, pH, and aging of the barium chloride solution are factors that must be carefully controlled to make this method of analysis reliable.

The integrating nephelometer [18] has been extensively studied as a tool in monitoring various parameters of the atmospheric aerosol. This instrument provides a direct measure of visibility degradation.

Other Methods. The conductometric method [93, 94, 95] depends upon the increase in conductivity of a sulfuric-acid-hydrogen-peroxide solution as sulfur dioxide is oxidized to sulfate. Thus, all soluble gases, particulates, and aerosols that yield electrolytes in solution or which increase or decrease the hydronium ion concentration are sources of interferences. Interferences are produced by nitrogen dioxide [61] and ozone [92]. The conductivity of the solution is also altered by changes in temperature.

Polarographic analysis [68] for SO_2 has been employed, but the complexity and stability of the instruments have limited their application mainly to the laboratory. The sensitivity of the method is about 0.02 ppm for a relatively small air sample. The main limitation is largely the ability to accurately measure the lines of the polarogram. Sulfur compounds and nitrites generally do not interfere with the analysis.

The potentiometric method [25, 29, 111] for sulfur dioxide involves the absorption of SO_2 from a flow-controlled air stream in an acidified bromide solution. Such materials as organic sulfides, H_2S, mercaptans, and disulfides interfere. Also, olefins and other unsaturates along with phenolic compounds interfere. The presence of these materials yields relatively high values. Low values result if chlorine, bromine, nitrogen dioxide, or ozone are present.

The lead peroxide candle [112] is sometimes used to measure the cumulative sulfation caused by gaseous sulfur dioxide in ambient air over a period of time. After exposure of the candle, composed of PbO_2 paste applied to a tube of gauze, the lead oxide containing sulfate is stripped off with aqueous sodium carbonate. The amount of sulfate formed is then weighed by a standard gravimetric procedure. Factors [79] that are known to influence measurements are wind velocity, wetness of the candle surface, temperature dependence, batch-to-batch candle variations, lead dioxide particle size, the presence of reduced sulfur compounds, and the use of different types of binders for the paste.

A long-path spectral procedure employing a molecular correlation spectrometer [10] has been applied to atmospheric analysis. This instrument compares the spectrum of the air sampled with an optical correlation mask, a procedure that continuously compares the "fingerprint" of the gas being measured with a reference. This instrument has been used for NO and NO_2.

Methods for ozone analysis include rubber cracking [72] and long-path ultraviolet [10] techniques. Long-path ultraviolet instruments for ozone have been too unstable for long-term atmospheric analysis. One concept uses total rubber crack [15] depth as a measure of ozone concentration. A more refined device [11, 26] utilizes the principle of creep under tension at constant stress. Rubber cracking is not completely specific for ozone since various free radicals interfere [23].

New reagent-less instruments are under development to sense and measure pollutants [64]. A modified flame photometric sulfur detector has been used to measure sulfur dioxide by gas chromatography. Fuel cells can be made for burning CO, CH_4, SO_2, and other pollutants. Likewise, semiconductors change resistance when exposed to certain pollutants. Microwave plasma detectors used in conjunction with spectrophotometers can analyze certain gases.

A method for measuring nitric oxide in photochemical smog research has been developed using the chemiluminescence from the rapid reaction between ozone and nitric oxide [63]. This method has been shown to detect levels of 0.01 ppm and monitor continuously with fast response. Good stability and selectivity of nitric oxide appear to characterize this approach.

4.4.3 Organic Gaseous Pollutants

Total Hydrocarbons. Organic substances are reactive in photochemical-type air pollution. Paraffinic hydrocarbons, benzene, and acetylene react slowly while olefins, aromatics, and aldehydes react at significant rates. Mass spectrometry [84, 98], dispersive and nondispersive infrared [52], and flame ionization [9] have been applied for total hydrocarbon analysis.

Atmospheric monitoring for total hydrocarbon usually entails the use of a flame ionization detector. With the addition of a selective combustor [47], an automatic system has been developed that records not only methane but methane plus all other hydrocarbons. A sensitivity of 0.004 ppm hydrocarbon with a response of 0.1 s has been reported for this approach.

Detailed Hydrocarbons. Since the levels of pollutants normally found in the atmosphere are below the limits of detection, detailed analysis of hydrocarbons in the atmosphere must use cold-trapping or absorptive techniques to concentrate the hydrocarbons present in a large sample of air. This concentrated sample is then generally analyzed using gas chromatography. By the use of suitable substrates on packed columns, the concentrated hydrocarbon sample can be separated into specific components and determined by flame ionization and electron-capture gas chromatographic techniques [7, 8, 36].

Aliphatic Nitrogen Compounds. Another important class of phytotoxic compounds are the peroxyacylnitrates (PAN). Infrared spectroscopy [71] and electron capture [24] have been used to identify and analyze members of this series. A gas chromatograph equipped with an electron-capture detector has been used for batch analysis of atmospheric PAN [91]. The sensitivity of the method is approximately 0.001 ppm.

Aliphatic Oxygenated Compounds. The methodology available for separating, identifying, and measuring oxygenates is neither specific nor sensitive.

The colorimetric methods for aldehydes are mentioned in Section 4.4.2, p. 219. Hydrazone derivatives of aldehydes have been analyzed by gas chromatograph to provide a method for measurement and identification of individual aldehydes [51].

Much less work has been done on ketones, acids, and alcohols than on aldehydes. Very few measurements of atmospheric ketones are available [22]. Formic acid has been determined chemically [55] but has not been detectable by long-path infrared [82]. Alcohols have been found in atmospheric cold-trapped samples analyzed by mass spectrometric methods [98], but could not be identified in the infrared spectra of samples [82].

4.4.4 Particulate Pollutants

The most commonly used definition of particulates is the material retainable on a filter which removes from the air at least 99% of the particles 0.3 μm and larger in diameter.

Physical Methods. Samples analyzed for particulate number and size require a different sampling procedure than those analyzed gravimetrically. In some cases, single samples are sufficient whereas in others numerous samples taken over an extended time period are required. Samples are usually collected by one of several procedures, including settlement, filtration, impingement, electrical precipitation, and thermal precipitation [30].

The physical measurements that can be made of atmospheric particulates include size distribution, weight concentration, number concentration, and optical characterization [27].

Chemical Methods. Chemical analysis entails that an initial separation of inorganic and organic material be made. This is usually done by extracting the residue with suitable organic solvents. Several fractionation schemes [39, 104] have been utilized to separate the organic fractions into classical groupings such as weak and strong acidic, basic and neutral fractions. These fractions are then subjected to a variety of analyses.

4.4.5 Instrumental Methods

Continuous Monitoring. Automatic and continuous samplers and analyzers have been built based on most of the physiochemical principles of atmospheric analysis discussed. These include colorimetric, conductometric, iodometric, chemiluminescence, gravimetric, volumetric, gas chromatography, and nephelometry [19].

Spectrophotometry. Applications of spectrophotometry to atmospheric pollutant analysis are discussed in Section 4.4.2, p. 221 [60, 88].

Ultraviolet absorption is not entirely satisfactory as a method for trace analysis of pollutants. Difficulties center on insufficient sensitivity and interference by coexisting materials in the atmosphere. However, a few specific gaseous pollutants including ozone, mercury (0.0001 ppm), tetraethyl lead (0.13 ppm), aniline, xylene, and monochlorobenzenes [35, 90] have been measured. The use of multiple reflection apparatus greatly increases the sensitivity of the method but sacrifices stability.

The direct field measurement of pollutants appears attractive, and a number of instruments have been developed to monitor certain industrial atmospheres [21]. Generally, this method lacks the sensitivity for trace analysis of atmospheric impurities.

Fluorescence is a useful technique for the analysis of a wide variety of materials, providing interfering substances are removed. The method has been applied for the analysis of beryllium, hydrogen fluoride, and uranium in atmospheric particulate matter [69, 70, 102].

Spectrographic Methods. Spectrographic methods include all techniques based on emission and absorption phenomena in the visible, ultraviolet, and infrared regions. All are useful for the analysis of material removed from the atmosphere.

Emission spectrography is widely used to establish the elemental composition of mixtures and for the exploratory examination of particulate matter [113].

The applicability of flame photometry is limited to approximately 30 elements over a wide range of sensitivities [20]. However, the higher sensitivity for detection of alkalies and the alkaline earths makes the flame method attractive for these elements.

X-Ray Diffraction. X-ray diffraction is a supplementary technique that is used in conjunction with elemental analysis. It is extremely useful for identifying crystalline materials in particulate matter and other materials. Of particular value is the fact that samples are available for subsequent analysis by other methods [50].

Polarographic Methods. Polarographic methods are applicable to the determination of any reducible or oxidizable substance and can, therefore, be used to analyze organic and inorganic compounds. Although simultaneous determination of a number of materials is possible, sample concentration techniques are required for maximum sensitivity and specificity of detection of atmospheric pollutants [97].

Microscopic Methods. The role of microscopy in advancing knowledge concerning the physical character of atmospheric particles has been reviewed [58, 59].

Interferometry and Thermal Conductivity. These methods are of little value for general air pollution work because of insufficient sensitivity and specificity. Interferometry has been employed for locating sources of pollution when only one pollutant is involved [85]. Thermal conductivity has been applied in special situations where stack effluents or near immediate sources of pollutants are of analytical interest [43].

Miscellaneous Methods. Very sensitive and specific analytical methods employing radioactivity [107], sonic absorption [33], and proton scattering [113] have been applied in characterizing a number of atmospheric samples.

References

1. Altshuller, A. P., and J. J. Bufalini, "Photochemical Aspects of Air Pollution: A Review," *Photochem. Photobio.* **4**, 97, 1965.
2. Altshuller, A. P., and J. J. Bufalini, "Photochemical Aspects of Air Pollution: A Review," *Environ, Sci. Tech.* **5**, 39–63, 1971.

3. Altshuller, A. P., and L. J. Leng, "Application of the 3-Methyl-2-benzothiazolone Hydrazone Method for Atmospheric Analysis of Aliphatic Aldehydes," *Anal. Chem.* **35**, 1541, 1963.

4. Altshuller, A. P., and S. P. McPherson, "Spectrometric Analysis of Aldehydes in Los Angeles Atmosphere," *J. Air Poll. Control Assoc.* **13**, 109, 1963.

5. Altshuller, A. P., et al., "Analysis of Aliphatic Aldehydes in Source Effluents and in the Atmosphere," *Anal. Chem. Acta.* **25**, 101, 1961.

6. Altshuller, A. P., D. L. Miller, and S. F. Sleva, "Determination of Formaldehyde in Gas Mixtures by the Chromotropic Acid Method," *Anal. Chem.* **33**, 621, 1961.

7. Altshuller, A. P., "Atmospheric Analysis by Gas Chromatography," in *Advances in Chromatography,* Vol. 5, Eds., J. C. Giddings and R. A. Keller, pp. 229–262, Marcel Dekker, New York, 1968.

8. Altshuller, A. P., "Organic Gaseous Pollutants," in *Air Pollution,* Ed. A. C. Stern, Academic Press, New York, 1968.

9. Andreatch, A. J., and R. Feinland, "Continuous Trace Hydrocarbon Analysis by Flame Ionization," *Anal. Chem.* **32**, 1021, 1960.

10. Barringer, A. R., and B. C. Newberry, "Molecular Correlation Spectromer for Sensing Gaseous Pollutants," paper 67–196, presented at the 60th Annual APCA Meeting, Cleveland, Ohio, June, 1967.

11. Beatty, J. R., and A. E. June, "A Simple Objective Method for Estimating Low Concentrations of Ozone in Air," *Rubber World* **131**, 232, 1954.

12. Beckman, A. O., J. D. McCullough, and R. A. Crane, "Microdetermination of Carbon Monoxide in Air," *Anal Chem.* **20**, 674–677, 1948.

13. Bertolocini, R. J., and J. E. Barney, "Colorimetric Determination of Sulfate Using Barium Chloranilate," *Anal. Chem.* **29**, 281, 1957.

14. Bovie, H. J., and R. J. Robinson, "Sodium Diphenylamine-sulfonate as an Analytical Reagent for Ozone," *Anal. Chem.* **33**, 1115–8, 1961.

15. Bradley, C. E. and A. J. Haagen-Smit, ."The Application of Rubber in the Quantitative Determination of Ozone," *Rubber Chem. Tech.* **24**, 750–755, 1951.

16. Byers, D. H., and B. E. Saltzman, "Determination of Ozone in Air by Neutral and Alkaline Iodide Procedures," *J. Am. Indust. Hyg. Assoc.* **19**, 251–7, 1958.

17. Chambers, L. A., J. F. Milton, and C. E. Cholak, "A Comparison of Particulate Loadings in the Atmosphere of Certain American Cities," presented at the Third National Air Pollution Symposium, Pasadena, California, 1955.

18. Charlson, R. J., N. C. Ahlquist, and H. Selvidge, "The Use of the Integrating Nephelometer for Monitoring Particulate Pollution," 10th Conf. on Methods in Air Pollution and Ind. Hygiene Studies, San Francisco, California, 1969.

19. Cholak, J., "Analytical Methods," in *Air Pollution Handbook,* Eds. P. L. Magill, F. R. Holden and C. Ackley, pp. 11–1 to 11–42, McGraw-Hill Book Co., New York, 1956.

20. Cholak, J., and D. M. Hubbard, "Spectrochemical Analysis with the Air-Acetylene Flame," *Ind. Eng. Chem.,* Anal. Ed. **16**, 728–35, 1944.

21. Cook, W. A., "A Review of Automatic Indicating and Recording Instruments for Determination of Industrial Atmospheric Contaminants," *Am. Ind. Hyg. Assoc. Quart.* **8**, 42–8, 1947.

22. Coulson, D. M., "Polarographic Determination of Semicarbozones," *Anal. Chem. Acta.* **19**, 284, 1958.
23. Crabtree, J., and B. S. Biggs, "Cracking of Stressed Rubber by Free Radicals," *J. Polymer Sci.* **11**, 280–281, 1953.
24. Darley, E. F., K. A. Kettner, and E. R. Stephens, "Analysis of Peroxyacyl-nitrates by Gas Chromatography with Electron Capture Detection," *Anal. Chem.* **35**, 589–591, 1963.
25. Dickinson, J. E., "The Operation and the Use of Titrilog and the Autometer," presented at the 49th Annual Meeting of the Air Pollution Control Assoc., Buffalo, New York, paper 56–39, 1956.
26. Doughty, R. V., and D. O. Erisman, "A Reliable Low Cost Instrument for Determining Atmospheric Oxidant Levels," *J. Air Poll. Control Assoc.* **11**, 428–430, 1961.
27. Germogenova, O. A., J. P. Friend, and A. M. Sacco, "Atmospheric Haze: A Review," Rept. PB 192–102, Bolt Beranek and Newman, NITS, U.S. Dept. of Commerce, Springfield, Virginia, March 31, 1970.
28. Gerraglio, F. P., and R. M. Manganelli, "Laboratory Evaluation of SO_2 Methods," *Anal Chem.* **34**, 675, 1962.
29. Giever, P. M., and W. A. Cook, "Automatic Recording Instruments as Applied to Air Analysis," *AMA Arch. Ind. Health* **21**, 233, 1960.
30. Giever, P. M., *Air Pollution,* Vol. II, Ed., A. C. Stern, pp. 249–253, Academic Press, New York, 1968.
31. Grant, W. N., "Colorimetric Determination of Sulfur Dioxide," *Ind. Eng. Chem. Anal. Ed.* **19**, 245, 1947.
32. Griffin, S. W., and W. W. Skinner, "Small Amounts of Sulfur Dioxide in the Atmosphere. Improved Methods for the Determination of Sulfur Dioxide when Present in Low Concentrations in Air," *Ind. Eng. Chem.* **24**, 862, 1932.
33. Gucker, F. T., Jr., "Determination of Concentration and Size of Particulate Matter by Light Scattering and Sonic Techniques", Proc. 1st Nat. Air Pollution Sym. Stanford Res. Inst., L. A. California 1949.
34. Haagen-Smit, A. J., and M. F. Brunelle, "The Application of Phenolphtalein Reagent to Atmospheric Oxidant Analysis," *Int. J. Air Poll.* **1**, 51–9, 1958.
35. Hanson, V. F. "Ultraviolet Photometer, Quantitative Measurement of Small Traces of Solvent Vapors in Air," *Ind. Eng. Chem.*, Anal. Ed. **13**, 119–123, 1941.
36. Harris, W. E., and H. W. Habgood, *Programmed Temperature and Gas Chromatography,* John Wiley and Sons, New York, 1966.
37. Hauser, T. R., and R. L. Cummins, "Increasing Sensitivity of 3-Methyl-2-Benzothiazolone Hydrazone Test for Analysis of Aliphatic Aldehydes in Air," *Anal. Chem.* **36**, 679, 1964.
38. Hodgeson, J. A., "Review of Analytical Methods for Atmospheric Oxidants Measurements," presented at the Conference on Methods in Air Pollution and Industrial Hygiene Studies, 11th, Berkeley, California, March 30–April 1, 1970.
39. Hueper, W. C., et al., *Arch. Pathol.* **74**, 89, 1962.
40. Jacobs, M. B. and S. Hockheiser, "Continuous Sampling and Ultramicro-determination of Nitrogen Dioxide in Air," *Anal. Chem.* **30**, 426–8, 1958.
41. Jacobs, M. B., *The Chemical Analysis of Air Pollutants,* Interscience, New York, 1960.

42. Jacobs, M. B., and L. Greenburg, "Sulphur Dioxide in New York Atmosphere," *Ind. Engr. Chem.* **48**, 1517, 1956.

43. Jacobs, M. B., *The Analytical Chemistry of Industrial Poisons, Hazards & Solvents*, 2nd ed., Interscience, New York, 1949.

44. Kanno, S., "The Colorimetric Determination of Sulfur Oxides in the Atmosphere," *Intern. J. Air Poll.* **1**, 231, 1959.

45. Katz, M., "Photoelectric Determination of Atmospheric Sulfur Dioxide Employing Dilute Starch-Iodine Solution," *Anal Chem.* **22**, 1940, 1950.

46. Keily, H. J., and L. B. Rodgers, "Nephelometric Determination of Sulfate Impurities in Certain Reagent Grade Salts," *Anal. Chem.* **27**, 759, 1955.

47. King, W. H., "The Monitoring of Hydrogen, Methane, and Hydrocarbons in the Atmosphere," *Environ., Sci. Tech.* **4**, 1136–41, 1970.

48. Larsen, R. I., "Determining Reduced-Emission Goals Needed to Achieve Air Quality Goals—A Hypothetical Case," *J. Air Poll. Control Assoc.* **17**, 823–8, 1967.

49. Leighton, P. A., *Physical Chemistry: Photochemistry of Air Pollution*, Vol. IX, Academic Press, New York, 1961.

50. Lennox, D., and J. Leroux, "Applications of X-ray Diffraction Analysis in the Environmental Field," *Arch. Ind. Hyg. Occup. Med.* **8**, 359–370, 1953.

51. Leonard, R. E., and J. E. Kiefer, "Qualitative Determination of Formaldehyde in Gross Impure Mixtures Containing Other Carbonyl Compounds," *J. Gas Chromat.* **4**, 142–143, 1966.

52. Littman, F. E., and J. O. Denton, "Infrared Spectrometric Method for Monitoring Gaseous Organic Substances in Atmosphere,"*Anal. Chem.* **28**, 945–949, 1956.

53. Lodge, J. P., et al., "Nitrogen Dioxide Interference in the Colorimetric Determination of Atmospheric Sulfur Dioxide," presented at the 148th National Meeting, American Chemical Society, Chicago, August 30–September 4, 1964.

54. Lyshkow, N. A., "A Rapid and Sensitive Colorimetric Reagent for Nitrogen Dioxide in Air," *J. Air Poll. Control Assoc.* **15**, 481, 1965.

55. Mader, P. O., G. R. Cann, and L. Palmer, "Effects of Polluted Atmospheres on Organic Acid Composition in Plant Tissue," *Plant Physiol.* **30**, 318, 1955.

56. Magill, P. L., F. R. Holden, and C. Ackley, Eds. *Air Pollution Handbook*, pp. 2–45, McGraw Hill Book Co., New York, 1956.

57. Mayrsohn, H., and C. Brooks, "The Analysis of PAN by Electron Capture Gas Chromatography," presented at the 1965 Am. Chem. Soc. Western Regional Meeting.

58. McCrone, W. C., "Morphological Analysis of Particulate Pollutants," *Air Pollution*, Vol. II, Ed. A. C. Stern, Academic Press, New York, 1968.

59. McCrone, W. C., R. G. Draftz, and J. G. Delly, *The Particle Atlas*, Ann Arbor Sci. Publishers, Inc., Ann Arbor, 1967.

60. Mellon, M. G., *Analytical Absorption Spectroscopy: Absorptometry and Colorimetry*, John Wiley and Sons, New York, 1950.

61. Moore, G. E., A. F. W. Cole, and M. Katz, "The Concurrent Determination of Sulfur Dioxide and Nitrogen Dioxide in the Atmosphere." *J. Air Poll. Control Assoc.* **7**, 25–28, 1957.

62. Nauman, R. V., P. W. West, and G. C. Gaeke, "Spectrophotometric Study of

the Shiff Reaction as Applied to the Quantitative Determination of Sulfur Dioxide," *Anal. Chem.* **32,** 1307, 1960.

63. Niki, H., A. Warnick, and R. R. Lord, "An Ozone-NO Chemiluminescence Method for NO Analysis in Piston and Turbine Engines," SAE 710072, Soc. Auto. Eng., Auto Eng. Congress, Detroit, Jan. 11–15, 1971.

64. O'Keefe, A. E., "Air Pollution Instrumentation. State of the Art," Annual Meeting, American Council of Independent Laboratories, Inc., Washington, D.C., October 20, 1968.

65. Otto, H. W., and R. H. Daines, "Genesis of Phytotoxic Air Pollutants and Their Identification and Measurement by Chemical and Phytological Methods," *IEEE Trans. Geosci. Elect.* **GE-8** (1), 59–69, 1970.

66. Parr, S. W., and W. D. Staley, "Determination of Sulfur by Means of the Turbidimeter," *Ind. Eng. Chem.* Anal. Ed. **3,** 66–7, 1931.

67. Patty, F. A., and G. M. Petty, "Nitrate Field Methods for the Determination of Oxides of Nitrogen," *J. Ind. Hyg. Toxicol.* **25,** 361, 1943.

68. Paulus, H. J., E. P. Floyd, and P. H. Byers, "Determination of Sulfur Dioxide in Atmospheric Samples. Comparison of a Colorimetric and a Polarographic Method," *Am. Ind. Hyg. Assoc. J.* **15,** 4, 1954.

69. Pringsheim, P., and M. Vogel, *Luminescence of Liquids and Solids and Its Practical Applications,* Interscience, New York, 1943.

70. Radley, J. A., and J. Grant, *Fluorescent Analysis in UV*, 3rd ed., D. Van Nostrand, New York, 1939.

71. Renzetti, N. A., and R. J. Byran, "Atmospheric Sampling for Aldehydes and Eye Irritants in Los Angeles Smog—1960," *J. Air Poll. Control Assoc.* **11,** 421–424, 1961.

72. Rugg, J. S., "Ozone Crack Depth Analysis for Rubber," *Anal. Chem.* **24,** 818–821, 1952.

73. Saltzman, B. E., and N. Gilbert, "Iodometric Microdetermination of Organic Oxidants and Ozone," *Anal. Chem.* **31,** 1914–20, 1959.

74. Saltzman, B. E., "Selected Methods for the Measurement of Air Pollutants," Method D-1, PHS 999-AP-11, 1965.

75. Saltzman, B. E., and A. F. Wartburg, Jr.. "Absorption Tube for Removal of Interfering Sulfur Dioxide in Analysis of Atmospheric Oxidant," *Anal. Chem.* **37,** 779, 1965.

76. Saltzman, B. E., "Colorimetric Microdetermination of Nitrogen Dioxide in the Atmosphere," *Anal. Chem.* **26,** 1949–55, 1954.

77. Saltzman, B. E., "Selected Methods for the Measurement of Air Pollutants," Public Health Service Publ. 999-AP-11, 1965.

78. Saltzman, B. E., "Colorimetric Microdetermination of Nitrogen Dioxide in the Atmosphere," *Anal. Chem.* **26,** 1949–55, 1954.

79. Saltzman, B. E., "Methods of Measuring and Monitoring Atmospheric Sulfur Dioxide," Natl. Air Pollution Control Admin., Cincinnati, Ohio, Publ. 999-AP-11, 1965.

80. Sawicki, E., et al, "The 3-Methyl-2-bensothiazolone Hydrozone Test," *Anal. Chem.* **33,** 93, 1961.

81. Scaringelli, F. P., B. E. Saltzman, and S. A. Frey, "Spectrophotometric Determination of Atmospheric Sulfur Dioxide," *Anal, Chem.* **39,** 1709–19, 1967.

82. Scott, W. E., E. R. Stephens, P. L. Hanst, and R. C. Doerr, "Further Develop-

ments in the Chemistry of the Atmosphere,"*Proc. Am. Petrol. Inst., Sect. III.* **37**, 171–83, 1957.

83. Shepherd, M., 'Rapid Determination of Small Amounts of Carbon Monoxide in Air," *Anal. Chem.* **19**, 77–81, 1947.

84. Sheppard, P. A., et al., "Isolation, Identification, Estimation of Gaseous Pollutants in Air," *Anal. Chem.* **23**, 1431–1440, 1951.

85. Stamm, R. F., and J. T. Whalen, "Calibration of the Rayleigh Refractometers Without Recourse to Gas Chambers," *J. Ind. Hyg. Toxicol* **29**, 203–207, 1947.

86. Stephens, E. R., "Photochemistry of Smog," *California Air Emissions,* Vol. I, Air Pollution Research Center, University of California, Riverside, April–June 1969.

87. Stern, A. C., Ed., *Air Pollution,* Vols. I, II, III, Academic Press, New York, 1968.

88. Stillman, J. W. "Bibliography of Photoelectric Spectrophotometric Methods of Analysis for Inorganic Ions," *Proc. Am. Soc. Testing Materials* **44**, 740–748, 1944.

89. Stratmann, H., "Microanalytical Methods for the Determination of Sulfur Dioxide," *Mikrochim. Acta* **6**, 668, 1954.

90. Strong, J., "A New Method of Measuring the Mean Height of the Ozone in the Atmosphere," *J. Franklin Inst.* **231**, 121–155, 1941.

91. Taylor, O. C., E. R. Stephens, and E. A. Cardiff, "Automatic Chromatographic Measurement of PAN," paper 68–70, presented at the 61st Annual APCA Meeting, 1968.

92. Terraglio, F. P., and R. M. Manganelli, "Laboratory Evaluation of Sulfur Dioxide Methods and Influence of Ozone-Oxides of Nitrogen Mixtures," *Anal. Chem.* **34**, 675–677, 1962.

93. Thomas, M. D., et al., "Automatic Apparatus for Determination of Small Concentrations of Sulfur Dioxide in Air," *Ind. Engrg. Chem.,* Ann. Ed. **4**, 253, 1932.

94. Thomas, M. D., et al., "Automatic Apparatus for Determination of Small Concentrations of Sulfur Dioxide in Air. Application to Hydrogen Sulfide, Mercaptans and other Sulfur and Chlorine Compounds," *Ind. Engrg. Chem.,* Ann. Ed. **15**, 287, 1943.

95. Thomas, M. D., J. O. Ivie, and T. C. Fitt, "Automatic Apparatus for Determination of Small Concentrations of Sulfur Dioxide in Air," *Ind. Engrg. Chem.,* Ann. Ed. **18**, 383, 1946.

96. Treon, J. F. and W. E. Crutchfield, "Rapid Turbidimetric Method for the Determination of Sulfates," *Ind. Engrg. Chem.,* Ann. Ed. **14**, 119, 1942.

97. Wawzonek, S., "Organic Polarography," *Anal. Chem.* **21**, 61–6, 1949.

98. Weaver, E. R., et al., "Interpretation of Mass Spectra of Condensates from Urban Atmospheres," *J. Res. NBS* **59**, 383–404, 1957.

99. West, P. W., and G. C. Gaeke, "Fixation of Sulfur Dioxide as Disulfitomercurate (II), Subsequent Colorimetric Estimation," *Anal. Chem.* **28**, 1816–19, 1956.

100. West, P. W., and F. Ordoveza, "Elimination of Nitrogen Dioxide Interference in the Determination of Sulfur Dioxide," *Anal. Chem.* **34**, 1324, 1962.

101. Whitby, K. T., and W. E. Clark, "Electro-Aerosol Particle Counting and

Size Distribution Measuring System for the 0.015 to one μm Size Range,"
Tellus **18,** 573–86, 1966.

102. White, C. E., "Fluorometric Analysis," *Anal. Chem.* **21,** 104–8, 1949.
103. Wilson, D., and S. L. Kopczynski, "Laboratory Experiences in Analysis of Nitric Oxide with Dichromate Paper," Paper 67–199, 60th Annual APCA Meeting, Cleveland, Ohio, 1967.
104. Wynder, E. L. and D. Hoffman, *J. Air Poll. Control Assoc.* **15,** 155, 1965.
105. Volmer, W., and F. Z. Frohlich, "Turbidimetric Determination of Sulfate," *Anal. Chem.* **126,** 414, 1944.
106. Vrone, P. F., and W. E. Boggs, "Acid-Bleached Fuchsin in Determination of Sulfur Dioxide in the Atmosphere," *Anal. Chem.* **23,** 1517, 1951.
107. Yanwich, P. E., "Radioactive Isotopes as Tracers," *Anal. Chem.* **21,** 318–21, 1949.
108. Zurlo, N., and A. M. Grinnini, "Measurement of Sulfur Dioxide Content of the Air in the Presence of Nitrogen and Heavy Metals," *Med. d. Lavoro.* **5,** 330, 1962.
109. "The Sources of Air Pollution and Their Control," Dept. of Health, Education and Welfare, USPHS #1548, prepared for 1966 National Conference on Air Pollutants, Washington, D. C.
110. Atmospheric Haze: A Review, CRC-APRAC Project CAPA-6-68, Natl. Tech. Info. Service, Rept. No. PB 192-102, U. S. Dept. of Commerce, Springfield, Virginia.
111. "ASTM Standards on Methods of Atmospheric Sampling and Analysis," Am. Soc. Testing Mater., 2nd ed., 1962.
112. Department of Scientific and Industrial Research, "The Investigation of Atmospheric Pollution," H. M. Stationery Office, London, England, 18th Rept., 1931–1932, 20th Rept., 1933–1934.
113. Stanford Res. Inst., 2nd Interim Rept., "The Smog Problem of L. A. County," Western Oil & Gas Assoc., Los Angeles, 1949.

Robert H. Daines

4.5 BIOLOGICAL SENSORS

Green plants sample and respond to constituents in the atmosphere. They respond adversely when exposed to toxic concentrations of gaseous or particulate materials. Some gaseous pollutants (fluorides, and sulfur from sulfur-containing gases) are absorbed [8, 27, 30] by leaves. Particulate pollutants are deposited and accumulated on leaf surfaces [35]. Analysis of such foliage can confirm the presence of these gaseous and particulate pollutants in the environment.

The susceptibility of plant species as measured by visible response varies with the phytotoxic pollutant, with each pollutant having its own spectrum of responding plant species [2]. In addition the age of the responding leaves and the pattern of leaf injury varies among the various pollutants. Therefore, it is frequently possible to identify pollutants present in toxic concentrations by a study of the vegetation in the area. The difficulties inherent in field surveys include frost, moisture stress, sunburn, fertility, insects, mites, plant diseases, and leaf age. These can produce symptoms that resemble those resulting from phytotoxic air pollutants. The ground pattern of plant response to a pollutant can aid in locating the source of an emission. Vegetation may be the first to announce the presence of a new pollutant.

The more important atmospheric pollutants that produce plant responses are sulfur dioxide (SO_2), fluorides, chlorine (Cl), hydrogen chloride (HCl), ethylene (CH_2CH_2) and the oxidants—ozone (O_3), peroxyacetyl nitrate (PAN), and nitrogen dioxide (NO_2). In addition the presence of phytotoxic particulates such as arsenicals, and probably nonphytotoxic airborne particles containing lead can be revealed by plant response or foliage analysis [35]. Of the pollutants listed, SO_2, fluorides, Cl, HCl, ethylene (usually from coal gas), and arsenicals have records of plant damage in limited areas, with the most severe response occurring near the point of origin of the pollutant.

Pollutants causing great concern are those that come from the combustion of, or additives to, fossil fuels. Such pollutants are SO_2, lead-containing particulates, unburned hydrocarbons and NO_x. Of these NO_2 absorbs energy from the ultraviolet rays of the sun (absorber) and transfers this energy to other molecules, thereby increasing their reactivity. In addition, NO_x also enters into chemical reactions occurring in the sunshine resulting in the occurrence of new compounds including ozone, PAN, and its analogs.

The meteorological parameters most effective in removing pollutants from their point of origin are winds and convection currents. If the wind direction is constant during emission from a single source, the ground pattern of plant injury is pie-shaped with the apex occurring at the point of the emission. Shifting winds result in an erratically shaped area of response, with the injury being most severe near the source of the pollutant. The area involved depends on meteorological parameters occurring at the time, the sensitivity of the vegetation, and the volume of the pollutant released. In areas where the volume of pollutants entering the air space (combustion) is relatively constant from day to day, visible plant response occurs sporadically. Such episodes occur during the periods of atmospheric stagnation, especially during periods of thermal inversions. When the depth of the mixing layer (height of the inversion) is not great, phytotoxic concentrations at ground level build up rapidly. The meteorological conditions that result in the accumulation of pollutants at ground level also result in warm humid periods. These latter parameters favor wide open stomata and increased plant sensitivity to phytotoxic pollutants [38].

4.5.1   Field Surveys

Many field surveys have been made using plants as indicators of air pollution. One of the most thorough surveys (SO_2) involved the areas around a large smelter, where studies involved extensive field inspections supported by controlled fumigations to check plant species susceptibility and symptoms expression [27, 36, 41]. These studies also considered the influence of meteorological parameters on the concentration or dispersion of the pollutants.

To determine the effect of SO_2 on plant growth, borings were made in the trunks of trees and measurements of the annular rings were made [28]. By comparing the diameter of the annual rings of a number of species of trees growing within and beyond the area affected by SO_2, and by considering the amount of wood produced annually before and after the smelter began operation, an indication of the area of response of each tree species and the impact of the pollutant on growth was secured.

Five criteria are used as a basis for selecting native indicator plant species [7]. The species selected should be sufficiently sensitive to the pollutant in question to respond at a level below the sensitivity of economically important plant species growing in the area; the species selected should be widely distributed through the area under study; the indicator plants should exhibit characteristic markings that are easily recognized; the species should be present and responsive throughout the growing season; and the species should grow from a terminal shoot through the year. The ground pattern of plant injury depends upon dispersion patterns which are conditioned by topography, ground cover, local weather conditions, and effective emitter heights.

In one survey, injury to economically important plants was noted when SO_2 concentrations equaled 0.95 ppm for 1 hr, 0.55 ppm for 2 hr, 0.35 ppm for 4 hr, or 0.25 ppm for 8 hr [15].

A survey conducted along the Houston ship channel showed that the sulfur content of the leaves of native elm, Arizona ash, and loblolly pine varied with proximity to the ship channel, with the highest amounts occurring near the channel [4, 33].

A marked increase in fluoride in the leaf tissue was observed after an aluminum factory began operations. Fluoride concentrations were correlated with distance from the factory and with wind direction [8].

In addition to field surveys using native species as indicators, gardens in strategic locations, with plant species that respond to low concentrations of selected air pollutants, are used as a means of monitoring the air. This method has been refined by growing plants under controlled nutrition in greenhouses using charcoal-filtered air, moving them out-of-doors to selected locations for a predetermined period of time, and observing the amount of injury that develops [37].

Since environmental parameters influence plant sensitivity, plant response may provide rather poor correlation with concentration of the pollutant.

Controlled studies have demonstrated a relationship between relative humidity, the width of the stomatal openings on tobacco and pinto bean foliage, and the concentration of ozone needed to evoke plant response. In these studies high humidity resulted in increased stomatal apertures and ozone fleck. At 95% relative humidity a 2-hr fumigation of 0.10 ppm ozone evoked plant response about equal to that resulting from a 2-hr exposure to 0.30 ppm at a relative humidity of 25% [38].

Since air pollutants enter the leaf almost exclusively through open stomata, parameters other than humidity that alter the size of the stomatal openings affect plant sensitivity. Such factors include soil moisture, soil texture, temperature, and light intensity [19, 27, 42, 43, 46, 51]. In addition, plants whose stomata close at night are much less responsive to fumigations that occur during periods of maximum stomatal closing than to exposures during the daytime.

Most of the controlled fumigation studies have involved the use of single pollutants. However, in nature, mixtures of pollutants are the general rule. Ozone-like symptoms developed on Bel W3 tobacco after 2-hr exposures to mixtures of ozone (0.037 ppm) and SO_2 (0.24 ppm) [34]. Plants exposed to these gases separately at the above concentrations did not develop visible injury symptoms. Mixtures of SO_2 and NO_2 have evoked considerably more phytotoxic response than would be expected from the two gases when used separately [20].

4.5.2 Phytotoxic Responses

Particulates. Particulates are nonspecific air pollutants whose common denominator is their physical characteristics. These solids are small (measured in mm) and can be sufficiently buoyant to be carried for a distance of a few miles from their point of emission in quantities sufficient to produce plant response. Such pollutants can result in direct toxicity, shading accompanied by reduced photosynthesis, objectionable residues on salable products, reduced fruit set, and changes in the pH of the soil. Under some conditions there can be sufficient alteration of pH to affect availability of essential nutrient elements.

Direct toxicity results when chemical particles that are sufficiently soluble to result in the occurence of toxic concentrations are deposited on leaves. Such toxicants as calcium arsenite and copper oxide produce necrotic, usually circular, spots on leaves, stems, and fruit. When dust from cement plant kilns is exposed to free moisture, it hydrates and forms a crust-like deposit on the upper leaf surface [9]. Rolling of leaf margins and killing of tissue located between the veins can result [13].

Cherry fruit set was reduced on the side of trees nearest a cement factory. Pollen germination was prevented by the dust deposited on the stigma [1].

Sulfur dioxide. Injury to leaves of plants from SO_2 toxicity has been classified as acute or chronic. Acute injury which results from the rapid absorption of a toxic dose of SO_2 is manifested by marginal or intercostal necrotic areas. The necrotic areas extend through the leaf and are visible on both surfaces. The intercostal spots occur between the veins while the areas immediately bordering the veins are seldom injured. These areas are characterized by few stomata and very limited intercellular spaces. Badly damaged leaves are often shed prematurely. The younger fully expanded leaves are most susceptible to SO_2, followed by the older leaves, while the enlarging leaves are the last to show acute injury.

The leaves of the monocotyledonous plants also vary in their response to toxic exposures of SO_2. Certain varieties of tulips, which are very sensitive to this pollutant, develop tip and marginal necrosis. The grain crops and grasses respond to toxic concentrations by the destruction of lengthwise areas between the veins. The current year's needles of conifers (pine, fir, spruce) that show acute SO_2 toxicity, exhibit a bright orange-red tip necrosis. The proportion of the needle to exhibit necrosis is a measure of the severity of the exposure. Although current-season needles are the most sensitive to SO_2 toxicity, older needles tend to become chlorotic and are shed prematurely. Such trees make limited growth and may die prematurely. Conifer needles absorb more SO_2 and are more sensitive to the gas during summer months than during the autumn and winter period [27].

Since the foliage of many plant species varies in susceptibility as the season progresses, plant species used as indicators of SO_2 should be changed with the season. Tulips, zinnia, and crabapple exhibit injury in the spring. Later in the year such crop plants as alfalfa, squash, or buckwheat are effective monitors of the gas. Trees that are very sensitive to SO_2 include trembling aspen, white birch, eastern white pine, and jack pine. Chickweed, mustard, and annual bluegrass are very susceptible to acute SO_2 toxicity. The bracken fern is also a good indicator. Other plants susceptible to SO_2 toxicity are bean, clover, cotton, soybean, bachelor's button, morning glory, sweetpea, larch, catalpa, American elm, mulberry, plantain, mallow, and sunflower.

Chronic injury results from a prolonged absorption of SO_2 at sublethal concentrations or from a short sublethal exposure to high-level concentrations of SO_2. SO_2 is absorbed on the moist surfaces of the mesophyll cells of the leaf where it is converted to the very phytotoxic sulphite. If the sulphite formed from SO_2 absorption is oxidized to sulphate at about the same rate as the gas is absorbed, acute injury will be avoided. However, if the sulphate (which is about 30 times less toxic than the sulphite ion) accumulates beyond the threshold value that the plant cells can tolerate, chronic injury occurs [49]. Chronic injury becomes manifest as a yellowing or chlorosis of the leaf, with these symptoms occasionally appearing first on the under surface of the leaf [6, 50]. On some plants the under-surface symptom may be a silvering which results from the destruction of mesophyll cells just above the lower epidermis. Perhaps the most common sulphate toxicity symptom is the development of yellowish-green areas between the veins, while the veinal area remains a normal green color. Such leaves are usually abscised prematurely.

Fluorides. Fluorides escape into the atmosphere mainly as hydrogen fluoride, silicone tetrafluoride, and particulates containing fluorine. Gaseous compounds such as hydrogen fluoride and silicone tetrafluoride are responsibile for fluoride injury to vegetation. Particulates, unless they are soluble, show little tendency to produce visible injury [39].

Fluorides are extremely toxic to a few plant species while other plants show considerable tolerance. Certain varieties of gladiolus, as well as Chinese apricots, prunes, blueberries, corn, ponderosa pine, and short-leafed and loblolly pines are very sensitive to fluorides. Ragweed, privet, and weeping willow can tolerate many times as much fluoride without visible response.

Experiments indicate that some plant species absorb and retain fluoride more efficiently than others. Corn and tomato fumigated together showed 39 ppm of fluoride in the dried corn and 78 ppm in the tomato foliage [30]. The old leaves on the tomato plants accumulated more fluoride than the more sensitive expanding foliage. In general the more sensitive plant species, and the most sensitive foliage (young) absorb less fluoride than do more resistant species or the less responsive older leaves on plants.

Acute fluoride injury symptoms occur as necrotic tissue, usually along the margins and tips of young leaves. Occasionally streaking or spotting occurs. Affected tissue first appears as water-soaked areas which become buff to reddish-brown on drying.

Of the monocotyledonous plants, some members of the lily family, such as gladiolus and tulip are among the most sensitive of plants. On these plants, injury is usually characterized by necrotic areas at or near the tip of the leaf blade. However, some gladioli develop considerable intercostal streaking. Fluorine is absorbed over the entire leaf surface, but it moves toward the tips and edges where toxic concentrations result in tip and marginal necrosis.

Among conifers, fluoride-induced needle blight is characterized by brown or reddish-brown necrosis which begins at the tip and progresses toward the base. Occasionally the injury appears as localized bands which are separated from the tip injury by bands of green tissue. Where injury is severe, needles tend to be shorter than usual and to drop prematurely. The needles are most sensitive while they are elongating.

Although leaves are considered to be the part of the plant that responds first to fluoride injury, damage to other plant organs has been reported. Cyclamen flowers are more susceptible to fluoride toxicity than the leaves of the same plant, with the injury symptoms occurring on the margins of the petals and sepals [42]. Soft red sutures and splitting of peach fruits along the sutures have been produced by fluoride fumigations and by the use of fluoride-containing sprays [3].

Fluoride-damaged pine needles exhibit collapsed areas only a few cells in advance of necrotic tissue. The upper and lower epidermal cells and the xylem tissue are most resistant and the paranchyma most sensitive to fluoride toxicity. Phloem cells are the first to be injured, with the phloem and xylem parenchyma becoming extended and distorted [45].

Chronic fluoride injury expresses itself as a loss of chlorophyll, resulting in the development of a chlorotic or mottled pattern on affected leaves. Citrus, oaks, maple, and poplar exhibit chlorotic areas along leaf margins which often extend inward between the veins, towards the midrib.

On corn, chlorosis from low dosages of HF occurs as small, irregularly shaped chlorotic spots appearing first at the tips and margins of the leaves. With increased exposure, the mottle may become visible on much of the leaf surface, and under more severe conditions, may develop a chlorotic zone along the leaf margins. When this marginal chlorosis occurs, much of the leaf surface will be a lighter green than normal and necrotic areas often develop along the margins and tips of such leaves.

Controlled fumigation at 3 ppm HF has resulted in the mottling of corn foliage, whereas gladioli develop necrotic areas from exposures to even lower concentrations [26].

Chlorine. Study has shown injury to radish and alfalfa foliage after a 2-hr exposure to 0.1 ppm of chlorine [5]. Drought increases resistance, and wet leaves do not alter susceptibility to chlorine.

Leaves of sugar maple and crabapple trees have been observed to be severely injured (dropping) whereas field corn only a few feet away remained undamaged. Injury on maple and horsechestnut is largely marginal whereas the tree of heaven, Virginia creeper, and blackberry exhibit intercostal markings. Certain varieties of tulip exhibit a bleach to the under-surface of older and middle-age leaves while surrounding plants remain undamaged. Foliage injury patterns from chlorine toxicity are variable. Marginal necrosis is a common symptom for many species; however, injury appears as scattered areas between or along the veins of some plants. Injury to onion leaves and pine needles occurs as white to tan or light brown necrotic tips.

Old to middle-aged leaves (current season) seem to be the most sensitive, with injury usually showing about equally on both surfaces. However, in some plants the upper surface, in others the lower surface, seems to be the most responsive to chlorine as a pollutant. On some plants exposure to a chlorine fumigation results in a bleaching of susceptible leaves.

Plants sensitive to chlorine include alfalfa, radish, mustard, johnny-jump-up, primrose, tulip, sugar maple, boxelder, horsechestnut, witch hazel, and blackberry [22].

Hydrogen Chloride. Viburnum and larch seedlings were killed in less than 2 days by exposures to 5 to 20 ppm hydrogen chloride [18]. HCl exposures resulted in the occurrence of bleached lesions and necrotic margins to foliage of deciduous trees.

Ethylene. Ethylene, acetylene, and propylene are olefins that, if used in sufficient concentrations, produce similar plant responses. Ethylene is 100- to 1000-fold more phytotoxic than the latter two compounds.

Perhaps the most costly plant damage resulting from ethylene is to greenhouse-grown orchids. The symptoms are dry sepal injury, failure of buds to open, sleepiness, and the occurrence of yellow leaves. The most sensitive structures in the cattleya orchid are the opening flower buds, which develop sepal injury when exposed to concentrations as low as 0.002 ppm ethylene [14]. Older flowers are more tolerant; they require about 0.04–0.05 ppm to produce sepal response, and if exposed to about 0.01 ppm an opening flower may close (sleepiness). It is believed that ethylene (the ripening gas) speeds life processes and causes symptoms of decline. Cut carnation blooms showed a reduced vase life when exposed to 0.05 ppm ethylene and the effect was overcome by increasing the carbon dioxide content to 2.2 percent of the ethylene air mixture [44]. The total life of the flowers was reduced below that of flowers held in ethylene-free air only when the carbon dioxide was less than 0.35 % of the ethylene-air mixture.

Ethylene acts as a growth hormone, and plant foliage exposed to it may show chlorosis, necrosis, or abscission. Epinasty, a common response to ethylene, has provided a convenient monitor for this pollutant. Where ethylene is suspected, tomato plants may be used as indicators since they respond to low concentrations by developing a downward bending (epinasty) of the leaf petioles.

4.5.3 Products of Photochemical Reactions in the Atmosphere

Ozone. A stipple to the upper surface of grape leaves has been attributed to ozone [40]. A light-colored flecking of the upper surface of tobacco leaves, called weather fleck, has been found to be due to ozone toxicity [23]. The plants known to be affected by this pollutant include forest trees, ornamental plants, vegetables, herbs, grains, forage and fruit crops, and weed plants [10]. Ozone probably causes more injury to vegetation in the United States than any other air pollutant. Plants and chemical monitors have recorded its occurrence in elevated concentrations over large areas situated near the more densely populated regions [17].

Plant responses to ozone toxicity can be grouped into four types of symptom expressions: pigmented lesions, surface bleaching, bifacial necrosis, and chlorosis [25]. The palisade cells (tightly appressed, elongated cells occurring just below the upper epidermis) are usually injured first [29]. These cells are absent in monocots. In some plants, especially in trees and shrubs, exposure to ozone results in a pigmentation of cell walls resulting in a colored lesion, usually visible on the upper leaf surface. Sorghum develops highly colored purplish-red areas from exposures to ozone. Such lesions usually do not cross even the small veins, but all the area bordered by the veins may be colored. Occasionally, pigmentation from these small areas diffuses beyond these spots, imparting coloration to otherwise normally appearing cells and even to small veins. The epidermal cells above these colored spots usually do not show visible evidence of injury.

Unpigmented necrosis, varying from small to relatively large areas, is a common symptom of ozone toxicity to the leaves of many herbaceous and woody plant species. Such areas involving the palisade and frequently the upper epidermal cells, often occur on the upper surface of dicotyledenous plants. Only infrequently do such symptoms develop on the under-leaf surface of such plants. A silvery appearance of the upper surface of spinach leaves often follows a leaf damaging exposure to ozone, but does not persist more than a day. On plants that lack palisade cells (monocots), these whitish necrotic areas develop about equally on both upper and lower leaf surfaces.

When injury is more severe, necrosis may extend through the leaf, producing a spot that is apparent on both surfaces. Such bifacial necrotic spots become paper thin on drying and in most species are white to off-white or tan in color. However, on some plant species, such necrotic spots are darker in color. Monofacial and bifacial necrosis may occur on a single leaf. Small areas of light green cells may occur primarily on the upper leaf surface following ozone fumigations. Such spots may coalesce forming a large, irregular, mottled appearance. These symptoms commonly occur on pine needles exposed to toxic levels of ozone. On some plants premature yellowing and dropping of the older leaves occurs.

From a single exposure of spinach plants that are in about the fifteenth leaf stage, depending on the severity of the exposure, three or four leaves will develop necrotic upper surface and bifacial lesions. The oldest of these responding leaves may exhibit necrotic spots over the entire leaf-blade with the spots being more concentrated toward the stem end. On the youngest responding leaf, the injury will be entirely or largely confined to the terminal portion. This same tendency to affect a tissue of a certain age is shown in the response of leaflets on a compound leaf, such as that of potato, where only lateral leaflets in a certain position along the petiole are affected.

Some plant species are more susceptible when they are young than when they are older. Spinach in about the 10–15 leaf stage is more likely to be injured by ozone than are older plantings.

Fully expanded leaves are the ones that usually exhibit symptoms of ozone toxicity. Plants with deficient or excess nitrogen nutrition are less susceptible than are plants receiving optimum nutrition [31]. At low and high concentrations of soluble sugar in the leaves, susceptibility to ozone is reduced and maximum response occurs at intermediate levels (2 to 4 mg/g fresh wt.).

Some plants that occasionally exhibit ozone toxicity symptoms are tobacco, lettuce, swiss chard, endive, chicory, escarole, spinach, cultivated dandelion, mustard, parsley, borage, onions, shallots, red beets, potatoes, tomatoes, beans, peas, eggplant, cucumber, watermelon, squash, sweet corn, barley, wheat, rye, oat, clover (red), grass (orchard), sweetpea, petunia, begonia, gladioli, ponderosa pine, white pine, boxelder, tulip tree, ash, and poplar.

In areas where ozone pollution is marking the foliage of many species of plants, concentrations of 0.05 to 0.10 ppm are not uncommon.

PAN and Its Analogs. PAN (peroxyacetyl nitrate) and its analogs (PPN-peroxypropionyl nitrate) (PBN-peroxybutryl nitrate) are highly phytotoxic [37]. PAN typically causes glazing and bronzing to the lower leaf surface of broad leafed plants [11].

PAN injures the spongy mesophyll cells just above the lower epidermis, but under more severe conditions it also destroys palisade cells. Such bifacial injured areas are difficult to distinguish from ozone-induced spots. These lesions may vary in color from light buff in spinach, chickory, or romaine lettuce, to grayish-brown in cultivated dandelion to a brownish color in the herb borage. Leaves that are just completing their expansion are the most sensitive to this pollutant.

Many grasses and small grains are sensitive to PAN and respond to phytotoxic exposures by the development of a narrow band of chlorotic or necrotic tissue across the leaf blade. When such injury occurs, the youngest responding leaf blade develops injury at the tip, while older responding leaves show the necrotic band lower down on the leaf blade [49]. The effect of age of leaf is also shown on broad leaves (petunia) and compound leaves (potato).

Some varieties of petunia (white) are especially sensitive to under-leaf surface injury, and some varieties of greenhouse-grown chrysanthemums may occasionally exhibit this type of injury. Other plants showing injury symptoms resembling those produced by PAN are Swiss chard, table beets, spinach, romaine and iceberg lettuce, cultivated dandelion, curly endive, potatoes, tomatoes, and annual bluegrass.

Nitrogen Dioxide. Continuous exposures to NO_2 concentrations ranging from 0.3 to 0.5 ppm caused a gradual change in appearance of pinto bean plants [47]. The symptoms produced were a downward cupping and a darker green leaf color than that of untreated plants. The plants exposed to the NO_2 fumigations exhibit growth suppression after 1 week to 10 days.

When concentrations sufficient to produce acute symptoms are used, irregular areas of tissue between the veins collapse and turn white to light tan. These symptoms closely resemble injury produced by sulfur dioxide. Sensitive areas of tissue closely correspond with those susceptible to ozone [48]. The apex of a recently expanded leaf is the area of response on that leaf. Older leaves exhibit injury progressively nearer the leaf base.

NO_2 concentrations of about 6 ppm during a 7-hr fumigation are required to produce markings on the foliage of the most sensitive plant species [48]. Plants are more susceptible when exposed to NO_2 in darkness, but even then the concentration required is much greater than those occurring in nature. In addition, studies have shown that a large percentage of NO_x occurring in nature is present as NO during the hours of darkness, but that NO_2 predominates during periods of sunshine.

References

1. Anderson, P. J., "The Effect of Dust from Cement Mills on the Setting of Fruit," *The Plant World*, **17**, 57–68, March 1914.
2. Benedict, H. M., and W. H. Breen, "The Use of Weeds as a Means of Evaluating Vegetation Damage Caused by Air Pollution," in *Proceedings of the Third National Air Pollution Conference*, pp. 177–190, 1955.
3. Benson, N. R., "Fluoride Injury or Soft Suture and Splitting of Peaches," *Pro. Am. Soc. Horticulture Sci.*, **74**, 184–198, Dec. 1959.
4. Bieberdorf, F. W., C. L. Shrewsbury, H. C. McKee, and L. H. Krough, "Vegetation as a Measure Indicator of Air Polution—Part 1. The Pine (Pinus-taeda)," *Bull. Torrey Botanical Club*, **85**, 197–200, May–June 1958.
5. Brennan, E., I. A. Leone, and R. H. Daines, "Chlorine as a Phytotoxic Air Pollutant," *Air Water Poll.*, **9**, 791–797, Dec. 1965.
6. Brisley, H. R., C. R. Davis, and J. A. Booth, "Sulphur Dioxide Fumigation of Cotton with Special Reference to Its Effect on Yield," *Agronomy J.*, **51**, 77–80, Feb. 1959.
7. Cole, G. A., "Air Pollution with Relation to Agronomic Crops: III. Vegetation

Survey Methods in Air Pollution Studies," *Agronomy J.*, **50**, 553–555, Sept. 1958.

8. Compton, O. C., L. F. Remmert, and W. M. Mellenthin, "Fluorine Levels in 1961 Crops of the Dalles Area," Oregon Agricultural Experimental Station Report 153, June 1963.

9. Czaja, A. Th., "The Effect of Dust with Alkaline Surface Reaction of the Particles, Especially of Lime-Kiln and Cement-Kiln Dusts on Plants," *First International Congress on Plant Pathology*, London, 1968.

10. Daines, R. H., I. A. Leone, and E. Brennan, "Air Pollution As It Affects Agriculture in New Jersey," New Jersey Agricultural Experimental Station Bulletin 794, March 1960.

11. Daines, R. H., I. A. Leone, and E. Brennan, "Air Pollution and Plant Response in the Northwestern United States," in *Agriculture and the Quality of Our Environment*, American Association for the Advancement of Science. Publication 85, pages 11–29, Plimpton Press. Norwood. Massachusetts. 1967.

12. Daines, R. H., "Sulfur Dioxide and Plant Response," *J. Occupational Med.*, **10**, 516–524, Sept. 1968.

13. Darley, E., "Symptomatology of Particulate Injury to Vegetation," in *Handbook of Effects, Assessment Vegetation Damage*, Center of Air Environment Studies, Pennsylvania State University, University Park, 1969.

14. Davidson, O. W., "Effects of Ethylene on Orchid Flowers," *Pro. Am. Soc. Horticulture Sci.*, **53**, 440–446. May 1949.

15. Dreisinger, B. R., "Sulfur Dioxide Levels and the Effects of the Gas on Vegetation near Sudbury, Ontario," *58th Annual Meeting of the Air Pollution Control Association*, Paper No. 65-121, Toronto, 1965.

16. Dugger, W. M., Jr., O. C. Taylor, E. Cardiff, and R. Thompson, "Relationship Between Carbohydrate Content and Susceptibility of Pinto Bean Plants to Ozone Damage," *Pro. Am. Soc. Horticulture Sci.*, **81**, 304–315, Dec. 1962.

17. Green, M. H., "New Jersey Air Monitoring Systems and Air Quality Data," New Jersey State Department of Health Technical Bulletin No. A-69-1, 1969.

18. Hasselhoff, E., and G. Lindau, *Die Beschädigung der Vegetation durch Rauch (Damage to Vegetation by Smoke)*, II. Kapitel, Chlor und Salzsäure (Chapter II, Chlorine and Hydrochloric Acid), pp. 203–256, Gebrüder Borntraeger, Leipzig, 1903.

19. Heck, W. W., J. A. Dunning, and I. J. Hindawi, "Interactions of Environmental Factors on the Sensitivity of Plants to Air Pollution," *J. Air Poll. Control Assoc.*, **15**, 511–515, Nov. 1965.

20. Heck, W. W., "Factors Influencing Expression of Oxidant Damage to Plants," *Ann. Rev. Phytopathology*, **6**, 165–188, 1968.

21. Heck, W. W., and A. S. Heagle, "Measurement of Photochemical Air Pollution with a Sensitive Monitoring Plant," *J. Air Poll. Control Assoc.*, **20**, 97–99, Feb. 1970.

22. Heck, W. W., R. H. Daines, and I. J. Hindawi, "Other Phytotoxic Pollutants. Section F. Recognition of Air Pollution Injury to Vegetation," Agricultural Committee of the Air Pollution Control Association Informative Report No. 1 TR-7, 1970.

23. Heggestad, H. E., and J. T. Middleton, "Ozone in High Concentrations as Cause of Tobacco Leaf Injury," *Science,* **129**, 208–210, 23 Jan. 1959.

24. Hendrix, J. W., and H. R. Hall, "The Relationship of Certain Leaf Characteristics and Flower Color to Atmospheric Fluoride-sensitivity in Gladiolus," *Pro. Am. Soc. Horticulture Sci.,* **72**, 503–510, Dec. 1958.

25. Hill, A. C., H. E. Heggestad, and S. N. Linzon, "Ozone. Section B. Recognition of Air Pollution Injury to Vegetation," Agricultural Committee of the Air Pollution Control Association Informative Report No. 1 TR-7, 1970.

26. Hitchcock, A. E., L. H. Weinstein, D. C. McCune, and J. S. Jacobson, "Effects of Fluorine Compounds on Vegetation, with Special Reference to Sweet Corn," *J. Air Poll. Control Assoc.* **14**, 503–508, Dec. 1964.

27. Katz, M., "Sulfur Dioxide in the Atmosphere and Its Relation to Plant Life," *Ind. Eng. Chem.,* **41**, 2450–2465, Nov. 1949.

28. Lathe, F. E., and A. W. McCallum, *Effects of Sulfur Dioxide on Vegetation,* Chapter 7, pp. 174–206, National Research Council of Canada, Ottawa, 1939.

29. Ledbetter, M. C., P. W. Zimmerman, and A. E. Hitchcock, "The Histopathological Effects of Ozone on Plant Foliage," *Contributions Boyce Thompson Inst.,* **20**, 275–282, Oct.–Dec. 1959.

30. Leone, I. A., E. Brennan, and R. H. Daines, "Atmospheric Fluoride: Its Uptake and Distribution in Tomato and Corn Plants," *Plant Physiology,* **31**, 329–333, Sept. 1956.

31. Leone, I. A., E. Brennan, and R. H. Daines, "Effects of Nitrogen Nutrition on the Response of Tobacco to Ozone in the Atmosphere," *J. Air Poll. Control Assoc.,* **16**, 191–196, April 1966.

32. Macdowall, F. D. H., E. I. Mukammal, and A. F. W. Cole, "Direct Correlation of Air-Polluting Ozone and Tobacco Weather Fleck," *Can. J. Plant Sci.,* **44**, 410–417, Sept. 1964.

33. McKee, H. C., "Use of Vegetation to Measure Air Pollution," Instituto Mexicano de Ingenieros Guimicos, Annual Technical Meeting, Mexico, 1961.

34. Menser, H. A., and H. E. Heggestad, "Ozone and Sulfur Dioxide Synergism: Injury to Tobacco Plants," *Science,* **153**, 424–425, 22 July 1966.

35. Motto, H. L., R. H. Daines, D. M. Chilko, and C. K. Motto, "Lead in Soils and Plants: Its Relationship to Traffic Volume and Proximity to Highways," *Environ. Sci. Tech.,* **4**, 231–237, March 1970.

36. National Research Council of Canada, "Effects of Sulfur Dioxide on Vegetation," Publication 815, 1939.

37. Noble, W. M., and L. A. Wright, "Air Pollution with Relation to Agronomic Crops: II. A Bio-Assay Approach to the Study of Air Pollution," *Agronomy J.,* **50**, 551–553, Sept. 1958.

38. Otto, H. W., and R. H. Daines, "Plant Damage by Air Pollutants: Influence of Humidity on Stomatal Apertures and Response to Ozone," *Science,* **163**, 1209–1210, 14 March 1969.

39. Pack, M. R., A. C. Hill, M. B. Thomas, and L. G. Transtrum, "Determination of Gaseous and Particulate Inorganic Fluorides in the Atmosphere," American Society of Testing Methods Special Technical Publication No. 281, pp. 27–44, 1960.

40. Richards, B. L., J. T. Middleton, and W. B. Hewitt, "Air Pollution With

Relation to Agronomic Crops: V. Oxidant Stipple of Grape," *Agronomy J.,* **50**, 559–561, Sept. 1958.

41. Scheffer, T. C., and G. G. Hedgcock, "Injury to Northwestern Forest Trees by Sulfur Dioxide from Smelters, U. S. Department of Agriculture Forest Service Technical Bulletin No. 1117, Washington, June 1955.

42. Seidman, G., I. J. Hindawi, and W. W. Heck, "Environmental Conditions Affecting the Use of Plants as Indicators of Air Pollution," *J. Air Poll. Control Assoc.,* **15**, 168–170, April 1965.

43. Setterstrom, C. and P. W. Zimmerman, "Factors Influencing Susceptibility of Plants to Sulphur Dioxide Injury. I," *Contributions Boyce Thompson Inst.,* **10**, 155–181, Jan.–March 1939.

44. Smith, W. H., and J. C. Parker, "Prevention of Ethylene Injury to Carnations by Low Concentrations of Carbon Dioxide," *Nature,* **211**, 100–101, July 2, 1966.

45. Solberg, R. A., D. F. Adams, and H. A. Ferchau, "Some Effects of Hydrogen Fluoride on the Internal Structure of Pinus Ponderosa Needles," *Proceedings of the 3rd National Air Pollution Symposium*, pp. 164–176, Pasadena, 1955.

46. Stoklasa, J., *Die Beschädigungen der Vegetation durch Rauchgase und Fabriks-exhalationen (Injuries to Vegetation Through Smoke Gases and Factory Exhalations)*, Urban und Schwarzenberg, Berlin, 1923.

47. Taylor, O. C., and F. M. Eaton, "Suppression of Plant Growth by Nitrogen Dioxide," *Plant Physiology,* **41**, 132–135, Jan. 1966.

48. Taylor, O. C., and D. C. MacLean, "Nitrogen Oxides and the Peroxyacyl Nitrates. Section E. Recognition of Air Pollution Injury to Vegetation," Agricultural Committee of the Air Pollution Control Association Informative Report No. 1 TR-7, 1970.

49. Thomas, M. D., and R. H. Hendricks, "Effects of Air Pollution on Plants," in *Air Pollution Handbook*, pp. 1–44, Eds. P. L. Magill, F. R. Holden, and C. Ackley, Section 9, McGraw-Hill Book Company, New York, 1956.

50. Thomas, M. D., "Effects of Air Pollution on Plants," *Air Pollution*, pp. 233–278, Columbia University Press, New York, 1961.

51. Wells, A. E., "Results of Recent Investigations of the Smelter Smoke Problem," *Ind. Eng. Chem.,* **9**, 640–646, July 1917.

Roscoe R. Braham, Jr.

4.6 CLOUD PARTICLE SENSORS

Cloud particle sensors are used in determining the shapes, concentrations, size distributions, mass per unit volume of air, and other properties of the water drops and ice particles which collectively constitute natural fogs and clouds.

Drops in natural clouds range in concentrations from about 50 to 2,000 cm^{-3}. In size they range from 1 to 2 to about 100 μm diameter. Most cloud-drop-size distributions are of the log normal form. The modal drop size for natural clouds appears to range from about 7 to 25 μm diameter. Drops larger than the modal size decrease in frequency with increasing drop size. Drops larger than about 100 μm are commonly called precipitation particles (drizzle and rain) because they are capable of surviving a fall from cloud base to the ground in most climates.

Ice particles in natural clouds may be single ice crystals (tens of μm to a few mm maximum dimension), aggregates of ice crystals (up to several mm), frozen drops, and semicompact irregular masses formed by freezing of cloud drops upon contact with crystals, crystal aggregates or frozen drops.

4.6.1 Cloud Water Content

Heated Wire. Water drops striking a heated wire cool it in proportion to the rate of impingement of water. A measurement of the temperature of the wire, or the power required to maintain a particular temperature, then becomes a measurement of the impinging liquid water. Knowledge of the cross section of the sensor wire, its collision efficiency and the true airspeed of cloud past the sensor, permits deduction of cloud liquid water content [36].

A device in most common use employs a single Ni-Fe sensing wire 0.5 mm diameter and 2.8 cm long, mounted perpendicular to the airstream and connected as one arm of a balanced ac bridge circuit. This wire is heated several hundred degrees by ac power. A second resistance wire, mounted parallel to the airstream and shielded from direct droplet impingement, is connected as an adjacent arm of the bridge to compensate for variations in cooling due to changes in airspeed and air density. The bridge is manually balanced in clear air; its imbalance in cloudy air, together with a scale adjustment for true airspeed gives an output calibrated directly in cloud water content [1]. Calibration of heated wire water sensors is difficult for lack of suitable standards [39, 51].

The most serious limitations of the heated wire device are its small sampling volume, its vulnerability to damage from snow pellets and hail, and a tendency for zero drift because of imperfect action of the compensation wire. The small sampling volume effectively restricts its use to cloud water sensing.

A variation of the heated wire instrument makes use of a cloud sensor consisting of 14 exposed heated wire loops each 4.1 cm long plus a precipitation sensor consisting of 236 cm of the same Ni-Fe resistance wire wound in a spiral-grooved forward facing cone about 5 cm in diameter [28].

Paper Tape. The conductivity of electrolyte impregnated paper is sensitive to small changes in water content. Using this principle a sensor for cloud water passes a 2.5-cm-wide coated paper behind a 3-mm-wide slit where it is exposed to the airstream and then across a pair of electrodes connected to an ac voltage source (ac to reduce polarization). The ac current is related to the rate of water interception at the slit [37, 59]. The relationship between water content and conductivity of the paper is somewhat nonlinear and temperature sensitive.

The paper tape has a slight advantage over a single-wire heated wire device in terms of sampling volume. However, in most situations this advantage is outweighed by its restriction to nonfreezing clouds, expendable paper stores, susceptibility of the paper to saturation and mechanical failure if exposed to rain while stopped, and the inherent difficulty of obtaining an accurate calibration.

Spinning Bowl. The spinning bowl [8] consists of a truncated conical cavity mounted with its 5-cm-diameter open truncated end facing forward into the airflow. The cavity spins at 3600 rpm about the flight axis. As water drops enter the cavity they are centrifuged to the rear outside edge where they collect to exit as a jet of water through a small radial hole. A small wire at the hole causes the exiting jet of water to deform into a thin web whose tangential dimension is proportional to the jet volume. Once every revolution of the cavity the jet passes through a stationary cone of light where a photodiode operates a counter to measure the transit time of the web and hence the rate of water interception of the cavity.

With a sampling area of 20.0 cm² the spinning bowl has a substantial sampling advantage over the conventional hot-wire and paper-tape instruments. The size of the cavity inlet results in low collision efficiencies for the smallest cloud droplets but this is usually not important since relatively little cloud water is contained in the smallest drops.

Cloud water content sensors are compared in Table 4.6-1.

4.6.2 Total Water Content

The total water density (vapor plus liquid and solid phases) of a parcel of cloudy air can be determined by measuring the dewpoint of the parcel

Table 4.6-1. Cloud Water Content Sensors

	Conventional Hot Wire	Paper Tape	Spinning Bowl
Output	#1. 27 mV dc full scale #2. 6. 0 Vdc full scale	100 μA dc full scale	Binary coded decimal
Dynamic range	0–6 g/m^3 at 60 m/s true air speed (TAS)	0–6 g/m^3 at 40 m/s TAS	0.2 to 4.5 g/m^3 at 85 m/s TAS
Transient response	1 s	0.1 s	0.3 s
Accuracy:			
Absolute	$\pm 10\%$ for drops $< 30\ \mu$m decreases rapidly for larger drops	Est $\pm 25\%$	Est $\pm 5\%$ full scale
Relative	$\pm 5\%$ for drops $< 30\ \mu$m	Est $\pm 10\%$	
Sampling area	0.14 cm^2	0.6 cm^2	20 cm^2
Overload effect	Nil	Saturates paper, paper may break	Nil
Response to other parameters	Ambient temp. and pres. changes may cause serious zero shift. Dynamic range depends upon TAS	Slightly sensitive to ambient temp.	Unknown
Environmental limitations	Usable in all natural water clouds, poor response to ice particles, wire may be broken by hail	Limited to nonfreezing clouds	Limited to nonfreezing clouds
Expendable stores limitations	None	20 min paper supply	None
Power requirements	0.8 A 115 V 400 Hz 15 A 28 Vdc	0.25 A 115 V 400 Hz 0.25 A 28 Vdc	

after heating it to evaporate the solid and liquid phases. The Lyman alpha total water probe operates on this principle [13, 41]. Cloudy air enters a forward facing sampling port directly into an evaporator unit designed for rapid evaporation while maintaining nearly isokinetic flow at the inlet. From the evaporator the air sample is divided, one portion is passed through a UV spectral-absorption cell while the other is passed through an optical dewpoint hygrometer. Using both sensors in the same system permits combining the advantages of high speed of response of the UV system (about 30 ms) with the much greater stability and more accurate calibration of the slower responding optical dewpoint hygrometer.

Combining the dewpoint temperature with the temperature and humidity of the original sample permits the calculation of the particle water content or condensate load. The absolute accuracies of available dew point instruments is about 0.5° above 0°C and about 1° below 0°C. Although this represents an error of only 3 to 8 percent in vapor density, it results in errors of 30 percent and more in particulate water content because the particulate load of natural clouds seldom exceeds 20 percent and is usually less than 10 percent of the saturated vapor content. A second source of error relates to sampled volume.

4.6.3 Cloud Particle Spectra

Slide Samplers. Most cloud particle spectra were obtained from small slides exposed to the airstream allowing particles to impact a suitable surface coating. Sampling slides are only a few mm wide to improve aerodynamic collision efficiencies and are exposed very briefly (a few ms) to obtain low cover fractions. A wide variety of materials has been used for the slide coatings. High-viscosity oils give excellent drop images, but require immediate photography to minimize evaporation losses. More permanent imprints can be obtained using MgO powder, carbon films, gelatin, or water soluble dyes.

The slide sampling technique suffers serious shortcomings, such as small sample volumes, restricted sample repitition rates, possibility of breakup of large particles and the need for sizing and counting particle images under a microscope. These disadvantages are frequently outweighed by the simplicity and reliability of slide sampling [9, 11, 17, 18, 31, 38, 53].

Continuous Formvar Replicator. Airborne Formvar replicators use a continuous 16 mm or 35 mm transparent tape to transport a ribbon of plasticized Formvar past a narrow slit exposed to the airstream. Cloud particles entering the slit impact into the soft plastic which subsequently hardens providing permanent replicas of small particles and impact craters of larger ones [29, 52].

Chloroform or ethylene dichloride solutions of Formvar (1–3%) are applied to the moving carrier tape close to the exposure slit. Slit widths (usually 1–5 mm) and tape speeds (up to 30 cm/s) are adjusted to prevent oversampling for a particular airspeed and anticipated particle concentration. A drying chamber is provided to insure complete hardening of the plastic before storage on a take-up reel.

The continuous replicator is a most useful device for obtaining data on cloud particle spectra and ice crystals along a flight path. It is subject to numerous problems such as condensation of water vapor upon the surfaces cooled by evaporating solvent, break-up of larger particles and aggregation to smaller ones after impact, flattening of drops in the shrinking plastic as it hardens, melting of crystals prior to hardening of the plastic, and spurious growth of frost in the soft plastic. Replicators seem to work best in clouds of small drops (10 to 30 μm) and low water content; it is frequently difficult to obtain good data in mixed-phase clouds [1, 60]. The Formvar replicator is compared with slide samplers in Table 4.6-2.

Continuous Gelatin Replicator. This continuously recording cloud-droplet replicator uses a thin ribbon of gelatin as the sampling surface. Excellent cloud-drop spectra have been obtained although the device "saturates" at approximately 0.32 gm/m^3 cloud water content [16].

In Situ Photography. In situ photography has the advantage of sampling

Table 4.6-2.   Cloud Drop Spectra Sensors

	Slide Samplers	Formvar Replicator
Output	Photographs or imprints of individual drops	Particle replicas in adhesive resin
Dynamic range	3 to 5 μm dia. drops Conc. $< 2000/cm^3$	3 to 50 μm dia. drops Conc. $< 3000/cm^3$
Accuracy:		
Drop size	Est $\pm 5\%$	Est $\pm 5\%$
Drop conc.	Poor from single slide due to small sample	Excellent
Resolution	Individual drops in sample	Individual drops in 0.01 s of flight path.
Sampling volume	1–5 cm^3 per slide	20 cm^3 per meter of path
Overload effect	Best data limited to 10% cover fraction	Best data limited to 10% cover fraction
Response to other parameters	Nil	Warm temp and high humidity promote spurious condensation. Can get frost growth below 0°C in mixed-phase clouds
Environment limitations	Usable in all tropospheric cloud conditions	Usable in all tropospheric cloud conditions
Expendable stores limitations	Supply of slides	Up to 4-hr record
Power requirements	Nil	7 A 28 Vdc

particles relatively undisturbed in their environment. Moreover it is the only known way to accurately distinguish between liquid and solid hydrometeors and, on the latter, to accurately determine shapes, sizes, and states of aggregation and accretion. The problems include particle size to platform velocity of $10^{-5} - 10^{-7}$ s.; low reflectance of drops and the angular dependent specular reflection from crystals; the requirement for a large in-focus volume; the difficulty of determining accurately which particles are in the in-focus volume, which will be a function of particle size; and the necessity for frame-by-frame data reduction.

Through the use of high-intensity pulsed illumination, combined in some instruments with mechanical motion of the lensing system to help compensate for drop motion, several investigators have successfully operated particle cameras from airplanes [10, 14, 27, 32].

Holographic Systems. Far-field (Fraunhofer) holography has been used to measure size spectra of naturally occurring fog drops from 5 to 100 μm diameter [47]. Larger drops could be handled by a demagnification technique. A Q-switched ruby laser was used to illuminate the sample volume and to stop the particle motion. Drop images were reconstructed by placing the hologram on a movable carriage and illuminating it with a He-Ne laser.

The reconstructions were imaged onto the face of a vidicon tube in a closed-circuit video system. Focused images were sized visually [23, 46, 58].

The most accurate method for holographic particle sizing, especially for resolving the details of solid particles such as snow flakes, is the two-beam Fraunhofer system [5]. The advantage of holographic systems is the ability to sense size and spatial distributions of particles in large volumes of undisturbed flow. The impossibility for a real-time read-out, and the tedious nature of image reconstruction are limitations.

4.6.4 Precipitation Particle Spectra

Optical Array. A linear array of illumination receptors is mounted normal to the local airflow and illuminated end-on with a collimated light of high intensity. In this arrangement the line of receptors serves as a size measuring scale of the shadows of particles passing through the sampling area and as a means for transmitting the event of a shadow passage to associated electronic circuitry for automatic sizing and counting [21, 22]. Optical arrays use either sub-millimeter solid state photodiodes or optical fibers individually coupled to photomultiplier tubes as sensor elements. When a passing particle casts its shadow on one or more of these sensors, it activates electronic circuitry to size and count the particle. Processing time for a single particle determination is less than 200 ns.

The optical-array instrument measures particle spectra directly in a way which lends itself to real-time display as well as to digital data recording. Moreover, by suitable choice of detector size and lensing, instruments can be built to cover any part of the cloud-drop to precipitation-particle size range. Computer processing of the recorded digital size information allows simultaneous calculation of water content and radar reflectivity.

Metal-Foil Impactor. Imprints resulting from impact of precipitation particles on strips of soft metal foils provide a means for sensing particle-size spectra from an airplane [4]. The minimum detectable particle size, and the ratio of imprint to particle size depends upon the type of particle, type of foil, impact speed, and geometry of the collecting point. A typical threshold value is 200 μm diameter for water drops impacting lead foil at 100 m/s.

In one continuous particle sampler [7] 2.5-cm-wide tapes of dead soft lead, 0.038 mm thick, attached to strips of 80-mesh wire cloth move past an exposure aperture. Another device uses somewhat heavier aluminium foil which can be handled as a continuous tape without backing material. This advantage is gained at the expense of a larger minimum detectable size [33]. Calibration of impression versus particle size is accomplished in wind tunnels or by the use of whirling arms [7, 44].

The major advantage of the foil instruments is their ability to sample precipitation particles in relatively low concentrations without interference from the many more numerous cloud droplets. Data reduction is tedious, usually requiring projection and sizing individual imprints. Ice pellets and snow pellets are readily detected and sized, raindrops are easily detected but are often difficult to size (especially large ones or those that have landed on the sides of the collecting surface) and snowflakes are frequently detected, but usually very little can be learned of their sizes or shapes.

Impact-Momentum Drop Spectrometers. This type of sensor is based upon transducing the impact momentum of drops striking a sensor plate [12, 50]. One sensor uses a piezoelectric force transducer to measure the force of raindrops impacting on the front end of a 5-cm-diameter cylinder exposed to the airstream [34]. The transducer output is proportional to the momentum of the impacting particle. A pulse analyzer in the electronic module sorts the pulses into nine adjustable class intervals. The system provides for immediate read-out as well as digital recording. The minimum resolvable particle is about 0.5 mm diameter; smaller particles cause a quasicontinuous (noise) signal rejected by the analyzer.

The momentum disdrometer (distribution of drops meter) combines a condenser microphone to measure the impact pressure while at the same time sucking the drop into a porous disk to measure its mass [20]. Geophones attached to an airplane windshield have been used to detect and size hailstones [45]. The impact momentum and optical array spectrometers are compared in Table 4.6-3.

Raindrop Spectrometers based Upon Light Scattering. Several precipitation particle spectrometers are based upon sensing the intensity of light scattered by particles passing through an illuminated sampling volume, or measuring

Table 4.6-3. Precipitation Particle Spectra Sensors

	Impact Momentum Spectrometer	Optical Array Spectrometer
Output	Binary coded decimal counts for each size class	Binary coded decimal counts for each size class
Dynamic range	0.5 to 7.0 mm dia. in nine adjustable class sizes	In principle from about 1 μm to 1 cm dia. However any given system limited by number of array elements times the element dia.
Accuracy	3% of drop size	$\pm \frac{1}{2}$ an array element dia.
Sampling frequency	500 Hz, one drop every 2 ms	5 MHz, one drop every 200 μs
Sample area	20 cm^2	Varies with drop size, typical value for particles $>$ 100 μm would be 10 cm^2
Response to other parameters	Must be carefully isolated from aircraft vibrations	Nil
Power requirements	15 W, 115 V 60 Hz	8 A 28 Vdc

the optical extinction caused by the particles. Pulses of scattered light (extinction pulses) are fed through a pulse-height analyzer and then summed into a number of class intervals. These devices have had limited success in part because of the different scattering properties of different sizes, shapes, and orientations. The technique is most frequently applied to sensing precipitation particles (D > 100 μm). The consequent requirement for a large sampling volume usually results in a large number of small particles simultaneously in the viewing volume. These are sensed as a continuous background noise [2, 6, 30, 48].

4.6.5 Cloud Condensation Nuclei Spectra

Cloud condensation nuclei (CCN) represent that small fraction (order 10^{-5}) of the total atmospheric population of condensation nuclei which are distinguished by their ability to become activated at the small supersaturation common to natural clouds (0.1 to 2 percent with respect to water). These particles can be counted in diffusion chambers operated at a desirably low supersaturation. Such a chamber is usually built with a top and bottom plate covered with wet filter paper and maintained at slightly different temperatures. The warm plate is put on top to maintain convective stability. In equilibrium, molecular conduction of heat and water molecules from the top to the bottom plate will establish very nearly linear gradients of temperature and vapor molecular concentration. However, since the saturated vapor concentration is not linear with temperature, the diffusion profiles of temperature and vapor concentration lead to supersaturation in the chamber, with the maximum occurring near the middle. If the air in the sample contains nuclei that can be activated at the supersaturation attained, they quickly nucleate and grow into small droplets that can be counted or photographed. Very important for understanding natural cloud structure is the spectra of CCN as a function of the degree of supersaturation. This requires operating the diffusion chamber at a number of saturation levels [24, 25, 40, 42, 43, 54, 55, 56].

4.6.6 Ice-Nuclei Spectra

Ice nuclei are the aerosol particles capable of initiating the ice phase inside clouds. The ice phase can be initiated by nucleation directly from vapor; by freezing of drops coming into contact with a suitable dry particle; and by freezing from particles which had been introduced into the drop at warmer temperatures. No single sensor can be expected to satisfactorily account for ice formation through all these ways. Among the techniques which have been found useful for certain aspects of ice-nuclei measurements are measuring the probability of drop freezing as a function of temperature [57], collection of all aerosol particles upon membrane filters which are subsequently exposed to cold moist air to activate any ice nuclei present [3, 35, 49], and passage of an air sample through a chamber containing supercooled air followed by counting of any ice particles present [15, 26].

References

1. Averitt, J. M., and R. E. Ruskin, "Cloud Particle Replication in Stormfury Tropical Cumulus," *J. App. Meteorology*, **6**, 88–94, Feb. 1967.
2. Bemis, A. C., "The Disdrometer, an Instrument for Measuring the Distribution of Raindrop Sizes Encountered in Flight," Massachusetts Institute of Technology, Department of Meteorology Technical Report No. 13, Cambridge, 1951.
3. Bigg, E. K., "Natural Atmospheric Ice Nuclei," *Sci. Progr.*, **49**, 458–475, July, 1961.
4. Bigg, E. K., F. J. M. McNaughton, and T. J. Methven, "The Measurement of Rain from an Aircraft in Flight," Royal Aircraft Establishment Technical Note 1, Mechanical Engineering 233, Farnborough, England, 1956.
5. Boardman, J., "Performance Testing of the Ground-Based and Snowflake Disdrometer Systems," Air Force Cambridge Research Laboratories Report 71-0164, Bedford, Massachusetts, 1971.
6. Borovikov, A. M., I. P. Mazin, and A. N. Nevzorov, "Large Particles in Clouds," in *Proceedings of the International Conference on Cloud Physics*, pp. 356–361, Toronto, Aug. 1968.
7. Brown, E. N., "A Continuous-Recording Precipitation Particle Sampler," *J. Meteorology*, **18**, 815–818, Dec, 1961.
8. Brown, E. N., "A Prototype Optical Flowmeter Applied to the Liquid Water Content Problem," National Center for Atmospheric Research TN/EDD-61, Boulder, Colorado, 1971.
9. Brown, E. N., and J. H. Willett, "A Three-Slide Cloud Droplet Sampler," *Bull. Am. Meteorological Soc.*, **36**, 123–127, March 1955.
10. Cannon, T. W., "High-Speed Photography of Airborne Atmosphere Particles," *J. Appl. Meteorology*, **9**, 104–108, Feb. 1970.
11. Clague, L. F., "An Improved Device for Obtaining Cloud Droplet Samples," *J. Appl. Meteorology*, **4**, 549–551, Aug. 1955.
12. Cooper, B. F., "A Balloon-Borne Instrument for Telemetering Raindrop-Size Distribution and Rainwater Content of Cloud," *Australian J. Appl. Sci.*, **2**, 43–55, March 1951.
13. E. G. and G., Inc., "Total Water Content Probe," Instruction Manual, Waltham, Massachusetts, 1969.
14. Elliott, H. W., "Cloud Droplet Cameras," National Research Council of Canada Report M1-701, Ottawa, Dec. 1947.
15. Fukuta, N., and G. K. Kramer, "A Fast Activation Continuous Ice Nuclei Counter," *J. Recherches Atmos.*, **3** (2nd year), 169–173, Apr.–June 1968.
16. Greiner, R. P., "An Airborne Hydrometeor Sampler," The Pennsylvania State University M. S. Thesis, University Park, 1969.
17. Houghton, H. G., and W. H. Radford, "On the Measurement of Drop Size and Liquid Water Content in Fogs and Clouds," in *Papers in Physics, Oceano-*

graphy and Meteorology, Vol 6, pp. 5–31, Massachusetts Institute of Technology, Cambridge, 1938.

18. Jiusto, J. E., "Cloud Particle Sampling," The Pennsylvania State University Department of Meteorology Report 6 (NSF G-24850), University Park, 1965.

19. Johnson-Williams Products, "Instruction Manual for J-W Liquid-Water-Content Indicator," Mountain View, California, 1968.

20. Katz, I., "A Momentum Disdrometer for Measuring Raindrop Size from Aircraft," *Bull. Am. Meteorological Soc.*, **33**, 365–368, Nov. 1952.

21. Knollenberg, R. G., "The Optical Array: An Alternative to Scattering or Extinction for Airborne Particle Size Determination," *J. Appl. Meteorology*, **9**, 86–103, Feb. 1970.

22. Knollenberg, R. G., and W. E. Neish, "A New Electro-Optical Technique for Particle Size Measurements," in *Proceedings of the Electro-Optical Systems Conference*, pp. 594–608, New York, Sept. 16–18, 1969, Industrial and Scientific Conference Management, Chicago.

23. Kunkel, B. A., "Fog Drop-Size Distributions Measured with a Laser Hologram Camera," *J. Appl. Meteorology*, **10**, No. 3, 482–486, June, 1971.

24. Laktinov, A. G., "Opredeleniye Kontsentratsii Oblachnikh Yader Kondensatsii (Determination of the Concentration of Cloud Condensation Nuclei)," *Dokl. Akad. Nauk SSSR (Proc. Acad. Sci. SSSR)*, **165**, No. 6, 1290–1293, 1965.

25. Langer, G., "Development and Evaluation of a New Cloud and Condensation Nucleus Counter," *Proceedings of the International Conference on Cloud Physics*, pp. 30–34, Toronto, Aug. 1968.

26. Langer, G., J. Rosinski, and C. P. Edwards, "Continuous Ice Nucleus Counter and its Application to Tracking in the Troposphere" *J. Appl. Meteorology*, **6**, 114–125, Feb. 1967.

27. Lavoie, R. L., et al., "Studies of the Microphysics of Clouds," The Pennsylvania State University Report 16 to National Science Foundation (GA-3956), pp. 53–96, University Park, Oct. 1970.

28. Levine, J., "Dynamics of Cumulus Convection in the Trades: A Combined Observational and Theoretical Study," Woods Hole Oceanographic Institution Reference 65-43, Woods Hole, Massachusetts, 1965.

29. MacCready, P. B., Jr., and C. J. Todd, "Continuous Particle Sampler," *J. Appl. Meteorology*, **3**, 450–460, Aug. 1964.

30. Mason, B. J., and R. Ramanadham, "A Photoelectric Raindrop Spectrometer," *Quart. J. Roy. Meteorological Soc.*, **79**, 490–495, Oct. 1953.

31. May, K. R., "Measurement of Airborne Droplets by the Magnesium Oxide Method," *J. Sci. Instr.*, **27**, 128–130, May 1950.

32. McCullough, S., and P. J. Perkins, "Flight Camera for Photographing Cloud Droplets in Natural Suspension in the Atmosphere," National Advisory Committee for Aeronautics Research Memo E 50K01a, 1951.

33. Meteorological Research, "Continuous Hydrometeor Sampler, Model 1220," technical data sheets, Altadena, California, 1968.

34. Metrodata Systems, *RP40 Raindrop Spectrometer, Technical Data*, Norman, Oklahoma, 1971.

35. Mossop, S. C., and N. S. C. Thorndike, "The Use of Membrane Filters for

Measurements of Ice Nucleus Concentration. I. Effect of Sampled Air Volume," *J. Appl. Meteorology*, **5**, 474–480, Aug. 1966.

36. Neel, C. B., Jr., "A Heated-Wire Liquid-Water Content Instrument and Results of Initial Flight Tests in Icing Conditions," National Advisory Committee for Aeronautics Research Memo A54123, 1955.

37. Newton, K. E., "Instrumentation of B-17 Airplane for Cloud Physics Research," Cloud Physics Laboratory Tech. Note 4, University of Chicago, Chicago, 1955.

38. Ohtake, T., "An Airborne Cloud-droplet Sampler," *The Science Reports of the Tohoku University*, Fifth Series, Geophysics, Vol. 15, pp. 59–65, March 1964.

39. Owens, G. V., "Wind Tunnel Calibrations of Three Instruments Designed for Measurements of the Liquid-Water Content of Clouds," Cloud Physics Laboratory Tech. Note No. 10, University of Chicago, Department of Meteorology, Chicago, 1957.

40. Radke, L. F., "An Automated Cloud Condensation Nucleus Counter," in *Proceedings of the International Conference on Cloud Physics*, pp. 35–37, Toronto, Aug. 1968.

41. Ruskin, R. E., "Hygrometer Developments at the U.S. Naval Research Laboratory," *Humidity and Moisture: Measurement and Control in Science and Industry*, Vol. 1, Ed. R. E. Ruskin, pp. 643–650, Reinhold Publishing Corp., New York, 1965.

42. Ruth, R. L., "A Cloud Nucleus Sampler for Aircraft Use," The Pennsylvania State University, Department of Meteorology, M. S. Thesis, University Park, 1968.

43. Saxena, V. K., J. N. Burford, and J. L. Kassner, Jr., "Operation of a Thermal Diffusion Chamber for Measurements on Cloud Condensation Nuclei," *J. Atmos. Sci.*, **27**, 73–80, Jan. 1970.

44. Schecter, R. M., and R. G. Russ, "The Relationship Between Imprint Size and Drop Diameter for an Airborne Drop Sampler," *J. Appl. Meteorology*, **9**, 123–126, Feb. 1970.

45. Schleusener, R. A., J. H. Hirsch, and L. B. Youngren, "Report on Operation of T-28 Armored Aircraft with Joint Hail Research Project at Greeley, Colorado, 21 July–1 August 1970," South Dakota School of Mines and Technology Report 70-13, Rapid City, 1970.

46. Silverman, B. A., B. J. Thompson, and J. H. Ward, "A Laser Fog Disdrometer," *J. Appl. Meteorology*, **3**, 792–801, Dec. 1964.

47. Silverman, B. A., G. B. Parrent, Jr., and B. J. Thompson, "Particle Size and Distribution Analysis Using Spatial Filtering Techniques," U. S. Patent 3,451,755, June 24, 1969.

48. Sing, C., B. Thompson, and A. Souza, "Protototype Development for Meteorological Instruments," Final Report on Contract AF 19(604)-6661, Technical Operations Research, Burlington, Massachusetts, 1964.

49. Stevenson, C. M., "An Improved Millipore Filter Technique for Measuring the Concentration of Freezing Nuclei in the Atmosphere," *Quart. J. Roy. Meteorological Soc.*, **94**, 35–43, Jan. 1968.

50. Smulowicz, B., "Analysis of the Impactometer, an Instrument for Measuring the Distribution of Raindrop Sizes Encountered in Flight," Massachusetts

Institute of Technology, Department of Meteorology Technical Report 19, Cambridge, May 16, 1952.

51. Spyers-Duran, P. A., "Comparative Measurements of Cloud Liquid Water Using Heated Wire and Cloud Replicating Devices," *J. Appl. Meteorology*, **7**, 674–678, Aug. 1968.

52. Spyers-Duran, P. A., and R. R. Braham, Jr., "An Airborne Continuous Cloud Particle Replicator," *J. Appl. Meteorology*, **6**, 1108–1113, Dec. 1967.

53. Squires, P., and C. A. Gillespie, "A Cloud-Droplet Sampler for Use on Aircraft," *Quart. J. Roy. Meteorological Soc.*, **78**, 387–393, July 1952.

54. Twomey, S., "The Nuclei of Natural Cloud Formation. Part I. The Chemical Diffusion Method and Its Application to Atmospheric Nuclei," *Geofisica Pure Applicata*, **43**, 227–242, 1959.

55. Twomey, S., "Measurements of Natural Cloud Nuclei," *J. Recherches Atmos.*, **1** (1st year), 101–105, July–Sept. 1963.

56. Twomey, S., and T. A. Wojciechowski, "Observations of the Geographical Variation of Cloud Nuclei," *J. Atmos. Sci.*, **26**, 684–688, July 1969.

57. Vali, G., "Estimates of Initial Cloud Glaciation from Nucleation Experiments," in *Proceedings of the Fifth Conference on Severe Local Storms*, St. Louis, pp. 154–160, American Meteorological Society, Boston, 1967.

58. Ward, J., "Laser Fog Disdrometer System," Final Report on Contract AF 19(628)-3813, Clearinghouse AD656487, Technical Operations, Inc., 1967.

59. Warner, J., and T. D. Newnham, "A New Method of Measurement of Cloud-Water Content," *Quart. J. Roy. Meteorological Soc.*, **78**, 46–52, Jan. 1952.

60. Yamashita, A., "Methods of Continuous Particle Sampling by Aircraft," *J. Meteorological Soc. Japan*, Series 2, **47**, 86–97, April 1969.

Forrest S. Mozer

4.7 ELECTRIC FIELD SENSORS

Electric fields play fundamental roles in the dynamics of astrophysical plasmas. The acceleration of charged particles creates the aurora and radiation belts in the magnetosphere, solar cosmic rays in the solar atmosphere, and galactic cosmic rays in the galaxy. The origin of these electric fields lies in the charge distributions and changing currents associated with the plasma response to these electric fields. Their theoretical understanding requires calculations of particle motions and fields, which is a plasma problem. Similar examples of enhanced particle acceleration and large electric fields arise in laboratory plasmas, which are generally difficult to study and understand because of limitations imposed by the small size of the plasmas, their short lifetimes and eddy currents in the walls. Measurements of electric fields in space contribute in an essential way to fundamental plasma physics as well as to an understanding of the origin of auroras, the radiation belts, solar cosmic rays, and galactic cosmic rays.

This understanding requires measurements of both dc or quasistatic electric fields and the ac electric fields in electromagnetic and electrostatic waves.

4.7.1 Potential Difference Measurements

The first quantitative measurements in laboratory plasmas were performed by Langmuir and Mott-Smith (1926), who measured and explained the voltage-current characteristics of small metal electrodes inserted into plasmas. These Langmuir probe techniques were soon extended to the determination of electric fields in laboratory plasmas through measurement of the potential differences between pairs of such electrodes [8, 20, 28]. The Langmuir double probe technique is the only developed method for making long-term in situ electric field measurements on balloons, rockets, or satellites.

4.7.2 Plasma Bulk Flow Measurements

A class of electric field measurements has been developed and performed that relies on observations of the bulk flow of the plasma. These measurements utilize the fact that a cold, collisionless plasma moves in crossed, time-independent, electric and magnetic fields with a velocity $v = E \times B/B^2$. Since the magnetic field, B, in the magnetosphere is known, the electric field, E, may be obtained from the measured plasma motion.

VLF Wave Propagation. The first successful measurement of magnetospheric plasma motion and of the equatorial, azimuthal, electric field in the magnetosphere was made by analysis of very-low-frequency (VLF) wave propagation [7]. VLF whistler mode waves are excited by lightning strokes and propagate through the magnetosphere along magnetic field lines on which there are slight enhancements of the plasma density, called ducts. The frequency dispersion in the waves gives information on the plasma density along the magnetic field line and on the equatorial magnetic field strength in the duct. As it moves with the rest of the plasma, the duct's equatorial magnetic field strength changes and its motion may be inferred from several hours of observation of its changing equatorial magnetic field through analysis of whistlers created by many lightning strokes.

Barium Cloud Releases. Artificial barium plasmas have been injected into the magnetosphere and electric fields have been deduced from ground observations of the motion of the clouds [12, 14, 15, 16, 30].

Radar Observation of Motions of Natural Ionospheric Irregularities. A general characteristic of the E- and F-regions of the ionosphere is the existence of enhancements of the plasma density, called irregularities. Radar observation of echoes incoherently or coherently scattered by these irregularities provides plasma drift velocities and electric fields from the Doppler effect [3, 4, 31, 32].

Other Plasma Flow Measurements. Particle detectors on satellites can, in principle, measure the bulk flow of the magnetospheric plasma and relate it to magnetospheric electric fields if pressure gradients in the plasma are small [6, 11].

A split Langmuir probe detector that measures differences of currents to two plates immersed in the ionospheric plasma has been flown on several rockets [5].

The motion of auroral arcs has been conjectured as due to an $\mathbf{E} \times \mathbf{B}$ drift of the magnetospheric plasma. Recent measurements of electric fields show that this idea is not valid on a small scale [15, 30] nor during the breakup phase of an active auroral display [18], but that the auroral motion is in general agreement with that expected from an $\mathbf{E} \times \mathbf{B}$ drift during quiet periods before or between substorms [18].

An effect of the $\mathbf{E} \times \mathbf{B}$ drift velocity is that the energy of a particle varies sinusoidally during its gyration around the magnetic field line because the $\mathbf{E} \times \mathbf{B}$ velocity is alternately parallel and antiparallel to the gyration velocity. An experiment has been flown on several sounding rockets to detect differences in energy of gyrating plasma ions incident on the detector from opposite directions.

Because all of the above-discussed techniques suffer disadvantages in tenuous plasmas such as those of the far side of the moon, the interplanetary medium, or the distant magnetosphere, a method involving the analysis of the trajectory of an artificial beam of particles has been studied [2, 21]. This concept involves ejection of a beam and observation of its trajectory, from which the electric field can be deduced if the magnetic field is known.

4.7.3 Theory of the Potential Difference Technique

A Langmuir double probe electric field detector consists of a pair of separated conductors whose measured potential difference divided by their separation distance is the component of electric field in the direction of the pair of conductors. Such detectors have measured electric fields of magnitudes between about one and 500 mV/m with probable errors generally less than $\pm$ 20 percent.

Consider the interaction of a single isolated conductor with the plasma in which it is immersed. Assume that the plasma contains a single species of ions, that these ions and the electrons are at the same temperature, T, that the conductor is at rest in the plasma, and that the conductor is large compared to the Debye length. Such a conductor acquires a net negative charge until it reaches a potential at which the current of electrons energetic enough to overcome the potential and strike the conductor is just equal, in magnitude, to the sum of the ion current, the current, I, to the measuring electronics, and the current, I_p, arising from nonthermal particles, photo-emission, and all otherwise neglected sources. This current-balance equation for an isotropic plasma in thermal equilibrium is

$$I + I_p + Ane \left(\frac{kT}{2\pi m_i} \right)^{1/2} = Ane \left(\frac{kT}{2\pi m_e} \right)^{1/2} \exp\left(\frac{eV}{kT} \right), \quad (4.7\text{--}1)$$

where V is the potential of the conducting surface with respect to the nearby plasma, m_i and m_e are the ion and electron masses, e is the electronic charge, k is Boltzmann's constant, n is the density of the plasma, and A is the surface area of the conducting body. From Eq. 4.7-1 the potential that a floating body (one that is isolated such that $I = 0$) reaches in such a plasma is

$$V = \frac{kT}{e} \ln \left[\left(\frac{m_e}{m_i} \right)^{1/2} + \frac{I_p}{Ane(kT/2\pi m_e)^{1/2}} \right]. \quad (4.7\text{--}2)$$

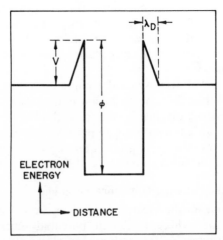

Figure 4.7-1 The electron potential energy as a function of position near an isolated conducting body immersed in a plasma.

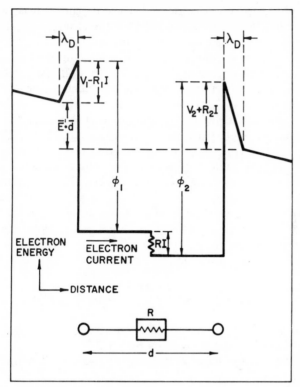

Figure 4.7-2 The electron potential energy as a function of position near a Langmuir double probe.

The value of V in the ionosphere, where I_p is negligible, is typically about -1 V.

The spatial dependence of an electron's potential energy near the conducting body discussed above is plotted in Figure 4.7-1. Far from the body, the potential is that of the plasma whose potential is assumed to be constant in the diagram of Figure 4.7-1. As the body is approached to within a distance of the order of a Debye length, λ_D, the electron potential energy increases until, at the surface of the conductor, the potential with respect to the nearby plasma is that given by Eq. 4.7–2. As the surface of the conductor is penetrated, the electron potential energy decreases by an amount equal to the work function of the conductor, and the potential becomes that of the sea of electrons inside the metal.

The electron energy-level diagram obtained when a pair of separated conductors are tied together through a resistance, R, representing the input circuit of a differential voltmeter is given in Figure 4.7-2. Far from either conductor the electron potential energy varies in space because of the assumed presence of an electric field. The difference of plasma potentials at the two conductors is $\mathbf{E'} \cdot \mathbf{d}$, where $\mathbf{E'}$ is the electric field strength and $\mathbf{d}$ is the separation distance of the two conductors. As either conductor is approached from the outside along a line containing the two conductors, the electron potential energy begins to differ from plasma potential within a distance the order of a Debye length from the conductor. At its surface, the maximum potential-energy level is reached, at which point the potential decreases as the surface of the conductor is crossed and the level falls to that of the sea of electrons inside the conducting body. The electron seas inside the two conductors are at different potentials because of the current, I, flowing through the resistance, R, connecting the two conducting bodies. Since this current flows to one conductor from the other, the surfaces of the two conductors are not at the same potential with respect to their nearby plasmas. The potential of the surface of either conductor with respect to its nearby plasma can be computed from Eq. 4.7–2. In a linear approximation, the effect of this current is to change the potential of the surface of either conductor by an amount $R_{\text{sheath}} I$ where

$$R_{\text{sheath}} = \frac{\partial V}{\partial I} = \frac{(2\pi m_e kT)^{1/2}}{Ane^2} \exp - \left(\frac{eV}{kT} \right). \qquad (4.7\text{–}3)$$

Thus, the potential of the surface of conductor 1 is $V_1 - R_1 I$, where V_1 is given by Eq. 4.7–2 and R_1 is given by Eq. 4.7–3.

A summation of the terms of the potential-energy diagram of Figure 4.7–2, gives

$$RI = \frac{\mathbf{E} \cdot \mathbf{d} + (V_1 - V_2) + (\phi_1 - \phi_2)}{1 + (R_1/R) + (R_2/R)}, \qquad (4.7\text{–}4)$$

where RI is the signal measured by the differential voltmeter and $(\phi_1 - \phi_2)$ is the difference in work functions of the two conductors, or their contact potential difference. While it is desired to measure $\mathbf{E'} \cdot \mathbf{d}$, the potential difference created by the external electric field, the differential voltmeter actually measures this quantity as well as the contact potential difference between the two conductors and the difference of floating potentials across the two sheaths. In addition, the entire measured quantity is attenuated by the resistance divider effect of R_1, R_2, and R. The relative importance of each of these terms depends on the conditions of the measurement (plasma density, temperature, field strength).

4.7.4 Rockets or Satellites in the Ionosphere

The Plasma Potential Difference Term. For a properly designed electric field experiment, the leading term on the right side of Eq. 4.7–4 should be the $\mathbf{E'} \cdot \mathbf{d}$ term arising from the electric field $\mathbf{E'}$ existing in the rocket coordinate system. The field, $\mathbf{E}'$, measured on a rocket or satellite is related to that in an earth-fixed frame of reference by

$$\mathbf{E}' = \mathbf{E} + \mathbf{v} \times \mathbf{B}, \tag{4.7-5}$$

where $\mathbf{E}$ is the electric field in the earth-fixed system, $\mathbf{v}$ is the vehicle velocity, and $\mathbf{B}$ is the intensity of the earth's magnetic field. To transform the measured data into the earth's frame of reference, knowledge of the vehicle attitude and trajectory is required.

The magnitude of the $\mathbf{v} \times \mathbf{B}$ term of Eq. 4.7–5 depends on the vehicle trajectory but is in the range of 5 to 50 mV/m for rockets and up to 500 mV/m for ionospheric satellites. Since typical auroral zone ionospheric electric fields have strengths of 10–100 mV/m, a careful analysis must be performed to insure accuracy.

The Work Function Term. The difference in average work functions between the two spheres $(\phi_1 - \phi_2)$ can be as large as several hundred millivolts if proper precautions are not taken.

A method of preparation of uniform carbon coated spheres has been developed to produce surfaces with work function variations less than a few millivolts.

The Sheath Resistance Terms. The magnitude of the decrease in measured signal caused by the voltage divider action of the sheath resistances, R_1 and R_2, with the electronic input resistance, R, has been measured on several rockets. These measurements gave sheath resistance values in the range of 10^4–10^6 Ω in the ionosphere. Since this value is much less than the electronic input resistance, R, the resistance terms in the denominator of Eq. 4.7–4 produce errors in the resultant field that are much less than 1 mV/m in the normal mode of detector operation.

The Sheath Potential Difference Terms. The sheath potential differences V_1 and V_2 depend on the currents collected by the two spheres so their magnitudes are dependent on photoemission, wake effects and energetic particle currents. Such currents are generally small in the ionosphere. Furthermore, any extra current source perturbs the electric field measurement only if the currents to the two spheres are different. Nonzero differences of sheath potentials may arise from spatial gradients in the plasma parameters or from geometric asymmetries between the two spheres. Magnitudes of the former effects are usually negligible [9] and geometric asymmetries are diminished if the conducting elements are spherical and the sphere geometry is symmetrized by addition of boom stubs on the outside ends of the spheres.

The flight of parallel pairs of conductors on a single rocket demonstrated that wake effects do not perturb the field measurement through the $(V_1 - V_2)$ term.

Two types of sheath potential differences associated with the geometric asymmetry from the presence of the rocket body have been identified in analyzed electric field data. The first occurs when the rocket is oriented relative to the magnetic field such that one of the spheres is on a field line that intersects the rocket body. Since the current collection by a sphere on this field line is perturbed by the presence of the rocket body, its sheath potential is altered and a pulse appears in the potential difference between the sphere of interest and any other sphere.

The second sheath potential difference is that arising from the photo-electric emission in daytime flights. If one sphere rotates into the shade of the rocket, the photocurrents from the two spheres will be unequal and a nonzero $(V_1 - V_2)$ term results. These asymmetries produce negligible errors in the field measurement, even for fields as small as a few millivolts/meter. The photoelectric emission from carbon coated spheres in unfiltered sunlight has been measured to be 1.5×10^{-9} A/cm^2.

Ionospheric electric fields can be measured with accuracies largely determined by uncertainties in the subtraction of the induced electric field due to the vehicle motion. For properly designed experiments, this accuracy is the order of or smaller than 10 percent of the electric field being measured.

4.7.5 Satellites in the Magnetosphere or Other Tenuous Plasmas

In much of the magnetosphere the plasma density is about one particle/cm^3, the temperature is about 1 keV, and the electric field magnitude is about 1 mV/m. The measurement of such electric fields in such plasmas by the Langmuir double probe technique is difficult.

Floating Potential Problems. The floating potential of a conductor immersed in a thermal plasma having a temperature of 1 keV is seen from Eq. 4.7–2 to be about -4 keV if the nonthermal currents, I_p, are zero. In measuring the few-millivolt potential difference between two such conductors due to the electric field, these large floating potentials cancel out in principle, since both conductors are at the same potential of $-4\,kT/e$ with respect to the nearby plasma. In practice, the cancellation of better than one part in 10^6 required to make meaningful measurements of the small electric fields is impossible to achieve because of error signals due to electronic misalignments (common mode errors) and sphere misalignments (different shapes, areas, orientations in the magnetic field, or orientations with respect to satellite components).

The photoemission current, I_0, of carbon coated spheres in sunlight is 1.5×10^{-9} A/cm^2, and has an energy spectrum that is approximately exponential with an *e*-folding potential, V_0, of about 1 V. A carbon conductor in sunlight in a rarefied plasma will have its positive floating potential determined by equating the high-energy tail of the photoemission to the electron saturation current of the plasma since all other currents in Eq. 4.7–1 are negligible. Thus, the current balance equation becomes

$$\frac{A}{4} I_0 \exp\left(-\frac{V}{V_0}\right) - Ane\left(\frac{kT}{2\pi m_e}\right)^{1/2} + I = 0. \tag{4.7-6}$$

The potential of the sphere is given by the solution to Eq. (4.7–6) as

$$V = V_0 \ln\left[\frac{4ne}{I_0}\left(\frac{kT}{2\pi m_e}\right)^{1/2} - \frac{4I}{AI_0}\right]. \tag{4.7-7}$$

The floating potential (V of Eq. 4.7–7 for $I = 0$) and the sheath resistance of a 15-cm-diameter sphere ($\partial V/\partial I$ obtained from Eq. 4.7–7) are listed in Table 4.7-1 for several sets of plasma parameters that might be encountered in the deep magnetosphere.

Table 4.7-1. Plasma Conditions, Floating Potentials, and Sheath Resistances

					Nighttime		Daytime	
			Electron					
			Saturation	Debye	Floating	Sheath	Floating	Sheath
	n	kT/e	Current	Length	Potential	Resistance	Potential	Resistance
Condition	(cm^{-3})	(V)	(A/cm^2)	(m)	(V)	(Ω)	(V)	(Ω)
Plasmaphere, just inside the plasmapause	3×10^2	1	3×10^{-10}	.3	-4	10^8	$+0.5$	5×10^6
Plasmasheet Condition 1	10	100	10^{-10}	20	-400	3×10^{10}	$+1.5$	10^7
Plasmasheet Condition 2	1	1000	3×10^{-11}	200	-4000	10^{12}	$+3$	5×10^7
Slot between plasmapause and plasmasheet, Condition 1	1	10	3×10^{-13}	60	-40	10^{12}	$+8$	5×10^9
Slot between plasmapause and plasmasheet, Condition 2	25	1	2×10^{-11}	1	-4	10^9	$+3$	7×10^7

Table 4.7-1 shows that the floating potentials of conducting spheres in sunlight are a few volts positive and vary by only a few volts over the range of plasmas that can be encountered. The magnitudes of the floating potentials measured on rocket flights in the ionosphere are similar and can be subtracted with sufficient accuracy to allow measurement of the small fields that may be present.

Since insulators have photoemission currents in sunlight that can be a few orders of magnitude smaller than those of conductors, the floating potentials of some insulated surfaces on magnetospheric satellites are similar to the nighttime calculations of Table 4.7-1 and can therefore be the order of kilovolts. Metallic surfaces on the shady side of the satellite are at similar potentials. Thus, there are electric fields the order of hundreds of volts/meter in the vicinity of the satellite. Since the Debye length is comparable to satellite dimensions (see Table 4.7-1), these fields are measured by the electric field detector. To avoid this, all surfaces of the satellite should be good photoemitters so that their floating potentials are a few volts positive and they should be tied together and to satellite ground or to potentials that are fixed with respect to satellite ground so that their floating potentials do not become large when they are in the shade of the satellite.

Large-Debye-Length Problems. From Table 4.7-1 it is seen that the Debye length of the plasma in the magnetosphere can be comparable to or greater than the dimensions of the electric field antennas or the satellite. This negates the convenient approximation made above that sheaths around the antennas or the satellite do not overlap and thus, that the presence of an object cannot affect the potential of another object.

The Debye-length calculations of Table 4.7-1 neglect the presence of photoemission currents which are, in fact, the major currents that flow. Their presence decreases the effective size of the sheaths around the spheres to the point where sheath-overlap problems are negligible.

The effect of overlapping sheaths, should they exist, can be understood qualitatively by considering the potential difference measured between two spheres when another object (the satellite) is between them, and the sheaths from all three objects overlap. The potential of each sphere is the sum of potentials from the electric field to be measured, its floating potential, the potential from the overlap of the sheaths of the sphere and the satellite, and the potential from the overlap of the sheaths of both spheres.

When the potential difference between the two spheres is obtained, all effects cancel except the desired terms arising from the electric field of interest. The above argument gives the desired conclusion that large sheaths or Debye lengths do not affect the dc field measurement, after making the implicit assumptions that the potentials induced on each sphere by the overlap potentials are identical and cancel out exactly, and the fields superimpose linearly to yield the resultant electric field.

Electronic Problems. From Table 4.7-1 it is seen that the sheath resistance of the photoelectrically biased spheres can be as great as 10^{10} Ω. The input impedance of the measurement electronics must be large compared to this value. A large error is made if amplifier drifts are not removed.

4.7.6 Balloons in the Atmosphere

The vertical electric field measured on balloons in the atmosphere arises from vertical currents that discharge the earth which is continually charged by worldwide thunderstorm activity. The horizontal electric field during periods of fair weather near the balloon is essentially equal to the one second, several hundred kilometer average of the ionospheric electric field [22, 27].

The balloon electric field detector also consists of pairs of separated conductors whose potential differences are measured.

The atmospheric plasma is isotropic since its collision frequency is much greater than its gyration frequency. Thus, the two conductors comprising the electric field sensor do not have to be rigorously symmetric. Flat plates or any other shape conductors may therefore be used in this experiment.

The negative charge carriers in the atmosphere are negative ions with masses comparable to those of the positive ions. Thus the sheath potential drop of Eq. 4.7-2 becomes zero (since $I_p = 0$ in the atmosphere) and there are no sheath problems in the measurement.

A complication in the measurement arises from the resistivity of the atmospheric plasma, which has been neglected in the consideration of the essentially fully ionized plasmas of the ionosphere and magnetosphere. The signal, V, arising from the electric field being measured on the balloon causes a current V/R to flow through the electronic input impedance, R. This current returns through the atmosphere along a path having resistance R_{atm}. Unless $R \gg R_{atm}$, the voltage divider effect of the two resistances causes most of the potential difference, V, to appear across R_{atm} and an attenuated signal is measured.

The atmospheric resistance can be computed by considering the potential drop necessary to cause a given current to be collected by a sphere of radius, r_0, when the current flow is impeded by ion-neutral collisions. The value thus obtained is

$$R_{atm} = \frac{m_n v}{4\pi n e^2 r_0},$$ (4.7-8)

where m_n is the mass of the neutral atoms, v is the ion-neutral collision frequency, n is the ion density, and e is the electronic charge. For a 30-cm-diameter sphere located near the surface of the earth, $R_{atm} \approx 10^{13}$ Ω, while at an altitude of 30 km, $R_{atm} \approx 10^{11}$ Ω. Thus the input impedance of the field measuring electronics must be greater than about 10^{14} Ω.

Problems associated with contact potential differences are unimportant in the balloon measurement of vertical electric fields because such fields are large, typically > 200 mV/m. Contact potential problems in the horizontal measurement are obviated by causing the payload to rotate about a vertical axis and analyzing only the horizontal signal occurring at the rotation frequency. Since this technique also eliminates errors from vertical fields present in the not-exactly-horizontal sensors, horizontal fields smaller than 1 mV/m have been accurately measured in the presence of large vertical fields and contact potentials.

References

1. Aggson, T. L., "Results of Magnetospheric Electric Field Measurements," invited talk at the Conference on Electric Fields in the Magnetosphere, Rice University, Houston, 1969.
2. Anderson, H. R., and R. H. Manka, "Electric Fields at the Lunar Surface, Sources and Methods of Measurement," in *Electromagnetic Exploration of the Moon*, Ed. W. I. Linlor, pp. 117–125, Mono Book Corporation, Baltimore, 1970.
3. Balsley, B. B., "Some Characteristics of Non-Two-Stream Irregularities in the Equatorial Electrojet," *J. Geophys. Res.* **74**, 2333–2347, May 1, 1969.
4. Balsley, B. B., "Nighttime Electric Fields and Vertical Ionospheric Drifts near the Magnetic Equator," *J. Geophys. Res.* **74**, 1213–1217, March 1, 1969.
5. Bering, E., M. C. Kelley, and F. S. Mozer, "Split Langmuir Probe Measurements of Currents, Electric Fields, Temperatures, and Densities in an Aurora," *Trans. Am. Geophys. Union* **51**, 404, April 1970.
6. Bogott, F. H., and F. S. Mozer, "The Magnetopause Electric Field Inferred from Energetic Particle Measurements on ATS-5," *J. Geophys. Res.* **76**, No. 4, 892–899, Feb. 1, 1971.
7. Carpenter, D. L., and K. Stone, "Direct Detection by a Whistler Method of the Magnetospheric Electric Field Associated with a Polar Substorm," *Planet. Space Sci.* **15**, 395–397, Feb. 1967.
8. Dittmer, A. F., "Experiments on the Scattering of Electrons by Ionized Mercury Vapour," *Phys. Rev.* **28**, 507–520, Sept. 1926.
9. Fahleson, U. V., "Theory of Electric Field Measurements Conducted in the Magnetosphere with Electric Probes," *Space Sci. Rev.* **7**, 238–262, Oct. 1967.
10. Fahleson, U. V., M. C. Kelley, and F. S. Mozer, "Investigation of the Operation of a d.c. Electric Field Detector," *Planet. Space Sci.* **18**, 1551–1561, Nov. 1970.
11. Freeman, J. W., Jr., "Observation of Flow of Low-Energy Ions at Synchronous Altitude and Implications for Magnetospheric Convection," *J. Geophys. Res.* **73**, 4151–4158, July 1, 1968.
12. Föppl, H., G. Haerendel, L. Haser, R. Lüst, F. Melzner, B. Meyer, N. Neuss, H. -H. Rabben, E. Rieger, J. Stöcker, and W. Stoffregen, "Preliminary Results of Electric Field Measurements in the Auroral Zone," *J. Geophys. Res.* **73**, 21–26, Jan. 1, 1968.
13. Gurnett, D. A., "Satellite Measurements of DC Electric Fields in the Ionosphere," in *Particles and Fields in the Magnetosphere*, Ed. B. M. McCormac, pp. 239–246, D. Reidel Publishing Co., Dordrecht, 1970
14. Haerendel, G., R. Lüst, and E. Rieger, "Motion of Artificial Ion Clouds in the Upper Atmosphere," *Planet. Space Sci.* **15**, pp. 1–18, Jan. 1967.
15. Haerendel, G., R. Lüst, E. Rieger, and H. Volk, "Highly Irregular Artificial Plasma Clouds in the Auroral Zone," in *Atmospheric Emissions*, Ed. B. M. McCormac and A. Omholt, pp. 293–303, Van Nostrand Reinhold, New York, 1969.

16. Haerendel, G., and R. Lüst, "Electric Fields in the Ionosphere and Magnetosphere," in *Particles and Fields in the Magnetosphere*, Ed. B. M. McCormac, pp. 213–228, D. Reidel Publishing Co., Dordrecht, 1970.

17. Kavadas, A., and D. W. Johnson, "Electron Densities and Electric Fields in the Aurora," in *Space Research, IV*, pp. 365–370, North-Holland, Amsterdam, 1964.

18. Kelley, M. C., R. Serlin, J. A. Starr, and F. S. Mozer, "The Relationship Between Magnetospheric Electric Fields and the Motion of Auroral Forms," *J. Geophys. Res., 76*, No. 22, 5269–5277, August 1, 1971.

19. Kellogg, P. J., and M. Weed, "Balloon Measurements of Ionospheric Electric Fields," Presented at the Fourth International Conference on the Universal Aspects of Atmospheric Electricity, Tokyo, 1968.

20. Langmuir, I., "Oscillations in Ionized Gases," *Natl. Acad. Sci. Proc.* **14,** 627–637, Aug. 15, 1928.

21. Melzner, F., and G. Volk, "Proposal for an Experiment for the ESRO Geostationary Satellite," ESRO Proposal S-329, 1970.

22. Mozer, F. S., "Balloon Measurements of Vertical and Horizontal Atmospheric Electric Fields," *Pure Appl. Geophys.* **84**, No. I, 32–45, 1971.

23. Mozer, F. S., "Power Spectra of the Magnetospheric Electric Field," *J. Geophys. Res.* **76**, No. 16, 3651–3667, June 1, 1971.

24. Mozer, F. S., and P. Bruston, "Electric Field Measurements in the Auroral Ionosphere," *J. Geophys. Res.* **72**, 1109–1114, Feb. 1, 1967.

25. Mozer, F. S., and U. V. Fahleson, "Parallel and Perpendicular Electric Fields in an Aurora," *Planet. Space Sci.* **18**, 1563–1571, Nov. 1970.

26. Mozer, F. S., and R. H. Manka, "Magnetospheric Electric Field Properties Deduced from Simultaneous Balloon Flights," *J. Geophys. Res., 76*, No. 7, 1697–1712, March 1, 1971.

27. Mozer, F. S., and R. Serlin, "Magnetospheric Electric Field Measurements with Balloons," *J. Geophys. Res.* **74**, 4739–4754, Sept. 1, 1969.

28. Penning, F. M., "Über die intermittierende Glimmentladung in Neon (Intermittent Glow Discharge in Neon)," *Physikalische Zeitschrift* **27**, 187–196, 1 April 1926.

29. Potter, W. E., and L. J. Cahill, Jr., "Measurements near an Auroral Electrojet," *Trans. Am. Geophys. Union* **49**, 650, Dec. 1968.

30. Wescott, E. M., J. D. Stolarik, and J. P. Heppner, "Electric Fields in the Vicinity of Auroral Forms from Motions of Barium Vapor Releases," *J. Geophys. Res.* **74**, 3469–3487, July 1, 1969.

31. Woodman, R. F., "Vertical Drift Velocities and East-West Electric Fields at the Magnetic Equator," *J. Geophys. Res.* **75**, 6249–6259, Nov. 1, 1970.

32. Woodman, R. F., and T. Hagfors, "Methods for the Measurement of Vertical Ionospheric Motions near the Magnetic Equator by Incoherent Scattering," *J. Geophys. Res.* **74**, 1205–1212, March 1, 1969.

Harold E. Dinger

4.8 ELECTROMAGNETIC FIELD SENSORS

The electromagnetic spectrum extends from audio frequencies up through radio, infrared, visible light, ultraviolet light, X-rays, and cosmic rays. The frequency ranges of the various portions of the spectrum are shown in Table 4.8-1.

Table 4.8-1. Regions of the Electromagnetic Spectrum

Type of Wave	Frequency (Hz)
Radio waves	Below about 3×10^{12}
Infrared waves	3×10^{12} to 4×10^{14}
Visible light	4×10^{14} to 8×10^{14}
Ultraviolet light	8×10^{14} to 7×10^{18}
X-Rays, gamma rays	Above about 7×10^{18}

Since all electromagnetic waves travel (in a vacuum) at a velocity of about 3×10^8 m/s, the wavelength in meters corresponding to a certain frequency can be found by dividing 3×10^8 by the frequency in hertz.

By international agreement the radio portion of the spectrum is divided into frequency bands with the designations shown in Table 4.8-2. Other radio bands are shown in Table 4.8-3.

Table 4.8-2. Radio Frequency Bands

Band Name	Abbrev.	Frequency
Very-low frequency	VLF	3 to 30 kHz (k = kilo- = 10^3)
Low frequency	LF	30 to 300 kHz
Medium frequency	MF	300 to 3000 kHz
High frequency	HF	3 to 30 MHz (M = mega- = 10^6)
Very-high frequency	VHF	30 to 300 MHz
Ultra-high frequency	UHF	300 to 3000 MHz
Super-high frequency	SHF	3 to 30 GHz (G = giga- = 10^9)
Extreme high frequency	EHF	30 to 300 GHz

Table 4.8-3. Unofficial Radio Frequency Bands

Band Name	Abbrev.	Frequency
Extremely low frequency	ELF	Below 3000 Hz
——	——	300 to 3000 GHz
——	——	3 to 30 THz (T = tera- = 10^{12})

There are a large number of applications for many kinds of electromagnetic sensors, each having different requirements and employing different techniques, depending on frequency considerations, type of platform, and the purpose for which the observations are desired. This section is restricted to basic electromagnetic concepts and a brief description of a typical ELF/VLF application.

Some electromagnetic field sensor applications are studies in various frequency ranges and under various geophysical conditions, radiation detection (cosmic, X-ray, optical, radio astronomy), electromagnetic distance measurements, electromagnetic navigation systems, time synchronization, animal tracking, monitoring air pollution, automated electronic highway systems, electrical storm detection and/or distribution, and radiation hazards (radio).

Electromagnetic fields in the radio portion of the spectrum (ELF to EHF) include ordinary radio communication signals (AM, FM, PM, PCM, TV), radio noise (atmospheric and man-made), extraterrestrial radio noise (from the sun, moon, planets, stars, quasars, pulsars), lightning stroke radiation (analysis and distribution), and whistlers, swishes, tweeks, chorus, hiss, and other VLF/ELF emissions.

4.8.1 Electromagnetic Waves

Stationary electric charges have an associated electric field. Electric charges in uniform motion, or steady electric currents, produce steady magnetic fields. When the current is varying with time (electron acceleration or deceleration), an electromagnetic field is produced. Such a field consists of energy in wave motion. Electromagnetic waves can be simple sine waves (unmodulated radio-frequency carriers), modulated waves (AM, FM, PCM), or highly complex combinations, such as some forms of radio noise. Electromagnetic fields are created by radio and radar stations; electric power lines; lightning discharges; ignition systems; electrical appliances; various processes occurring on the sun, planets, and stars and other terrestrial and extraterrestrial manifestations (such as chorus, hiss, and other VLF emissions).

Electric field sensors are described in Section 4.7 and magnetic field sensors in Section 4.15. Other allied topics are discussed in Sections 4.14, 4.16, 5.3, 5.4, 5.5, and 5.7, and Chapters 6, 7, and 8.

For an electromagnetic wave in free space, the electric and magnetic energies are equally dvided, and either type of field can be measured. Poynting's vector for direction and rate of energy flow is

$$\mathbf{P} = \mathbf{E} \times \mathbf{H}, \qquad (4.8-1)$$

and the free-space path impedance is

$$Z = \frac{\mathbf{E}}{\mathbf{H}} \quad (= 120\pi\Omega \text{ when } \mathbf{E} \text{ and } \mathbf{H} \text{ are in MKS units}), \quad (4.8-2)$$

where $\mathbf{P}$ is the power density in watts per square meter, $\mathbf{H}$ is the magnetic field strength in ampere turns per meter, and $\mathbf{E}$ is the electric field strength in volts per meter.

Whenever transmission path conditions change, the path impedance can change. For measurement purposes it is difficult in practice to approach free-space conditions because the insertion of a probe into the field changes the impedance of the space in the neighborhood of the probe. Whenever the path impedance changes, the waves are subjected to refraction or reflection with consequent changes in the $\mathbf{E}$ and $\mathbf{H}$ vectors.

The $\mathbf{E}$ and $\mathbf{H}$ vectors are perpendicular to each other. If the $\mathbf{E}$ vector is perpendicular to the earth (or other point of reference) the magnetic vector is horizontal. The wave is then said to be vertically polarized as the electric field is usually used to designate the plane of polarization. Under certain conditions the vectors rotate as the wave is propagated, in which case they are said to be elliptically (or perhaps, circularly) polarized. Some radiation, such as that from most sources of ordinary visible light, is normally randomly polarized.

Some of the methods (modes) of wave propagation are ground wave, sky wave, beam, waveguide, underwater, knife-edge diffraction, scatter (forward-, back-, ionospheric-, tropospheric-, transequatorial-, meteortrial-), auroral reflection, and sporadic-E.

The field strength of an electromagnetic wave radiated from a short vertical antenna over perfectly conducting ground has a theoretical value of

$$E = 300 P_T^{1/2} \quad \text{mV/m}, \qquad (4.8-3)$$

at a distance of 1 km, where P_T is the power radiated in kilowatts.

For practical radiators the field strength can vary from near the theoretical value down to about one-tenth of that value. The value of E at distances greater than 1 km will depend on path conditions, such as refraction, reflection, and absorption. Where ground attenuation can be neglected, another expression for E is

$$E = 377 \frac{Ih_e}{d_{km}} \quad \text{mV/m}, \qquad (4.8-4)$$

where I is the antenna current in amperes, d_{km} is the distance from the antenna in kilometers, and h_e is the effective height of the transmitting antenna in meters. The power density represented by this field strength is

$$P = \frac{E^2}{120\pi} \quad \text{W/m}^2. \tag{4.8-5}$$

4.8.2 The Ionosphere

The D-, E-, F_1-, and F_2-ionospheric layers surrounding the earth greatly influence radio-wave propagation throughout a large part of the radio spectrum. The amount of the influence depends on signal frequency, angle of wave incidence upon the layers, and the electron density or state of ionization of the layers. The latter depends on the angle of the sun with respect to the ionosphere and on the magnetic activity on the sun. As the signal frequency decreases, the bending effect of the ionosphere on the wave increases. At the higher frequencies the wave penetrates with little or no change in direction. At lower frequencies part of the energy in the wave is usually refracted enough to return to earth.

Some of the energy penetrating the ionosphere is also absorbed. This effect is also greater at the lower frequencies. In certain frequency ranges there is a region beyond where the ground wave is detectable and before the reflected wave appears in which no signal can be heard. This is called the *skip zone*. Time variations of ionospheric characteristics can cause signal fading.

One ionospheric radio-wave propagation method makes use of the magneto-ionic mode. It is responsible for whistlers, chorus, hiss, and other VLF emissions.

Other radiofrequency investigations and services for which a type of electromagnetic sensor may be required are sudden ionospheric disturbances (SID's), polar blackouts, ionizing radiation from solar flares, cosmic-ray storms, Faraday rotation, diversity reception, vertical ionospheric soundings, oblique soundings, topside soundings, lightning flash counting, lightning flash analysis, radio noise, cosmic noise, absorption effects, reflection properties, ionospheric drift, spread-F, and laser beam techniques.

4.8.3 Radio Noise

Radio noise can be the signal to be measured (atmospheric or cosmic radio noise, for example), or it can be a background interference (either internal or external to the receiver itself) in which the signals are immersed. Radio noise can be classified by origin as natural (atmospheric, cosmic) or man-made (ignition, switching) or by waveform as fluctuation (tube noise, thermal noise), impulsive (ignition noise), or both combined (background atmospheric noise).

Noise studies include the radiation from individual lightning strokes or from the hundreds of strokes occurring each second throughout the world, corona discharge, and precipitation static. The radio-noise waveform varies with each different source and is modified by the receiver. Theoretically, a single radio frequency passes through a good receiver with only an amplitude modification. If the frequency is modulated, the bandwidth of the receiver must be wide enough to accommodate the additional frequencies comprising the signal. Most radio noise requires large bandwidths to permit its passage relatively unchanged in form. Fluctuation noise consists of overlapping pulses closely spaced in time and has a very broad frequency spectrum, hence the name "white" noise. Impulse noise refers to separated impulses of noise, such as those from automobile ignition systems, or those from a radar transmitter. Combination fluctuation and impulsive noise takes many forms and is represented by background radio static and motor noise.

An excellent example of fluctuation noise is the thermal noise which originates in all bodies above absolute zero in temperature. This is caused by thermal agitation of the electrons and its voltage is given by

$$E^2 = 4RkTB, \qquad (4.8\text{--}6)$$

where E is the rms. value of the radio noise voltage, R is the resistance in ohms between the points of measurements, k (Boltzmann's constant) is 1.38×10^{-23} J °K^{-1}, T is the temperature in degrees Kelvin, and B is the bandwidth of the measuring device (in hertz).

The power represented by this noise is given by

$$P = kTB \qquad (4.8\text{--}7)$$

Thus, noise power is proportional to bandwidth. The noise voltage appearing at the output of a receiver can be a complicated function of input voltage, depending on the effects of the receiver circuits on that particular type of noise. The measurement of radio noise is difficult compared to the measurement of more simple signals.

The parameters measured in radio-noise studies are noise power, effective noise power (dB relative to kTB), average envelope voltage, average logarithm of voltage, envelope crossing rate, quasipeak amplitude, slide-back-peak amplitude, amplitude probability distribution, or other statistical time distributions.

Radio astronomy work usually involves "white" noise and specialized receivers called radiometers (see Section 5.7). Other emissions from space are highly impulsive in character, such as some radiation from Jupiter and pulsar signals.

4.8.4 Whistlers

If a high-fidelity audio amplifier is connected to an antenna, the VLF/ELF electromagnetic energy from an individual lightning stroke will be received in the daytime as a click or crash consisting of a broad spectrum of radio frequencies arriving at exactly the same time. At night, because of the presence of a large number of ray paths, the lower frequencies arrive slightly later, producing a somewhat different sound called a *tweek*. If the height of the ionosphere is known, careful measurements of the time-frequency characteristics of a tweek permit the distance to the source lightning stroke to be calculated.

Under certain conditions, using the same amplifier, musical or semi-musical sounds (whistlers) consisting of a broad band of descending tones can be heard [16].

Some early observations made in England led to the following conclusions [37]:

1. Whistlers fall into two general categories—long and short.
2. Long whistlers were usually preceded by an atmospheric click (spheric).
3. Long whistlers were more numerous during the English summer.
4. Short whistlers apparently were not preceded by an identifiable spheric.
5. Short whistlers were more numerous during the English winter.
6. All loud spherics were followed by a long whistler in some of these early observations.

The whistler theory developed from these observations depends on the dispersion equation [12, 16, 37] given by

$$D = T_d f^{1/2} \quad (\mathrm{s}^{1/2}), \tag{4.8-8}$$

where f is the component frequency and T_d is the time delay for that frequency.

Some of the VLF energy from a lightning flash is guided along the earth's magnetic field lines through the ionosphere. This path is called a magneto-ionic duct. Below the *nose* frequency (the frequency of minimum time delay) which varies with the electron gyro frequency, the whistler frequencies decrease with time; above the nose frequency they increase with time.

The path is between the northern and southern hemispheres. A lightning stroke occurring in the northern hemisphere produces a short (one-hop) whistler at or near the magnetic conjugate point in the southern hemisphere. Reflection can occur at the end of the path allowing the whistler energy to travel back to the northern hemisphere where it would be received as a long (two-hop) whistler, having about twice the dispersion. Under good propagation conditions (low absorption, sufficient ionization) the whistler can make several round trips (as many as forty) between the hemispheres before becoming too weak to detect. The frequencies become more dispersed with each trip.

Multiple lightning strokes can cause multiple whistlers, usually having identical dispersions. At times, energy from the same stroke may be channelled into two or more ducts having slightly different path characteristics resulting in multiple whistlers with slightly different dispersions.

4.8.5 Hiss, Chorus, and Other VLF Emissions

The same receiving equipment used for receiving whistlers is also suitable for receiving VLF emissions known as *hiss* and *chorus* that appear to result from particle interaction in the magneto-ionic duct. Hiss sounds very much like fluctuation noise. Its intensity may remain quite constant for minutes or vary almost continuously. Chorus is a semimusical, chirpy noise that may occur with or without hiss accompaniment. Both occur more readily at times of high magnetic activity.

Other VLF emissions are sometimes referred to as discrete emissions and called *rising tones, hooks, falling tones,* or perhaps combinations of these. There are also periodic emissions, both dispersive and non-dispersive, usually having periods similar to those of whistlers.

Some emissions can be triggered by other emissions. Hiss may be triggered by strong whistlers. Sometimes a strong spheric occurring during a whistler will change the nature of the whistler [10, 11]. The above discussed phenomena have all been naturally occurring emissions. Studies have also been made of magneto-ionic mode signals excited by VLF communication signals [16].

4.8.6 Instrumentation

In its most simple concept, an electromagnetic field sensor consists of a sensitive, low-noise radio receiver of the desired frequency range connected to a suitable antenna system and followed by an appropriate display or read-out device. Accurate calibration, timing, and switching arrangements are generally required, especially if the measurements are to be correlated with other geophysical processes or events, or with similar measurements being made on a synoptic basis. For extremely weak signals in noise, special techniques, such as auto-correlation, may be required in order to retrieve the information.

A typical receiving arrangement is shown in block form in Figure 4.8-1. Depending on the particular service for which it is to be used, additional circuits that may be required are filter, time constant, phase detector, dc amplifier, integrator, heterodyne oscillator, and mixer. The output may be audio, visual, mechanical, or electrical.

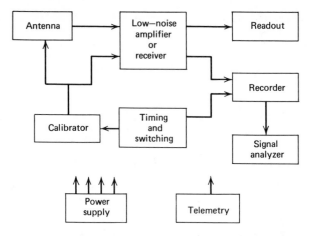

Figure 4.8-1   Basic electromagnetic sensing equipment.

Some of the considerations in receiver design are bandwidth requirements (both rf and af), fixed or variable frequency operation, wide-band operation, signal-to-noise ratio, sensitivity and selectivity.

For the analysis of VLF emissions the receiver should have a frequency range from about 500 Hz to at least 30 kHz, although most of the energy is contained in the lower part of this range. It is often necessary to use a high-pass filter to exclude harmonics of the power line frequency and a low-pass filter to prevent overloading of the first amplifier stage by strong local radio signals. Either a loop or a long-wire antenna can be used. A tape recorder with accurate timing signals is essential. Signal analysis is usually done with an audio spectrograph.

4.8.7   Antennas

Antennas can take various forms such as isotropic (a theoretical form), linear (single wire or beam), loop, slot, parabolic reflector (dish) and interferometer array (usually some combination of one or more of the above).

Measurements require the knowledge of certain electrical characteristics, such as effective height (or length), gain, or effective area.

The effective height of a loop (area small compared with the wavelength) is

$$h_e = \frac{2\pi N A}{\lambda},$$ (4.8-9)

where N is the number of turns, A is the area of loop per turn, and λ is the wavelength.

The effective height of a straight vertical antenna (less than $\lambda/4$ in actual height, h) is

$$h_e = \frac{\lambda}{\pi \sin(2\pi h/\lambda)} \sin^2(\pi h/\lambda). \qquad (4.8-10)$$

The power density in free space is related to the electric field intensity by Eq. 4.8–5. The power density at a distance d from an isotropic radiator is

$$P = W/4\pi d^2 \quad W/m^2, \qquad (4.8-11)$$

where W is the power radiated. The gain of an antenna is

$$G = \frac{4\pi A_e}{\lambda^2}, \qquad (4.8-12)$$

where the effective areas, A_e, for typical antennas are
isotropic: $A_e = \lambda^2/4\pi = 0.8\lambda^2$
short dipole: $A_e = \frac{3}{8}\lambda^2$
half-wave dipole: $A_e = 0.13\lambda^2$
dish: $A_e = 0.54 \times$ area of aperture [45].
If the gain is known from measurement, the area can be found from

$$A_e = G\lambda^2/4\pi. \qquad (4.8-13)$$

When the effective height of a receiving antenna is known, a receiver calibrated in volts connected to the antenna measures radio field strength in volts per meter since

$$V_r = E_s h_e, \qquad (4.8-14)$$

where E_s is the field strength, V_r is the received voltage (determined by signal generator), and h_e is the effective height (in meters). This field method has been widely used on the lower, medium, and high frequencies. At frequencies where dish antennas become practical, measurements are usually made in terms of power instead of field strength.

References

1. Allen, John L., "Array Antennas: New Applications of an Old Technique," *IEEE Spectrum* **1** (11), 115–130, 1964.
2. *Annals of the International Geophysical Year*, Vol. III, Pergamon Press, New York, 1957.
3. Bohn, Erik V., *Introduction to Electromagnetic Fields and Waves*, Addison-Wesley, Reading, Mass., 1968.
4. Brown, G. M., Ed., *Space Radio Communication*, Elsevier Publishing Co., New York, 1961.

5. Brown, J., Ed., *Electromagnetic Theory, Parts 1 and 2* (International series of monographs in electromagnetic waves), Pergamon Press, New York, 1967.
6. Chamberlain, J. W., *Physics of the Aurora and Airglow*, Academic Press, New York, 1961.
7. Collin, R. E., and Francis J. Zucker, Eds., *Antenna Theory, Parts 1 and 2*, McGraw-Hill Book Company, New York, 1969.
8. Corliss, William, *Space Probes and Planetary Exploration*, D. Van Nostrand, Princeton, New Jersey, 1965.
9. Darney, Carl H., and Curtiss C. Johnson, *Introduction to Modern Electromagnetics*, McGraw-Hill Book Company, New York, 1969.
10. Dinger, H. E., "Whistling Atmospherics," NRL Report 4825, Navy Research Laboratory, Washington, D. C., 1956.
11. Dinger, H. E., "Sifflements Atmospheriques Exceptionnels," *Onde Elec.* **37** (362), 526–34, 1957.
12. Eckersley, T. L., "Musical Atmospherics," *Nature* **135**, 104–5, 1935.
13. Evans, John V., and Tor Hagfors, Eds., *Radar Astronomy*, McGraw-Hill Book Company, New York, 1968.
14. Finley, John W., "Antennas and Receivers for Radio Astronomy" in *Advances in Radio Research*, Vol. 2., J. A. Saxton, Ed., Academic Press, New York, 1964.
15. *Geophysics and the International Geophysical Year*, American Geophysical Union, Washington, D.C., 1958.
16. Helliwell, Robert A., *Whistlers and Related Ionospheric Phenomena*, Stanford University Press, Palo Alto, 1965.
17. Horner, F., Ed., *Radio Noise of Terrestrial Origin*, Proc. of Comm. IV, 13th General Assembly of URSI, London, 1960, Elsevier Publishing Co., New York, 1962.
18. Horner, F., "Radio Noise from Thunderstorms" in *Advances in Radio Research*, Vol. 2., J. A. Saxton, Ed., Academic Press, New York, 1964.
19. Horner, F., Ed., "Radio Noise of Terrestrial Origin," *Progress in Radio Science*, Vol. IV, Elsevier Publishing Co., New York, 1965.
20. Jennison, R. C., *Introduction to Radio Astronomy*, McGraw-Hill Book Company, New York, 1967.
21. Joint Technical Advisory Committee (JTAC), *Radio Spectrum Utilization*, McGraw-Hill Book Company, New York, 1964.
22. Kraus, John D., "Recent Advances in Radio Astronomy," *IEEE Spectrum* **1** (9), 78–95, 1964.
23. Kraus, John D., *Radio Astronomy*, McGraw-Hill Book Company, New York, 1966.
24. Liepart, J. P., R. W. Zeek, L. S. Bearce, and E. Toth, "Penetration of the Ionosphere by Very-Low-Frequency Radio Signals—Interim Results of the LOFTI I Experiment," *Proc. IRE* **50** (1), 6–17, 1962.
25. Letzer, Seymour, and Norman Webster, "Noise in Amplifiers," *IEEE Spectrum* **7** (8), 67–75, 1970.
26. Levine, Morton A., "Plasmas in Space," *IEEE Spectrum* **3** (11), 43–47, 1966.
27. Moore, Richard K., "Radiocommunication in the Sea," *IEEE Spectrum* **4** (11), 42–51, 1967.
28. National Research Council; National Academy of Science, *Remote Sensing* (with special reference to agriculture and forestry), 1970.

29. Nussbaum, Allan, *Electromagnetic Theory for Engineers and Scientists,* Prentice-Hall, Englewood Cliffs, New Jersey, 1965.

30. Pierce, E. T., "The Monitoring of Global Thunderstorm Activity," in *Proc. of the Fourth International Conference on Universal Aspects of Atmospheric Electricity, Tokyo, May 1968.* Gordon and Breach Science Publishers, New York, 1968.

31. Radio Astronomy Issue, *Proc. IRE* **46** (1), 1958.

32. *Reference Data for Radio Engineers,* 5th ed. ITT, Howard Sams and Co., Indianapolis, Indiana.

33. Ross, D., *Lasers, Light Amplifiers and Oscillators,* Academic Press, New York, 1969.

34. Ross, Monte, *Laser Receivers, Devices, Techniques, Systems,* John Wiley & Sons, Inc., New York, 1966.

35. Saxton, J. A., Ed., *Advances in Radio Research,* Vol. 2, Academic Press, New York, 1964.

36. Schonland, B. F. J., *The Flight of Thunderbolts,* Oxford University Press, New York, 1964.

37. Storey, L. R. O., "An Investigation of Whistling Atmospherics," *Phil. Trans. Roy. Soc.* (Lond.) A, **246,** 113–141, 1953.

38. Storey, L. R. O., "Whistlers," *Sci. Am.* **194** (1), 34–37, 1956.

39. Thomas, H. A., and R. E. Burgess, *Survey of Existing Information and Data on Radio Noise over the Frequency Range 1–30 Mc/s,* Department of Scientific and Industrial Research, London, Her Majesty's Stationery Office, 1947.

40. Tyras, George, *Radiation and Propagation of Electromagnetic Waves,* Academic Press, New York, 1967.

41. Uman, Martin A., *Lightning,* McGraw-Hill Book Company, Inc., New York, 1969.

42. Weeks, Walter L., *Antenna Engineering,* McGraw-Hill Book Company, Inc., New York, 1968.

43. Whitten, R. C., and I. G. Poppoff, *Physics of the Lower Ionosphere,* Prentice-Hall, Englewood Cliffs, New Jersey, 1965.

44. Williams, Donald J., and Gilbert D. Mead, Eds., *Magnetospheric Physics,* American Geophysical Union, Washington, D.C., 1969.

45. Wolff, Edward A., *Antenna Analysis,* John Wiley and Sons, New York, 1966.

Wilfred K. Klemperer

4.9 PRECISION GEODETIC SENSORS

4.9.1 Precision Optical Methods

Observatory Instruments. Traditionally, the high-precision instruments for astronomic position and time determinations are maintained in observatories. A number of improvements have been made to these instruments, and new portable instruments for precision geodesy (based on those very refined observatory models) are being developed. An example is the portable zenith camera. Instruments which have evolved include the Danjon astrolabe [11] and the Photographic Zenith Tube (PZT) [15]. Both latitude and longitude can be determined with such an instrument from a single set of observations (30 or more stars) in one night. The precision with which positions can be determined from groups of selected stars should approach ± 0.1 arc second for a single night's observation. In practice, the actual errors are larger due to the passage of weather fronts (atmospheric wedge refraction), poor seeing conditions, and errors in the star positions (proper motion, etc.).

Lunar Retroreflector Experiments. Laser ranging of the lunar surface was demonstrated in 1961 [29]. Early in 1965, a scientific team proposed that the first Apollo astronauts place a (passive) retroreflector on the lunar surface [1]. The great advantage of this laser ranging technique (over that of using the natural lunar surface as a reflector) is the capability of making range measurements of very high precision. The major part of the energy of the return pulse is not spread out over ten or more microseconds by the terrain but, instead, returns from a well-defined point. An experimental package of 100 fused silica corner cubes was placed on the moon in 1969. An identical array of corner cubes was set up in the Fra Muro region in 1971. The Russians had a French retroreflector package on their first (unmanned) Lunar roving vehicle [3, 20] which landed in 1970.

The Lunar Ranging Experiment (LURE) [28] transmits a 4-ns-wide pulse from a four-stage, Q-switched ruby laser through a 272-cm telescope aimed at the lunar retroreflector. Approximately 10^{18} photons impinge on a circle about 5 km in diameter (depending on atmospheric seeing) on the lunar surface and, if the telescope has been aimed properly, a weak but similar 4-ns-wide pulse is returned to the earth in the direction from which it came. An area around the transmitter about 5 km in diameter is illuminated by this pulse, which arrives after a delay of about $2\frac{1}{2}$ s from the time the laser fired. The telescope collects about 10 to 20 photons and (typically) one photoelectron is counted every ten laser shots. With the precision of the timing circuits used, the LURE can routinely measure the relative time delay of the average returned laser pulse to within $\pm$ 2 ns ($\pm$ 30 cm in range).

One considerable advantage of the lunar retroreflector technique over radio frequency techniques is that atmospheric water vapor has a much smaller effect at optical than at radio frequencies. Disadvantages are that the lunar ephemeris is not yet known to better than 100 m, coordinated measurements with other observing sites are difficult since any given optical pulse can be received only within 5 km of the laser transmitter, laser transmitters of the required power represent a substantial engineering achievement, and until much larger retroreflectors are placed on the moon, only large telescopes can be used in this work.

Optical Satellite Methods. From camera and laser satellite geodesy programs, the positions of certain worldwide control points are now known to $\pm$ 10 m [14]. Laser satellite tracking promises to improve the precision to $\pm$ 1 m.

4.9.2   Precision Radiofrequency Methods

Radiofrequency Satellite Methods. Several methods for determining positions from satellite transmissions have been developed. One example is the TRANSIT satellite system. TRANSIT 1-B was successfully launched in 1960 and has been followed by many others. In practice, the Doppler shift of a fixed frequency beacon (originally near 50 MHz) is recorded, and the minimum slant-range or closest approach to the satellite determined from the maximum rate-of-change of Doppler. Refinements include receiving (from coded satellite transmissions) updated information about the satellite orbit. Using the satellite TRANSIT 1968–12A, the position of the earth's pole of rotation could be determined to $\pm$ 1 m [2]. See Section 3.8.

Another scheme is the radar transponder for the GEOS-B satellite capable of $\pm$ 2 to $\pm$ 5 m accuracy in range determination [13]. Equipment known as "Secor" for *se*quential *co*leration in *r*ange has been used on various satellites [6] to obtain positions as good as $\pm$ 10 m. The range determinations are made simultaneously from a number of ground stations (the positions of three or more being known). At a later time (for a second position of the satellite), the experiment is repeated. The positions of unknown ground stations are then determined from the known positions, ranges, and accurate information about the satellite orbit.

New systems could improve radiosatellite position determinations by an order of magnitude [4].

Very-Long-Baseline Interferometry (VLBI). Interferometry of celestial sources at optical frequencies dates from the work of Michelson in 1890 [18, 19]. Consequently the simplest form of radio interferometer is known as Michelson's interferometer. A number of improvements to the radio interferometer have been made [8, 17, 24]. Very successful efforts to extend the baseline (using microwave links) were carried out [21] in the 1960's. VLBI with independent atomic clocks was first demonstrated in 1967 [5, 7]. The various interferometers are shown schematically in Figure 4.9-1.

Theory. The response of a two-element, correlation interferometer as a function of θ, the direction of a plane (monochromatic) radio wave of radian frequency ω_0, is

$$P(\theta) = \cos\left(\frac{\omega_0}{c} D \sin\theta\right), \tag{4.9-1}$$

where D is the antenna separation and c the velocity of light. This is identical in form to the equation for a very-long-baseline interferometer used to observe a distant "point" source of radio noise, provided the receiver bandwidth is a small fraction of the operating radian frequency ω_0.

If the system bandwidth is *not* a small fraction of ω_0, fringes well away from the central maximum are reduced in amplitude because of destructive interference. This can be described by rewriting Eq. 4.9–1 as

$$P(\theta) = \Gamma_{1,2} \cos\left(\frac{\omega_0}{c} D \sin\theta\right), \tag{4.9-2}$$

where ω_0 is now the *mean* radian frequency and $\Gamma_{1,2}$ is the mutual-coherence function. When observing a distant "point" source of radio noise with omnidirectional antennas, a wide-bandwidth system exhibits directivity, and fringes are obtained over only a narrow range of θ. This can be shown by

$$\Gamma_{1,2}(\boldsymbol{r}, \tau) = \langle V_1(\boldsymbol{r}_1, \tau_1) V_2^*(\boldsymbol{r}_2, \tau_2) \rangle, \tag{4.9-3}$$

which expresses mutual-coherence as the correlation of the wave field at the antennas 1 and 2 shown in Figure 4.9-2.

Unambiguous determination of the angle θ to a point radio source is equivalent to the problem of determining that delay time $\tau = D \sin\theta/c$ which maximizes the correlation (fringe amplitude). For the simple case of the signal incident on antenna 1 the same as that incident on antenna 2 (except for the time delay), the cross-correlation function is identical to the auto-correlation:

$$\zeta(\tau) = \langle V_1(t) V_1^*(t + \tau) \rangle, \tag{4.9-4}$$

and by the Wiener-Kinchine theorem

$$\zeta(\tau) = \int_{-\infty}^{\infty} F(\omega) \cos \omega\tau \, d\omega, \tag{4.9-5}$$

where $F(\omega)$ is the received *power spectrum*.

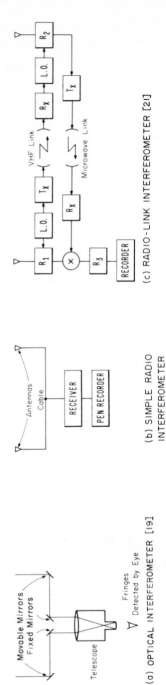

(a) OPTICAL INTERFEROMETER [19]

(b) SIMPLE RADIO INTERFEROMETER

(c) RADIO-LINK INTERFEROMETER [21]

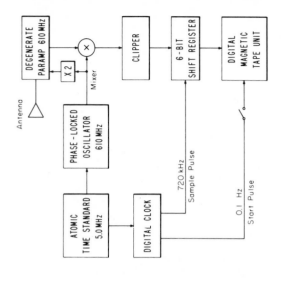

(e) NRAO MK.1 VLBI SYSTEM
(Identical System at Other Site)

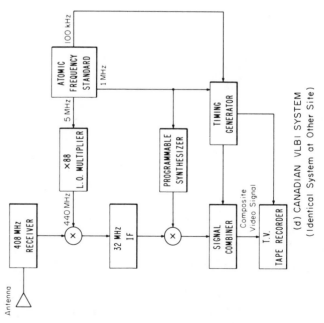

(d) CANADIAN VLBI SYSTEM
(Identical System at Other Site)

Figure 4.9-1 Interferometers.

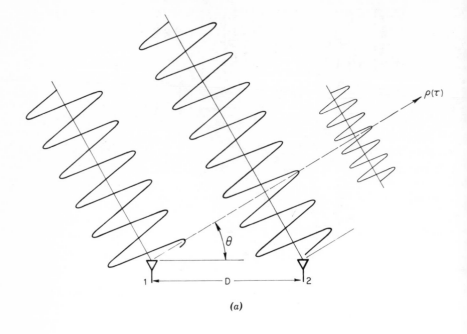

(a)

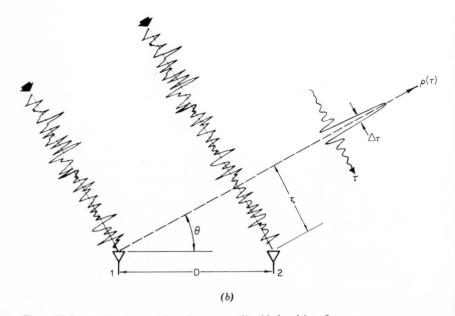

(b)

Figure 4.9-2   (*a*) Narrow-band interferometer; (*b*) wide-band interferometer.

The Fourier transform relationship of Eq. 4.9–5 shows that a wide-band interferometer has a narrow angular response function. With a system operating over a sufficiently wide frequency range, delay time can be unambiguously* determined. Practical methods of realizing wide-band VLBI have been developed [10, 22, 23]. It is unnecessary to record all of the bandwidth at one time; portions of the spectrum ("windows") suffice.

The character of the fringe pattern can be shown by expressing the system response in terms of the Celestial Equator Reference Frame, where δ is the declination of the radio source, H is the hour angle of the radio source, d is the declination of the baseline pole, h is the hour angle of the baseline pole, and D is the distance between antennas. Then

$$P(H, \delta) = \Gamma_{1,2} \cos \left\{ \frac{\omega_0}{c} D(\sin d \sin \delta + \cos d \cos \delta \cos (H - h) + \alpha \right\},$$

(4.9–6)

where α is a phase constant which depends on the initial adjustment of the interferometer.

The fringe pattern expressed by the cosine term in Eq. 4.9–6 moves across the sky at the sidereal rate. The "fringe frequency" is the rate at which fringes cross the source and is given by the derivative of Eq. 4.9–6 as

$$f = \frac{\omega_0 D}{2\pi c} \frac{\cos d \cos \delta \sin (H - h)}{13751} \text{ cycles per sidereal second.} \quad (4.9–7)$$

In very-long-baseline work, this rate can be substantial. If δ and H are both zero (source at zero declination and central meridian transit), the East-West baseline is 3000 km, and the wavelength (λ) is 3 cm (X-band), then $D = 10^8 \lambda$ and $f = 7.26 \times 10^3$ cycles per sidereal second or about 7 kHz. To facilitate the data processing, it is customary to offset one of the local oscillators to remove most of the fringe rate.

For the fringe pattern due to a given source, there are six unknowns and only three observable quantities [9, 26]. Taking data three times a day on three separate sources (or more, to overdetermine the system) permits solutions for all quantities including the unknown time difference between the station clocks.

Atomic Frequency Standards. The development of highly stable local oscillators made very-long-baseline interferometry possible. The stable oscillator used at each site must have good short-term stability. Its multiplied output must exhibit less than one radian of random-phase jitter for

---

*Note that *a priori* knowledge of the baseline parameters is not required. The existence of a well-defined maximum in the cross-correlation function also avoids the ambiguity problem common to most geodemeters and similar fringe-counting systems.

the duration of an integration period (typically 160 s). That is,

$$160\omega\sigma_{\Delta t/t} \leqq 1, \qquad (4.9\text{-}8)$$

where ω is the local oscillator (LO) frequency in radians per second and $\sigma_{\Delta t/t}$ is a measure of the stability. Accordingly, for LO operation at X-band $(6 \times 10^{10} \text{ rad/s})$ the stability must be better than one part in 10^{13}. Rubidium-vapor atomic clocks are perfectly adequate for work below S-band $(2 \times 10^{10} \text{ rad/s})$. Although some data on source size have been obtained with such clocks even at 22 GHz [9], work of geodetic quality at X-band and above generally requires hydrogen maser oscillators (or their equivalent) with stability of parts in 10^{15}.

Tape Recorders. In order to store the information obtained simultaneously at each antenna location for later processing, magnetic tape is most convenient. One system uses analog TV video recorders and a sophisticated, one-of-a-kind analog processor. The bandwidth is about 4 MHz, and data runs 1 hr in length can be recorded without interruption. Another system uses standard 1.25 cm digital computer tape drives and digital processing. This is convenient, since any large computer can be used to process the data. A disadvantage is that the available bandwidth is only 360 kHz (the limitation being the serial 720 kb/s maximum recording rate of the tape decks). A later digital system uses wide-band portable color television recorders. This system formats a 4 Mb/s data stream (2 MHz bandwidth) on 5 cm.-wide tape along with clocking information for synchronization with a second tape on playback. Over 4 hr of uninterrupted data can be stored on a standard 1600 m reel of tape. This wide-band, long-playing digital equipment permits great flexibility in carrying out precise experiments over longer periods with several stations simultaneously. Note that the number of possible baselines for N stations is $N \cdot (N - 1)/2$, so that for as few as 9 stations there are 36 possible baselines which could all run simultaneously by doing all the correlations required at a later time. Experiments of this sort are useful for eliminating systematic errors.

Other Instrumentation. Other components of the VLBI system are standard low-noise receivers, programmed frequency synthesizers (required at the higher frequencies to slow down the apparent fringe rate and for frequency windowing), and suitable calibration equipment. A good calibration scheme uses identical pseudo-random binary code generators at each site synchronized by the local clock [25, 32].

Operational VLBI systems do not require better than state-of-the-art equipment. As atomic frequency standards improve, longer integration times can be attempted at higher frequencies and less computing time will be required to analyze the data.

Errors. The greatest uncertainty in VLBI measurements is the correction for atmospheric and ionospheric phase path. Other errors such as clock offsets and antenna deflections can be made negligible. Earth tides, which can amount to 10–20 cm, can be corrected for since they are periodic and predictable. Ionospheric (plasma) effects vary as λ^2, a strong dependence, so that measurements at two widely separated frequencies can be used to estimate the amount of correction required. Another method for estimating the ionospheric correction is to measure the relative phase-delay difference of the two magneto-ionic modes at a single frequency. VLBI can also be carried out at frequencies above 5 GHz for which the relative ionospheric correction becomes practically negligible. For the neutral atmosphere at radio frequencies, there is (in the zenith direction) an excess delay of about 8 ns over what it would be in vacuum. This is equivalent to a 2.5-m excess path length and varies as the secant of the zenith angle. The amount of delay varies with water vapor content and, to a smaller degree, with atmospheric pressure. Calculations using model atmospheres [31] for VLBI experiments [16] have been carried out. Geodesy to parts in 10^7 can be attained by VLBI using ordinary ground-based weather data. To attain accuracies of parts in 10^8 will require radiosonde data and perhaps other techniques such as water-vapor line radiometry looking along the same path to the radiosource [27].

References

1. Alley, C. O., et al., *J. Geophys. Res.* **70**, 2267, 1965.
2. Anderle, R. J., and L. K. Benglass, *Bull. Geodes.* **96**, 125–41, 1970.
3. Aviation Week, p. 19, 7 December 1970.
4. Aviation Week, p. 52, 22 February 1971.
5. Bare, C., et al., *Science* **157**, 189, 1967.
6. Blackband, W. T., Ed., *Advanced Navigational Techniques* Techivision Services, Slough, England, 1970.
7. Broten, N. W., et al, *Science* **156**, 1592, 1967.
8. Brown, R. H., H. P. Palmer, and A. R. Thompson, *Phil. Mag.* **46**, 857, 1955.
9. Burke, B. F., "Long Baseline Interferometry," *Phys. Today* **22**, No. 7, 54–63, 1969.
10. Burke, B. F., et al, "Studies of H_2O Sources by Means of a Very-Long-Baseline Interferometer," *Astrophys. J. Ltrs.* **160**, L63–68, 1970.
11. Danjon, A., "The Impersonal Astrolabe," in *Stars and Stellar Systems I.*, Ed. Kuiper, G. P. and B. M. Middlehurst, University of Chicago Press, Chicago, 1960.
12. Hinteregger, H. F., "A Long Baseline Interferometer System with Extended Bandwidth," *1968 NEREM Convention Record* **10**, 66–67, 1968.
13. Leitao, C. D., et al., NASA Report X-16-68-1, 1968.
14. Mancini, A., "Satellite Geodesy: Data Acquisition," *Trans Am. Geophys. Union* **52**, IUGG 34–37, 1971.

15. Markowitz, W., "The P.Z.T. and the Dual Rate Moon Camera," in *Stars and Stellar Systems I.*, Ed. by Kuiper, G. P. and B. M. Middlehurst, University of Chicago Press, Chicago, 1960.
16. Mathur, N. C. et al., *Radio Sci.* **5**, 1253, 1970.
17. McReady, L. L., J. L. Pawsey, and R. Payne-Scott, *Proc. Roy. Soc. (Lond.)* A **190**, 357, 1947.
18. Michelson, A. A., *Phil. Mag.* **30**, 1, 1890.
19. Michelson, A. A., and F. G. Pease, *Astrophys. J.* **53**, 249, 1921.
20. *Nature* **228**, 1017–18, 12 Dec. 1970.
21. Palmer, H. P., et al., *Nature* **213**, 789, 1967.
22. Rogers, A. E. E., *Radio Sci.* **5**, 1239, 1970.
23. Rogers, A. E. E., and J. M. Moran, "VLBI as a Means of Worldwide Time Synchronization", MIT Lincoln Labs Report, Feb. 1969.
24. Ryle, M., and D. D. Vonberg, *Proc. Roy. Soc. (Lond.)* A **193**, 98, 1948.
25. Sargent, H. H., and W. K. Klemperer, "A Decametric Long Baseline Interferometer System," *Radio Sci.* **5** (10), 1283, 1970.
26. Shapiro, I. I., and C. A. Knight, "Geophysical Applications of Long-Baseline Interferometry," in *Earthquake Displacement Fields and the Rotation of the Earth*, Eds. L. Mansinha et al., pp. 284–301, D. Reidel Publishing Co., Dordrecht, Holland 1970.
27. Schaper, L. W., D. H. Staelin, and J. W. Waters *Proc. IEEE* **58** (2), 272–3, Feb. 1970.
28. *Science*, **170**, 1289, 18 Dec. 1970.
29. Smullin, L. D., and G. Fiocco, *Proc. IRE* **50**, 1703, 1962.
30. Stanford University, "Final Technical Report on a Preliminary Design of a Drag-Free Satellite and its Application to Geodesy," NASA Contract NAS 12–695, May 1969.
31. Thayer, G. D., "Atmospheric Effects on Multiple-Frequency Range Measurements," ESSA Tech. Report IER 56-ITSA 53, 1967.
32. Yen, J. L., "The Canadian L.B.I. System," IEEE Northeast Electr. Res. Eng. Rec. **10**, 64–65, 1968.

J. C. Harrison

4.10   GRAVITY SENSORS

Gravity values are needed in standards laboratories for establishing fundamental metrological units. An absolute accuracy of one part in a million suffices for this purpose.

Maps of the spatial variations of gravity are used to deduce density variations within the earth and the shape of the geoid. Only gravity differences need be measured, but the sensitivity required may be as high as 0.01 mGal (1 Gal = 1 cm/s^2) in precise surveys and 0.1 mGal in routine geophysical prospecting.

Tidal variation of gravity with time at a given location gives information about the rigidity of the earth and about the ocean tides. The amplitude of this tidal variation attains about 0.2 mGal. A sensitivity of 1 μGal is required and the instrumental calibration factor should be known to better than 1 percent. Gravimeters which respond to ground accelerations are superior to conventional seismographs in the period range of the earth's free oscillations (4–54 min); a sensitivity of at least 0.1 μGal is required.

Questions about the nature of gravitation itself have prompted attempts to detect gravitational radiation and aroused interest in searching for possible variations in the value of the Newtonian constant. These experiments require the utmost possible sensitivity.

Three principles have been widely used in the design of gravity sensors. These are the pendulum, direct measurement of the acceleration of a freely falling body, and force balances (gravity meters) in which the gravitational force on a proof mass is balanced against some other force. A fourth principle, that of exciting resonant oscillation of a system by gravitational forces and measuring the amplitude of oscillation, has been used recently in antennas designed to detect gravitational radiation and in gradiometers.

Traditionally, gravity at a site has been determined by a combination of relative pendulum and gravity meter measurements to establish differences between sites. By starting from one site at which gravity is known, it is possible to establish values at a network of other sites [5, 42]. All gravity values to date have been based on the absolute determination at Potsdam using reversible pendulums [7] published in 1906. Although the Potsdam value has been known to be about 14 mGal too high for several decades, revision of the standard has been delayed until general agreement can be reached as to the exact correction needed.

The accuracy of absolute pendulum measurements is limited to 1 mGal by the pivot, that of relative measurements to about 0.1 mGal by the reproducibility of the pendulum periods [19, 37, 42]. Accurate absolute gravity determinations have been made by free-fall methods, and portable instruments can measure absolute gravity more accurately than the relative pendulum apparatus can measure differences.

4.10.1 Free-Fall Methods

Free-fall methods include laboratory experiments and tracking of space-craft in free-fall orbits [13]. Two trajectories have been used in laboratory experiments—the straightforward drop and the symmetrical free motion or "up-and-down" path. In the former it is necessary to time the falling object through at least three horizontal planes; in the latter the intervals between upwards and downwards passage through two planes are timed.

The main considerations are defining and measuring the separation of the horizontal reference planes, timing the passage of the object through these planes, ensuring true free fall of the object which means eliminating air resistance and magnetic and electrostatic forces, eliminating or correcting for movements of the reference planes during the fall, and eliminating effects of rotation of the object during its fall. The planes are defined using optical interference methods with highly stable continuous-wave lasers that produce monochromatic fringes.

The optical system of the Hammond-Faller Apparatus [10, 16] is illustrated in Figure 4.10–1. Light from the laser is collimated and directed onto the beam splitter of a Michelson interferometer employing corner cube reflectors. Beams returned from the falling and reference cubes are combined at the beam splitter, one of the exit beams being directed onto a photomultiplier tube where the fringes are observed. The lower mirror may be moved away, allowing the mercury pool to be used for vertical alignment of the apparatus. The dropping distance is 1 m.

Fringes are observed at the photomultiplier tube following the release of the falling corner cube, each fringe corresponding to the motion of the reflector through a distance of one-half wavelength. Two scalars, gated on and off by a precise time standard, count the fringes in two time intervals starting simultaneously. The durations and starting time of these intervals relative to the release of the corner cube can be preset. The small delays between the gate opening and passage of the first fringe, and between the gate closing and the passage of the next fringe, are measured. The gating intervals, corrected for these small delays, yield the times of passage through three planes separated by known integral numbers of half wavelengths and thus provide sufficient data for the determination of accelerations.

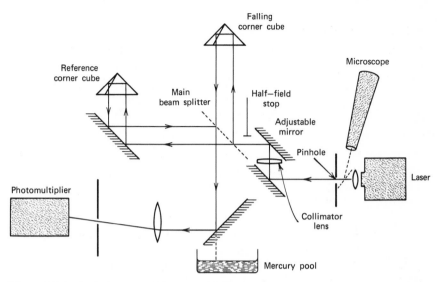

Figure 4.10-1   Optical system of Hammond-Faller free-fall absolute gravity apparatus [16].

The standard deviation of results from several hundred individual drops was about 0.1 mGal. The 70-percent confidence limits of the complete determination was about 0.05 mGal, the largest items in the error budget arising from the effect of solenoid current on the reference cube suspension spring, and uncertainty in the wavelength of light from the laser.

Other recent determinations used the symmetrical free motion of a glass sphere which focussed light from one slit onto another slit to generate a signal on passing through the reference plane [8, 9] or used white light fringes and symmetrical free motion of a corner cube reflector [30, 34]. The comparison in Table 4.10-1 indicates good agreement [16].

Table 4.10-1.   Gravity Measurements

Reference Site	Author	Value Transferred to Site		
Commerce Building	Tate [34]	980	104.77	± 0.45 mGal
Washington, D. C.	Hammond–Faller	980	104.24	± 0.08 mGal
British Fundamental	Cook [9]	981	181.81	± 0.13 mGal
Station, Teddington	Hammond–Faller	981	181.865	± 0.06 mGal
Sèvres, France	Sakuma [30]	980	925.965	± 0.006 mGal
Site A	Hammond–Faller	980	925.965	± 0.05 mGal

4.10.2 Spring Balances

Gravity meters, the most commonly used instruments for measurements on land, are spring balances in which the gravitational attraction on a proof mass is balanced against elastic forces in a spring. The equation of motion of a mass M, supported by a vertical spring with constant k, and damped with a viscous force $D\dot{x}$, is

$$M\ddot{x} + D\dot{x} + kx = Mg, \qquad (4.10\text{--}1)$$

where x is the spring extension relative to its unstressed length. The static deflection is Mg/k and the sensitivity to changes in gravity $\partial x/\partial g$ is (M/k). The static deflection cannot be much greater than 10 cm in an instrument of convenient size, and the sensitivity is thus limited to about 10^{-5} cm/mGal. This is not large enough for a simple optical arrangement to be used for detecting the displacement, and an astatic suspension is normally employed to increase the sensitivity [18, 35].

The suspension of one instrument shown in Fig. 4.10-2a consists of a spring FC of small unstrained length L and spring constant k, supporting a beam DW which is pivoted at D. The lengths a, b, d, r and angles θ, β, Δ are defined in the figure. Both $\delta = \pi/2 - \theta$ and Δ are small. Then [23] the equilibrium condition is

$$g = \frac{kab(1 - L/r)\sin\theta}{Md\sin(\theta + \Delta)}, \qquad (4.10\text{--}2)$$

and the meter sensitivity $\partial\theta/\partial g$, correct to first-order terms in the small

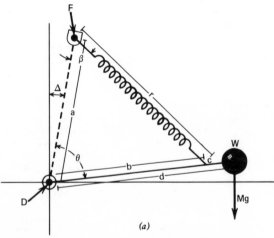

(a)

Figure 4.10-2 (*a*) Basic geometry of gravity meter suspension.

quantities L, δ, Δ is

$$\frac{Md}{kab}\left(\Delta + \frac{Lab}{r^3}\right)^{-1}. \tag{4.10-3}$$

In practice, L is made zero and the equilibrium condition is

$$g = \left[\frac{kb}{Md}\ \frac{\sin\theta}{\sin(\theta + \Delta)}\right]a, \tag{4.10-4}$$

and

$$\frac{\partial\theta}{\partial g} = \frac{Md}{kab\Delta}. \tag{4.10-5}$$

The meter sensitivity is set to a convenient value by adjusting Δ. In order to read the meter, a is changed by moving the upper end of the spring F until the beam is brought to the horizontal, this position being designated by the "reading line" on the eyepiece scale of the telescope through which a fiducial wire on the beam is observed. Under this condition the quantities in the square brackets on the right-hand side of Eq. 4.10–4 are constants, so that changes in a are proportional to changes in g.

The adaptation of this suspension to a gravity meter is shown in Figure 4.10-2*b*. The upper end of the spring is moved by a measuring screw operating through a double reduction lever system. The line of action of the main spring acts through the center of gravity of the mass, so that

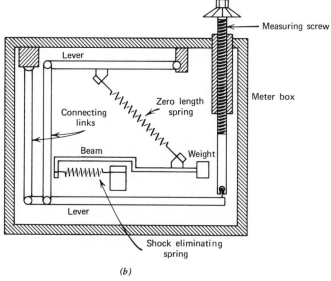

(b)

Figure 4.10-2 *(b)* Construction of gravity meter.

the reaction of the pivot is horizontal and can be supplied by horizontal springs, eliminating the need for a mechanical pivot.

The gravity meter of Figure 4.10-2 with its carrying case and battery weighs about 8.5 kg. It has a range of 7000 mGal, can be read to 0.01 mGal and has a drift rate of less than 1 mGal per month. Repeated measurements usually indicate standard deviations between 0.02 and 0.05 mGal for a single measurement.

4.10.3 Measurements from Moving Platforms

It is necessary to correct gravity measured on a moving platform for its horizontal speed over the curved rotating earth. This Eötvös correction is given by [17]

$$E = 2V\omega_e \cos\phi \sin\alpha(1 + h/a) + V^2/a$$
$$\cdot\{1 + h/a - \varepsilon[1 - \cos^2\phi(3 - 2\sin^2\alpha)]\}, \qquad (4.10\text{–}6)$$

where V is the ground speed of the vehicle, α its track azimuth, h its height, and ϕ its geographic latitude; a is the earth's semimajor axis, ε its flattening, and ω_e its rotation rate.

It is also necessary to consider the vertical motion of the platform, for the instantaneous gravity value sensed is $(g + \ddot{z})$, where $\ddot{z}$ is the vertical acceleration of the platform. If the sensor averages instantaneous values over a period T, it measures

$$\langle g \rangle + [\dot{z}]_0^T/T, \qquad (4.10\text{–}7)$$

where the first term is the mean value of gravity, and the second is the change in vertical velocity per unit time during the observation. Ship motions have periods that are short compared with the time intervals over which it is normally desired to measure, so it is satisfactory merely to filter the output. In aircraft, however, it is necessary to correct continuously for changes in height, which seriously limits the shortness of the interval over which an accurate reading can be obtained.

Sea gravity meters have been made to operate on gyrostabilized platforms which maintain the sensor vertical. A block diagram of the kind of platform used is shown in Figure 4.10-3. The platform is slaved to gyros mounted on the table itself; although the gyros tend to maintain a constant direction in space, the rotation of the earth, movement of the ship, and imperfections in the gyros would make it slowly tilt over in practice. The gyros are therefore precessed by the outputs of horizontal accelerometers also mounted on the platform in a manner designed to keep the average horizontal accelerometer output nulled. The loop parameters have an important influence on the over-all performance [25, 33].

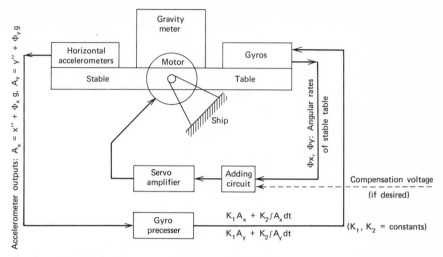

Figure 4.10-3 Block diagram of stabilized platform [25].

Two gravity meter sensors are used most at sea [14, 15, 24, 25, 31]. One is a modified version of the land meter with heavy air damping, a suspension stiffened against unwanted degrees of freedom, a photoelectric beam position read-out and the angle Δ of Eq. 4.10–5 made equal to zero. The other meter has an aluminum beam, restraining filaments to prevent pivot motion, and horizontal cylindrical springs to balance the gravitational couple. Eddy-current damping is provided by a permanent magnet. Cross-coupling corrections must be made (see Figure 4.10-4).

Accelerometers designed for inertial navigators have been used in place of gravity meters on stable platforms [26, 32]. Two such accelerometers have been used in sea gravity meters. One is a single-axis, pendulous force rebalance instrument [22, 36]. The mass consists of a circular coil suspended in a magnetic field and forming the center plate of a three-plate differential capacitor in a position-sensing bridge circuit, the output of which controls the coil current to maintain the coil at the null position. The other device employs the natural frequency of transverse vibration of a string supporting a mass [41]. The string is placed in a magnetic field, and the emf generated by its movement is amplified and fed back to maintain the oscillations.

An accuracy of about 1 mGal can be obtained with modern shipboard instruments, and navigational uncertainties are often the most important source of error in gravity anomaly determinations. Airborne measurements have been made on an experimental basis. The measurement problems are correcting for the changing vertical velocity of the airplane and determining the very large Eötvös corrections.

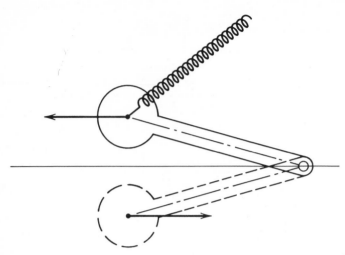

Figure 4.10-4 The gravity meter beam pivots about its axis in response to the vertical component of wave acceleration. If the inertial forces due to the horizontal wave accelerations are correlated with this motion a systematic couple will act on the beam resulting in an erroneous gravity measurement. This cross-coupling error is removed by a special-purpose computer whose inputs are beam position and component of horizontal acceleration parallel to the beam.

4.10.4   Tidal and Seismic Gravity Meters

A great advance in earth-tide instruments was the development of a gravity meter similar to that described in Section 4.10.2 but with increased measuring screw sensitivity (and decreased range), in which the beam position is sensed photoelectrically and the measuring screw driven by a servomotor to maintain the beam on the reading line [6]. Recent developments have included capacitance methods for sensing the beam position and for nulling it electrostatically [2, 27, 40]. The sensitivity of the capacitance position sensor is high enough to dispense with an astaticized spring balance, and yet retain adequate over-all sensitivity to gravity [2].

The drift inherent in spring balances owing to aging of the spring has so far prevented accurate measurement of the fortnightly and monthly tides and of possible secular changes in gravity. An attempt was made to improve long-term stability by supporting a superconducting ball in a magnetic field produced by currents in superconducting coils. The field gradient acts as a spring supporting most of the weight, and the ball position is sensed and held fixed by a capacitative bridge sensor and electrostatic feedback arrangement [29]. Drift has been reduced below 1 mGal per day. Additional information on tide sensors is given in Section 4.26.

4.10.5 Lunar and Planetary Measurements

The first extraterrestrial gravity measurement came from the inertial measurement unit on the Apollo 11 lunar landing module using pulsed integrating pendulous accelerometers [28]. There is no serious difficulty in adapting spring gravity meters of the type described in Section 4.10.3 for use on the moon (or Mars), and it should be possible to transfer the terrestrial calibration. The difficulties of performing free fall determinations on the moon with apparatus of the same general type as that described in Section 4.10.2 are operational rather than instrumental.

A technique for sensing the gravity field from lunar or planetary orbiters is desirable. Some information can be obtained by tracking the orbiter, but this method does not yield all the information needed, particularly about the far side of the moon [20]. Gravity cannot be measured by instruments on the satellite, but it is possible to measure gradients of gravity. A newly developed dynamic gradiometer offers advantages of sensitivity and discrimination against nongravitational effects. This gradiometer is a development arising from research into the detection of gravitational radiation [39]. Gravitational forces at twice the rotational frequency excite a scissors-like oscillation in a sensor consisting of dumbbell masses mounted on flexural pivots. This mode of oscillation, driven at its resonant frequency, produces an electrical output by means of piezoelectric transducers mounted on the central pivot. Many mechanical problems are alleviated in a free-fall environment and, if performance limited by Brownian noise can be achieved, ultimate sensitivities of 10^{-11} s^{-2} with 30-s integration times are possible [1, 11, 12].

References

1. Bell, C. C., R. L. Forward, and H. P. Williams, "Simulated Terrain Mapping with the Rotating Gravity Gradiometer," in *Advances in Dynamic Gravimetry*, Instrument Society of America, Pittsburgh, 1970, pp. 115–128.
2. Block, B., and R. D. Moore, "Measurements in the Earth Mode Frequency Range by an Electrostatic Sensing and Feedback Gravimeter," *J. Geophys. Res.* **71,** 4361–4375, September 1966.
3. Block, B., and R. D. Moore, "Tidal to Seismic Frequency Investigations with a Quartz Accelerometer of New Geometry," *J. Geophys. Res.* **75,** 1493–1505, March 1970.
4. Browne, B. C., "The Measurement of Gravity at Sea," *Mon. Not. Roy. Astron. Soc., Geophys.* Suppl. **4,** 271-279, September 1937.
5. Browne, B. C., "Memorandum on the Technique of Measuring Relative

Gravity Values on Land by Means of Pendulums," *Bull. Geod.* No. 64, 146–168, June 1968.

6. Clarkson, H. N., and L. J. B. LaCoste, "Improvements in Tidal Gravity Meters and their Simultaneous Comparison," *Trans. Am. Geophys. Union* **38**, 8–16, February 1957.

7. Cook, A. H., "The Absolute Determination of the Acceleration due to Gravity," *Metrologia* **1**, 84–114, July 1965.

8. Cook, A. H., "A New Absolute Determination of the Acceleration due to Gravity at the National Physical Laboratory, England," *Phil. Trans. Roy. Soc.*, Ser. A, **261**, 211–252, March 1967.

9. Cook, A. H., and J. A. Hammond, "The Acceleration due to Gravity at the National Geophysical Laboratory," *Metrologia* **5**, 141–142, October 1969.

10. Faller, J. E., "The Precision Measurement of the Acceleration of Gravity," *Science*, **158**, 60–67, October 1967.

11. Forward, R. L., "Rotating Tensor Sensors," *Bull. Am. Phys. Soc.* **9**, 711, December 1964.

12. Forward, R. L., C. C. Bell, D. Berman, T. D. Beard, and L. R. Miller, "Detection of Static, Gravitational Force Gradient Fields with a Rotating Mass Sensor," *Bull. Am. Phys. Soc.* **12**, 1127, December 1967.

13. Gaposchkin, E. M., and K. Lambeck, 1969 Smithsonian Standard Earth (II), Smithsonian Astrophysical Observatory Special Report 315, May 1970.

14. Graf, A., "Das Seegravimeter," *Z. fur Instrumkde* **60**, 151–162, September 1958.

15. Graf, A., and R. Schulze, "Improvements on the Sea Gravimeter Gss 2.," *J. Geophys. Res.* **66**, 1813–1821, June 1961.

16. Hammond, J. A., "A Laser-Interferometer System for the Absolute Determination of the Acceleration of Gravity". Joint Institute for Laboratory Astrophysics Report 103, University of Colorado, Boulder, February 1970.

17. Harlan, R. B., "Eötvös Corrections for Airborne Gravimetry," *J. Geophys. Res.* **73**, 4675–4679, July 1968.

18. Heiskanen, W. A., and F. A. Vening Meinesz, *The Earth and its Gravity Field*, Chapter 4, McGraw-Hill Book Company, New York, 1958.

19. Jackson, J. E., "The Cambridge Pendulum Apparatus," *Geophys. J.* **4**, 375–388, 1961.

20. Kaula, W. M., "The Gravitational Field of the Moon," *Science*, **166**, 1581–1588, December 1969.

21. Kittel, C., *Elementary Statistical Physics*, pp. 152–153, John Wiley & Sons, New York, 1958.

22. Klingenmeier, P., M. Meldnum, and D. Stubbs, "Testing a High Performance Digital Gravity Meter—Some Problems, Some Solutions", in *Proceedings of the Fourth Inertial Guidance Test Symposium held at Air Force Missile Development Center, New Mexico, 6–8 November 1968, Vol. 1*, Missile Development Center—Technical Report 68–76, 1968.

23. LaCoste, L. J. B., "A Simplification in the Conditions for the Zero-length-Spring Seismograph," *Bull. Seis. Soc. Am.* **25**, 176–179, April 1935.

24. LaCoste, L. J. B., U.S. Patent 2,977,799, 1961.

25. LaCoste, L. J. B., "Measurement of Gravity at Sea and in the Air," *Rev. Geophys.* **5**, 477–526, November 1967.

26. Macomber, G. R., and M. Fernandez, *Inertial Guidance Engineering*, Chapter 4, Prentice-Hall Inc., Englewood Cliffs, New Jersey, 1962.
27. Moore, R. D., and W. E. Farrell, "Linearization and Calibration of Electrostatically Fedback Gravity Meters," *J. Geophys. Res.* **75**, 928–932, February 1970.
28. Nance, R. L., "Gravity: First Measurement on the Lunar Surface," *Science* **166**, 384–385, October 1969.
29. Prothero, W. A., and J. M. Goodkind, "A Superconducting Gravimeter," *Rev. Sci. Instr.* **39**, 1257–1262, September 1968.
30. Sakuma, A., "Etat actuel de la nouvelle détermination absolut de la pesanteur au Bureau International des Poids et Mesures," *Bull. Geod.* No. 69, 249–260, September 1963.
31. Schulze, R., "Automation of the Sea Gravimeter Gss 2," *J. Geophys. Res.* **67**, 3397–3401, August 1962.
32. Slater, J. N., *Inertial Guidance Sensors*, part B, Reinhold Publishing Corp., New York, 1964.
33. Talwani, M., *The Sea*, Vol. IV, Interscience-John Wiley, New York, 1970.
34. Tate, D. R., Acceleration Due to Gravity at the National Bureau of Standards, *N.B.S. Monograph 107*, U.S. Government Printing Office, Washington, D.C., June 1968.
35. Tomascheck, R., "Tides of the Solid Earth," in *Encyclopedia of Physics*, Vol. XLVIII, pp. 799–806, Springer, Berlin, 1957.
36. U.S. Naval Oceanographic Office, "Report on Prototype Gravity Measuring System" in *Proceedings of the First Marine Geodesy Symposium, September 28–30, 1966*, pp. 189–194, U.S. Government Printing Office, Washington, D.C., 1967.
37. Valliant, H. D., "The Canadian Pendulum Apparatus, Design And Operation", *Publ. Earth Phys. Branch*, **41**, 47–66, Dept. Energy, Mines, and Resources, Ottawa, 1971.
38. Vening Meinesz, F. A., *Theory and Practice of Pendulum Observations at Sea*, Publication of the Netherland Geodetic Committee, J. Waltman, Delft; Part I, 1929; Part II, 1941.
39. Weber, J., "Observation of the Thermal Fluctuations of a Gravitational-Wave Detector," *Phys. Rev. Letters*, **17**, 1228–1230, December 1966.
40. Weber, J., and J. V. Larson, "Operation of La Coste and Romberg Gravimeter at Sensitivity Approaching the Thermal Fluctuation Limits," *J. Geophys. Res.* **71**, 6005–6009, December 1966.
41. Wing, C. G., "MIT Vibrating String Surface-Ship Gravimeter," *J. Geophys. Res.* **74**, 5882–5894, November 1969.
42. Woollard, G. P., and J. C. Rose, *International Gravity Measurements*, Society of Exploration Geophysicists, Madison, Wisconsin, 1963, Part 4.

Delvin S. Fanning and Kenneth A. Rayburn

4.11 GROUND CONSTITUENT SENSORS

4.11.1 Constituent Characteristics

Ground constituents are solid materials (rocks and minerals) which occur primarily in consolidated forms (rocks) or in unconsolidated forms (soils and sediments). They also occur to a lesser extent suspended in liquids and gases. While they are primarily inorganic in form, the organic forms (e.g., coal) can be very important from an economic and environmental point of view. In terms of abundance, silicates are the most important group of minerals composing ground constituents of the earth and moon —and presumably of the other planets and their satellites in the solar system. Other important groups of minerals, in terms of abundance as grouped by their anions, are halides, sulfides, sulfates, carbonates, and oxides and hydroxides. In some instances, metals also occur.

Man is interested in the ground constituents from two main points of view. First, some of these materials are of economic value and important in promoting and maintaining man's physical well being. Secondly, the ground constituents are of value from a more purely scientific standpoint. Since the minerals and rocks that form and are stable depend upon environmental conditions, these materials are important keys to the history of the earth, the solar system, and the universe.

There are a large number of chemical and mineralogical forms that exist in geologic deposits, and a large number of techniques are available for their identification and analysis. Competent field earth scientists (geologists, soil scientists, mineralogists) are better than most mechanical devices for identifying rocks, soils, and minerals in the field and for directing the sampling of these materials for laboratory analyses.

The kinds of information about ground constituents that instruments are used to obtain can be grouped into two main and several peripheral categories. The two main categories are the chemical and the mineralogical composition of the material. Chemical composition refers to the kinds and amounts of chemical elements (and in some instances the isotopes of the elements) present. Mineralogical composition refers to the kinds and amounts of minerals present. Minerals are specified not only by the chemical elements present, but more importantly by the arrangement of elements in unique crystal structures. Minerals composed of the same element, or elements, can have markedly different properties (e.g., graphite versus diamond) and modes of origin.

Other ground constituent characteristics of interest include particle size (either discrete particles in unconsolidated materials or mineral grains in solid rocks), particle shape, particle surface features, and the arrangement of particles. Also the stratigraphic sequences of rocks in geologic columns and the profile distribution of soil characteristics can be of considerable interest.

Although some progress has been made in developing instruments to study ground constituents in situ [2], most studies of these materials continue to be made on samples that are brought to laboratories, where more precise conditions exist and better analyses can be made. The need to bring samples of moon rocks and soil back to earth laboratories for accurate chemical and mineralogical analyses exemplifies this situation.

4.11.2 Chemical Analysis Instruments

The most frequently used instrumental techniques for chemical analysis are shown in Table 4.11-1. The theoretical background and detailed discussion of the instruments employed with each technique are adequately covered in at least one of the references cited.

Table 4.11-1. Instrumental Techniques for Chemical Analysis

Technique	Elements Analyzed	Qual.	Quantitative Range	Field Use	Applications	Ref.
X-Ray emission	Na-U	Yes	ppm–100%	No	Solids, liquids	1, 5, 17, 21
non-dispersive	Al-U	Yes	0.1–100%	Portable Remote	Solids, liquids	18, 31
Optical emission	Li-U[a]	Yes	ppb–100%	Portable	Solids, liquids	3, 4
Atomic absorption	Li-U[b]	No	ppb–10%	Portable	Solutions only	13, 31, 32
Flame emission	Li-U[b]	No	ppb–10%	Portable	Solutions only	12, 13, 31, 32
Electron microprobe	B-U	Yes	ppb–100%	No	Solids, microscale analysis	6, 14
Spark source mass spectroscopy	H-U	Yes	ppb–10%	No	Isotope ratio	10, 30, 34
Alpha particle scattering	Li-U	Yes	Range ests. only	Remote	Surface analysis Surveyor expts.	37, 39, 41

[a] Except N, O, halogens.
[b] Except C, N, O, halogens and noble gases.

The columns denoting element range covered and normal quantitative range should not be taken as absolute since these are subject to considerations of matrix effect and sample size problems. The field-use column indicates the following: "No"—no field or remote units currently available; "Portable"—smaller field units available but require on-site personnel for operation; "Remote"—automated, or automatic operation by remote control units available or designed. All of the technique types noted are commercially available in laboratory versions.

Additional instrumental techniques other than the main ones listed in Table 4.11-1, which are of limited use, are Mössbauer spectroscopy [33] for determination of the oxidation state of iron and electron spectroscopy techniques such as electron spectroscopy for chemical analysis (ESCA) or Aüger for determination of binding energies, oxidation state, and bonding studies of surfaces [40]. Neutron activation analysis is a powerful technique for trace analysis—but beyond the reach of most laboratories and certainly beyond portable or remote use at the present state of the art [23].

One of the best sources of information on instrumentation and technique applications is the Fundamental Reviews edition of *Analytical Chemistry*, published in alternate years. The latest reviews cover nearly all analytical techniques and their applications, with references to the latest texts covering the theoretical background of the techniques of interest [44].

4.11.3 Mineralogical Analysis Instruments

Chemical analysis data help in identifying and in giving quantitative estimates of the minerals present in a sample. Some procedures for allocating the elements from chemical analysis data to minerals (qualitatively identified by mineralogical techniques such as X-ray diffraction) have been developed [16, Chapter 11]. However, to truly identify minerals it is necessary to use instruments that measure or reflect (in a broad sense) the atomic arrangement of the elements. The two main instrumental techniques that have been used by mineralogists to do this have been X-ray diffraction (powder cameras, diffractometers) and the petrographic microscope (petrography) (Table 4.11-2).

X-ray diffraction is based upon the Bragg law, $n\lambda = 2d \sin\theta$, which describes the condition for obtaining a diffracted beam of X-rays from a crystal where n is an integral number, λ is the wavelength of the X-rays, d is the distance between equally spaced parallel planes of atoms (or ions) in the crystal, and θ is the angle between the impinging beam of X-rays and the planes.

A set of diffraction peaks, or reflections, is obtained for powdered specimens of minerals that are examined over a range of θ angles. The pattern obtained is unique for each mineral and serves as a "fingerprint" by which it can be identified. X-ray diffraction patterns of known minerals and chemical compounds have been accumulated and classified by various workers and a file of these patterns is available from the American Society

Table 4.11-2. **Instrumental Techniques for Mineralogical Analysis**

Technique	Use	Limitations	Ref.
X-ray diffraction Powder techniques utilizing:			36
Random orientation specimens	Mineral identification using ASTM card file system	Sample must be a powder. Identification problems with mixtures of minerals.	43
Parallel orientation specimens	Identifying and studying clay minerals	Do not get general reflections from clay minerals	9, 16
Single crystal techniques	Crystal structure (atomic arrangement) studies	Structures for minerals of complex crystal systems are difficult to solve. Cannot be done on clay-size particles	15, 43
Petrographic microscope			7
Grain mounts	Mineral identification	Limited particle size range. Cannot identify minerals of clay-size particles. Particle size fractionation usually necessary	(Optical properties of different minerals)
Thin sections	Mineral identification	Thin section preparation required	
	Fabric studies	Unconsolidated materials must be impregnated before sectioning	8 (Soil fabric)
Thermal methods			29
Differential thermal analysis (DTA)	Study temperatures of mineral transformations	Not diagnostic for mineral identification in most instances	27, 29
Thermal gravimetric analysis (TGA) (Thermal-balance)	Qualitative and quantitative mineral analysis in some instances	Not diagnostic for mineral identification in most instances	29
Infrared (IR)	Bonding and functional group studies	Not diagnostic for mineral identification in most instances	25, 26
Magnetic separators	Separating magnetic minerals	Only use with nonmagnetic minerals is for purification with respect to magnetic minerals	35
Electron microscope	Study size, shape, surface features of small particles and surfaces	Not diagnostic by itself for mineral identification in most instances	19
Electron diffraction	Study structure of materials amorphous to x-rays	Patterns too complex for most identification work	20, 28
Electron microprobe	Study chemical composi- tion, which may give indication of mineral- ogy, of micron-size selected areas; study of weathering, etc.	Mineral identification is only by inference from chemical composition	14, 22

for Testing and Materials (ASTM). Thus the pattern from an unknown mineral can be compared to standard patterns until a match is obtained [43]. X-ray diffraction can be used on clay-size materials as well as on coarse particles, which normally require grinding before they can be used for identification by the powder techniques (Table 4.11-2).

Mineral identification employing the petrographic microscope is based upon differential light transmitting and reflecting properties of minerals [7]. These techniques have a longer history than X-ray diffraction and are still used by mineralogists. They are limited, compared to X-ray diffraction, in that they are of little use in identifying minerals of clay size although they can be used to study the orientation of clay bodies in thin sections [8]. Although these light microscopes are quite mobile and can be used in the field, they are delicate and are normally used in a laboratory because of the considerable sample preparation and accessory equipment (oils, etc.) needed for good work.

The other instrumental techniques listed in Table 4.11-2 are seldom used independently for routine mineralogical analysis, although well-equipped mineralogical laboratories often have many of these instruments. Some are quite useful for specific kinds of minerals (e.g., magnetic separators), whereas others are used primarily to give supplemental information to that given by X-ray diffraction and the petrographic microscope.

4.11.4 Instruments for Particle Size, Shape, and Arrangement Analysis

To have a complete characterization of ground constituent material, in addition to chemical and mineralogical analysis data, it is necessary to know about the size, the shape, and the arrangement of the particles or grains composing the material. Also information about the nature of particle surfaces may be needed. These features can help in determining the conditions of deposition or formation of the material as well as weathering transformations.

Particle-size analysis of unconsolidated materials is normally done by sieving to separate the coarser particle sizes (sand or coarser) [11]. Finer particles (silt and clay) are normally measured by pipette analysis or by hydrometer [11]. The actual fractionation of these finer particle sizes so that samples of them can be examined is normally done by centrifugation [16].

The size, shape, and arrangement of particles can also be studied with microscopes. Thin sections of "undisturbed" samples are normally employed with petrographic microscopes to study these characteristics [8]. Electron microscopes are employed to study the shapes (or habits) of clay-size particles [19]. Scanning electron microscopes are especially good for examining three-dimensional surface features in detail [24]. Surface features are also studied with normal electron microscopes by means of the replica technique [19].

References

1. Adler, I., *X-Ray Emission Spectrography in Geology*, Elsevier, New York, 1966.
2. Adler, I., and J. I. Trombka, *Geochemical Exploration of the Moon and Planets*, Springer-Verlag, New York, 1970.
3. Ahrens, L. H., and S. R. Taylor, *Spectrochemical Analysis*, 2nd ed., Addison-Wesley, Reading, Massachusetts, 1961.
4. American Society for Testing and Materials, *Methods for Emission Spectrochemical Analysis*, 5th ed., ASTM Committee E-2, American Society for Testing and Materials, Philadelphia, Pennsylvania, 1968.
5. Bertin, E. P., *Principles and Practices of X-Ray Spectrochemical Analysis*, Plenum Press, New York, 1970.
6. Birks. L. W., *Electron Probe Microanalysis*, Interscience, New York, 1963.
7. Bloss, F. D., *An Introduction to the Methods of Optical Crystallography*, Holt, Rinehart, and Winston, New York, 1961.
8. Brewer, R., *Fabric and Mineral Analysis of Soils*, John Wiley and Sons, New York, 1964.
9. Carroll, D., "Clay Minerals: A Guide to Their X-ray Identification," Geol. *Soc. Am. Spec.* Paper 126, 1970.
10. Carver, R. D., and P. G. Johnson, "Use of a Spark Source Mass Spectrograph for the General Analysis of Geological Samples," *Appl. Spectr.* **22**, 431, 1968.
11. Day, P. R., "Particle Size Fractionation and Particle Size Analysis," in *Methods of Soil Analysis, Part 1*, C. A. Black, editor-in-chief, pp. 545–567, Agronomy Monographs No. 9, American Society of Agronomy, Inc., Madison, Wisconsin, 1965.
12. Dean, J. A., *Flame Photometry*, McGraw-Hill Book Company, New York, 1960.
13. Dean, J. A., and T. C. Rains, *Flame Emission and Atomic Absorption Spectrometry: Theory*, Vol. 1, Marcel Dekker, New York, 1969.
14. Heinrich, K. F. J., Ed., "Quantitative Electron Probe Microanalysis". NBS Spec. Publ. 298, National Bureau of Standards, Washington, D.C., 1968.
15. Henry, N. F. M., H. Lipson, and W. A. Wooster, *The Interpretation of X-ray Diffraction Photographs*, Macmillan and Co., London, 1960.
16. Jackson, M. L., *Soil Chemical Analysis—Advanced Course*, Published by the author, Department of Soils, University of Wisconsin, Madison, Wisconsin, 1956.
17. Jenkins, R., and J. L. DeVries, *Practical X-Ray Spectrometry*, 2nd ed., Phillips Technical Library, Springer-Verlag, New York, 1970.
18. Karttunen, J. O., and W. R. Harmon, "Determination of Uranium in Ores and in Solution Using a Portable Non-Dispersive X-Ray Spectrograph," *Spectrochim. Acta* **24B**, 301, 1969.
19. Kittrick, J. A., "Electron Microscope Techniques," in *Methods of Soil Analysis, Part 1*, C. A. Black, editor-in-chief, pp. 632–652, Agronomy Monographs No. 9, American Society of Agronomy, Inc., Madison, Wisconsin, 1965.
20. Kittrick, J. A., "Electron-Diffraction Techniques for Mineral Identification," in *Methods of Soil Analysis, Part 1*, C. A. Black, editor-in-chief, pp. 652–670,

Agronomy Monographs No. 9, American Society of Agronomy, Inc., Madison, Wisconsin, 1965.

21. Liebhafsky, H. A., H. G. Pfeiffer, E. H. Winslow, and P. D. Zemany, *X-Ray Absorption and Emission in Analytical Chemistry*, John Wiley and Sons, New York, 1960.

22. Long, J. V. P., Electron Probe Microanalysis, in *Physical Methods in Determinative Mineralogy*, Ed. J. Zussman, Academic Press, London, 1967.

23. Lutz, G. J., R. J. Boren, R. S. Maddock, and W. W. Meinke, "Activation Analysis: A Bibliography," NBS Tech. Note 467, Parts 1 and 2, National Bureau of Standards, Washington, D.C., 1968.

24. Lynn, W. C., and R. B. Grossman, "Observations of Certain Soil Fabrics with the Scanning Electron Microscope," *Soil Sci. Soc. Amer, Proc.* **34**, 645–648, 1970.

25. Lyon, R. J. P., "Infrared Absorption Spectroscopy," in *Physical Methods in Determinative Mineralogy*, Ed. J. Zussman, Academic Press, London, 1967.

26. Lyon, R. J. P., "Infrared Analysis of Soil Minerals," in *Soil Clay Mineralogy*, Ed. C. I. Rich and G. W. Kunze, University of North Carolina Press, Chapel Hill, 1964.

27. Mackenzie, R. C., Ed., *The Differential Thermal Investigation of Clays*, Mineralogical Society, London, 1957.

28. McConnell, J. D. C., "Electron Microscopy and Electron Diffraction," in *Physical Methods in Determinative Mineralogy*, Ed. J. Zussman, Academic Press, London, 1967.

29. McLaughlin, R. J. W., "Thermal Techniques," Chapter 9 in *Physical Methods in Determinative Mineralogy*, Ed. J. Zussman, Academic Press, London, 1967.

30. Margrave, J. L., Ed., "Mass Spectrometry in Inorganic Chemistry," Advances in Chemistry Series, Vol. **72**, American Chemical Society, Washington, D.C., 1968.

31. Mavrodineanu, R., "Bibliography on Flame Spectroscopy: 1800–1966," NBS Misc. Publ. 281, National Bureau of Standards, Washington, D.C., 1967.

32. Maxwell, J. A. *Rock and Mineral Analysis*. Interscience, New York, 1969.

33. May, L., Ed., *Introduction to Mössbauer Spectroscopy*, Plenum Press, New York, 1971.

34. Morrison, G. H., and A. T. Kashuba, "Multielement Analysis of Basaltic Rock Using Spark Source Mass Spectrometry," *Anal. Chem.* **41**, 1842, 1969.

35. Muller, L. D., "Laboratory Methods of Mineral Separation," in *Physical Methods in Determinative Mineralogy*, Ed. J. Zussman, Academic Press, London, 1967.

36. Parrish, W., Ed., *Advances in X-Ray Diffractometry and X-Ray Spectrography*, Centrex Publishing Company, Eindhoven, The Netherlands, 1962.

37. Patterson, J. H., "Chemical Analysis of Surfaces Using Alpha Particles," *J. Geophys. Res.* **70**, 1311, 1965.

38. Rhodes, J. R., and T. Furuta, "Applications of A Portable Radioisotope X-Ray Fluorescence Spectrometer to the Analysis of Minerals and Alloys," *Advan. X-Ray Anal.* **11**, 249, 1968.

39. Turkevich, A., K. Knolle, R. A. Emmert, W. A. Anderson, J. H. Patterson, and E. Franzgrote, "Instrument for Lunar Surface Chemical Analysis," *Rev. Sci. Instr.* **37**, 1681, 1966.

40. Siegbahn, K., C. Nordling, A. Fahlman, R. Nordberg, K. Hamnn, J. Hedman, G. Johansson, T. Bergmark, S. E. Karlsson, I. Lindgren, and B. Lindberg, *ESCA-Atomic, Molecular, and Solid State Structure Studied by Means of Electron Spectroscopy*, Almquist and Wiksells, Upsala, 1967.

41. Turkevich, A., K. Knolle, E. Franzgrote, and J. H. Patterson, "Chemical Analysis Experiment for the Surveyor Lunar Mission," *J. Geophys. Res.* **72**, 831, 1967.

42. Winchell, A. N., and H. Winchell, *Elements of Optical Mineralogy, Part II. Description of Minerals*, John Wiley and Sons, New York, 1951.

43. Zussman, J., "X-Ray Diffraction," in *Physical Methods in Determinative Mineralogy* Ed. J. Zussman, Academic Press, London, 1967.

44. —, "Analytical Reviews: Fundamentals 1970," *Anal. Chem.* **42**, No. 5, 1970.

J. L. Hieatt, W. N. Keller, and R. K. Schisler

4.12 HORIZON SENSORS

Earth horizon sensors are used in spacecraft to provide a means for indicating the local vertical. Their applications include utilization as sensory components for active control systems (attitude control) and as measuring instruments to provide correlation between the orientation of experiments, antennas, etc., and the local vertical (attitude determination). The two functions are very similar and the same instruments are often used for both; the primary difference is that attitude control must be done in real time while attitude determination often permits more leisurely ground processing.

Considerable effort has been directed in the design of horizon sensors to the determination of the best spectral region for defining the space-to-earth discontinuity and providing the greatest immunity to unwanted radiation. Most horizon sensors use the earth's radiation in the infrared spectrum (2 to 30 μm). It should be noted that the location of the observed horizon in this spectral range does not coincide with the conventional visible horizon observed at the discontinuity between the atmosphere and the earth's surface, but instead is located in the upper regions of the atmosphere. These sensors utilize long-wavelength thermal detectors operating at room temperature such as thermistors, metal bolometers, or thermopiles.

4.12.1 Operational Concepts and Applications

Horizon sensors can be divided into four basic types: active scanners (self-contained scan), passive scanners (spacecraft scan), edge trackers, and radiance balancers. The spacecraft application and operational constraints dictate the appropriate type or types of sensor which can be used.

Active scanners are instruments in which a narrow instantaneous field of view is made to move in space across portions or all of the earth's image. Various scan mechanisms are employed to cause the motion. The signals from the resulting energy measurements are compared with spacecraft axes information to compute the relative position of the earth in one or more axes. Passive scanners are very similar except that the motion of the spacecraft itself is used to replace the scan mechanism.

Edge-tracking earth sensors are somewhat similar in appearance to active scanners, also using a scan mechanism to move a small instantaneous field of view. In the edge-tracker the average position of the field of view tracks the horizon. The angular position of the track point is a direct measure of the position in spacecraft coordinates of that horizon; this may be used alone or in combination with several such measurements at different points along the horizon to compute the local vertical.

Radiometric balance sensors compare ("balance") the radiant energy received from different portions of the earth. When the radiance from opposite sides of the earth is equal, the optical axis is assumed to be pointing at the center.

4.12.2 Design Considerations

Nature of the Earth's IR Emission. The infrared radiation from the earth is essentially the graybody emission from the earth due to its nonzero temperature and the transmission, absorption, and re-emission of energy by the earth's atmosphere. Figure 4.12-1 indicates the computed spectral radiance of the earth and its atmosphere for several blackbody temperatures and experimental data from balloon flights. Since the infrared radiation from the earth is affected by the atmosphere, seasonal variations and weather conditions cause variations in the spectral radiance. The ultimate accuracy that can be obtained from horizon sensors is limited by these variations. For this reason the selection of spectral passband for a horizon

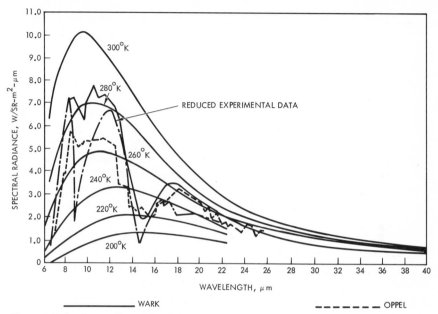

Figure 4.12-1   Comparison of theoretical and measured spectral radiance [1].

sensor is largely dictated by the magnitude of radiance variations due to seasonal and weather changes.

Spectral Selection. The spectral bandpass of a horizon sensor is based on several considerations including horizon radiation stability, magnitude of available radiance, and sensitivity to competing radiation sources, (sun, moon, etc.).

The earliest horizon sensors were designed based only on optimizing the signal-to-noise ratio of the instrument by allowing the total spectral input that the optical materials utilized would permit. This led to serious problems due to solar reflections from clouds. To solve this problem, the spectral passband was limited to wavelengths beyond 5 μm. This led to the discovery of the relative horizon radiance instability that can occur over any wide passband which includes atmospheric "windows." The variation in transmission of the atmosphere as a function of weather conditions is dramatically demonstrated for the visible atmospheric window during storms and on cloudy days.

The primary trade-off parameters that are considered in the choice of spectral bandpass are accuracy and stability requirements and size, weight, and power. A narrow spectral passband appropriately selected can optimize horizon radiance stability, but at the expense of relatively low radiance. This could result in reduced accuracy due to a low signal-to-noise ratio or in larger size, weight and perhaps power requirements than would result from a choice of a wider passband. Since the accuracy and stability requirements are generally specified in terms of angles, the operational altitude can be an important factor in the spectral passband selection.

Typical choices of spectral passband are 14 to 16 μm, approximately 13 to 18 μm, and 13 to 40 μm.

The 14- to 16-μm band is a spectral region in which the atmospheric carbon dioxide (CO_2) strongly absorbs radiation and re-emits as a gray body. Since the CO_2 mixing ratio is approximately constant throughout the atmosphere, there is CO_2 absorption up to the maximum altitudes where the sensible atmosphere exists. At these extreme altitudes weather effects almost cease to exist and seasonal variations are minimized. Figure 4.12-2 indicates extreme horizon radiance profile variations for the 14- to 16-μm band.

The 13- to 18-μm bandwidth is often the choice for high-altitude missions (on the order of synchronous altitude or greater). At these altitudes the horizon stability is almost unaffected by the increased bandwidth but the radiance levels are approximately three times larger.

The 13- to 40-μm band has been chosen for some missions. This band utilizes the CO_2 and rotational water absorption bands from 20 to 40 μm. The selection of this band results in considerably increased radiance but the variation in the water vapor content with altitude results in larger variations in radiance and poorer horizon stability than the other two passbands.

Estimated horizon radiance profiles are given in Ref. 2 through 5.

Competing Radiance Sources. One of the most important design considerations is the effect of radiance sources other than the earth and its atmosphere on horizon sensor performance. Primary sources of competing radiance are direct sun and moon interference, radiance from portions of the spacecraft or reflections from spacecraft structures and internal reflectances in the horizon sensor. Other stellar objects are generally not a problem because the sensor field of view is usually large (on the order of perhaps 1° or greater in diameter) compared to the angular subtense of stellar objects. Since the horizon sensor is designed such that the horizon fills the field of view, the signal level from a very small angular source is generally much smaller than the earth signal.

Proper design of the horizon sensor and care in mounting on the spacecraft so that the spacecraft structure cannot get into the field of view of the sensor can eliminate most of the problems from competing radiation except direct radiance from the sun and/or the moon. Avoidance of problems due to radiance from the sun and moon involves the use of auxiliary sun sensors with auxiliary horizon sensors or horizon sensors with multiple or movable fields of view to avoid scanning the detected sun, or the utili-

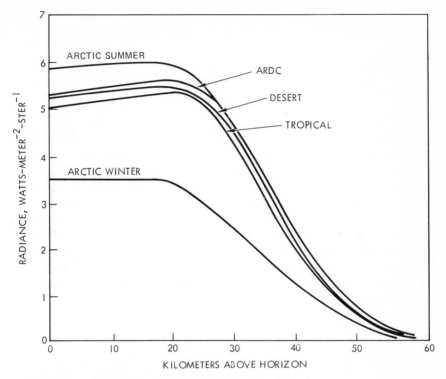

Figure 4.12-2 CO_2 horizon radiance profiles (14 to 16 μm) [2].

zation of logic to reduce sun and moon effects in scanning horizon sensors based on angular subtense sensing.

Interfaces. The primary interfaces which must be considered in the design of a horizon sensor can be grouped into several categories: optical, electrical, mechanical and thermal.

Optical interfaces often include the requirement to have the horizon sensor field of view or alignment axis boresighted with some other instrument on board the spacecraft. In addition, the relation of the horizon sensor field of view to the spacecraft structure is of importance. Alignment mirrors, surfaces, and pins usually satisfy this requirement. For some applications a window may have to be provided.

Electrical interfaces include the electrical power requirements for the earth sensor, the available power from the spacecraft, signal input and output parameters, and electromagnetic interference.

Mechanical interfaces include mounting and alignment provisions on the spacecraft, size, weight, and shape of the horizon sensor, and the available mounting space. Other important mechanical interfaces are the consideration of vibration and shock to the earth sensor which can be greatly influenced by the mounting location and spacecraft structural parameters.

The thermal interfaces include all heat transfer between the earth sensor and the spacecraft and between spacecraft components and the external environment. Thermal environment can have a great effect on earth sensor performance and reliability.

These interfaces are not independent; compromises must be made to provide the best system which satisfies mission requirements.

Detector Selection. The choice of an infrared detector for a horizon sensor depends upon spectral passband requirements, signal responsivity and noise performance, complexity resulting from cooling requirements, and of course on cost and availability.

The most sensitive detectors available for wavelengths of 14 μm and above are photosensitive detectors such as doped germanium photoconductors. These detectors, in order not be limited by internal thermal effects, must be operated at temperatures below about 20°K. Because of the reliability, cost, weight, and size implications associated with detector cooling, these detectors have not been used for horizon sensors utilized primarily for spacecraft control.

The usual choice of detector for a horizon sensor is based on thermal rather than photo effects. Examples of thermal detectors are bolometers and thermopiles.

The bolometer is constructed of a material with a temperature-dependent parameter, usually resistance. A typical bolometer consists of two elements in series with a bias source, with one of the elements shielded from and the other exposed to the infrared radiation to be sensed. Any change in the temperature of one element with respect to the other is sensed as a change in voltage distribution across the elements. Figure 4.12-3 shows a typical bolometer circuit for a resistive bolometer. Because of the difficulty in making the two elements identical, the dc voltage level from the bolometer varies with ambient temperature. For this reason it is usually utilized for scanning-type horizon sensors where the signal of importance is time modulated and the output is ac coupled to eliminate the usually slowly varying output caused by ambient temperature variations.

The thermopile consists of a number of thermocouples connected in series. These thermocouples are arranged so that incoming infrared radiation heats one junction of the thermocouples with respect to the other, causing a voltage to be generated by thermoelectric effect.

All of these detectors are usually limited in practice by preamplifier noise, resulting in the best signal-to-noise ratio for the detector with the highest responsivity. The most commonly used detectors in order of

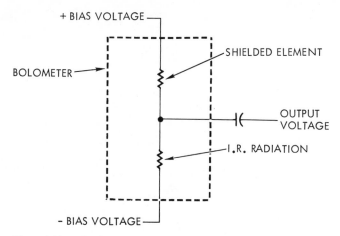

Figure 4.12-3 Typical bolometer circuit.

decreasing responsivity are thermistor bolometer, thermopile, and metal bolometer. The thermistor bolometer is often used with an immersion lens resulting in an even larger responsivity.

The detector chosen for a scanning type of horizon sensor is usually a thermistor bolometer because of the relatively large responsivity and short time constant. For specific sensors where detector stability is required, such as a radiance balancing horizon sensor, either the thermopile or metal bolometer can be chosen.

4.12.3 Active Horizon Scanners

Horizon scanners are sensors that mechanically or electronically scan a large volume of object space with a small instantaneous field of view. The majority of active horizon sensors built to date have been mechanical scanners. Scan patterns are generally either conical or linear.

Conical Scanners. In the case of an active control system application, the axis of the scan pattern is fixed relative to the vehicle control axis. Figure 4.12-4 illustrates a simple scanner with a conical scan pattern. In this example the axis of scan rotation lies in the plane containing the vehicle roll and yaw axes.

The scan pattern can be generated with either refractive or reflective optical elements which cause the instantaneous field of view to be off-set from the sensor axis. A motor rotates the optical elements, causing the instantaneous field of view to sweep out a cone centered about the sensor axis.

The IR detector output signal collected by the instantaneous field of view in an ideal situation would be a square wave with width proportional

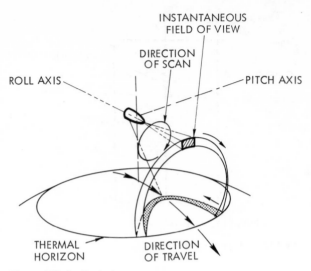

ROLL AXIS

INSTANTANEOUS
FIELD OF VIEW

DIRECTION
OF SCAN

PITCH AXIS

THERMAL
HORIZON

DIRECTION
OF TRAVEL

Figure 4.12-4 Conical scan pattern.

to earth-crossing time, or dwell time. The sensor generates an internal reference pulse which indicates when the instantaneous field of view passes through a reference plane.

If the earth signal is centered about the reference pulse, the vehicle pitch axis must be normal to the local vertical; this is the condition for zero roll angle. Roll attitude is proportional to the phase of the earth pulse relative to the reference. Pitch attitude can be obtained in a similar manner from another sensor mounted 90° away in yaw. One important element in the accuracy of measuring local vertical with a scanner is relative duty cycle at the earth pulse. For this reason conical scanners are generally used in relatively low-altitude circular-orbit spacecraft applications.

Linear Scanners. A second method of implementing a scanning earth sensor is to generate a linear scan pattern. Instead of causing the instantaneous field of view to rotate continuously in a single direction, as in the conical scanner, it is made to oscillate back and forth. The resulting fan-shaped field of view scans across the earth in a straight line. Mechanizations of this type of sensor usually have a reflective optical element and the rotor portion of a motor suspended on flexure pivots or the equivalent. The angular position of the instantaneous field of view is measured by a digital encoder or analog pickoff, and this measured angle is compared with the processed signal from the IR detector to permit computation of the local vertical or horizon angle. Figure 4.12-5 illustrates the operation of such a sensor. These instruments have the advantage of avoiding motor bearings, and the ability to scan in both directions permits an increase in accuracy

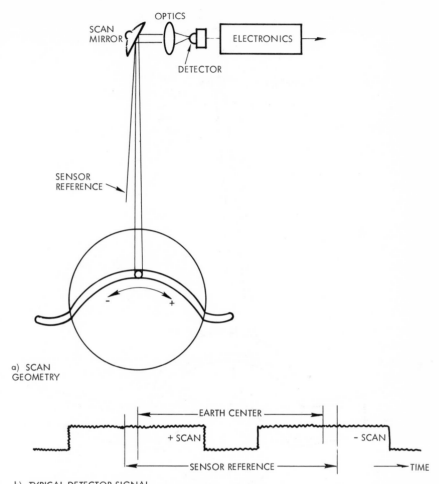

a) SCAN
GEOMETRY

b) TYPICAL DETECTOR SIGNAL

Figure 4.12-5 Horizon scanner—linear scan.

by cancellation of electronic delays. Because of the limits on angular position caused by the flexure mechanism, these instruments are generally used only at high altitudes (i.e., synchronous) where the subtended angle of the earth is relatively small, or in circular orbits, where several such instruments may be used, each scanning across an earth edge without reaching the other side.

4.12.4 Passive Horizon Scanners

In the case of spin-stabilized spacecraft, the rotation of the vehicle itself may be used to provide the scanning motion. This concept has been used for a

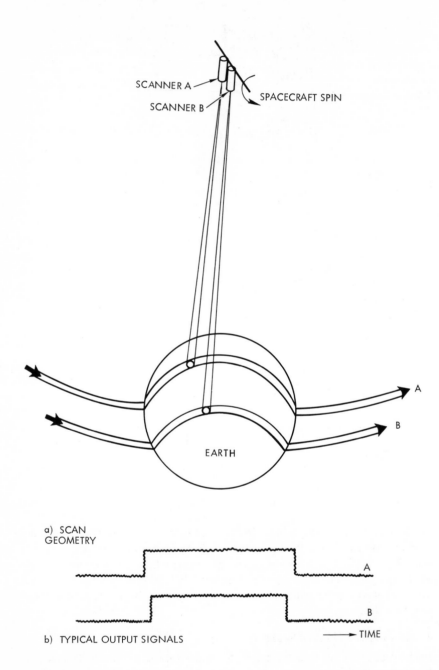

a) SCAN
GEOMETRY

b) TYPICAL OUTPUT SIGNALS

Figure 4.12-6   Horizon scanning using two passive scanners with spinning spacecraft.

large number of despun communications satellites, in which the spin axis of the spacecraft is oriented such that the spacecraft equatorial plane intersects the earth. The earth sensor in this case simply consists of a fixed telescope, a detector, and relatively simple electronics. As the field of view crosses the earth, a pulse is generated. The center of this pulse, after correcting for delays, represents the time of crossing the center of the earth. The usual approach is to use two such sensors to provide two-axis attitude information. The two fields of view are offset in opposite directions from the earth center as shown in Figure 4.12-6.

Relative width of the pulses from the two sensors is a measure of spin axis inclination, while the center of the pulses provides a redundant indication of crossing of the center of the earth.

4.12.5 Edge-Tracking Sensors

Edge tracking is achieved by driving a detector field of view to a particular location relative to the horizon. The field of view is dithered across the horizon by a servomechanism that derives its error signal from the detector waveform (bolometer detector) or held at grazing incidence to the horizon (thermopile detectors).

An example of the dithering field of view edge tracker employs a horizon point-tracking scheme for tracking the horizon at four points around the horizon circle as illustrated in Figure 4.12-7. Attitude information is derived from the simple geometric relationships between the four lines of sight to the horizon (i.e., the angle between a reference in the sensor and the horizon). Note that only three sight lines are required. The sensor

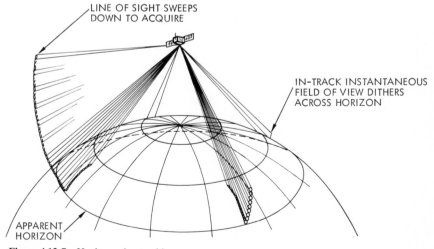

Figure 4.12-7 Horizon edge tracking geometry.

produces four sight lines; three lines are used in primary operation and the fourth is redundant. In the signal-processing design, a threshold on a fixed percentage of the peak detector output signal is used to generate a rectangular waveform. The oscillating field of view is controlled so that the average value of this waveform is zero. Thus the dither is centered on a threshold point, with a fixed relationship to the horizon.

Because of the tracking capability of the edge tracker, it is continuously observing the horizon and wastes little information time observing space or earth. It is thus capable of relatively wide system bandwidth (often useful for control system applications) and is well suited to spacecraft missions which have eccentric orbits with large variations in spacecraft altitude.

4.12.6 Radiance-Balancing Sensors

Radiance-balancing sensors employ image-plane rather than object-plane scanning of a detector field of view. The radiance balancing technique assumes that when a balance in radiance is achieved, the sensor optical axis is pointing at the center of the illuminated or self-emitting target.

One form of this sensor type consists of four individual detectors with large fields of view that cover much of the earth. The outputs of the two detectors viewing opposite horizons are differenced to provide a signal proportional to earth offset angle. A second form drives a small detector field of view to grazing incidence on the horizon. A third version utilizes an electronically scanned array of sensors and has no moving parts. An example of the first form is illustrated in Figure 4.12-8.

Radiance-balancing sensors are usually designed to utilize the thermal discontinuity at the edge of an optically formed earth image without modulating the incoming energy. Consequently, the detectors employed may be thought of as strictly temperature sensors. Their output is a function of the temperature distribution of the image which they are viewing. This temperature distribution is influenced not only by the flux density collected and focused by the optics but also by radiation and conduction from sources

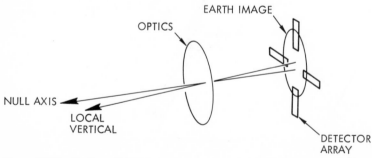

Figure 4.12-8 Radiance balance sensor.

within the sensor itself. The detector cannot differentiate between temperature differentials caused by the image and those caused by the local environment of the sensor. Thus the primary design task is to assure that thermal gradients caused by the sensor's local environment are of a magnitude sufficiently less than those caused by the earth's image so that the required accuracy can be achieved.

The application of a radiance balance sensor is limited by the constraints of image plane scanning unique to the particular implementation. For example, in the concept illustrated in Figure 4.12-8, the complete earth image is formed on the detector array. This implementation is limited to mission altitudes consistent with achievable wide field of view IR optics.

4.12.7 Performance

Performance, particularly accuracy, is primarily limited by the variations and non-uniformities of the earth IR emission.

The effect of internal noise and horizon radiance variation on earth sensor performance can be illustrated by considering a typical scan-through earth sensor. Since this type of sensor is designed so that the field of view of the sensor is small compared to the apparent earth size, the peak-to-peak radiance signal, neglecting second-order effects, is

$$S = A\Omega \int_0^\infty N(\lambda)\, T(\lambda)\, R(\lambda)\, d\lambda, \qquad (4.12\text{--}1)$$

where S is the peak-to-peak signal in V, Ω is the sensor field of view in sr, A is the sensor effective aperture area in cm^2, $N(\lambda)$ is the spectral radiance of the earth and its atmosphere in $W\text{-}cm^{-2}\text{-}sr^{-1}\text{-}\mu m^{-1}$, $T(\lambda)$ is the spectral transmission of the sensor optics, $R(\lambda)$ is the spectral responsivity of the sensor detector in $V\text{-}W^{-1}$ and λ is the wavelength in μm.

The noise for a typical sensor is determined by both the detector and preamplifier noise contributions. Typical sensor designs result in worst case signal-to-noise ratios on the order of $50:1$.

The simplest radiance signal processing technique is the use of a fixed threshold level for determination of a horizon crossing. Figure 4.12-9 shows the voltage waveform and indicates the effect of noise on the waveform. Under conditions where multiple threshold crossings cannot occur (this is the usual condition which can be guaranteed by design techniques, such as the utilization of hysteresis in the threshold circuit) the noise caused error is given by

$$\alpha = \frac{1}{S/N \cdot s'}, \qquad (4.12\text{--}2)$$

where α is the rms error in detected horizon location in degrees, S/N is the signal to rms noise ratio in the threshold circuit, and s' is the normalized

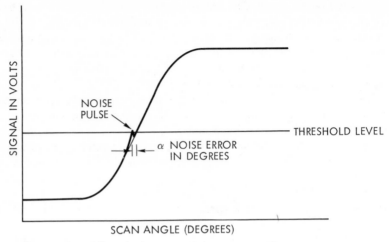

SCAN ANGLE (DEGREES)

Figure 4.12-9   Effect of noise on sensor output.

signal slope at the threshold level in volts per degree divided by the peak signal in volts.

The effect of horizon radiance variation on the detected horizon location is shown in Figure 4.12-10. This error can be considerable for large variations in radiance level. The amplitude of this error as well as that for

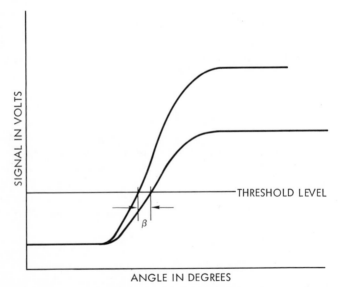

ANGLE IN DEGREES

Figure 4.12-10   Effect of horizon radiance variation on sensor output.

noise caused errors depends upon specific sensor design parameters such as the sensor processing electronics bandwidth and the specified spectral passband. The errors due to horizon radiance variations can be reduced by more sophisticated electronic processing techniques. Overall accuracies in the range of $\pm 0.02°$, 3σ, are currently achievable at synchronous and higher altitudes and in the range of $\pm 0.05°$ to $0.1°$, 3σ, at lower altitudes.

References

1. Duncan, J., W. Wolfe, G. Oppel, and J. Burn, "IRIA State-of-the-Art Report, Infrared Horizon Sensors," Report 2389-80-T, April 1965.
2. Hanel, R., W. Bandeen, and B. Conrath, "The Infrared Horizon on the Planet Earth," Report X-650-62-164, August 1962.
3. Kirk, R. J., B. F. Watson, E. M. Brooks, and R. O. Carpenter, "Infrared Horizon Definition," NASA CR-722, April 1967.
4. Thomas N. L., "Earth's Radiance for CO_2 Absorption Band Horizon Sensor," LMSC/687266, 9 April 1969.
5. Wark, D., J. Alishouse, and G. Yamamoto, "Calculation of the Earth's Spectral Radiance for Large Zenith Angles," Meteorological Satellite Laboratory Report No. 21, October 1963.

Elias J. Amdur

4.13 HUMIDITY SENSORS

Humidity measurement is important because atmospheric moisture is a major factor in the operation of the atmospheric heat engine and affects the properties of materials.

4.13.1 Humidity Reporting

A factor which differentiates industrial practice from geophysical practice is the different humidity reporting forms. Humidity is a chemical or material variable that can be reported in several distinct forms, some of which are not interconvertible without the use of outside information. Humidity sensors yield information in one of these forms.

At least nine humidity reporting forms are in use. They can be divided into six classes, which can be reclassified into two groups—those which are quantitative expressions in which the water weight or volume enters and those which quantitatively report observed data which can be related to one of the first group through charts or tables:

I. Expressions based on water substance
    A. vapor density
    B. relative humidity
    C. humidity to gas ratios
       1. specific humidity (mixing ratio)
       2. weight fraction-wet basis
       3. mole fraction
II. Phenomenological forms
    A. vapor pressure
    B. dew point
    C. psychrometric forms
       1. wet and dry-bulb readings
       2. wet-bulb depression and dry-bulb reading

Vapor Density. Chemists express the concentration of solutions in terms of the weight of a material in a unit volume of the solution. Alternative forms are (1) the weight per unit volume of solute*, (2) the molarity, which is the number of gram molecular weights of solute in a liter of volume of solution, and (3) the molality, which is the number of gram molecular weights of solute in a kilogram of solvent. In gaseous mixtures, the only one of these forms currently used is the simple ratio of weight of water per unit of wet volume. This ratio is called the vapor density. The vapor density of the water in a space is an "absolute humidity" reporting form. It is useful because rates of evaporation, diffusion and reaction are related to differences in concentration of the constituents of a system.

Vapor density is rarely used, except in the gas industry where moisture

---

*The gravimetric determination of water in a gas yields the data in the form, "weight of water per volume of dry gas" at the temperature of the experiment. It is the only instance of the use of this particular form. At low temperatures the value converges to the vapor density. At temperatures above about $-20°C$ a correction must be made for the contribution of water vapor to the volume of the mixture. Precise computation of this correction is involved, but approximate correction by means of the gas laws is adequate for most purposes.

content is expressed as pounds per million cubic feet of gas, because the numerical value for a sample varies with temperature and pressure in "open" systems* ordinarily encountered.

Instrumentation for the absolute humidity forms is not as readily available as for relative humidity. Infrared hygrometry and dielectric constant methods yield data primarily as vapor density.

Relative Humidity. The relative humidity (RH) form is widely used in comfort air-conditioning and the storage and processing of materials. The moisture content of materials (more exactly, the activity of the contained moisture) is a function of RH and is almost independent of temperature over moderate ranges of temperature. For example, control of the swelling and contraction of wooden or skin-covered musical instruments, the moisture content of flour or tobacco, and the registration of multicolor printing on paper require consideration of relative humidity.

The RH form is not conservative with respect to either temperature or pressure, in closed or open systems. Since relative humidity may be defined as the ratio of the moisture content of a space to the saturation moisture content†, it is affected by temperature which sets the reference saturation moisture content. Changes in the pressure of a given sample affect the moisture concentration.

Relative humidity is not readily used in computational procedures but is easily handled as an analog and does not complicate instrumentation as long as this form is desired as the final display result. Computations involving the changes in relative humidity with variations in temperature and pressure require conversion to another form.

Because relative humidity affects the properties of materials, property changes form a reproducible analog of this reporting form which is used in most humidity instruments.

Humidity-to-Gas Ratios. At least three methods of expressing humidity as a ratio are in use, but only one is sufficiently useful to be retained. The ratio of the weight of water vapor to a unit weight of coexisting dry gas is widely used in engineering and meteorology under the designation "specific

---

*An "open" system is connected to the ambient, like an unstoppered bottle. A closed system, like a securely stoppered bottle, is isolated and the weight of water in it is fixed and independent of temperature. A form which yields numerical values which do not vary with temperature, pressure, or other influences is known as a "conservative" form with respect to those influences.

†The definition preferred by the author is the weight ratio of the water vapor in a given volume to that which would exist in the same volume at saturation, which is the ratio of vapor densities. Wexler [11] uses the ratio of mole fractions. While there are no theoretical objections to its use, this form causes computational complexities. The ratio of vapor pressures is also common. This definition is objectionable on theoretical grounds except where water vapor is the only gas present. The difference between the relative humidity values calculated according to the two acceptable definitions is nearly 0.5% RH at 25°C and 50% RH. It is negligible below 0°C.

humidity" or "humidity mixing ratio." Other ratios are the percent wet basis, and the mole fraction.*

Specific humidity, humidity mixing ratio $= \dfrac{\text{weight of water vapor}}{\text{weight of dry gas}}$

Percent wet basis $= \dfrac{\text{weight of water vapor}}{\text{total weight of wet gas}} \times 100$

Mole fraction $= \dfrac{\text{moles water vapor}}{\text{total moles wet gas}}$

The dry-basis ratio is used because it is a completely conservative form and lends itself readily to computations where the dry-gas analysis does not change. Wet-basis forms are equally conservative but difficult to use in computations. None of the ratio forms can be used when water vapor is the only gas in the system. In such cases the relative humidity, vapor pressure, or dewpoint forms, must be used.

The humidity mixing ratio is preferred in meteorology. Unfortunately, no widely available sensing system exists for this form although some methods are possible. By conducting measurements on a gas sample with a relative-humidity, vapor-pressure, or dewpoint sensor in a system held at constant gas temperature and pressure, one can obtain a signal which can be interpreted as a humidity mixing ratio. The system can be calibrated to read out directly in this form.†

Vapor Pressure. When gas-free water is introduced into an evacuated thermostated space, the pressure rises to a value which depends upon the temperature of the water. This observed pressure is called the "vapor pressure" of the water. Both thermodynamic considerations and actual analytical determinations can be used to relate the vapor pressure to the vapor density in a space.

The vapor pressure of water is an absolute humidity form that is preferred to vapor density because it is more conservative. A numerical value for vapor pressure varies with total pressure in open systems but is not affected by temperature. Since most humidity systems are open systems with practically constant total pressure, the vapor pressure is essentially a constant value for a particular humid gas sample.

The vapor pressure is simple to handle in computational processes and to convert to several other reporting forms. Where water vapor is the

---

*Weight and volume ratios are also expressed as 'parts per million', either on a dry or wet basis. They are readily derived from the above listed forms by the use of conversion factors and, thus, are not different reporting forms.

†The form "percent of saturation" is used to describe the ratio of the mixing ratio at a particular condition to the mixing ratio at saturation at the same temperature. It is a derived form and is not the same as the relative humidity although it approximates that form at low temperature. No instrumentation has been devised to measure humidity and to report it in this form.

only gas in a system, the vapor pressure can be directly measured with an absolute pressure gauge. The partial pressure of water vapor in a system containing other gases is not easily measured.

Dewpoint. The water vapor pressure in a system which consists of a vapor space and a plane water surface in isothermal equilibrium is known as the "saturated" vapor pressure of water at that temperature. The saturation vapor pressure of water as a function of temperature is very accurately known, so that it is theoretically possible to determine the vapor pressure in a space by determining the temperature of a plane water surface which is in equilibrium with it.

Dewpoint determination is an experiment designed to approximate the saturation temperature of the vapor in a space. A surface is cooled until dew forms on it and the temperatures at which the dew just forms and just evaporates from the surface are determined. The mean of these two temperatures is regarded as the dewpoint. Other procedures involving cooling of the gas by expansion until a cloud of condensate forms, and methods utilizing hygroscopic salts are in use. The dewpoint measurement yields a useful approximation of the saturation temperature, is an alternative method of expressing absolute humidity and is useful in itself because it indicates the temperature at which moisture condenses or clouds form in a space. This type of measurement is independent of the original temperature of the gas being investigated. Since compression or expansion of a gas affects its vapor pressure, the dewpoint is also affected by total pressure changes. The dewpoint and vapor pressure are therefore equally conservative.

There are a variety of dewpoint sensing systems including the self-regulating lithium-chloride electrical dewpoint hygrometer [8]. The dewpoint form is second to relative humidity in popularity.

Wet- And Dry-Bulb Psychrometry. The wet- and dry-bulb sling psychrometer is a widely used humidity instrument. The simplified basic equation for this instrument shows that it yields absolute humidity:

$$(e_w - e_a) = C(T_a - T_w), \tag{4.13-1}$$

where e_w is the vapor pressure of the water on the wick, e_a is the desired ambient vapor pressure, T_a is the dry-bulb temperature, and T_w is the wet-bulb temperature. The relative humidity can be obtained by dividing e_a by e_w. In practice, tables, charts, or accepted empirical equations of RH versus the wet- and dry-bulb temperatures are used. Psychrometric charts that present adiabatic saturation lines rather than wet-bulb temperature lines are frequently used although the wet bulb does not operate under conditions of adiabatic saturation.*

*Adiabatic saturation hygrometers have been developed [6, 10].

The constant C is the ratio of the sensible heat transfer coefficient of the bulb used to the vapor transfer coefficient, converted to its equivalent value in terms of heat of evaporation. Since this ratio is not a constant unless the air velocity exceeds 4.5 m/s, a motorized blower drawing* the air sample over the bulbs yields more satisfactory data than are obtained with a sling instrument.

In practical instruments of this type the sensible heat of the water brought to the bulb, stem conduction errors, radiation errors, and other factors such as incomplete wetting of the wick or impure water result in high values for the observed wet-bulb temperature.

The true form for direct reporting of psychrometric data is wet- and dry-bulb readings since no fully accepted conversion tables or formulas exist.

Wet-Bulb Depression. Wet- and dry-bulb methods utilizing separate channels for each bulb are most accurate when the difference between the readings is great enough to minimize the calibration errors of the thermometers and is not great enough to cause large errors of the type which cause high wet-bulb readings. The best readings are obtained between 25 and 80% RH. The requirement for testing at high RH values has brought methods for direct reporting of the wet- and dry-bulb difference into prominence.

Instrumentation for observation of the wet-bulb difference or depression is the same as that used for ordinary psychrometric readings, except that two matched linear calibration resistance thermometers are used, one dressed in the wet-bulb wick and the other exposed directly to the air. The two resistance thermometers form legs of an electrical bridge network in a manner that yields an emf proportional to the temperature difference between the two thermometers. Separate instrumentation is used to determine the dry-bulb reading.

At high RH the relationship between the wet-bulb depression and the relative humidity is insensitive to the dry-bulb temperature [16]. This permits the use of crude dry-bulb temperature instrumentation, and direct conversion between the wet-bulb depression and the relative humidity.

The accuracy of the interconversion of wet-bulb depression and relative humidity depends on the accuracy of existing tables which retain the uncertainty of psychrometric formulas. Thus the wet-bulb depression retains in itself the characteristics of a reporting form.

4.13.2 Humidity Sensing Methods for Geoscience

Humidity measurements of interest in geoscience are made in three types of environments: surface or near surface conditions in the natural outdoors

---

*A fan or blower heats the air which passes through it. The use of a fan to blow rather than to draw air over the assembly is equivalent to an increase in dry-bulb temperature.

setting, simulated surface environments produced in the laboratory, and sonde or upper-air conditions. The requirements for sensing methods differ for these environments and certain methods suitable for laboratory simulations are not practical in the field.

The collection of data for climatology does not require rapid-response instrumentation. This permits the consideration of a wider variety of instrumentation than is possible for micrometeorological work, where a time response of a fraction of a second is desirable. Most available humidity instrumentation has a time constant of a fraction of a minute to several minutes.

The Psychrometer. The most widely used humidity instrument for surface meteorological and climatological work is the manual wet- and dry-bulb psychrometer. The measurement cycle requires five or six minutes.

Remote reporting automatic psychrometric sets have been used for industrial applications such as high-temperature ovens. These devices have fan aspiration, automatic water feed, and resistance thermometer or thermocouple temperature reporting. Their use is limited to cases where maintenance is readily available for renewal of the distilled water supply and the changing of wicks, and where the wick and water supply will not be subjected to freezing temperatures.

Mechanical RH Sensors. Mechanical RH hygrometers have been satisfactory for indoor work. One sensor uses a human hair bundle with a spring and a position transducer to form a relative humidity transducer. Mechanical hygrometer elements do not operate well over the wide range of relative humidity which prevails outdoors.

Electrical Conductance and Capacitance RH Sensors. The only fairly successful RH sensors have been those using electrical conductivity sensors [2]. These consist of a very thin layer of sensitive material on a plastic base fitted with noble-metal electrodes. Flat plastic plates with interlocking grid electrodes and cylinders with bifilar windings have been used successfully. As shown in Figure 4.13-1 these sensors are available in a number of types each of which covers a relatively narrow range of relative humidity.

The exact range of each sensor, roughly defined as the relative humidity span corresponding to extreme values of 10,000 Ω and 4 MΩ, is a function of the ratio of lithium chloride to polyvinyl alcohol in each. Because of the extremely great change in resistance with relative humidity and the absence of hysteresis, they are very useful in fixed-humidity rooms but not (individually) for the recording of RH data over a wide range of values.

Figure 4.13-2 shows how a composite sensor is constructed of a series of individual element sensors to give a straight line characteristic of relative humidity versus conductivity. This characteristic is displaced upward or downward by temperature variations (3.6% RH per 10°C). The accuracy of composite sensors is $\pm 1.5\%$ RH, and no hysteresis can be detected.

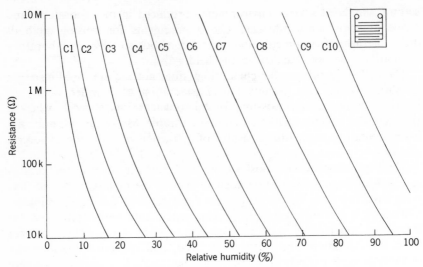

Figure 4.13-1 25°C characteristic of 10 lithium chloride-PVA relative humidity sensors.

Because of the use of several sensors instead of one, the failure or drift of one element is not catastrophic.

These composite sensors are fairly rapid in response but subject to deterioration or temporary offset at humidity ambients near saturation. A heavy protective coating is applied to composite sensors destined for outdoor exposure. The resulting sensor is slower in response but more

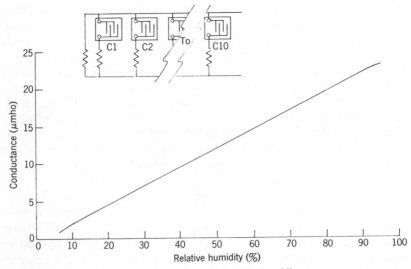

Figure 4.13-2 25°C characteristic of composite relative humidity sensor.

stable. An electrical conductivity sensor based on another principle is one prepared from surface sulfonated polystyrene screened with a silver electrode grid. This sensor shows hysteresis of several percent RH, has a wider initial tolerance, and a more rapid drift of calibration than a protective coated composite sensor of lithium chloride elements. Because of its lower initial cost, it finds use largely in nonmeteorological applications.

Electrical capacitance humidity sensors appear to offer theoretical advantages in their resistance to contamination-induced drift and in properties of lesser importance in outdoor humidity sensing, but the relatively low signal-to-base-capacitance ratio and certain design problems have retarded development.

Protection of electrical humidity sensors from contact with water or particulate contaminants is necessary to minimize errors and calibration drift with time. Shelters with white exteriors and insulation are used to minimize temperature deviations at the sensor.

Cold Mirror Dewpoint Hygrometers. Cold mirror dewpoint hygrometers have been developed which duplicate the manual dewpoint measuring process on a continuous basis. A rod conductor is cooled continuously by refrigeration or by Peltier cooling so that a mirror at its end is brought below the dew or frost point.

The thickness of dew is sensed by optical means and the signal used by an electronic control system to control heat which is added to the rod conductor so that the dew can be reevaporated. At equilibrium a thin layer of water droplets is stabilized on the mirror and the temperature of the mirror is determined by a thermocouple, thermistor, or resistance thermometer buried in the rod near the mirror. Dewpoint instruments of this type cannot distinguish between supercooled water and ice, leading to some ambiguity in results below 0°C.

There are several variations of the classical automatic dewpoint instrument. In one version [9] the cold surface is a nonconductor with a grid electrode system. As the dewpoint is approached moisture is absorbed by ionic impurities on the nonconductor surface and a rapid change in conductivity is used to control the temperature of the cold surface. This type of detector is sensitive to the concentration and hygroscopicity of adventitious ions. The presence of varying amounts and types of ionic contamination leads to a variation in the deviation from the true dewpoints.

Electrical Self-Heating Dewpoint Hygrometer. This hygrometer consists of an inert cloth layer over an insulating tube wound with bifilar electrodes of noble metal. This element is treated with a dilute lithium chloride solution which is dried to form an evenly distributed electrolyte. A 20 to 30 V alternating current is applied to the bifilar winding. The temperature of the interior of the base tube is measured with an inserted thermometer. If the ambient RH is less than 11 % the thermometer reads the ambient dry-bulb temperature. If the RH exceeds 11% the thermometer indicates a higher temperature that is a function of the ambient dewpoint [8]. When used in applications where slow achievement of equilibrium is satisfactory, dewpoints as low $-45°C$ can be measured. The upper limit, governed by the choice of materials, generally exceeds $+75°C$ dewpoint. Because of the presence of practically dry salt there is no signal ambiguity due to the difficulty in distinguishing between water and ice.

Self-heating hygrometers have been widely applied in airport applications. The sheltered sensor is located near the runway and the telemetered information is utilized in aircraft load computations.

Simulated Surface Environments. In the laboratory a wider range of sensors can be used since it is possible to limit accidental conditions, tolerate shorter exposures, and provide more maintenance. A simple bimorphic sensor made of a metal strip and paper may suffice if it is frequently calibrated. Sophisticated instruments include infrared, microwave resonance, and pneumatic bridge hygrometers [7, 12, 14]. Another approach is to utilize methods which produce atmospheres precisely and directly through physical procedures whereby measurement of the humidity variable is bypassed. Such methods are the two pressure, two temperature and split stream humidity atmosphere procedures [1, 14] which are often used for calibration of hygrometers and can provide precisely known conditions within a test chamber.

Sonde Hygrometers. Upper-air measurements can be made by means of instruments carried in a free balloon, dropped with a parachute (dropsonde), or mounted in an aircraft. In all of these cases the instrument passes rapidly through a series of air samples. The air frequently contains water in the form of rain, fine mist, or snow. Criteria for a satisfactory sonde humidity system include a rapid response over the entire range of atmospheric temperatures down to stratospheric lows and a rapid recovery from wetting or a suitable protection system to prevent contact with liquid water.

References

1. Amdur, E. J., and R. W. White, "Two Pressure Relative Humidity Standards," in Wexler Ref. [13] Vol. 3, pp. 445–454.
2. Dunmore, F. W., "An Electrical Hygrometer and Its Application to Radio Meteorography," *J. Res. N.B.S.* **20**, 723, 1938.
3. Dunmore, F. W., "An Improved Electrical Hygrometer," *J. Res. NBS* **23**, 701, 1939.
4. Dunmore, F. W., "Humidity Measuring," U.S. Patent 2,295,570, 1942.
5. EG & G Cambridge Systems, Bulletin SFC-4 (100K).
6. Greenspan, L., and A. Wexler, "An Adiabatic Saturation Psychrometer," *J. Res. NBS.,* **72 C**, No. 1, 33–47, 1968.

7. Lück, W., *Feuchtigkeit*, R. Oldenbourgh, München and Wien, 1964.
8. Nelson, D. E., and E. J. Amdur, *The Mode of Operation of Saturation Temperature Hygrometers Based on Electrical Detection of a Salt-solution Phase Transition* in Wexler Ref. [13] Vol. 2, pp. 617–626.
9. Singer VAP-AIR Division, Action Reports Vol. 3, No. 3.
10. Wentzel, J. D., "An Instrument for Measurement of the Humidity of Air," *ASHRAE J.* **66,** 67, 1961.
11. Wexler, A., "Calibration of Humidity Measuring Instruments at the National Bureau of Standards," *ISA Trans.* **7,** No. 4, 356–362, 1968.
12. Wexler, A., "Electric Hygrometers," N.B.S. Circular 586, 1957.
13. Wexler, A., Ed., *Humidity and Moisture Measurement in Science and Industry*, Vol. 1–4, Reinhold Publishing Co., New York, 1965.
14. Wexler, A., and W. G. Bombacher, "Methods of Measuring Humidity and Testing Hygrometers," N.B.S. Circular 512, 1951.
15. Wexler, A., Personal communication.
16. Zimmerman, O. T., and I. Lavine, *Psychrometric Tables and Charts*, 2nd ed. Industrial Research Service, Dover, New Hampshire, 1964.

George R. Carignan

4.14 ION AND ELECTRON SENSORS

4.14.1 The Ionosphere

At altitudes greater than about 60 km above the earth's surface, a significant fraction of the atmosphere exists in the charged state, principally positive ions and electrons. The region of the earth's atmosphere between approximately 60 and 1000 km has been termed the ionosphere. The ionosphere is produced by the interaction of solar radiation and energetic particles with the neutral atmosphere. Ionized particles ultimately recombine as the result of encounters between ions and electrons. During their lifetime charged particles are acted on by forces quite different from those acting on neutral particles. Because of the nature of the source and loss mechanisms and the constraints of electric and magnetic fields, charged particles are distributed in their own characteristic way.

Early experimental study of the ionosphere involved remote radio techniques. Its existence was proved experimentally in 1925 [1] and a year later the ionosonde was devised [5]. In situ measurements of the ionosphere began in the late 1940s when rockets became available for scientific research [15].

4.14.2   Measurement Techniques

Virtually all instruments that are used for in situ measurement of ions and electrons in the ionosphere take advantage of the intrinsic charges of the particles. Either electric or magnetic fields, generated by the instrument, are used to deflect the particles along trajectories that enable measurement of the desired parameter.

Langmuir Probes and Ion Traps. The current-voltage characteristic of an electrode immersed in a plasma can be used to determine the number density of the ions and electrons and their energy distributions [12]. When a floating body is immersed in a plasma, it assumes a potential that causes the net current to it to be zero. Since electrons at the same temperature as ions have a much greater thermal velocity due to their lighter mass, a probe initially collects more negative than positive charge. The probe ultimately reaches a negative potential just large enough to cause the net current to it to be zero. This potential is usually referred to as the floating potential.

If a voltage is applied to the probe, the net current to it is no longer zero. The volt-ampere curve can conveniently be divided into three distinct regions (see Figure 4.14-1): the ion saturation region (AB), which corresponds to a probe highly negative with respect to the plasma, the electron saturation region (CD), which corresponds to a probe positive with respect to the plasma, and the intermediate electron retarding region (BC). In the region labeled AB, the probe potential is sufficiently negative to repel essentially all electrons, and the current reaching the probe is a function of the ion parameters and the velocity and orientation of the probe with respect to the plasma. Equations relating the measured current to the ion density for a moving probe have been derived [10].

In the electron saturation region the current reaching a stationary cylindrical probe is

$$I_e = AN_e e \left(\frac{kT_e}{2\pi m_e} \right)^{1/2} \frac{2}{\pi^{1/2}} \left(1 + \frac{eV}{kT_e} \right)^{1/2}, \qquad (4.14-1)$$

where A is the area of the probe, N_e is the electron density, e is the unit charge, k is Boltzmann's constant, T_e is the electron temperature, m_e is the electron mass, and V is the probe to plasma potential.

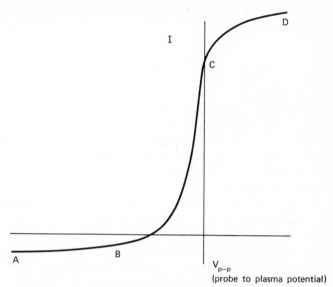

Figure 4.14-1 Langmuir probe volt-ampere curve showing the three characteristic regions: AB, ion saturation; BC, electron retardation; and CD, electron saturation.

Note that V is the voltage of the probe with respect to the plasma. The plasma potential is determined by locating the inflection point of the current characteristic, and the electron density can be determined directly from the known and measured parameters.

Between the extremes of ion and electron saturation lies a region where the electron current predominates and is an exponential function of the voltage as given by

$$I_e = AN_e e \left(\frac{kT_e}{2\pi m_e} \right)^{1/2} \exp(eV/kT_e). \qquad (4.14-2)$$

The ion current is subtracted from the total current measured giving I_e, from which T_e can be explicitly determined. An extrapolation of the ion saturation region to the point where I_e is to be determined is most often used to determine the ion current to be subtracted.

Impedance Probes. Electron densities were first measured in the ionosphere by the impedance probe in 1957 [9]. If a pair of electrodes are immersed in an ionized medium and a suitable radio-frequency signal is applied between them, then the rf current flowing between the electrodes is comprised of a constant displacement current due to the in vacuo electrode to electrode capacitance plus an opposite current proportional to the electron density in the local plasma [11]. The dielectric constant (K) of an ionized medium is given by the simplified Appleton-Hartree equation

$$K = 1 - \frac{4\pi N_e^2}{m_e \omega^2}, \qquad (4.14\text{–}3)$$

where ω is the probing frequency.

This simplified form of the equation is valid for only very specialized conditions where the probing frequency is high compared to the electron collision frequency and the gyro frequency. The effects of the sheath about the probe also complicate the implementation of the measurement.

Resonance Probes. If a small radio-frequency voltage is applied in series with a Langmuir probe circuit, the probe current is increased in all regions of the probe characteristic. This is a result of rectification through the nonlinearity of the plasma. For frequencies greatly below the plasma frequency, the increase in current is independent of frequency and is dependent only on the electron temperature. At frequencies in the region of the plasma frequency a large increase in current is observed. In early interpretations the frequency of the maximum increase was taken to be the plasma frequency, which led directly to electron density through the equation

$$\omega_n = e\left(\frac{N_e}{\varepsilon_0 m} \right)^{1/2}, \qquad (4.14\text{–}4)$$

where ω_n is the plasma frequency and ε_0 is the free space permittivity.

Subsequent work showed that the frequency corresponding to the peak in current was a lower one than the plasma frequency, a circumstance resulting from the sheath about the probe [7]. The following equation has been derived [8] to relate the measured resonance frequency ω_r and the plasma frequency as a function of the probe radius R and the sheath thickness S:

$$\omega_r = \frac{\omega_N}{\cdot (1 + R/S)^{1/2}}. \qquad (4.14\text{–}5)$$

Thus, if the sheath thickness can be determined, the ambient electron density can be deduced from the frequency of resonance.

Mass Spectrometers. Many techniques are available to separate the constituents of a mixed plasma. Three of these (the radio-frequency mass spectrometer [3], the magnetic mass spectrometer, [13], and the quadrupole mass filter [14]) have been extensively developed for space flight use. Particles are selected according to their charge-to-mass ratio either by differential deflection in a magnetic field or through stability (or instability) in an alternating electric field. Particles of a selected charge to mass ratio are directed into a detector system for measurement. By varying the parameters of the separation field the mass range of interest is scanned, providing a measure of the relative abundance of the plasma constituents.

4.14.3 Sensors

The common denominator of sensors for ions and electrons is the requirement to present to the plasma an electrode of controlled potential in a region that, to the extent possible, is unperturbed by the presence of the instrument platform or spacecraft. Several classes of probes often utilize cylindrical protrusions from the instrument platform which serve as the sensor electrodes. These include the resonance and impedance probes and the Langmuir probe. Ion traps and ion mass spectrometers commonly are flush mounted to a spacecraft skin with a small orifice for introducing particles into the instrument. Behind the orifice suitable electrodes perform the particle analysis.

Langmuir Probe. The essential functions of the Langmuir probe measurement of electron temperature and density are the application of a known sweep voltage between a conducting probe and a reference electrode, and

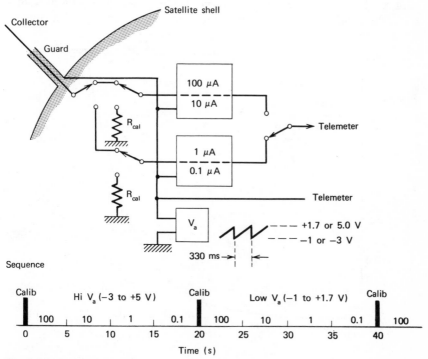

Figure 4.14-2 Block diagram of a Langmuir probe. Time line shows the sequence of operations [4].

the measurement of the resultant current to the probe. A typical arrangement is shown in Figure 4.14-2.

In the illustration, a cylindrical collector is mounted radially to a satellite shell. A sweep voltage V_a is applied between the shell and a guard electrode which is used to remove the collector from the influence of the satellite skin. The same voltage is applied to the collector through linear current detectors so that the current to the collector as a function of the applied voltage is measured. The value of V_a and the output of one of the sequentially selected current detectors are telemetered and constitute the measurement. To cover the dynamic range of densities and temperatures encountered in a typical application, two ranges of V_a and four detector sensitivities are employed, time-shared, as illustrated in the sequence diagram. Also, an in-flight current calibration is periodically performed.

In Figure 4.14-3, an actual volt-ampere curve measured on a satellite [4] is shown. The electron temperature as determined from this curve is 1900°K and the density, although not shown, is about 5×10^4 electrons/cm^3. More advanced versions of the Langmuir probe instrument perform an in-flight analysis of the volt-ampere curve that permits automatic selection of the proper current and voltage ranges and reduces the telemetry bandwidth required for a given spatial resolution. Table 4.14-1 gives typical values for engineering parameters and measurement capabilities of the instrument.

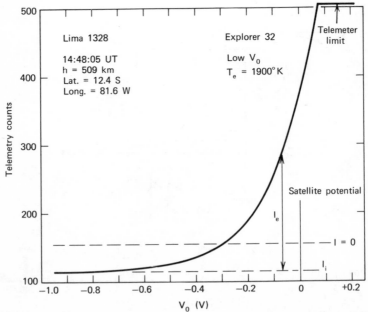

Figure 4.14-3 Volt-ampere curve of a typical Langmuir probe. A smooth exponential has been drawn through the raw data points [4].

Table 4.14-1. Langmuir Probe

Parameters measured	Electron temperature
	Electron density
Dynamic range	Temperature 300 to 10,000°K
	Density 10^2 to 10^7 electrons/cm^3
Accuracy	Temperature $\pm 5\%$
	Density $\pm 10\%$
Weight	0.9 kg
Power	2 W

Ion Trap. The ion trap or retarding potential analyzer is comprised of a sensor head and an electronic system to provide electrode voltages and to measure the collected current. One of the most successful instruments is shown in Figure 4.14-4 [6].

Grids 1 and 2 are maintained at vehicle potential and grid 3 is swept over a voltage range of $+ 20$ to $- 2$, corresponding to the case where all ions are retarded to a case where all are collected. Grid 4 is maintained at $- 9$ V to suppress photoemission from the collector and to exclude thermal electrons.

A plot of the collector current versus retarding voltage is shown in Figure 4.14-5. The data analysis technique used involves curve fitting an equation containing density, temperature, and ionic mass as free parameters to the measured points. The least-squares fit provides the values tabulated on the figure.

Table 4.14-2. Planar Ion Trap

Parameters measured	Ion temperature
	Ion density
	Ion composition
Dynamic range	Temperature 300 to 10,000°K
	Density 5 to 5 $\times$ 10^6 ions/cm^3
Accuracy	Ion temperature $\pm 5\%$
	Ion density $\pm 10\%$
Weight	1.4 kg
Power	2 W

Salient features of the instrument are tabulated in Table 4.14-2. No accuracy is assigned to the ion composition measurement since accuracy depends crucially on the relative abundances and the ionic mass of the constituents. Qualitatively, ions of widely separated mass are well resolved with accuracy decreasing as mass separation decreases.

Impedance Probe. The impedance probe requires that two electrodes be immersed in the plasma. In one instrument [11], two 10-cm discs are mounted 10 cm apart on booms to remove them from the influence of the rocket. The two disc electrodes are connected to a bridge circuit, shown in Figure 4.14-6, which is mounted in the boom near the electrodes. The bridge circuit is driven by a crystal-controlled 10-MHz sine wave of 3.0 V rms. A sawtooth waveform is also applied so that the probe is swept through the plasma potential. To improve the electric field in the vicinity of the electrodes the boom is also driven with the sawtooth, which necessitates the isolation transformer shown in the bridge circuit.

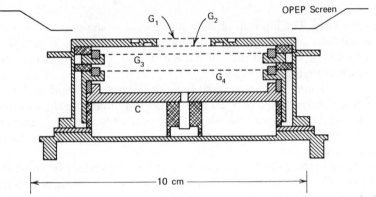

Figure 4.14-4 Schematic cross section of the RPA sensor head, to scale. Cross hatched sections are insulators [6].

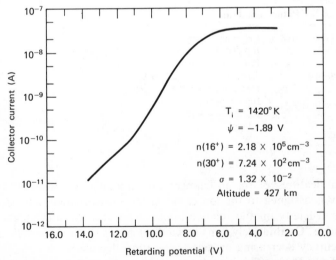

$T_i = 1420°\,K$

$\psi = -1.89\ V$

$n(16^+) = 2.18 \times 10^5\,cm^{-3}$

$n(30^+) = 7.24 \times 10^2\,cm^{-3}$

$\sigma = 1.32 \times 10^{-2}$

Altitude = 427 km

Figure 4.14-5 Plot of ion current versus retarding potential. Solid points are the observed data and the line, the least squares fit to the points [6].

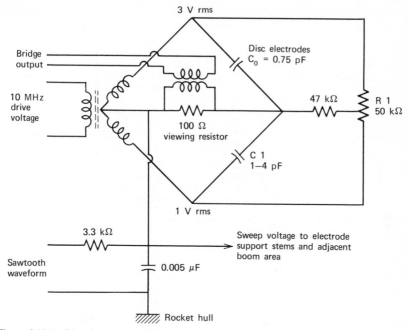

Figure 4.14-6 Impedance probe bridge circuit. The two disc electrodes form one arm of the bridge [11].

A block diagram of the over-all system is shown in Figure 4.14-7. The off-balance output voltage from the bridge is frequency shifted, amplified, detected, and telemetered. A subsidiary amplifier with a gain of four is used to increase the dynamic range. A complementary output and phase comparator permits an in-flight calibration.

A telemetry record showing the raw data is shown in Figure 4.14-8. The data analyzed according to the equations given in Section 4.14.2, p. 332, yield the value of electron density.

Resonance Probes. The implementation of the resonance probe measurement can be inferred from the discussion of the physical basis of the experiment and the foregoing description of the Langmuir probe and the impedance probe. The basic requirements for the measurement are that a small variable frequency rf signal be applied to a Langmuir probe and that the frequency corresponding to enhancements in the probe current be observed. Experiments have been conducted to perform in-flight comparisons of the resonance probe and impedance probe measurements [2].

Ion Mass Spectometer. The radio-frequency mass spectrometer has been used extensively for composition measurements of the earth's ionosphere. A simplified diagram that illustrates its operation is shown in Figure

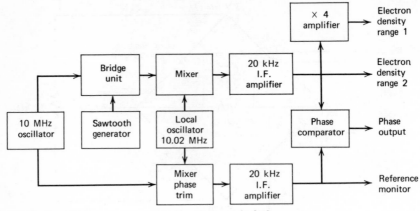

4.14-7 Overall block diagram of the impedance probe [11].

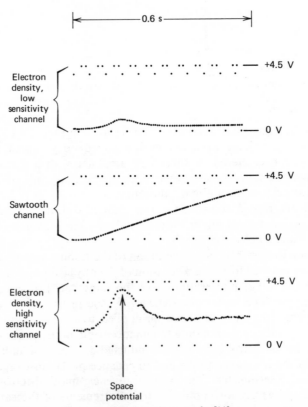

Figure 4.14-8 Telemetered outputs of the impedance probe [11].

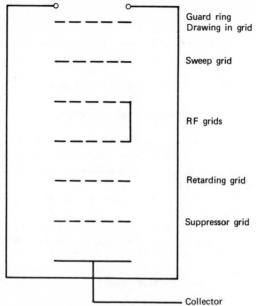

Figure 4.14-9 Functional diagram of the rf ion mass spectrometer.

4.14-9. Ions enter the instrument through an orifice surrounded by a guard ring which is intended to present a uniform field in the entry region. A "drawing-in" grid, usually with a programmable negative potential, draws ions into the spectrometer. Ion drift velocity is established by an accelerating potential on the sweep grid. For each ion mass there is a "resonant velocity." Ions traversing the tube at the resonant velocity gain energy from the rf field and are able to overcome the retarding potential and reach the collector. Nonresonant ions, having gained no energy, are rejected. Thus, a range of masses can be swept by varying the value of the voltage on the sweep grid. Usually several analyzer stages are placed in series to improve resolution. A supressor grid returns any secondary electrons to the collector.

The performance of a three-stage instrument of this type [16] is given in Table 4.14-3.

Table 4.14-3. Radio-Frequency Ion Mass Spectrometer

Parameters measured	Ion composition
	Ion density
Dynamic range	10^2 to 10^5 ions/cm^3
Mass range	1 to 36 amu
Resolution	1 in 25 amu
Weight	3 kg
Power	2 W

References

1. Appleton, E. V., and M. A. F. Barnett, "Local Reflections of Wireless Waves from the Upper Atmosphere," *Nature* **115**, 333, 1925.
2. Baker, K. D., A. M. Despain, and J. C. Ulwick, "Simultaneous Comparison of RF Probe Techniques for Determination of Ionospheric Electron Density," *J. Geophys. Res.* **71**, No. 3, 935–944, 1966.
3. Bennett, W. H., "Radiofrequency Mass Spectrometer," *J. Appl. Phys.* **21**, 143–149, 1950.
4. Brace, L. H., G. R. Carignan, and J. A. Findlay, "Experimental Evaluation of Ionosphere Electron Temperature Measurements by Cylindrical Electrostatic Probes," Presented at the XIIIth Plenary Meeting of COSPAR held in Leningrad, U.S.S.R., 1970.
5. Breit, G., and M. A. Tuve, "A Test of the Existence of the Conducting Layer," *Phys. Rev.* **28**, 554–575, 1926.
6. Hanson, W. B., S. Sanatani, D. Zuccaro, and T. W. Flowerday, "Plasma Measurements with the Retarding Potential Analyzer on OGO 6," *J. Geophys. Res.* **75**, No. 28, 5438–5501, 1970.
7. Harp, R. S., and G. S. Kino, "Measurement of Fields in the Plasma Sheath by an Electron Beam Probing Technique," *Proc. 6 th Int. Conf. Ioniz. Phenom. Gases,* Paris, 1963, **3**, pp. 45–50.
8. Harp, R. S., and F. W. Crawford, "Characteristics of the Plasma Resonance Probe," *J. Appl. Phys.* **35**, No. 12, 3436–3446, 1964.
9. Jackson, J. E., and J. A. Kane, "Measurement of Ionospheric Electron Densities using an RF Probe Technique," *J. Geophys. Res.* **64**, 1074–1075, 1959.
10. Kanal, M., "Theory of Current Collection of Moving Cylindrical Probes," *J. Appl. Phys.* **35**, No. 6, 1697–1703, 1964.
11. MacKenzie, E. C., and J. Sayers, "A Radio Frequency Electron Density Probe for Rocket Investigation of the Ionosphere," *Planet. Space Sci.* **14**, No. 8, 731–740, 1966.
12. Mott-Smith, H. M., and I. Langmuir, "The Theory of Collectors in Gaseous Discharges," *Phys. Rev.* **28**, 727–763, 1926.
13. Nier, A. O., "A Mass Spectrometer for Routine Isotope Abundance Measurements," *Res. Sci. Inst.* **11**, 212–216, 1940.
14. Paul, W., H. P. Reinhard, and V. Von Zahn, "Das elektrische Massenfilter als Massenspektrometer und Isotopentrenner," *Zeitschrift für Physik* **152**, 143–182, 1958.
15. Reifman, A., and W. G. Dow, "Theory and Application of the Variable Voltage Probe for Exploration of the Ionosphere," *Phys. Rev.* **15**, 1311A, 1949.
16. Taylor, H. A., Jr., H. C. Brinton, and C. R. Smith, "Positive Ion Composition in the Magnetoionosphere Obtained from the OGO-A Satellite," *J. Geophys. Res.* **70**, No. 23, 5769–5781, 1965.

Thomas L. Skillman

4.15 MAGNETIC FIELD SENSORS

4.15.1 Characteristics of the Earth's Magnetic Field

One of the planetary properties of the earth is its generation of a magnetic field [13, 15, 26, 34, 46, 64]. This field originates from sources internal and external to the solid earth. The internal source (thought to be a self-maintaining dynamo in the fluid core) is the principal contributor to the main field and is described to a first approximation by the field of a magnetic dipole placed at the earth's center. The main field is distorted by crustal anomalies of magnetized rock in the top 30 km of the earth's surface, by sources due to ionospheric currents, by the effects of plasmas in the magnetosphere, and by distortion of the magnetospheric magnetic field due to the solar wind.

The earth's magnetic field is constantly changing both in direction and intensity. These changes are sometimes systematic and sometimes variable. The secular variation, which takes place on the time scale of years is the largest of the systematic changes and is believed to be caused by the earth's fluid core. Other systematic changes are known as the daily variations which occur mainly during the daylight hours and are controlled by the sun and the moon. The solar heating of the atmosphere generates a system of electric currents at heights roughly between 95 and 130 km (E-region of the ionosphere). Concentrated ionospheric currents along the magnetic equator on the dayside of the earth produce the equatorial electrojet whose effect varies daily, seasonally, and according to solar activity. The tidal oscillation of the atmosphere due to the gravitational force of the moon generates an electric current in the ionosphere whose magnetic field is observable at the earth's surface. This lunar geomagnetic variation is about one-tenth the magnitude of the solar heating effect.

Superimposed on these systematic variations are irregular changes recordable over the entire earth's surface and primarily concentrated at high latitudes along oval-shaped belts. These auroral belts are believed to be caused by ionospheric currents tied to magnetic field lines extending out into the magnetospheric regions. Concentrated currents flowing in these belts are known as auroral electrojets and are primarily Hall currents. These currents are driven by horizontal electric fields in the region of the ionosphere where energetic electrons precipitate from the magnetosphere and increase the ionization and hence the conductivity. Enhanced disturbances as observed in the horizontal component of the magnetic field directly beneath the auroral electrojets are designated magnetic bays. These bays usually last for 1 or 2 hr and often recur at intervals of several hours.

Worldwide magnetic disturbances lasting one to several days are called magnetic storms. Magnetic storms often, but not always, begin with a sudden commencement which is the result of an increase or decrease of the solar wind pressure on the boundary surface between the magnetosphere and interplanetary space. Sudden impulses have the same source but are not accompanied by a subsequent magnetic storm. Magnetic effects are observed from ring currents caused by the motion of charged particles about the earth.

Currents are induced within the earth by any of the external time-varying fields mentioned above due to the electrical conductivity of the earth. In the case of the daily variations, approximately two-thirds of the variance is caused by the currents in the ionosphere and approximately one-third from currents induced in the earth.

There are also variations of shorter periods and smaller amplitudes. These more or less continuous variations are called geomagnetic micropulsations and have oscillation periods ranging from minutes to fractions of a second.

The magnitude of the earth's magnetic field is approximately 30,000 γ at the geomagnetic equator and 65,000 γ at the poles. The amplitude of field variations is very dependent on the location of the surface measurement, principally due to latitude effects. The range of these variations is from approximately 1 γ (small micropulsations) to several thousand γ (large magnetic storms) observed at high geomagnetic latitudes.

The earth's magnetosphere and magnetospheric tail which is confined by the solar wind plasma is shown in Figure 4.15-1. The solar corona is continuously expanding outward and the resulting plasma flow is referred to as the solar wind. Solar magnetic fields that extend beyond the solar corona are swept outward into interplanetary space by the solar wind and have been found to co-rotate with the sun and to exhibit a sector structure with the field direction alternately pointing toward or away from the sun. The earth's magnetic field is confined by this solar wind plasma forming the region known as the magnetosphere and forming a standing bow shock on the upstream side of the magnetosphere. The area between the bow shock and the magnetosphere boundary is called the magnetosheath. The magnetosphere is compressed on the side nearest the sun and is stretched out on the opposite side to form the magnetosphere tail. The magnetopause is the boundary between the magnetosheath and the magnetosphere.

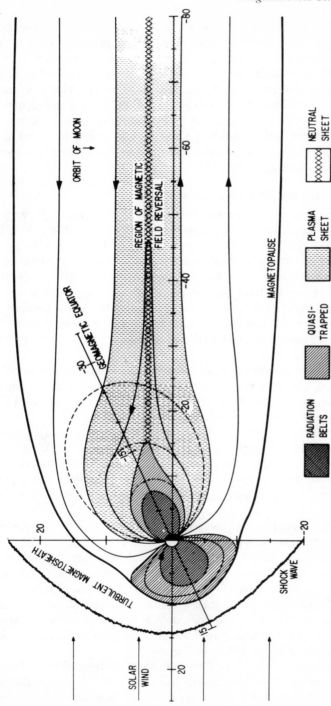

Figure 4.15-1 Magnetosphere and tail.

A uniformly magnetized sphere can be described mathematically at the sphere's surface. The field has a horizontal intensity $H = H_0 \sin \theta$, a vertical intensity $Z = 2H_0 \cos \theta$, and total field intensity $B = (H^2 + Z^2)^{1/2}$, where θ is the polar angle (geomagnetic colatitude), and H_0 is the horizontal intensity at the equator (0.30953 G). This has been done to a first approximation for both a centered dipole and an eccentric dipole [45]. The earth's main field can be described, outside its source region, by a scalar magnetic potential (γ) and expressed in a spherical harmonic series. The Gaussian coefficients needed for this description have been determined experimentally [14].

4.15.2 Magnetic Field Measurements

Classification of Instruments. Numerous magnetometers have been developed to satisfy specific requirements [8, 33, 35, 38, 40, 45, 48, 59]. These instruments can be classified in many overlapping ways. One division is made between mechanical and electrical instruments. Mechanical instruments are based on the balancing of the external force on a magnet against a restoring force. Electrical instruments are based on a variety of principles. Another division of magnetometers is made between analog and digital devices. Analog intruments have an output of current, voltage, or power that is proportional to the measured field, while digital instruments are frequency devices.

Still another division is made between component and total field magnetometers. Component instruments measure the field along one axis of the sensor whereas the total field instrument measures the total intensity of the magnetic field.

Magnetic instruments are also classified as magnetometers, variometers, and gradiometers. Magnetometers measure magnetic fields for specific time intervals. Variometers make continuous measurements relative to a base line that is controlled by absolute measurements with a magnetometer. Gradiometers measure the difference in the magnetic field between two points.

Table 4.15-1 is a listing of the more common considerations in the selection of a magnetometer design.

Noise sources should contribute a signal less than the intrinsic noise sensitivity of the sensor (including data output quantization errors if present). The attenuation of higher-frequency fluctuations by the bandpass of the magnetometer or by time-averaging discrete samples at a uniform time interval should be considered. Aliasing occurs when a frequency higher than the critical cutoff frequency (Nyquist frequency) appears equivalent to a much lower frequency.

Calibration of a mechanical system can be performed by placing a permanent magnet at a known distance from the center of the recording magnet or by applying a known field from a current flowing through a coil.

Table 4.15-1. Considerations in Selection of Sensor

Type of measurement	
Component	— Total field
Relative	— Absolute
Scalar	— Vector
Analog	— Digital
High field	— Low field
Wide range	— Low range
Steady field	— Rapid fluctuations
Environment of the measurement	
Location	— Land, sea, air, space
Orientation	— Stabilized, spinning, towed, on boom
Thermal	— Temperature-controlled, fluctuating
Artificial	— ac and dc, magnitude, and rate of
fields	— occurrence
Timing	— Timing available for required accuracy

Instrument characteristics	
Accuracy	Frequency response
Precision	Data sample rate (aliasing)
Sensitivity	Power
Resolution	Weight
Calibration	Volume
Type output	Complexity of design
Maintenance	Method of recording
Temperature sensitivity	

Measurement Terminology. The measurement of the magnetic field consists of determining the direction and intensity of the field. Measurements on the earth's surface use two reference planes, the plane of the horizon and the true meridian which is defined by the great circle through the geographic poles. True bearing or true azimuth of an object is the angle between the true meridian at the point of observation and the vertical plane containing the object and the point of observation. The magnetic meridian at a point is the vertical plane fixed by the direction of the magnetic lines of force (the direction taken by a perfect compass needle). The magnetic declination D (see Figure 4.15-2) is the angle from the true meridian to the magnetic meridian, and is east or west declination depending on magnetic north being east or west of true north. Declination and variations of the compass are synonymous. Dip or inclination, I, is the angle which the lines of force make with the plane of the horizon. It is north (positive) or south depending on which end of the compass needle dips below the horizontal. In rectangular coordinates, X is the component along the horizontal direction in the geographical meridian and is positive when pointing north, Y is the horizontal component at right angles to the geographical meridian and is positive when pointing east, and Z is the vertical component and is positive in the northern hemisphere.

The magnitude of the field is called total intensity, F. The tesla is the quantity of magnetic flux density (Wb/m^2). One gamma equals 1 nanotesla, and 1 gauss equals 100,000 gamma (γ).

Figure 4.15-2 illustrates the various components (elements) of the earth's magnetic field. The geomagnetic coordinate system using local horizontal (H), vertical (Z), and declination (D) is most desirable for measurements in the magnetosphere where the internal geomagnetic field dominates. For studying the deformed geomagnetic field, solar magnetic coordinates are used. In studies dealing with the solar wind interaction with the earth, it has been found useful to use the solar magnetospheric coordinate system. This system takes into account the 23.4° angle of the earth's rotation axis and the normal to the ecliptic plane and also the angle between the non-axial dipole axis, M, of the geomagnetic field and the earth's rotational axis. The

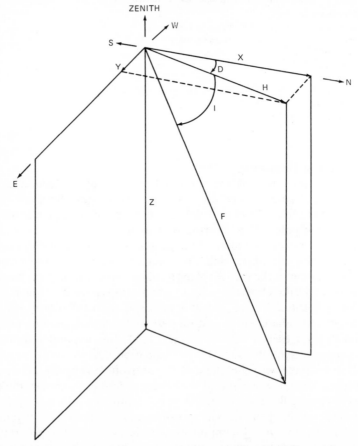

Figure 4.15-2   Components of the geomagnetic field vector in the geodetic coordinate system.

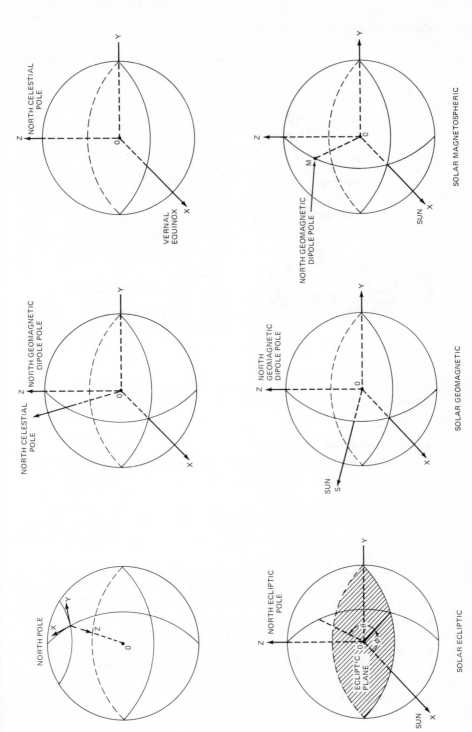

Figure 4.15-3 Coordinate systems.

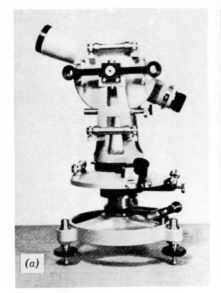

(a)   (b)

Figure 4.15-4 (*a*) Theodolite attachment on theodolite base; (*b*) magnetometer attachment on theodolite base.

most common system used when studying the interplanetary magnetic field is the solar ecliptic coordinate system. These coordinate systems are illustrated in Figure 4.15-3.

4.15.3 Mechanical Instruments

Theodolite Magnetometer. Absolute measurements are commonly made by using a theodolite magnetometer for the D and H elements and an earth inductor for determination of the angle of dip, I. The vertical field is usually computed directly from the H and I measurements. The true meridian (used to determine declination) is obtained from observations of the sun (or other star), which give the azimuth of a prominent object called a mark.

Figure 4.15-4 shows a magnetometer of this type, together with the theodolite which is used for astronomical observations and for angle measurements between marks. A hollow, octagonal magnet hangs in a horizontal position in a stirrup suspended by a fiber of gold or phosphor-bronze. One end is closed with a disk of plain glass on which two perpendicular lines are engraved, while the other end is a collimating lens which permits pointing on the intersection of the lines with the telescope when the latter has been focused for distant objects. A scale in the telescope permits quick measurement of small motions of the magnet. The whole assembly is mounted on the theodolite base so that the angle between the pointings at the magnet and at the mark can be measured with the required accuracy. So that the fiber will not turn the magnet slightly from the magnetic meridian, the torsion in the fiber must be removed, the plane of detorsion being determined by suspending a nonmagnetic weight in place of the magnet. The effect of differences between magnetic and optic axes is eliminated by making readings with the magnet first in an erect position and then inverted.

To measure *H* (horizontal intensity) with this instrument, two types of observations on the same magnet are combined. The product *HM* is measured by oscillations where *M* is the magnetic moment of the magnet. The magnet (called a "long magnet") is suspended as for measuring *D*, and is disturbed momentarily with a small auxiliary magnet. It then oscillates in a horizontal plane. The time of one oscillation depends primarily on *H*, *M*, and *K* (*K* = moment of inertia). Increase of either *H* or *M* decreases the time of an oscillation so that (knowing *K*) *HM* can be derived.

The quotient *H/M* is determined by deflections. Another magnet (the "short magnet") is suspended in the stirrup, while the long magnet is placed horizontally on a deflector bar that extends magnetically east and west of the short magnet (Figure 4.15-4). This turns the short magnet out of the magnetic meridian by an angle (μ) depending upon *H/M*.

A mechanical instrument that measures *H* conveniently and rapidly is the Quartz Horizontal Magnetometer (QHM). It is a direct method of measurement of *H* by determining the torque acting on a magnet when it is deflected from the magnetic meridian. It is a relative instrument since it requires a knowledge of the magnetic moment of the magnet and the torque is proportional to the product *HM*.

Earth Inductor. In order to measure the total vector field, it is necessary to measure the angle of inclination (*I*), and this is done with an earth inductor (Figure 4.15-5).

The earth inductor is a small dynamo, whose operation is based on the fact that when a coil of wire forming a closed circuit is rotated in a magnetic field, an electrical current is generated in the coil except when the axis of rotation of the coil is parallel to the lines of force of the field. This null position is determined by observing a light spot position on a scale.

A copper coil on a cylindrical core is rotated with the axis of rotation in the magnetic meridian. The inclination of the axis of rotation is adjusted until no current is induced in the coil.

Magnetic Field Balance. The magnetic field balance, a mechanical instrument for measuring relative values of the vertical intensity, is used principally in geophysical prospecting. A drawing of the magnet system is shown in Figure 4.15-6.

The magnet system consists of two hardened, magnetized, and gold-plated steel lamellae (*A*, *A'*) properly aged, with equal magnetic moments,

Figure 4.15-5 Earth Inductor.

and joined by a piece of square tubing (*B*) carrying a mirror on the upper face. A quartz knife edge (*K*) passes centrally through the lamellae and the tubing. Projecting horizontally from two ends of the tubes are metal spindles of different material carrying small weights (*L*, *T*) which are adjustable. The metal spindles are used for temperature compensation (*T*) and for balancing the intensity of the vertical field (*L*). The underside of the iron center piece is provided with a stud carrying a bronze weight (*S*) which can be locked in position. The adjustment of this weight makes it possible to vary the sensitivity of the magnet system.

The mirror is viewed from above with a telescope which has a scale in the eyepiece and the observer sees this scale and its reflection. The relation of the two indicates the change in position of the magnet system.

Variometers. Variometers contain a recording magnet that is free to move about a specified axis. The magnet turns in response to changes in intensity normal to the axis of rotation and to the magnetic axis of the magnet. A mirror attached to the magnet reflects light from a light source to the photographic paper, which is wrapped around a revolving drum in the recorder. Each variometer produces a straight "base line" (by reflection

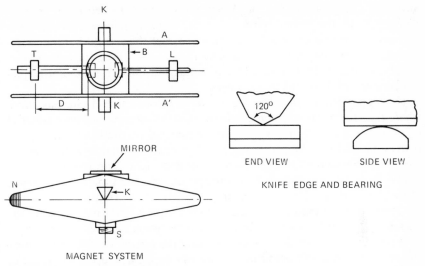

Figure 4.15-6 Magnetic field balance.

from a fixed mirror) to give a definite point from which to measure ordinates. Each variometer is compensated by one of several methods: mechanical compensation, optical compensation, or, the most satisfactory, magnetic compensation. Magnetic compensation is obtained by using one or more small control magnets set in a calculated position so that as the temperature changes around the variometer its reaction with the recording magnet is varied with the changing temperature.

The recording magnet of the *D* variometer is suspended by a fine quartz fiber and adjusted so that the magnet lies in the magnetic meridian. The fiber prevents the magnet from following absolutely the changes of declination, but the effect is very slight and is determinable [36].

The *H* variometer has a recording magnet which is also suspended by a quartz fiber but the torsion head is turned until the magnetic axis of the magnet lies magnetically east-west. In this position the mechanical couple of the fiber balances the opposing magnetic couple of the magnet. Slight changes in *H* cause the system to move until the couples are again in balance.

The recording magnet of a *Z* variometer is mounted like that of a magnetic field balance (see Figure 4.15-6).

4.15.4 Electrical Instruments

Fluxgate Magnetometer. The fluxgate magnetometer (saturable core) is a magnetically saturable transformer [4, 12, 44, 52, 53]. The sensor is a high-permeability core with a primary and secondary winding. The primary is driven at kilohertz frequencies which in turn magnetically drives the core into plus and minus saturation. When the sensor is placed in an external magnetic field, the second harmonic of the drive frequency is picked up in the secondary, and the magnitude and phase of the second harmonic is a measure of the amplitude and direction (+ or − along the sensor axis) of this external field. Figure 4.15-7 is a diagram illustrating the principles of operation of fluxgate magnetometers.

The upper left of the drawing shows a normal *B-H* hysteresis curve where H_c is the point at which the driving field H saturates the core material ($\pm B_s$). ΔH is the offset (bias) due to the external magnetic field and $\pm H_D$ is the amplitude of the primary drive. The resultant magnetic flux, B_R, is linked to the secondary as shown at the top right. The resulting signal output from the secondary, V_s, is proportional to the time rate of change of the flux (dB/dt). The asymmetric waveform of the output as shown in the lower right of the figure contains both even and odd harmonics of the frequency as ΔH is nonzero. The second harmonic is $\pm 90°$ out of phase with the primary with the sign determining the direction of the field along the core axis, and the amplitude is linearly dependent upon the value of $\Delta H/H_D$.

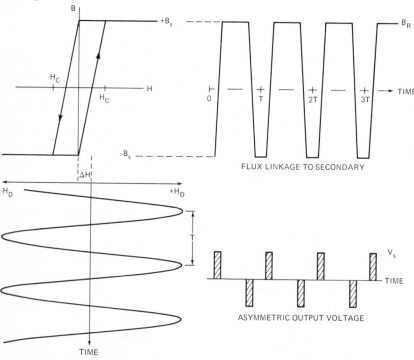

Figure 4.15-7 Fluxgate operating principles.

The primary drive must be absolutely free of second harmonic distortions. Also, the core should be driven into complete saturation in both the plus and minus directions to help eliminate any residual magnetism that would affect the zero level of the instrument.

Several fluxgate sensors have been developed. These include a single high-permeability core with both primary and secondary coils [67, 71], two parallel rods with the primary wound in series opposition on the cores and the secondary wound about both cores, and a closed loop for the excitation field with a ring core inside a secondary winding [28, 29, 31, 32, 41].

These are parallel fluxgate sensors that have the excitation field parallel to the direction of the field to be measured. In another sensor the excitation field is applied perpendicular to the field to be measured. This principle is orthogonal gating and is due to changes in $B/(H - H_0)$ while parallel gating is due to changes in dB/dH as taken from the normal B-H curves for magnetic materials [57, 58].

A block diagram of a simplified fluxgate magnetometer is shown in Figure 4.15-8.

Induction Magnetometer. One type of induction magnetometer, the earth inductor, is discussed in Section 4.15.3, p. 351, because it is partly mechanical [37, 39]. Another type, the induction coil, can have either an air core, which is very large, or a ferromagnetic core. The coil can be used in a stationary position where the ambient field furnishes the varying magnetic field or can be spun either by a direct drive or by attachment to a spinning base such as a spinning satellite. Spinning coils are used primarily to study fluctuations in the magnetic field. An emf is generated in the coil by the varying magnetic field which changes the flux linkage. Only the component which is parallel to the axis of the coil is effective. The induction magnetometer does not respond to dc or slowly varying fields, but its sensitivity increases linearly with frequency, and therefore is very useful for measuring rapid fluctuations.

Proton Magnetometer. The free precession magnetometer measures the total field vector, F, of the earth's magnetic field by measuring the precession frequency of protons in this field [5, 43, 50, 60, 68]. This precession frequency is given by the relationship $\omega = \gamma_p H$, where ω is the angular frequency, H is the value of the magnetic field, and γ_p is the gyromagnetic ratio of the proton determined by fundamental atomic constants for the proton [7]. In a field of 1 G the precession frequency is 4258 Hz. Since this precession represents a time-dependent variation of magnetic moment, it induces a voltage in a coil surrounding a sample containing protons (i. e., water, alcohol, hexane). The frequency of this induced voltage can be measured to determine the value of the magnetic field. The induced voltage is also proportional to the net polarization of the sample, that is, the orientation

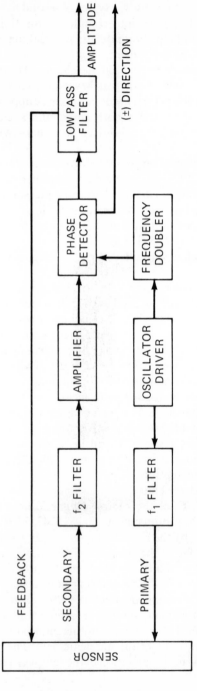

Figure 4.15-8 Simplified block diagram of single-axis fluxgate magnetometer.

of the magnetic moment of the individual protons. This polarization is proportional to applied fields and is given as $M = \chi H$, where M is the polarization, H the field, and χ is the nuclear susceptibility.

In operation, a strong polarizing field is applied at right angles to the earth's magnetic field. After a time T_1 this polarizing field is turned off very rapidly to leave the induced polarization in a direction approximately perpendicular to the earth's magnetic field. In this condition the nuclear polarization precesses at the Larmor frequency and induces a voltage in the coil. The frequency of this induced voltage is independent of orientation of the instrument relative to the earth's magnetic field, although the signal amplitude does depend to some extent on the angle between the polarizing field and the earth's field. The precession frequency is measured to obtain the field.

Alkali-Vapor Magnetometer. Alkali-vapor magnetometers have been operated using such metals as Rb-85, Rb-87, and Cs-133 [25, 47, 49, 51, 54, 61, 62, 65, 66]. These magnetometers are based on the technique of optical pumping, a process of aligning atomic spin orientations so they become preferentially oriented into one of the Zeeman levels of the ground state [6, 10, 11]. This is accomplished by passing circularly polarized light from an alkali metal lamp through an absorption cell containing alkali metal vapor and a buffer gas. By the absorption of optical photons the atoms in the absorption cell are raised from the ground state to the first excited state. The atoms then immediately decay and return with equal probability to the various Zeeman levels in the ground state. As the atoms in the highest Zeeman level (or lowest, depending on sense of polarized light) cannot absorb additional energy, the repetition of rising to one state ($\Delta m = 1$) when excited, and falling back to all states when returning to the ground state, causes the majority of atoms eventually to end in the nonabsorbing state called the pumped state. In this state all the atoms have their spin axes aligned, and the maximum amount of light is transmitted through the cell. This transmitted light is monitored by a photocell and indicates the degree of pumping (spin orientation) within the cell [19].

When the atoms in the pumped state are irradiated with radio waves at the frequency corresponding to the energy transitions of this state, some of these atoms are disoriented and less light passes through the cell. This decrease in transmitted light is an indication of a unique absorption line in the vapor. Since the magnetic field is simply related to the observed transition frequency, this absorption line provides a convenient method of measuring the magnetic field. The frequency corresponding to the Zeeman splitting of the energy levels has a second-order splitting effect which is proportional to the square of the magnetic field intensity.

Optically pumped magnetometers can be classified into the straight transmitted light [61, 62] and the cross-beam type instruments (Table 4.15-2, Figure 4.15-9).

Table 4.15-2. Optically Pumped Magnetometers

	Straight Transmitted Light	Cross-Beam
Light source	Single light path	Two light paths at right angles to one another
Direction of light axis to magnetic field	Parallel	Light beam monitored for field measurement is perpendicular to the field
Output from detector	dc voltage level indicating occurrence of resonance frequency	Larmor frequency (ac output)

The self-oscillating magnetometer is a variation of the cross beam. A single light path, which is at an angle to the magnetic field, is used to obtain a Larmor frequency output from the photocell. The output acts as the rf source on the absorption cell and closes the loop of the self oscillator (Figure 4.15-10) [54].

The dual gas cell magnetometer eliminates the orientation errors due to the asymmetric resonance curve of the smeared Zeeman sublevels. In the highest plus state the peak of the resonance curve is shifted off-center in one direction, and if the lowest state is used as the pumped state, its center is shifted in the opposite direction. Using two cells with the output modulation of one detector fed back to furnish the rf drive for the opposite gives a summed frequency that has its peak in the center of the Zeeman levels.

Table 4.15-3 lists the Larmor frequency in hertz for some of the metals used in alkali vapor magnetometers [9, 17, 23, 24, 55, 69].

Helium Magnetometer. Helium magnetometers have been built for both high (earth field) and low field measurements [1, 16, 27, 56]. The advantages of this instrument favor earth field measurements and specifically gradiometer applications.

Variations in transmitted light through an absorption cell are used to make the measurement of the field in a manner similar to the alkali-vapor magnetometers. Although unpolarized optical resonance radiation can be used in this instrument, normal usage employs circularly polarized light. Atoms in the 2^3S_1 metastable state when radiated with pumping

Table 4.15-3. Conversion Constants

Frequency (hertz) = constant × magnetic field (gamma)

f_p(proton)   = 0.04257586(32) × H

f_{85}(rubidium) = 4.667370(35)   × H ⎫

f_{87}(rubidium) = 6.995746(52)   × H ⎬ Center of structure [55]

f_{133}(cesium) = 3.498577(26)   × H ⎭

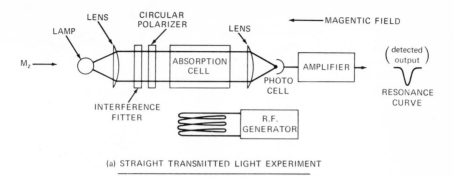

(a) STRAIGHT TRANSMITTED LIGHT EXPERIMENT

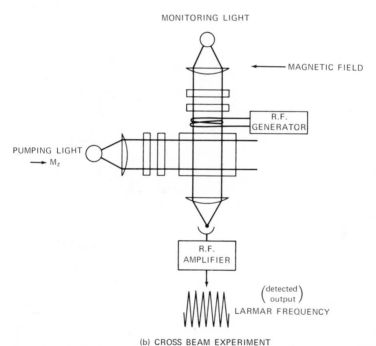

(b) CROSS BEAM EXPERIMENT

Figure 4.15-9 Optical pumping experiments.

light assume an unbalanced population density which is then redistributed by external electronic devices, for example, an rf field corresponding to the Larmor precession frequency. The transmitted light is detected and monitored for the field measurement.

4.15.5 Recent Advancements

Narrow-Line Magnetometer. The narrow-line magnetometer is an alkali vapor instrument which takes advantage of the rapid response of

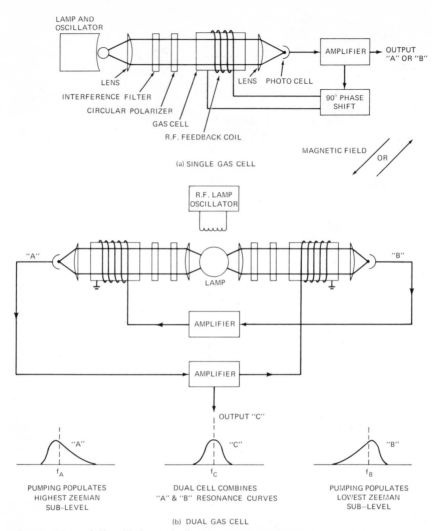

(a) SINGLE GAS CELL

(b) DUAL GAS CELL

Figure 4.15-10 Self-oscillating optically pumped magnetometers.

the self-oscillator and the ability to lock-on to a separate transition component inherent in a straight transmitted light magnetometer [3] (Figure 4.15-11). The narrow-line reference device is used to observe this narrow resonance by using low light intensity, weak rf intensity, and vapor with a long relaxation time. A low-frequency modulation is applied along the magnetic field direction and the phase difference from the output determines the frequency of the self-oscillator. An instrument of this type has both absolute accuracy and rapid response to variations.

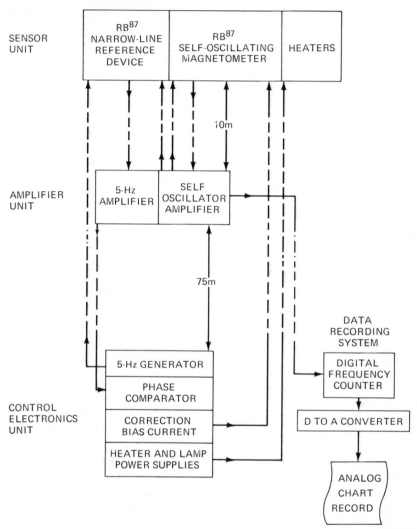

Figure 4.15-11 Narrow-line magnetometer system.

Superconducting Magnetometer. Extremely sensitive magnetometers have been built using the unique properties due to the microscopic quantum nature of superconductors [17, 18, 21, 22, 42]. One of these properties, quantized magnetic flux, has led to the superconducting magnetometer based on the use of Josephson junctions of both the tunnel and point contact type devices (Figure 4.15-12). The basic circuit is usually a super-conducting loop containing one or two Josephson junctions. A current is

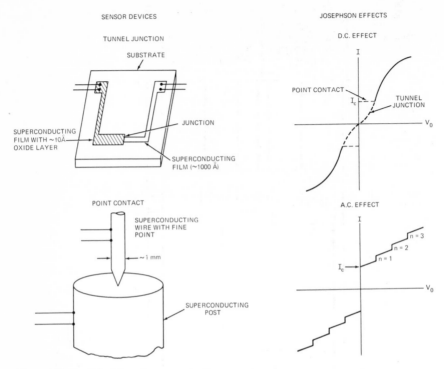

Figure 4.15-12   Principles of superconducting magnetometers.

fed through the loop which is coupled to the magnetic field being measured. Change in critical current (I_c) is related to change in magnetic flux density. As shown in Figure 4.15-12 a Josephson junction has a characteristic maximum zero-voltage current I_c which depends on the junction material and the magnetic flux density. I_c is of the order of a few milliamperes and V_0 is usually a few millivolts. These instruments are used in medical research for measuring magnetic fields from the heart and brain.

References

1. Aldrich, R.G., "A Helium Magnetometer That Utilizes Light-Modulation," Thesis from the United States Naval Post-Graduate School, 1962.
2. Alldredge, L. R., "A Proposed Auromatic Standard Observatory," *J. Geophys. Res.* **65**, 3777–3786, 1960.
3. Allen, J.H., and P.L. Bender, "Narrow Line Rubidium Magnetometer for High Accuracy Field Measurement," private communication, November 1971.

4. Aschenbrenner, H., and G. Goubau, "Eine Anordnung zur Registrierung rascher magnetischer Störungen," *Hochfrequenz-technik u. Electroakustik*, **47**, 177–181, 1936.

5. Bacon, F.W., Jr., "Adaptation of a Free Precession Magnetometer to Measurements of Declination", Thesis from the United States Naval Post-Graduate School, Monterey, California, 1955.

6. Bell, W.E., and A.L. Bloom, "Optical Detection of Magnetic Resonance in Alkali Metal Vapor," *Phys. Rev.* **107**, 1559, 1957.

7. Bender, P.L., and R.L. Driscoll, "A Free Precession Determination of the Proton Gyromagnetic Ratio," *IRE Trans. Instr.* **7**, 176–180, 1958.

8. Bender, P.L., "Measurement of Weak Magnetic Fields by Optical Pumping Methods," *Compt. Rend. de 9ᵉ Colloq. Ampère*, p. 621, Librairie Payot, Geneva, 1960.

9. Bender, P.L., "Comparison of the Rb 87 and Proton Zeeman Transition Frequencies in the Earth's Magnetic Field," *Phys. Rev.* **128**, 2218–2221, 1962.

10. Bloom, Arnold L., "Optical Pumping," *Scientific American*, October 1960.

11. Bloom, Arnold L., "Principles of Operation of the Rubidium Vapor Magnetometer," *Appl. Optics*, **1**, No. 1, 61–68, Jan, 1962.

12. Bozorth, R.M., *Ferromagnetism*, D. Van Nostrand, New York, 1951.

13. Cahill, L.J., "The Geomagnetic Field," in *Space Physics*, Ed. D.P. LeGalley and A. Rosen, John Wiley & Sons, New York, pp. 301–349, 1964.

14. Cain, J.C., and R.A. Langel, "Geomagnetic Survey by the Polar Orbiting Geophysical Observatories OGO-2 and OGO-4 1965–1967," in *World Magnetic Survey 1957–1969*, Ed. A. J. Zmuda, Paris, IAGA Bull. 28, 1971.

15. Chapman, Sidney, and Julius Bartels, *Geomagnetism*, Oxford University Press, London, 1940.

16. Colegrave, F.D., and P.A. Franken, "Optical Pumping of Helium in the 3*S* Metastable State," *Phys. Rev.* **119**, 680, 1960.

17. Dayem, A.H., and C.C. Grimes, "Microwave Emission from Superconducting Point-Contacts," *Appl. Phys. Letters* **9**, 47, 1966.

18. Deaver, B., and W. Fairbanks, "Experimental Evidences for Quantized Flux in Superconducting Cylinders," *Phys. Rev. Letters* **7**, 43, 1961.

19. Dehmelt, H.G., "Modulation of a Light Beam by Processing Absorbing Atoms," *Phys. Rev.* **105**, 1924–1925, 1957.

20. DeVuyst, A.P., "Operation of an Automatic Digital Magnetic Observatory," *Inst. Roy. Met. de Belgique* 1968.

21. Doll, R., and M. Nabaver, "Experimental Proof of Magnetic Flux Quantization in a Superconducting Ring," *Phys. Rev. Letters* **7**, 51, 1961.

22. Doyle, Owen, "Josephson Junctions," *Electronics*, March 1, 1971.

23. Driscoll, R.L., and P.L. Bender, "Proton Gyromagnetic Ratio," *Phys. Rev. Letters* **1**, No. 11, 1958.

24. Dristoll, Raymond L., "Electronic *g* Factor of Rubidium," *Phys. Rev.* **136**, No. 1A, A54–A57, 1964.

25. Farthing, W.H., and W.C. Folz, "Rubidium Vapor Magnetometer For Near Earth Orbiting Spacecraft," *Rev. Sci. Instr.* **38**, No. 8, 1023–1030, August 1957.

26. Fleming, J.A., "Physics of the Earth," *Terrestrial Magnetism and Electricity,* Vol. 8, McGraw-Hill Book Company, New York, 1939 (Reprinted with corrections in 1949 by Dover Publications, New York).

27. Franken, P.A., and F.D. Colegrave, "Alignment of Metastable Helium Atoms by Unpolarized Resonance Radiation," *Phys. Rev. Letters* **1**, 316, 1958.

28. Geyger, W.A., "The Ring-Core Magnetometer—A New Type of Second-Harmonic Fluxgate Magnetometer," *AIEE Trans.* **81**, pt. 1, Comm. and Elec., 65, March 1962.

29. Geyger, W.A., *Non-Linear Magnetic Control Devices,* Chapters 13 and 14, McGraw-Hill Book Company, New York, 1964.

30. Gillis, H.J., "A General Computer Data Processing System: Documentation of the ATS-5 Ground Station Magnetometer Program," Goddard Space Flight Center Document X-645-70-458, November 1970.

31. Gordon, D.I., et al., "Factors Affecting the Sensitivity of Gamma-Level Ring-Core Magnetometers," *IEEE Trans. Magnetics* **MAG-1,** No. 4, 330, December 1965.

32. Gordon, D.I., R.H. Lundsten, R. A. Chiarodo, and H. H. Helms, Jr., "A Fluxgate Sensor of High Stability for Low Field Magnetometry," *IEEE Trans. Magnetics* **MAG-4,** 397–401, 1968.

33. Hazard, Daniel L., "Directions for Magnetic Measurements," U. S. Dept. of Commerce Serial No. 166, U. S. Government Printing Office, Washington, D.C., 1930 (corrected 1957).

34. Heppner, J.P., *Physics of Geomagnetic Phenomena,* Vol. 2, Academic Press, New York, 1968.

35. Hood, P., "Magnetic Surveying Instrumentation," *A Review of Recent Advances, Mining and Groundwater Geophysics,* Geological Survey of Canada Economic Report No. 26, pp. 3–31, 1967.

36. Howe, H.H., "On the Theory of the Unifilar Variometer," *Terr. Mag.* **42,** 29–42, Baltimore, 1937.

37. Inouye, G.T. and D.L. Judge, "The Optimized Iron Cored Induction Magnetometer," TRW Report 9821-6001-RUOO, May 18, 1964.

38. International Conference, "Low Magnetic Fields of Interest in Space and Geophysics, Abstracts," National Center of Space Studies, Paris, pp. 20–23, May 1969.

39. Judge, D.L., M.G. McLeod, and A.R. Sims, "The Pioneer I, Explorer VI, and Pioneer V High Sensitivity Transistorized Search Coil Magnetometer," *IRE Trans. Space Electron. Telemetry* **SET-6,** 114–121, 1960.

40. Ledley, B.G., "Magnetometers for Space Measurements Over a Wide Range of Field Intensities," *Rev. Phys. Appl.* **5,** 164–168, 1970.

41. Ling, S.C., "A Fluxgate Magnetometer for Space Application," Cornell University Center for Radiophysics and Space Research, Ithaca, New York, Quart. Status Report to NASA, CRSR 124, NASA Contract NASR-46, June 1, 1962.

42. London, F., *Superfluids,* John Wiley and Sons, New York, 1950.

43. Mansir, Dolan, "Magnetometer at Work in Outer Space," *Radio-Electron.* **31,** 38–41, 1960.

44. Marshall, Stanley V., "An Analysis of the Fluxgate Magnetometer," University of Missouri, Columbia, Thesis, 1967.
45. Matsushita, S., and Wallace H. Campbell, Ed. Papers, "International Geophysics Series," *Physics of Geomagnetic Phenomena*, Vol. I and II, Academic Press, New York, 1967.
46. McComb, H.E., "Magnetic Observatory Manual," U.S. Dept. of Commerce Special Publication 283, U.S. Government Printing Office, 1952.
47. Meilleroux, J.L., "Progrès Récent Sur Le Magnétomètre à Vapeur de Césium Type 'Asservi'," *Rev. Phys. Appl.* **5**, 121–130, 1970.
48. Ness, Norman F., "Magnetometers For Space Research," *Space Sci. Rev.* **11**, 459–554, 1970.
49. Osgood, C., "Design and Use of a Gradiometer Connected Rubidium Magnetometer," *Rev. Phys. Appl.* **5**, 113–118, 1970.
50. Packard, N.E., and R.H. Varian, "Free Nuclear Induction in the Earth's Magnetic Field," *Phys. Rev.* **93**, 941, 1954.
51. Parsons, L.W., and Z.M. Wiatr, "Rubidium Vapor Magnetometer," *J. Sci. Instr.* **39**, 292–300, 1962.
52. Primdahl, F., "Bibliography of Fluxgate Magnetometers," Vol. 41, No. 1, Dept. of Energy, Mines, and Resources, Ottawa, Pub. Earth Physics Branch, 1970.
53. Primdahl, Fritz, "The Fluxgate Mechanism," *IEEE Trans. Magnetics* **MAG-6**, No. 2, 376–383, June 1970.
54. Ruddock, K.A., "Optically Pumped Rubidium Vapor Magnetometer for Space Experiments," *Proc. Intern. Space Sci. Symp.*, Vol. **2**, p. 692.
55. Sarles, Lynn R., "Conversion Constants," private communication, October 1966.
56. Schearer, L.D., *Advances in Quantum Electronics*, Ed. J.R. Singer, p. 239, Columbia University Press, New York, 1961.
57. Schonstedt, E.O., and H.R. Irons, "Airborne Magnetometer for Determining All Magnetic Components," *Trans. Am. Geophys. Union* **34**, No. 3, 363–378, 1953.
58. Schonstedt, E.O., "Saturable Measuring Device and Magnetic Core Therefor," U. S. Patents 2, 916, 696 and 2, 981, 885, December 8, 1959 and April 25, 1961.
59. Serson, Paul H., "Bibliography of Magnetometers" International Association of Geomagnetism and Aeronomy, Commissions I and IX, 1972.
60. Shapiro, I.R., J.D. Stolarik, and J.P. Heppner, "The Vector Field Proton Magnetometer for IGY Satellite Ground Stations," *J. Geophys. Res.* **65**, 913–920, 1960.
61. Skillman. T.L., and P.L. Bender, "Measurement of the Earth's Magnetic Field with a Rubidium Vapor Magnetometer," *J. Geophys. Res.* **63**, No. 3, 513–515, 1958.
62. Skillman, Thomas L., "Rubidium Vapor Magnetometer," *Intern. Hydro. Rev.* **37**, No. 1, January 1960.
63. Snare, R.C., F.R. George, and R.F. Klein, "Data Logging System for Monitoring Geophysical Phenomena," Publication Number 743, Institute of

Geophysics and Planetary Physics, University of California, Los Angeles, March 1969.

64. Sugiura, M., and J.P. Heppner, "The Earth's Magnetic Field," *Introduction to Space Science,* Eds. W.N. Hess and G.D. Mead, Chapter 1, Gordon and Breach, Science Publishers, New York, 1965.

65. Unterberger, R.R., "Direct Recording of Small Geomagnetic Fluctuations," *J.Geophys. Res.* **65,** 4213–4216, 1960.

66. Usher, M.J., W.F. Stuart, and J.H. Hall, "A Self-Oscillating Rubidium Magnetometer for Geophysical Measurements," *J. Sci. Instr.* **41,** 544–547, 1964.

67. Vacquier, V.V., and Gulf Research and Development Co., U.S. Patent No. 2, 406, 870, 1946.

68. Waters, G.S., and P.D. Francis, "A Nuclear Magnetometer," *J. Sci. Instr.* **35,** 88–93, March 1958.

69. White, G.W., W.M. Hughes, G.S. Hayne, and H.G. Robinson, "Determination of g-Factor Ratios for Free Rb[85] and Rb[87] Atoms," *Phys. Rev.* **174,** 23–32, 1968.

70. Wienert, K.A., "Notes on Geomagnetic Observatory and Survey Practice," UNESCO, Van Boggenhoudt, Brussels, 1970.

71. Wyckoff, R.D., "The Gulf Airborne Magnetometer," *Geophys.* **13,** No. 12, April 1948.

Joseph B. Reagan

4.16 PARTICLE RADIATION SENSORS

Particle radiation sensors have been used extensively in space physics in the investigation of such phenomena as the solar wind, energetic solar particle events, cosmic radiation, the aurorae, and geomagnetically trapped particles. The primary particles encountered in these events are protons, electrons, and alpha particles although neutrons and heavier nuclei have also been measured. In many of these phenomena all three particle types are present simultaneously. The range of energies encountered is extremely broad, extending from less than 10^3 eV in the solar wind to 10^9 eV in solar particles events and up to 10^{20} eV in cosmic radiation. Flux intensities range from a few particles/cm^2-s in cosmic radiation to greater than 10^9 particles/cm^2-s in auroral precipitation. A variety of sensors and sensor configurations have been used to detect, discriminate, and analyze these particle radiations. Three of the most important of these sensors are the scintillation detector, the semiconductor detector, and the electron-multiplier detector.

4.16.1 Scintillation Detectors

Principles of Operation. The scintillation detector consists of a material in the form of a solid, liquid, or gas that scintillates, that is, produces flashes of light upon excitation by particle radiation [5, 18, 19, 33]. The scintillating material, or scintillator, is optically coupled to a photomultiplier tube which converts the incident light flashes into electrical pulses whose amplitude is proportional to the number of photons in the scintillation. The latter, in turn, is proportional to the total energy deposited by the incident particles.

The general arrangement of the scintillator and photo-multiplier and the basic processes involved in the detection and measurement of a charged particle are shown in Figure 4.16-1. The entire system must be maintained light-tight by a suitable enclosure with opaque windows and apertures as appropriate. An incident charged particle of energy E impinges on the scintillator where it dissipates its energy in the ionization and excitation of the molecules. A fraction of this energy is converted into photons P which are radiated in all directions. The scintillator is usually surrounded by a reflector to maximize the number of these photons which fall on the photocathode P'. A fraction of these photons cause the emission from the semitransparent cathode of photoelectrons which are then accelerated by the potential applied between the cathode and the first dynode of the photomultiplier.

The ratio of the photoelectrons emitted per incident photon, referred to as the quantum efficiency ε, has a maximum value between 0.10 and 0.30 and depends on the type of photocathode material used. N electrons ($N = \varepsilon P'$) strike the first dynode and each ejects further electrons by secondary emission. If δ is the average number of secondary electrons per incident electron which are ejected from the dynode and are collected by the next dynode, then δN electrons impinge on the second dynode. Values of δ range from 3 to 6 for conventional dynode materials such as Ag-Mg-O-Cs and Cu-Be-O, to 30 to 50 in first dynodes employing GaP as the secondary emitter. This electron multiplication process is repeated at subsequent dynodes, each of which is at a higher potential than the preceding one. If there are n dynodes, each with a multiplication factor δ, the

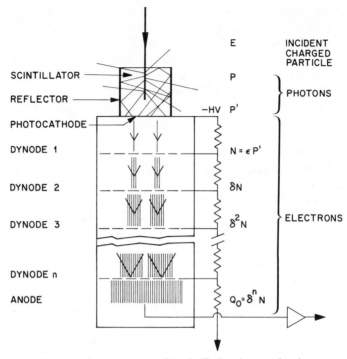

Figure 4.16-1 The general arrangement of a scintillation detector for the measurement of charged particles.

number of electrons finally emerging from the last dynode and collected at the anode is $Q_0 = \delta^n N$. The number of dynodes in a typical tube varies from 6 to 12, and δ^n, the overall electron gain of the tube, ranges from 10^4 to 10^8. This electron avalanche produces a voltage pulse in the output capacitance at the anode terminal of the photomultiplier.

The anode pulses can be analyzed in a variety of ways to obtain the desired information on the incident radiation. The pulses above a given energy level can be counted to obtain an integral measure of the radiation intensity or they can be analyzed in terms of pulse height to provide detail on the energy spectrum of the radiation. If the intensity of the radiation is so high that counting techniques are not practical, a measurement of the average current from the photomultiplier is often utilized to obtain the total energy deposition.

Solid Scintillators. The density of most solid scintillators is such that charged particles with energies up to 100 MeV can be stopped in material only a few centimeters thick. The decay time of the light flashes (the time for the light to fall to $1/e$ of its peak value) in typical solid scintillators is from a few nanoseconds to a few microseconds. Since photomultiplier tubes are capable of responding to these fast light pulses, the solid scintillation detector is extremely valuable where high-intensity radiation is to be analyzed or where fast coincidence between particle detectors is to be performed. An advantage of solid scintillators is that virtually any size or shape detector can be easily and inexpensively fabricated.

Scintillator Characteristics. Solid scintillators are classified into two groups, inorganic and organic. The inorganic scintillators most widely used are the pure and activated alkali iodides NaI, NaI (Tl), LiI, LI (Eu), CsI (Na), CsI, and CsI (Tl). Of these, the first five are hygroscopic and must be sealed in air-tight containers for general use. Other types of inorganic crystals which have been found to scintillate are the tungstate group consisting most noticeably of $CdWO_4$ and $CaWO_4$ and the sulfide group consisting of $ZnS(Ag)$ and $ZnS(Cu)$. The recent development and commercial availability of CaF_2 (Eu) crystals for charged particle detection offers many of the advantages of the alkali iodides without the associated inconveniences.

The organic materials that are efficient scintillators generally fall in the class of aromatic hydrocarbons, both pure and substituted. In contrast to the alkali halides, no activator is added to organic crystals. The most widely used solid organic scintillators are anthracene, trans-stilbene, and plastic, an organic scintillator dissolved in either polystyrene or polyvinyltoluene.

The physical events responsible for the light pulses generated by charged particles in organic and inorganic scintillators are different. In the case of an activated alkali halide crystal the luminescence is caused primarily by the presence of the activator in small concentration. The charged particle in slowing down in the crystal loses its energy primarily to electrons of the crystal, resulting in ionization and excitation of the constituents of the crystal lattice. The light emitted generally decays in time with an exponential (or nearly exponential) dependence $I = I_0 e^{-\lambda t}$. The transition probability λ is a characteristic of a particular type of crystal, its activator, and crystal temperature. The quantity $1/\lambda$ is called the decay time of the scintillator.

The scintillation process in organic crystals is primarily a molecular phenomenon. The luminescence arises from the de-excitation of the weakly bound molecules from their first excited state. The details of the energy transfer in organic scintillators is very complex and probably involves a number of processes including electron-hole pair migration, quantum-mechanical resonances, and photon emission and reabsorption.

The scintillation process in organic compounds exhibits distinct differences from that in inorganic crystals. The luminescence emission from organic materials occurs on a much faster time scale than in inorganic

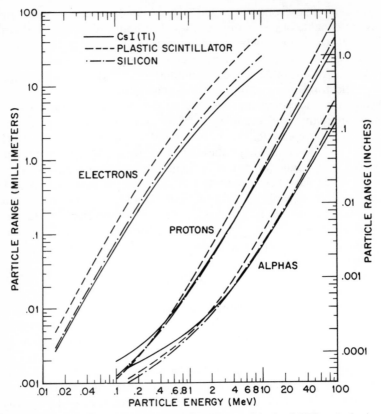

Figure 4.16-2   Range of protons, electrons and alpha particles in CsI(Tl) and plastic scintillators and in silicon semiconductor detectors.

materials. The conversion efficiency (the efficiency with which charged-particle energy is converted to light) of organic scintillators is generally less than that of inorganic crystals. The difference in light output per unit energy deposited from various charged particles is significantly greater for organic scintillators. Organic crystals have a nonlinear relationship between pulse height and energy for heavy charged particles over a wide range of energy. The characteristics of several frequently used solid scintillators for the measurement of charged particles in geophysical applications are shown in Table 4.16-1.

Particle Range and Efficiency.   In selecting a scintillator size and shape for an application, the range of the particle or particles of interest in the scintillator is of prime consideration. In Figure 4.16-2 the range of protons [12], electrons [4], and alpha particles [28] in CsI (Tl), plastic scintillator and silicon materials are shown over the energy range 10 keV to 100 MeV.

Table 4.16-1. **Characteristics of Commonly Used Scintillators in Geophysical Applications**

Scintillator	Wavelength of Maximum Emission (Å)	Density (g/cm³)	Decay Time (μs)	Relative Light Output for Electron Excitation	Remarks
NaI(Tl)	4100	3.67	0.3	1.00	Highest light output Wide variety of crystal sizes available Hygroscopic; requires air-tight seal
CsI(Tl)	4200–5700	4.51	α 0.5 β 0.8	0.26–0.40	Available as large, clear crystals Relatively nonhygroscopic and convenient to handle Variation in decay time for different particles permits particle discrimination [11, 32] Exhibits long-lived light decay following intense excitation by particle fluxes [22]
CaF$_2$(Eu)	4350	3.18	0.90	0.50	Inert Nonhygroscopic High resistance to thermal and mechanical shock Low vapor pressure Does not exhibit long-lived light decay under particle radiation [15] Light output comparable to CsI(Tl) and anthracene Low electron backscatter characteristics
Anthracene	4400	1.24	0.03	0.48	Available in a variety of shapes and sizes up to several centimeters Widely studied
Plastic (polyvinyl-toluene)	4100	1.02	~ 0.002	0.21–0.33	Available in wide variety of shapes and sizes; can be easily machined and formed Nonhygroscopic Fastest decay time Low electron backscatter characteristics No long-lived phosphorescences Relatively low cost

In most cases, the scintillator is made somewhat thicker than the range of the maximum particle of interest. Figure 4.16-2 illustrates one of the main advantages of the solid scintillator, namely, the ability to stop energetic particles in relatively short distances.

The next important consideration is the conversion efficiency of a given scintillator material as a function of particle type and energy. Organic scintillators generally yield a lower absolute light output than inorganic scintillators for a given particle type and energy as shown in Table 4.16-1. The organic scintillators also exhibit a greater relative variation in light output for different particles of the same energy as shown in Figure 4.16-3. The light response of plastic scintillator is a linear function of electron energy down to energies below 100 keV [24, 33]. In contrast, the response to protons and alpha particles is distinctly nonlinear.

Inorganic scintillators do not exhibit such pronounced differences and nonlinearities although variations in response to different particles do exist [3, 16].

Photomultipliers. The parameters by which a photomultiplier is evaluated for scintillation applications are as follows [23, 29].

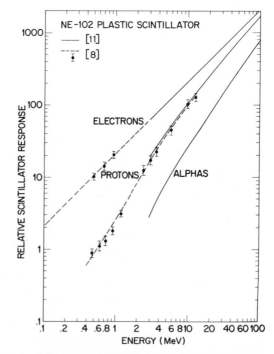

Figure 4.16-3 Relative light output of NE-102 plastic scintillator for protons, alpha particles and electrons over the energy range 0.1–100 MeV.

Quantum Efficiency. This is the number of photoelectrons emitted from the cathode per incident photon of energy. It is expressed as a percentage with values ranging between 10 percent for the Cs_3Sb (S-11) cathodes to 30 percent for the multi-alkali cathodes such as Na-K-C-Sb³ (S-20) and K-Cs-Sb² (RCA-115) at the wavelengths of maximum scintillation emission. The higher quantum efficiencies are desirable since the ultimate output signal amplitude and quality are proportional to the initial number of photoelectrons.

Cathode Sensitivity. This is a measure of the photocurrent emitted per lumen of incident light flux ($\mu A/lm$) for an emission spectral response corresponding to a 2854° K blackbody or as the photocurrent per watt of incident light flux (mA/W) at the wavelength of interest, but usually stated at the wavelength of maximum spectral response. In Figure 4.16-4, the emission characteristics of NaI (Tl), CaF_2 (Eu), and Pilot-B scintillators are superimposed on the relative response curves of several of the popular photocathode materials chosen for scintillation applications.

Gain. This is the ratio of the anode signal current to the photocathode signal current at a stated electrode potential difference. The current gain of a photomultiplier is given by

$$G = \delta^n = (KV_s)^n, \qquad (4.16\text{--}1)$$

where δ is the mean secondary emission factor of the dynodes, n is the number of dynodes, V_s is the potential difference between successive dynodes, and K is the straight line slope of the secondary emission factor as a function of dynode voltage. To illustrate the sensitivity of the photomultiplier gain to supply voltage variations, differentiate Eq. 4.16-1 with respect to V_s to give

$$\frac{dG}{dV_s} = nK(KV_s)^{n-1}, \qquad (4.16\text{--}2)$$

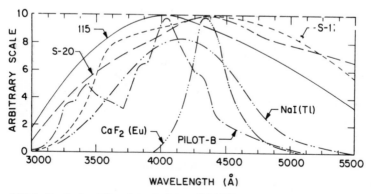

Figure 4.16-4 Spectral emission characteristics of NaI(Tl), CaF_2 (Eu) and Pilot-B scintillators compared to the response curves of three photocathode types.

and divide by Eq. 4.16–1 to give

$$\frac{dG}{G} = n\frac{dV_s}{V_s}. \tag{4.16-3}$$

Hence, if a gain stability of 1 percent is desired and a ten-stage multiplier ($n = 10$) is chosen, the dynode voltages must be stable to 0.1 percent to achieve this result. This requirement necessitates the use of a very high-quality high-voltage power supply. Miniature power supplies capable of providing such stability at low power consumption for satellite applications are available [31, 32].

Linearity. This is the extent and limit to which there is a direct proportional correspondence between illumination and output current. Since the scintillation illumination is often linearly proportional to the incident energy, it is important for spectrometer applications to establish the range of anode currents over which linearity exists within an acceptable limit. In general, a linear relationship exists over several orders of magnitude from the dark current to space charge saturation limits.

Frequency Response. This is the ability of the photomultiplier to respond to fast scintillation pulses. In most modern photomultipliers the rise time (the time that the electrical output takes to rise from 10 percent to 90 percent when the illumination is instantaneously increased) is of the order of a few nanoseconds and well matched to typical scintillator decay times.

Resolution. This is the ability of the scintillator-photomultiplier combination to discriminate between two closely spaced monoenergetic inputs. Resolution is usually measured by obtaining a pulse-height distribution of the pulses from a monoenergetic source of charged particles. The pulse-height resolution η is defined as the ratio of the full-width in pulse-height of the resulting distribution as measured at the half-intensity point to the pulse height at the maximum intensity point, i.e., $\eta = \Delta V_{1/2MAX}/V_{MAX}$. Therefore, the lower the pulse-height resolution, the greater the ability of the photomultiplier to discriminate between pulses of nearly equal height or energy. Pulse-height resolution is a function of the light output of a scintillator at a given energy and the quality of the photocathode. For scintillators, the resolution is approximately inversely proportional to $\sqrt{E}$. Typical values of pulse-height resolution for 1-MeV electrons in plastic scintillators are 10–15 percent of which 5–8 percent is attributable to the photomultiplier.

Stability. This is the variation in overall sensitivity as evidenced by shifts and drifts. Shifts are characterized by a rapid change in gain immediately following a change in luminous flux. Drifts are characterized by a change in gain as a function of time without any change in luminous flux. Both types of instabilities can be minimized to the order of a few percent through selection criteria. Shifts can occur as a result of rapid variations in counting rate.

Dark Current and Noise. Dark current is the anode current observed in the absence of any incident illumination. It sets a limit to the lowest intensity of continuous light which can be detected and therefore should be minimized. A reduction by a factor of 10 in the total dark current can be obtained by cooling a tube to 0°C. Noise is a consequence of the discrete nature of the incident light quanta and the statistical fluctuations of resulting electrons. Noise can be reduced through the selection of scintillators with high conversion efficiency. Noise exists in both the signal current and in the dark current.

Mechanical Integrity. The sensor must survive the acceleration, vibration, and shock of rocket and satellite launches without significant performance degradation. Ruggedization of glass envelopes and introduction of shorter and heavier lead lengths have resulted in a class of high-quality photomultipliers capable of surviving the severest of launch environments.

Temperature Effects. The influence of temperature on a photomultiplier is predominantly attributable to changes in the spectral sensitivity of the photocathode due to volume and surface effects in the cathode semiconductor. The average escape depth of the electrons increases as the temperature decreases while the resistance of the photocathode increases rapidly as the temperature decreases. These effects result in temperature coefficients that vary widely as a function of wavelength, cathode type and even from tube to tube of the same type. Figure 4.16-5 shows the relative response of several photomultipliers of the same type having S-11 photocathodes as a function of temperature at a wavelength of approximately 4100 Å. Also shown is the relative light output of three common scintillators as a function of temperature. In general, variations in scintillator response due to temperature are far less than those experienced in photomultipliers. The plastic scintillator is especially notable for its stability to temperature changes over a wide range. In practice, corrections for the temperature variations of the scintillator-photomultiplier combination are usually accomplished by the addition of temperature-compensating components in the high-voltage power supply.

Applications. Plastic scintillation detectors in conjunction with evaporated aluminum windows have been used extensively on polar-orbiting satellites as threshold detectors for measuring both low-energy (few-keV) electrons and protons in the auroral zones of the earth [26]. In this application the energy threshold is established by the thickness of the evaporation, and the total energy deposited in the scintillator above this threshold is measured. Protons are separated from electrons by the application of a sweep magnet and by the difference in energy threshold for the two particles to penetrate the aluminum window.

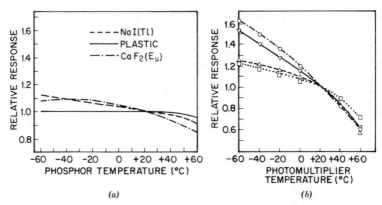

Figure 4.16-5 (*a*) Relative light output of NaI(Tl), plastic and CaF$_2$ (Eu) scintillators as a function of temperature; (*b*) relative response of several photomultipliers at 4100 A (S-11 photocathodes) as a function of temperature.

In a recent space application [17], an evaporated CsI(Tl) layer approximately 500–15000 Å thick was deposited on a plastic scintillator to discriminate and measure both low-energy protons (> 170 keV) and electrons (> 18 keV) on the basis of the particle pulse heights in both scintillators. At higher energies, extensive use has been made of the stopping power of solid scintillators.

Plastic scintillation detectors have been used to measure the high-energy protons in the inner radiation belt [31] as well as to measure the flux and spectra of energetic protons and electrons in solar-flare particle events [25].

4.16.2 Semiconductor Detectors

Principles of Operation. Semiconductor radiation detectors are solid-state ionization chambers in which solid, semiconducting silicon or germanium is used as the detection medium. When a charged particle enters a silicon semiconductor detector, it creates one free electron-hole pair for each 3.6 eV of energy deposited in the detector. The equivalent rate for germanium is one electron-hole pair for each 2.9 eV of deposition. Silicon detectors are predominantly utilized in charged particle spectroscopy while germanium detectors are used extensively for gamma-ray spectroscopy.

The charge-production rate in silicon is essentially independent of particle energy and specific ionization density (energy loss per unit path length) over a wide range of energies. Therefore, the detector response is linear over an equally wide energy range provided that the active depth (depletion region) of the detector exceeds the range of the particle and that the electric field across the depletion region is sufficiently large to separate the charge carriers before they recombine. The advantages of a solid, silicon material for stopping energetic charged particles were illustrated in Figure 4.16-2 along with two scintillators.

Figure 4.16-6 shows a cross-sectional view of one popular type of semi-conductor detector, the surface-barrier detector. This type of detector is essentially a large-area diode with a thin layer of *p*-type silicon covering the sensitive face of an *n*-type wafer. Electrical contacts for applying the depletion bias are made through a thin evaporated film of gold (40 μg/cm^2) over the *p*-type material and aluminum (40 μg/cm^2) over the rear surface of the *n*-type silicon wafer. The active area of the device is defined by the diameter D, while W is the depth of the depletion region and L is the total physical thickness of the wafer. Devices with active areas up to 10 cm^2 and with depletion thicknesses ranging from 10 to 1000 μm are commercially available.

The detector is sensitive only to the depth of the depletion region where an electrical field exists resulting from the back bias (V_b) applied to the diode. Particle radiation that enters the depletion region produces free charge carriers which are separated by the electrical field to produce output current pulses proportional in amplitude to the energy deposition. In most applications these current pulses which flow through the load resistor, R_L, are amplified by a charge-sensitive preamplifier.

The signal at the output of the semiconductor detector is much smaller than the signal from the scintillation detector. Using the best scintillation phosphor in terms of light output [NaI(Tl)] and a typical S-11 photoca-

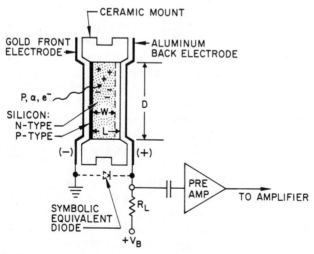

Figure 4.16-6   Cross-sectional view of the surface-barrier semiconductor detector.

thode sensitivity of 60 μA/lm, some 200 eV of particle energy is required to produce a single photoelectron from the cathode. This value is directly comparable to the 3.6-eV requirement per ion-electron pair in a silicon detector. The silicon detector is therefore 50 to 60 times more efficient in generating signal charge than the scintillation detector for the same energy input. Since the pulse-height resolution of the final signal is directly proportional to the square root of the initial number of electrons (the distribution is Gaussian), the silicon detector possesses an inherent resolution 7 to 8 times better than the scintillation detector. As mentioned in Section 4.16. 1, p. 374, a plastic scintillator-photomultiplier combination might realize a resolution of 100 keV at an electron energy of 1 MeV. A comparable thickness silicon detector might show a resolution of 12 keV at the same energy.

Space Applications. For space applications the semiconductor detector has several advantages over other types of radiation sensors. The detectors are small and can be stacked to obtain greater range for detecting higher energy particles or for performing complex particle identification [14, 25]. Bias for the detectors is usually below a few hundred volts so that high-voltage power supplies are not required. A high degree of voltage regulation is also not required, as in the case of the photomultiplier. Semiconductor detector response is essentially independent of particle type and energy over a wide range in marked contrast to organic scintillators. Semiconductor detectors have essentially 100 percent efficiency for the detection of protons, alpha particles, and electrons and have been used to measure particles as low as tens of keV and as high as several hundred MeV in energy. The energy resolution is also significantly better than scintillation detectors at comparable energies. Semiconductor detectors are, however, much more susceptible to radiation damage than scintillation detectors. Notable degradation in surface-barrier detectors is observed at proton fluxes of 10^{10} protons/cm^2. Lithium-drifted silicon detectors show radiation damage effects at fluxes as low as 10^8 protons/cm^2. In long-lived, earth-orbiting space missions, degradation can be anticipated as a result of exposure to protons in the inner and outer radiation belts of the earth and in solar-flare produced particle events. Operation of the detectors with the back aluminum window exposed to the incident radiation significantly increases the resistance to low-energy radiation damage [6].

Semiconductor detectors with proper mounting and design considerations are also capable of meeting the requirements for temperature, shock, vibration, light, and high vacuum encountered in space flight.

Characteristics. To determine the detector thickness necessary for a given space experiment, the range in silicon of the particles of interest must first be obtained from Figure 4.16-2 or from the nomogram shown in

Figure 4.16-7 [20]. To stop a 6-MeV proton, a 23 MeV alpha particle or a 250-keV electron, Figure 4.16-7 shows that a totally depleted detector of 300 μ thickness is required. A straight line from the specific resistivity (Ω cm) of the silicon material used in the detector through the depletion depth point gives the required bias voltage. The nomogram also gives the specific capacity of the detector in pF/mm^2. The detector capacitance along with cable and stray capacitance at the input to the preamplifier are important parameters in establishing overall system noise and hence energy resolution. When detectors are stacked to achieve greater depletion depth, the total capacitive load to the preamplifier is obtained by summing the detector capacitance and the stray capacitances. The overall noise in a stacked system is obtained by the mean square process,

$$N_T^2 = N_1^2 + N_2^2 + \ldots = \sum_i N_i^2, \qquad (4.16\text{--}4)$$

where N_i is the individual noise contribution from each detector. Operating temperature is another important consideration in the design of semiconductor detectors for space applications. Commercially available detectors are capable of operating over the temperature range -78 to $+45°C$. Degradation of performance is, however, experienced at elevated temperatures. Increasing the operating temperature causes the detector leakage current to increase by a factor of 3 for each 10°C rise, resulting in a noise width increase of approximately 1.7 per 10°C rise.

While decreasing the operating temperature of the detector reduces the junction noise and leakage current, the capacitance of the device does not change but remains a limiting parameter of instrument noise. The ultimate limit in low-temperature operation is the expansion coefficient of the silicon, the lavite ring in which the detector wafer is mounted and the bonding epoxy.

When a silicon surface-barrier detector is exposed to light, it produces a marked increase in current and noise. For space applications where it may be necessary for the detector to view sunlight, the detector can be made light-tight by reverse mounting and by the application of an evaporated layer of aluminum (~ 140 μg/cm^2) to the surface.

4.16.3 Electron-Multiplier Detectors

Low-energy charged particles are extremely important contributors to the outer radiation belt of the earth and to aurora phenomena. In fact, the major portion of the total energy input to aurora is supplied by charged particles having energies below 40 keV. In this energy region it is critically important that the amount of material surrounding the active volume of the detector be minimized or eliminated entirely. Scintillation detectors having only a very thin evaporated coating of aluminum over the scintilla-

380 *In Situ and Laboratory Sensors*

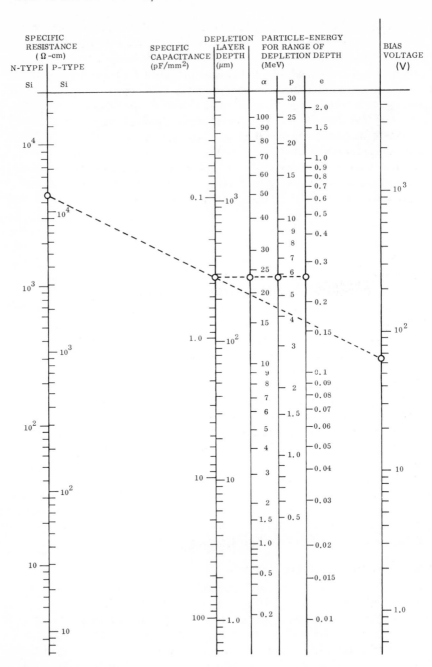

Figure 4.16-7 Silicon detector parameters nomogram.

tor to reduce light sensitivity have been effectively used to detect electrons with energies as low as a few keV. The pulse height in the scintillator at such energies is, however, too low for effective pulse analysis. The total anode current rather than individual pulses is usually measured as representative of the total incident energy flux. This technique, however, is not very sensitive to low particle fluxes. A similar situation exists for semiconductor detectors, but since the detection efficiency is higher than in scintillators, pulse-height analysis techniques for energies in the tens of keV region are possible.

Bare electron multipliers utilizing the dynode structure of a photomultiplier tube (see Figure 4.16-1) have also been used to directly detect charged particles. The absence of a photon-electron conversion process permits the use of a material with a high work function, thereby reducing thermal noise. The detector response extends to extremely low-energy particles because of the complete elimination of an entrance window. The anode current or pulses from such a detector provide a measure of the incident particle flux. The output pulse-height distribution of such a device is usually quite broad and the stability of the dynode structure, particularly when exposed to air, is variable.

A device that has many of the desirable features for low-energy detection in space is the continuous-channel electron multiplier [34]. The channel multiplier, as shown in Figure 4.16-8, is a curved glass capillary tube having dimensions such that the length-to-inside-diameter ratio is 50 to 100. The inside diameter is from less than a tenth of a millimeter to more than a millimeter. A layer of semiconducting material having secondary electron emission characteristics similar to tungsten is deposited on the

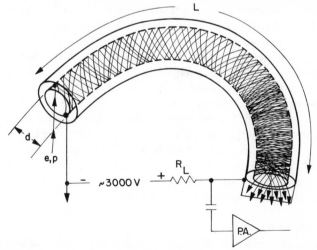

Figure 4.16-8   Cross-sectional view of a channel-multiplier sensor.

interior surface of the tube. When a potential difference is applied between the ends of the tube, an axial electrical field is established down its length. Any electron ejected from the inner surface by photoelectric or secondary emission processes will thus be accelerated down the tube. Simultaneously, the electron will drift across the tube with the lateral velocity acquired in the ejection process. The electron multiplication occurs when the potential difference and tube dimensions are such that these free electrons gain enough energy from the electric field between encounters with the surface that, on the average, more than one secondary electron is generated at each encounter. In this way, a single electron ejected at the low potential input end of the tube can produce an electron cascade at the high potential or output end of the channel. Electron gains of 10^8 are realized in currently available units. The units are simple, inexpensive, small, rugged, and fairly stable with respect to changes in the secondary emission ratio when exposed to air. Channel multipliers exhibit pulse-height resolution of approximately 50 percent when operated at sufficiently high voltage to produce saturated pulses.

The efficiency of the input material for the detection of low-energy particles is quite high. Recent measurements [1, 21] indicate the absolute efficiency for the detection of low-energy electrons is near unity at 1 keV and between 0.5 and 0.8 at 10 keV. The efficiency for detecting low-energy protons is not as well established as for electrons. Reported proton detection efficiencies at an incident energy of 1 keV vary between 0.27 [8] and 0.8 [25].

Several measurements on the fatigue of channel multipliers when exposed to a variety of conditions such as certain vapors, high counting rates and potential gradient have been reported [13, 27]. When properly processed and selected, channel multipliers have demonstrated sufficiently high stability for operation in space over several years.

Since the channel multiplier inherently provides no discrimination between particles of different energies, some external method of particle type and energy selection must be employed. One type of low-energy particle spectrometer [27] employs analyzing magnets and thin foils to establish energy thresholds. An electron flux enters an analyzing magnet that bends selected energies into the channel multiplier aperture. A second detector for the measurement of protons is included. This measurement utilizes a sweep magnet to deflect electrons away from the aperture. Protons having energies greater than the range of the thin foil mounted over the channel multiplier aperture are detected.

4.16.4   Nucleonics

The common characteristic of the three types of particle radiation detectors discussed in this section is that a pulse of electrical charge, Q, results when a charged particle is detected. The magnitude, rise time and duration of this charge pulse, however, varies significantly with the type of detector and the energy of the charged particle. The scintillation detector and the electron multiplier produce charge pulses at their respective anodes that are from 10^2 to 10^3 times greater than the charge pulse from a semiconductor detector for an equivalent energy input. While a charge-sensitive preamplifier is normally used in all three cases to couple this charge to subsequent electronic circuitry, the requirements on quality are drastically different.

In the case of the scintillation and electron-multiplier detectors, the prime role of the preamplifier is to convert the charge at the high-impedance anode into a voltage signal with a low-output impedance capable of driving a coaxial cable and other circuits. The preamplifier acts essentially as an impedance transformer. In many cases, no additional amplification is necessary prior to performing subsequent analysis. Since the signal is large from these detectors, electronic noise in the preamplifier circuitry is usually not an important parameter. The inherent resolution of the scintillator-photomultiplier combination and the electron multiplication process become the limiting factor of performance and not the noise of external electronics. Simple, low-cost preamplifiers can therefore be utilized with these types of detectors.

Since the charge pulse from semiconductor detectors is significantly smaller than from the previous detectors, the requirements of the preamplifier-amplifier combination are much more severe. Also, since the inherent resolution of the detector is substantially superior to either the scintillation or electron-multiplier detectors, the electronic noise must be maintained well below the detector noise if maximum overall system performance is to be achieved. A sophisticated charge-sensitive preamplifier is therefore required for this application in which state-of-the-art, low-noise field-effect transistors are used as the input elements. Selection of critical components and high-quality devices is a necessity. Amplification is usually incorporated into the preamplifier and in addition a main amplifier always follows. Shaping of the preamplifier pulses is performed in the main amplifier to obtain optimum signal-to-noise performance [9].

Once signal amplitudes in the several-volt range are obtained, the pulses can be analyzed by a variety of means. The simplest counting system involves an integral discriminator and a means of accumulating pulses over a period of time such as a digital counter or an analog ratemeter. All pulses above a selected voltage (i.e., energy) level are counted in this scheme. In space applications, this simple technique has been employed extensively. The next degree of sophistication involves the use of differential discriminators where pulses falling within selected levels or "windows" are accumulated in separate counters or the average rate within each "window" is converted to an analog voltage such as occurs in the operation of the rate meter.

The ultimate degree of sophistication involves the use of pulse-height analyzers [2] and particle identification circuitry [14] to obtain substantial detail on the incoming particles. In the pulse-height analyzer technique, the pulses are sorted into many small energy bins to reproduce the spectrum of the incoming particles. The content or the "address" of each energy bin is transmitted for subsequent sorting on the ground. In many space applications, coincidence and anticoincidence requirements must be satisfied between multiple detectors before the pulse-height analysis is performed.

References

1. Archuleta, R.J., and S.E. DeForest, "Efficiency of Channel Electron Multipliers for Electrons of 1–50 keV," *Rev. Sci. Inst.* **42**, 89, 1971.
2. Bakke, J.C., J.B. Reagan, R.D. Reed, W.L. Imhof, and J.D. Matthews, "A High-Sensitivity Particle Spectrometer for the Measurement of Polar-Cap-Absorption Events," *IEEE Trans. Nucl. Sci.* **NS-17**, 91, 1970.
3. Baskin, S., R.R. Carlson, R.A. Douglas, and J.A. Jacobs, "Response of CsI (Tl) Crystals to Energetic Particles," *Phys. Rev.* **109**, 434–436, June 1958.
4. Berger, M., and S.M. Seltzer, "Tables of Energy Losses and Ranges of Electrons and Positrons," NASA Report SP-3012, 1964.
5. Birks, J.B., *The Theory and Practice of Scintillation Counting,* Macmillan Co., New York, 1964.
6. Coleman, J.A., D.P. Love, J.H. Trainor and D.J. Williams, "Low-Energy Proton Damage Effects in Silicon Surface Barrier Detectors," *IEEE Trans. Nucl. Sci.,* **NS-15**, No. 1, 482, 1968.
7. Craun, R.L., and D.L. Smith, "Analysis of Response Data for Several Organic Scintillators," *Nucl. Inst. Meth.* **80**, 239–244, 1970.
8. Egidi, A., R. Marconero, G. Pizzella, and F. Sperli, "Channel Fatique and Efficiency for Protons and Electrons," *Rev. Sci. Instr.* **40**, No. 2, 88, 1970.
9. Fairstein, E., and J. Hahn, "Nuclear Pulse Amplifiers—Fundamentals and Design Practice," *Nucleonics* **23**, No. 7, 56, July 1965; **23**, No. 9, 81, Sept. 1965; **23**, No. 11, 50, Nov. 1965; **24**, No. 1, 54, Jan. 1966; No. 3, 68, Mar. 1966.
10. Gooding, T.J., and H. G. Pugh, "The Response of Plastic Scintillators to High-Energy Particles," *Nucl. Inst. Meth.* **7**, 189–192, 1960.
11. Hrehuss, G., "A New Method of Mass Discrimination," *Nucl. Inst. Meth.* **8**, 344–347, 1960.
12. Janni, J.F., "Calculations of Energy Loss, Range, Pathlength, Straggling, Multiple Scattering, and the Probability of Inelastic Nuclear Collisions for 0.1-to 1000-MeV Protons," Air Force Weapons Laboratory Report AFWL-TR-65-150, September 1966.
13. Klettke, B.D., N.D. Krym, and W.G. Wolber, "Long Term Stability Characteristics of Commonly Used Channel Electron Multipliers," *IEEE Trans. Nucl. Sci.* **NS-17**, No. 1, 72–80, Feb. 1970.

14. Lanzerotti, L.J., H.P. Lie, and G.L. Miller, "A Satellite Solar Cosmic Ray Spectrometer with On-Board Particle Identification," *IEEE Trans. Nucl. Sci.* **NS-16**. No. 1. 343. 1969.

15. Menefee, J., C.F. Swinehart, and E.W. O'Dell, "Calcium Fluoride as an X-Ray and Charged Particle Detector," *IEEE Trans. Nucl. Sci.* **NS-13**, 1, 720 724, Feb. 1966.

16. Meyer, A., and R.B. Murray, "Scintillation Response of Activated Ionic Crystals to Charged Particles," *IRE Trans. Nucl. Sci.* **NS-7**, 22–24, June 1960.

17. Mozer, F.S., F.H. Bogott, and C.W. Bates, Jr., "Development of a Double-Layered Scintillator for Separating and Detecting Low-Energy Protons and Electrons," *IEEE Trans. Nucl. Sci.* **NS-15**, 144–146, June 1968.

18. Murray, R.B., "Scintillation Counters," *Nuclear Instruments and Their Uses —Volume 1*, Chapter 2, Arthur H. Snell, Ed. John Wiley and Sons, New York, 1962.

19. Neiler, J.H., and P.R. Bell, "The Scintillation Method," *Alpha-, Beta-, and Gamma-Ray Spectroscopy*, Vol. 1, pp. 245–302, Amsterdam, North Holland, 1965.

20. ORTEC Catalog 1002, 28, 1970.

21. Paschmann, G., E.G. Shelley, C.R. Chappell, R.D. Sharp, and L.F. Smith, "Absolute Efficiency Measurements for Channel-Electron-Multipliers Utilizing a Unique Electron Source," *Rev. Sci. Instr.* **41**, 1706, 1970.

22. Peterson, L.E., "Radioactivity Induced in Sodium Iodide by Trapped Photons," *J. Geophys. Res.* **70**, No. 7, 1762–1765, 1965.

23. "Photomultiplier Manual," RCA Technical Series PT-61, 1970.

24. Pleitz, W., "Light Yield of the Plastic Scintillator NE-102A for Electrons from 3-11 keV," *Z. Phys.* **190**, No. 4, 446–454, 1966.

25. Reagan, J.B., J.C. Bakke, J.R. Kilner, J.D. Matthews, and W.L. Imhof, "A High-Resolution, Multiple Particle Spectrometer for the Measurement of Solar Particle Events,'. *IEEE Trans. Nucl. Sci.* **NS-19**, No. 1, 554, 1972.

26. Reagan, J.B., D.L. Carr, J.D. McDaniel, and L.F. Smith, "Satellite Instrumentation for the Measurement of Auroral Phenomena," *IEEE Trans. Nucl. Sci.* **NS-11**, No. 3, 441, 1964.

27. Reed, R.D., E.G. Shelley, J.C. Bakke, T.C. Sanders, and J.D. McDaniel, "A Low-Energy Channel-Multiplier Spectrometer for ATS-E," *IEEE Trans. Nucl. Sci.* **NS-16**, No. 1, 359, 1969.

28. Rich, M., and R. Madey, "Range Energy Tables," University of California Positron Laboratory Report UCRL-2301, 1954.

29. Schonkeren, J.M., *Photomultipliers*, Eds. H. Kates and L.J. Thompson, Amperex Electronics Co., April 1970.

30. Shute, D.S., "A Hybrid Cockroft-Walton Multi-Dynode Photomultiplier Supply for Space Applications," *IEEE Trans. Nucl. Sci.* **NS-17**, 130, Feb. 1970.

31. Smith, R.V., J.B. Reagan, and R.A. Alber, "Use of Scintillation Counters for Space Radiation Measurements," *IRE Trans. Nucl. Sci.* **NS-9**, 386, June 1962.

32. Storey, R.S., W. Jack, and A. Ward, "The Fluorescent Decay of CsI (Tl) for Particles of Different Ionization Density," *Proc. Phys. Soc.* **72**, 1, 1958.

33. Swank, R.K., "Characteristics of Scintillators," *Ann. Rev. Nucl. Sci.* **4,** 111–140, 1954.
34. Wiley, W.C., and C.F. Hendee, "Electron Multipliers Utilizing Continuous Strip Surface," *IRE Trans. Nucl. Sci.* **NS-9,** 103, 1962.

S. Ben-Yaakov and I. R. Kaplan

4.17 pH SENSORS

Since the introduction of the pH concept in 1909, pH measurements have become crucial to all studies involving aqueous solutions [3, 17]. The pH concept is based on the assumption that when an acid (HA) or base (B) is introduced into water it reversibly ionizes or dissociates liberating a hydronium or hydroxyl ion as follows:

$$HA + H_2O \rightleftharpoons A^- + H_3O^+$$

or

$$B + H_2O \rightleftharpoons B^+ + OH^-.$$

For historical reasons and for simplicity, the hydronium ion is generally referred to as the hydrogen ion, H^+. As the concentration of the hydrogen ion, $^cH^+$, is very low (10^{-6} to 10^{-9} g/liter) for most natural waters, pH was originally defined in terms of the logarithm of hydrogen ion concentration. Later the concept of pH was redefined in terms of activity ($^aH^+$), which more accurately expresses the chemical state of the hydrogen ion in solution:

$$pH = -\log {}^aH^+. \tag{4.17-1}$$

In real terms, pH = 7 at 25°C, is accepted as being neutral, solutions having pH values below 7 increase in acidity toward pH = 0 (unit normality of acid) and increase in basicity toward pH = 14 (unit normality of an alkaline solution).

In nature, values for pH range from 1.0 to 10.5 [1]. Lower values occur most frequently in volcanic areas where halogens and oxides of sulfur are emitted and hydrolyzed to acid, and in acid mine waters. In the latter case, acids form through the microbiological oxidation of metal sulfides, in particular iron sulfide (pyrite). The bacteria which perform these oxidations (Thiobacillus spp and Ferrobacillus spp) are able to proliferate under such highly acid conditions. At the other end of the scale, very high pH values (10.5) have been measured in the soda lakes of the Mojave Desert, California. The pH here is controlled by the dissociation of sodium carbonate.

A knowledge of pH is important to earth scientists for two reasons. First, most organisms live within a narrow pH range or can tolerate only a narrow range. Thus, pH is an important parameter in ecological control [1]. Second, most elements, complexed or free, undergo reactions in the presence of water leading to ion formation. These reactions are controlled by pH which can determine the nature of the products [15]. It is possible to predict the nature of the ion and its maximum concentration in simple aqueous solutions from a knowledge of the thermodynamic dissociation constants of hydroxides or other complexes and from a measurement of pH.

4.17.1 The Practice of pH Measurements

Two methods have been used for pH determination [2, 5, 10, 17, 19]: colorimetric and electrochemical. The colorimetric method uses organic compounds which undergo a color change as a function of pH. Indicators can be added to the test solution, and the color attained by the mixture will be a function of pH. Alternately, a piece of paper previously treated with an indicator (e.g., litmus paper) can be exposed to the unknown solution to produce a color change. The actual pH value is determined by comparing the indicator's color to a calibration chart, or by measuring the optical absorbance with a spectrophotometer. The resolution of the colorimetric method is poor (approximately 0.1 pH unit for maximum resolution).

The electrochemical [1, 3] technique uses electrochemical cells which produce an electrical potential shift proportional to pH. Two half-cell electrodes, a reference and an indicator, are required. The indicator electrode is sensitive to $^aH^+$ and produces a proportional potential, which, when added to the fixed reference electrode potential completes an electrochemical cell. This output potential (emf) of a pH electrode assembly, the sum of the potentials of the reference and indicator electrode, can be measured by an electronic instrument. Under ideal conditions the potential of the reference electrode can be considered constant. In practice the reference electrode potential can undergo changes which introduce uncertainties in the pH determination.

The hydrogen electrode [18] was the first electrochemical system to be used as a pH indicator, since it is reversible (in the thermodynamic sense) to hydrogen ions. This electrode consists of a thin platinum sheet over which hydrogen gas is bubbled. The assembly is placed in the test solution and the potential of the platinum metal is compared with the potential of a reference electrode inserted into the solution. This method is extremely cumbersome, and the stream of hydrogen gas through the solution can introduce errors by stripping dissolved gases from the solution.

The glass membrane electrode [14] has considerably simplified and increased the precision of pH measurements. Most pH measurements are made with a glass electrode which produces an electrical potential proportional to the pH of the test solution.

The Glass and Reference Electrodes. A schematic representation of a pH sensor composed of a glass membrane electrode and a reference electrode is shown in Figure 4.17-1. The glass membrane is made of a specially composed thin glass bulb which develops an electrochemical potential as a function of hydrogen ion activities on both sides of the membrane. A linear function is obtained, [14] relating electrode potential to hydrogen ion activity with a slope that approaches, in most cases, the theoretical Nernst slope, which can be expressed by*

$$E_m = -2.303 \frac{RT}{F} \log \frac{{}^aH_i^+}{{}^aH_e^+} + E_{as}, \qquad (4.17\text{--}2)$$

where E_m is the potential across the glass membrane, R is the gas constant (8.3144 J/°K/mole), T is the absolute temperature in °K, F is the Faraday number (96493.1 C/g equivalent), ${}^aH_i^+$ is the activity of hydrogen ion in internal solution, ${}^aH_e^+$ is the activity of hydrogen ion in external solution and E_{as} is the asymmetry potential. The asymmetry potential, E_{as}, originates from the difference in response of the two sides of the membrane. This asymmetry results in a residual potential, even when the external solution is made equal to the internal one.

The potential generated by the glass electrode is thus a linear

---

*The sign of this equation was assigned by assuming that the potential is measured with respect to the external solution. Although this is opposite to the sign convention normally accepted by physical chemists [18], it is consistent with the practical pH measurement procedure.

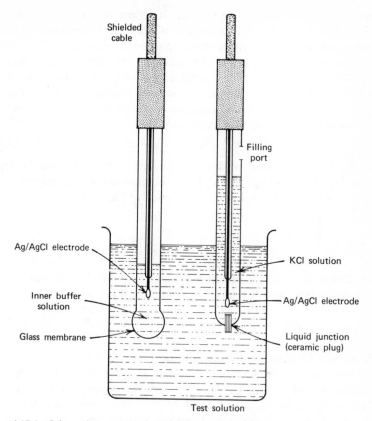

Shielded cable

Filling port

Ag/AgCl electrode

KCl solution

Inner buffer solution

Ag/AgCl electrode

Glass membrane

Liquid junction (ceramic plug)

Test solution

Figure 4.17-1 Schematic representation of a glass and a reference electrode.

function of the pH of the external solution, as shown in Eq. 4.17–3, derived from Eqs. 4.17–1 and 4.17–2.

$$E_m = -\frac{2.303RT}{F}\text{pH}_e + \frac{2.303RT}{F}\text{pH}_i + E_{as}, \qquad (4.17\text{–}3)$$

where pH_i is the known pH of the internal solution and pH_e is the measured pH. There is no way to directly measure E_m, since any attempt to measure the emf generates unknown electrochemical potentials between the solution and the measuring electrodes. The common practice is to use electrodes which generate a reproducible potential that can be determined during standardization of the pH sensor. This is achieved by using reference electrodes such as calomel or silver–silver–chloride which are reversible to the chloride ion and generate a constant reproducible emf when placed in a solution of constant chloride ion activity. The external reference cell cannot be placed directly in the test solution since the chloride ion activity of this solution is unknown. Instead it is placed in a solution of known concentration, generally a concentrated potassium chloride (KCl) solution, which communicates with the test solution through a liquid junction (Figure 4.17-1). However, such a junction between two different solutions generates an emf. The potential measured between the glass and the reference electrode (E_s) is therefore composed of the sum of emf's generated by the various electrochemical cells and is

$$E_s = E_{rr} + E_j + E_m + E_{gr}, \qquad (4.17-4)$$

where E_{rr} is the potential of the internal cell of the reference electrode, E_j is the liquid junction potential, E_m is the glass membrane potential and E_{gr} is the potential of the internal reference of the glass electrode. Combining Eqs. 4.17-3 and 4.17-4 gives

$$pH_e = -\frac{F}{2.303RT}E_s + pH_i + \frac{F}{2.303RT}[E_{gr} + E_{rr} + E_j + E_{as}].$$
$$(4.17-5)$$

Proper design of a pH sensor can make pH_i, E_{gr}, and E_{rr} constant for any given temperature, and for this condition Eq. 4.17-5 becomes

$$pH_e = -\frac{F}{RT2.303}(E_s - E_j - E_{as}) + K_t, \qquad (4.17-6)$$

where K_t is a temperature-dependent constant.

The two interfering potentials (E_j and E_{as}) are inherent in any pH measurement and are taken into account by proper calibration. The practical pH scale makes use of standard (buffer) solutions to calibrate the pH glass and reference electrode assembly. Long-term stability of the pH sensor is directly related to the constancy of the inherent parameters E_{as}, E_j, K_t, and the stability of the glass membrane potential slope. The stability of the asymmetry potential (E_{as}) is the limiting factor.

Equation 4.17-6 indicates that the sensitivity of the pH sensor ($\Delta pH/\Delta E_s$) at a given temperature is constant and equal to 16.9044 $\Delta pH/V$ or 59.156 mV per pH unit at 25°C. The theoretical sensitivity at any other temperature can be found by evaluating $F/2.203RT$. The output potential of the pH sensor is a function of temperature. In practice the sensor is standardized in a buffer at the same temperature as the tested solution (see Section 4.17.3). Special calibration procedures are required when the standardizing buffer and the test solution are not at the same temperature. An example of such a procedure is given in Section 4.17.3, p. 397.

Several pH sensors (glass and reference electrodes) with a variety of characteristics are available to meet particular applications. Combination electrodes, which contain both a reference and a glass electrode in one construction are available.

Instrumentation. Special instrumentation is required for measuring pH with a glass and reference electrode [3]. The main difficulties are the extremely high internal impedance of the glass electrode and the low dc output potential. The internal impedance of the sensor is a function of the thickness of the glass membrane and its composition and it may range between 10^7 and 10^{10} Ω. This necessitates that the input stage, which should be a low-drift dc amplifier, must have an extremely high input impedance. Another requirement is that the input current to the amplifier be very low to decrease the voltage drop in the sensor. The specific requirements of the amplifier clearly depend on the sensor to be used. A typical pH meter for general use has an input impedance of 10^{13} Ω, input current of 10^{-12} A, and short-term stability of 0.1 mV.

The pH meters can be divided into potentiometric and direct-reading instruments. Most instruments are direct-reading. A block diagram of a direct reading pH meter is given in Figure 4.17-2. The input stage of a direct-reading meter should conform with the requirements outlined above, which have been realized by using electrometer amplifiers, chopper amplifiers, FET, MOS FET, and in special applications vibrating reed electrometers. The buffered, and generally amplified voltage, is then converted to pH units by a resistor network that is adjusted (either manually or automatically by a thermistor placed in solution) to give the proper voltage-pH relation at any given temperature. The temperature adjustment control is part of a resistive attenuator and is adjusted to obtain the desired volt-pH ratio. The "standardizing" control is an offset voltage control, by which the pH is adjusted to read the pH of the buffer during calibration. Prior to measurement, the sensor and meter are standardized in a calibration (buffer) solution by adjusting the meter with a bucking voltage control to read the pH of the buffer.

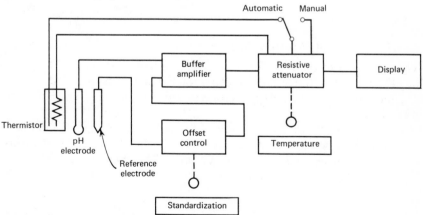

Figure 4.17-2   Block diagram of a direct-reading pH meter.

4.17.2 The Practical pH Scale

An error due to liquid junction potential is inherent in any practical pH measurement which uses a reference electrode with a liquid junction. The magnitude of this emf is negligibly small when the test solution has a low ionic strength (< 0.1 mole/liter). Since there is no way to directly measure the liquid junction potential, all pH measurements are comparative rather than absolute.

The practical pH value is defined operationally [2, 4] in terms of the observed difference of the potential output of the pH sensor when the standard reference solution is replaced by a test solution:

$$pH_x = pH_s + K(E_s - E_x),\tag{4.17-7}$$

where pH_x is the pH of the test solution, pH_s is the pH of the standard solution, E_x is the output of the pH cell sensor when placed in the test solution, E_s is the output of the pH sensor when placed in the standard solution, and K is a constant approximately equal to $F/2.303RT$.

This operational definition has been adopted universally and serves as the basis for electrometric pH determinations. The United States National Bureau of Standards* has assigned pH values to several buffer solutions used as standards for pH determinations [4]. Maximum relative accuracy of these standards is 0.01 pH unit.

The standard solutions can also be used to measure K (Eq. 4.17–7) for a given sensor. This is done by monitoring the output of the sensor when placed in two standard solutions and using the relation

$$K = \frac{pH_{s_1} - pH_{s_2}}{E_{s_2} - E_{s_1}}.\tag{4.17-8}$$

Information on glass electrode errors is available [2, 21].

4.17.3 Oceanic pH Measurements

The pH is a controlling factor in ecology and ionic distribution. Changes in pH within sea water are indicative of biological (metabolic) and inorganic processes. Thus, pH is a property which helps identify a particular water body.

The pH range in open ocean water [20] is between 8.4 and 7.5. Surface waters tend to be most alkaline, probably due to CO_2 (or H_2CO_3) consumption by algae, and show a minimum of 7.5–7.6 at a depth of a few hundred meters as a result of excretion of carbon dioxide during the

---

*The British Standards Institution has also established a pH scale that is somewhat different from the NBS scale.

decomposition of organic matter below the photic zone. The deep sea (> 4 km) appears to show a slight downward decrease.

The processes controlling the pH of the ocean are not entirely understood. Two mechanisms are most widely accepted. The first involves equilibrium among the carbonate (calcium carbonate)-bicarbonate and carbon dioxide systems:

$$CO_3^= \underset{-H^+}{\overset{+H^+}{\rightleftharpoons}} HCO_3^- \underset{-H^+}{\overset{+H^+}{\rightleftharpoons}} H_2CO_3 \rightleftharpoons H_2O + CO_2. \qquad (4.17\text{-}9)$$

The second involves equilibrium between clays, entering the ocean from rivers or on the ocean floor, and the overlying water, which results in an exchange of hydrogen ions for metal cations, as shown in Eq. 4.17-10 during the conversion of kaolinite to chlorite:

$$Al_2Si_2O_5(OH)_4 + SiO_2 + 7H_2O$$
$$+ 5Mg^{++} \rightleftharpoons Mg_5Al_2Si_3O_{10}(OH)_8 + 10H^+. \quad (4.17\text{-}10)$$

In the first case, a change in pH of the ocean could drastically affect the CO_2 content of the atmosphere. In the second case hydrogen ion reactions involving clays may be the most important control mechanism for a steady-state (long-term) composition of ocean waters. A comprehension of the pH controlling mechanism and its consequences is a significant reason for measuring and understanding the variation of pH in the ocean.

Until recently, most pH measurements on bodies of water were made by bringing samples to the surface in bottles, allowing them to warm to some temperature, measuring the pH relative to a buffer solution and making pressure-temperature corrections [22]. Such measurements suffered from alteration occurring from reactions between the collecting vessel and the water and through loss of volatiles during collection, temperature change, and measurement.

In Situ Glass and Reference Electrodes. The glass membrane electrode can be used for in situ oceanic pH measurements because the glass membrane maintains its sensitivity to pH at high hydrostatic pressures when both sides of the membrane are exposed to the same pressure [11–13]. Figure 4.17-3 is a cross section of a glass electrode that has been successfully used in oceanic pH measurements to a depth of 3 km [6].

The glass electrode [9] consists of two compartments, a silicone fluid compartment and the internal solution compartment. The sections are pressure equalized by a flexible diaphragm which is made of plastic tubing extending into the silicone fluid compartment. Another flexible tubing equalizes the internal pressure of the electrode with the external hydrostatic pressure to eliminate any differential pressure across the thin glass membrane. The potential of the glass electrode is measured against a silver–silver-chloride (Ag/AgCl) reference electrode (Figure 4.17-4) com-

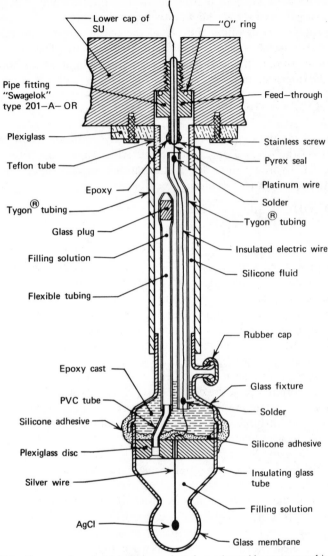

Figure 4.17-3 Cross section of a high-pressure glass electrode used in oceanographic measurements.

posed of two compartments, the upper silicon fluid compartment and the lower KCl filling solution compartment. The KCl solution is compressed by a rubber bulb to maintain a hydrostatic pressure higher than the outside pressure to ensure an outflow of the KCl through a nylon wick tip which serves as the liquid junction.

In Situ Instrumentation. The high internal resistance and the low dc output of the pH sensor demands in situ signal conditioning. The buffered signal can then be transmitted to the surface via an electrical cable, or recorded in situ by a suitable recording unit.

Direct analog transmission over long electrical cables can result in appreciable errors due to attenuation along the line. A more reliable approach is to convert the signal to a format which does not depend on amplitude measurement [8]. Figure 4.17-5 is a block diagram of a probe using this method. The output signal of the pH sensor is buffered and converted to a frequency signal by an analog-to-frequency converter [7]. This signal is either sent to the surface or recorded in place by a magnetic tape recorder unit. A more detailed schematic of the buffer amplifier is given in Figure 4.17-6. The high input impedance is achieved by an operational amplifier. The dc drift of the operational amplifier is monitored by shorting relay B and measuring the output current. Relays A and B are special dry reed relays with an insulation resistance of 10^{15} Ω.

Results. The practical standardization of the pH sensor (Section 4.17.2) assumes that both the standard buffer and test solution are at the same temperature. The response of the high pressure electrode is within 1 to

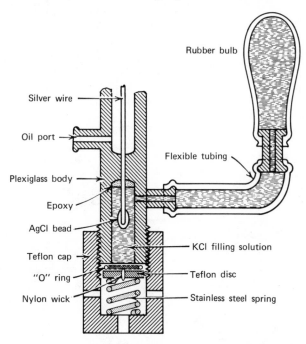

Figure 4.17-4 Cross section of a high-pressure reference electrode used in oceanographic measurement.

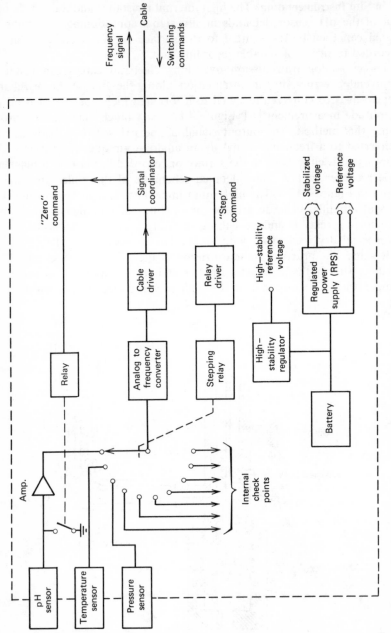

Figure 4.17-5 Block diagram of an oceanographic probe designed for in situ pH measurement.

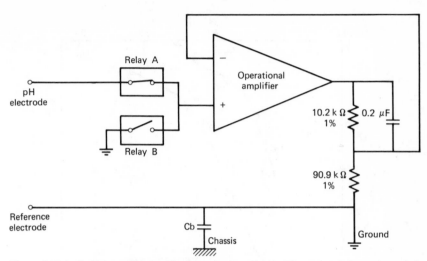

Figure 4.17-6 Buffer amplifier used in the probe described in Figure 4.18-5. Operation amplifier is model 311 manufactured by Analog Devices Co. (U.S.A.).

2% of the theoretical slope. Hence, if the standard solution is within 1 pH unit of the expected pH, the error is only 0.01–0.02 pH units if the slope is assumed to be ideal $F/2.303RT$. The pH sensor can be calibrated by the following procedure: A buffer with a pH value close to the expected pH range in which the pH sensor is immersed, is cooled down to 0°C and allowed to warm slowly. The sensor output is recorded every 1°C until the maximum expected temperature is reached. These data, the pH of the buffer and Nernst's slope at the in situ temperature range, are then used to calculate the true pH from the in situ measurement by applying Eq. 4.17-7.

Asymmetry potential changes with time are monitored by periodically inserting the pH sensor in the buffer solution. The glass electrode asymmetry potential shifts with pressure in the order of 1 mV per 10,000 psi.

A typical pH profile obtained twenty miles west of San Diego is shown in Figure 4.17-7. The insert in the figure demonstrates the ability of the probe to follow small changes in pH. The reproducibility of the pH measurement was ± 0.02 pH units and the relative accuracy estimated to be 0.03 pH units. The response time of the pH sensor is two seconds, but the thermal time constant is 1 min.

Special Application. The oceanographic pH sensor has been used to perform in situ carbonate saturometer experiments [8]. The pH is measured before and after a sample of sea water has been equilibrated with a solid phase of $CaCO_3$ [23]. The pH shift is used to calculate the amount of $CaCO_3$ that has been precipitated from, or dissolved in, the sea water [8]. Figure 4.17-8 is a block diagram of the in situ carbonate saturometer. A glass

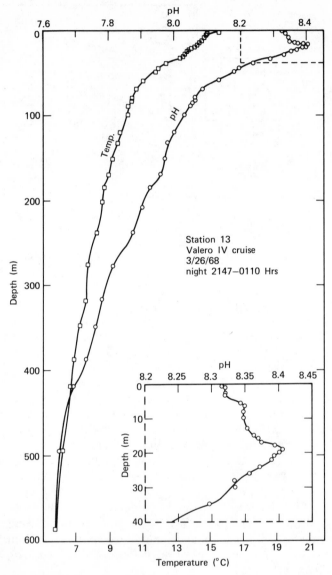

Figure 4.17-7 A pH profile measured on the continental borderland of Southern California with oceanographic probe.

electrode is inserted in a plexiglass cup filled with coarse CaCO₃ and connected through a hose to a pump and a solenoid-operated valve. Sea water is pumped through the cell, and the pH electrode then registers the pH of the sea water. During the off period the trapped sea water reacts with

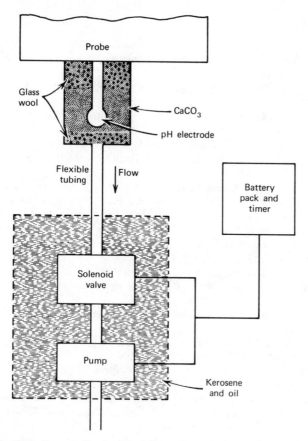

Figure 4.17-8 Block diagram of an in situ calcium carbonate saturometer.

the $CaCO_3$, moving toward equilibrium with the solid phase. A pH change occurs according to Eq. 4.17–11 depending on whether the water is undersaturated or supersaturated with respect to calcium carbonate.

$$2CaCO_3 + 2H^+ \rightleftharpoons 2HCO_3^- + 2Ca^{++}. \qquad (4.17-11)$$

Information on the degree of saturation of $CaCO_3$ in the oceans is important for understanding the mechanisms controlling the amount of $CaCO_3$ in oceanic sediment and the carbon dioxide and calcium cycles in nature.

Acknowledgment

The instrumentation studies reported in this section have been supported by a contract from the AEC Division of Biology and Medicine, Contract No. AT (04-3)-34 PA 178.

References

1. Baas-Becking, L. G. M., I. R. Kaplan, and D. Moore, "Limits of the Natural Environment in Terms of pH and Oxidation Reduction Potentials". *J. Geology* **68**, 243–284, May 1960.
2. Bates, R. G., *Determination of pH, Theory and Practice*, John Wiley and Sons, New York, 1964.
3. Bates, R. G., *Electrometric pH Measurements*, p. 312, John Wiley and Sons, New York, 1954.
4. Bates, R. G., "Meaning and Standardization of pH Measurements," *Symposium on pH Measurements*, Am. Soc. Testing Material Publication No. 190, p. 1–11, 1956.
5. Bates, R. G., "The Glass Electrode," in *Reference Electrodes, Eds.* D. J. G. Ives and G. J. Janz, pp. 231–269, Academic Press, New York, 1961.
6. Ben-Yaakov, S., and I. R. Kaplan, "A pH-Temperature Profile in Ocean and Lakes Using an In Situ Probe," *Limn. Oceangr.* **4**, 688–693, October, 1968.
7. Ben-Yaakov, S., "Analog to Frequency Converter is Accurate and Simple," *Electronic Design* **14**, 96–98, July 4, 1968.
8. Ben-Yaakov, S., "An Oceanographic Instrumentation System for In Situ Measurement," Ph.D Thesis in Engineering, University of California, Los Angeles, 1970.
9. Ben-Yaakov, S., and I. R. Kaplan, "High Pressure pH Sensor for Oceanographic Applications," *Rev. Sci. Instr.* **39**, 1133–1138, August 1968.
10. Britton, H. T. S., *The Measurement of Hydrogen Ion Concentration*, Vol. 1, Chapman and Hall, London, 1955.
11. Disteche, A., "Electrochemical Measurements at High Pressure," *J. Electrochem. Soc.* **109**, 1084–1092, November 1962.
12. Disteche, A., "Nouvelle Cellule à Electrode de Verre Pour la Mesure Directe du pH aux Grandes Profondeurs Sous-marines," *Bull. Inst. Oceanogr. Monaco* **64**, (1320), 1–10, November 1964.
13. Disteche, A., "pH measurements With Glass Electrode Withstanding 1500 kg/cm³ Hydrostatic Pressure," *Rev. Sci. Instr.* **30**, 474–478, June 1959.
14. Eisenman, G., Ed., *Glass Electrodes for Hydrogen and Other Cations, Principles and Practice*, Marcel Dekker, New York, 1967.
15. Garrels, R. M., *Mineral Equilibria*, Harper & Bros., New York, 1960.
16. Garrels, R. M., "Silica: Role in the Buffering of Natural Waters," *Science* **148**, 69, January 1965.
17. Gold, V., *pH Measurements*, John Wiley and Sons, New York, 1956.
18. Ives, D. J. G., and G. J. Janz, *Reference Electrodes*, Academic Press, New York, 1961.
19. Mattock, G., *pH Measurement and Titration*, Heywood, London, 1961.
20. Park, K., "Deep-sea pH," *Science* **154**, 1540–1542, December 23, 1966.
21. Smith, W.H., and D.W. Hood, "pH Measurements in the Ocean: A Sea Water Secondary Buffer System," in *Ken Sugawara Festival Volume, Recent Researches in the Field of Hydrosphere, Atmosphere, and Nuclear Geochemistry*, pp. 183–202, Eds. Y. Miyake and T. Koyama, Maruzen Co., Tokyo, 1964.
22. Strickland, T. D. H., and T. R. Parson, "A Manual of Sea Water Analysis," Fisheries Research Board of Canada Bull. 125, Ottawa, 1965, pp. 31–33.
23. Weyl, P.K., "The Carbonate Saturometer," *J. Geol.* **69**, 32–43, January, 1961.

Ralph Bernstein

4.18 POSITION SENSORS

Position sensors provide relative position or displacement information. Applications include positioning a camera a precise distance over the ocean bottom, sensing inertial acceleration or velocity to determine platform position, sensing vibratory displacement, and measuring the strain on a structure by displacement (strain) sensing. Positioning systems which use some of the sensors discussed in this section are described in Section 3.8.

4.18.1 Position Sensor Types

Position sensors can be categorized as translational and rotational. Each of these can be further subcategorized as position, velocity, acceleration, and jerk sensors. Examples are given in Table 4.18-1. Velocity and acceleration sensors are included because position is frequently derived from integration of velocity or acceleration data.

Where the information is in an analog form, analog integrators are used to form the result

$$x(t) = \int^t v(\tau)\, d\tau = \int\int^t a(\tau)\, d\tau d\tau. \qquad (4.18\text{-}1)$$

If the information is in a digital form, counters or computers are used to form the result

$$x(nT) = \sum^n v(iT) = \sum\sum^n a(iT). \qquad (4.18\text{-}2)$$

Some position sensors are combined to achieve a particular output form. Examples are a rate gyro with an induction potentiometer transducer to convert angular precession rate to a time-varying voltage, or a mass displacement accelerometer with a differential transformer to provide an electrical output voltage proportional to mass displacement.

4.18.2 Sensor Calibration

Distance and time standards are used to calibrate position sensors. The meter was defined in terms of the wavelength of a krypton-86 lamp in 1960 [13] and is equal in length to 1,650,763.73 wavelengths in vacuum corresponding to the transition between the energy levels $2P_{10}$ and $5d_5$ of krypton-86. The reproducibility [17] of the standard is about 2

Table 4.18-1. Typical Translational and Rotational Sensors

Translational position sensors	Rotational position sensors
Linear potentiometer	Potentiometer
Linear capacitor	Rotary capacitor
Displacement encoder	Rotary encoder
Strain gage	Synchro
Differential transformer	Rotary differential transformer
Variable inductance pickup	Induction potentiometer
Variable permeance pickup	Pendulum
Piezoelectric sensor	Gyro (rate-integrating)
Vibrating wire depth sensor	
Translational velocity sensors	Rotational velocity sensors
Time measurement	Pulse generator
Velocimeter	Tachometer
Electromagnetic log	Drag cup
Moving coil sensor	Gyro (rate)
Moving magnet sensor	Stroboscopic lamp
Pitot-static tube	Flyball pick-up
Savonius rotor current meter	Pendulum
Pendulum velocity meter	
Translational acceleration sensors	Rotational acceleration sensors
Mass displacement accelerometer	Piezoelectric sensor
Piezoelectric accelerometer	Pendulum
Pendulum accelerometer	Gyro
Vibrating wire	Mass rotation accelerometer

parts in 10^8 and can be used at this precision level to measure lengths in the range of 10^{-9} to 1 m.

Since position is commonly derived from velocity and acceleration measurements, a time standard also is needed. In 1964 the second was defined as the interval of time corresponding to 9,192,631,770 cycles of the atomic resonant frequency of cesium-133 [13]. Repeatability of these resonators are a few parts in 10^{11}. The measurement of time is discussed in Section 4.27.

Devices that are used to calibrate position sensors include displacement-time measuring systems, shake tables, tilt tables, ballistic pendulums, drop test devices, and centrifuges.

4.18.3 Displacement Potentiometer

A displacement potentiometer consists of a linearly constructed resistive element with a movable contact. The contact is mechanically engaged to the device whose position is being measured, and the voltage, V, sensed between the movable contact and the terminal of the resistance is proportional to displacement (see Table 4.18-2). The small nonlinearity due to the finite internal resistance of the voltmeter or recorder connected to the potentiometer terminal can be derived as

$$\frac{V}{e} = \frac{1}{x/x_{FS} + (R/R_M)\left[1 - x/x_{FS}\right]},\qquad(4.18\text{--}3)$$

where e is the excitation voltage, x is the displacement, R is the potentiometer resistance, x_{FS} is the full-scale contact displacement, and R_M is the voltmeter internal resistance. For sufficiently high meter resistance and sufficiently low potentiometer resistance, the nonlinearity can be made very small. Potentiometers are very reliable, but are subject to wear and noise effects.

4.18.4 Displacement Encoder

A displacement encoder consists of a strip of material with conducting, magnetic or optically opaque areas alternating with nonconducting, nonmagnetic, or optically transparent areas in a particular coding scheme. Table 4.18-2 shows four bands arranged in a binary sequence. A detector array (electrical contact, magnetic, or light detector) is positioned over the encoded strip, and the output binary values represent the displacement or position of the detector array, relative to the encoded strip.

There are a number of coding schemes that can be used [7, 12, 16], such as the reflected binary or Gray code, which are designed to minimize the possibility of reading errors or to avoid ambiguous readings. For a binary sequence, n channels, and a maximum displacement range of x_{max} the resolution, Q is

$$Q = \frac{x_{max}}{2^n}.\qquad(4.18\text{--}4)$$

The error can generally be held to $\pm\frac{1}{2}$ of the least significant bit or

$$e = \pm\frac{1}{2}\frac{x_{max}}{2^{n-1}} = \pm\frac{x_{max}}{2^n}.\qquad(4.18\text{--}5)$$

Encoders with as many as 20 channels are available, providing a high degree of resolution and accuracy.

4.18.5 Resistance Strain Gage

Strain gages are used to measure small displacements resulting from applied forces and pressures. A conductor of uniform cross-sectional area A, length L, and resistivity ρ, has a resistance

$$R = \frac{\rho L}{A}.\qquad(4.18\text{--}6)$$

Table 4.18-2. Sensor Properties

Sensor Name	Schematic	Relation
Displacement potentiometer		$E = Kx$ where $K = \dfrac{E}{x_{FS}}$
Displacement encoder		$N = \displaystyle\sum_{j=0}^{J} C_j 2^j$ where $C_j = f(x)$ $= 0$ or 1
Strain gage		$\Delta R = f(x)$ where $R = \dfrac{E}{I}$
Differential transformer		$E_0 = Kx$
Variable inductance pickup		$E_0 = Kx$

Principle	Performance		Comments
Voltage drop sensed by linearly displaced potentiometer wiper provides measure of displacement.	Displacement: Resolution: Linearity:	0.25 to 50 cm 0.0025 to 0.005 cm 0.05 to 0.10% FS	Will become noisy with wear. Should minimize R and maximize voltmeter resistance for minimum errors.
Binary sequence developed by the use of conducting, magnetic, or optical segments sensed by electrical contact, magnetic, or optical detector.	Displacement: Resolution: Accuracy: Total Count: No. of Channels:	x (variable) $x_{max}/2^n$ $\pm x_{max}/2^n$ 2^n n	Special codes available to eliminate ambiguity and reduce errors.
Resistance change is due to length and area change and piezo-resistance effect due to applied strain (displacement). Displacement can be determined from resistance change.	Displ. Sens: Strain: Accuracy: Thermal Sens:	0.004 cm 10^{-7}–10^{-1} cm/cm 0.15% FS 0.01%/°F	Gages exhibit temperature sensitivity. Requires compensation to achieve maximum accuracy.
Voltage induced in two secondary coils is proportional to iron core position. Series opposition connection of coils results in output voltage proportional to core position	Displacement: Nonlinearity: Sensitivity:	0.01 to 10 cm 0.1 to 0.5% 0.2 mV to 0.6 V/cm	Excitation frequency should be at least ten times displacement frequency
Change in position of iron core causes a proportional increase and decrease in inductance of the two coils. The bridge circuit output voltage is proportional to the change in inductance and thus to displacement.	Displacement: Nonlinearity: Sensitivity:	0.25 to 500 cm 0.02 to 1% 2 to 16 V/cm	Can be made to measure large displacements (up to 500 cm) with good accuracy.

Table 4.18-2. **(continued)**

Sensor Name	Schematic	Relation
Capacitance pickup		$E_0 = Kx$ where $K = K'CE$
Time measurement		$x = \frac{1}{2}(T_1 + T_2)C$ where C = velocity of propagation
Frequency measurement, vibrating wire		$f = \left(\dfrac{T + Kx}{mL} \right)^{1/2}$ where T = Tension of string Kx = Change in tension at depth X m = Mass of unit length of wire L = Length of wire

Principle	Performance		Comments
Capacitance of feedback capacitor is varied by separating capacitor plates by amount of unknown displacement. Output voltage is proportional to displacement.	Displacement: Resolution: Accuracy:	0.0025–1.25 cm 1.2×10^{-5} cm 2% FS	Various capacitors used for different displacement ranges.
Travel time of acoustic, electromagnetic or light pulse measured, and distance determined by multiplication by propagation velocity.			Error in determining distance dependent on accuracy of knowledge of velocity of propagation.
Natural vibration of wire changes due to change in tension of wire caused by depth change. Actuator, A, excites wire, and frequency detected by sensor, S.	Accuracy:	$\pm 0.25\%$ of depth	Requires periodic calibration.

Table 4.18-3. Sensor Characteristics

Sensor Name	Schematic	Relation
Time measurement		$\bar{v} = \dfrac{x}{T_X}$ where $T_X = T_2 - T_1$
Frequency measurement velocimeter		$\Delta f = \dfrac{v^2}{2cx}$ where c = propagation velocity
Moving coil velocity sensor		$E_0 = Kv$
Moving magnet velocity sensor		$E_0 = Kv$
Pendulous velocity meter (pendulous integrating gyro accelerometer)		$\Theta_P = \dfrac{P}{H} v$

Principle	Performance	Comments
Average velocity determined by precise time measurement over fixed distance.	$\Delta \bar{v} = \dfrac{\Delta v}{T_x}$, $\quad \Delta \bar{v} = \dfrac{x \Delta T_x}{T_x^2}$ where $\Delta \bar{v}$ = error in v Δx = uncertainty in x ΔT_x = uncertainty in T_x	Accuracy dependent upon counter resolution, timing accuracy and distance over which measurement made.
Velocity determined by measurement of frequency change due to differing propagation times in "Sing-Around" circuit.	$\Delta v = \dfrac{cx}{v} \Delta f_e$ where Δf_e = error in measurement of Δf	Accuracy dependent upon knowledge of propagation velocity and frequency change measurement error.
Vibratory velocity determined from voltage induced in coil moving in permanent magnet field.	Displacement: 0.25 cm Sensitivity: 0.06 V/(cm/s) Nonlinearity: $\pm 1\%$	Used for sensing vibratory velocities.
Vibratory velocity determined from voltage induced in coil by moving magnet.	Displacement: 1.3–23 cm Sensitivity: 0.04–0.25 V/(cm/s) Nonlinearity: $\pm 1\%$	Used for sensing vibratory velocities.
Pendulous mass attached to gyro gimbal senses linear accelera-tion. Resulting torque causes gyro precession which when detected and used in feedback loop results in precession angle proportional to velocity.	Bias: 10^{-5} g Sensitivity: 2×10^{-7} g Range: ± 20 g Linearity: 5×10^{-6} g/g^2	Precise temperature control needed.

Table 4.18-4. Sensor Comparisons

Sensor Name	Schematic	Relation
Accelerometer (suspended mass)		$x = \dfrac{m}{K} a$
Accelerometer (piezoelectric)		$E_0 = Ka$ where K = accel. scale factor
Accelerometer (pendulum)		$i = \dfrac{P}{K} a$ where i = torquer current K = torquer scale factor p = pendulousity
Redundant strapdown inertial reference unit (SIRU)		$\bar{F} = m\bar{a}$ $\bar{T} = \bar{H}X\bar{\omega}_p$

$\alpha = 31°\ 43'\ 2.9''$

Principle	Performance (Typical)		Comments
pring constrained mass is displaced in the presence of an acceleration by an amount proportional to the acceleration.	Sensitivity: Linearity: Range: Bias:	10^{-6} to 10^{-5} g 0.01 % to 0.1 % FS $\pm$ 30 g 10^{-5} g	Requires precise alignment (10 arc s). Requires periodic calibration.
rystal under high compression provides a near-linear output to an acceleration along sensitive axis due to piezoelectric effects.	Sensitivity: Linearity: Range: Freq. Range: Temp. Drift:	50 mV/g 1 % FS 0.03–10000 g 20–4kHz($\pm$5%) $\pm$ 10%(0 to 360°C)	Used commonly for shock and vibration analysis.
endulous element sensitive to acceleration rotates in presence of acceleration. Rotation detector senses rotation and provides torque signal to torquer to maintain angular position of pendulum. Amplifier current is proportional to acceleration.	Sensitivity: Linearity: Range: Freq. Range: Scale Factor Stability:	5×10^{-7} g 0.005–0.02 % FS $\pm$ 100 g 0–100 Hz $\pm$ 0.01–0.02 % /yr	Commonly used as a level sensor.
ertial sensing of rotation and acceleration using gyros and accelerometer in a nonorthogonal arrangement of six gyros and six accelerometers. Computer used to compute and to transfer data to principal axes.	Dependent upon components.		Redundancy arrangement of sensors provides maximum failure independence and and detection (2 gyro-accelerometer failures cause no increase in error, 3 still allow operation). Also, systematic errors are minimized.

If the conductor is stretched or compressed, its resistance changes due to dimensional changes and a piezoresistance effect which represents the dependence of resistivity on the mechanical strain. The change in resistance due to a mechanical strain is [2, 11, 13]

$$dR = \frac{\rho dL(1 + 2v)}{A} + \frac{Ld\rho}{A},$$ (4.18–7)

where v is Poisson's ratio. The gage factor is

$$\frac{dR/R}{dL/L} = 1 \qquad\qquad +2v \qquad\qquad + \frac{d\rho/\rho}{dL/L}$$ (4.18–8)

| Resistance change due to length change | Resistance change due to area change | Resistance change due to piezoresistance effect |

Thus, measurement of dR/R allows the determination of the strain dL/L which allows the determination of the change in length dL.

Common strain gages use metallic materials such as Advance and Iso-Elastic or semiconductor materials. The advantage of semiconductor strain gages is their very high gage factor (typically 130) relative to metallic strain gages (2 to 3.5). For metallic materials, most of the resistance change is due to dimensional changes; for semiconductor materials most of the resistance change is due to piezoresistance effects.

Strain gages can be either bonded (cemented) to the surface of the device under strain or unbonded. Bonded strain gages generally use metallic wire or foil in a grid construction. Unbonded strain gages use flexure plates holding the strain wires under stress. Both bonded and unbonded strain gages measure motions due to tension and compression. Unbonded gages are primarily used to measure force, pressure, and acceleration. Bonded strain gages are very effective in measuring small displacements due to strains. Amplifiers and Wheatstone bridge circuits are commonly used.

Temperature compensation is generally required. In some applications, such as beam displacement transducers, four gages can be used in a bridge circuit compensating manner to minimize temperature effects.

4.18.6 Differential Transformer

A differential transformer provides displacement information by the use of a variable inductance. A cylindrical 3-coil transformer, with a soft iron core whose position is variable and proportional to the displacement to be measured, causes the voltage induced in two identical secondary windings to change in relation to the core position from null.

The secondary coils have induced in them a voltage whose amplitude varies with the position of the iron core. With the secondaries connected in series opposition, a null occurs at $x = 0$ where the induced voltages cancel each other out.

The open-circuit output voltage is [2]

$$\frac{E_0}{E}(j\omega) = \frac{(M_1 - M_2)/R_p}{\left[\left(\omega \dfrac{L_p}{R_p}\right)^2 + 1\right]^{1/2}} e^{j\phi}, \tag{4.18-9}$$

where

$$\phi = 90° - \tan^{-1} \omega \frac{L_p}{R_p}, \tag{4.18-10}$$

M_1 and M_2 are the mutual induction of the secondary coils, L_p is the primary coil inductance, R_p is the primary coil resistance, and ω is the excitation frequency. The net mutual inductance, $(M_1 - M_2)$, varies linearly with core motion. Thus for a relatively large range of x, $E_0 = Kx$ or a proportional relation exists. The excitation frequency ω is adjusted such that the zero cross over occurs at $x = 0$. The excitation voltage is in the frequency range 60 Hz to 20 kHz. Suitable phase sensitive demodulators and low pass filters are used to convert the variable phase, amplitude modulated output voltage to a dc voltage proportional to x.

The dynamic response is primarily limited by the excitation frequency, and should be more than ten times the displacement frequency.

4.18.7 Variable Inductance

The variable inductance sensor operates on a principle similar to the differential transformer, with only two coils used. When the iron core is positioned at null, both coils have the same inductance, the bridge circuit is balanced, and the output voltage is zero. A change in the position of the iron core results in a proportional increase in inductance of one coil and a decrease in the other which results in an output voltage. The output voltage can be made close to a linear function of displacement.

4.18.8 Capacitance

A number of capacitance displacement sensors are commonly used. One that uses a high-gain amplifier to achieve good linearity and sensitivity is known as a feedback-type capacitive pickup. A capacitor with variable plate separation is used. One plate of the capacitor is connected to the device whose displacement is to be measured. The displacement of the capacitor plate results in a change in capacitance that is used in the high-gain feedback amplifier circuit. For an amplifier with gain $A \gg 1$.

$$E_0 = \frac{C}{C_x} E. \qquad (4.18\text{-}11)$$

The capacitance C_x is related to the displacement x by

$$C_x = \frac{K'}{x}, \qquad (4.18\text{-}12)$$

where K' is related to the capacitance area and dielectric constant. By substitution for C_x,

$$E_0 = Kx. \qquad (4.18\text{-}13)$$

Thus, a linear relationship exists between the output voltage and displacement of the capacitor plate.

Since the excitation voltage, E, is an ac voltage, the resultant output is an amplitude modulated carrier. The displacement can be determined by the detection of the output voltage envelope. Various capacitor plates are used, depending upon the displacements to be measured. It is possible to detect motion as small as 2 μm using suitable capacitive sensors.

4.18.9   Propagation Time Measurement

Distance can be determined accurately by the measurement of the travel time of an acoustic [1, 4, 15] or electromagnetic energy pulse over the unknown distance. This principle has been used to accurately position an underwater camera a fixed distance above the ocean bottom. A transducer located with the camera emits a short burst of 12 kHz energy at 1-s intervals. The transmitted pulse travels directly to the surface and is also reflected from the ocean bottom and then travels to the surface. A sonar transducer on a surface ship detects the direct and reflected sonar pulses, and the difference in travel time is determined and is a measure of the distance that the camera is above the ocean bottom. The same principle is commonly used to determine the ocean depth beneath a ship. The travel time of an acoustic pulse from the ship, to ocean bottom, and back to the ship is a measure of the ocean depth. The primary source of error in these types of systems is the uncertainty in the propagation velocity of the sonar signal and background noise effects. Frequently, average values (such as 1.5 km/s for the speed of sound in water) are used, resulting in an error proportional to the distance being measured. Accuracy can be increased by periodically determining the speed of sound in water as a function of depth, and computing an average speed of sound. An additional source of error is reflection of the sonar energy from off-axis higher bottom surfaces.

The same principle is employed in radar range and bearing systems and radar altimeters, the basic difference being the use of electromagnetic rather than acoustic energy and transducers.

A display such as a facsimile recorder or cathode-ray tube is used for man-monitoring of the data and to provide an integration effect to the signal in the presence of the noise background.

4.18.10 Vibrating-Wire Frequency Measurement

A vibrating wire can be used to measure ocean depth. A fine tungsten wire is stretched in a magnetic field. The wire vibrates due to excitation by the field at a frequency which is determined by the length, tension, and mass per unit length of the wire. As the device is lowered into the ocean, the pressure change causes a change in the length of the case which causes the tension of the wire to change, resulting in a different natural vibration frequency. A suitable sensor detects the vibration frequency.

For a wire of length L, with tension T, and mass m, per unit length, the frequency is

$$f = \frac{1}{L} \sqrt{T} \frac{1}{(m/L)^{1/2}} = \left(\frac{T}{mL} \right)^{1/2}. \qquad (4.18\text{--}14)$$

A change in pressure on the device results in a proportional change in string tension (increase or decrease depending upon the design) so that

$$T_x = T + Kx, \qquad (4.18\text{--}15)$$

where K is a constant which depends upon the device construction. By substitution, the frequency-depth relation is

$$f_x = \left(\frac{T + Kx}{mL} \right)^{1/2}. \qquad (4.18\text{--}16)$$

4.18.11 Travel Time Measurement

Velocity can be determined by measuring the device or platform travel time between two points whose separation is precisely known. Many methods exist for this type of measurement, varying from the use of simple photocell circuits to radar or laser systems. The basic components of the system are a source of energy, a set of detectors, a clock, and a precisely known course or separation. The detectors provide a start and stop signal to the clock, from which the duration time is determined. The average velocity, $\bar{v}$, is determined from duration time and displacement by

$$\bar{v} = \frac{x}{T_x}, \qquad (4.18\text{--}17)$$

where x is the detector separation and T_x is the time separation between detector signals. This measurement is illustrated in Table 4.18-3.

The errors inherent in this type of measurement are caused by errors in timing and an error in knowledge of detector separation. Equation 4.18–17 yields the error equations

$$\Delta v = \frac{\Delta x}{T_x}, \qquad \Delta v = \frac{x \Delta T_x}{T_x^2}, \qquad (4.18-18)$$

where the Δ's correspond to the measurement errors or uncertainties of the variables. It can be seen that maximizing the time (separation) results in minimum errors, but causes an average, as opposed to instantaneous, velocity determination to be made.

4.18.12 Frequency Measurement by Velocimeter

The measurement of frequency can be used as a measure of velocity in some applications. A velocimeter provides both oceanographic platform velocity and sound propagation velocity [3, 4]. A transmitter at station 1 sends a pulse of energy towards a receiver at station 2. The receiver triggers a transmitter at station 2, which sends a pulse of energy which is received at station 1, which triggers the station 1 transmitter. In this "sing-around" mode of operation, the frequency of the loop transmission is

$$f = \frac{1}{T_1 + T_2} = \frac{c}{2x}, \qquad (4.18-19)$$

where T_1 and T_2 are the propagation times of transmissions 1 and 2, respectively, c is the velocity of propagation in water, and x is the separation between stations 1 and 2. For a stationary velocimeter, the frequency provides a measure of velocity of propagation.

If the sensor moves at a constant velocity, V, in the direction of the transmissions, the apparent propagation velocity in each direction changes, resulting in different propagation times. The propagation time of the first transmission is

$$T_1' = \frac{x}{c + v}, \qquad (4.18-20)$$

and the second transmission is

$$T_2' = \frac{x}{c - v}, \qquad (4.18-21)$$

so that the new loop transmission frequency is

$$f_v = \frac{1}{T_1' + T_2'} = \frac{c^2 - v^2}{2cx}, \qquad (4.18-22)$$

and

$$\Delta f = f - f_v = \frac{v^2}{2cx}. \qquad (4.18\text{--}23)$$

The change in frequency is thus an indication of the velocity of the sensor in the direction of transmission. Clearly, the separation, x, and the propagation velocity, c, must be well known, and the frequency f_v precisely measured in order to provide accurate results.

4.18.13 Moving Coil or Magnetic Velocity Sensor

When a coil moves in a magnetic field, a voltage is induced in the coil that is proportional to the relative velocity of the coil and the magnetic field. This voltage is

$$E_0 = BLv, \qquad (4.18\text{--}24)$$

where B is the flux density, L is the length of coil, and v is the relative velocity of coil and magnet.

This type of sensor is commonly used to measure vibratory velocities, and is accurate to about $\pm 1\%$. Increased sensitivity can be achieved by the use of additional length of coil and higher flux densities.

A moving magnet velocity sensor operates on the same principle as the moving coil sensor, but with the field provided by a permanent magnet, and the magnet connected to the device whose velocity is to be measured. Vibratory strokes as long as 25 cm can be measured.

4.18.14 Pendulum Velocity Meter

A pendulous (or unbalanced) gyro-accelerometer provides inertial velocity information without the need for integrating acceleration. The device consists of a constrained pendulum accelerometer and an integrating gyro. The gyro gimbal case has a viscous constraint (damping) and a pendulous mass attached to the gyro gimbal. In the presence of an acceleration perpendicular to the gyro spin axis and gimbal axis, the pendulous mass causes a torque about the gimbal axis which causes the gyro to precess.

By the use of gimbal angle detectors and torquing of the outer case of the device by an amount proportional to the gimbal angle, the pendulous mass maintains its orientation and the outer case has a precession angle. The precession angle of the gyro is proportional to the integral of the acceleration and is

$$\theta_p = \frac{P}{H} \int a \, dt = \frac{P}{H} v, \qquad (4.18\text{--}25)$$

where a is the linear acceleration, v is the linear velocity, P is the pendulosity of mass unbalance, and H is the gyro angular momentum. Thus, the outer case rotation provides an accurate measure of linear velocity.

4.18.15 Suspended-Mass Accelerometer

If an acceleration, a, acts on a mass, m, then a force,

$$F = ma, \tag{4.18-26}$$

results according to Newton's second law. If the mass is constrained in the direction of acceleration by springs of stiffness, k, the mass is displaced, x, an amount proportional to the acceleration since the spring force and force due to acceleration are equal and opposite. Thus

$$F = kx = ma, \tag{4.18-27}$$

or

$$x = \frac{m}{k} a, \tag{4.18-28}$$

and the mass displacement is a measure of the applied acceleration. The accuracy of position sensing by double integration of acceleration depends upon the magnitude of the uncompensated bias of the accelerometer and the stability of the gain terms. This sensor is illustrated in Table 4.18–4.

4.18.16 Piezoelectric Accelerometer

Piezoelectric accelerometers are commonly used to measure accelerations resulting from shock and vibrations. A piezoelectric crystal is preloaded to a high stress, which results in the crystal being in a more linear operating range and also assures that the crystal is always in compression for all applied accelerations. The additional load due to acceleration causes an output voltage proportional to the shock or vibration. Due to charge leakage, a sustained acceleration does not cause a constant output.

4.18.17 Pendulum Accelerometer

This accelerometer uses a pendulum suspended from a very low friction support and damped in a viscous fluid. In the presence of an acceleration along the sensitive axis, the pendulum is displaced in relation to the applied acceleration. Displacement sensors detect the displacement of the pendulum from the null position and a proportional electrical signal is applied to electromagnetic torquers in a direction to balance the force due to acceleration and return the pendulum to a null condition. The torquer current is proportional to the sensed acceleration, and double integration of the current yields position information. The accelerometers can be used in acceleration environments of ± 20 g (up to 100 g possible) with non-linearity errors less than 5×10^{-6} g/g^2.

4.18.18 Redundant Strapdown Inertial Reference Unit

This inertial platform [14] uses six gyros and accelerometers mounted so that their sensitive axes are at an angle of 31° 43′ 2.9″ to the principal axes (*X, Y, Z*). A computer processes the gyro and accelerometer data and transforms the data to determine angular motion around and linear acceleration along the principal axes. Table 4.18-4 shows the mounting arrangement. The 12 pentagon faces form a dodecahedron array. Each principal axis is served by four of the gyros and accelerometers providing inherent redundancy and failure detection and minimizing geometric error amplification. As many as three of the six gyro/accelerometer modules can fail without seriously affecting accuracy.

References

1. Alexiou, A. G., "Conducting Underwater Surveys with a Multipurpose Instrument," *ISA J., 8*, Dec. 1961.
2. Doebelin, E. O., *Measurement Systems: Application and Design*, McGraw Hill Book Company, New York, 1966.
3. Dunlap, G. D., and H. H. Shufeldt, Eds. *Dutton's Navigation and Piloting*, 12th ed., United States Naval Institute, Annapolis, Maryland, 1969.
4. Edgerton, H. E., "Uses of Sonar in Oceanography" *Electronics, 33*, 93, June 24, 1960.
5. Fernandez, M., and G. R. Macomber, *Inertial Guidance Engineering*, Prentice-Hall, Englewood Cliffs, New Jersey, 1962.
6. Glasstone, S., *Sourcebook on the Space Sciences*, D. Van Nostrand, Princeton, New Jersey, 1965.
7. Grabbe, E. M., S. Ramo, and D. E. Wooldridge, *Handbook of Information, Computation and Control*, Vol. 1—Control Fundamentals, and Vol. 3—Systems and Components, John Wiley and Sons, New York, 1958.
8. Harris, C. M., and C. E. Crede, Eds., *Shock and Vibration Handbook*, Vol. 1, pp. 16–35, McGraw-Hill Book Company, New York, 1961.
9. Hix, C. F., Jr., and R. P. Alley, *Physical Laws and Effects*, John Wiley and Sons, New York, 1958.
10. Kearfott Technical Data for Precision Instrumentation, Kearfott Div. General Precision Aerospace, Little Falls, New Jersey.
11. Lion, K. S., *Instrumentation in Scientific Research*, McGraw-Hill Book Company, New York, 1959.
12. Lippel, B., "A Systematic Survey of Codes and Coders", *IRE Convention Record Pt. 8,* Information Theory, pp. 109–119, 1953.
13. McNish, A. G., "Fundamentals of Measurement", *Electro-Technology, 71*, No. 5, p. 113, New York, May 1963.
14. "MIT Develops Revolutionary Gyro System" *EDN/EEE,* page 18, 1 June 1971.
15. Snodgrass, J. M., and J. W. Cawley, Jr., "Temperature and Depth Circuitry

in a Telerecording Batherometer," Scripps Institution of Oceanography, Contributions, 1956, pp. 813–821, Feb. 1958.
16. Tou, J. T., *Digital and Sampled Data Control Systems*, McGraw Hill Book Company, New York, 1959.
17. Wildhack, W. A., "NBS—Source of American Standards," *ISA J.*, **8**, No. 2, 45–50, February 1961.

Eugene A. Mueller

4.19 PRECIPITATION SENSORS

4.19.1 Rainfall Quantity

The quantity of rainfall is of primary interest in rainfall measurement. A bowl was used as a raingage in 400 B.C. [4] and little improvement on this measuring technique has been accomplished in the intervening years. A nonrecording raingage consists of an orifice of known cross-sectional area to collect the falling rainfall and a container to store the water. The stored water volume or weight can then be measured. In this gage as well as most others there is a question as to the representativeness of the collected water. The area of the orifice, the height of the orifice from the ground, and the shape and size of the supporting structure affect the ratio of the amount of water collected to the amount which actually falls in an undisturbed area of the earth.

Standard Climatic Raingage. The standard climatological raingage in the United States is a nonrecording raingage that has a circular orifice 20 cm in diameter [1]. The orifice is located between 0.75 and 1 m from the ground in an open area. The amount of water collected is generally measured once a day by inserting a graduated measuring stick.

Tipping-Bucket Raingage. The tipping bucket raingage was one of the early types of recording rain sensors [7]. This sensor has a fixed volume that is systematically filled and emptied. Each time the volume is emptied a signal is available for recording, thus indicating the time at which the instrumental volume is reached. The tipping bucket is usually set to tip with 0.25 mm of rainfall. This sensor is capable of tipping with less water by increasing the orifice area. However, as the orifice becomes larger overfilling of the bucket before tipping is more likely at the higher rainfall rates.

This sensor has several disadvantages. At high rainfall rates the tipping mechanism does not operate sufficiently rapidly, and underestimation of total water occurs. This is particularly true if a large orifice area is chosen for great resolution in light rains. The record is a series of steps indicating times that certain amounts have been reached. Thus the resolution for low rainfall rates is limited. It is generally unsuitable for winter work where freezing rain or snow is encountered.

Weighing Raingage. One of the parameters of precipitation which is easily measured is the weight of the water captured by the orifice. Most modern recording raingages measure weight. The water from the orifice is funneled into a bucket which rests on a platform which is suspended by either a spring or a counterbalance. Water entering the bucket displaces the platform a distance proportional to the weight of the water, and this movement is recorded by a pen on a moving chart. An alternate recording method uses punched paper tape.

One of the major disadvantages of this type of gage is that it cannot be left unattended for indefinite periods of time as a practical method of emptying the accumulation of water in the buckets has not been developed.

4.19.2   Rate of Rainfall

Information on the quantity of rainfall for periods of an hour or more are not useful in determining the rate of rainfall. On the other hand, for very short time periods and for small areas, rainfall rate is a discontinuous function. Either a raindrop is contacting the sensor during the observation interval or it is not. Thus, strictly speaking, instantaneous rate of rainfall is not an appropriate term.

Any sensor which is responsive to the total quantity of rainfall can be used for rainfall rate if the time resolution is sufficiently good. The hydraulic time lag from the time water enters the orifice until it is resident in the collector is important in a raingage used for rainfall rate measurements. The tipping bucket is effective at moderate to high rainfall rates. At low rates the sensor suffers from the lack of sufficiently frequent tips, and average rainfall rates over long periods of time are the only calculable quantity. At the very high rates there is a tendency to undermeasure the volume of water. If the sensor is to be used as a rate recorder, this error can be calibrated and thereby reduced in the calculated rainfall rate. Relatively good time recording accuracy is required.

The weighing-bucket raingage can also be used as a rate sensor by observing the amount of water added in a given length of time. Time periods of about 1 min are as small as practical with existing gages because of the recording methods used.

Float Rainfall Rate Gage. Most sensors which respond directly to the rainfall rate depend on a calibrated orifice, tube, or weir. In one gage, a valve which controls an exit port from the rain collector is controlled by a float [2]. The higher the rain rate, the higher the float rises and the faster the water discharges. The float position is recorded by a mechanical linkage on a moving chart.

This sensor is subject to clogging by debris blown into the collector.

Capacitor Gage. Another rain rate recorder directs the water from the collector orifice through a channel with a copper wire on each side [6]. These two wires are plates of a condenser. The height of water in the channel is related to the rate of rainfall. The capacity between the two wires is in turn related to the height of the water.

At moderate to high rainfall rates, the sensor works well but at low rates water collects by virtue of surface tension in the collector and is released through the channel in spurts. Since the relationship between capacity and rainfall rate is nonlinear, these pulses or spurts are difficult to interpret. The time response of the sensor is considerably less than 10 s.

4.19.3   Size and Number of Raindrops

The earliest method was to allow the raindrops to fall on ruled slate and to measure the size of the resulting splash [5]. This method may be categorized as the measurement of the trace of a drop after impingement on a flat surface. The surfaces which have been employed include dyed filter paper, blue print paper, treated photographic film, and coated nylon screen. These methods are simple and require a minimum of equipment. They are calibrated by dropping drops of known size on the surface and measuring the resultant traces. As a result of drop splashes, the number of small drops can be overestimated. Also, for the calibration to be valid, the surface must be maintained perpendicular to the path of the drops. Under windy conditions this is impossible. The data must be reduced manually. Despite these objections, the filter paper method remains one of the most popular methods for obtaining drop size measurements.

Photographic Methods. Photographic methods have been employed to obtain images of raindrops. Since the occurrence of a raindrop in space is a relatively infrequent event, large volumes of space must be photographed in order to obtain a representative spectrum. Obtaining large sample volumes with high photographic resolution is difficult. The photographic method also suffers from the severe handicap that the data must be reduced manually. Thus, each image or trace is manually measured, recorded and the spectrum calculated. One method replaces the film with an image orthicon. In an associated special-purpose computer, the number of intersects of the vertical sweep with the drop image are tabulated and interpreted as the vertical dimension of the drop.

Momentum Measurement. One sensor which responds to the momentum of the raindrop has a platform suspended by springs that serves as the drop receiver. There is a coil attached to the platform and an associated coil on the fixed member. Any deflection or movement of the platform is electronically amplified and returned to the fixed coil to oppose the mechanical movement. Thus, very small mechanical movements of the platform result. This small movement reduces the sensitivity variation produced by impingements at different points of the receiver. The magnitude of the pulse from the amplifier is related to the momentum of the raindrop. Under the assumption that the raindrop is traveling at terminal velocity, the momentum can be related to the drop size. Automatic data processing of the amplitudes of the signal is available.

4.19.4 Snowfall Parameters and Sensors

The standard method for snow measurement is the use of snow stakes [1]. These stakes are made from wood about 5 cm square and secured in an upright position. Appropriate graduations indicate the depth of the snow. The stakes should be on level ground and their values averaged to obtain depth of snow.

Most devices which measure rainfall are used to measure the equivalent water from a snowfall. The major difficulty is to obtain a representative amount of the snow in a catchment orifice. The measuring orifice presents an obstacle to the flow of air, and since snow has a relatively low terminal velocity, the snow can be carried away from the orifice.

A more accurate measure of the amount of water can be obtained by sectioning a portion of snow, removing, and either weighing or melting and measuring volume.

To improve the ability of a raingage to catch snow, a wind shield is frequently employed. One shield consists of a ring 2 m in diameter with loosely hanging slats. These swinging slats do not permit snow to accumulate on them. The usage of this type of shield increases the amount of water captured in the gage by about 80%.

When raingages are being used for snow it is frequently necessary to heat the gages to melt the snow. Thus with the tipping-bucket gages heating is necessary.

Another method of measuring snow depth is to bury an innocuous radioactive source at ground level, and monitor the radiation from a mast overhead. As snow builds up over the radioactive source the radioactivity is effectively shielded from the monitor. Thus, the depth of snow (or equivalent water content) is measured by the amount of radioactive reduction.

4.19.5 Hail

Until recently there have been no serious attempts to measure size or intensity of hailfalls. At present there are two ways of obtaining information on hail.

The simplest method consists of covering styrofoam blocks with aluminum foil. Either the depth of the trace or the diameter can be measured and related to the initial diameter of the hailstone.

A second means utilizes a geophone and relates the change in momentum of the impacting hailstone to the signal from the geophone.

References

1. Department of Commerce, Weather Bureau, Circular B "Instructions for Climatological Observers", June 1962.
2. Jarde, Ramon, *Un Pluviograph d' intensitats,* Barcelona, 1921.
3. Joss, J., and A. Waldvogel, "Ein Spektograph für Niederschlagstropten mit Automatischer Auswertung," *Pure Appl. Geophys.* **68,** 1967.
4. Kurtyka, J. C., "Precipitation Measurements Study," Report of Investigation 20, Water Survey, 1953.
5. Lowe, E. J., "Raindrops," *Quart J. Roy. Meteorological Soc.* 1892.
6. Semplak, R. A., "Gauge for Continuously Measuring Rate of Rainfall," *Rev. Sci. Instr.* **37,** No. 11, 1554–1558, Nov. 1966.
7. Sprung, A., "Ueber Fern-Registrierung des Regens," Meterologische Zeitschrift, October 1897.

H. Dean Parry

4.20 PRESSURE SENSORS

4.20.1 Definition and Units

Pressure is force per unit area. Pressure values are specified in a variety of units, most consisting of a unit of force per unit of area. Historically, pressure was measured by a barometer or manometer, a circumstance that has resulted in measuring pressure also in terms of the height of the column of mercury which the pressure will support [17]. The accepted standard of atmospheric pressure at sea level is also used as a unit of pressure. Pressure units most commonly used, the definition of each and various conversion factors are listed in Table 4.20-1. Gage pressure is defined as pressure above atmospheric pressure. Absolute pressure is total pressure including the pressure of any ambient air.

4.20.2 Pressure as a Geoscience Parameter

Measurement of pressure is possible and meaningful only in fluids. Meteorology, limnology, oceanography and perhaps vulcanology and glaciology are branches of geoscience which deal with fluids and which use pressure measurements. Pressure in a fluid at rest in which there are no vertical accelerations except gravity varies with the depth according to the hydrostatic equation

$$\int_{P_1}^{P_2} dP = \int_{Z_1}^{Z_2} \rho g \, dZ, \tag{4.20-1}$$

where P is the pressure at any depth Z, ρ is the density at depth Z, and g is the acceleration due to gravity. The subscripts 1 and 2 refer to the upper and lower surfaces of any layer, respectively.

Liquids are essentially incompressible; hence, for a given homogenous liquid with no vertical temperature gradients, ρ is constant for all practical purposes. The density of fresh water is 1.00 g/cc and the average density of sea water is 1.025 g/cc [11]. The absolute pressure P at any depth Z is given to a good approximation by

$$P = P_0 + 0.1004(Z - Z_0) \quad \text{for sea water,} \tag{4.20-2}$$

and

$$P = P_0 + 0.098(Z - Z_0) \quad \text{for fresh water,} \tag{4.20-3}$$

Table 4.20-1. Pressure Units

Name of Unit	Definition, (Conversion)
1 pascal	1 newton/meter^2 (10 microbars)
1 bar	10^6 dynes/cm^2 (1.01325 atmosphere)
1 kilobar	10^3 bars
1 millibar	10^3 dynes/cm^2 (0.0295300 in. Hg)
	(0.750062 mm/Hg)
1 microbar	1 dyne/cm^2 (1×10^{-6} bars),
	(1/10 pascal)
1 cm Hg	The difference between the pressure at the base and top of a layer of mercury which occupies the space between two horizontal planes 1 cm apart crossing a vertical column of mercury having a density 13.5951 g/cm^3 and subjected to a gravitational acceleration of 980.665 cm/s^2 (13.33224 mb). (Note: The above density is for pure mercury at 0°C.)
1 in. Hg	Same as centimeter of Hg except the planes 1 (standard) in. (2.54 cm) apart. (33.86389 mbar)
1 atmosphere	1013.250 millibars (760,000 mm Hg), (29.92127 in. Hg)
1 lb/in.^{2}	1 (standard) lb per 1 (standard) in.^{2} (68.944 millibars)
1 torr	1/760 of one atmosphere (1.333224 mb) (Note: The torr is an obsolete unit.)
1 barye	1 dyne/cm^2 (1.000 microbar) (Note: Earlier usage specified other definitions now obsolete.)

where P_0 and Z_0 are pressure and depth, respectively, at some level above Z. When measuring from the water-air interface Z_0 is zero and P_0 is atmospheric pressure. In the above relationships P and P_0 are in bars and Z and Z_0 are in meters.

In the atmosphere ρ is not constant. Pressure-height relationships for many practical purposes are referred to a standard atmosphere which specifies pressure and temperature at sea level as 1013.25 mb and 286.16°K, respectively, and, in the first 11 km, temperature linearly decreasing with height at the rate of 0.0065°C/m [9]. Applying Eq. 4.20–1 to the standard atmosphere, invoking the perfect gas law, neglecting the effect of moisture on density, and reversing the sign of the vertical ordinate to give altitude instead of depth gives

$$P_A = P \left(\frac{286.16}{286.16 - 0.0065H} \right)^{- \frac{g}{0.0065R}}, \qquad (4.20–4)$$

where P_A is the altimeter setting in millibars, P is the pressure in millibars prevailing at any height H (meters), R is the gas constant for air, and g is the acceleration due to gravity in cm/s². The standard pressure-altitude relationship can be obtained by putting $P_A = 1013.25$ and computing P as a function of H. Pressure-altitude and altimeter setting are both useful in geoscience. The altimeter setting is frequently used as an approximation to atmospheric pressure reduced to sea level especially in automatic weather stations which have limited computational capability.

For nearly all geoscience purposes static pressure is desired but dynamic effects can disturb the measurement. A wind outside will alter the pressure inside a building; the magnitude and sign of the alteration is a function of the strength of the wind and the design and exposure of the building. Ventilating and air conditioning fans produce serious errors in pressure measurements for meteorology.

4.20.3 Measurement of Pressure

Over 400 pressure sensors are commercially available [19]. Most of these sensors fall into one of three categories:

1. A sensor in which the pressure is measured by the height of the vertical column of liquid it will support (e.g., mercurial barometer).

2. A sensor in which the fluid whose pressure is to be measured is brought in contact with an area of a flexible solid. Pressure is measured by measuring strain in the solid material (e.g., an aneroid barometer).

3. A sensor which employs the principle that a fluid boils when its vapor pressure equals the total ambient pressure (e.g., hypsometer).

In addition to these very commonly used concepts there are several other, highly ingenious ones which have been proposed but which are not used often in geoscience because of inherent inaccuracy or high cost.

4.20.4 Liquid Barometers

Figure 4.20-1 illustrates the liquid barometer. By Pascal's law the air pressure on the exposed mercury surface is transmitted to every surface in the fluid. Consider the surface within the tube which is at the level of the mercury in the cistern. Transmitted pressure is normal to this surface and hence directed upward. The weight of the column of mercury produces a downward pressure at this surface. If the mercury is stationary these two pressures must be equal and opposite. Mercury pressure is a function of the density of the mercury, the height of the column, and the magnitude of the gravitational force. Pressure can be measured by height of the mercury if gravity and density are standardized. In a real barometer instrumental errors are minimized by improving the vacuum, standardizing mercury density, and correctly measuring the height under known gravity.

The vacuum in a real instrument never has zero pressure. Air dissolved in the mercury escapes into the evacuated space. Some air enters the vacuum because of outgassing from the glass and air may also filter very slowly

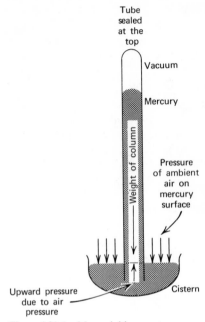

Figure 4.20-1   Mercurial barometer.

through the walls of the tubing. In some primary standard barometers the vacuum is constantly monitored by a McLeod gage and maintained by a high vacuum pump.

Density of the mercury is standardized by beginning with pure mercury rather than an amalgam, dispelling disolved air and keeping the mercury clean. As the pressure, and with it the column of mercury in the tube rises or falls, there is a compensating fall or rise in the cistern. The height of the column to be measured is the height of the mercury in the tube above the upper surface of the mercury in the cistern.

The Fortin barometer cistern (Figure 4.20–2a) has an ivory fiduciary point *A* at the zero point of the measuring scale. The mercury level in the cistern is raised or lowered by a screw mechanism *B* which raises or lowers a leather bag *C* serving as the bottom of the cistern. The height of the mercury surface in the cistern can be set very exactly by watching (through glass wall *D*) the ivory point and its reflection in the mercury surface.

In the Kew-type cistern (Figure 4.20-2b) the rise and fall of the mercury surface in the cistern is compensated by graduating the height scale, not in standard centimeters or inches but in $A/(A + S)$ length units, where S is the area of the tube at the level of the mercury and A is the area of the fixed-diameter cistern. An engine-divided scale or, in some very refined instruments, gage blocks are used to obtain accurate measurement of the column.

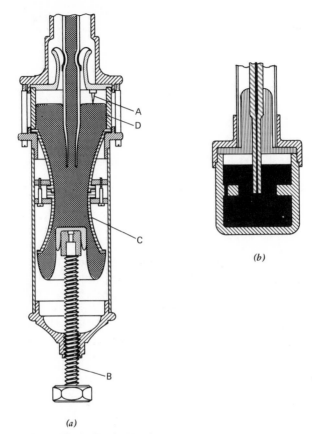

Figure 4.20-2   Fortin-Kew barometer.

Change of length of the mercury column produced by temperature variations might be considered as a change in density—the physical cause. However, it is conventional to express the conversion to standard barometeric height in terms of the linear coefficient of expansion of the scale, α, and the volume coefficient of expansion of the mercury, β.

If the pressure is to be given in terms of the height of a mercury column, it is necessary to normalize the height of the mercury as read to standard conditions of both mercury density and gravity. Density is standardized to that of pure mercury (13.5951 g/cm³) at 0°C. Heights are normalized to "standard" gravity of 980.665 cm/s². The normalized height in millimeters for a Fortin barometer is [18]

$$H_{\text{STD}} = \frac{g_{\phi z}}{g_{\text{STD}}} \left\{ H_T \left[1 - \frac{T(\beta - \alpha)}{1 + \beta T} \right] \right\}, \qquad (4.20\text{-}5)$$

where $g_{\phi z}$ is the local gravity, g_{STD} is the standard gravity, T is the temperature of both the mercury and the scale in °C (temperature of both is assumed to be the same), H_T is the height in millimeters read from the scale, and α and β are as defined above in appropriate metric units.

For purposes of barometry a value of 980.665 cm/s^2 has been adopted as standard gravity by both the International Committee on Weights and Measures and The World Meteorological Organization. A relationship has been proposed [9] which gives the acceleration of sea level gravity as a function of latitude ϕ as

$$g_{\phi,0} = 980.616(1\text{-}0.0026373 \cos 2\phi + 0.0000059 \cos^2 2\phi) \text{ cm/s}^2. \quad (4.20\text{-}6)$$

Gravity also decreases with height in the free air (the so-called free-air correction) at the rate of 0.0003086 cm/s^2/m. Because of terrain effect and other anomalies in the gravity field, files of local gravity measurements should be consulted [23].

Although the mercurial barometer normally is read manually, the reading has been automated in a number of ingenious ways. The weighing barometer is shown in Figure 4.20-3. The only function of the mercury is to produce a vacuum in the top of the tube. Since the mercury does not adhere to the glass walls, its weight does not directly add to the weight on the scale arm. The scale arm is supporting the weight of the glass tube and the downward pressure on the top of the tube which is not compensated by upward pressure because of the vacuum inside the tube [18, pp. 23–27].

Automation has been achieved electronically. The tube is constructed of paramagnetic stainless steel. A magnetic slug is floated on top of the mercury column. A differential transformer is moved up and down the tube by a servo-motor to a position where the transformer's opposing coils are null-balanced by the slug. In this manner the height of the mercury is determined. Usually a Kew-type cistern is used to obviate the need for setting a fiduciary point. The temperature correction is made constant by mounting the whole instrument in a constant temperature oven.

The height of the mercury column in the tube has also been measured by a pulse of hypersonic sound. The pulse originates from a transmitter submerged in the cistern mercury and travels in the tube through the mercury to its upper surface, which reflects the sound. A small microphone next to the transmitter picks up the reflected pulse. The time between the transmission of the pulse and its return is measured by a phase comparator. The height of the surface of the mercury in the tube is computed from this time and the known speed of sound in the mercury.

4.20.5 Elastic Pressure Sensors

A large group of sensors measure pressure by measuring the strain in a material subjected to a pressure-induced force. These sensors are not ad-

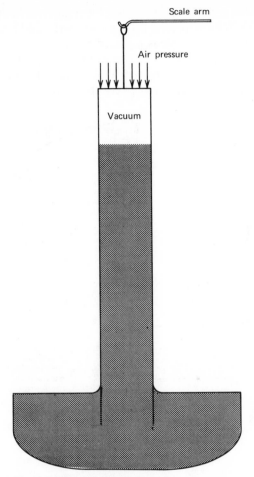

Figure 4.20-3 Weighing barometer.

versely affected by motion or small accelerations. A minimum of maintenance is required. It is possible to design sensors to cover the diverse ranges required by the geosciences. Operational meteorology requires a range from a few millibars to 1050 mb and oceanography for the deepest part of the ocean requires a range from 1 bar to more than 1100 bars. Elastic sensors are used to make virtually all operational pressure measurements in geoscience.

Perhaps the most commonly used elastic sensor is the aneroid cell (Figure 4.20-4b). Modifications of it are used in measuring atmospheric pressure at the earth's surface and to considerable heights. Other forms are used in measuring the extremely high pressure encountered in ocean depths.

The aneroid cell is read out either visually or automatically. If the pointer shown in Fig. 4.20-4a is placed before a properly calibrated scale, visual readout is possible. If the scale is replaced by graduated paper on a clock-driven drum or disc and a pen is substituted for the pointer, a mechanically recording pressure sensor is obtained. This is the form of the barograph in common use for measuring atmospheric pressure.

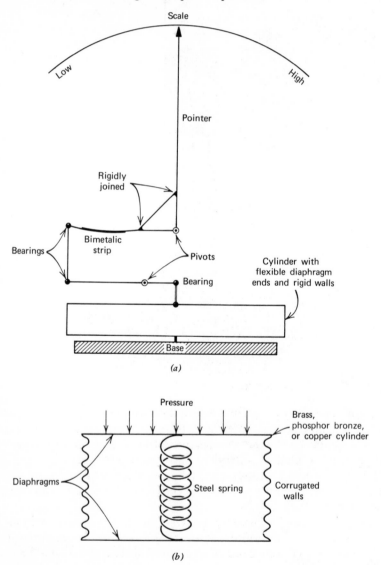

Figure 4.20-4 Aneroid cells.

Figure 4.20-5   Baroswitch.

The altimeter used in aircraft and by surveyors is an aneroid barometer whose scale is calibrated in height units corresponding to pressures in the standard atmosphere given by Eq. 4.20–4.

A common automatic pressure readout is the baroswitch, a contacting arm moved across contacts by the aneroid diaphragm. The radiosonde baroswitch shown in Figure 4.20-5 multiplexes a single communications channel to transmit temperature, humidity, and pressure. Pressure readout is obtained by recognizing the terminal contacted by the moving arm. For higher pressure ranges, the motion of the diaphragm is read out by strain gages. Automatic readout of the aneroid cell can also be accomplished by electrically insulating the opposite diaphragms to form a capacitor [24]. In another method the position of a ferromagnetic slug attached to the diaphragm is determined by a differential transformer.

Still another automatic readout technique is accomplished by a vibrating wire. In one form of this device an aneroid cell is attached by the center of one diaphragm to a frame. The frame has an arm which extends around the cell to a point several centimeters above the center of the other diaphragm. A fine wire is connected, under some tension, between the end of this arm and the center of the second diaphragm as shown in Figure 4.20-6. The wire is driven to oscillate in a magnetic field. The frequency of oscillation is a function of the tension in the wire which is, in turn, a function of the distance between the diaphragms. This readout has the advantage that it is a frequency which can be converted to a digital value by counting oscillations over a specific time interval.

A variation of the vibrating wire technique is the use of an aneroid cell with a mechanically unloaded diaphragm. It is made to vibrate by an electrical transducer system mechanically coupled to the diaphragm and driven by an oscillator. The mechanical resonant frequency of the diaphragm changes when pressure is applied because of the variable spring rate designed into it. The oscillator follows the resonant frequency of the diaphragm giving pressure as a frequency of oscillation [1].

Another technique employs a photocell (photoresistive type). Light from a source of constant intensity is passed through a shutter onto the photocell. The shutter has one plate attached by a small rod to the center of the aneroid diaphragm. Another plate is stationary. Both plates have holes in them so arranged that motion of the diaphragm changes the position of the holes with

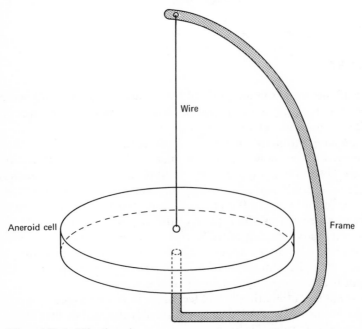

Figure 4.20-6   Vibrating wire.

respect to each other and thereby the size of the shutter aperture. This change in aperture size changes the amount of light falling on the photocell and its resistance [15, 20].

Modern aneroids used to measure air pressure near sea level, if not exposed to great temperature extremes, will measure pressure for weeks at a time without drifting from their average calibration by more than 0.15 mb.

The Bourdon Tube. The Bourdon tube is a circularly curved tube of strongly elliptical cross section which is completely evacuated or which contains a specific amount of dry inert gas. The radius of curvature of the tube changes with changing pressure (see Figure 4.20-7). There is no exact analytical expression relating these changes as a function of the physical properties of the tube. A useful approximation [14] is

$$dS = \frac{\lambda^4 LR\Delta P}{ESh^3},$$ \hspace{1cm} (4.20–7)

where dS is the displacement of the free end of the tube with the other end fixed to a reference frame and with a pressure change of ΔP; λ is the tube radius along the longer axis of the elliptical cross section, and δ is the radius along the shorter axis; h is the thickness of the tube wall; L is the total length of the tube; E is the modulus of rigidity of the tube material; and

$$R = \left[1 + \frac{\sin^2\alpha}{\alpha^2} - \frac{2\sin\alpha\cos\alpha}{\alpha} \right]^{1/2},$$ \hspace{1cm} (4.20–8)

where $\alpha = \phi/2$ and ϕ is the central angle subtended by length L. Another instrument uses a helical tube [6]. The Bourdon tube must be evacuated or contain dry inert gas at low pressure to measure pressure.

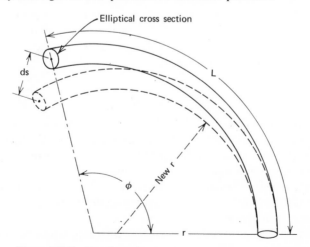

Figure 4.20-7 Bourdon tube.

Piezoelectric Effect Transducers. Certain crystals exhibit a change in dielectric constant when subjected to a stress. In one version of pressure sensor employing this effect two identical quartz crystals control identical oscillators. One crystal is isolated from ambient pressure, remaining under constant pressure. The other, exposed to ambient pressure, experiences a change of stress and hence a change in dielectric constant. This produces a change in frequency of the associated oscillator circuit. The shielded oscillator is beat against the exposed oscillator and the beat frequency is a measure of difference in pressures acting on the shielded and exposed crystals.

There is an obvious problem of keeping the pressure on the shielded or reference crystal constant. Temperature affects both crystals but if identical crystals are exposed to the same temperature, all temperature effects are compensated. This concept works best for large pressures and pressure ranges [13].

A *p-n* junction changes resistance when subjected to mechanical strain [7] and can be used to measure pressure [8, 22, 26]. Piezotransistors have been built. One version of this device is a silicon planar NPN transistor with stress-sensitive emmitter-base junction. Being both transistor and transducer the device is dual-purpose. Another instrument uses four interconnected strain gage resistors infused into the silicon diaphragm [5].

Errors in Elastic Sensors. Unless properly designed, all elastic sensors react strongly to temperature In one technique a bimetallic strip (Figure 4.20-4a) is placed in the mechanical readout linkage so that motion under changing temperature is equal and opposite to the temperature error of the sensor.

In a second technique, applicable only to aneroid cells and Bourdon tubes, enough gas is left inside the cell or tube so that the expansion of the gas exactly balances the decrease of stiffness of the sensor as the temperature increases. The residual pressure, R, to be left inside for exact compensation should be at a particular pressure

$$R = \frac{-\beta}{\alpha + \beta} P, \qquad (4.20\text{--}9)$$

where P is the pressure to be measured, α is the coefficient of expansion of gas, $1/273$ (which is 0.00366), and β is the temperature coefficient of the modulus of rigidity (shear modulus) of the sensing element.

In a third technique materials which have essentially zero temperature coefficients are used. When the walls of an aneroid cell are made thin so that the entire load is borne by an internal spring made of a material with a small β, the temperature error is very small.

A fourth technique is to process the signal with temperature compensating electronic circuitry.

The temperature error problem can be circumvented by putting the pressure sensor in a constant-temperature oven. This system works quite well with a pressure sensor since it is possible to control the temperature to which the sensor is exposed without disturbing the pressure being measured.

Other errors to which all elastic sensors are subject are hysteresis, shift of calibration from thermal shock [12] and long-term shift of calibration. An error is also produced if there is any loading of the sensor by the indicator mechanism. In mechanical indicators loading is due to friction or gravity. In electromagnetic indicators the loading can be due to magnetic or electrostatic forces.

4.20.6 Hypsometers

A liquid boils when its vapor pressure equals the total ambient pressure. Since a specific temperature corresponds to each vapor pressure, the ambient pressure can be measured by measuring the temperature of the vapor from a boiling liquid. The above description is quantified by the Clapyron-Clausius equation [10]

$$L = T\left(\frac{dP}{dT}\right)\Delta V, \qquad (4.20-10)$$

where L is the heat for change of state, ΔV is the change in volume for change of state, T is the absolute temperature, and dP is the change in pressure. Neglecting the volume of the liquid, using the ideal gas law, and integrating gives

$$P = \exp(-L/RT). \qquad (4.20-11)$$

4.20.7 Other Pressure Sensors

A specialized pressure sensor is used on the ocean bottom for detecting and measuring tsunamis, tides, and seiches. These phenomena involve relatively low amplitude, long waves. In contrast to the 12-hr tide period, the tsunami period is usually between 5 and 60 min, but can be longer. Its amplitude in open water can be as little as 0.1 m but can be 30 m at the coast line. The wavelength is several hundred kilometers. Such a wave is invisible and all but imperceptible in the open ocean. It might be detected by a pressure sensor capable of measuring pressure changes of 1 or 2 mb in a pressure environment of more than 100,000 mb. To do this the sensor would be placed on the bottom of the ocean, and it would monitor the total hydrostatic pressure including the atmospheric pressure. The ocean itself filters out the short waves (wind and swell waves on the ocean surface) so that the long-period tsunami would show as small and long-period variations in the high pressure prevailing at the ocean floor. Pressure sensors for this purpose include the electrokinetic transducer [2; 15, p. 106] and a Mossbauer effect detector that detects the motion of a membrane by sensing the Doppler effect of gamma rays [16].

Several pressure transducers have been used as tide gages including the piezoelectric crystal sensor, the vibrating wire aneroid, and a dual helix Bourdon tube sensor.

Another special situation for pressure transducers is the atmosphere at great heights where the pressure is too low to be sensed by standard operational instruments. Beyond the range of the hypsometer, most pressure measurements are made by converting the drag on a satellite or freely falling sphere to density and then converting to pressure.

Thermal gages employ a heated wire which is cooled by thermal conduction of the gas whose pressure is being measured. At pressures less than about 13 mb, heat conduction away from the wire by the gas molecules becomes a linear function of pressure, which then can be measured by applying a constant power to the heating of the wire and measuring the wire temperature (Pirani gage) or by measuring the power required to maintain the wire at constant temperature. In its usual form the device has a range from about 1 to 10^{-3} mb. Claims for much lower pressure limits [4] and for a much larger range have been made [25].

The radioactive ionization gage can also be used to measure pressures in the very high atmosphere. In one form the gage consists of an electronic circuit measuring the current between a housing and an ion collector as the gas is ionized by an α source. The range of the technique is from about 1300 mb to 10^{-3} mb [15]. The thermionic ionization gage employs essentially the same principle except that a thermionic (electron-emitting) filament is substituted for the α source.

4.20.8 Sensors of Pressure Variation

In day-to-day meteorological operations, rate of pressure change is usually read visually from a sensitive barograph. Pressure change meters, called variometers, measure the rate of change of pressure, dP/dt, and work on the same principle as an aircraft rate of climb indicator. The instrument is simply an aneroid pressure cell with a small hole in it. If the pressure is not changing, air leaks slowly through the hole to balance pressure inside and outside the cell. If the pressure outside changes, air cannot flow through the small hole fast enough to maintain equilibrium, and the diaphragm of the cell is deflected.

Pressure pulses having a frequency of less than about 20 Hz are generated in the atmosphere by several meteorological phenomena and are best detected by large chambers containing microphones and having properly designed and placed apertures [3]. 1000-Hz sinusoidal pressure variations of the order of 3×10^{-5} μb have been detected and measured using a multi-element array of loud speakers as the transducer [21].

References

1. Blanchard, W. C., "Study Phase of Radiosonde Pressure Sensor," Final Report on Contract No. 0-35214 to ESSA, Weather Bureau, Equipment Development Laboratory, Silver Spring, Md., July, 1970 (unpublished manuscript).
2. Collins, J. L., *Sohon Electrochemical Devices, Marine Sciences Instrumentation*, Vol. 2, p. 163, Plenum Press, New York, 1963.
3. Cook, R. K. and A. J. Bedard, Jr., "On the Measurement of Infrasound," *Geophys. J. Roy. Astron. Soc.* **26**, 5–12, Dec. 1971.
4. Ellet, A., and R. M. Zabel, *Phys. Rev.* **37**, 1102, 1931.
5. Fairchild Controls, Sunnyvale, Calif., Reports # 1 and # 2 to ESSA, Weather Bureau on Contract No. 0-35237 (unpublished manuscript).
6. Filloux, J. H., "Bourdon Tube Deep-sea Tide Gages," in *Proceedings of the International Symposium on Tsunamis and Tsunami Research,* International Union of Geodosy and Geophysics, University of Hawaii, East-West Center Press, Honolulu, 1970.
7. Hall, H., J. Bardeen, and G. Pearson, "The Effects of Pressure and Temperature on the Resistance of *p-n* Junctions of Germanium," *Phys. Rev.* **84**, 129–132, Oct. 1951.
8. Hardesty, C., private communication, NASA, Langley, 1972.
9. Harrison, L. P., "Manual of Barometry," U.S. Dept. of Commerce, Weather Bureau, Washington, D.C., 1963.
10. Hix, C. G., Jr. and R. P. Alley, *Physical Laws and Effects,* John Wiley and Sons, New York, 1958.
11. Hodgman, C. D., Ed., *Handbook of Chemistry and Physics,* Chemical Rubber Publishing Co., Cleveland, Ohio, 1960.
12. Iseley, C. W., NOAA, NOS Engineering Development Laboratory private communication, Feb. 1972.
13. Karren, H. E., and J. A. Leach, "Quartz Resonator Pressure Transducer," *IEEE Trans. Indust. Control Instr.,* July 1969.
14. Kleinschmidt, E., *Handbuch der Meteorologischen Instrumente*, Julius Springer, Berlin, 1935.
15. Lion, K. S., *Instrumentation in Scientific Research,* McGraw-Hill Book Company, New York, 1959.
16. McCue, J. C., "A Radioisotope Deep Ocean Aide and Tsunami Sensor," (unpublished manuscript).
17. Middleton, W. E. K., *The History of the Barometer*, The Johns Hopkins Press, Baltimore, 1964.
18. Middleton, W. E. K., and A. F. Spilhaus, *Meteorological Instruments,* University of Toronto Press, 1953.
19. Minnar, E. J., Ed., *I.S.A. Transducer Compendium,* Plenum Press, New York, 1963.
20. Ortman, G. C., "Photo Electric Pressure Sensing," *I.S.A.J.* **9**, No. 12, 63–64, Dec. 1962.
21. Parry, H. D. and M. J. Sanders, Jr., "The Design and Operation of an Acoustic Radar," *IEEE Trans. Geosci. Electron.* **GE-10,** No. 1, 1972.

22. Parker, C. D., "A Feasibility Study of a Miniature Solid-State Pressure Transducer," NASA Contractor Report NASA CR-1366, National Aeronautics and Space Administation, Washington, D.C., 1969.
23. Rice, D. A., NOAA, NOS, Gravity and Astronomy Branch, Rockville, Maryland, private communication, Feb. 1972.
24. Utterback, N. G., and T. Griffith, Jr., "Reliable Submicron Pressure Readings with Capitance Manometry," *Rev. Sci. Instr.* **37**, No. 7, July 1966.
25. Weise, E., *Zeitsch. Tech. Physik,* **24**, 66, 1947.
26. Wortman, J. J., "Effect of Mechanical Strain on *p-n* Junctions," NASA Contractor Report NASA CR-275, National Aeronautics and Space Administration, Washington, D.C., 1965.

Neil Brown

4.21 SALINITY SENSORS

The salinity of sea water can only vary as a result of direct interaction with some other part of the environment. For example, the salinity can be increased or decreased by evaporation or precipitation due to interaction with the atmosphere. Variations in salinity also occur when different kinds of water masses mix. Consequently, a complete understanding of the dynamics of the oceans requires a detailed knowledge of the spatial and temporal distribution of temperature and salinity.

4.21.1 Measurement Techniques

Chemical Methods. The titration method involves the precipitation of the halides, mainly the chlorides, by titration against a standardized solution of silver nitrate. The relative proportion of dissolved solids is fairly constant throughout the oceans. Consequently, a measurement of one of the main components such as the chlorides results in a measure of the total. This technique is of limited precision and requires extreme care by the operator. The further disadvantage is that chemical methods cannot be readily automated for continuous measurements from an in situ instrument.

Physical Methods. Salinity [22] can be determined by measuring density, sound velocity, refractive index, or electrical conductivity. The most highly developed method uses electrical conductivity. All the physical methods require a very precise knowledge of both temperature and pressure to obtain a useful accuracy in the determination of salinity. For example, to obtain a salinity accuracy of $\pm$ 0.01 ppt (parts per thousand), temperature must be known with an accuracy of at least 0.01°C.

In all cases the variation of the physical quantity being measured due to temperature variations is usually greater than the variation due to the normal range of salinity in the open ocean. Furthermore, with density, sound velocity, and refractive index, the variation from pure water to normal sea water is relatively small. For a change from pure water to normal sea water at 0°C, the variation in sound velocity is approximately 3%, the variation in density is approximately 2.7%, and the change in refractive index is approximately 0.5%.

4.21.2 Conductivity Sensors

Conductivity is approximately proportional to salinity and is essentially zero for pure water. Consequently, salinity can be determined to much greater accuracy and sensitivity by the use of electrical conductivity measurements than any of the other physical methods mentioned above. For in situ measurements where telemetry methods are required, electrical conductivity is a convenient measurement to make and is widely used.

Factors to be considered in sensor design include electrical, mechanical, and chemical effects on the sensor of pressure, temperature, fouling, and corrosion.

Electrode-Type Conductivity Sensors. Electrode-type sensors fit into two basic catagories, the two-electrode cell [2, 10, 17, 21, 23] and the four-electrode cell [12]. The two-electrode cell is not suitable for extended measurements of the highest accuracy because the electrochemical reactions that occur at the electrode-electrolyte interface have the effect of introducing an impedance between the electrode and the seawater which cannot be predicted. The magnitude of this impedance depends on operating frequency, electrode area, electrode surface cleanliness, and the geometry of the complete cell.

One cell design [21] minimizes these effects and results in measurements of highly acceptable accuracy. However, the design has extremely poor flushing characteristics and is unsuitable for an in situ instrument. Figure 4.21-1 is a schematic of the cell. The cell has two electrodes of very large area and a very long small-diameter tube. This gives a ratio of cell polarization

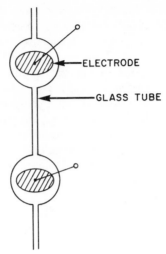

Figure 4.21-1 Conductivity cell [21].

impedance to overall cell resistance that is extremely low, thus resulting in high accuracy.

Four-Electrode Cell. The four-electrode cell [12] essentially eliminates the problems of polarization impedance. This cell, shown in Figures 4.21-2*a* and 2*b*, utilizes one pair of electrodes to inject a known current into the sea water circuit and a second pair of electrodes to measure the resulting voltage. This cell is analogous to the four-terminal technique used in

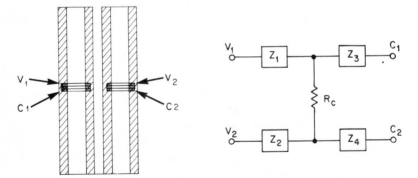

V_1, V_2 = VOLTAGE ELECTRODES $Z_1 \rightarrow Z_4$ = POLARIZATION IMPEDANCES

C_1, C_2 = CURRENT ELECTRODES R_C = CELL RESISTANCE

(*a*)   (*b*)

Figure 4.21-2 (*a*) Four terminal conductivity cell physical arrangement; (*b*) four terminal conductivity cell electrical equivalent.

resistance measurement to eliminate the effect of lead resistance. The resistance of the cell is defined as the ratio of the open-circuit voltage measured at the voltage electrodes to the current flowing through the current electrodes. This resistance is independent of the polarization impedance that occurs at each of the four electrodes.

One typical method of using this cell is as shown in Figure 4.21-3. Negative feedback via the cell automatically adjusts the alternating current through the cell such that E_c is equal to the applied input reference voltage E_r. If the open-loop gain of A is infinite, the input current flowing through Z_1 and Z_2 is zero, and the voltage across the sea water circuit is equal to E_r. That is,

$$I_c R_c = E_r, \tag{4.21-1}$$

so that

$$I_c = \frac{E_r}{R_c}, \tag{4.21-2}$$

where I_c is the cell current, R_c is the cell resistance, and E_r is the reference voltage (see Figure 4.21-3). Also,

$$E_0 = I_c R_0 = \frac{E_r}{R_c} R_0. \tag{4.21-3}$$

The cell resistance is

$$R_c = \frac{K}{\sigma}, \tag{4.21-4}$$

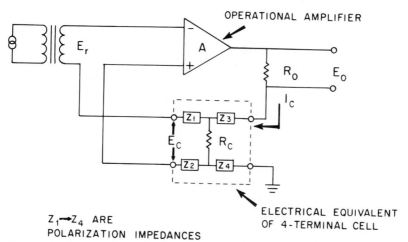

$Z_1 \rightarrow Z_4$ ARE
POLARIZATION IMPEDANCES

Figure 4.21-3 Four electrode conductivity cell and operational amplifier circuit.

where K is the cell constant and σ is the conductivity of seawater in the cell. Therefore

$$E_0 = \frac{E_r R_0 \sigma}{K}, \qquad (4.21-5)$$

and

$$\frac{E_0}{E_r} = \sigma \, \frac{R_0}{K}. \qquad (4.21-6)$$

Inductively Coupled Conductivity Sensors. Figure 4.21-4 shows the mechanical details of a typical inductively coupled conductivity sensor [4, 8, 9, 14, 15, 16]. Sea water forms a conducting single-turn loop which is common to the two toroidal coaxial transformers in the sensor. The voltage applied to the input winding of one transformer induces a current flow in the sea water loop which in turn impresses a magnetomotive force (mmf) on the second transformer. The induced current, and consequently the conductivity, can be measured by reducing the mmf to zero by passing an equal and opposite current through a single-turn winding on the second transformer. Balance is determined by observing a voltage null on a detector winding on the second transformer.

This sensor can be used at audio frequencies with optimum results and since it has no electrodes, it is not subject to the same kind of degradation as the simple two-electrode type. Fouling due to the growth of marine organisms particularly in the center hole is a problem only if it becomes so gross as to significantly change the effective geometry of the sensor.

Any conductivity method senses the conductance of the particular geometrical configuration of the conducting medium, not the specific conductivity. Thus any type of sensor suffers from the effects of gross fouling. Since the inductivity coupled types have better electrical characteristics if the center hole is large, the effective fouling is proportionally less with more favorable geometries.

4.21.3   Salinity, Temperature, Pressure Relationships

Temperature Effect on Conductivity. The change in conductivity of sea water as a function of temperature [5] is large, doubling from 0 to 30°C. For a salinity 35 ppt the conductivity (σ_t) of sea water is given by

$$\sigma_t = \sigma_{15} f(T), \qquad (4.21-7)$$

where σ_{15} is the conductivity at 15°C,

$$f(T) = a + bT + cT^2 + dT^3 + eT^4, \qquad (4.21-8)$$

$a = 0.67652$, $b = 0.201317 \times 10^{-1}$, $c = 0.99887 \times 10^{-4}$, $d = 0.19426 \times 10^{-6}$, $e = 0.67249 \times 10^{-8}$, and T is the temperature (°C) [5].

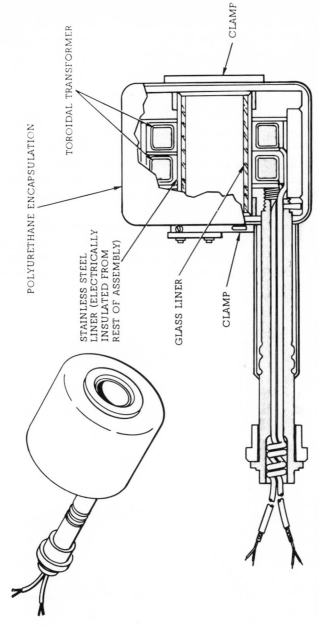

Figure 4.21-4 Inductively coupled conductivity cell, mechanical arrangement.

The Effect of Pressure on Conductivity. The effect of pressure on conductivity of sea water is shown in Figure 4.21-5 [1, 3, 18]. For any one particular value of salinity in the range of 30 to 40 ppt, the fractional change in conductivity due to an increase in pressure is given fairly accurately by

$$\frac{\Delta\sigma}{\sigma} = (aP + bP^2)(K_1 e^{c/T} + K_2), \qquad (4.21\text{-}9)$$

where a, b, c, K_1 and K_2 are constants, σ is the conductivity at atmospheric pressure, P is the pressure, and T is the temperature in $^\circ$K.

The fractional change in conductivity depends slightly on salinity, but in practice the salinity dependence is not important because at depths where the pressure effect is large the salinity converges to a value close to 35 ppt in the larger oceans of the world.

The curves in Figure 4.21-5 show that the pressure factor is large and highly dependent on temperature. Usually at extreme depths ocean temperatures converge to a value of about 2 or 3°C permitting simplifications in the expressions or in the design of pressure compensating circuits.

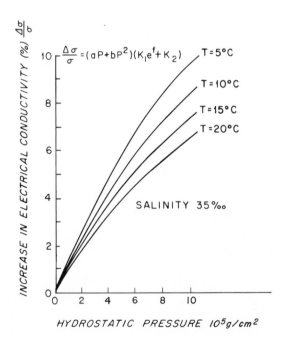

Figure 4.21-5 Change of conductivity of sea water (S = 35.0 ppt) with hydrostatic pressure [3].

The Salinity-Conductivity Ratio Relationship. The relationship between conductivity ratio and salinity over the range of 10 to 40 ppt is given by [5]

$$S = a + bR + cR^2 + dR^3 + eR^4 + fR^5, \qquad (4.21-10)$$

where

$$R = \frac{\text{conductivity of sample at } 15°\text{C}}{\text{conductivity of 35 ppt sea water at } 15°\text{C}},$$

$a = -0.2189$, $b = 29.8196$, $c = 7.9555$, $d = -3.8860$, $e = 1.5653$, and $f = -0.2347$. Where the conductivity ratio is measured at a temperature other than 15°C,

$$R = R_T + (R_T - 1)(jR_T + kR_T^2)(mT + nT^2 - 1), \qquad (4.21-11)$$

where

$$R_T = \frac{\text{conductivity of sample at temperature } T}{\text{conductivity of 35 ppt sea water at temperature } T},$$

T is the temperature (°C), $j = 0.01750$, $k = -0.00450$, $m = 0.080$, and $n = -0.00089$ [5].

4.21.4 Salinity Measuring Systems

An In Situ Salinometer. One instrument in wide use takes advantage of the normal distribution of temperature and salinity with depth to permit some simplifications in the description of the conductivity, temperature, salinity, and pressure relationships, which simplify the hardware required to implement them [6].

The salinity bridge which is the vital part of this instrument is shown in Figure 4.21-6. It consists of an inductively coupled conductivity sensor, a primary temperature compensation circuit, a second-order temperature compensation circuit, and a pressure compensation circuit.

Conductivity Sensor. The mechanical details of this sensor are shown in Figure 4.21-4. It consists of two coaxial toroidal transformers electrostatically and magnetically shielded from each other and housed inside a pressure-proof housing. The housing is insulated from sea water by means of a plastic coating. It consists of an outer and inner cylindrical section and two end caps. The inner cylinder is electrically insulated from both the sea water and the two end caps.

Temperature Compensation Circuits. Many techniques have been used to provide temperature compensation for direct-reading salinometers. These have included use of a thermistor resistor circuit adjusted to have the same temperature coefficient of conductance as the temperature coefficient of conductivity of sea water [9, 13, 17, 19]. Another technique is to use a compensating conductivity cell, with a sample of standard sea water

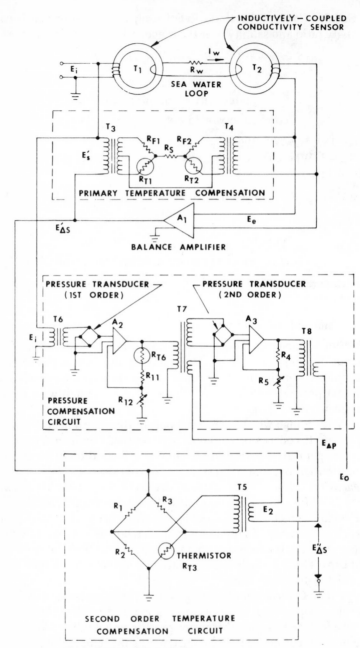

Figure 4.21-6 Salinity bridge.

permanently sealed in the cell [8]. It is fitted with a flexible membrane so that the sea water in the cell tends to come to thermal and pressure equilibrium with the surrounding sea water. The variations in conductivity of such a cell with temperature and pressure match the variations of conductivity of the surrounding sea water. However, it is extremely difficult to construct a cell sufficiently small to have a fast thermal response time and yet still remain stable with time.

Figure 4.21-6 shows how the temperature compensation circuit is used in conjunction with the inductively coupled conductivity sensor and a balance amplifier to form a basic salinity bridge. An input voltage E_i, which is common to the bridge input toroid (T_1) and the input of the temperature compensation network induces current in two circuits. One is in the sea water loop current (I_w) and the other is the output current of the temperature compensation network. Both these currents will induce an mmf on the output toroid (T_2). If the balance amplifier (A_1) output is short circuited, the output voltage E_e is then a function of the difference between these two mmf's. If they are initially equal, then changes in the sea water current (I_w) due to temperature-induced changes in conductivity are reflected by proportional changes in the output of the temperature compensation network so that an initial balance is not disturbed by changes in temperature.

Second-Order Temperature Effects. The salinity error signal (E'_{As}) is slightly temperature dependent because the primary temperature compensation is adjusted to give accurate compensation for one value of salinity (usually 35 ppt), but the temperature coefficient of conductivity of sea water varies slightly with salinity [11]. This means that as the salinity differs increasingly from 35 ppt, the temperature dependence increases. Figure 4.21-6 shows a Wheatstone bridge circuit with a thermistor in one arm to compensate for salinity changes. The bridge error signal is a function of both salinity and temperature because its input E'_{As} is a function of salinity and the thermistor resistance varies with temperature.

Pressure-Compensation Circuit. Figure 4.21-6 shows a schematic of the pressure-compensation circuit which consists of two pressure transducers, a temperature sensor, and two amplifiers. An input signal to the first pressure transducer results in an input to the first amplifier A_2, which is proportional to the sea water pressure. The gain of A_2 is determined by a negative feedback circuit consisting of the thermistor R_{T6} and the resistors R_{11} and R_{12}. The resistance of a thermistor R_t at any temperature is given by

$$R_t = R_0 \exp\left(B(1/T - 1/T_0)\right), \qquad (4.21\text{--}12)$$

where R_0 is the resistance at temperature T_0, T is the temperature of the thermistor in $°$K, and B is a constant. Therefore

$$R_t = K_3 R_0 \exp(B/T), \qquad (4.21\text{--}13)$$

where

$$K_3 = \exp(-B/T_0). \tag{4.21–14}$$

A series circuit consisting of a thermistor and fixed resistor (R_F) has a resistance R_s given by

$$R_s = R_0 K_3 \exp(B/T) + R_F = R_0 \left(K_3 \exp(B/T) + K_4\right), \tag{4.21–15}$$

where

$$K_4 = \frac{R_F}{R_0}. \tag{4.21–16}$$

By the appropriate choice of thermistor type and the resistance values of R_{11} and R_{12}, the constants B, K_3, and K_4 in Eq. 4.21–15, can be made proportional to the constants c, K_1, and K_2 in Eq. 4.21–9. Consequently, if such a series circuit is used as one of the elements of a negative-feedback circuit of a high-gain amplifier, then the gain of this amplifier can be made to vary with temperature in the same way as $\Delta\sigma/\sigma$. Therefore, the output of A_2 is

$$E_A = K_5 P\left(K_1 \exp(c/T) + K_2\right). \tag{4.21–17}$$

Now, since part of the output of A_2 is applied to the second pressure transducer, the output of this transducer is proportional to the product of the output of A_2 and the sea water pressure applied to the transducer. That is, its output is proportional to pressure squared. Therefore the output of A_3 is given by

$$E_3 = K_6 P^2\left(K_1 \exp(c/T) + K_2\right). \tag{4.21–18}$$

The output of this transducer is then amplified in a second amplifier A_3, and the output of this amplifier added to the output of the first amplifier, (A_2), using isolated windings on the two coupling transformers $(T_7$ and $T_8)$. By the correct phasing of the transformer windings and the correct adjustment of the gain of the two amplifiers, the constants K_5 and K_6 can be made equal to a and b, respectively, in Eq. 4.21–9. Thus the pressure-compensation circuit accurately simulates fractional increases in conductivity with pressure at any temperature.

Complete Salinity Bridge. Figure 4.21-6 shows a circuit of the complete salinity bridge. It consists of the conductivity sensor, the primary temperature-compensation circuit, balance amplifier, second-order compensation circuit, and the pressure-compensation circuit.

The total error due to temperature compensation, pressure compensation, and second-order compensation is less than ± 0.02 ppt in salinity for a salinity range of 30–40 ppt. a temperature range of 0–25°C and pressure range of 0–700 kg/cm².

Conductivity, Temperature, and Pressure (CTP) Systems. Attempts have been made to develop CTP systems using digital sensors and digital processing to compute salinity. Techniques are currently being developed which will permit digitizing to 16-bit accuracy (past devices were limited to 12-bit accuracy). These techniques along with modern low-cost minicomputers will permit salinity to be determined to much higher accuracies.

References

1. Adams, L. H., and R. E. Hall, "The Effect of Pressure on Electrical Conductivity," *J. Phys. Chem.* **35**, 2145, 1931.
2. Bradshaw, A. L., and K. E. Schleicher, "A Conductivity Bridge for the Measurement of Salinity of Sea Water," Woods Hole Oceanographic Institution Tech. Report, Ref. No. 56-20, 1956.
3. Bradshaw, A. and K. E. Scheicher, "The Effect of Pressure on the Electrical Conductance of Sea Water," *Deep Sea Res.* **12**, 151–162, 1965.
4. Brown, N. L., "A Proposed In Situ Salinity Sensing System," *ISA Marine Sci. Instr.* **2**, 19, Nov. 1962.
5. Brown, N. L., "Salinity, Conductivity and Temperature Relationships of Seawater Over the Range 0 to 50 ppt," Report published under ONR Contract NOnr-4290 (00), March 1966.
6. Brown, N. L., "An In Situ Salinometer for Use in the Deep Ocean," *ISA Marine Sci. Instr.* **4**, 563, Jan. 1968.
7. Brown, N. L., "The PARALOC—A Precise Telemetry Subcarrier Oscillator," submitted for publication to 4th National ISA Marine Sciences Instrumentation Symposium, January 1968.
8. Brown, N. L., A. L. Bradshaw, and K. E. Schleicher, "A Recorder for In situ Measurement of Salinity by the Inductive Method, Temperature, and Depth," unpublished work under Office of Naval Research Contract No. 2196-7 at Woods Hole Oceanographic Institution.
9. Brown, N. L., and B. V. Hamon, "An Inductive Salinometer," *Deep Sea Res.* **8**, No. 1, 65–75, June 1961.
10. Cox, R. A., "The Thermostat Salinity Meter," National Institute of Oceanography, Internal Reporter No. C2, 1958.
11. Cox, R. A., "Temperature Compensation in Salinometers," *Deep Sea Res.* **9**, 504–506, 1962.
12. Dauphinee, T. M. "In Situ Conductivity Measurement Using Low Frequency Square Wave A. C.," *ISA Marine Sci. Instr.* **4**, 555 1968.
13. Dorestein, R. R., *Kon Nederlands Meteor. Inst.,* **59**, 387, 1954.
14. Esterson, G. L., "The Induction Conductivity Indicator," The Johns Hopkins University, Chesapeake Bay Inst. Tech. Rept. XIV, Ref. No. 57–3, 1957.
15. Esterson, G. L., and D. W. Pritchard, "C.B.I. Salinity-Temperature Meters," Proc. Conf. Coastal Engrg. Instruments, Berkeley, Calif. (1955). Univ. of Calif. Council on Wave Research of the Engineering Foundation, Berkeley, Calif., pp. 260–271, 1956.

16. Gupta, S. R., and G. J. Hills, "A Precision Electrodeless Conductance Cell for Use at Audio prequencies," *J. Sci. Instr.* **33**, 313–314, 1956.

17. Hamon, B. V., "A Portable-Temperature Chlorinity Bridge for Estuarine Investigations and Sea Water Analysis," *J. Sci. Instr.* **33**, 329, 1956.

18. Hamon, B. V., "The Effect of Pressure on the Electrical Conductivity of Sea Water," *J. Mar. Res.,* **16**, 83–89, 1958.

19. Hamon, B. V., and N. L. Brown, "A Temperature-Chlorinity-Depth Recorder for Use at Sea," *J. Sci. Instr.* **35**, 452–458, 1958.

20. Hinkelmann, H., "Temperature Compensation for Sea Water Conductivity," *Keiler Meer, sf.* **17**, 25–31, 1961.

21. Jones, G., and G. M. Bollinger, "The Measurement of the Conductance of Electrolytes III. The Design of Cells," *J. Am. Chem. Soc.* **53**, 411–451, 1931.

22. Sverdrup, H. V., M. W. Johnson, and R. H. Fleming, *The Oceans,* Prentice-Hall, Englewood Cliffs, New Jersey, 1942.

23. Wenner, F., E. H. Smith, and F. M. Soule, "Apparatus for the Determination Abroad Ship of the Salinity of Sea Water by the Electrical Conductivity Method," *Natl. Bur. Std. J. Res.* **5**, 711–732, Research Paper No. 223, 1930.

W. P. Johnson III

4.22 SEISMIC SENSORS

Seismic is derived from the word meaning to shake, and usually relates to the earth. The shake may be due to natural causes or to man-made energy sources. Magnitudes on earth vary from a possible low displacement of 10^{-10} m. The upper limit is not clearly defined but dislocations of more than 10 m can be encountered. The frequency band is from 0.01 to 500 Hz. In exceptional cases this bandwidth has been exceeded. The seismic-wave sensor is part of an instrument system called a seismograph used to measure the earth shake.

4.22.1 Principle of Operation

Seismic sensors are based on a mass (m) attached to a spring or other restoring force which is connected to the instrument case in a manner that isolates the mass from the case by the restoring force. Damping is necessary to control response in the vicinity of resonance and is viscous so that the damping force is proportional to velocity. A schematic of the essential elements is shown in Figure 4.22-1. The physical principles which determine the relation between case motion and mass motion are well defined in mechanics, assuming the mass to be restricted to linear motion without rotation and without Coulomb friction. Otherwise general equations to describe the sensor would only be approximations.

4.22.2 General Mechanical Equations

The general equation for the sensor in the absence of case motion is

$$m \frac{d^2 x}{dt^2} + C \frac{dx}{dt} + kx = 0, \tag{4.22-1}$$

where C is the damping coefficient and k is the spring constant.

If simple harmonic motion is applied to the case, the relation between mass and case motion called the magnification factor can be ascertained. Neglecting the phase angle between mass and case, the equation in dimensionless form is

$$M = \frac{n^2}{[(1 - n^2)^2 + 4b_t^2 n^2]^{1/2}}, \tag{4.22-2}$$

where

$$M = \frac{\text{case-to-mass motion}}{\text{case motion}}, \tag{4.22-3}$$

$$n = \frac{\omega}{\omega_0} = \frac{\text{case motion frequency}}{\text{undamped suspension frequency}}, \tag{4.22-4}$$

$$b_t = \frac{C_0}{C} = \frac{\text{actual damping force}}{\text{damping force required for critical damping}}. \tag{4.22-5}$$

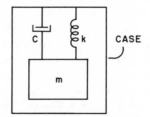

Figure 4.22-1 Seismic sensor schematic diagram [7].

This equation is used as the basis for determining the response of a sensor at a number of simple harmonic frequencies. These points are connected by a line on graph paper to obtain the steady-state sinusoidal response. Near resonance the response can become infinite, for the theoretical case of no damping. The phase relation is given by

$$\phi = -\arctan \frac{2b_t n}{1 - n^2} \cdot \tag{4.22-6}$$

The damping force is converted to a dimensionless factor for convenient use. The unit factor is critical damping, which occurs when the smallest possible damping force is present to prevent any oscillation of the mass after the reception of a transient or step-function force. The three damping states are under-critical, critical, and over-critical. Except in rare cases the damping is at or below critical. The most common damping factor, 0.7071, produces a response curve which rapidly approaches a magnification factor of unity at frequencies above resonance without exceeding the value of unity throughout the frequency spectrum. The equation for determining under-critical damping is

$$b_t = \frac{\log_e(A_1/A_2)}{\{\pi^2 + [\log_e(A_1/A_2)]^2\}^{1/2}}, \tag{4.22-7}$$

where A_1 is the amplitude of one oscillation decrement, and A_2 is the amplitude of the succeeding half-cycle.

The above equations are used for comparative purposes in selecting and using the sensor. However, the sensor is seldom subjected to simple waveforms. One type of waveform, the wavelet, has been analyzed and proven useful. Other possible waveforms can also be analyzed and the sensor response determined.

4.22.3 Practical Seismic Sensors

No practical sensor is available to satisfy all requirements of the general equation. All sensors have a lateral natural frequency, usually higher than the desired natural frequency, so that two degrees of freedom are always encountered. This second degree of freedom should be well beyond the desired passband or the response to vector forces on the case will be too difficult to interpret. The undesirable effect can be reduced considerably by the use of balanced mechanical and transducer construction.

Distortion is generally proportional to the case-to-mass motion. Distortion is caused by nonlinearity due to a nonlinear spring constant or a mass which moves in an arc or rotates. The mechanical distortion must be added to the transducer distortion to obtain the complete sensor rating.

4.22.4 Transducers

The most common transducer has a coil of wire moving at right angles to a magnetic flux path. The transduction constant relating output voltage to velocity can be expressed as

$$G = \frac{E}{V} = \bar{B}l, \qquad (4.22-8)$$

where E is potential in volts, $V = \dot{x}$ is velocity, $\bar{B}$ is the average flux density across the coil, and l is the length of conductor in coil. Most seismic sensor and transducer units are CGS and $\bar{B}l$ is multiplied by 10^{-8}.

With current flow in the coil a restraining force, or damping force, is present which is proportional to velocity. This form of damping may be due in part to a metallic coil form with a transduction constant G_0 producing a damping force ratio b_0. The total damping force ratio b_t is the arithmetic sum of $b_0 + b_c$.

Another form of transduction consists of a variable magnetic gap coupled to a coil in the iron circuit. This is a very nonlinear system with a large value of negative incremental k and a large value of coil self-inductance. The system is not in general use. Other transducers use air-core differential transformers or variable capacitors or inductors. These have nonlinear spring constants and must be restricted to very low power levels. The damping in these cases is best supplied by a coil or vane in a magnetic flux path.

Transducers are not actually necessary particularly in the very-low-frequency region. Many older seismometers contain mechanical-advantage devices connected to a stylus for direct writing on a moving surface. Some use a mirror to reflect a light beam to an actinically sensitized surface.

4.22.5 Noise in Seismic Sensors

Practical sensors must have sufficient inertial mass to reduce thermal agitation noise. This is not always possible, especially at natural frequencies at the lower end of the spectrum. The noise can be calculated by mechanical equations, but in the case of electromagnetic damping an electrical analog is used for convenience. The noise voltage is

$$E_n = [4kT\bar{R}B]^{1/2}, \qquad (4.22-9)$$

where E_n is the noise in RMS volts, k is $1.37 \cdot 10^{-23}$ J/$^\circ$ K, T is the temperature in degrees Kelvin, B is the frequency bandwidth in hertz and $\bar{R}$ is the average parallel value of shunt and coil impedance. This is a variable dependent on T due to conductor temperature coefficient.

This noise increases by a factor of 3 periodically and rarely attains a level of 10 times this value. In temperature-controlled sensors convection air currents cause slight motions of the mass that appear as noise.

Stray electrical field pickup may be present and is reduced by dual transducer construction and electromagnetic and electrostatic shields. When long lines are anticipated, the transducer must have very low capacitance and resistive leakage to the case to reduce ground pickup.

4.22.6 Long-Period Seismometers

These sensors are different topographically from Figure 4.22-1 because 0.04-Hz sensors require a suspension with about 39 m deflection at the surface of the earth. Because such dimensions are inconvenient, mechanical-advantage methods related to the spring constant are employed. The deflection of a vertical seismometer is

$$A_s = \frac{g}{\omega_0^2}, \tag{4.22-10}$$

where A_s is the mass-to-case displacement in meters, g is the acceleration (about $9.8 \, \text{m/s}^2$ on earth), and ω_0 is 2π times the undamped natural frequency.

Most vertical sensors utilize the LaCoste slant spring [4]. This consists of a horizontal beam attached to the case through spring hinges with very little restoring force. The other end is attached to the principal mass and transducer. In the simple form a negative-length helical spring is attached to the mass by a ligature and the opposite end is attached to the case through a ligature above the hinges. The spring is normally slanted at 45° to support the mass in the earth's gravitational field. This sensor has no theoretical low-frequency limit.

Horizontal sensors consist of a pendulum forced to operate at some angle to the gravitational field. When the sensor is horizontal the field has no effect and the frequency is determined by the rotational inertia of the mass beam combination with the restoring force supplied by the hinges.

4.22.7 Short-Period Seismometers

Short-period seismometers have natural resonant frequencies from about 0.3 to 2 Hz. The slant spring is not used because of rather large frequency changes with case tilt. Two three-element spiders are used to center and guide the coil-mass assembly and a lift spring is used to counteract gravitational forces. Geophone-type construction may also be used with the spiders formed upward so that no lift spring is necessary. The terms seismometer or geophone are used in this frequency region depending on the use of the sensor.

4.22.8 Exploration Geophones

Geophone sensors are manufactured primarily for the seismic method of oil prospecting and for intrusion detection. In these sensors the transducer and physical components become indistinguishable. The mass consists of the coil assembly and the case is part of the magnetic circuit. Sensors are available with natural frequencies from 2 to 80 Hz. The transducer is a coil moving perpendicular to a magnetic field. These sealed sensor units are generally placed in a plastic housing with a stud bottom for mounting to different types of planting attachments including flat bases, spikes, or tripods.

All geophones should withstand random drops of 0.5 m to a heavy steel plate once per second for at least four hours. Some units will withstand shock impacts of more than 20,000 g combined with large values of rotational acceleration.

4.22.9 Horizontal Seismic Sensors

All sensors are somewhat sensitive to tilt, both statically and dynamically. At long to short periods the sensitivity is so great that the horizontal response may be obscured. It is important to plant the sensors so they are level and do not drift with time. The displacement due to tilt may be expressed as

$$A_n = \sin \theta A_s, \qquad (4.22\text{--}11)$$

where A_n is the case-to-mass motion due to tilt from horizontal, θ is the tilt angle, and A_s is the case-to-mass motion due to vertical acceleration (see Eq. 4.22–10).

4.22.10 Seismic Sensors in Space

Seismic sensors cannot operate alone in space since energy cannot be transmitted to the case, but they can be used on space platforms or in space vehicles. In a circular orbit around a large body such as the earth, or moon, the sensor will be of the horizontal type regardless of spacecraft attitude. In an elliptical orbit accelerations are present which cause case-to-mass motion that can be computed from Eq. 4.22–2.

4.22.11 Stability of Seismic Sensors

The stability factors can be grouped into irreversible and reversible categories. The magnetic circuit and the spring have both.

The irreversible effects of the magnetic circuit can be removed during manufacture by continuous cycling over any expected temperature range, typically − 50° to 85°C. The magnet must not be remagnetized after this operation.

The suspension system drops relative to the case due to spring creep. This can be minimized by temperature cycling the complete sensor. Ageing until the creep reaches a plateau position must be done on all seismometers or geophones with responses to about 3 Hz. The springs are then adjusted for proper support of the mass.

Several reversible effects are present. The magnetic flux has a temperature coefficient from 0.01 to 0.015 %/°C. The spring modulus of elasticity changes very slightly with temperature and can cause large changes in spring deflection with a consequent change of spring constant. It is therefore necessary to use constant-modulus alloys for short-period seismometers. Long-period seismometers must have this provision in addition to a fairly constant temperature environment.

In addition the coil resistance changes with temperature about $0.039\%/°C$ for copper or aluminium wire. The metallic coil form has a similar temperature coefficient.

4.22.12 Typical Sensors

Virtually all sensors are active units with the transducer consisting of a coil moving relative to a magnetic field with an electromagnetic damping force. The transduction is based on a center value of coil resistance; which can be changed (due to standard wire sizes) by increments of 1.26^2, with transduction increments of 1.26. Resolution is limited only by noise due principally to thermal agitation.

Geophones are available from 1.4 cm in diameter by 1.6 cm high to 5.4 cm in diameter by 5.4 cm high. Seismometers are generally cased and vary from 7.5 cm in diameter by 15 cm high to units with a volume of 10^5 cm^3 at the longer periods. See Table 4.22-1.

Table 4.22-1. Typical Sensors

Type	Geophone	S.P. Seis.	L.P. Seis.
Size(cm)	3.2 dia × 3.3	11 dia. × 35	40 × 30 × 60
Construction	Balanced	Balanced	Slant spring
Transduction factor	0.37 V/cm/s	1.9 V/cm/s	2.1 V/cm/s
Coil resistance	380 Ω	500 Ω	1130 Ω
Max. case to coil motion	0.4 cm total	1.25 cm total	2 cm total
Guarantee period	3 yr	3 yr	1 yr
Natural frequencies	4.5, 8, 10, 14, 20, 30 Hz	0.5, 1.0 Hz	Adjustable 10–30 s
Lateral frequency range	80–300 Hz	170 Hz	—
Relative response to lateral motion at lateral frequency	−6 to −30 dB	−20 to −40 dB	—
Reliability	99.5%	99.5%	—
Availability	2 weeks	6 weeks	—

4.22.13 Special Sensors

Special sensors are units for which there is no standard tooling and few stock parts. An example is a geophone which produces a very small amount of external magnetic flux and can be used in close proximity to instruments quite sensitive to magnetic fields. This sensor consists of two single coil geophones mounted face to face with the coils in intimate contact. The magnetic field is therefore enclosed in soft steel. The stray field is greatly reduced and high-premeability low-saturation alloys can be used to reduce the stray field strength to any desired level with a minimum of size and weight. Actually the two coils become one and the magnetic circuit is made in two parts for magnetization.

Another example is a very-high-output seismometer or geophone which requires a damping factor of over ten times critical with an undamped natural frequency of about 1.9 Hz. The open-circuit transduction is 11.5 V/cm/s.

4.22.14 Other Sensors Which Can Be Used For Seismic Detection With Restrictions

Vibration sensors may be used in the range between 0.25 mm/s peak velocity to 50 g of acceleration limited to 1.25 cm peak-to-peak case motion. The transduction is about 0.04 V/cm/s. The lateral response is less than 2 percent. The units are fluid filled and have a thermal coefficient of transduction of 0.03%/°C. These are self-generating devices and have a typical coil value of 940 Ω. Piezoelectric sensors can also be used at the higher frequencies or very high levels of excitation. These accelerometers have basic sensitivities of 3 to 10 mV/g with an internal impedance which is capacitive and equal to 400 to 700 pF.

References

1. Dennison, A. T., "The Design of Electromagnetic Geophones," *Geophys. Prospecting* **1**, No. 1, 3–28, 1953.
2. Freberg, C. R., and E. N. Kemler, *Elements of Mechanical Vibration*, Chapter 3, John Wiley and Sons, New York, 1943.
3. LaCoste, Jr., L. J. B., "A New Type Long Period Vertical Seismograph," *Physics*, **5**, 178–180, July 1934.
4. Mechtly, E. A., *The International System of Units*, NASA SP-7012, U.S. Government Printing Office, Washington, D.C.
5. O'Brien, P. N. S., "Geophone Distortion of Seismic Pulses and Its Compensation," *Geophys. Prospecting* **13**, No. 2, 283–305, June 1965.
6. Parker, R. J. and R. J. Studders, *Permanent Magnets and Their Application*, John Wiley and Sons, New York, 1962.
7. Ricker, Norman, "Wavelets," *Geophys.* **5**, No. 4, 348–386, October, 1940, also **9**, No. 3, 314–323, July, 1944.
8. Symon, Keith R., *Mechanics*, Chapter 2, Addison-Wesley, Reading, Massachusetts, 1964.

Robert H. Lee

4.23 SUN SENSORS

A wide variety of sun sensors is available, depending on the spectral region of interest. The earth's atmosphere has only a few spectral "windows" where transmission is high enough to make solar radiation measurements feasible. One of these includes the radio wavelengths from a few millimeters to a long-wavelength cutoff caused by the earth's ionosphere. The second major window extends from about 0.35 μm to a few μm. This is the visible and near infrared portion of the spectrum.

4.23.1 Purposes

Solar Astronomy. The study of solar physics requires the measurement of many parameters including the solar constant, flares, prominences, magnetic fields (by sensing the Zeeman splitting of emission or absorption lines), solar temperature at various heights in the sun's atmosphere, and the solar corona. Nearly all of these measurements require the use of narrow optical bandwidths. In the visible portion of the spectrum, photography is commonly used, both for direct photographs (usually through optical band-narrowing filters) and for recording spectra. Various television techniques have been used successfully.

In the visible region the multiplier phototube (MPT) is most commonly used where high precision of measurement is needed or where differencing or rapid scanning (spectral or spatial) can be used to detect a small change against a strong background. The PIN diode is replacing the MPT in many applications. A variety of sensors are in use in the infrared spectral region.

Direction Sensing. In applications such as navigation and platform orientation it is necessary to know the direction of the sun. Solar telescopes on platforms use pointing or guiding mechanisms with feedback servomechanisms. Silicon photovoltaic cells are well suited to this application when used with a simple lens for focusing an image of the sun, or with a shadowing edge or disc. The energy received is adequate for the silicon cell, which has advantages of simplicity, ruggedness, and stability. A high sensitivity exists because the spectral sensitivity of the cell is matched to the solar output, and a wide optical bandwidth can be used.

Intervening Medium. The sun can be used as an energy source for the measurement of some parameter of the medium between the sun and the sensor. Parameters that can be measured include cloud cover, aerosols, precipitable water, and atmospheric constituents. With the exception of simple cloud cover (sunshine) recorders, the sensors must be used with auxiliary devices such as light-gathering equipment (mirror or telescope, with tracking capability) and equipment for isolating the spectral regions of interest (spectrograph or narrow-band interference filter).

Table 4.23-1. Transmittance from ∞ to h for Normal Incidence [7]

Wavelength (μm)	Sea Level	11.6 km
0.34	0.212	0.845
0.36	0.351	0.889
0.38	0.476	0.910
0.40	0.539	0.928
0.45	0.635	0.953
0.50	0.691	0.960
0.55	0.718	0.952
0.60	0.737	0.946
0.65	0.777	0.972
0.70	0.805	0.985
0.80	0.829	0.992
0.90	0.847	0.997
1.06	0.860	0.998
1.26	0.869	0.999
1.67	0.882	0.999
2.17	0.897	1.0
3.5	0.915	1.0
4.0	0.923	1.0

4.23.2 The Source

Figure 4.23-1 shows the solar spectral irradiance incident upon the earth's atmosphere [7]. It can be approximated by a blackbody at 6000°K.

Solar irradiance is a complex function of altitude [2]. Figure 4.23-2 shows measurements made from an aircraft at 11.6 km and at sea level with the same instrument [7]. Table 4.23-1 shows the calculated transmittance through the atmosphere to sea level and to 11.6 km at several wavelengths. Beyond 0.90 μm the values are given for optical windows in the atmosphere. At visible wavelengths attenuation is caused primarily by molecular and aerosol scattering. As the wavelength increases from the visible, molecular absorption becomes increasingly important, with water vapor being the major contributor [3, 5].

Some measurements are disturbed by radiation from the sky [1]. Thermal radiation from the sky peaks around 10 μm, decreasing rapidly toward shorter wavelengths. Sunlight scattered by the atmosphere has a spectral characteristic which peaks in the blue region and decreases rapidly toward

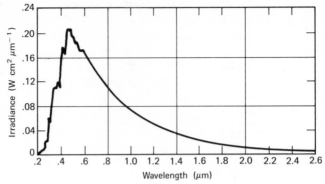

Figure 4.23-1 Solar spectral irradiance incident on atmosphere [7].

longer wavelengths. This results in a minimum sky brightness in the spectral region around 3 μm.

4.23.3 Sensor Types

Two classes of photosensors are in widespread use in the visible portion of the spectrum. One class, the vacuum photoemissive sensor, includes vacuum, gas-filled, and multiplier phototubes. The multiplier phototube is the most useful for measurement purposes. The other class, the solid-state photocell, includes photoconductivity bulk effect, photovoltaic, and photoconductive junction type sensors. These devices generate a current which is linear with respect to received radiant power.

In the infrared region, a third class of sensors uses the thermal effect. In these devices, the absorption of radiation results in an increase in tempera-

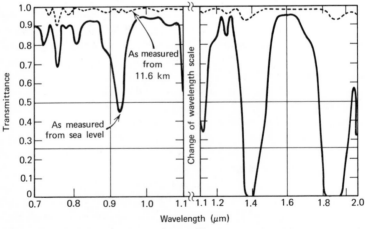

Figure 4.23-2 Atmosphere transmittance [7].

ture in the detector, producing a measurable change in some detector property. These sensors include bolometers in which the resistance changes, and thermocouples, thermopiles, and pyroelectric devices which generate a voltage change. Thermal detectors usually have a fairly constant output over a wide range of wavelengths since they detect energy and not photons. Their response extends into the far infrared regions, but they have a slow response time and high noise level.

Photoelectric Theory. The maximum energy of an emitted photoelectron is proportional to the energy of the light quanta (hv) less the energy of the work function (ϕ) which must be given to an electron to allow it to escape into the surrounding vacuum. This can be written as

$$E = \frac{mv^2}{2} = hv - \phi. \qquad (4.23-1)$$

The photoelectric effect for each metal is characterized by a value (ϕ) expressed in electron volts. If the energy received by an electron is just sufficient to allow it to escape ($hv_0 = \phi$), then v_0 is the threshold frequency of the exciting radiation and is related to the long-wavelength cutoff (λ_0) and the velocity of light (c) by

$$\lambda_0 = c/v_0. \qquad (4.23-2)$$

This can be rewritten as

$$\lambda_0 = \frac{1.24}{\phi} \mu m. \qquad (4.23-3)$$

The photoelectric yield (quantum efficiency) of pure metals is very low, less than 10^{-3} electron per incident photon.

The most useful photoemitters are semiconductors. In a semiconductor the highest filled energy band for electrons is called a valence band. Above the valence band is an energy gap, called the forbidden gap, where no electron energy states exist. Above the forbidden gap is a band of permitted energy called the conduction band. At ordinary temperatures it contains few electrons. The potential barrier from the bottom of the conduction band to the vacuum surface is called the electron affinity. Photoconductivity may result when radiation of sufficient energy excites an electron in the valence band and causes it to move to the conduction band. Photoemission does not take place until the electron escapes the barrier to the vacuum. Thus the minimum photon energy necessary to produce electron emission is the sum of the energy required to move from the valence band to the conduction band and that required to overcome the electron affinity. For a multi-alkali photocathode these values are 1.0 and 0.55 eV, respectively [4].

Semiconductors used as photoemitters have high quantum efficiencies, with yields up to 30 percent compared with 0.1 percent for metals. In a semiconductor, a photoexcited electron has a lower probability of losing its energy in an electron-electron collision than in a metal where many free electrons are present. The photoelectric responses of several materials are shown in Figures 4.23-3 through 4.23-5.

In solid-state photocells, the absorption of an incident photon of sufficient energy raises an electron from the valence band to the conduction band, creating a hole-electron pair. In the photovoltaic junction, space charge in the junction separates the hole-electron pairs, giving rise to a photovoltage. The pair separation is in the direction to forward bias the junction.

In a photoconductive semiconductor there are two excitations used. In intrinsic excitation, hole-electron pairs are generated when a photon raises an electron from the valence band to the conduction band. In extrinsic excitation, current carriers are produced by photons exciting electrons from or to impurity levels which lie within the forbidden gap of the semiconductor material. In both cases photocurrent is made to flow by means of a reverse bias across the junction.

Multiplier Phototubes (MPT). An MPT can be conveniently subdivided into a photocathode, an electron-optical input system, a secondary emission multiplier, and an anode.

The photocathode can be either semitransparent or opaque. The semitransparent photocathode is used in end-on phototubes, with the photosensitive material deposited on the inside of the tube window. The incident radiation passes through the window, is absorbed in the photocathode, and electrons are emitted into the vacuum within the tube. The cathode

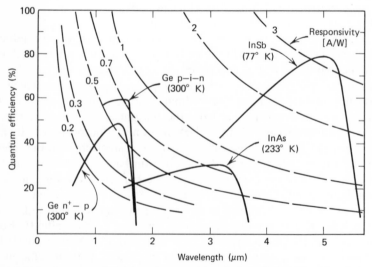

Figure 4.23-3 Quantum efficiency and responsivity of photodiodes that operate between 1 and 6 μm [4].

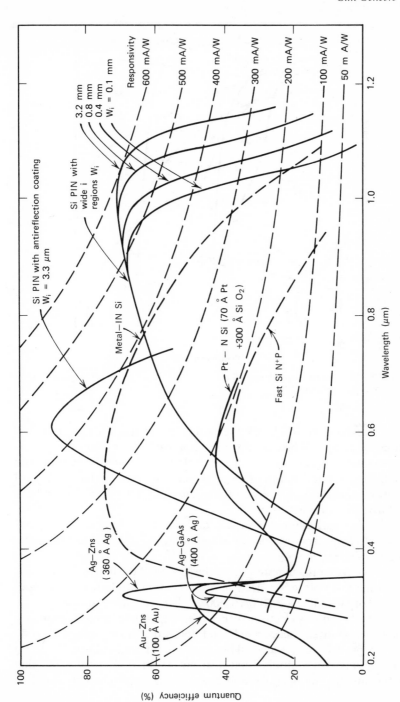

Figure 4.23-4 Wavelength dependence of quantum efficiency and responsivity for several high-speed photodiodes [4].

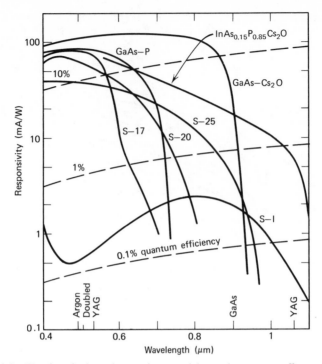

Figure 4.23-5 Wavelength dependence of responsivity and quantum effiency for several high-efficiency photosurfaces [4].

must be thick enough to absorb the radiation but thin enough to permit electrons to escape. When an opaque photocathode is used, the photo-sensitive material is deposited on a metal plate or film inside the tube, and electrons are emitted from the side on which the radiation falls. Standard designations for the relative spectral response of photo devices have been defined from S-1 through S-25.

The electron-optical input system is designed to collect, accelerate, and focus the electrons from the photocathode onto the first element (dynode) of the multiplier section.

The secondary emission multiplier is the section of the tube where current amplification takes place. An electron from the photocathode is accelerated by the potential difference (100 to 300 V) between the cathode and the first dynode. When this electron strikes the first dynode, several secondary electrons are emitted, accelerated, and focused onto a second dynode where the process is repeated. This process continues through several dynodes or stages of multiplication. Current gain is 10^3 to 10^7 depending upon the number of dynodes and the voltage applied. Gain stability, susceptibility to magnetic fields, resistive leakage, and transit time spread are factors affecting the choice of dynode type.

Solid-State Photodiodes. The solid-state photodiode is often easier to use than a multiplier phototube. The silicon photodiode is useful over the entire visible range and into the near infrared to about $1.1\mu m$.

Two types of photodiodes are the PIN and PN. The PIN photodiode has a heavily doped P region and a heavily doped N region separated by a nearly intrinsic (I) region with a high resistivity compared to the P and N regions. When a reverse bias is applied to the diode a depletion region develops whose depth is dependent upon the resistivity of the I region and the voltage applied across it. This depth is the distinguishing difference between a PN and a PIN device. A PN device has a depletion layer only a few μm thick, but a PIN device can have a depletion region more than a mm thick.

A photon absorbed by the PIN diode produces a hole and an electron. If this absorption occurs in the depletion region, the hole and the electron are separated by the electric field, thus contributing to current flow. For high quantum efficiency, the P region is thin to avoid absorbing photons before they reach the depletion region, which is sufficiently thick to assure absorption of the photon.

A PIN photodiode normally is used as a photoconductive device with reverse bias applied. Operated in this manner it is a current generator. Advantages of this mode are fast response time, high responsivity, good linearity (better than 2 percent over more than eight decades) and high output at longer wavelengths.

The major disadvantage of the photoconductive mode is dark current with its associated shot noise. Dark current is a combination of bulk leakage and surface leakage. Surface leakage can be reduced by a "guard ring" to shunt surface currents away from the signal circuit. Leakage current increases as reverse bias is increased. Dark current is temperature dependent, with every $10°C$ temperature increase causing the dark current to approximately double.

By operating the PIN diode as a photovoltaic (self-generating) device dark current is greatly reduced. If it is operated into a low impedance, good linearity is maintained. If it works into a high-impedance load, the voltage output is a quasi-logarithmic function of the irradiance. In the photovoltaic mode, responsivity is lower than it is in the photoconductive mode. This reduction varies with manufacturing technique, but the responsivity may be reduced to about one-half that of the reverse bias case.

Active areas can be as small as 1.3×10^{-6} cm^2 and as large as 5 cm^2, and of almost any shape. Arrays of diodes with densities of more than 40 per linear cm can be manufactured. The spectral response of the cell can also be modified, within limits, by choice of resistivity of the bulk silicon.

For example, for 3000 Ω/cm material the peak responsivity is about 0.95 μm and for 10 Ω/cm material the peak is about 0.65 μm.

A large number of photon detecting materials are available in the infrared region. A number of them approach the theoretical maximum detectivity when cooled to liquid nitrogen or liquid helium temperatures [4]. Standard relative spectral response curves S-26 through S-40 have been assigned to cover most of the elemental infrared photodetecting materials.

4.23.4 Noise

When making photoelectric measurements at either very low light levels, or at higher light levels with high precision, sensitivity or precision or both are inevitably limited by noise. Noise induced from external sources can be reduced or eliminated by shielding, filtering, and careful attention to signal returns (grounds) and other common paths which the desired signal shares, perhaps unintentionally, with some disturbing signal. See Section 8.2.

A special case is magnetic disturbance within a multiplier phototube. A magnetic field can distort the trajectories of electrons as they progress from the photocathode down the dynode chain. A varying magnetic field (frequently induced at powerline frequencies) can modulate the gain of an MPT. A steady field can reduce the gain, or, if it prevents electrons emitted by the photocathode from reaching the first dynode, degrade the signal-to-noise ratio.

Johnson Noise. This is the frequency-independent noise arising from thermal fluctuations in the motions of electrons in a resistive element. This RMS noise voltage is

$$V_J = (4kTR\,\Delta f)^{1/2}, \tag{4.23-4}$$

where k is Boltzman's constant (1.38×10^{-23} J/$^{\circ}$K), T is the absolute temperature ($^{\circ}$K) and Δf is the effective noise bandwidth of the measuring system. Since this noise is temperature dependent, it can be reduced by cooling the resistive elements which are the sources.

Shot Noise. This results from statistical fluctuations in the number of electrons leaving or arriving at a surface (i.e., electrons emitted from a photocathode or entering or leaving a semiconductor contact). The RMS current fluctuation is given by

$$i_n = (2ei\,\Delta f)^{1/2}, \tag{4.23-5}$$

where i is the average current and e is the electron charge (1.6×10^{-19} C). This relationship can be used to determine the noise from dark current, or "noise-in-signal," the shot noise arising from the fluctuations in the photocurrent caused by the "signal" flux input. In many applications, changes in input flux are to be measured rather than the total flux. In this case the total flux contributes shot noise but does not contribute signal. This means that unnecessary "background" flux should be excluded from the sensor whenever possible.

Shot noise due to dark current can be reduced by cooling the photo device. Shot noise due to "signal" or "background" currents is not significantly affected by temperature. The signal-to-noise ratio can be improved by using devices with high quantum efficiency since the signal current increases directly with quantum yield, while noise current increases as the square root.

For a multiplier phototube, the RMS current fluctuation (caused by shot noise) at the anode is

$$i_{n(a)} = \mu(2ei_t F \, \Delta f)^{1/2}, \qquad (4.23-6)$$

where μ is the gain of the multiplier, i_t is the total collected photocathode current (dark current plus photo current), and F is the noise factor of the multiplier (1.1 to 1.5 for most tubes).

It is usually advantageous to modulate the signal component of the radiation by a chopper or other means and to use a narrow bandpass amplifier tuned to the modulation frequency. A modulation factor, M, can be defined as the smoothed peak-to-peak change in photo current caused by the modulator, divided by the smoothed peak photo current contributed by the signal and background flux. Then the signal current at the anode is

$$i_s = \mu \, WMR, \qquad (4.23-7)$$

where W is the flux in lumens or watts and R is responsivity of the photocathode in amperes per lumen or per watt.

If the signal current is small compared to the signal plus the background photo current and the modulation of the signal component is sinusoidal, the signal-to-noise ratio at the anode of an MPT is

$$\frac{S}{N} = \frac{\text{RMS modulated signal current}}{\text{RMS noise current}} = \frac{M}{4} \left(\frac{i_t}{eF \, \Delta f} \right)^{1/2}. \qquad (4.23-8)$$

For most solar applications this method of determining noise is more realistic than calculations involving only dark current. Equivalent noise input (ENI) or noise equivalent power (NEP) are measurements of the noise caused by dark current. ENI is the radiant flux which produces an RMS signal current equal to the noise current in a bandwidth of 1 Hz. NEP is a measure of the radiant flux required to generate a signal-to-noise ratio of unity when detector noise is referred to a 1-Hz bandwidth. NEP is commonly used as a figure of merit for infrared detectors, but the conditions of measurement must be specified to make comparisons meaningful. ENI and NEP are most useful when referenced to a specific wavelength of radiation, in which case they are similar measurements.

ENI for good MPT's under favorable conditions is on the order of 10^{-16} W. NEP for silicon PIN diodes at their peak response is about 10^{-14}W/Hz$^{1/2}$.

Flicker Noise. This is frequently called "$1/f$ noise" since the power fluctuations vary roughly as the reciprocal of the frequency. It is usually detectable at frequencies below 1000 Hz, and generally becomes the dominant noise below 100 Hz. It can be reduced by device selection and by modulating and amplifying at frequencies of 100 Hz or higher.

Multiplier phototubes do not have this noise but do have small drifts which can cause similar results, especially at frequencies below 1 Hz. Since multiplier gain is a steep function of applied dynode voltage, noise can be introduced in the multiplication process by small power supply fluctuations which can have a $1/f$ form.

4.23.5 Signal Conditioning

When using a multiplier phototube, the most important requirement is a highly stable dynode voltage supply to prevent gain changes. The anode load resistor must be small enough so that capacitance of the anode, signal leads and external circuitry does not extend response time beyond tolerable limits. Conventional MPT's with electrostatic focusing have bandwidths up to 100 MHz, far in excess of the requirements for solar sensors.

A problem is caused by slow drifts in output level of an MPT brought about by insulator charging, fluorescence, and other effects. The effect of these drifts can be reduced by modulating or chopping the signal and using ac amplification and synchronous (phase-sensitive) detection. If the modulator alternates between the signal flux and a stable comparison source of comparable flux level, the effects of gain variations due to multiplier fatigue or dynode supply variations are reduced.

Low-noise amplifiers are not necessary with MPT's because of the gain in the multiplier. This is not true with photovoltaic and photoconductive cells. If the low light level measurement capability of these devices is to be realized, extreme care must be taken to keep amplifier noise at a minimum.

Chopping or modulating frequencies should be at least 100 Hz if practicable. For photodiodes, choice of the modulating frequency is a compromise between reducing low-frequency ($1/f$) noise, and keeping the response time short enough to avoid attenuating the modulation frequency in the photodiode output.

References

1. Bell, E., L. Eisner, J. Young, and R. Oetjen, "Spectral Radiance of Sky and Terrain at Wavelengths Between 1 and 20 Microns. II. Sky Measurements," *J. Opt. Soc. Am.* **50**, 1313–1320, Dec. 1960.

2. Elterman, L., "UV, Visible, and IR Attenuation for Altitudes up to 50 km," AFCRL Environmental Research Paper No. 285, April 1968.
3. Gates, D. M., "Near Infrared Atmospheric Transmission to Solar Radiation," *J. Opt. Sos. Am.* **50**, 1299–1304, Dec. 1960.
4. Melchior, H., M. Fisher, and F. Arams, "Photodetectors for Optical Communications Systems," *Proc. IEEE* **58**, 1466–1486, October 1970.
5. Murcray, D., J. Brooks, F. Murcray, and W. Williams, "Atmospheric Absorptions in the Near Infrared at High Altitudes," *J. Opt. Soc. Am.* **50**, 107–112, Feb. 1960.
6. Oliver. B. M., "Thermal and Quantum Noise," *Proc. IEEE* **53**, 436–454, May 1965.
7. Thekaekara, M. P., Ed., "The Solar Constant and the Solar Spectrum Measured From a Research Aircraft," NASA TR-R-351, Oct. 1970.

Harold W. Smith

4.24 TELLURIC CURRENT SENSORS

It has been known for more than a century that time varying electric currents resulting from natural sources are continuously flowing in the earth and that these "telluric currents" are correlated with the time-varying magnetic field at the surface of the earth. The various theories on the relationship between these electric and magnetic fields were reviewed in 1940 [4]. Later several investigators began to recognize the electromagnetic nature of these fields and their relation to the electrical resistivity structure of the earth [2, 11, 14]. Although this section is primarily devoted to the measurement of telluric currents, it is the basic electromagnetic nature of the process which requires that magnetic variations be discussed as well if an adequate background is to be developed.

Time-varying magnetic disturbances, often called geomagnetic micropulsations, and associated telluric currents are now known to be due almost entirely to sources external to the solid earth if those components due to man-made excitations are ruled out. Comprising the near dc to several kilohertz frequency range of the electromagnetic spectrum, these natural micropulsations (the term micropulsation will be used when both geomagnetic and telluric current signals are being considered) have sources of excitation in the atmosphere, ionosphere, and the earth's magnetosphere.

The early and continued success of satellites which resulted in the discovery of the solar wind and the magnetosphere renewed interest in micropulsation phenomena as it became increasingly evident that they were related to disturbances on the boundary of the magnetosphere. It had long been observed that micropulsations were associated with solar activity, including sun spots and flares, but until this increased knowledge was gained, the generating mechanisms and the modes of propagation were not well understood.

Interest in the measurement of telluric currents and/or the related geomagnetic micropulsations stems from two main areas of research. The first is the study of their relation to the solar wind and other phenomena within the boundary of the magnetosphere. Characteristics of certain micropulsation signals are believed to be an important source of information relative to plasma instabilities and other conditions indicative of the state of the magnetosphere. It is well established, for example, that high-altitude nuclear detonations within the boundary of the magnetosphere produce anomalous micropulsation signals over an extensive area at the surface of the earth.

The second area of interest is based on the information which can be gained relative to the electrical resistivity structure of the earth from the surface to depths in the order of several hundred kilometers for the extremely low-frequency variations. This resistivity information is useful in determining the structure and composition of the earth's crust and upper mantle.

It is feasible to induce telluric currents in the earth using man-made sources for determining the electrical resistivity structure and for underground signaling and communication.

4.24.1 Sources

Micropulsation sources can first be separated between those of natural and man-made origin. There are two distinct natural sources which are effective in different portions of the frequency spectrum, although there is some overlap. The primary source of those currents below about 5 Hz has been mentioned briefly. It involves disturbances on the boundary of magnetosphere created by the solar wind. These disturbances are believed to propagate as hydromagnetic waves in the plasma region extending from the magnetospheric boundary to the ionosphere where they manifest themselves as current systems within the ionosphere. These current systems may then be viewed as antenna systems exciting electromagnetic waves which propagate through the lower atmosphere and into the earth itself. The frequency of oscillations arising from hydromagnetic wave excitation is limited to 5 Hz or less, which is an effective cutoff imposed by absorption of electromagnetic energy in the ionosphere. In effect, the ionosphere is transparent to propagation frequencies below this cutoff and is opaque until very much higher frequencies are reached.

The source of micropulsations with frequencies above this value is due almost entirely to lightning discharges from thunderstorm activity or "spherics," as they have been called. Activity is widespread at any given time over many parts of the earth with definite centers of this activity in certain locations depending upon the seasons. Such activity can produce micropulsations at frequencies below 5 Hz if the distance is not great, but most of the measurable excitation extends upward in frequency into the kilohertz range. One very interesting source mechanism is the excitation of the cavity between the earth and ionosphere, both of which are good electrical conductors, by these electrical discharges creating the so-called "Schumann resonances" of this cavity. The fundamental resonance mode is centered near 7 Hz with higher-order modes near 15, 22, and 28 Hz. These resonances are often relatively "high-Q" or narrow band, and exhibit shifts in the center frequency as the cavity conditions change with ionospheric densities. These resonances are almost always in evidence but vary widely in amplitude as conditions change.

Man-made sources have a variety of forms depending on the particular application, but the objective in each case is to induce telluric currents in the earth in a controlled fashion. Often these systems take the form of either an electric or magnetic dipole. The applied excitation can be single or repetitive impulses, a continuous signal at a number of discrete frequencies, or other known forms.

Other man-made excitations from industrial sources such as electric railways and power systems act as undesirable noise for natural or controlled experiments. These can be locally severe to the extent that useful measurements are difficult or impossible.

4.24.2 Classification and Characteristics of Natural Signals

Through the years a classification system has evolved for natural micropulsation signals of hydromagnetic wave origin. The system is based on morphological properties (frequency, amplitude, duration), rather than a more general classification based on the mechanism of generation or connection with other types of phenomena. A broad distinction is made between continuous and irregular pulsation trains. Continuous types are designated by the symbol Pc and consist of a series of quasisinusoidal oscillations lasting up to several hours. Irregular types are designated Pi and frequently consist of well damped series of oscillations usually lasting less than one hour.

Continuous or Pc micropulsations are divided into subgroups according to their periods as shown in Table 4.24-1.

Table 4.24-1. Classification of Pc Micropulsations

Designation	Period Range (s)
Pc 1	0.2–5
Pc 2	5–10
Pc 3	10–45
Pc 4	45–150
Pc 5	150–600

Irregular or Pi micropulsations are divided into two subgroups, again according to period range as shown in Table 4.24-2.

Table 4.24-2. Classification of Pi Micropulsations

Designation	Period Range (s)
Pi 1	1–40
Pi 2	40–150

Both Pc and Pi micropulsations have been studied extensively by many investigators over the past several years leading to a vast accumulation of detailed characteristics relative to each of the subgroups [5, 6].

Pc 1 micropulsations or "pearls" or hydromagnetic emissions as they have variously been called have probably been studied more extensively than any other type. They are sinusoidal oscillations with periods mainly between 0.2 to 5 s and often occur in the form of separate bursts or "pearls" lasting from tens of minutes to hours. In mid latitudes they may occur a few times a month mainly after midnight and on into the early morning hours. However, they are primarily a daytime phenomenon in the auroral zone. They exhibit a fine structure including a rising-tone of frequency dispersion and hold the most promise as an indicator of the state of the magnetosphere. They are frequently observed simultaneously at conjugate point stations and at times over one hemisphere.

Pc 2, 3, 4 micropulsations are primarily a daytime phenomenon reaching their maximum frequency of occurrence before or near local noon. Most common are a series of pulsations lasting several hours with periods usually in the range from 10 to 60 s. There is a dependence of oscillation period of this group on planetary magnetic activity. Amplitudes increase sharply with latitude, often up to an order of magnitude in the auroral zone. They are frequently global in distribution on the dayside and can be global at times for the whole earth.

Pc 5 micropulsations with periods ranging from about 150 to 600 s can have extremely large amplitudes and last from a few minutes to several hours. There is a broad latitude distribution of Pc 5's decreasing rapidly in amplitude with distance from the auroral zone and showing some evidence of equatorial enhancement. Their distribution is not greatly extended in longitude.

Irregular micropulsations of the Pi 1 and Pi 2 types show a close relationship with magnetic storms and auroral zone phenomena, although they can occur during periods of very quiet magnetic activity. Their intensity is strongly dependent on latitude, reaching amplitudes an order of magnitude greater in the auroral zone. Their maximum frequency of occurrence is near local midnight. Pi 2 pulsation trains are generally distributed over the night hemisphere and often over the entire globe.

Natural signals above the hydromagnetic wave cutoff of approximately 5 Hz have received far less attention and except for the earth-ionospheric cavity resonances have not been subjected to any detailed classification. This is a natural consequence of the fact that these signals are highly irregular in both amplitude and frequency distribution. Characteristics of telluric currents in the frequency range extending into the kilohertz region are more uncertain because of the wide variation in signal amplitude and the greater exposure to industrial noise. Also lacking has been the stimulation which is generated by the presence of specific applications of measurements in this portion of the spectrum. Nevertheless, interest is on the increase, primarily in the area of determining the electrical resistivity of earthen structures or ore bodies at the shallower depths of penetration of these signals. There is also an interest in the possibility of communication systems operating in this frequency range.

Micropulsation Amplitudes. In a consideration of the expected range of amplitudes of natural micropulsation signals a careful distinction must be made between the nature of geomagnetic micropulsations and telluric currents, aside from the fact that different sensors are required. This is because the amplitudes of the geomagnetic micropulsations are little affected by the local resistivity of the earth whereas telluric current amplitudes can vary more than an order of magnitude due to changes in the local geological structure alone.

Geomagnetic micropulsation amplitudes vary widely with the class of event, with geomagnetic activity and with latitude. On very quiet days (magnetically) narrow bandwidth recordings of Pc 1 micropulsations have amplitudes as small as a few milligammas. Pc, 2, 3, 4 continuous pulsations are characteristically in the 0.1 to 1 γ range and the Pc 5 oscillations and some of the large Pi 2 events associated with magnetic storms can reach amplitudes as high as several hundred γ in the auroral zones.

Telluric current amplitudes are by custom expressed in terms of the potential difference between electrode pairs per unit separation distance, most frequently in millivolts per kilometer. Except as modified by the different frequency responses of sensors and the local resistivity structure, the range of telluric current amplitudes behaves in much the same way as the geomagnetic micropulsations for the respective micropulsation types. The total range of telluric current amplitude variations can be quite large; from a few microvolts/km to several hundred millivolts/km under the most extreme conditions of geomagnetic activity and local geology. Obviously, such a large dynamic amplitude range can create problems in measuring and recording these signals.

Frequency Spectrum. Figure 4.24-1 shows the bounds of power density spectra for micropulsation signals in the frequency range from 10^{-4} to 10 Hz which were encountered in a series of seven locations in sedimentary deposits in central Texas. The telluric currents (E spectrum) would be expected to be larger in highly resistive granite or shield areas and, along with the geomagnetic micropulsations (H spectrum), would be expected to increase with latitude. Above this frequency range the spectra normally rise slowly on the average, but are highly variable. However, the most striking feature in Figure 4.24-1 is the steep negative slope of these spectra with increasing frequency. The range of amplitudes virtually requires that this frequency spectrum be covered with measurements in two or more bands. It also makes it highly desirable to use a considerable degree of "pre-whitening" within each band by shaping the sensor response with filters to flatten or at least compress the wide dynamic range of the recorded variations. If the amplitude and phase response of the pre-whitening filters are accurately known, they can be removed from the recorded signals. This process was used to obtain the spectra shown in Figure 4.24-1.

4.24.3 Telluric Current Measuring Systems

A system for measuring telluric currents can have a variety of forms and and can utilize a variety of individual components depending upon the application. For example, the frequency range from 10^{-4} to 10^4 Hz extends over eight full decades and requires a fairly elaborate system to measure and record signals in several frequency bands. On the other hand, a rather simple system can measure the large amplitude telluric currents in the lowest two or three decades with accuracy and ease.

A typical telluric current measuring system for the frequency range from

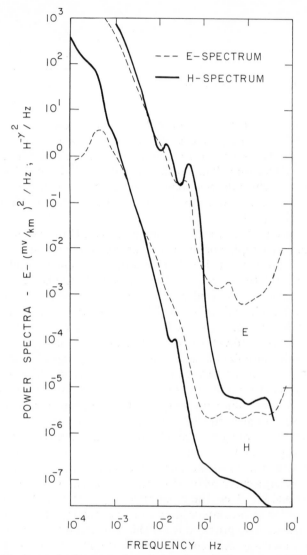

Figure 4.24-1 Power density spectra bounds for micropulsations at seven stations.

about 10^{-4} to 20 Hz which has been used by a number of investigators is described in some detail in the following sections. It is shown in block diagram form in Figure 4.24-2.

Electrodes. Electrodes commonly take the form of buried sheets of lead or one of the nonpolarizing types. An example of the latter is shown in Figure 4.24-3 and consists of a Pyrex filter crucible filled with a saturated

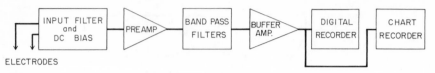

ELECTRODES

Figure 4.24-2 Block diagram of a telluric current measuring system.

cadmium chloride buffer electrolite in which a pure cadmium rod is suspended. Contact with the ground is made through the bottom of the crucible, a fine ceramic filter. When possible, electrodes are installed deep enough to reach ground moisture. Water is added to some of the local soil to produce a thin slurry into which the electrode is inserted. A plastic cover is provided for protection against excessive drying, sudden inrush of water during rain, and small animals.

Electrodes are placed in a cross configuration 200 or 300 m apart and oriented along geographic or magnetic north-south and east-west lines so that the electrodes are located on the four corners of a square. Each electrode potential is cabled to the center of the square via the center conductor of a coaxial cable. At the center each electrode cable is connected to a multiconductor cable with individually shielded conductors to the instrument van some 150 to 200 m away. The shields are left open at the electrode ends and are carried individually to the instrument panel.

This electrode installation has a number of operational advantages: (1) a measurement of the leakage resistance from cable shield to ground provides a fast check for cable damage without the necessity of disconnecting the electrode; (2) the cables are heavy and flexible enough to lie flat on the ground making burial normally unnecessary; (3) the cross configuration also provides two orthogonal sets of parallel lines spaced some distance apart. Comparison of these telluric signals lends some insight into the degree of inhomogeneity in the immediate vicinity of the measuring site as well as providing a check on the electrodes in operation.

Input Filter and DC Bias. The input filter is a notch filter for 60 and 120 Hz rejection prior to preamplification to avoid preamplifier saturation by power frequencies. Also provided is a dc biasing circuit to counter the electrode contact potential difference, typically 10 millivolts or more. The dc

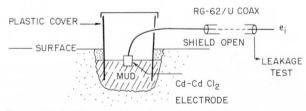

Figure 4.24-3 Nonpolarizing electrode installation.

offset voltage, read on a meter at the preamplifier output, is minimized to avoid preamplifier saturation. A resistance pad is provided in the input circuit to match the electrode resistance to the input resistance of the notch filter. A standard calibration signal can be connected to the system input in place of the electrode line.

Preamplifier. A low-noise chopper carrier preamplifier with differential input has been found to be the most satisfactory arrangement. Voltage gains of 40 to 60 dB are adequate and good input common mode rejection is essential.

Bandpass Filter and Buffer Amplifier. Adjustable bandpass filters utilizing operational amplifiers and RC elements have proven to be successful in covering the wide frequency range of interest. Typically, these filters have adjustable three pole transfer functions with mid-band gains of unity. After each filter a variable gain buffer amplifier with five gain steps allows for gain change without perturbing the system filters which can take the better part of an hour to recover from a switching transient. The buffer amplifiers also provide a low impedance isolated output for driving tape recorders or other recording media.

Recording System. The recording system can be simple or elaborate depending on the particular application. Included in Figure 4.24-2 are both an analog and a digital recorder. For some applications a simple strip chart recorder is sufficient, but if quantitative results are required, as for magnetotelluric analysis of subsurface resistivity profiles, both are almost essential.

4.24.4 Field Measurements

Certain precautions in installation and measuring techniques are necessary for the acquisition of high-quality telluric current data. The nonpolarizing type of electrode needs to be installed at least several hours prior to recording to allow the stabilization of the electrolyte with the earth fluids. The principal disadvantage of lead electrodes is that an even longer time is usually required, in some instances several days or more.

The first test after the system is put into operation is to connect the electrodes so that parallel telluric currents can be examined. If these signals are not virtually identical, there is either an abnormality somewhere in the system or there is a locally severe inhomogeneous condition in the earth. Upon successful completion of this test the electrodes are then connected to measure orthogonal telluric current components. Electrode condition should be checked at frequent intervals by measuring electrode resistance and bias potential.

Telluric current signals should be monitored for a time to estimate the proper gain settings and to adjust the filters to pre-whiten the signal spectra in the frequency band to be measured. Even so, sudden commencements of magnetic storms require system gain reductions.

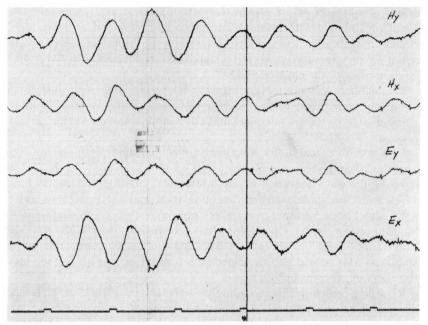

Figure 4.24-4   Daytime Pc 3 oscillations.

Careful system calibration by insertion of test signals and accurate knowledge of the total amplitude and phase characteristics of the whole system are required for quantitative analysis of the recorded signals.

Sample micropulsation recordings are shown in Figures 4.24-4 through 4.24-6. In each figure E_x and E_y are, respectively, the north-south and east-west telluric current components; H_x, H_y, and H_z are, respectively, the north-south, east-west, and vertical geomagnetic micropulsation components. Figure 4.24-4 shows an example of daytime continuous oscillations of the Pc 3 type. The rectangular pulses on the bottom trace are spaced at 1-min intervals. It should be noted that the orthogonal E and H pairs, E_x and H_y, and E_y and H_x are strikingly similar (except that the polarity convention used causes the E_x and H_y components to be out of phase). This is a typical situation for a location where the geology is reasonably uniform laterally.

Figure 4.24-5 shows an example of Pc 1 oscillations. It is an unusual case where simultaneous recordings in Austin, Texas and La Grande, Oregon have pulsation envelopes which are highly correlated. Furthermore, there is a consistent time shift between envelope peaks of about 2 s with the Oregon station leading in time.

Figure 4.24-6 shows a typical case of nighttime Pi 2 activity without the E_y trace.

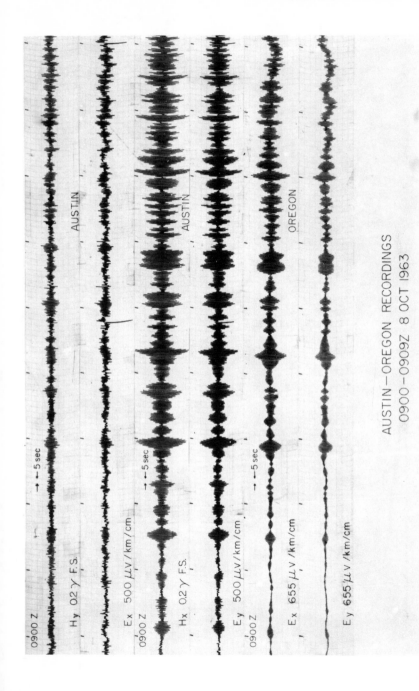

Figure 4.24-5 Pc 1 oscillations.

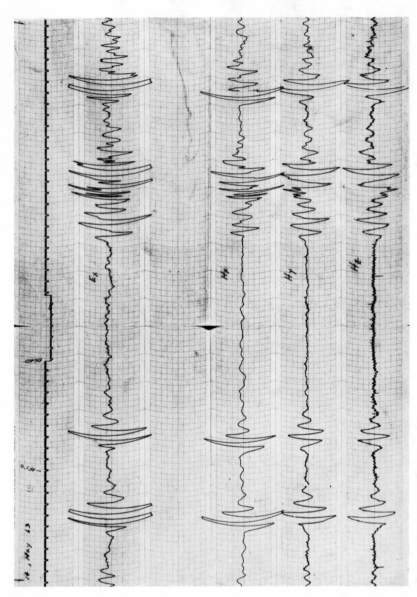

Figure 4.24-6 Nighttime Pi 2 activity.

References

1. Bostick, F. X., Jr., and H. W. Smith, "Investigation of Large-Scale Inhomogeneities in the Earth by the Magnetotelluric Method," *Proc. IRE* **50**, No. 11, 2339–2346, 1962.
2. Cagniard, L., "Basic Theory of the Magneto-Telluric Method of Geophysical Prospecting," *Geophys.* **18**, 605–635, 1953.
3. Cantwell, T., and T. R. Madden, "Preliminary Report on Crustal Magnetotelluric Measurements," *J. Geophys. Res.* **65**, 4202–4205, 1960.
4. Chapman, S., and J. Bartels, *Geomagnetism,* Oxford University Press, London, 1940.
5. Jacobs, J. A., *Geomagnetic Micropulsations,* Springer-Verlag, Berlin, 1970.
6. Jacobs, J. A., and K. Sinno, "World-Wide Characteristics of Geomagnetic Micropulsations," *Geophys. J.* **3**, 333, 1960.
7. Keller, G. V., L. A. Anderson, and J. I. Pritchard, "Geological Survey Investigations of the Electrical Properties of the Crust and Upper Mantle," *Geophys.* **31**, No. 6, 1078–1087, December 1966.
8. Orange, A. S., and F. X. Bostick, "Magnetotelluric Micropulsations at Widely Separated Stations," *J. Geophys. Res.* **70**, 1407, 1965.
9. Srivastava, S. P., J. L. Douglass, and S. H. Ward, "The Application of the Magnetotelluric and Telluric Methods in Central Alberta," *Geophys,* **28**, No. 3, 426–446, June 1963.
10. Swift, C. M., Jr., "A Magnetotelluric Investigation of an Electrical Conductivity Anomaly in the Southwestern United States," Ph.D. thesis, Massachusetts Institute of Technology, 1967.
11. Tikhonov, K., N. V. Lipskaya, and B. M. Yanovsky, "Some Results of the Deep Magneto-Telluric Investigations in the USSR," *J. Geomagn. and Geoelec.* **15**, No. 4, 275–279, 1964.
12. Vozoff, K., R. M. Ellis, and M. D. Burke, "Telluric Currents and Their Use in Petroleum Exploration," *Bull. Am. Assoc. Petr. Geol.* **48**, No. 12, 1890–1901, December 1964.
13. Wait, James R., "Theory of Magneto-Telluric Fields," *J. Res. NBS* **66D**, No. 5, 509–541, September–October 1962.
14. Wait, J. R., "On the Relation Between Telluric Currents and the Earth's Magnetic Field," *Geophys.* **19**, 281–289, 1954.

Frank S. Castellana

4.25 TEMPERATURE AND HEAT FLOW SENSORS

4.25.1 Thermodynamic and Practical Temperature Scales

The modern definition of temperature is based on a detailed analysis of reversible cycles and equilibrium. The result of this analysis is the Kelvin thermodynamic temperature scale defined by the relation

$$\frac{T_1}{T_2} = \frac{Q_1}{Q_2}, \qquad (4.25-1)$$

where Q_1 and Q_2 are the heats rejected and absorbed by a body carrying out a Carnot cycle between the temperatures T_1 and T_2 [12]. This is a relative scale that can be fixed either by assigning a definite temperature to a particular physical state, or by assigning a definite numerical value to the temperature difference between two physical states. Once this is done, all other temperatures on the scale are established. In October 1960, the International Committee on Weights and Measures fixed the Thermo-dynamic Temperature Scale by assigning the triple point of water a temperature of exactly $273.16°K$ [2].

Practical difficulties inherent in direct application of Eq. 4.25–1 have led to adoption of an International Practical Temperature Scale (IPTS) based on the numerical definition of reproducible fixed points, and the assignment of standard instruments calibrated at these points to interpolate between regions of the scale. The defining fixed points and interpolating instruments of the 1968 IPTS are shown in Table 4.25-1 [2]. For temperatures below the triple point of hydrogen, various vapor pressure and magnetic scales are in use [6, 11].

4.25.2 Principles of Temperature Measurement

Any property of a substance that changes with temperature in a well-defined monotonic fashion can serve as a thermometric indicator and be the basis of a temperature sensor. The sensor can be the body whose temperature is to be measured, a separate element inserted into the body, or a device located at some distance from the body. In the first case, the indicator is a temperature-sensitive characteristic of the body; in the second situation, the sensor and the body must achieve thermal equilibrium; and in the last case, the sensor detects emitted thermal radiation.

Table 4.25-1.   The 1968 International Practical Temperature Scale

Defining Fixed Points	Temperature (°K)	Interpolating Instrument (Interpolating Relation)
Triple point of hydrogen	13.81	
Boiling point of hydrogen (25/76 atmospheres)	17.042	
Boiling point of hydrogen	20.28	Platinum resistance thermometer (defined reference function [2])
Boiling point of neon	27.102	
Triple point of oxygen	54.361	
Boiling point of oxygen	90.188	
Freezing point of water[a] ————	273.15 ————	
Triple point of water	273.16	
Boiling point of water	373.15	Platinum resistance thermometer (modified Callendar Equation)
Freezing point of zinc	692.73	
Freezing point of antimony[a] ————	903.89 ————	
Freezing point of silver	1235.08	Platinum, 10% rhodium to platinum thermocouple (parabolic equation)
Freezing point of gold ————————	1337.58 ————	
		Optical pyrometer (Planck's law)

[a] Secondary fixed point.

Optimal selection of a sensor requires careful evaluation of all factors related to the measurement problem. The more important of these are absolute temperature level, required measurement accuracy, required time response, physical state of the body, and chemical nature of the body. The first three factors have a strong influence on choice of the sensor, while the last two relate more closely to the method of thermal connection and the necessary environmental protection.

Complete evaluation of any sensor requires a thorough knowledge of its range, sensitivity, accuracy, and time response. Range refers to the maximum and minimum temperatures, sensitivity to the minimum detectable change in temperature, accuracy to the difference between the indicated temperature and the true temperature, and time response to the rate at which a change in temperature is followed. The last three parameters must be considered in the light of the overall measuring system, since even a sensor capable of the greatest precision and the fastest time response is subject to limits of the readout instrumentation. In almost all cases, however, a careful selection of this instrumentation usually results in sensor constraints being limiting.

The response time τ is precisely defined as the time required for the sensor to come to within $1/e$ of equilibrium after application of a step change in temperature as described in Section 4.1. The response time is determined by design of the sensor and by the manner in which it is applied to the system.

The flow of heat q into the sensing element is related to its mean temperature rise by the relation

$$q = \rho V C_p \frac{dT_e}{dt}, \tag{4.25-2}$$

where ρ, V, T_e, and C_p are, respectively, the element's density, volume, temperature, and heat capacity. With the resistance to heat flow from the body at temperature T_b to the element defined as

$$R = \frac{(T_b - T_e)}{q}, \tag{4.25-3}$$

it can be shown [1] that the response time is given by

$$\tau = \rho V C_p R. \tag{4.25-4}$$

The mass ρV, and the heat capacity C_p are intrinsic characteristics of the sensor, whereas the resistance R is determined by the degree of thermal connection.

Based on these requirements a number of practical devices have been developed, each achieving optimal utilization under different conditions or constraints. They are divided into two classes, those that depend on electrical phenomena and those that rely on the physical change of a solid, liquid, or gas. In the former group are thermocouples, resistance thermometers, and pyrometers. In the latter group are liquid and gas-filled systems, bimetallic elements and surface indicators [1, 5, 12].

4.25.3 Thermocouples

The thermocouple is perhaps the most versatile and widely used temperature measuring instrument. Its operation is based on the fact that an electric current flows continuously in a closed circuit formed from two dissimilar conductors when the junctions of the conductors are maintained at different temperatures. Usually the conductors are metal wires that make contact at two points. One of the contact points is a reference junction maintained at a reference temperature. The emf developed by the thermocouple is then a function of the temperature of the other junction. Since this measuring junction can be made arbitrarily small, the thermocouple is ideally suited to rapid response point measurement, or to situations where introduction of a larger element would alter the body's temperature field.

The laws that govern thermocouple design have been stated as follows [12]:

1. An electric current cannot be sustained in a circuit of a single homogeneous metal, however varying in section, by the application of heat alone.

2. The algebraic sum of the thermoelectromotive forces in a circuit composed of any number of dissimilar metals is zero if all of the circuit is at a uniform temperature.

3. The thermoelectromotive force developed by any thermocouple of homogeneous metals with its junctions at any two temperatures T_1 and T_3, is the algebraic sum of the emf of the thermocouples with one junction at T_1 and the other at any other temperature T_2, and the emf of the same thermocouple with its junctions at T_2 and T_3.

A device for measuring the thermally induced emf can be introduced into the circuit at any point without effectively altering the circuit if the added junctions are all at the same temperature. If a temperature-emf calibration with a standard reference is available, the calibration with any other reference at a different temperature can be calculated.

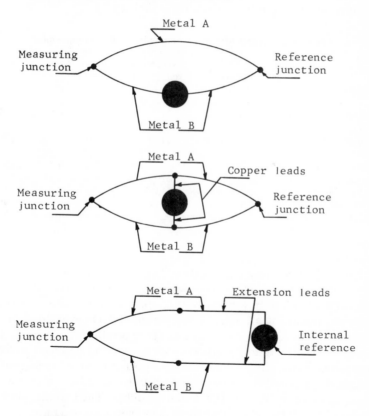

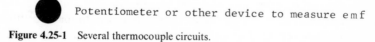

Potentiometer or other device to measure emf

Figure 4.25-1   Several thermocouple circuits.

Typical thermocouple circuits are shown in Figure 4.25-1. The first two circuits are commonly employed in laboratory work where a reference bath is available, while the third is used with instrumentation having a reference junction compensator to account for changes in instrument reference temperature.

Metal Combinations in General Use. A thermocouple can theoretically be formed from any two dissimilar metals. Because of sensitivity limitations, corrosion resistance, and reproducibility, only a few combinations have found wide application. These are summarized in Table 4.25-2, together with the temperature ranges for general use, the maximum allowable operating temperatures, and the emf's generated at $0°$, $100°$ and $400°C$ [4, 13]. The high-temperature limit is a function of alloy melting point, resistance to oxidation, and thermoelectric stability, and the low-temperature limit is usually determined by the thermoelectric sensitivity.

Table 4.25-2. The Characteristics of Some Commonly Used Thermocouples

Thermocouple Type	Usual Temperature Range (°C)	Maximum Temperature (°C)	Generated emf (mV) at		
			0°C	100°C	400°C
90% platinum—10% rhodium to platinum	0 to 1450	1700	0.000	0.643	3.250
87% platinum—13% rhodium to platinum	0 to 1450	1700	0.000	0.646	3.398
Copper to constantan	− 200 to 300	600	0.000	4.276	20.865
Iron to constantan	− 200 to 760	1000	0.000	5.40	22.07
Chromal-P to alumel	− 200 to 1100	1350	0.000	4.10	16.39
Chromal-P to constantan	− 100 to 1000	1000	0.000	6.32	28.95
Tungsten—5% rhenium to tungsten—26% rhenium	to 2500	2500	0.000	1.41	6.71

Selection of the best metal combination depends on the operating range, accuracy, cost, and environment. The platinum to platinum-rhodium combination is the most stable, especially at temperatures in excess of 400°C. Its main disadvantage is a relatively low sensitivity. The iron-constantan thermocouple can be used over a wide range of temperatures, is mechanically strong, has a moderately low thermal conductivity, and performs well in both oxidizing and reducing atmospheres. Copper-constantan finds widest application at temperatures below 100°C. At temperatures in excess of 300°C, oxidation of the copper is a serious problem. The chromel-alumel combination has a relatively longer lifetime at elevated temperatures than the iron-constantan, especially in oxidizing atmospheres, but its resistance to reducing atmospheres is poor, and it is more subject to metal brittleness if contamination occurs. For ultra-high-temperature applications, in the region of 2500°C, the tungsten-5% rhenium to tungsten-26% rhenium metal combination has been used successfully [7].

Calibration Requirements and Other Considerations. Since the thermoelectric properties of a metal depend in part on physical parameters such as grain structure, all thermocouples of one type do not necessarily have identical temperature-emf relationships or agree exactly with standard tables. Accuracies of the order of ±0.1°C are usually attainable.

Thermocouples are subject to errors from metal inhomogeneity and the presence of parasitic voltages. Nonuniformity in the thermocouple material can result in the generation of an extraneous emf if the thermocouple is located in the region of a sharp temperature gradient (the Thomson effect). Corrections for the resulting error are extremely difficult, especially if temperature distributions are not known [9]. Parasitic voltage effects can be significant in the presence of moisture or inadequate insulation. Precautions must be taken to eliminate conduction errors which can be minimized by using small-diameter wire, adequate insulation, low-conductivity metals, and an adequate depth of immersion.

The reliability and lifetime of a thermocouple can be extended by protecting the junction and wires with glass or with metal-sheathed ceramic covers. This results, however, in a corresponding reduction of the instrument's time response.

4.25.4   Resistance Thermometers

The resistance offered by a conducting material to the passage of an electric current is a function of temperature and forms the basis of precision resistance thermometry. The resistance thermometer consists of a suitable metallic or nonmetallic sensing element, a precision circuit for measuring electric resistance, and an appropriate temperature-resistance calibration curve.

Numerous circuit arrangements are available for measurement of the electric resistance. The most common is the simple Wheatstone bridge circuit subject to the null-balance relation (Figure 4.25-2)

$$\frac{r_{se}}{r_b} = \frac{r_1}{r_2}. \tag{4.25-5}$$

For $r_1 = r_2$, the resistance of the sensing element is identical to the resistance of the variable balancing resistor r_b. For work involving high precision, modifications of this basic circuit have been developed to account for factors such as lead resistance and contact resistance [1].

All resistance thermometers require use of a finite electric current to effect temperature measurement. The resulting heating can cause errors in high-accuracy situations. This limitation can be overcome by extrapolating the resistance at several values of the current to zero current.

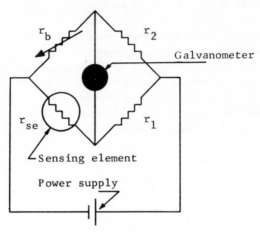

Figure 4.25-2 A simple Wheatstone bridge circuit.

Metallic Elements. While in theory almost any metal is acceptable as a sensing element, high-purity platinum is used almost exclusively in precision work. In a fully annealed and strain-free state, platinum has an especially stable and reproducible temperature-resistance relationship. This relationship is usually given by the Callendar-Van Dusen interpolation formula [12]

$$T = \frac{R_T - R_0}{\alpha R_0} + \delta\left(\frac{T}{100} - 1\right)\frac{T}{100} + \beta\left(\frac{T}{100} - 1\right)\left(\frac{T}{100}\right)^3. \quad (4.25\text{-}6)$$

where T is the temperature in $^\circ$C, R_T the element resistance at T°C, R_0 the element resistance at 0°C, and α, δ, and β characteristic constants of the sensor. For $T > 0^\circ$C, $\beta = 0$. The value of R_0 is established by direct observation of R_T at 0°C, and α is found by measuring R_T at 100°C and by using the relation

$$\alpha = \frac{R_T - R_0}{100 R_0}. \quad (4.25\text{-}7)$$

Two additional fixed-point measurements provide values of δ and β. Calibrated platinum sensors can be as accurate as $\pm 0.001^\circ$C, and can operate successfully within the temperature range of approximately -260°C to 1100°C. At higher temperatures sensor stability is affected. For applications where cost is a factor, copper, nickel, tungsten, and other less expensive metals are used.

In general, these elements are made from wire mounted as a coil on a suitable ceramic base. They can also be formed as a vapor deposit of the metal on a ceramic substrate if increased strength and high resistance are important.

Thermistors and Other Nonmetallic Elements. Thermistors are semiconductors with an extremely high, but nonlinear, negative temperature coefficient of resistance. Calibration stability is usually poorer than most metal resistance elements, but the sensitivity can be significantly greater over a small temperature range. To assure maximum reproducibility, thermistors are usually formed from substances whose electrical properties are the least sensitive to impurity content. Common examples are solid solutions of Fe_3O_4, $MgCr_2O_4$, and $MgAl_2O_4$, and sintered mixtures of NiO, Mn_2O_3, and Co_2O_3. In manufacture, the mixture of compressed oxides is formed into the desired shape and then fired in a controlled atmosphere. Usual sensor configurations are beads, disks, washers, and rods. Calibration stability can usually be improved by sealing the thermistor in a constant atmosphere glass envelope and aging for a period of several weeks at an elevated temperature.

Ordinary thermistors usually conform to standard calibration curves to within $\pm$ 1°C, while preaged and sealed thermistors used at low temperatures can have an accuracy an order of magnitude greater. The useful temperature range is from about -100°C to 300°C. For very-low-temperature applications (less than 15°C) doped germanium and carbon sensors have been successfully employed [12].

4.25.5 Radiation Pyrometry

Radiation pyrometry as a temperature measuring technique is based on the functional dependency of a body's spectral radiance on its temperature. The method is especially attractive for elevated temperature measurement, since no part of the sensing instrumentation must necessarily be in close proximity to the subject body.

The spectral radiance of a blackbody—defined as the energy radiated per unit time, per unit wavelength, per unit projected area, per unit solid angle—is related to its temperature by the Planck radiation equation [12]

$$N_{b\lambda} = \frac{C_1}{\lambda^5 (e^{C_2/\lambda T} - 1)\,\pi}. \qquad (4.25\text{–}8)$$

$N_{b\lambda}$ is the blackbody spectral radiance at the wavelength λ, T is the thermodynamic temperature, and C_1 and C_2 are constants.

The direct use of this equation for temperature determination demands absolute radiation measurements which are difficult to make. For this reason, the spectral radiance is usually measured relative to a standard spectral radiance, and the ratio of the unknown to the standard used as the measure of temperature. The standard spectral radiance selected is that at the equilibrium of solid and liquid gold (1064.43°C). Since any observable radiation consists of a finite spectral band, the comparison is made in terms of a photometric brightness given by the integral of the spectral radiance over all wavelengths within a particular band.

Use as a temperature sensor requires first a relative brightness measurement, and then relation of this relative brightness to temperature through appropriate calibration. The temperature obtained in this manner is called an apparent temperature, and is exactly equal to the actual temperature only when the body's emissivity is unity.

The emissivity within a given wavelength range is a function of the source surface condition, geometry, and temperature, and must be accounted for in precise measurement. Care must also be exercised to eliminate all sources of extraneous radiation, and any intervening matter such as dust or smoke that might absorb radiation.

Two classes of sensing instruments exist: total-radiation pyrometers and spectrally selective pyrometers. Total radiation pyrometers detect incident radiation over a wide bandwidth, while the generally more accurate spectrally selective pyrometers restrict the radiation to a relatively narrow bandwidth [1]. In the latter group there are three main types: optical, photoelectric, and two-color.

The total radiation pyrometer is comprised of an optical system for focusing the radiated energy and a temperature-sensitive detecting element such as photocell, bolometer, or thermopile (Figure 4.25-3). The electrical output of the sensing element is related to the source temperature through calibration factors. These devices can measure temperatures from 100°C to several thousand degrees, but with relatively limited accuracy.

The optical pyrometer uses the human eye as its radiation sensor. The source of radiation is projected as an image onto the heated tungsten filament of a pyrometer lamp. The lamp current is adjusted to match the brightness of the filament to the brightness of the source, and the temperature is determined from an appropriate lamp current versus temperature calibration curve. If the radiation source is at a temperature higher than can be feasibly accommodated by filament current adjustment, a suitable

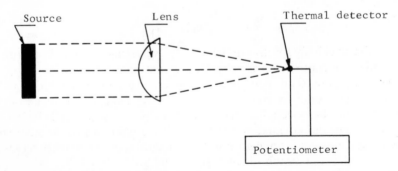

Source   Lens   Thermal detector

Potentiometer

Figure 4.25-3 Schematic of a radiation pyrometer.

filter is used to reduce brightness to within the workable temperature range. With a calibrated instrument, accuracies of ± 1 to 2°C, over a range of about 700°C to several thousand degrees can readily be expected.

The optical pyrometer is limited by the sensitivity of the human eye. This limitation is overcome by the photoelectric pyrometer which replaces the human eye with a sensitive phtomultiplier circuit [12]. The resulting increased precision in brightness matching can improve the sensitivity and accuracy by an order of magnitude. The photomultiplier circuit can be replaced by a thermopile or resistance thermometer with some sacrifice in sensitivity and accuracy.

The two-color pyrometer measures the intensity of radiation in two selected wave bands, computes the ratio of the energies, and converts this ratio into a temperature indication. If the subject emissivity is identical at the two wavelengths, the measurement is free from emissivity errors [1, 3].

4.25.6 Liquid-Filled Sensors

Liquid-filled sensors are based on the thermal variation of fluid expansion and are usually divided into two classes: liquid-in-glass thermometers and liquid-pressure thermometers. Instruments in both groups have the advantages of being direct reading, reliable, and compact. With various working fluids, ranging from alcohol and mercurcy to gallium, measurable temperatures span the range of − 200°C to 1000°C.

Liquid-in-Glass Thermometers. The principal features of solid-stem liquid-in-glass thermometers are shown in Figure 4.25-4. The thermometric fluid contained in the bulb of the thermometer moves in the glass capillary tube in response to temperature changes. The expansion chamber prevents buildup of excessive pressures in gas-filled thermometers, and the contraction chamber is provided to prevent contraction of the liquid column into the bulb.

The thermometer can be calibrated as a total-immersion thermometer, a partial-immersion thermometer, or a complete-immersion thermometer. A total-immersion instrument indicates the correct temperature when the bulb and entire liquid column are in direct contact with the fluid; a partial-immersion instrument when the bulb and a specified part of the stem (to the immersion line) are in contact with the fluid; and a complete-immersion instrument when the entire thermometer is in contact with the fluid.

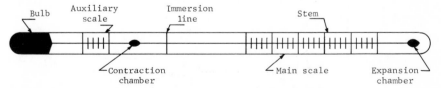

Figure 4.25-4 Principal features of the solid-stem liquid-in-glass thermometer.

The accuracy of this sensor is limited by its construction characteristics, the most important being glass instability, bore nonuniformities, and inaccurate scale graduation. With calibrated instruments, accuracies of $\pm\,0.1°C$ or better can be expected. Since the span of measurement is inversely proportional to sensitivity, any one instrument is usually of a limited range.

Liquid Pressure Sensors. Liquid pressure sensors are usually formed from a bulb connected by capillary tubing to a flexible spiral element (Figure 4.25-5). The liquid in the bulb generates a pressure of sufficient magnitude to partially uncoil the spiral when exposed to a temperature within the instrument's range span. The resulting rotary movement is transmitted through a shaft to a pointer which indicates temperature on a graduated scale. Instruments of this design can transmit signals over a distance, but their accuracy is usually limited to about $\pm\,1°C$.

4.25.7 Other Sensors

Bimetallic elements are formed from two metals having different thermal expansion coefficients. The two metals are bonded together and formed into a spiral shaped sensor. Rotary movement of the sensor induced by temperature changes is transmitted through a shaft to a pointer. For every value of temperature within the range span, there is a corresponding pointer position which is read as temperature on a graduated scale. These instruments are generally used at temperatures below $600°C$, and while their accuracy and sensitivity are limited, they have a wide range and a fairly low cost.

Gas-filled thermometers are based on the known variation of gas pressure with temperature, and in their common form are similar in construction and use to the flexible liquid-filled thermometer. They are generally used over ranges between $-80°C$ and $400°C$.

Vapor-pressure thermometers differ from gas-filled thermometers in that the bulb system is partially filled with a volatile liquid such as methyl chloride, ether, or hexane, and the indicating pressure is determined by

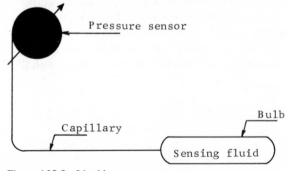

Figure 4.25-5 Liquid pressure sensor.

the liquid's temperature-vapor pressure relationship. Temperatures from about $-40°C$ to $400°C$ can be measured depending on the nature of the liquid.

Pyrometric cones are small pyramids made from a mixture of oxides that rely as indicators on the temperature dependency of solid fusion. They are useful for approximate measurement within the range of 540 to 2000°C.

Surface indicators are temperature-sensitive chemical materials that are applied as a surface film. Available for a temperature range of about 40° to 1500°C, they indicate by a visual change in condition such as a sudden molten appearance.

4.25.8 Comparison of Sensor Characteristics

The most widely used and versatile of the sensors described in the preceding sections are compared in Table 4.25-3. The indicated numerical values are only representative and can vary considerably depending on sensor design and construction. The response time is a function of sensor design and the thermal resistance between the body and the sensor (see Eq. 4.25–4).

Table 4.25-3.   A Comparison of the Characteristics of Several Important Sensors

Characteristic	Thermocouples	Platinum Resistors	Thermistors	Pyrometers	Filled System Thermometers
Active or passive	Active	Passive	Passive	Active	Active
Output form	Electrical	Electrical	Electrical	Visual and electrical	Visual and mechanical
Range (°C)[a]	−200 to 2500	−260 to 1100	−90 to 300	100 to 3000 +	−100 to 1200
Time response[b]	Fair to excellent	Fair to good	Fair to excellent	Usually less than 0.10 s	Poor
Transfer function	Relatively linear	Approximately linear	Nonlinear	Nonlinear	Approximately linear
Accuracy	As good as ±0.10°C	Better than ±0.10°C	±1°C or better	±1°C or better	Less than ±1°C[c]
Sensitivity	Better than 0.10°C	Better than 0.05°C	Better than 0.01°C	Better than 1°C	Better than 1°C[c]
Stability	Good	Excellent	Good	Good	Good
Power consumption	None	Moderate	Moderate	Moderate	None
Relative cost	Low to moderate	Moderate	Low	High	Low to moderate

[a] Any single sensor of a particular class will not necessarily cover the entire range.
[b] Can vary considerably depending on sensor construction (see Table 4.25-4).
[c] Except for calibrated liquid in glass thermometers which can be better than ±0.10°C.

The response of a thermocouple improves as the wire diameter decreases and is better in a liquid medium than in a gas. Some representative time constants are indicated in Table 4.25-4.

The reliability and useful life of a sensor depends on its construction and on the conditions under which it is used. Most sensors serve reliably for

Table 4.25-4.   A Comparison of Time Constants for Several Sensors

Sensor Description	Sensor Environment	Response Time (s)
Bare iron-constantan thermocouple	Still air	240
(1.6 mm wire diameter) [1]	Air, 350 m/min	8
	Water, 18 m/min	< 1
Bare chromel-alumel thermocouple	Still air	60
(0.9 mm wire diameter)	Air, 350 m/min	3
	Water, 18 m/min	< 1
Platinum resistance thermometer	Still air to still	
in stainless steel sheath	water–immersion	
2.1 mm O.D.	velocity 1 m/s	1.4
3 mm O.D.		1.9
3 mm O.D.		5.5
Tungsten resistance element	Air, 30 m/s	3×10^{-4}
(0.005 mm O.D. wire)		
Mercury-in-glass thermometer	Still air	
9 mm O.D.		500
1.25 cm O.D.		680
Mercury-in-glass thermometer	Air, 60 m/min	
9 mm O.D.		240
1.25 cm O.D.		330

relatively long periods of time when used within their recommended temperature range. Outside that range, especially in the presence of corrosive atmospheres, the useful life can be considerably shortened.

4.25.9   Heat Flow

Heat is defined as energy transferred across a boundary between two systems as the result of a temperature difference between the systems. The transfer can take place by conduction or by radiation, depending on whether or not the systems are in physical contact.

Measurement of conduction heat flow requires the independent determination of a steady-state temperature gradient dT/dy, and the effective thermal conductivity k of the material across which the gradient is measured. The heat flow q, in terms of a flux per unit area, is related to these quantities by Fourier's law of heat conduction (see Fig. 4.25-6)

$$q = -k\frac{dT}{dy}. \tag{4.25-9}$$

If temperature differences can be precisely measured and the thermal conductivity of the body is known, Eq. 4.25-9 can be used directly. The thermal conductivity can be determined independently by taking a material

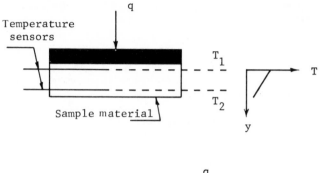

$$k = \frac{q}{(T_1 - T_2)/(y_2 - y_1)}$$

Figure 4.25-6 A simple device for measuring thermal conductivity.

sample of known geometry, applying a known heat flux, and measuring the resultant temperature gradient. Although probe construction is generally adapted to the particular measurement problem [8, 10], some commercial probes in the form of flat plates using miniature thermopiles to measure temperature differences are available. These sensors can operate within the temperature range of -185 to $1100°C$, can have sensitivities as great as 0.1 J/hr cm² mV, and can measure heat flux densities as high as several hundred thousand J/hr cm².

References

1. Baker, H. D., E. A. Ryder, and N. H. Baker, *Temperature Measurement in Engineering*, Vol. I, John Wiley and Sons, New York, 1953; Vol. II, John Wiley and Sons, New York, 1961.
2. Benedict, R. P., "International Practical Temperature Scale of 1968," *Leeds and Northrup Technical Journal*, No. 6, pp. 2–12, 1969.
3. Brender, B. B., and H. W. Newkirk, "Multicolor Pyrometry," Atomic Energy Commission Research and Development Report HW-56505, Sept. 1, 1958.
4. Gray, D. E., Ed. *American Institute of Physics Handbook*, McGraw-Hill Book Company, New York, 1963.
5. Herzfeld, C. M., Editor-in-Chief, *Temperature, Its Measurement and Control in Science and Industry*, Vol. III, Pt. 1, Reinhold Publishing Corp., New York, 1962; Pt. 2, Reinhold Publishing Corp., New York, 1962.
6. Hudson, R. P., and R. S. Kaeser, "Cerous Magnesium Nitrate: A Magnetic Temperature Scale 0.002–2°K," *Physics* **3**, 95–113, Feb. 1967.
7. Kuhlman, W. C., "Status Report of Investigation of Thermocouple Materials for Use at Temperatures Above 4500°F," Society of Automotive Engineers Paper 750D, presented at meeting Sept. 23–27, 1963.
8. Langseth, M. G., Jr., A. E. Wechsler, E. M. Drake, G. Simmons, S. P. Clark, Jr.,

and J. Chute, Jr., "Apollo 13 Lunar Heat Flow Experiment," *Science* **168**, 211–217, 10 April 1970.

9. Lorrain, J., and C. F. Bonilla, "Cross Conduction Errors in Thermocouples. Correction of Long Swaged Thermocouples at High Temperatures," *Nucl. Engr. Design* **8**, 251–272, Aug. 1968.

10. Runcorn, S. K., *Methods and Techniques in Geophysics*, Vol. I, pp. 1–61, Interscience Publishers, New York, 1960.

11. Sherman, R. H., et. al., "The 1962 He Scale of Temperatures IV. Tables," *J. Res. Natl. Bur. Std.*, Section A, **68A**, No. 6, 1964.

12. Swindells, J. F., Ed., "Precision Measurement and Calibration—Selected National Bureau of Standards Papers on Temperature," National Bureau of Standards Special Publication 300, Vol. 2, pp. 7–16, 121–158, 213–232, 363–390, and 391–403, Aug. 1968.

13. Weast, R. C., Ed., *Handbook of Chemistry and Physics,* 50th Edition, The Chemical Rubber Company, Cleveland, 1969–1970.

R. Lawrence Swanson

4.26   TIDE SENSORS

4.26.1   Relationship of Tide to Environment

Tide is the periodic daily or semidaily fluctuation of the sea surface. The degree of vertical displacement is variable and not visually perceptible in some portions of the world. In the fifth century B.C. Herodotus documented the oldest known reference to the tides, those in the Red Sea [7]. In the next century, Pytheas recorded that the motion of the moon and the tide were related.

The tides have always been important as an aid to navigation. High tide increases the extent of navigable waters and low water is used to expose the hull of a beached vessel for maintenance. Tidal currents aid the navigator in many estuaries by providing a favorable current for entrance or departure.

Knowledge of the tides is essential because wharves, buildings, and other structures have to be built with the ever-changing water level in mind. With the increasing use of offshore structures, a knowledge of the ocean tides is rapidly becoming important.

Information on estuarine and coastal circulation is needed to lessen the stress placed on our ecological system through inadequate waste management procedures. The tides play an important role in determining the rates of dilution, mixing, and flushing of the coastal waters.

The offshore oil industry has created the necessity of precisely defining boundaries. Since coastlines constantly undergo changes, boundaries are difficult to delineate. Tidal datum planes appear to be the most effective method to date. Such datum planes as mean low water, mean lower low water, mean high water, mean higher high water, mean tide level, and mean sea level can be determined relative to a specified epoch or time period. Long series of tidal observations are needed for such datum determinations.

4.26.2 Elementary Tidal Theory

Equilibrium Tide. The effect of the moon on tides remained a mystery until Isaac Newton stated his laws of gravity in 1687. Newton's work and that of Daniel Bernoulli in 1740 led to the equilibrium theory of tides, the basis for understanding simple tidal generation [2].

To help visualize how the tide is produced only the influence of the moon will be considered. Newton's law of gravitation states that two bodies are attracted in direct proportion to the product of their masses and inversely proportional to the square of the distance between them. The earth and moon are not on a collision course because the gravitational attraction between them is balanced by the centrifugal force due to the rotation of the earth and moon about the centroid of the earth-moon system. The centroid is within the interior of the earth, 4600 km from the earth's center [2]. The two opposing forces exactly balance each other at the earth's center, but at the surface of the earth the two forces are slightly out of balance. The centrifugal force is slightly less than the gravitational attraction on the side of the earth nearest the moon and greater on the opposite side.

However, the gravitational attraction of the earth on the particle of water at the earth's surface is considerably greater than that due to the mass of the moon. At points on the earth's surface other than the line connecting the centers of the two bodies, the gravitational attraction on a particle due to the moon can be resolved into components normal to the earth's surface and tangent to the surface. The component normal to the

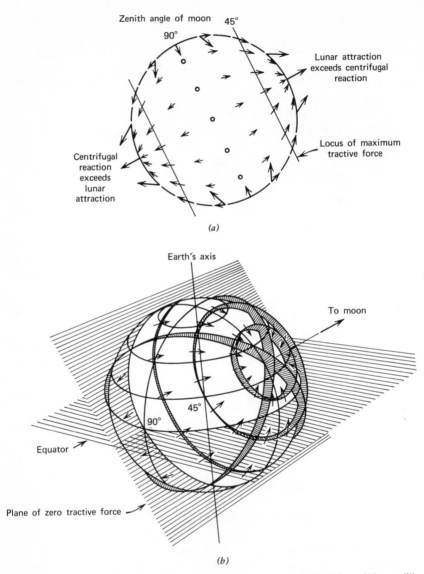

(a)

(b)

Figure 4.26-1 (a) Configuration of the tractive forces, and (b) configuration of the equilibrium tide [19].

surface is negligible in producing the tides, but due to the ocean's fluid characteristics, the tangential component is sufficient to generate the tidal bulges as illustrated in Figure 4.26-1.

A similar tide-generating force is caused by the other heavenly bodies, but the sun is the only other body that produces a significant force. The sun has considerably greater mass than the moon but the distance to the sun is also greater so that the force due to the sun is only 46 percent of that of the moon.

The range of the tide (difference between successive high and low waters) fluctuates with time primarily as a result of the changing positions of the sun and moon with respect to the earth.

Semimonthly the range is increased, with the high tides higher and the low tides lower. These are known as spring tides and are associated with new and full moon. At this time the tractive forces of the sun and the moon are aligned to increase the tidal bulge.

Between full and new moon the sun and moon are at quadrature resulting in a decreased tidal range with the high tides lower and the low tides higher. These conditions are referred to as neap tides.

Celestial orbits are elliptical; the distance between the satellite and the body around which it revolves varies. The moon encircles the earth relative to the sun approximately once a calendar month so that it passes through its apsides twice a month. Consequently, the tide range is increased monthly when the moon passes through perigee. The range is decreased with the same periodicity at apogee.

A similar situation occurs as the earth moves about the sun, passing through perihelion and aphelion. The effect is less pronounced than the counterparts of the moon's motion and occurs on a yearly basis instead of monthly.

The changing declinations of the moon and sun also play an important role in modifying the tide. The periodicity of the moon's declinational change is approximately $27\frac{1}{3}$ days with maximum values of nearly 28.6° north and south of the equator. At maximum declination the attractive force due to the moon is uneven with respect to the equator, causing a difference in the heights of succeeding high waters and succeeding low waters at a given location. This difference, known as diurnal inequality, is generally a maximum when the maximum declination occurs producing the tropic tides. Equatorial tides occur when the moon is over the equator and the diurnal inequality is a minimum. Tropic and equatorial tides occur semimonthly [11].

Variation of Real Tide from Equilibrium Theory. In the equilibrium theory the earth is assumed to be a smooth surface completely covered by a hydrosphere that is in hydrostatic equilibrium. The effects of friction, inertia, depth of the ocean, and Coriolis force are neglected. If the equilibrium assumptions were valid, the tide would respond instantaneously to the tide-producing forces. Instead, the time of high tide varies considerably throughout the world's oceans with respect to the moon's passage over the local meridian. The height of tide also cannot be explained by the simplified theory. The equilibrium theory does not explain the observed tidal phenomena, but provides insight into the causes and fluctuations [3].

Prediction of Tide. Tide prediction capability is due to the fact that the tide is driven by heavenly bodies whose movements are well ordered in both time and space. The resulting tidal time series is a continuous periodic function that readily lends itself to curve-fitting procedures.

A modified harmonic analysis is generally performed on a time series of tidal heights. In this type of analysis a function of the form

$$H(t) = H_0 + A_i \cos(n_i t - \alpha_i) \tag{4.26-1}$$

is fitted to the data, where $H(t)$ represents the height of tide at any time t, H_0 is the height of mean level above a reference datum such as mean low water, A_i is the amplitude of the tidal constituents, n_i is the speed of the constituents, and α_i is the phase of the constituents at $t = 0$. Since the n's are known speeds from the motions of the astronomic bodies, it is possible to solve for the amplitudes and phases of the various tidal constituents [10].

Long series of observations are necessary to accurately determine the amplitudes and phases of the constituents. Fifteen days of hourly data is a minimum and a month is preferred. Data are collected for a year when it is desired to have the long term periodicities separated in the harmonic analysis.

4.26.3 Types of Tide

A time series of tidal heights (marigram) is distinctive to a specific location. There are, however, general characteristics of the tides throughout the world that permit the establishment of a classification system. Illustrative examples at several locations about the United States are shown in Figures 4.26-2 and 4.26-3.

A tide is diurnal if during the period of a lunar day of $24^h 50^m$ there occurs only one high water and one low water. Diurnal tides are found in the northern Gulf of Mexico and in Southeast Asia.

The semidiurnal tide is most common. It is characterized by the occurrence of two high waters and two low waters in the lunar day. The elevations of succeeding high waters and succeeding low waters are nearly the same. A semidiurnal tide is found along the east coast of the United States.

In the mixed tide, there are two high waters and two low waters in a lunar day. However, succeeding high waters and succeeding low waters are generally different in height. These differences are known as diurnal inequality. Mixed tides are common to the west coast of the United States and to Alaska and Hawaii.

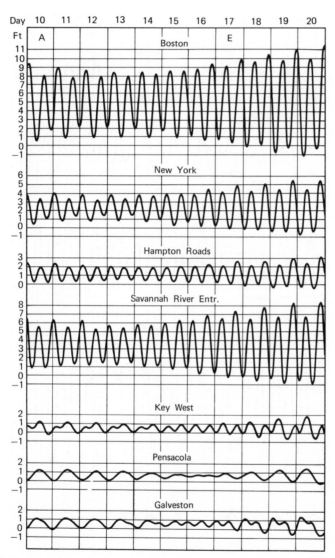

Figure 4.26-2 Typical tide curves for United States ports. Lunar data: max. S. declination, 9th; apogee, 10th; last quarter, 13th; on equator, 16th; new moon, 20th; perigee, 22nd; max. N. declination, 23rd [15].

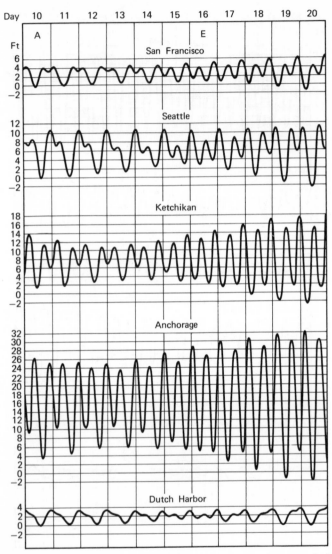

Figure 4.26-3 Typical tide curves for United States ports. Lunar data: max.S. declination, 9th; apogee, 10th; last quarter, 13th; on equator, 16th; new moon, 20th; perigee, 22nd; max. N. declination, 23rd [17].

4.26.4   Phenomena Associated with Tides

Phenomena often associated with tides are those of the meteorological tide, seiche, and tide current. The first two often distort the normal tide pattern of the marigram.

Meteorological Tide. The water column responds to any external forces that are applied to it such as wind and barometric pressure. The change due to these two forces is known as the meteorological tide (storm surge).

In shallow coastal areas the wind interacts with the water surface and moves water from one area to another. The water body response depends on the wind field and other complicating factors such as topography and the stage of the tide. However, in coastal areas the water surface usually responds directly to the wind. Thus an onshore wind raises the water level on the coast and an offshore wind lowers the level.

The water surface responds to changes in barometric pressure like an inverted barometer, tending to decrease the elevation of the surface with an increase in pressure. The process is complicated, depending upon the size of the barometric disturbance and the speed with which it moves in addition to the characteristics of the body of water.

The meteorological tide is particularly noticeable with the passage of large storms such as hurricanes. Strong winds and low pressure can raise the water level considerably. If this increase in water level is superimposed on the astronomic high tide, extreme flooding is often the result.

There is a periodicity in the meteorological tide with the change of seasons. If the astronomic tide is small, the random meteorological disturbances often represent a significant portion of the total change in water level.

Seiches. A seiche is a stationary wave oscillation whose period depends on the dimensions of the local body of water. These oscillations (Figure 4.26-4) can have periods of only a few minutes up to many hours.

A seiche is generated by an external force, often one of the same forces that generate a storm surge. After the force has been removed, the body of water responds by oscillating at its natural frequency. The seiche can have one or more nodes [18].

As the seiche period approaches that of the tide, the range of tide can be considerably affected. Some of the great tidal ranges in the world can be attributed to this interaction of the tide of the open ocean and the seiche of a semienclosed body of water. The large tidal fluctuations in the Bay of Fundy exemplify this interaction [9].

Tidal Current. A tidal current is a horizontal flow of water generated by the tide-producing forces [11]. Tidal currents, like the tides themselves, are periodic, and they can be analyzed and predicted by harmonic tech-

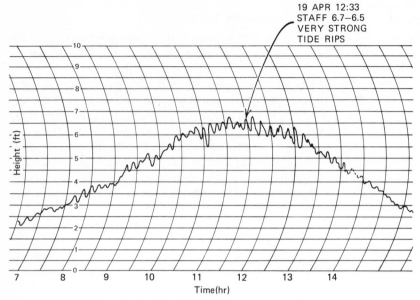

Figure 4.26-4 Seiche at Sachem Head, Conneticut, April 19, 1969.

niques. The procedure is more complicated because the tidal current is two dimensional.

There are some cases, however, in which the movement of the waters is confined to one dimension. In these instances the water flows alternately toward (flood) and away (ebb) from the land.

In less restricted waters the deflecting force of the earth's rotation causes the flow direction to change continually with time. The locus of points of the current vectors roughly describes an elliptical path and the flow is often called rotary.

Generally, maximum flood is associated with a high tide and maximum ebb with a low tide. In theory this relationship holds for a progressive wave. In a standing wave the maximum currents occur at mean tide level. However, the simple theory seldom holds in nature. Interaction of numerous complicating factors such as friction, topographic effects, and river flow cause the horizontal and vertical tidal motions to be unique for any specific locality.

4.26.5 Tidal Sensors

Little work has been done to determine the response characteristics of tide gages. Efforts have been made to dampen high frequencies, hopefully leaving the lower frequencies with a nearly linear response. Hourly values or averages of tidal heights are usually used in tide studies.

Staff. The tide staff is the simplest tide gage. It consists of a graduated scale marked in units of length.

The staff is constructed of wood with painted graduations. The adverse effects of the marine environment rapidly deteriorate the painted markings. Notches in the staff at the foot or meter marks help to facilitate recovery of the entire scale [13]. An available vitrified scale is much easier to maintain. It can be cleaned of marine growth periodically and is not readily subject to corrosion. This staff is made in short sections that can be individually replaced in case of damage.

A staff gage should be installed with a self recording tide gage to permit comparison observations for quality control. The staff allows the tidal observations to be referenced to bench marks on the ground to determine and monitor the various tidal datum levels.

Since the tide staff is a direct-reading instrument, the limitations on its use lie solely with the observer. Under ideal conditions it is difficult to read the staff closer than the nearest 3 cm. For tidal studies the observer is generally required to record the staff readings only once an hour. If a short series (up to three days) of tidal data is all that is required, it is difficult to find a system that is more cost effective.

Mechanical Float-Actuated Tide Gage. A gage of this type depends upon the movement of a float in a stilling well being transmitted via a wire, cable, or beaded chain to a recording device. The motion of the wire is transferred to a float wire wheel and a counterpoise weight is attached to the end of the wire. This weight keeps tension in the wire with a rising tide but is not heavy enough to cause the float to "hang-up" on a falling tide. Protection of the counterpoise from the elements is accomplished by using the float well, a section of pipe, a special tide house, or an existing structure.

Water enters the float well near the bottom and comes to equilibrium with the water surface exterior to the well (Figure 4.26-5). The degree of damping depends upon the size of the opening to the sea and the location of the orifice in relation to the surface.

The lag between the rate of rise of the water inside the well compared to that outside the well can be approximated by means of Torricelli's theorem, which states that the velocity of water through an orifice is

$$v = (2gh)^{1/2}, \qquad (4.26-2)$$

where g is the acceleration of gravity and h is the difference in water level on either side of the orifice. Modifying the equation for friction, it has been shown that for the ratio

$$\frac{\text{diameter of orifice}}{\text{diameter of float well}} = 0.1. \qquad (4.26-3)$$

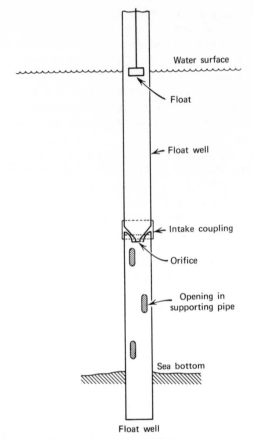

Figure 4.26-5 Mechanical float-actuated tide gage [14].

The time for the water inside the well to come to equilibrium with that outside (due to change in the tidal elevation) is a few seconds [4].

The effects of surface waves can be reduced considerably by placing the orifice near the bottom. In simple classical wave theory a wave is considered to be a deepwater wave if the ratio of depth of the water to wavelength (h/λ) is greater than 1/2. The radius of the orbital motion of the water particles is

$$r_d = Ae^{-kd}, \tag{4.26-4}$$

where d is the depth of interest, A is the amplitude of the surface wave and k is equal to $2\pi/\lambda$. Thus at a depth of $\lambda/2$ the radius of the orbital motion is 1/23 that at the surface. For shallow water waves, the vertical motion decreases even more rapidly as the orbital path becomes elliptical with the semi-major axis parallel to the bottom [19].

Gas-Purging Tide Gage. A gas-purging or "bubbler" tide gage works on the principle that the rate of gas leakage through an orifice at the base of a water column is inversely proportional to the pressure head.

A gas is bled off a supply source (Figure 4.26-6) through a reducing valve. Normally nitrogen gas is used, since it is inert, dry, relatively inexpensive, and available [14]. A differential regulator in the line assures that there is a minimum pressure available to prevent water from backing up in the line through the orifice. The gas is then led through a needle valve and a bubble chamber that permits visual inspection of the flow rate which varies with the range of tide at different locations. From the bubble chamber, the gas flows to the orifice which is securely anchored in the water column. The pressure in the tubing is measured by a transducer bellows of a standard pressure-type strip-chart recorder [8].

The proper adjustment of the bubble rate corresponding to a given range of tide is extremely important. The use of a low bubble rate and an improperly secured orifice are the most common mistakes encountered with the gas-purging installation. There can be errors due to variations in the weight of the gas, friction of the gas in long tubing, variations in the density of sea water, extreme fluctuations due to wave action, and variations in barometric pressure [8].

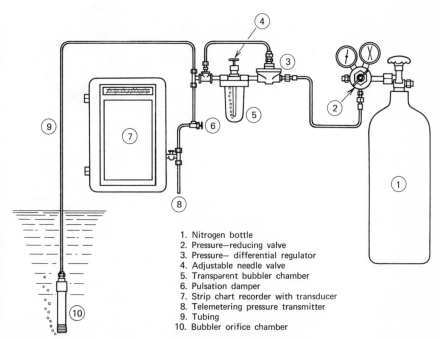

1. Nitrogen bottle
2. Pressure—reducing valve
3. Pressure— differential regulator
4. Adjustable needle valve
5. Transparent bubbler chamber
6. Pulsation damper
7. Strip chart recorder with transducer
8. Telemetering pressure transmitter
9. Tubing
10. Bubbler orifice chamber

Figure 4.26-6   Gas-purging pressure tide gage, functional diagram [8].

Sea Floor Sensors. Progress has been made in developing instrumentation for sea floor tide sensors, but the state of the art is still rudimentary. Sea floor sensors measure the pressure resulting from a change in the head of water to indicate the change in tide level. A diaphragm mounted on one end of a pressure case transmits the fluctuations to a vibrating wire [12] or a Bourdon tube [5]. The frequency of the vibrating wire or the degree of flexing of the Bourdon tube is recorded as an analog or a digital signal.

Unique Methods. Often tide observations are required when time, money, or equipment are not readily available and it is necessary to improvise. Two common methods are shipboard observations and sounding from a marked position.

A depth recorder can be used as a tide gage. In areas of rapidly changing bottom, the vessel is anchored from the bow and the stern to prevent the

Table 4.26-1. Tide Sensors

Parameter	Tides			
	Sensor Type			
Characteristic	Tide Staff (Vitrified)	Float Actuated	Gas-Purging Pressure	Sea-Floor Pressure
Active or passive	Passive	Passive	Passive	Passive
Output type	—	Analog or digital	Analog	Analog or digital
Output form	—	Strip chart or paper tape	Strip chart	Strip chart
Accuracy, absolute	—	3cm	1 % full scale	0.5 cm
Resolution	—	3 mm	3 mm	—
Power consumption	—	For digital $7\frac{1}{2}$ V dc source	—	Variable depends on other sensors in package
Environmental limits				
operating	None	8 days to 1 month	8 days	Variable
storage	None	1 month	6 weeks	Variable
Reliability	Excellent	Excellent	Good	Unknown
Life	10 yr	10 yr	10 yr	Unknown
Availability	Readily	Readily	Readily	Scarce

ship from swinging over variable bottom topography. A continuous trace can be maintained, if desired. If the ship cannot be double anchored and the bottom is flat, it is also possible to obtain usable information.

Periodic lead line observations at a fixed position (such as next to a piling) are often an adequate means of determining the change in water level.

References

1. Barbee, Wiliam D., *Tide Gages in the U.S. Coast and Geodetic Survey*, Coast and Geodetic Survey, Rockville, Maryland, 1965. Prepared for distribution at the X Pan American Consultation on Cartography, Guatemala City.
2. Defant, A., *Physical Oceanography*, Vol. 2., Chapters 7, 8, 9, Pergamon Press, New York, 1961.
3. Dietrich, Günter, *General Oceanography, An Introduction*, Chapter 9, Interscience Publishers, New York, 1963.
4. Doodson, A. T., and H. D. Warburg, *Admiralty Manual of Tides*, corrected reprint, Chapter 10, Her Majesty's Stationery Office, London, 1966.
5. Filloux, J. H., "Deep Sea Tide Gauge with Optical Readout of Bourdon Tube Rotations," *Nature* **226**, 935–937, June 6, 1970.
6. Lennon, G. W., "The Evaluation of Tide Gauge Performance Through the Van De Casteele Test," *Cahiers Oceanographiques* **20** (10), 867–877, December, 1968.
7. Marmer, H. A., *The Tide*, Chapter 2, D. Appleton and Company, New York, 1926.
8. Poling, Austin C., and Wesley M. Butler, "The Gas Purging Pressure Tide Gauge," *Internatl. Hydrographic Rev.* **50** (1), 103–111, January 1963.
9. Redfield, A. C., "The Analysis of Tidal Phenomena in Narrow Embayments," *Papers in Physical Oceanography and Meteorology*, **11** (4), 1–36, 1950.
10. Schureman, Paul, *Manual of Harmonic Analysis and Prediction of Tides*, United States Government Printing Office, Washington, D. C., 1941.
11. Schureman, Paul, *Tide and Current Glossary*, United States Government Printing Office, Washington, D. C., 1963.
12. Snodgrass, Frank E., "Deep Sea Instrument Capsule," *Science* **162**, 78–87, October 4, 1968.
13. The Hydrographer of the Navy, *Admiralty Manual of Hydrographic Surveying, Volume Two, Tides and Tidal Streams*, Part 1, The Hydrographer of the Navy, Taunton, Somerset, England, 1969.
14. United States Coast and Geodetic Survey, *Manual of Tide Observations*, United States Government Printing Office, Washington, D.C., 1965.
15. United States Coast and Geodetic Survey, *Tide Tables: High and Low Water Predictions, East Coast of North and South America, Including Greenland, 1970*, United States Government Printing Office, Washington, D. C., 1970.
16. United States Coast and Geodetic Survey, *Tide Tables: High and Low Water Predictions, Europe and West Coast of Africa, Including Mediterranean Sea, 1970*, United States Government Printing Office, Washington, D. C., 1970.

17. United States Coast and Geodetic Survey, *Tide Tables: High and Low Water Predictions, West Coast North and South America, Including the Hawaiian Islands, 1970*, United States Government Printing Office, Washington, D. C., 1970.
18. Von Arx, William S., *Introduction to Physical Oceanography*, Chapter 3, Addison-Wesley, Reading, Mass., 1962.
19. Williams, Jerome, *Oceanography*, Chapter 16, Little Brown and Company, Boston, 1962.

William Markowitz

4.27 TIME SENSORS

4.27.1 Basic Concepts

Epoch and Time Interval. Time measurement involves two quantities: *epoch*, which specifies when a single, instantaneous event occurs, and *time interval* (TI), which specifies the duration of a continued event or the elapsed time between two instantaneous events. Epoch and TI are analogous to position and length in linear measurement.

Epoch is normally given in *universal time* (UT) and TI in terms of the *second*, which is defined by a specific number of electromagnetic oscillations of the cesium-beam atomic clock. UT is based on the rotation of the earth about its axis.

Atomic time (AT) is defined by adopting an initial epoch for the time generated by an atomic clock. The AT defined by any combination of atomic clocks will differ slightly from that defined by another combination. These differences are about 1 part in 10^{12} in frequency, which means a divergence of 1 μs in 10 days.

Ephemeris time (ET) is defined by the orbital motion of the earth about the sun. ET is the independent variable in the equations of dynamics.

Within the errors of measurements, ET and AT provide the same measure of time interval. UT, which is rotational time, is variable with respect to both ET and AT.

The angular position of the earth about its axis is needed for celestial navigation, geodetic surveying, satellite tracking, and for civil use. For these reasons, UT is in worldwide use to define epoch in the form called *Coordinated Universal Time* (UTC). Time signals transmit UTC, which is indicated by clocks in scientific laboratories. Public clocks indicate zone time, which differ from UTC by an integral number of hours; minutes, seconds, and fractions are the same.

The epoch of UTC is kept close to that defined by the rotation of the earth (its angular position) but the interval between successive seconds is that of the atomic clock. *Leap seconds* are used to keep UTC close to rotational time.

Thus, the time system is a dual system; epoch is based on UT and time interval on the atomic second.

Geodetic, geophysical, and satellite tracking operations are recorded in UTC, but are generally required in some other form of epoch, namely, some form of UT or ET. AT is often used as an intermediary.

Coordinated Universal Time. Local mean solar time is obtained from observations of stars at an observatory. UT, in the form denoted UT0, is obtained by adding a constant λ_o, called the conventional longitude. Because of polar motion the instantaneous longitude is $\lambda_o + \Delta\lambda$, with $\Delta\lambda$ different for each observatory. UT1, defined as UT0 + $\Delta\lambda$, is the same for all stations except for errors of observation. UT1 is smoothed by adding an empirical periodic correction for yearly and semiyearly variations in speed of rotation of the earth, ΔSV; UT2 = UT1 + ΔSV.

UT2 still contains variations with respect to ET and AT, and coordinated universal time (UTC) was introduced. UTC is obtained from an atomic clock whose epoch is shifted exactly 1 s, as necessary, so as to remain, generally, within 0.5 s of UT. The maximum difference allowed is 0.95 s.

The shift is made by introducing *leap seconds*. Normally, the last second in each calendar day is 23h 59m 59s and the next second is 0h 00m 00s of the next day. However, if clocks are to be retarded then the second which follows 23h 59m 59s is denoted 23h 59m 60s of the *same* day, and the next second denoted 0h 00m 00s of the *next* calendar day.

A leap second may be introduced on the first day of any month. Some years may contain no leap seconds, and others contain one, two, or more. Currently, about 1 retardation per year is expected.

The Bureau International de l'Heure (BIH) coordinates time and frequency information, issues corrections $\Delta\lambda$ and ΔSV, and selects the leap second dates.

Conversions to UT1, UT2, A.1, and ET. Only UTC is in common use, as indicated by clocks, for scientific purposes. UT1, UT2, AT, and ET are obtained from UTC by applying the appropriate differences. See Section 4.27.4.

The atomic time scale of the BIH is called *International Atomic Time* and is denoted by TAI.

4.27.2 Timekeeping Equipment

General Items. Equipment used in timekeeping includes radio-frequency receivers, clocks, oscilloscopes, decimal counters, and chart recorders.

Precise time is commonly obtained from radio transmissions. An oscilloscope compares received time pulses with pulses generated by a local clock. The clock provides time continuously. A decimal counter compares clocks to the microsecond or higher accuracy. Chart recorders record differences in phase (epoch) or in frequency.

Communications Receivers. High-frequency (HF) radio time signals provide UTC accurate to about 1 ms when corrected for propagation time and receiver delay. Time signals on HF are transmitted from numerous stations about the world on allocated frequencies, such as WWV and WWVH, and on other frequencies such as CHU. The WWV frequencies are 2.5, 5, 10, 15, 20, and 25 MHz. Time signals are also broadcast from communications stations about the world.

For many purposes an accuracy of 1 ms is sufficient. For such applications, little more is required than a communications receiver, a clock, and an oscilloscope. Even when very high precision is required and other equipment is used, such as Loran-C, it is useful to have a communications receiver for checking the second of UTC. This ensures the minimum chance of error. An expensive receiver is not required.

Loran-C Receivers. Loran-C is a radio navigational system which operates through the reception, via ground wave, of pulses transmitted from chains of stations on 100 kHz. The pulse repetition rates are different for each chain. Loran-C provides both epoch and time interval (frequency) with high precision, in a simple manner, and at relatively low cost. The waveform is shown in Figure 4.27-1.

Two types of receivers which extract time information from the navigational pulses are available, visual and coherent-phase detection. A repetition synthesizer must be used with the visual receiver to trigger the oscilloscope at the pulse repetition rate. Frequency can be calibrated to about 1 part in 10^{12} in one day with either receiver.

Epoch can be obtained with high precision by correcting for ground wave travel time, which can be computed with an accuracy of 1 or 2 μs. The coherent receiver identifies the third cycle of each pulse train. With the visual receiver the wrong cycle may be identified, leading to a constant error, a small multiple of 10 μs. The cost of a synthesizer and visual receiver is about one-fifth that of a coherent receiver.

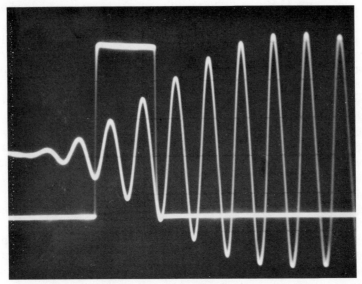

Figure 4.27-1 Loran-C pulses photographed with a dual-trace oscilloscope at Marquette University, Milwaukee, Wis. Pulse from station at Dana, Indiana, distance 350 km, and time marker from local clock.

VLF and Omega. Very-low-frequency (VLF) communications transmissions (14 to 25 kHz) and *Omega* navigational transmissions (10 to 14 kHz) provide frequency calibration of high precision. A diurnal phase shift occurs from day to night but this can be eliminated by making measurements at 24-hr intervals, preferably daylight to daylight. Frequency can be calibrated to about 1 part in 10^{11} in one day. Phase tracking receivers are available which track both VLF and Omega. With a chart recorder it is possible to record the difference in phase between a local clock and the incoming signal, generally controlled by an atomic clock. If the epoch of a local clock has been determined (e.g., via a portable atomic clock), it is possible to determine the phase of the local clock thereafter through use of the VLF or Omega record. If there is an interruption, however, the phase information is destroyed, which is not the case with Loran-C.

Time signals are emitted by some VLF stations such as NBA, Canal Zone, and GBR, Rugby. Because of the slow buildup of VLF signals, a precision of only about 1 or 2 ms in epoch is obtained.

4.27.3 Clocks

General; Quartz. Clocks used for precise timekeeping are electronic. They consist of (a) an oscillator and (b) a counting device. Part (b) indicates hours, minutes, and seconds, either by digital lights or a dial mechanism, and generates precise time markers (pulses) at various rates, e.g., 1, 10,

100, 1,000, or 10^6 per second. The stability of rate is governed by the oscillator.

Quartz crystal clocks are generally reliable, small, and relatively inexpensive. However, a quartz crystal changes frequency with age. Also, abrupt changes can occur. The drift rate is about 1 part in 10^8 per day for a quartz-crystal chronometer. Laboratory oscillators can be obtained with drift rates of about 1 part in 10^{10} to 1 part in 10^{11}. The latter is equivalent to a change in rate of about 1 μs per day per day.

Atomic Clocks. An *atomic clock* is a device which produces coherent electromagnetic radiation through a quantum transition and which counts the cycles produced. The three types in general use are cesium-beam, rubidium gas cell and hydrogen maser.

Cesium. The second is defined in the SI system as 9, 192, 631, 770 periods of oscillation of the radiation produced by a specified transition of cesium-133.

It is practicable to construct small cesium beam clocks with high-frequency stability and reliability. The average deviation in frequency from the absolute is about 4 parts in 10^{12}. The frequency of each such clock, as manufactured, is independent; it is not adjusted with respect to some standard.

The Naval Observatory atomic time scale, A.1, is based on a system of 16 atomic clocks. The frequency of the mean is constant to 2 parts in 10^{14} for intervals of 5 to 100 days. The mean has probably drifted less than 1 part in 10^{13} in 2 yr. Portable cesium clocks carried by airplane show an average deviation of about 1.0 μs after a 15-day round trip.

The mean time between failures of a cesium clock is about 2 yr. Approximate dimensions and weight are $42 \times 40 \times 22$ cm; 27 kg. A separate power supply and battery are required.

Rubidium Gas Cells. The spectral purity of a rubidium oscillator exceeds that of a cesium oscillator, and for intervals up to 1 day the frequency stability of rubidium exceeds that of cesium. However, for longer intervals cesium is more stable. A rubidium clock must be set in frequency according to some frequency standard, and may drift 1 part in 10^{11} within a week. It is therefore not in the same class as cesium in providing a primary reference frequency. The weight of a rubidium clock is about two-thirds that of a cesium clock.

Hydrogen. Hydrogen masers provide high spectral purity and, for intervals up to 1 day, the highest frequency stability. However, they are inferior to the best cesium oscillators for intervals over 10 days.

Hydrogen masers are large, require considerable auxiliary equipment, and are not readily available commercially. They have been used in very-long-baseline interferometry experiments but are not well suited for general use.

Selection of Clock. The cost of an atomic clock relative to cesium is: rubidium, $\frac{1}{2}$; hydrogen, 10. If the purchase of an atomic clock is justified, then it is generally advisable to buy a cesium clock in preference to rubidium because of its freedom from drift.

Portable Atomic Clocks; Relativity. The U. S. Naval Observatory transports cesium-beam atomic clocks by airplane to synchronize atomic clocks about the world to within 1 μs.

Clocks in airplanes are affected by relativistic changes in frequency compared to clocks fixed on the ground. The frequency increases by 1.1×10^{-16}/m of height. The change due to motion depends on the speed relative to the ground, direction of motion, and latitude; it can be positive or negative. Detection of the relativistic effects is difficult; stability of clock rate is affected by transportation. However, experiments were made in 1971. The gain on a round-the-world westward flight was about 0.27 μs and loss on the eastward flight was about 0.06 μs. These results are in reasonably good agreement with theory.

Satellite Techniques. The two-way exchange of time pulses via communications satellites from ground stations is used to synchronize clocks to 0.1 μs. The one-way transmission from a clock on board a satellite, even from a synchronous satellite, provides a lesser precision because both the orbital position of the satellite and the rotational position of the receiving station must be known. Special equipment is required for receiving signals from satellites. For the general user, satellite techniques are not as useful as radio transmissions from stations at fixed sites.

4.27.4 Information on Transmissions

The U.S. Naval Observatory distributes information on time determinations, corrections to the epoch and frequency of radio transmissions, and differences between the various kinds of time described in Section 4.27.1. It also distributes information for BIH.

Franklin S. Harris, Jr.

4.28 VISIBILITY SENSORS*

4.28.1 Quantities Used in Visibility

The ability to detect or distinguish objects is characterized by a quantity called visibility. It is affected by a large number of factors including the distribution of illumination, the reflective properties of the object to be detected, the properties of the medium, the detection system and the properties of the eye [21, 45, 46]. Because of differences in terminology and symbols as used in the literature, the quantities used are defined below with their symbols and the elementary relationships between them [1, 47]. There are other widely used systems in oceanography [21, 35, 63, 67]. The definitions are radiometric with photometric equivalents given in parenthesis.

ϕ Radiant (luminous) flux. Time rate of flow of radiant (luminous) energy. Unit: watt (lumen).

I Radiant (luminous) intensity (of a source in a given direction). The radiant (luminous) flux emitted by a source, or element of a source in an infinitesimal cone containing the given direction divided by the solid angle. $I = d\phi/d\omega$. Unit: watt/steradian (candela).

L Radiance (luminance). Radiant (luminous) flux per unit solid angle per projected area of a surface. $L = dI/(dA \cos\theta \, d\omega)$. Unit: watt/meter^2-steradian (candela/meter^2).

L^* Path radiance (luminance). Radiance (luminance) in line-of-sight of length r caused by ambient radiation (light) scattered into the direction of the sight path. Unit: watt/meter^2-steradian (candela/meter^2).

L_p^* Path function. Radiance (luminance) per unit length in the direction of the line-of-sight generated by the radiation (light) scattered into the beam, or the scattered radiance (light) of the beam. Unit: watt/meter^3-steradian (candela/meter^3).

E Irradiance (illuminance, illumination). The radiant (luminous) flux on an infinitesimal element of surface containing the point under consideration, divided by the area of that element. $E = d\phi/dA$. Unit: watt/meter^2 (lumen/meter^2 = lux).

E_d Downward irradiance (illuminance). The radiant (luminous) flux incident on an infinitesimal element of the upper face ($0°-180°$) of a hori-

---

* Publication partly supported by NASA Contract No. NAS1-9438-28.

zontal surface containing the point considered, divided by the area of the element. $E_d = d\phi_d/dA$. Unit: watt/meter^2 (lux).

E_u Upward irradiance (illuminance). The radiant (luminous) flux incident on an infinitesimal element of the lower face ($180°–360°$) of a horizontal face containing the point being considered, divided by the area of that element. $E_u = d\phi_u/dA$. Unit: watt/meter^2 (lux).

ρ Reflectance. The ratio of the reflected radiant (luminous) flux to the incident flux. $\rho = \phi_r/\phi_0$.

τ Transmittance (transmissivity). The ratio of the transmitted radiant (luminous) flux to the incident flux ϕ_t/ϕ_0.

τ_b Beam transmittance. The transmittance for a beam of diameter small compared to its length. For a homogeneous medium $\ln \tau = -\sigma r$ for a path length r.

α Absorptance. The ratio of the radiant flux lost from a beam by means of absoprtion to the incident flux. $\alpha = \phi_\alpha/\phi_0$.

β Scatterance. The ratio of the radiant flux scattered from a beam to the incident flux. Term used in oceanography. $\beta = \phi_B/\phi_0$.

μ Attenuance. The ratio of the radiant flux lost from a beam by means of absorption and scattering to the incident flux. $\mu = \alpha + \beta = 1 - \tau$.

a Absorption coefficient. The internal absorptance of an infinitesimally thin layer of the medium normal to the beam, divided by the thickness (Δr) of the layer. $a = -\Delta\phi/(\phi\Delta r)$. Unit: meter^{-1}.

$\beta'(\theta)$ Volume scattering function. The radiant intensity (from a volume element in a given direction) per unit of irradiance on the volume and per unit volume. $\beta'(\theta) = dI(\theta)/(E\,dv)$. Unit: meter^{-1}.

b (Total) scattering coefficient. The internal scatterance of an infinitesimally thin layer of the medium normal to the beam, divided by the thickness (Δr) of the layer. $b = -\Delta\phi/(\phi\Delta r) = \int_{4\pi}\beta'(\theta)\,d\omega$. Unit: meter^{-1}.

σ Attenuation (extinction) coefficient. The internal attenuance of an infinitesimally thin layer of the medium normal to the beam divided by the thickness (Δr) of the layer. $\sigma = (\alpha + \beta)/\Delta r = -\Delta\phi/(\phi\Delta r)$. Unit: meter^{-1}.

σ_k Diffuse attenuation coefficient. The internal attenuance of an infinitesimally thin layer of the medium for downward or upward diffused radiance (as in the ocean). $\sigma_k = \Delta E_d/(E_d\Delta r)$. Unit: meter^{-1}.

d Attenuance (attenuation) length. The distance in which the non-attenuated irradiance is reduced to $1/e$ or about 0.37 of the initial amount. Unit: meter.

V Visual range. The distance at which a given object can just be detected against its background. Unit: meter.

V' Meteorological range. The distance a black target can be seen with a threshold contrast ε. $V' = (1/\sigma)\ln(1/\varepsilon)$. For $\varepsilon = 0.02$ use V'_2, for $\varepsilon = 0.05$, use V'_5. Unit: meter.

C Contrast. The difference between object luminance and background luminance divided by the background luminance. $C = (L - L')/L'$.

RVR   Runway visual range. The maximum distance along a runway at which the runway lights are visible to a pilot after touchdown or the distance he can see dark targets against the horizon sky or cloud background, whichever distance is greater.

SVR   Slant visual range. The slant distance at which a pilot sees the required visual segment of the precision approach lights of a runway.

4.28.2   Visibility, the Eye and Instrumentation

Various kinds of visibility are used. In meteorology [31] it is the greatest distance in a given direction at which it is just possible to see and identify with the unaided eye a prominent dark object against the sky at the horizon, in the daytime and at night, a known, preferably unfocused, moderately intense light source. After visibilities have been determined around the entire horizon circle, they are resolved into a single value of *prevailing visibility* (the greatest horizontal visibility prevailing throughout at least half of the horizon circle) for reporting purposes.

There are inherent difficulties with the requirement that visibility markers be recognized. The more rigorously defined concept of the *visual range* avoids reference to recognition. Daytime estimates of visibility are subjective evaluations of atmospheric *attenuation of contrast*, while nighttime estimates represent attempts to evaluate something quite different, attenuation of *flux density*.

The difference between detection and recognition is great in sea water. Due to deterioration in the image because of the variation in the index of refraction, the contrast varies approximately inversely as the third power of the distance [19]. This deterioration or smearing of the image results from light scattered at small angles by objects large compared to the wavelength of light which have indices of refraction little different from water. Therefore beam attenuation, of importance in image formation and photography, is about 2.7 times that of diffuse attenuation. This effect results in an underwater lamp disappearing as an object before the light goes out [19, p. 221].

When the eye is the detector, consideration must be given to differences in individual eyes, the size of the target, whether the eye is dark-adapted and the luminance background. The results of extensive experiments on these factors have been summarized and nomographs given of the parameters and meteorological range [6, 45, 46]. The eye is also sensitive to color differences and motion.

There are four "laws" which are useful. Bouguer's law describes the rate of decrease of flux in a plane-parallel beam, strictly for one wavelength, as it penetrates a medium which both scatters and absorbs:

$$L = L_0 e^{-\sigma r}. \tag{4.28-1}$$

The law is not valid for high attenuation coefficients [25, p. 92]. For a black body with $\varepsilon = 0.02 = e^{-\sigma r}$, the meteorological range is $V'_2 = 3.91/\sigma$. Since there is usually little absorption in the atmosphere, σ can be replaced by b.

Koschmieder's law is

$$L = L_0 e^{-\sigma r} + L_h (1 - e^{-\sigma r}), \tag{4.28-2}$$

where L_h is the luminance of the horizon sky, and the second term is the apparent luminance of a black object due to skylight scattered into the path between the object and the eye. This equation can be expressed in the form used in oceanography [2] as

$$L_r = L_0 \tau_r + L_r^*, \tag{4.28-3}$$

where L_0 is again the inherent luminance at the object, τ_r is the beam trans-mittance through a distance r, and L_r^* is the path luminance of length r and direction θ caused by ambient light scattered into the direction of the sight path, and L_r is the apparent luminance of point of regard after passage through path length r in direction θ.

If the light source is a point, the spreading of the spherical wave must also be included:

$$E = I r^{-2} d^{-\sigma r}. \tag{4.28-4}$$

There is an empirical relation between the extinction coefficient and the wavelength of light λ called Ångström's law:

$$\sigma = k_1 \lambda^{-4} + k_2 \lambda^n. \tag{4.28-5}$$

The first term gives the attenuation due to Rayleigh scattering, the second is the usual form of Ångström's law [33, 37], where k_1 and k_2 are constants. Actually n varies from about 0 for fogs (not wavelength dependent) to about -4 for small particles (Rayleigh) [46, p. 258]. Ångström's value was about -1.3. Therefore, since there is not a simple universal relationship between σ and λ, there is not one between the meteorological range and the spectral attenuation curve either [15].

There have been extensive studies of backscattering characteristics of a medium to establish relationships for predicting the visibility from back-scattering measurements. There is no simple relationship because there are too many variables which influence the scattering return. These include absorption, size distribution of the particles, and variations in the index of refraction [13, 23]. Experimental measurements have shown limited success using a laser source for backscattering [5, 8, 27, 30, 64], and better results with near backscattering using white light to predict visibility [41, 65].

Tables have been constructed to predict visibility giving the relations between extinction coefficient and transmissivity, extinction coefficient for daytime or nighttime (with point sources of light) [40], and (when the atmospheric model is assumed) the attenuation and transmittance [4; 22; 36; 41, pp. 23, 36; 44; 45].

Television systems [3] and image converters using infrared with a GaAs source have been developed to give improved visibility over the eye [38].

There are three principal instrumental techniques which can be used to improve the visibility [43; 67, pp. 80–83]: range gating in which the source is pulsed and only light whose return time is such that it comes from the volume of interest is examined [16, 17]; crossbeams in which the object is illuminated by a source separated laterally from the viewing point so that regions in front of and behind the object are not illuminated [10], and polarization discrimination in which the source is polarized and a polarization analyzer in front of the receiver distinguishes an object against the background by the different effects on polarization of the light.

Values of the visibility parameters vary widely. For a standard atmosphere the visibility due to Rayleigh scattering alone is 336 km [22, p. 2]. In the Antarctic the ratio of total air scattering to Rayleigh scattering is 1.24 to 1.36 [24]. A visibility of 100 km has about equal amounts of Rayleigh and aerosol scattering [37, p. 143]. The visibility transition from clear air to haze occurs at about 15 km, haze to fog at about 1.2 km [27, pp. 2–3], and in shower clouds 10–20 m depending on the water content [59]. The value of clear air has been given for σ as 1.750×10^{-5} m^{-1} [46, p. 255]. Atmospheric values of $\sigma = 2.6 \times 10^{-5}$ m^{-1} and 60×10^{-5} m^{-1} with visibility greater in winter than summer [14] have been measured. There is also an improvement with altitude [54]. Snow causes 10 to 15 times the attenuation of light as rainfall [49]. A detailed experimental and theoretical summary has been made of the reduction of visibility and attenuation of solar radiation [68, 69]. The major variable which influences the scattering for a given amount of material is the size distribution [48, 50].

Underwater the angular distribution of irradiance of the proper amount and its direction at the object with given reflectance characteristics determine the initial contrast, while the corresponding natural background at the observer and light scattered into the field of view along the transmission path determine the contrast which must be detected [19]. Swimmers lose sight of each other, with ample daylight, at a distance very close to $4/\sigma$ [18].

The open-sea attenuation lengths are from two-thirds to one-seventh those of distilled water in the 0.440-μm wavelength [58]. In the blue wavelength of 0.480 μm, about 60 percent of the attenuation of clear blue water is due to scattering and 40 percent to absorption [19, p. 218]. About 1 percent of the sunlight is needed for photosyntheses. This is reached at a 100 m depth for clearest ocean water, 10 to 30 m in coastal water and less than 3 m for very turbid harbors or estuaries [12]. Undersea submersibles have detected light penetration to depths of 700 m [9]. At 10 m depth the spectral distribution of available daylight is about the same as the response curve of the human eye, while at great depths the residual light peaks at 0.465 μm [39, p. 43]. Approximately 95 percent of the sunny weather sunlight of 1 kW/m² irradiation is absorbed beneath the surface. Half of the radiation is infrared, which is absorbed within the first meter of the surface [20, p. 226]. In the clearest ocean water the total scattering by water molecules is about 7 percent of the total scattering [20, p. 65]. Scattering in the sea is predominantly due to transparent biological organisms ranging from bacteria to jelly fish and particles large compared to the wavelength of light [19, p. 217]. When there is little scattering, the E_d/E_u is about 300 compared to the usual ocean ratio of downwelling to upwelling irradiance of 50 [58, p. 110].

4.28.3 Principles of Operation of Instruments

Transmission and Attenuation Transmissometers. A transmissometer consists of a light source whose power after traversing a given path is measured by a detector. The variation in the power received, for a given distance, compared to no loss, is the transmission. The transmissivity is $\tau = L/L_0$, but this is also $e^{-\sigma r}$ from which the attenuation coefficient is obtained. If the absorption is negligible then the attenuation coefficient reduces to the scattering coefficient. Optical alignment between source and detector must be maintained. The instrument must be located so that its measurement is representative of the desired region. Usually several instruments are necessary to take into consideration path or regional inhomogeneities or patchiness. The instrument can have a folded path in which a corner reflector returns the beam back to the same position as the source. The instrument can have two paths of different lengths to handle a wider range of visibilities. In the measurement of diffuse attenuation, distinction must be made, particularly in oceanography, between loss from a collimated beam, and small angle forward scattering.

Meters for Total and Angular Scattering Measurements. The power scattered from the beam can be measured directly by collecting most of the 4π solid angle scattering. The scattering into the region of the continuing beam, and the backscattering near the incident beam are not included and must be taken into consideration by suitable calculation and calibration. In practice some of the total scattering meters (integrating nephelometers) collect the scattering from one side of the beam only. The scattering functions depend on the relative index of refraction, the particle shape, the total number and the distribution of sizes, and the scattering angle [50]. Particles below 0.1 μm radius do not contribute much to the scattering [51]. Polar nephelometers can obtain more information by measuring the scattering as a function of angle. Some meters measure the scattering at a particular angle: near-forward direction, a side angle, near backward, or 180° backscattering (as in conventional lidar).

Absorption Meters. Because it is difficult to measure the absorption alone directly, it is more convenient to measure the attenuation and scattering and compute difference to obtain the absorption.

Continuous Versus Pulsed Source Systems. Continuous sources change with time and require recalibration or circuit compensation. Many systems modulate the continuous source to discriminate against light background and electrical noise by measuring the detector signal only at the modulation frequency. A pulsed source has several advantages. It is easy to distinguish between the pulsed light and background, the duty cycle is short with high power for short periods of time, and the systems can have high reliability. If the source is a laser with monochromatic light, a filter can give effective spectral discrimination against background light. A further advantage of a pulsed source is the application of range-gating to measure light only from the volume of interest.

4.28.4 Instruments for Atmospheric Measurements

Considerable information is available on visibility measurements [26], and the analysis of visibility observation methods [28, 45, 46].

Transmissometers. The typical airport tranmissometer is usually two-ended with source and receiver separated on a fixed baseline of about 165 m. Both source and receiver are shielded against stray light and the weather. For airports it has been found that three transmissometers are needed, one at touch-down, one at mid-point, and one at roll-out position on the runway [56]. A helium-neon laser source can be used with a silicon photodetector in the receiver [28, 41, 53]. Instead of black targets, a system of cavities can be used in the measuring of the visual range [29]. Such transmissometer measurements combined with computation systems give the visibility along the ground, but for airport use the SVR is needed because of the complications of horizontal and vertical variations in the scattering material which influence the incoming pilot's slant range vision.

Solar Transmissometer. The simple solar transmissometer utilizes the sun as the source and the transmission is measured through the atmosphere by a selenium photocell. The zenith angle is measured and the transmission at 0.500 μm wavelength measured in a field-of-view of 1°. From nomographs the decadic (base 10 instead of e) turbidity is determined [24, 66].

Integrating Nephelometer. Several instruments have been developed that integrate the scattering from a small volume of air [11, 52, 55]. The light source, such as a xenon lamp, is pulsed frequently to illuminate a volume of air, either open to the atmosphere or closed with the air drawn through. A phototube accepts light scattered out of the beam over a wide angle. The system has been used for general meteorological purposes, visibility, and to obtain mass loading in air pollution. It can be airborne.

Angular Scattering Meters. Many instruments have been designed to take advantage of scattering measurements to predict visibility [26, 28, 45]. They vary from near forward angles of 10° to 40° to near backscattering and backscattering at 180°. Ultraviolet is used in some instruments so they can be used in daylight. Forward-scatter instruments have the advantage of increased scattering at small angles and less index of refraction effect. Visible radiation is commonly used but some instruments use near infrared. For most purposes broad-band or white light has advantages over mono-chromatic light [60]. Detailed tests have been made on near-backscattering instruments which have demonstrated satisfactory reliability and cor-relation with estimates in visual observations [41, 65].

Pulsed Laser Systems. The advantages of high-power continuous or extremely high-power pulsed operation with narrow beams and mono-chromatic radiation make lasers an exceptional light source. The first applications were made with lidar (*l*ight *d*etection *a*nd *r*anging). Tests have been made with ruby lasers in pulsed systems to study visibility, and good correlations with visibility predictions from backscattering measurements have been obtained [5, 8, 27, 32, 64]. The beam can be sent in any direction, including a desired slant path at an airport. The scattering at two angles can be used to obtain the slant transparency [42]. For eye safety, lasers in the near infrared have advantages over the visible.

4.28.5 Instruments for Oceanographic Measurements

One of the oldest and simplest means of determining visibility is to lower a Secchi disk 30 cm in diameter into the water until it disappears, which is the depth of visibility. It is easy to use but great care is required for reliable measurements [61; 67, pp. 72–78]. Radiance distribution under water is important for image contrast, but it is difficult to measure. One technique is to use a 180° field-of-view lens and photographic film [57].

Transmissometers. Underwater transmissometers have been more diffi-cult to develop than the atmospheric counterparts because of the require-ments of water immersion and the different optical characteristics of the medium. An instrument should measure only the beam transmission for beam attenuation, in contrast to the diffuse attenuation, which is easier to determine [2; 19; 20; 35, p. 48; 39; 67, pp. 99–106].

Scatterometers. Though there are some scatterance meters which use a fixed angle such as 30°, 45°, or 90° to measure the scattering, most designs measure the light scattered by scanning through somewhat less than a half-circle, and hence could be called polar nephelometers. Integrating nephelometers have been designed. Since small-angle scattering is of especial importance in the sea, special instruments have been designed for these measurements [6; 7, pp. 107–108; 35, chap. 2; 62].

Absorption Meters. While it is difficult to measure absorption directly and easier to measure the attenuation and scattering to derive the absorption, absorptances of dissolved organic substances have been measured [35, p. 55].

Acknowledgment

Helpful suggestions have been given by Frederick C. Hochreiter, National Weather Service, Sterling, Virginia, John C. Ludwick, Institute of Oceanography, Old Dominion University, Norfolk, Virginia, and M. Patrick McCormick, NASA Langley Research Center, Hampton, Virginia.

References

1. *American National Standard Nomenclature and Definitions for Illuminating Engineering*, RP-16 (USAS z7.1–1969). Approved August 16, 1967 by USA Standards Institute, Illuminating Engineering Society, New York, 1969.
2. Austin, R. W., "Assessing Underwater Visibility," *Optical Spectra,* **4**, No. 5, 34, May, 1970.
3. Backer, M. C. C. de, "Videometry and Air Transport Safety," *Flight Safety* **3**, No. 4, 23–25, February, 1970.
4. Beck, R. H., "A Pilot Looks at Visibility," *Aerospace Safety,* **25**, No. 7, 4–7, July, 1969.
5. Bertolotti, M., L. Muzii, and D. Sette, "On the Possibility of Measuring Optical Visibility by Using a Ruby Laser," *Appl. Opt.* **8**, 117–120, 1969.
6. Blackwell, H. R., "Contrast Thresholds of the Human Eye," *J. Opt. Soc. Am.* **36**, 624–643, 1946.
7. Boileau, A. R., "Atmospheric Measurements in the Vicinity of Crater Lake, Oregon," *Appl. Opt.* **7**, 1907–1911, 2252–2258, 1968.
8. Brown, R. T., Jr., "Backscatter Signature Studies for Horizontal and Slant Range Visibility," Research Center, Sperry Rand Corp., Sudbury, Massachusetts, Final Report No. RD-67-24, Project No. 450-402-O1E, May, 1967. [AD 659 469, National Technical Information Service, Springfield, Virginia].
9. Busby, R. F., "Undersea Penetration by Ambient Light, and Visibility," *Science* **158**, 1178–1179, 1967.
10. Cap, S., "Vision in the Ocean," *Sperry Rand Eng. Rev.* **23**, No. 2, 2–10, 1970.
11. Charlson, R. J., N.C. Ahlquist, H. Selvidge, and P.B. MacCready, Jr., "Monitoring of Atmospheric Aerosol Parameters with the Integrating Nephelometer," *J. Air Poll. Control Assn.* **19**, 937–942, 1969.
12. Clarke, G. L., and E. J. Denton, "Light and Animal Life," in *The Sea,* Ed. M. N. Hill, Vol. 1, pp. 456–468, Interscience Publishers, New York, 1962.
13. Conner, W. D., and J. R. Hodkinson, "Optical Properties and Visual Effects of Smoke-Stack Plumes," Public Health Service Publication No. 999-AP-30, U. S. Department of Health, Education and Welfare, Public Health Service, Cincinnati, Ohio, 1967.
14. Crosby, P., and B. W. Koerber, "Scattering of Light in the Lower Atmosphere," *J. Opt. Soc. Am.* **53**, 358–361, 1963.
15. Curcio, J.A., "Evaluation of Atmospheric Aerosol Particle Distribution from Scattering Measurements in the Visible and Infrared," *J. Opt. Soc. Am.* **51**, 548–551, 1961.

16. Donati, S., and A. Sona, "Further Results on Range Gating Technique: Visibility in Sea Water," *Optoelectronics* **1**, No. 3, 155–159, August, 1969.
17. Donati, S., and A. Sona, "Optical Range Gating to Extend Visibility in the Fog," *Alta Frequencia* **39**, 202, 1970.
18. Duntley, S.Q., "Underwater Visibility," in *The Sea*, Ed. M. N. Hill, Vol. 1, pp. 452–455, Interscience Publishers, New York, 1962.
19. Duntley, S. Q., "Light in the Sea," *J. Opt. Soc. Am.* **53**, 214–233, 1963.
20. Duntley, S. Q., "Visibility in the Oceans," *Optical Spectra*, pp. 64–69, Fourth Quarter, 1967.
21. Duntley, S.Q., J.I. Gordon, J.H. Taylor, C.T. White, A.R. Boileau, J.E. Tyler, R. W. Austin, and J. L. Harris, "Visibility," *Appl. Opt.* **3**, 549–598, 1964.
22. Elterman, L., "Vertical-Attenuation Model with Eight Surface Meteorological Ranges 2 to 13 Kilometers," Air Force Cambridge Research Laboratories, Bedford, Massachusetts, Environmental Research Paper No. 318, AFCRL 70-0200, March, 1970.
23. Fenn, R. W., "Correlation between Atmospheric Backscattering and Meteorological Visual Range," *Appl. Opt.* **5**, 293–295, 1966.
24. Fischer, W. H., "Atmospheric Aerosol Background Level," *Science* **171**, 828–829, 1971.
25. Germogenova, O. A., J. P. Friend, and A. M. Sacco, "Atmospheric Haze: A Review," Bolt, Beranek and Newman, Inc., Cambridge, Massachusetts, Report No. 1821, 31 March 1970 [PB 192 102 National Technical Information Service, Springfield, Virginia].
26. Grimes, Annie E., "An Annotated Bibliography on Methods of Visibility Measurement, 1950–1969," ESSA, Atmospheric Sciences Library, ASL-2, Silver Spring, Maryland, August, 1969 [PB 188 652 National Technical Information Service, Springfield, Virginia].
27. Hill, E.T., "Lasers for Measuring Slant Visual Range," NATO Advisory Group for Aerospace Research and Development, Conference Proceedings for Norway meeting, Sept. 24–Oct. 3, 1969, pp. 35-1 to 10.
28. Hochreiter, F.C., "Analysis of Visibility Observation Methods," ESSA Technical Memorandum WBTM T and EL 9, System Development Office, Test and Evaluation Laboratory, Sterling, Virginia, October, 1969 [PB 188 327 National Technical Information Service, Springfield, Virginia].
29. Hood, J. M., Jr., "A Two-Cavity Long-Base Mode Meteorological Range Meter," *Appl. Opt.* **3**, 603–608, 1964.
30. Horman, M. H., "Measurement of Atmospheric Transmissivity Using Back-Scattered Light from a Pulsed Light Beam," *J. Opt. Soc. Am.* **51**, 681–691, 1961.
31. Huschke, R.E., Ed., *Glossary of Meteorology*, American Meteorological Society, Boston, Massachusetts, 1959.
32. Inaba, H., T. Kobayashi, T. Ichimura, M. Morihisa, and K. Taira, "Measurement of Propagation Characteristics of Optical Beams in the Atmosphere and Analysis of Returned Echo Signals Using a Laser Radar System with A-Scope Representation," *Elect. Comm. Japan* **51B**, 45–52, 1968.
33. Irvine, W. M., and F. W. Peterson, "Observations of Atmospheric Extinction from 0.315 to 1.06 Microns," *J. Atmos. Sci.* **27**, 62–69, 1970.

34. Ito, H., "Time and Space Variation in Meteorological Elements in the Aerodrome and Its Vicinity," *Aeronautical Meteorology,* pp. 142–157, WMO No. 227 TP121, Geneva, Switzerland, 1969.

35. Jerlov, N.G., *Optical Oceanography,* Elsevier, New York, 1968.

36. Jones, R.F., "Time and Space Variations of Visibility and Low Cloud within the Approach Control Area," *Aeronautical Meteorology,* pp. 97–101, WMO No. 227 TP 121, Geneva, Switzerland, 1969.

37. Junge, C.E., *Air Chemistry and Radioactivity,* p. 142, Academic Press, New York, 1963.

38. Krumreich, P., "Das Problem der Sicht bei der Schlechtwetterlandung in der Civilluftfahrt," Arbeits- und Forschungsgemeinschaft Graf Zeppelin, Stuttgart (1969), 75p. [N70-10793 National Technical Information Service, Springfield, Virginia].

39. Lankes, L.R., "Optics and the Physical Parameters of the Sea," *Optical Spectra,* pp. 42–49, May, 1970.

40. List, R.J., *Smithsonian Meteorological Tables,* 6th ed., revised, pp. 452–478, Smithsonian Institution, Washington, D.C., 1963.

41. Lomer, L. R., "Fog Detectors for Unmanned Aids to Navigation," United States Coast Guard Field Testing and Development Center, Report No. 512 (July, 1970) [AD 710 086 National Technical Information Service, Springfield, Virginia].

42. Markelov, V.A., "A Means of Measuring Inclined Transparency with a Laser," Scientific Research Institute of Hydrometeorological Instrument Manufacture, Leningrad, Informative Materials on Hydrometeorological Instruments and Observation Methods, No. 18 (1968). Translation: FTD-HT-23-337-69, pp. 87–106 [AD 704 033 National Technical Information Service, Springfield, Virginia].

43. Marquedant, R., and H. Hodara, "Advanced Optical Devices for DSSP," *Marine Sci. Instr.* **4,** 241–251, 1968 (Plenum Press, New York).

44. McClatchey, R.A., R.W. Fenn, J.E.A. Selby, J.S. Garing, and F.E. Volz, "Optical Properties of the Atmosphere," Air Force Cambridge Research Laboratories, Bedford, Massachusetts, Environmental Research Paper, No. 331, AFCRL 70-0527, 22 September 1970.

45. Middleton, W.E.K., *Vision Through the Atmosphere,* University of Toronto Press, Toronto, 1952.

46. Middleton, W.E.K., "Vision Through the Atmosphere," in *Handbuch der Physik,* Ed. S. Flügge, Springer-Verlag, Berlin, **48,** 254–287, 1957.

47. Nicodemus, F.E., "Optical Resource Letter on Radiometry," *J. Opt. Soc. Am.* **59,** 243–248, 1969.

48. Noll, K.E., P.K. Mueller, and M. Imaga, "Visibility and Aerosol Concentration in Urban Air," *Atm. Envir.* **2,** 465–476, 1968.

49. O'Brien, H.W., "Visibility and Light Attenuation in Falling Snow," *J. Appl. Meteor,* **9,** 671–683, 1970.

50. Pilat, M.J., and D.S. Ensor, "Plume Opacity and Particulate Mass Concentration," *Atm. Envir.* **4,** 163–173, 1970.

51. Pueschel, R.F., and K.E. Noll, "Visibility and Aerosol Size Distribution," *J. Appl. Meteor,* **6,** 1045–1052, 1967.

52. Rae, J.B., and J.A. Garland, "Stabilized Integrating Nephelometer for Visibility Studies," *Atmos. Envir.* **4**, 219–223, 1970.
53. Rossler, J., "Transmissionsmessungen mit Laserlicht (6328 Å) am Meteorologischen Observatorium Aachen," *Meteor. Rundschau* **21**, 26–28, 1968.
54. Rozenberg, G.V., "Optical Investigations of Atmospheric Aerosol," *Sov. Phys. Uspekhi* **11**, 354–380, 1968.
55. Ruppersberg, G.H., "Registrierung der Sichtweite mit dem Streulichtsschreiber," *Beitr. Physik Atm.* **37**, 252–263, 1964.
56. Schlatter, E.E., "Evaluation of Several Multi-Transmissometer Systems," ESSA, Weather Bureau, Systems Development Office, Test and Evaluation Laboratory, Atlantic City, New Jersey, SRDS Report No. RD-67-2, December 1967 [N68-22228 National Technical Information Service, Springfield, Virginia].
57. Smith, R.C., R.W. Austin, and J.E. Tyler, "An Oceanographic Radiance Distribution Camera System," *Appl. Opt.* **9**, 2015–2022, 1970.
58. Spiess, F.N., in *Hydronautics,* Eds. H.E. Sheets and V.T. Boatright, Jr., p. 107, Academic Press, New York, 1970.
59. Tverskoi, P.N., *Physics of the Atmosphere,* p. 465, Israel Program for Scientific Translations, Jerusalem, 1965 [NASA TT F-288, TT 65-50114 National Technical Information Service, Springfield, Virginia].
60. Twomey, S., and H.B. Howell, "The Relative Merits of White and Monochromatic Light for the Determination of Visibility by Back-Scattering Measurements," *Appl. Opt.* **4**, 501–505, 1965.
61. Tyler, J.E., "The Secchi Disc," *Limnology and Oceanography* **13**, 1–6, 1968.
62. Tyler, J.E., and R.W. Austin, "A Scattering Meter for Deep Water," *Appl. Opt.* **3**, 613–620, 1964.
63. Tyler, J.E., and R.W. Preisendorfer, "IV Transmission of Energy within the Sea, Chapter 8, Light," in *The Sea,* Ed. M.N. Hill, Vol. 1, Interscience Publishers, New York, 1962.
64. Viezee, W., E.E. Uthe, and R.T.H. Collis, "Lidar Observations of Airfield Approach Conditions: An Exploratory Study," *J. Appl. Meteor.* **8**, 274–283, 1969.
65. Vogt, H., "Visibility Measurement Using Backscattered Light," *J. Atmos. Sci.* **25**, 912–918, 1968.
66. Volz, F., "Photometer mit Selen-Photoelement zur spektralen Messung der Sonnenstrahlung und zur Bestimmung der Wellenlängenabhängigkeit der Dunsttrübung," *Arch. Met. Geophy. Biokl.* **B10**, 100–131, 1959.
67. Williams, J., *Optical Properties of the Sea,* United States Naval Institute, Annapolis, 1970.
68. Zuev, V.E., *Atmospheric Transparency for Visible and Infrared Radiation,* Izdatel'stvo "Sovetskoe Radio," Moscow, 1966.
69. Zuev, V.E., *Propagation of Visible and Infrared Waves in the Atmosphere,* Izdatel'stvo "Sovetskoe Radio," Moscow, 1969.

P. G. Brewer, R. A. Horne, and T. R. S. Wilson

4.29 WATER CONSTITUENT SENSORS

Water has a central role in the interactions between atmosphere, lithosphere, hydrosphere, and biosphere. The two most important chemical characteristics of natural waters are salinity and pH, which are discussed in Sections 4.21 and 4.17.

4.29.1 Chemical Composition of Natural Waters

Sea Water. Sea water is roughly a 0.5M NaCl solution, about 0.05M in $MgSO_4$, and contains at least a trace of almost everything else. The major sea water constituents are given in Table 4.29-1. The amounts of suspended inorganic particulate material and suspended and/or dissolved organic material are always relatively negligible. The total suspended particulate matter in sea water is rarely greater than 2 mg/kg and of this the organic portion can be as great as 60% but is usually less than 30%. The dissolved organic material is usually equivalent to less than 2 mg (of carbon/kg [26]).

Table 4.29-1. The Major Chemical Constituents of Sea Water (g/kg)

Salinity (%)	Na^+	Mg^{++}	Ca^{++}	K^+	Cl^-	$SO_4^=$	HCO_3^-
10	3.074	0.370	0.118	0.111	5.530	0.775	0.041
20	6.148	0.739	0.236	0.221	11.059	1.550	0.081
30	9.222	1.109	0.354	0.332	16.589	2.324	0.122
35	10.759	1.294	0.413	0.387	19.354	2.712	0.142
40	12.296	1.479	0.472	0.442	22.118	3.094	0.162

The oceans are well if slowly mixed; thus the ratios of the major constituents remain constant (Marcet's principle, 1819) even though the total salt concentration or salinity is variable. The slowness of the mixing processes (mean water resident times range from 10 to 1000 yr [10]) means that it is relatively easy to produce serious local pollution problems.

One important class of inorganic solutes in sea water are an exception to Marcet's principle, the so-called "nutrients," including phosphate, silicate, nitrite, nitrate, and ammonia. These substances are involved in the life cycle of organisms and they are subject to perturbations depending on local biological activity [14].

Another important class of inorganic solutes in sea water is the dissolved gases: carbon dioxide (CO_2), nitrogen (N_2), oxygen (O_2), and the noble gases. The quantities are very small (less than 40 ml (STP)/l) but highly significant. CO_2 dissolves to become involved with the carbonate equilibria in the marine environment, while O_2, a product of photosynthesis, is strongly dependent on local biological activity.

The chemistry of sea water is largely controlled by four factors: the ionic strength, the concentration of the complexing chloride ion, the acidity expressed as pH (which is in turn controlled by carbonate and silicate equilibria), and the concentration of dissolved oxygen.

Fresh Waters. The dissolved salts of terrestrial stream and lake waters are derived from three sources. Weathering of rocks in the catchment area constitutes the largest natural contribution. The total dissolved salt content from this source varies with the rock type, being generally below 50 parts per million (ppm) for igneous and metamorphic rocks and up to 200 ppm for sedimentary types. A subsidiary natural source of dissolved salts is provided by geochemical recycling of sea salt via the atmosphere.

Other dissolved material in fresh water streams is introduced by the activities of man. Agricultural, industrial, and municipal discharges add both organic and inorganic material. As long as the capacity of the stream to absorb this material is not exceeded, most of the organic compounds are quickly broken down, mainly by bacterial degradation [52]. The inorganic material, however, remains unchanged [22].

In general, levels of dissolved salt in the range over 200 ppm are due to human activities. Such activities supply chloride, sodium, and sulphate ions, while natural run-off contributes most of the carbonate and alkaline earths [16, 22]. The levels of other inorganic ions are usually very low (< 1 ppm) in unpolluted waters. An unpolluted stream in a state of natural equilibrium contains typically only about 10 ppm of organic material, mainly refractory material such as humic acids [37, 44, 45]. These compounds may complex inorganic cations [45].

"Nutrient salts," such as phosphate, nitrate, and silicate, occur at low levels in natural run-off. They are also added directly by human activity, and can be produced by the bacterial degradation of dissolved and particulate organic matter. They are vital to the well-being of the plant community in a stream. Excess nutrients unbalance the natural system and can lead to unnaturally high production of organic material by plants and algae. An unpolluted stream is normally in equilibrium with the atmospheric gases, of which oxygen is of the most consequence. This oxygen is consumed by the respiration of living organisms in the stream, and in a balanced situation is replaced by the oxygen produced by photosynthesizing plants. In an unbalanced situation excessive respiration, usually by heterotrophic bacteria engaged in the breakdown of organic material, causes the dissolved oxygen level to drop to zero. In this anoxic situation only specialized bacteria can survive; their metabolic processes are powered in part by the reduction of sulphate to sulphide providing the characteristic unpleasant hydrogen sulphide odor.

The meeting of river and sea is an ecologically important area. The biological productivity is high because of the continual supply of nutrients carried by the inflowing river [51]. The estuarine environment influences the productivity of surrounding offshore waters, since spawning and juvenile stages of deep water species often occur inshore. The mixing of salt and fresh water can occur in several different modes [39].

The large variation in total dissolved salts between different rivers, and with time in any given river, requires that a sensor have a wide dynamic range. The possible variations in the matrix are also large, especially in estuaries, and a sensor must be very specific for the desired species in order to reject these variations. There are other factors of importance, notably reliability, power requirements and operating costs.

Table 4.29-2 shows the composition of natural waters.

4.29.2   pE or E_h Sensors

Just as acid-base equilibria in natural water systems are controlled by hydrogen ion activity, oxidation-reduction reactions are controlled by electron activity, and it is convenient to treat the electron as a reagent, analagous to the proton, and to define a quantity, pE:

$$pE \equiv - \log[e^-], \qquad (4.29-1)$$

Table. 4.29-2.   Composition of Natural Waters from Various Sources (Concentrations in ppm)

	Unpolluted			Marine Aerosol Influence Mean of 10 Lakes Nova Scotia	Saline Lake Water Bad Water Death Valley California	Mean Worlds' Rivers	Ocean Salinity [40]
	Igneous rock Saguenay R. Quebec	Sedimentary Rock Couchichingl Ontario	Polluted Cuyahoga R. Ohio				
HCO_3^-	6.1	126.1	120	0.0	187	58.4	142
SO_4^{2-}	3.1	16.8	82	6.2	4,960	11.2	2,712
Cl^-	0.5	3.4	32	7.5	21,400	7.8	19,354
F^-	—	—	0.2	—	4.9	—	1.3
NO_3^-	0.6	0.5	4.3	—	—	1	1
Ca^{2+}	3.6	36	54	1.1	1,230	15	413
Mg^{2+}	1.5	4.8	12	4.7	148	4.1	1,294
Na^+	—	—	21	5.0	14,100	6.3	10,759
K^+	—	—	3.4	4.2	594	2.3	387
Fe	0.2	—	0.6	—	—	0.67	—
SiO_2	3.0	4.8	6.7	—	49	13.1	0–4
Total	18	198	336	29	42,700	120	35,000

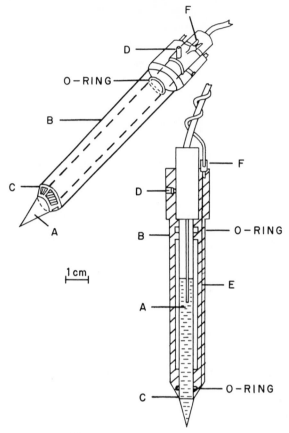

Figure 4.29-1 Compound electrode for measuring E_h. A, spear type glass electrode; B, perspex holder; C, metal foils (Pt or Au for E_h measurements); D, securing screw; E, lead wire; F, connector.

analagous to the pH [46]

$$pH \equiv - \log [H^+], \qquad (4.29\text{-}2)$$

where the bracketed terms represent the activities of electrons and protons, respectively. Inert metal electrodes (usually platinum) are used in conjunction with a reference electrode (Figure 4.29-1). After the observed cell emf has been corrected for liquid junction potentials, the redox potential, E_h is related to pE by the expression

$$pE = \frac{E_h}{R\,Tn\,F - 1 \ln 10} + \log K_{ref}, \qquad (4.29\text{-}3)$$

where R is the gas constant, T is absolute temperature, F is the Faraday constant and K_{ref} is the reference electrode constant. If a scale is chosen based on the standard hydrogen reference electrode

$$H^+ + e^- \rightleftharpoons \tfrac{1}{2} H_2(g), \qquad \log K \equiv 0, \qquad (4.29\text{–}4)$$

then Eq. 4.29–3 becomes

$$pE = \frac{E_h}{RTnF - 1\ln 10}. \qquad (4.29\text{–}5)$$

At 25°C the value of $(RTnF - 1\ln 10)$ is 59.155 mV. Thus, inserting log $P_o^2 = -0.68$, $\log[H^+] = 8.1$, and $\log[H_2O] = -0.01$ into the equilibrium constant expression

$$\tfrac{1}{2}O_2(g) + 2H^+ + 2e^- \rightleftharpoons H_2O(l), \quad \log = 41.55, \qquad (4.29\text{–}6)$$

the pE for sea water in equilibrium with air at 25°C is 12.5 [46].

The measurement of E_h and pE is difficult [4, 53]. Some of the principal sources of uncertainty are shown in Figure 4.29-2. The E_h and pH character-

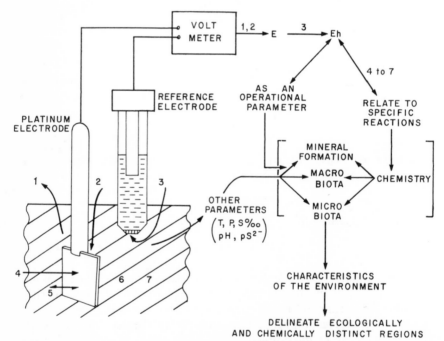

Figure 4.29-2 Problems associated with E_h measurement [53]. 1, release of gases (e.g., H_2S); 2, introduction of air; 3, liquid junction effects: (*a*) suspension effect, (*b*) sulfide content; 4, direct attack on Pt; 5, trace component controlling potential; 6, whole system out of equilibrium; 7, microenvironments can be important.

istics of sea water are shown in Figure 4.29-3 and those of other natural water and biological systems are shown in Figure 4.29-4. Oxidation-reduction potential probes also find use in monitoring industrial chemical processes [29].

4.29.3 Dissolved Oxygen Sensors

With the exception of certain species of bacteria, every living thing requires oxygen for the maintenance of its life processes. Oxygen in water is traditionally determined by the Winkler method [55] in one of its several modifications [11, 38]. This method is not easily automated. More successful approaches for continuous and remote measurements use polarographic methods. The simple dropping mercury electrode [9, 48] is subject to chemical interference from easily reduceable ions and is not mechanically reliable. Membrane electrodes have been much more widely applied. These were first used to minitor blood oxygen levels [15] and later adapted for environmental work [12, 35]. A typical design is shown in Figure 4.29-5. A potential of 0.6 V is applied across the electrode, which quickly becomes polarized. The current is governed by the supply of oxygen through the membrane to the interior of the electrode. This is a function of the external dissolved oxygen concentration and ambient temperature. An integral thermistor temperature sensor is used to minimize errors caused by temperature changes. The membrane is subject to long-term drift and external fouling.

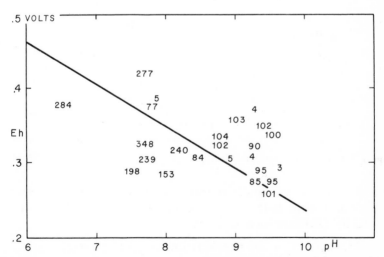

Figure 4.29-3 Influence of oxygen saturation on the E_h of sea water at different pH values [4]. The straight line corresponds to $E_h = 0.810 - 0.058$ pH. The numbers are percent oxygen saturation.

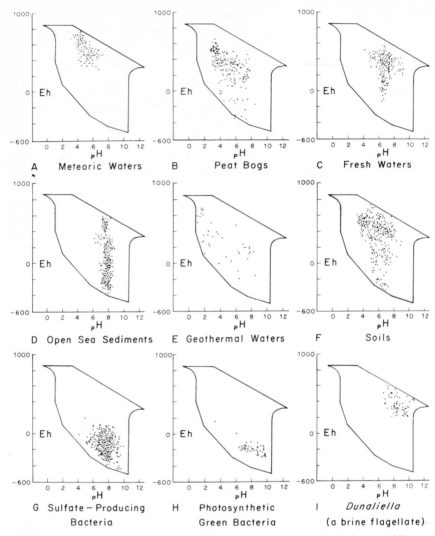

Figure 4.29-4 E_h -pH characteristics of various natural water and biological systems [4].

Another method for the determination of dissolved oxygen relies on the behavior of thallium in the presence of oxygen and water. The thallium is oxidized by a reaction such as

$$4\,Tl + O_2 + 2H_2O \rightleftharpoons 4\,Tl\,(OH) \rightleftharpoons 4\,Tl^+ + 4\,OH^-. \quad (4.29\text{--}7)$$

In a flowing system the thallous ion concentration in the electrode region is proportional to the dissolved oxygen level. The potential developed by the

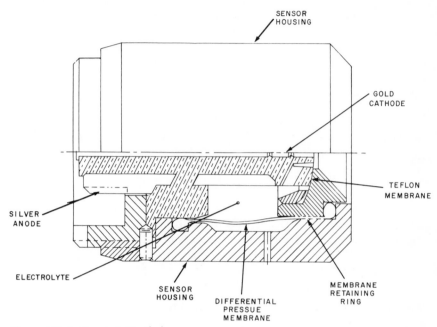

SENSOR
HOUSING

GOLD
CATHODE

TEFLON
MEMBRANE

SILVER
ANODE

ELECTROLYTE

SENSOR
HOUSING

DIFFERENTIAL
PRESSUE
MEMBRANE

MEMBRANE
RETAINING
RING

Figure 4.29-5 Practical Clark-type electrode for in situ oxygen measurements to 3000 m depths. Courtesy of Beckman Instruments, Inc.

cell is logarithmically dependent on thallous ion concentration close to the thallium surface. Temperature compensation is applied in a manner similar to that used for the membrane-type electrodes. The advantages of this sensor over the membrane designs include freedom from fouling (a fresh thallium surface is continually exposed and the thallous ion is very toxic) and a lower long-term drift. It is not, however, suited for marine applications, since at low temperatures (< 10°C) in sea water the thallium surface "passivates."

4.29.4 Ion Selective Electrodes

True sensors for monitoring dissolved ions are electrodes whose response varies with the parameter of interest. Such devices obey the Nernst equation

$$\Delta E = E_0 + \frac{RT}{nF} \Delta \ln a \qquad (4.29\text{--}8)$$

where ΔE is the output voltage change, a is the activity of the species considered, R, T, and F are the gas constant, absolute temperature, and Faraday, constant, respectively, and n is the ionic charge. E_0 is a constant of the electrode system. If the output responds to changes in activity of a single species, the device is a specific ion electrode.

The most widely used electrode of this type is the glass-membrane pH electrode which is highly selective for hydrogen ions [21]. Other available electrodes are specific for a wide variety of ions [19] (Table 4.29-3). All these electrodes utilize a membrane to separate the test solution from an inner reference solution in which the activity of the ion of interest is constant.

Non-ideal behavior of the membrane is a source of error. No membrane yet devised is truly selective for only one ion. Consequently the output of the real system can be described by

$$\Delta E = E_0 + \frac{RT}{nF} \Delta \ln a + \sum_i k_i b_i^{n_a/n_b}, \qquad (4.29\text{-}9)$$

where k_i and b_i are the selectivity ratio and the activity of the ith interfering ion. The evaluation of k_i is difficult since it is not a true constant but varies with the activities of the interfering species, which are themselves subject to changes.

Table 4.29-3. Commercially Available Specific Ion Electrode Characteristics

Ion	Membrane	Concentration Range (M)	pH Range	Interfering Ions
		Solid State Electrodes		
F^-, La^{+++}	LaF_3	$10^0 - 10^{-6}$	0–8.5	OH^-
Cl^-	$AgCl/Ag_2S$	$10^0 - 5 \times 10^{-5}$	0–14	Br^-, I^-, $S^=$, NH_3, CN^-
Br^-	$AgBr/Ag_2S$	$10^0 - 5 \times 10^{-6}$	0–14	I^-, $S^=$, NH_3, CN^-
I^-	AgI/Ag_2S	$10^0 - 5 \times 10^{-8}$	0–14	$S^=$, CN^-
SCN^-	$Ag/SCN/Ag_2S$	$10^0 - 5 \times 10^{-5}$	0–14	Br^-, I^-, $S^=$, NH_3, CN^-
$S^=$, Ag^+	Ag_2S	$10^0 - 10^{-7}$	0–14	Hg^{++}
CN^-	AgI, Ag_2S	$10^{-2} - 10^{-6}$	0–14	I^-, $S^=$
Cu^{++}	CuS/Ag_2S	$10^0 - 10^{-8}$	0–14	Hg^{++}, Ag^+
Pb^{++}	PbS/Ag_2S	$10^0 - 10^{-7}$	2–14	Hg^{++}, Ag^+, Cu^{++}
Cd^{++}	CdS/Ag_2S	$10^0 - 10^{-7}$	1–14	Hg^{++}, Ag^+, Cu^{++}
		Liquid-State Electrodes		
Ca^{++}	$(RO)_2PO_2^-$	$10^0 - 10^{-5}$	5.5–12	Zn^{++}, Fe^{++}, Pb^{++}, Mg^{++}, Ba^{++}
Cl^-	NR_4	$10^{-1} - 10^{-5}$	2–11	I^-, NO_3^-, Br^-, HCO_3^-, $SO_4^=$, F^-
BF_4^-	$NiL_3(NO_3)_2{}^a$	$10^{-1} - 10^{-5}$	2–12	I^-, NO_3^-, Br^-, OAc^-, HCO_3^-, Cl^-
NO_3^-	$NiL_3(NO_3)_2{}^a$	$10^{-1} - 10^{-5}$	2–12	I^-, Br^-, NO_2^-, Cl^-, $CO_3^=$, ClO_4^-, F^-
ClO_4^-	$FeL_3(NO_3)_2{}^a$	$10^{-1} - 10^{-5}$	4–11	I^-, NO_3^-, Br^-, F^-
K^+		$10^0 - 10^{-5}$	1–12	Cs^+, NO_3^-
M^{++b}	$(RO)_2PO_2^-$	$10^0 - 10^{-8}$	5.5–12	Zn^{++}, Fe^{++}, Cu^{++}, Ni^{++}, Ba^{++}, Na^+

^{a} L represents a substituted phenanthroline ligand.
^{b} Divalent cation electrode (Ca^{++} + Mg^{++}).

Since all these electrodes obey the Nernst equation, the output voltage corresponding to a 10-fold change in activity of the sample ion is approximately 59.2 mV for a monovalent ion and approximately 29.6 mV for a divalent ion at 25°C. Although the impedance of the electrode itself can be $10^7 \, \Omega$ [5], the current is also limited by the chemical changes induced in the half-cell by ion transport. Thus the output must be measured using a high impedance ($10^8 - 10^{12} \, \Omega$) device, especially if long service life is required.

Electrode membranes [41] are either liquid or solid. Liquid membranes consist of low dielectric constant, high molecular weight fluids in which the specific charge carriers are dissolved. The fluid is supported by an inert solid phase to provide mechanical rigidity. The charge carriers are large organic molecules which specifically bond to the ion of interest. Ideally all other ions are excluded from the membrane by the low polarity of the solvent. Solid-state membrane electrodes depend on ionic conduction through a crystal lattice. These are semiconductors in which conduction of the ion through the crystal phase proceeds by a lattice defect mechanism. Since movement through the lattice is restricted to ions of specific size and charge, these devices are highly selective. If the activity of the mobile ion in the test solution is controlled by solubility equilibrium with another ionic species, then the electrode responds to changes in activity of this species.

The mechanical properties of the solid-state membranes are superior. Liquid membranes have a much greater susceptibility to mechanical and thermal shock. In addition, the liquid membrane solvent has a low but finite solubility and must be replenished at intervals of less than one month.

The specific ion electrode and reference electrode are immersed in the sample solution and the voltage of the cell formed by the electrode pair is measured [25]. Calibration is carried out by comparison with standard solutions at the same temperature. Chemical analysis gives the concentration of a species; electrode systems give a measure of the activity. The two are related by the equation

$$a = \gamma C, \tag{4.29-10}$$

where C is the concentration in moles, γ is the ionic activity coefficient, and a is the activity. The activity coefficient is controlled by both the total ionic strength of the medium and by specific interactions between ions. Thus, for the estimate of concentration it is important that the ionic composition of the standard solution matches that of the sample solution as closely as possible [23].

Other procedures which have been applied to analysis using specific ion electrodes are capable of higher precision than the simple potentiometric approach described above. However, these procedures which include the method of standard additions and the use of electrodes as titration endpoint indicators [50] are usually applied to the analysis of discrete samples rather than to continuous monitoring.

4.29.5 Automatic Analysis

Automatic analyses can be divided into two types: those in which each sample is manipulated in its own container during serial addition of reagents, mixing, and the like, and those in which the operations are carried out during continuous flow. An example of the latter consists of a rotating sample table which presents discrete samples sequentially to a peristaltic pump system. The pump induces flow through flexible tubes of varying diameter. The differing volumes of sample and reagent thus pumped are combined and the discrete character of the sample is maintained by the admittance of air bubbles. These bubbles prevent smearing of the sample along the walls of the tube and loss of resolution. The sample and reagents are mixed in helical coils, and colorimetric determinations are made in the continuous flow cell of an interference filter photometer. The precision obtained from this procedure is approximately that of careful manual analysis [8].

Phosphate has been determined by reduction of phosphomolybdate to a molybdenum blue, either by an ascorbic acid [13], or by stannuous chloride [2]. Nitrate has been determined by reduction to nitrite by a heterogeneous reductor such as cadmium or zinc in the presence of various buffers [2, 6], and nitrite is determined by the well known Bendschneider and Robinson method to form a red azo dye. Silicate has been determined by molybdenum blue methods using both stannous chloride [2] and a p-methylaminophenol sulphate/sodium sulphite reagent [8].

For the automatic determination of dissolved organic carbon, inorganic carbon is removed from the sample by acidification and stripping with nitrogen gas. The sample is mixed with potassium peroxidisulphate and irradiated with UV light (900 W, 252 nm) in a quartz coil to oxidize dissolved organic carbon to carbon dioxide. The carbon dioxide is stripped from the liquid phase, purified, and absorbed in 0.01-N sodium hydroxide which flows through a conductivity cell. The monitored output of the conductivity cell is proportional to the amount of carbon dioxide dissolved. The method is very sensitive and as little as 100 μg of dissolved organic carbon per liter can readily be detected [20].

4.29.6 Carbon And The Carbonate System

The carbonate system provides pH control in natural waters by means of the following series of interrelated equilibria:

$$CO_{2(g)} \rightleftharpoons CO_{2(diss)}, \qquad (4.29\text{--}11)$$

$$CO_2 + H_2O \rightleftharpoons H_2CO_3, \qquad (4.29\text{--}12)$$

$$H_2CO_3 \rightleftharpoons H^+ + HCO_3^-, \qquad (4.29\text{--}13)$$

$$HCO_3^- \rightleftharpoons H^+ + CO_3^=, \qquad (4.29\text{--}14)$$

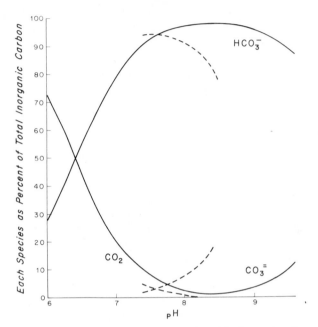

Figure 4.29-6 Distribution of various carbonate species with change in pH at 16°C [43]. Solid line, fresh water; dashed line, 19 ppt chlorinity sea water.

Reactions 4.29–11 and 4.29–12 are slow to reach equilibrium, while 4.29–13 and 4.29–14 are virtually instantaneous. The thermodynamic relations within these species, and between them and other ions present in natural waters, are complex [26, 47]. The theoretical relationships between pH and various carbonate species in sea water are summarized graphically in Figure 4.29–6. In addition, pH is discussed in Section 4.17.

The partial pressure of carbon dioxide in equilibrium with a water sample (P_{CO_2}) can be determined continuously by means of a flow system in which the water is scrubbed by a counter-current of CO_2-free gas. The gas flow is dried and passed to the infrared CO_2 monitor. For sea water the relationship between the change in total CO_2 in a sea water sample and the resultant change in the partial pressure of CO_2 in equilibrium with the sample at constant carbonate alkalinity is given by [47]

$$\frac{\delta P_{CO_2}}{P_{CO_2}} = 12.5 \frac{\delta \sum CO_2}{\sum CO_2}. \qquad (4.29-15)$$

Precise monitoring of the change in P_{CO_2} is a very sensitive method of estimating changes in total CO_2. Natural waters are monitored for total organic carbon (TOC), or chemical oxygen demand (COD) by systems which rely on the oxidation of organic carbon, either by wet [27] or dry [1] methods, and the subsequent determination of the CO_2 produced. This latter can be performed by electrochemical methods or by infrared measurement in the gas phase. At levels below that measured by total organic carbon monitors only specialized and highly sensitive methods, usually based on gas liquid chromatography (GLC), suffice.

References

1. Anon., "Continuous Total Organic Carbon Monitor," Automated Environmental Systems, Inc., Application Data Sheet No. B-101, 1969.
2. Armstrong, F. A. J., and E. C. LaFond, "Chemical Nutrient Concentrations and Their Relationship to Internal Waves and Turbidity off Southern California," *Limnol. Oceanogr.* **11**, No. 4, 583–547, 1966.
3. Armstrong, F. A. J., C. R. Stearns, and J. D. H. Strickland, "The Measurement of Upwelling and Subsequent Biological Process by Means of the Technicon Autoanalyzer and Associated Equipment," *Deep-Sea Res.* **14**, 381–389, 1967.
4. Baas-Becking, L. G. M., I. R. Kaplan, and D. Moore "Limits of the Natural Environment in Terms of pH and Oxidation-Reduction Potential," *J. Geol.* **68**, 243–284, 1960.
5. Brand, M. J. D., and G. A. Rechnitz, "Novel Impedance Measurements on Ion Selective Electrodes," *Anal. Chem.* **41**, 1185, 1969.
6. Brewer, P. G., and J. P. Riley, "The Automatic Determination of Nitrate in Sea Water," *Deep-Sea Res.* **12**, 765–772, 1965.
7. Brewer, P. G., and J. P. Riley, "The Automatic Determination of Silicate-Silicon in Natural Water with Special Reference to Sea Water," *Anal. Chim. Acta.* **35**, 514–519, 1966.
8. Brewer, P. G., and J. P. Riley, "A Study of Some Manual and Automatic Procedures for the Determination of Nitrate and Silicate in Ocean Water," *Deep-Sea Res.* **14**, 475–477, 1967.
9. Briggs, R. and G. Knowles, "Developments in the Use of the Wide-base D.M.E. for Determining Dissolved Oxygen and Oxygen Gases," *Analyst* **86**, 604–608, 1961.
10. Broecker, W. S., R. D. Gerard, M. Ewing, and B. C. Heezen, "Geochemistry and Physics of Ocean Circulation," in *Oceanography,* Ed. M. Sears, Pub. No. 67, Amer. Assoc. Adv. Sci., Washington, D.C., 1961.
11. Carpenter, J. H., "New Measurements of Oxygen Solubility in Pure and Natural Water," *Limnol. Oceanogr.* **11**, 264–277, 1966.
12. Carritt, D. E., and J. W. Kanwisher, "An Electrode System for Measuring Dissolved Oxygen," *Anal. Chem.* **31**, 5–9, 1959.
13. Chan, K. M., and J. P. Riley, "The Automatic Determination of Phosphate in Sea Water," *Deep-Sea Res.* **13**, 467–471, 1966b.
14. Chow, T. J., and A. W. Mantyla, "Inorganic Nutrient Anions in Deep Ocean Waters," *Nature,* **206**, 383–385, 1965.
15. Clark, L. C., R. Wold, D. Granger, and Z. Taylor, "Continuous Recording of Blood Oxygen Tensions by Polarography," *J. Appl. Physiol.* **6**, 189–193, 1953.

16. Conway, E. J., "Mean Geochemical Data in Relation to Oceanic Evolution," *Proc. Roy. Inst. Acad.* **48B,** 119–159, 1943.
17. Cowgen, P. A., V. E. Lyons, G. D. Barns, and F. G. Hall, "Conductivity Measurements Monitor Waste Streams," *Environ. Sci. Technol.* **4,** 116–121, 1970.
18. Crosby, N. T., A. L. Dennis, and J. G. Stevens, "An Evaluation of Some Methods for the Determination of Fluoride in Potable Waters and Other Aqueous Solutions," *Analyst* **93,** 643, 1968.
19. Durst, R. A., Ed., "Ion Selective Electrodes," National Bureau of Standards Special Publication 314, Washington, D.C., 1969.
20. Ehrhardt, M., "A New Method for the Automatic Measurement of Dissolved Organic Carbon in Sea Water," *Deep-Sea Res.* **16,** 392–397, 1969.
21. Eisenman, A., *Glass Electrodes for Hydrogen and Other Cations: Principles and Practice,* Marcel Dekker, New York, 1967.
22. Evans, R. L., "Addition of Common Ions from Domestic Use of Water," *J. Am. Water Works Assn.* **60,** 315–320, 1968.
23. Grant, M. S., and J. W. Ross, "The Use of a Total Ionic Strength Activity Buffer for Electrode Determination of Fluoride in Water Supplies," *Anal. Chem.* **40,** 1169, 1968.
24. Grummett, R. E. R., and I. G. McIntosh, "Occurrence of Arsenic in Soils and Waters of the Waiotapn Valley and Its Relation to Stock Health," *New Zealand J. Sci. Technol.* **21A,** 137–145, 1939.
25. Harwood, J. E., "The Use of an Ion Selective Electrode for Routine Fluoride Analysis on Water Samples," *Water Res.* **3,** 273, 1969.
26. Horne, R. A., *Marine Chemistry,* Wiley-Interscience, New York, 1969.
27. Ickes, J. H., E. A. Gray, N. S. Zaleiko, and M. H. Adleman, "Operating Experience with a COD Instrument in Industrial Wastes," *Automation in Analytical Chemistry, Technicon Symposium,* pp. 351–356, Mediad, New York, 1967.
28. Ives, D. J. G., and G. J. Janz, *Reference Electrodes, Theory and Practice,* Academic Press, New York, 1961.
29. Jones, R. H., "Oxidation—Reduction Potential Measurement", *Instr. Soc. Am. J.* **13,** 40–44, 1966.
30. Kanwisher, J., "pCO$_2$ in Sea Water and Its Effect on the Movement of CO$_2$ in Nature," *Tellus,* **12,** 209–215, 1960.
31. Kanwisher, J., "Effect of Wind on CO$_2$ Exchange Across the Sea Surface," *J. Geophys. Res.* **68,** 3921–3927, 1963a.
32. Kanwisher, J., "On the Exchange of Gases Between the Atmosphere and the Sea," *Deep-Sea Res.* **10,** 195–207, 1963b.
33. Keeney, D. R., B. H. Byrnes, and J. J. Genson, "Determination of Nitrate in Waters With the Nitrate-Selective Ion Electrode," *Analyst* **94,** 383, 1970.
34. Loomis, W. F., "Direct Method of Determining Carbon Dioscide Tension," *Anal. Chem.* **30,** 1865–1868, 1958.
35. Mackereth, J. J. H., "An Improved Galvanic Cell for the Determination of Oxygen Concentrations in Liquids," *J. Sci. Inst.* **41,** 38–41, 1964.
36. Manahan, S. E., "Fluoride Electrode as a Reference in the Determination of Nitrate Ion," *Anal. Chem.* **42,** 128. 1970.

37. Midwood, R. B., and G. T. Felbeck, Jr., "Analysis of Yellow Organic Matter for Fresh Water," *J. Am. Water Wks. Assn* **60**, 357–366, 1968.
38. Montgomery, H. A. C., N. S. Thom, and A. Cockburn, "Determination of Dissolved Oxygen by the Winkler Method, and the Solubility of Oxygen in Pure Water and in Sea Water," *J. Appl. Chem.* **14**, 280–296, 1964.
39. Prichard, D. W., *The Movement and Mixing of Contaminates in Tidal Estuaries,* Proc. First Intern. Conf. on Waste Disposal in the Marine Environment, Pergamon Press, London, 1960.
40. Riley, J. P., "Analytical Chemistry of Sea Water," in *Chemical Oceanography,* Eds. J. P. Riley and G. S. Skirrow, pp. 295–424, Academic Press, New York, 1965.
41. Ross, J. W., "Ion Selective Electrodes," NBS Special Publication No. 314, Washington, D. C., 1969.
42. Ross, J. W., "Solid State and Liquid, Membrane Ion-Selective Electrodes," in *Ion Selective Electrodes,* Chapter 2, N.B.S. Publication 314, Washington, D. C., 1969.
43. Saruhashi, K., "On the Equilibrium Concentration Ratio of Carbonic Acid Substances Dissolved in Natural Water," *Papers Meteorol. Geophys. (Tokyo)* **3**, 202–206, 1955.
44. Shapiro, J., "Chemical and Biological Studies of the Yellow Organic Acids of Lake Water," *Limnol. Oceanog.* **2**, 161–179, 1957.
45. Shapiro, J., "Yellow Acid-Cation Complexes in Lake Water," *Science* **127**, 702–704, 1958.
46. Sillen, L. G., "Master Variables and Activity Scales," in *Equilibrium Concepts in Natural Water Systems,* Ed. W. Stumm, Adv. Chem. Ser. No. 67, Am. Chem. Soc. Washington, D. C., 1967.
47. Skirrow, G., Chapter 7, in *Chemical Oceanography,* Eds. J. P. Riley, and G. Skirrow, Academic Press, London, 1965.
48. Sower, D. T., R. S. George, and C. H. Rhodes, "Polarography of Gases-Quantitative Studies of Oxygen and Sulphur Dioxide," *Anal. Chem.* **31**, 2–5, 1969.
49. Strickland, J. D. H., and T. R. Parsons, *A Manual of Sea Water Analysis,* Fish Res. Bd. Canada, Publication 125 (2nd ed.), Ottawa, 1965.
50. Tackett, S. L., "Automatic Titoation of Calcium with EDTA Using a Calcium Selective Electrode," *Anal. Chem.* **41**, 1703, 1969.
51. Teal, J., and M. Teal, *Life and Death of the Salt Marsh,* Little, Brown and Company, Boston, 1969.
52. Vallentyne, J. R., "The Molecular Nature of Organic Matter in Lakes and Oceans with Lesser Reference to Sewage and Terrestrial Soils," *J. Fish Res. Bd. Canada* **14**, 33, 1957.
53. Whitfield, M., "E_h as an Operational Parameter in Estuarine Studies," *Limnol. Ocean.* **14**, 547–558, 1969.
54. Williams, L. M., "Chemical Aspects of Fluoridation," *Tetracol.* **8**, 41, 1970.
55. Winkler, L. W., "Die Bestimmung des in Wasser gelosten Sauerstoffes," *Ber. dtsch. chem. Ges.* **21**, 2843–2845, 1888.

Vincent Cushing

4.30 WATER CURRENT SENSORS

Knowledge of the transport of heat, momentum, chemical concentration, and pollutants is important in geoscience. Water in motion is perhaps the most important vehicle of transport. The continual cycle of water circulation, evaporation, convection, precipitation, and return is the *hydrologic cycle* [12].

The largest measurement scale involves measuring steady average currents with a large instrument response time or small bandwidth. Next is the seasonal variation of these currents, then diurnal and semidiurnal variations. Resolution of water motion throughout these daily cycles requires instrument response times of one-half to one hour.

Eddies shed from streams within the ocean can require measurement response times ranging from several hours to a few minutes. Water surface waves associated with wind forces produce orbital velocity in the water where the period of alternation can be several seconds to a large fraction of a minute requiring instrument response times the order of 1 s. The measurement of turbulent velocities can require an instrument bandwidth of 10 to 100 Hz. High-frequency components of turbulence are associated with small dimensional scale so that turbulent velocity measuring instruments face the additional difficult constraint of small size to resolve the velocity of a very small volume of water.

Although water currents are mostly horizontal, vertical transport is important, and its measurement is difficult because it is small, relative to the horizontal motion.

The greatest use of water velocity instruments has been the metering of streams and rivers. Measurement of water current in the ocean is more difficult because speed as well as direction must be measured, the ocean is a comparatively hostile environment for long-term maintenance-free operation, access to the instrument for maintenance is difficult, operability at great depth and hydrostatic pressure must be provided for, and motion-free mounting of the instrument is virtually impossible so that measurement of ocean current relative to an earth-fixed coordinate system requires correction for motion of the instrument.

4.30.1 Requirements

Stream Flow. For stream flow measurement, the interest is in measuring volume transport. Time-averaged information is required, so that a rapid-response instrument is not necessary. An attendant measures the water velocity at several points throughout the cross-sectional area of a stream and sums the information to determine the overall volume transport of water [31]. The total volume transport correlates well with stream level, so that long-term flow information in streams is generally inferred from long-term records of water level.

The dynamic range requirement for stream flow measurement is from 0 to 4 m/s. On rare instances near a waterfall or rapids this could be extended to 10 m/s. An accuracy of 2 percent of reading is usually adequate.

For stream pollution studies total stream flow information may not be adequate and local stream velocity in the neighborhood of outfalls for wastes is important.

Near-Surface Ocean. In the shallow or near-surface ocean, fouling by marine plants and animals is a serious environmental problem [4, 24]. If the instrument is immersed for weeks or months, corrosion is also a serious problem. Operating temperatures range from $-5°C$ to $40°C$. Salinity gradients can be severe in estuaries where fresh water mixes with sea water [27].

Gradients in velocity can also be large. One study showed a horizontal sheer of 4 cm/s per meter [2]. The data also indicated that a difference of less than 1 cm/s in the surface layer velocity resulted in a change in net volume transport from approximately 30 m³/s to zero.

Orbital velocities associated with surface waves range from 5 to 50 cm/s with periods from 10 to 15 s [24]. By comparison, the relatively steady-state flow in the near-surface ocean can range from 2 to 10 cm/s, with periodicities the order of 100 to 1000 s and semidiurnal and diurnal periodicities. These superposed large orbital velocities put a strain on the linearity requirement for a water velocity transducer. Since water velocity is a vector quantity, the need for linearity implies that the transducer be linear in a vector sense. For example, in horizontal flow, if $\mathbf{n}$ is a horizontal unit vector fixed in the transducer's coordinate system, and $\mathbf{V}$ is the water velocity in earth-fixed coordinates, then as the transducer is rotated—that is, as the unit vector $\mathbf{n}$ rotates—the transducer should sense a magnitude of $\mathbf{n} \cdot \mathbf{V} = |\mathbf{V}| \cos \theta$, where θ is the angle between $\mathbf{n}$ and $\mathbf{V}$.

Deep Ocean. In the deep ocean, hydrostatic pressure is a principal environmental factor. If the interior of a water current meter can operate in high hydrostatic pressure, it is useful to fill the instrument with a fluid and maintain pressure equilibrium between the fluid and seawater. If the interior cannot be pressurized, the instrument must be a pressure vessel. Temperature in the deep ocean is much less variable than elsewhere, usually within a few degrees above freezing.

It is assumed that water velocity at depth is quite steady, so that high frequency response of a water current meter is unnecessary and satisfactory measurements can be made with a single-component velocity meter, an instrument that trails along the flow streamlines and measures the speed along these streamlines.

It is desirable to mount a current meter rigidly relative to an earth-fixed coordinate system. For measurements in remote regions of the ocean, such a fixed mounting is not available and the instrument must be suspended along a mooring line which is anchored at one end and supported by a buoy at the other. The coordinate frame of a water current meter is then translating and rotating in a manner determined by the forces acting on the buoy and mooring line [8]. The meter then senses water velocity relative to its own coordinate frame. A correction or coordinate translation must be made to determine the water velocity in the usually desired earth-fixed coordinate system. Vector linearity of the instrument is imperative in this situation.

Flow Interference. Emplacement of the measuring instrument disturbs the free-field flow pattern to be measured. A water current meter is often attached to or placed near a larger subsurface instrument package whose presence seriously affects the flow field. It is important that the meter be calibrated in the same configuration in which it is employed.

4.30.2 Sensor Types

Drag Force. The drag force method uses strain gauges to measure the hydrodynamic drag on a fixed object mounted on a stiff suspension. It is also available in a pendulum suspension where the angular deflection of the object is a measure of water velocity. The instrument measures magnitude and direction of water current and provides cosine response to changes in flow direction. Its output is essentially a quadratic function of water speed so that it is suitable for steady flow but unsuitable when a substantial alternating flow is superposed.

For steady-flow, the vector drag force $\mathbf{F}_D$ on an object is

$$\mathbf{F}_D = \frac{C_D A \rho |\mathbf{V}| \mathbf{V}}{2}, \qquad (4.30\text{--}1)$$

where C_D is the object's drag coefficient (dimensionless), A is the cross-sectional area of the object perpendicular to the flow, ρ is the fluid density, and $\mathbf{V}$ is the vector velocity of the fluid relative to the object.

Flow transverse to a circular cylinder has been extensively investigated, and this shape is generally employed in a drag-type water current meter. The drag coefficient for a circular cylinder is determined experimentally as a function of the dimensionless Reynolds number R_d defined by

$$R_d \equiv \frac{\rho V d}{\mu} = \frac{V d}{v}, \qquad (4.30\text{--}2)$$

where d is the diameter of the circular cylinder, μ is the fluids's viscosity, and $v(=\mu/\rho)$ is the fluid's kinematic viscosity. The drag coefficient for a circular cylinder as a function of Reynolds number R_d is shown in Figure

4.30-1 [13]. The figure shows that C_D varies as R_d^{-1} for R_d less than 1. Equation 4.30–1 shows that the drag force is a linear function of flow velocity V in this region. This is called the creeping flow regime. The higher-velocity region where C_D is constant extends from a Reynolds number of 10^3 to the critical Reynolds number of about 3×10^5. The drag force is a quadratic function of the flow velocity in this region.

The expression for drag force is more complicated when the fluid is accelerating relative to the object [14], making the method much less attractive for use in nonsteady flow [24].

Tracer Schemes. The tracer scheme, or Lagrangian method, involves tagging a packet of water particles and measuring the time of transit over a known distance. Classical tags have consisted of drift bottles, radio buoys, patches of dye or radioactive waste, and drogues—all used to determine water velocity at the surface.

Drogues, in the configuration shown in Figure 4.30-2, have also been used to measure subsurface currents [16]. The objective is to make the drag force F_D on the subsurface sail or drogue large compared with the drag force F_B on the supporting buoy (plus the distributed drag force on the supporting line). Neglecting the drag force on the line, these forces are described by

$$F_D = \frac{C_D A_D \rho (V_D - V)^2}{2},\qquad (4.30\text{--}3)$$

and

$$F_B = \frac{C_B A_B \rho (V_B - V)^2}{2},\qquad (4.30\text{--}4)$$

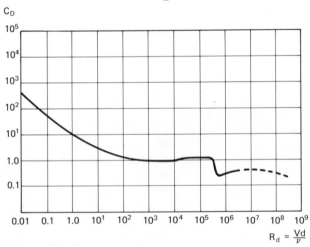

Figure 4.30-1 Drag coefficient C_D for a circular cylinder as a function of Reynolds number R_d [13].

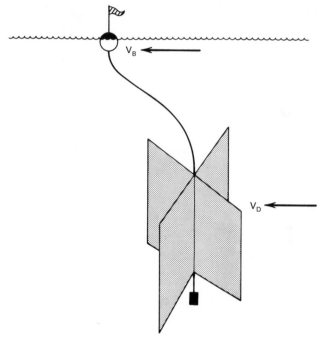

Figure 4.30-2 Subsurface drogue for measuring current velocity at depth d, supported from a surface buoy where the surface current is V_B.

where C_D and C_B are the drag coefficients for the drogue and buoy, respectively, A_D and A_B are the effective areas of the drogue and buoy, respectively, V_D and V_B are the water velocities at the drogue and buoy, and V is the net velocity of the tied-together combination. These equations can be combined to give

$$V = V_D + (V_B - V)\left(\frac{C_B A_B}{C_D A_D}\right)^{\frac{1}{2}}. \qquad (4.30\text{--}5)$$

The velocity V of the tied-together combination is sufficiently close to the water velocity V_D at the drogue depth when $(C_B A_B / C_D A_D)$ is sufficiently small.

The surface buoy and supporting line can be eliminated by using a neutrally buoyant weighted float made of material less compliant or compressible than seawater [33]. This subsurface drogue equilibrates at a depth where its preset average density is equal to the local density of seawater and is tracked by acoustic means.

A further variation [29] of the convecting drogue is shown in Figure 4.30-3. The method provides the average water velocity $\overline{V}$ between the surface and the bottom depth D. A projectile with buoyant force F_B is ballasted with a weight $2F_B$. If the projectile has a drag coefficient C_p and

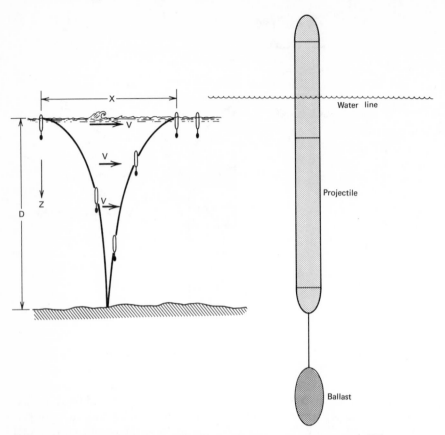

Figure 4.30-3   Falling projectile method of measuring water transport [29].

a cross-sectional area A_p, it falls with a velocity V_F given by

$$F_B = \frac{A_p C_p \rho V_F^2}{2}.$$ (4.30–6)

When the ballasted projectile hits bottom, the ballast is released. If the drag coefficient for the projectile is the same for both directions, it rises at the same velocity V_F.

During the fall and rise, the projectile convects horizontally with the local water velocity V. If the projectile disappears from the surface at time zero, reappears at time t, and travels a horizontal distance X the average horizontal water current velocity $\bar{V}$ is

$$\bar{V} = \frac{X V_F}{2D} = \frac{X}{t},$$ (4.30–7)

where D is the bottom depth. The volume transport per unit width T is

$$T = \frac{XV_F}{2}. \tag{4.30-8}$$

Heat Transfer. If a fluid passes over a hot wire, the rate at which heat is transferred from the wire to the fluid depends on the fluid velocity [20]. The technique has not had much use in water current metering because of nonlinearities and fragility. However, the sensitive area can be made very small, and it has found use for measuring very small-scale high-frequency turbulent velocities [26].

The sensor consists of a fine wire W strung between two terminals which project into the fluid stream as shown in Figure 4.30-4. The fine wire is usually made of platinum, tungsten, or nickel with a diameter of 0.01 to 0.1 mm and a length as small as a fraction of a millimeter. The thermal inertia of the system depends on the dimensions of the wire. A fine wire has high frequency response but is concomitantly delicate.

The resistance R_w of the fine wire depends on the wire temperature T_w and is

$$R_w = R_0[1 + \alpha(T_w - T_0)], \tag{4.30-9}$$

where R_0 is the wire resistance at temperature T_0, and α is the temperature coefficient of electrical resistance. The rate of heat loss P from the heated wire to the fluid stream is [15]

$$P = I^2 R_w = (C_1 + C_2 V^{\frac{1}{2}})(T_w - T_f), \tag{4.30-10}$$

where I is the electric current in the wire, C_1 and C_2 are constants for a given wire, V is the fluid velocity, and T_f is the temperature of the fluid.

In a constant-current meter, I is maintained constant and T_w is a measure of V. T_w is determined measuring R_w and using Eq. 4.30–9. In a constant-temperature meter T_w is maintained constant, a servo loop varies I so that R_w is maintained constant, and I is a measure of V.

The hot-wire technique provides the only configuration with a small enough size to measure very small-scale high-frequency fluid velocity variations.

Acoustic. There are two basic kinds of acoustic water current meters, those that are sensitive to the transit time of a sound pulse through a moving

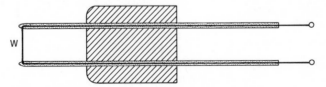

Figure 4.30-4 Hot-wire sensor for insertion into a fluid stream.

fluid, and those that measure the Doppler shift in acoustic energy scattered back from foreign particles convecting with the water [3, 17, 19, 32].

The transit time sensor block diagram is shown in Figure 4.30-5. The projector P_A initiates an acoustical pulse which is later sensed by hydrophone H_A. If the water velocity V is in the direction shown, the transit time T_A for the sound pulse to travel from P_A to H_A is

$$T_A = \frac{L}{c + V},$$ (4.30–11)

where L is the separation between P_A and H_A, and c is the speed of sound in water.

Similarly, a sound pulse initiated at the projector P_B is propagated in the opposite direction to the hydrophone H_B with a transit time T_B of

$$T_B = \frac{L}{c - V}.$$ (4.30–12)

The transit-time water current meter eliminates the sound speed dependence and ascertains the water velocity as

$$V = \frac{L(T_A - T_B)}{2T_A T_B}.$$ (4.30–13)

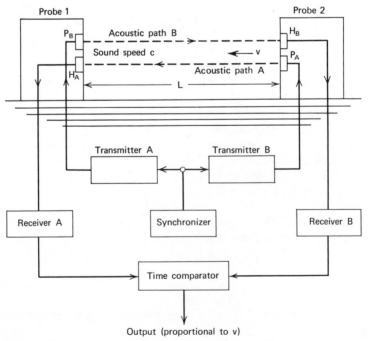

Figure 4.30-5   Principle of operation of transit-time acoustic water current meter.

Output linearity of the over-all instrument is better than 1 percent. Under conditions of thermal gradients, instantaneous velocity errors up to 100 mm/s can occur [19, 24]. Typical dynamic ranges are a few millimeters per second to several meters per second. The instrument requires stable, sophisticated electronics because the transit-time difference $(T_A - T_B)$ is generally less than a microsecond.

Pairs of probes can be installed so that the acoustic paths are at right angles to each other, and outputs from the two channels are then representative of the two orthogonal components of water velocity [22].

The Doppler acoustic meter is illustrated in Figure 4.30-6. The projector P radiates an acoustic beam at a frequency between 3 and 10 MHz with a beamwidth of approximately 4°. A similarly constructed hydrophone H also has a beamwidth of approximately 4°.

The radiated sound beam reflects or scatters from any inhomogeneities (gas bubbles or impurities) in the fluid. The energy backscattered from the inhomogeneities in the intercepted volume shown in Figure 4.30-6 is received by the hydrophone. If V is the water velocity, 2θ is the angle between the projector and receiver beam axes, and f_0 is the frequency of the radiated beam, the backscatter acoustic energy is Doppler shifted to a frequency f given by [18]

$$f = f_0 \left[\frac{1 + (V/c) \cos \theta}{1 - (V/c) \cos \theta} \right], \qquad (4.30-14)$$

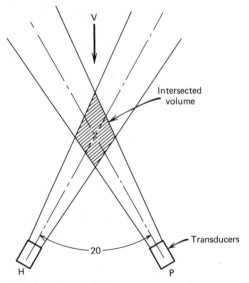

Figure 4.30-6 Principle of operation of Doppler type acoustic water current meter.

which is adequately approximated to be linear if $V/c \ll 1$. That is,

$$\nabla f \equiv f - f_0 = \frac{2f_0 V \cos \theta}{c}. \tag{4.30-15}$$

The quantity Δf is known as the Doppler shift in frequency, and V is

$$V = \frac{c\Delta f}{2f_0 \cos \theta}. \tag{4.30-16}$$

The instrument can be in error if the sound speed varies.

Impeller. The instrument most utilized for measurement of stream flow in the United States is a small cup type meter with a vertical axis of rotation [12, p. 119]. In turbulent water it tends to overregister. It provides a pulse every revolution or every fifth revolution of the meter wheel. Its velocity dynamic range extends from approximately 30 mm/s to 7 or 8 m/s.

The meter is not meant for unattended long-term operation. When properly maintained and operated, it provides an accuracy of 1 percent of value. The pivot bearing of the vertical axis is set in a deep inverted cup in which air is trapped to prevent entrance of water and dirt that might affect the frictional characteristics of the bearing.

In Europe, the most widely used current meter for stream flow measurement is a screw type with the axis of rotation pointed into the stream [12, p. 119].

A common ocean current meter is shown in Figure 4.30-7. The water speed sensor (a Savonius rotor) is shown in the lower part of the figure and the direction sensor (vane) directly above it. The Savonius rotor is divided in two by a horizontal mid-plane. The cross section of the Savonius rotor is shown in Figure 4.30-8. Moving water generates an unbalanced drag force and produces a torque about the axis.

At high speeds, the instrument is linear. At low speeds it has a stiction threshold of about 20 mm/s [9]. It is a high-inertia transducer that responds more rapidly to accelerating flow than to decelerating flow. With alternating flow superposed on the steady flow, the instrument tends to read a mean current that is higher than the actual steady flow. It senses current transverse to the axis of rotation, but the end plates are not adequate to eliminate response to vertical flow. The device therefore should not be used in the near-surface where surface waves can provide alternating orbital velocity or where up and down surges on the instrument's supporting line can cause excessive error.

Magnetic. An electrical current is induced in a conductor moving through a magnetic field. The induced electric field strength is

$$\mathbf{E} = \mathbf{v} \times \mathbf{B}, \tag{4.30-17}$$

where $\mathbf{v}$ is the velocity of motion, and $\mathbf{B}$ is the magnetic induction (flux density).

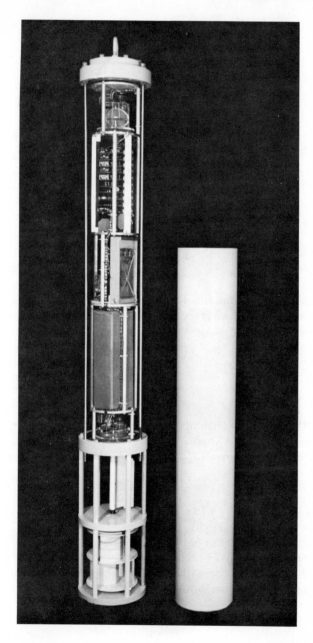

Figure 4.30-7 Photograph of Savonius rotor current meter.

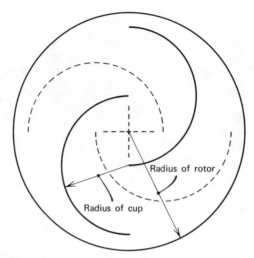

Figure 4.30-8 Cross section of a Savonius rotor.

Motion of a fluid in the presence of a magnetic field induces a voltage V throughout the fluid which is the solution to the classical Poisson equation [5]

$$\nabla^2 V = q, \qquad (4.30\text{--}18)$$

where

$$q = Z\,\mathbf{div}(\mathbf{v} \times \mathbf{B}). \qquad (4.30\text{--}19)$$

The quantity Z is effectively unity for fresh water and sea water.

For fluid velocities and conductivities in the range of interest for water current metering, **curl B** is zero, and Eq. 4.30–19, using a vector identity, becomes

$$q = \mathbf{B} \cdot \mathbf{curl}\, \mathbf{v}; \qquad (4.30\text{--}20)$$

curl v is the vorticity of the water flow field, and q is the source of the flow induced electrical potential field.

The magnetic induction can be obtained artificially from an electromagnet in the sensor or naturally as the earth's magnetic field. If B in the water is due to a dipole of moment **m**, the flow induced electric potential V about a sphere of radius R at the position **r** is [6]

$$V = \frac{3\mu_0}{8\pi|\mathbf{r}|^3}\,\mathbf{v}_0 \times \mathbf{m} \cdot \mathbf{r}, \qquad (4.30\text{--}21)$$

where $\mathbf{v}_0$ is the free field flow velocity, and μ_0 is the permeability of free space. In SI units V is in volts, **r** is in meters, $\mathbf{v}_0$ is in meters per second, **m** is in ampere-meters^2, and μ_0 is $4\pi \times 10^{-7}$ H/m. Equation 4.30–21 shows that the flow induced potential is a linear function (in a vector sense) of the free-field flow velocity v_0.

If two detection electrodes are placed diametrically opposite on the surface of a sphere at radius R, v_0 is perpendicular to the magnetic field's axis of symmetry, the potential difference ΔV sensed by the detection electrodes is

$$\Delta V = \frac{3\mu_0 \left|v_0\right| m \cos \theta}{4\pi R^2}, \tag{4.30–22}$$

where θ is the angle between the flow direction and the perpendicular to the line connecting the two detection electrodes.

If the earth's magnetic field is used to provide the induction, Eq. 4.30–20 shows that water flow behavior at all fluid flow boundaries contributes to the flow-induced potential. The geomagnetic electrokinetograph (GEK), which employs the earth's magnetic field for its induction [36] can be employed only in open water, far from the sea bottom, shorelines, or other boundaries of the flow field.

If the magnetic field is deactivated in an electromagnetic water velocity meter, it senses spurious voltages due to thermoelectric effects and electrochemical action at the interface between the electrode and the electrolyte (water). Alternating magnetic inductions as low as 10 Hz substantially eliminate the effects of thermoelectric and electrochemical noises [6, 25, 35].

4.30.3 Direction Sensor

When direction as well as speed must be sensed, a vane of the type shown in Figure 4.30-7 which trails along the water flow streamlines, is often used. The vane, which can move through 0–360°, is coupled to a microtorque potentiometer, and the output voltage is an indication of direction. The vane sensor has had its greatest use in conjunction with the Savonius rotor.

The vane is best suited to deep ocean work where changes in direction occur at a very slow rate. It is unsuitable near the surface if large orbital water velocity is superposed on steady flow, requiring rapid changes in vane position.

Component type velocity sensors such as the two-axis drag-force type, the multi-axis acoustic type, and the multi-axis electromagnetic type described above sense the Cartesian components of water velocity to determine direction.

4.30.4 Compass and Tilt Auxiliaries

The water current sensor provides water velocity (speed and direction) information in a coordinate system fixed to the sensor, so it is necessary to ascertain the orientation of the sensor relative to earth-fixed coordinates.

This added information can be derived from a gimballed magnetic compass with an accuracy of $\pm 2°$.

References

1. Bowden, K. F., and L. A. Fairbairn, "Measurements of Turbulent Fluctuation and Reynold's Stresses in a Tidal Current." *Proc. Roy. Soc. (Lond.). Series A,* **237,** 422–438, 6 Nov. 1956.
2. Cannon, G. A., "Observations of Motion at Intermediate and Large Scales in a Coastal Plain Estuary," Chesapeake Bay Institute of the Johns Hopkins University Technical Report 52, March 1969.
3. Chalupnik, J. D., and P. S. Green, "A Doppler-Shift Ocean-Current Meter," *Marine Sciences Instrumentation,* Vol. 1, Ed, R. D. Gaul, D. D. Ketchum, J. T. Shaw, and J. M. Snodgrass, pp. 194–199, Plenum Press, New York, 1962.
4. Crosby, R. M., F. H. MacDonald, T. R. Livermore, and W. Gumma, "The Survival Environment for Oceanographic and Meteorological Sensors," Texas Instruments Report No. 59009–2, Dallas, May 1970.
5. Cushing, V., "Induction Flowmeter," *Rev. Sci Instr.* **29,** 692–697, Aug. 1958.
6. Cushing, V., "Electromagnetic Water Current Meter," *Rev. Sci. Instr.* 1971.
7. Farady, M., "Part I, Experimental Researches in Electricity," pp. 125–162, "Part II, same," pp. 163–193, *Phil. Trans. Roy. Soc. Series A,* 1832.
8. Fofonoff, N. P., "A Technique for Analysing. . . Buoy System Motion," *Geo-Marine Tech.* **1,** 10–13, July 1965.
9. Fofonoff, N. P., and Y. Ercan, "Response Characteristics of a Savonius Rotor Current Meter," Woods Hole Oceanographic Institution Report No. 67–33. Woods Hole, Massachusetts, June, 1967, unpublished manuscript.
10. Fofonoff, N. P., "Current Measurements from Moored Buoys. 1959–1965," Woods Hole Oceanographic Institution Report No. 68–30, Woods Hole, Massachusetts, May 1968, unpublished manuscript.
11. Grant, H. L., R. W. Stewart, and A. Moilliet, "Turbulence Spectra from a Tidal Channel," *J. Fluid Mech.* **12,** 241–268, Feb. 1962.
12. Grover, N. C., and A. W. Harrington, *Streams Flow,* Chapter 1, Dover Publications, New York, 1966.
13. Hoerner, S. F., *Fluid-Dynamic Drag,* pp. 3–9, published by the author, Brick Town, New Jersey, 1965.
14. Keulegan, G. H., and L. H. Carpenter, "Forces on Cylinders and Plates in an Oscillating Fluid," *J. Res. Natl. Bur. Std.* **60,** 423–440, May 1958.
15. King, L. V., "On the Convection of Heat from Small Cylinders in a Stream of Fluid: Determination of the Convection Constants of Small Platinum Wires with Applications to Hot-Wire Anemometry," *Phil. Trans. Roy. Soc. Series A,* **214,** 373–432, 1914.
16. Knauss, J. A., *The Sea,* Ed. M. N. Hill, Vol. 1, Chapter 14, Interscience Publishers, New York, 1962.
17. Kocy, F. F., M. Kronengold, and J. Loewenstein, "A Doppler Current Meter," *Marine Sciences Instrumentation, Vol. 2, Proceedings of the Symposium on Transducers for Oceanic Research,* Ed. R. D. Gaul, pp. 127–134, San Diego, Nov. 8–9, 1962, Plenum Press, New York, 1963.

18. Landau, L. D., and E. M. Lifshitz, *Fluid Mechanics*, p. 260, Addison-Wesley, Reading, Massachusetts, 1959.
19. Lester, R. A., "High Accuracy, Self-Calibrating Acoustic Flow Meters," *Marine Sciences Instrumentation, Vol. 1*, a Collection of Instrumentation Papers Presented at the Marine Sciences Conference Held September 11–15, 1961, at Woods Hole, Massachusetts, pp. 200–204, Plenum Press, New York, 1962.
20. Lion, K. S., *Instrumentation in Scientific Research*, pp. 127–132, McGraw-Hill Book Company, New York, 1959.
21. Malkus, J. S., *The Sea*, Ed. M. N. Hill, Vol. 1, pp. 88–92, Interscience Publishers, New York, 1962.
22. Multer, R. H., "Measuring Directional Velocity in Water Waves with an Acoustic Flowmeter," Coastal Engineering Research Center Technical Memorandum No. 31, U.S. Army Corps of Engineers, Washington, April 1970.
23. Neumann, G., and W. J. Pierson, Jr., *Principles of Physical Oceanography*, Chapter 14, Circulation and Stratification of the Oceans, pp. 422–478, Prentice-Hall, Englewood Cliffs, New Jersey, 1966.
24. Olson, J. R., Flowmeters in Shallow-Water Oceanography," Naval Undersea Warfare Center Report No. NUWC TP 5, San Diego, Sept. 1967.
25. Olson, J. R., "Component Electromagnetic Flowmeter," Naval Undersea Warfare Center Report TN 193, San Diego, Oct. 1968.
26. Patterson, A. M., "Development of a Hot-Wire Instrument for Ocean Turbulence Measurements," Pacific Naval Laboratory TM 57–2, 1957.
27. Pritchard, D. W., and E. A. Pearson, *Wastes Management Concepts for the Coastal Zone*, pp. 26 et seq., National Academies of Sciences and Engineering, Washington, D. C., 1970.
28. Richardson, W. S., P. B. Stimson, and C. H. Wilkins, "Current Measurements from Moored Buoys," *Deep-Sea Res. Oceanogr. Abstracts* **10**, pp. 369–388, Oct. 1963.
29. Richardson, W. S., A. R. Carr, and H. J. White, "Description of a Freely Dropped Instrument for Measuring Current Velocity," *J. Marine Res.* **27**, 137–153, Jan. 1969.
30. Smith, C. G., and J. Slepian, "Electromagnetic Ship's Log," U.S. Patent No. 1,249,530, Dec. 11, 1917.
31. Smoot, G., and C. E. Novak, "Measurement of Discharge by the Moving-Boat Method," *U.S. Geological Survey Book 3*, 1968.
32. Suellentrop, F. J., A. E. Brown, and E. Rule, "An Acoustic Ocean-Current Meter," *Marine Sciences Instrumentation, Vol. 1*, a Collection of Instrumentation Papers Presented at the Marine Sciences Conference Held September 11–15, 1961, at Woods Hole, Massachusetts, pp. 190–193, Plenum Press New York, 1962.
33. Swallow, J. C., "A Neutral-Buoyancy Float for Measuring Deep Currents," *Deep Sea Res. Oceanogr. Abstracts* **3**, 74–81, Oct. 1955.
34. Takenouti, A. Y., *International Marine Science*, Vol. IV, UNESCO Publications Center, New York, Oct. 1966.
35. Tucker, M. J., N. D. Smith, F. E. Pierce, and E. P. Collins, "A Two-component Electromagnetic Ship's Log," *J. Inst. Navigation* **23**, 302–316, July 1970.
36. von Arx, W. S., "An Electromagnetic Method for Measuring the Velocities of Ocean Currents from a Ship Under Way," *Papers in Physical Oceanography*

and Meteorology, Vol. XI, pp. 1–62, Cambridge and Woods Hole, Massachusetts. 1950.

37. Wollaston, C., "Earth Currents," A. J. S. Adams, *J. Soc. Telegraph Engineers and Electricians,* **10**, 34–43, Feb. 10, 1881.

38. Young, F. B., H. Gerrard, and W. Jevons, "On Electrical Disturbances due to Tides and Waves," *Phil. Mag. 6th Series,* **40**, 149–159, July 1920.

Mace T. Miyasaki

4.31   WATER WAVE SENSORS

4.31.1   Nature of the Ocean Surface

The ocean surface consists of three general kinds of structure: capillary waves, gravity waves, and swell. These are characterized by differences in height, wavelength, and shape [4, 11, 25].

Capillary waves are the millimeter-sized variations in surface height that form in the presence of a light breeze. The formation of these wavelets is resisted by the surface tension of the water so that when the wind ceases, the capillary waves die out and the surface becomes smooth again.

Capillary waves, under the influence of a driving wind, can build up to the point where they contain sufficient energy to continue to exist after the driving wind has ceased. The cross-over point seems to be when the capillary waves reach a wavelength of about 1.73 cm, with a corresponding wave speed of about 24 cm/s. These waves, which are commonly known as gravity waves, continue to grow under the influence of a driving wind.

The speed of a wave depends upon its height. The superpositioning of many different waves of different heights results in a complex distribution of heights. Since each wave travels at a different speed, the structure changes continuously with time.

The maximum height a wave achieves under a given wind condition depends on the wind speed, duration, and fetch. Wind speed determines the amount of force in the wave-creating process. Duration is the period of time the wind has been blowing, and fetch is the distance over which it has maintained its speed and duration. For a given speed and fetch, a minimum duration is required to achieve a maximally (fully) developed sea after which only an increase in the wind conditions can generate a higher sea. As the seas rise in height, the waves become more peaked, and as the winds increase above about 10 m/s, the crests of the waves start to break. At higher wind speeds, more foaming action takes place, and wind streaks form. Table 4.31-1 describes the necessary conditions for a fully developed sea.

Eventually, either the driving winds diminish or the gravity waves travel beyond the influence of the generating winds. As they continue to propagate, the waves become smoothly rounded off and become what is commonly known as swell. Swell can propagate for thousands of miles.

Quantitative descriptions of the surface conditions of the ocean can be based on different parameters using the following definitions:

Height: The elevations of the crest above the trough.
Length: The distance from one crest to another.
Period: Length of time for two successive crests to pass one point.
Fetch: Horizontal distance over which a constant wind has been blowing.
Duration: The period for which a constant wind has been blowing.

Although average wave height is often used, a more meaningful statistic is the significant wave height, the arithmetic average of the one-third highest waves.

The Beaufort Sea State scale, shown in Table 4.31-1, is a descriptive scale based on a fully arisen sea for a given wind condition. This description is valid only if the wind has been blowing for the minimum duration over the required fetch.

The International Sea State scale, shown in Table 4.3-1, is based on the significant height of the waves and can therefore be applied to a sea in an intermediate stage of development.

The World Meteorological Organization has established a descriptive code (Table 4.31-2) used in the reporting of weather conditions at sea.

4.31.2 Mechanical Sensors

Float Gage. The simplest method of measuring the water level is the use of a float driving an indicator as shown in Figure 4.31-1. This instrument is not responsive to high-frequency waves because of the mechanical inertia present in the moving components.

Pressure Sensors. Pressure measurements are made from the ocean bottom or from a fixed position below the water surface. The weight of the column of water above the sensor and the pressure of the atmosphere above combine to produce an absolute pressure at the submerged sensor nominally

Table 4.31-1. Wind waves at sea. This table applies only to waves generated by the local wind and does not apply to swell originating elsewhere. Presence of swell makes accurate wave observations exceedingly difficult. The height of waves is arbitrarily chosen as the height of the highest 1/3 of the waves. Occasional waves caused by interference between waves or between waves and swell can be considerably larger. Only lines 7, 8, and 9 are applicable to swell as well as waves.

#	Quantity	Scale values
1	Wind velocity (m/s)	2 2.5 3 4 5 6 7 8 9 10 15 20 25 30 35
2	Beaufort wind and description	1 Light air · 2 Light breeze · 3 Gentle breeze · 4 Moderate breeze · 5 Fresh breeze · 6 Strong breeze · 7 Moderate gale · 8 Fresh gale · 9 Strong gale · 10 Whole gale · 11 Storm
3	Required fetch (km) — Fetch is the distance a given wind has been blowing over open water	100 200 300 400 500 600 700 800 1000 1500
4	Required wind duration (hr) — Duration is the time a given wind has been blowing over open water	5 20 25 30 35

When the fetch and duration are as great as indicated above, the following wave conditions exist. Wave heights can be up to 10% greater if fetch and duration are greater.

#	Quantity	Scale values
5	Wave height crest to trough (m)	0.3 0.5 (White caps form) 1 2 3 4 5 6 8 10 15 20
6	International sea state number and description	1 Smooth · 2 Slight · 3 Moderate · 4 Rough · 5 Very rough · 6 High · 7 Very high · 8 Precipitous
7	Wave period (s)	1 2 3 4 5 6 7 8 9 10 12 14 16 18 20
8	Wave length (m)	6 10 20 30 50 70 100 150 200 300 500
9	Wave velocity (m/s)	1 2 3 4 5 6 8 10 15 20 25 30
10	Particle velocity (m/s)	0.3 0.4 0.5 0.6 0.8 1 1.5 2 2.5 3 4 5
11	Wind velocity (m/s)	2 2.5 3 4 5 6 7 8 9 10 15 20 25 30 35

Table 4.31-2. Descriptive Code of the World Meteorological Organization

WMO WAVE CODE No. 42

Symbolic Form

1. $3\,P_wP_wH_wH_w$ For Sea

      where

        3—indicator

      P_wP_w—is period in whole seconds

      H_wH_w—height of waves in accordance with Table 18

      Note. The direction of sea waves is assumed to be the direction of the surface
             wind. The wave period is measured to the nearest second.

2. $d_wd_wP_wH_wH_w$ For Swell

      where

      d_wd_w—is direction in 10's of degrees from which swell is coming.

      P_w—period in accordance with Table 19.

      H_wH_w—swell height in accordance with Table 18.

CODE TABLE 18
H_wH_w—Mean Maximum Height of the Waves

Code Figure	Height in Feet	Code Figure	Height in Feet
00	1	11	17 or 18
01	1 or 2	12	19 or 20
02	3 or 4	13	21
03	5	14	22 or 23
04	6 or 7	15	24
05	8	16	25 or 26
06	9 or 10	99	Height not observed
07	11 or 12		
09	14 or 15		
10	16		

CODE TABLE 19
P_w—Period of the Swell Waves

Code Figure	Period (Seconds)
5	5 or less
6	6
7	7
8	8
9	9
0	10
1	11
2	12
3	13
4	14

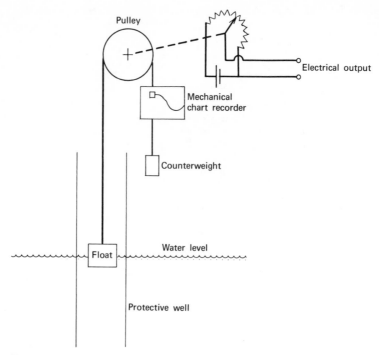

Figure 4.31-1 Float gage.

equal to 1 kg/cm^2 plus 1.04 g/cm^2 per cm of water depth. This pressure changes slowly with weather patterns that affect the atmospheric pressure and with tides that affect the mean water column height. The wave profile is superimposed on these slow pressure fluctuations.

Because the pressure transducer is usually mounted 5 to 15 m below the surface, higher-frequency waves are lost in the integrated pressure measured by the transducer. The pressure transducer is insensitive to wavelengths less than twice the transducer depth.

Pressure transducers used in wave measuring applications include the strain gage, the vibrating wire, and a relatively new type of sensor, the tunnel diode [31].

Strain Gage. A strain gage is a very thin conductor of electricity which decreases in cross-sectional area when stretched. This increases the resistance of the gage which can be measured electrically. Changes occurring under most strains are only on the order of a few milliohms, so the strain gage is usually part of a Wheatstone bridge circuit (Figure 4.31-2). Figure 4.31-3 shows a typical mechanical configuration.

The diaphragm used as the compliant element in Figure 4.31-3 is open to the pressure to be measured on one side and open to a controlled pressure

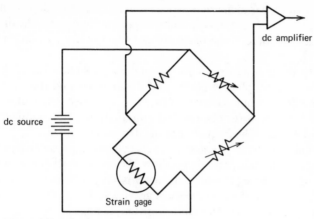

Figure 4.31-2 Strain gage circuit.

on the other. A pinhole is made in the controlled pressure side, exposing the top side of the diaphragm to a very low-pass filtered ambient pressure. The transducer thus responds only to wave height variations, reducing the dynamic range required of both the transducer and the recording equipment.

A typical strain-gage wave meter responds to wave periods from 4 to 300 s with sensitivity ranges of $\pm$ 175 g/cm^2 (3-m waves) and $\pm$ 350 g/cm^2 (6-m waves).

Vibrating Wire. The Vibrotron is a pressure transducer that transforms pressure into frequency through the vibrations of a taut wire. The wire, fixed at one end and attached at the other end to a diaphragm exposed to the

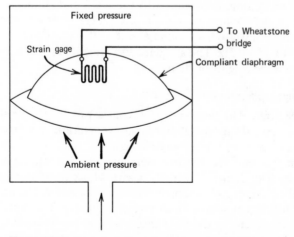

Figure 4.31-3 Typical strain gage pressure transducer.

ambient pressure, has a characteristic resonant frequency depending on the tension impressed upon it. As the ambient pressure increases, the diaphragm moves under the force of the pressure, and there is less tension on the wire [15, 20, 21].

Normally, the taut wire does not vibrate of its own accord, so an exciting force is provided by placing permanent magnets around the wire creating magnetic fields normal to the axis of the wire, and by passing an alternating current through the wire. The resultant interaction of the two fields causes the wire to vibrate at the frequency of the driving current.

Accelerometers. Accelerometers have been installed on buoys to sense the motions of the buoys as they follow the motions of the waves [2, 16, 27]. Since the response of an accelerometer is a function of the relative angle between the sensitive axis and the local gravity vector, some means of determining the local vertical of the accelerometer package is necessary.

One system is a spherically shaped buoy, about 60 cm in diameter, with a self-contained accelerometer, double-integrator, transmitter package. With a battery package to power the instrument for 9 months, the system weighs about 90 kg.

Accelerations are detected using a rod with one end clamped and the other end in contact with a conductive fluid. Stationary electrodes create a stationary electrical field in the fluid. The rod acts as a spring so that accelerations perpendicular to the rod move the free end, and the voltage at the rod changes.

The accelerometer is suspended in a fluid of almost equal specific gravity, and a pendulous suspension keeps the sensitive axis vertical. The natural frequency of the accelerometer, as a result of the suspension and mounting systems, is about 45 Hz.

Electrical integrators are tailored to the mass and stiffness of the buoy and accelerometer. A high-pass filter of 0.03 Hz gives an error of about 3 % in the range of 0.6 to 0.8 Hz and about a 30 % error over the increased limits to 0.03 and 1.0 Hz. The standard instrument is capable of operation in maximum wave heights of 20 m.

4.31.3 Electrical Sensors

Resistance Staff. The resistance staff consists of a resistive element suspended on a vertical support. Water height is determined by measuring the change in resistance of the sensing element.

In one instrument, the resistive element is suspended vertically in the sea water [1]. The element is usually a wire whose resistance is linear with length. A return conductor is submerged in the sea water. The wire is excited by a constant current source at an audio frequency. The voltage between the non-immersed end of the resistance wire and the return conductor is a direct function of the amount of the resistance element that is not submerged since the submerged section is electrically short-circuited by the conducting sea water. Alternating current reduces electrolytic deterioration of the parts of the system exposed to sea water.

In another instrument, discrete resistors are used instead of a continuous sensing element. The junctions of equal resistance resistors connected in series are exposed to the sea water at increments equal to the desired measuring resolution. As the water rises, successive resistors are short-circuited, providing an incremental record of wave height. This instrument is self-calibrating. Regardless of changes in current, voltage, or any modifying amplifiers in the instrument, the height of the incremental step is always known by the mechanical placement of the resistor junctions.

A third instrument is a variation of the step-resistance staff just described. In this instrument, the shunt resistance of the submerged resistors is measured. The resistances are carefully chosen to produce a linear shunt resistance, and the connection between the resistance staff and the return conductor is made on the immersed side of the staff.

A fourth resistance staff overcomes the necessity of the sensing elements coming into direct electrical contact with the fluid. This decreases errors due to corrosion, oil, and physical damage from debris floating on the surface. This sensor is a resistance element enclosed in a sealed jacket. Figure 4.31-4 shows an exploded view of a typical length of the sensing element. The precision wound element produces 75 to 150 contacts per meter. The sensor functions by compressing the jacket so that the internal elements

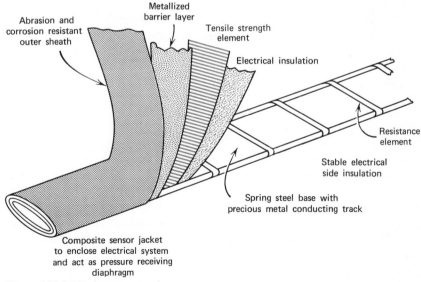

Figure 4.31-4 Metritape sensor element.

come into electrical contact with the conducting bus strip. progressively contacting the bending points of the resistance element. All contact points below the liquid level are shorted together so the resistance of the element above the liquid is a measure of the wave height. The incorporation of a tensile strength element allows the tape to be used in a one-point suspension system with a heavy weight to keep the sensor vertical and taut.

The accuracy and resolution of all resistance staffs depend on the stability and resolution of the current sources, the voltage reading and recording devices used, and the diameter of the sensing element.

Capacitance Staff. A capacitance probe shaped like a rod and placed vertically in a liquid increases in capacitance as the water level increases because air surrounding the probe is displaced by the liquid which has a higher dielectric constant. The probe is connected in an arm of a capacitance bridge. The change in capacitance of the probe created by the changing water level causes a corresponding imbalance voltage to appear in the bridge. This voltage is a direct analog of the water level [5, 26].

The probe consists of an insulated uniformly shaped cylindrical conductor and a return conductor in electrical contact with the liquid. The insulated cylindrical conductor can be an insulated wire, a Teflon-coated rod, or a mercury-filled glass tube. The two parts of the probe act like a coaxial capacitor, one plate being formed by the insulated conductor and the other formed by the conducting liquid. As the liquid rises, the capacitance increases in direct proportion to the liquid level.

A capacitance probe can be made very small in diameter to reduce the total wetted area and thereby reduce inaccuracies caused by surface tension of the water causing a slight difference in the capacitance measurement when the water is rising or falling. A miniscus forms around the probe, and if capillary waves are being measured with a sizable probe, appreciable errors can result. Probes consisting of number 28 enameled wire suffice in many cases.

4.31.4 Microwave Sensors

Radar Profiling. This technique uses an airborne radar altimeter with sufficient resolution to observe the surface wave motion relative to an aircraft [18].

One profiler is an FM-CW radar operating at a frequency of 4.3 GHz [30]. The carrier is modulated at 25 MHz. At a height of 150 m, the system illuminates a spot on the ocean about 5 m in diameter. Because of the large size of the spot, only wave periods greater than about 3 s can be measured, and it is necessary to analyze at least a 2.5 min. sample to provide statistically reliable results. This sensor requires compensation for vertical and horizontal motion of the aircraft.

Another radar profiler is the AN/SPN-37 Radar Wave Height Sensor. This AM-CW radar weighs 10 kg and is about the size of an attaché case.

Scatterometer. Section 5.6.4 discusses the use of a radar scatterometer and the response of its signals from the sea surface. For wave measurements, the scatterometer is an airborne CW radar system that propagates a fan-beam, 120° by 3°, in the direction of flight. Surface reflectivity is measured simultaneously at all angles of incidence. Reflectivity data are derived from the Doppler shift at the associated look angles. Figure 4.31-5 shows reflectivity measurements made at 13.3 GHz.

Radiometer. The theory and operation of microwave radiometers is discussed in Section 5.7.

Few quantitative data exist on the apparent temperature of the sea as a function of the wave conditions. One isolated observation shows that, in a sea state estimated to be 3, the ocean appeared 8°K warmer than a corresponding specular sea surface using horizontally polarized radiation at a frequency of 35 GHz. A study has been made of the measurement of the radiation at 19.4 GHz flying a radiometer at an altitude of 1 km over water whose temperature is 290°K [24]. Radiometers have sufficient sensitivity to detect the temperature changes predicted for differing wind speeds.

Figure 4.31-6 shows the temperature of horizontally polarized radiation

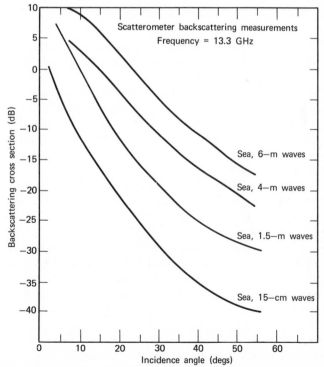

Figure 4.31-5 Scatterometer reflectivity as a function of wave height. Data: Ryan Aeronautical Co. Report No. 29172-146.

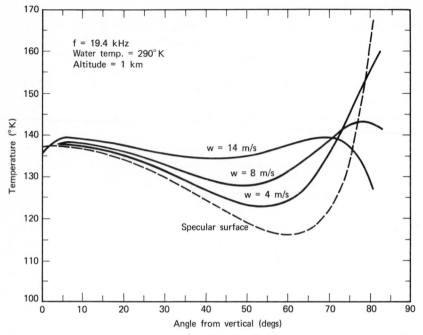

Figure 4.31-6 Temperature of horizontally polarized radiation as a function of angle (upwind case) for different wind speeds [24].

as a function of angle in the upwind case. Similar measurements in a cross-wind direction would yield an increase in temperature of a few degrees for observation angles near 0° and beyond 70°, and a decrease in temperature at intermediate angles. Vertically polarized observations do not vary more than a fraction of a degree with varying wind speeds when viewed at an observation angle of 50° at an altitude of 1 km.

A radiometer flown at altitudes of 1 or 30 km would indicate the temperatures shown in Fig. 4.31-7. A sea state scale based on the Beaufort Scale is superimposed on the abscissa to give quantitative sea state readings to the radiometer measurements.

The study [24] assumed a horizontally stratified atmosphere with no rain or clouds. If measurements are to be made from high altitudes where water vapor absorption has an effect on the observed temperature, additional measurements have to be made at other frequencies to eliminate this source of uncertainty.

4.31.5 Optical Sensors

Infrared Wave Profiler. An infrared wave profiler has been developed for use where a pier, tower, or ship is available for the instrument package support. A beam of light is amplitude modulated, and the reflected signal is phase compared to the transmitted modulating frequency to determine the height of the sea under the instrument. An electroluminescent diode is used as the source of noncoherent light in the infrared region (9600 Å). This is modulated at a frequency of 1 MHz. The resolution of this instrument depends on the light beam modulating frequency and the performance of the phase comparator. Comparison of the instrument output with an adjacent wave staff shows measurement capabilities of at least 2.5 cm when mounted 7 m above mean sea level. The instrument has operated successfully from the bow of a ship at distances of 15 m above the sea surface.

Laser Profiler. The laser profiler makes possible the surveillance of larger

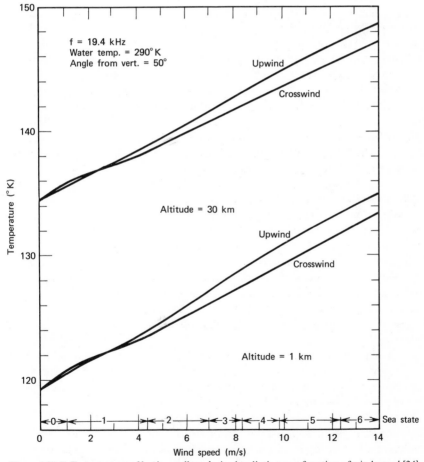

Figure 4.31-7 Temperature of horizontally polarized radiation as a function of wind speed [24].

areas in a short period of time when measuring ocean wave profiles from an airborne platform.

One instrument uses a helium-neon cw laser with a peak power output of 15 mW [19]. This beam is amplitude modulated and projected down to the surface of the water. The reflected light is collected by a 7.5-cm reflecting telescope. After passing through a bandpass filter, the reflected light amplitude is measured by a photomultiplier tube which produces an output voltage proportional to the amplitude of the reflected signal. This signal is phase compared with the original modulating signal to determine the measurement range.

The parameters of the instrument determine the measurement range and accuracy. The laser has a wavelength of 632 Å and a modulating frequency of 49.17 MHz. The resulting measuring range per 360° of phase change is 3.05 m. The output of the phase detector represents a total height variation of 1.5 m. Because the data are recorded with a voltage-controlled oscillator having a 1% accuracy, the accuracy of the wave height measurement is on the order of 1.5 cm. At an aircraft altitude of 200 m, this instrument illuminates a spot on the water about 3.4 mm in diameter.

Figure 4.31-8 shows a block diagram of the instrument. In the detection part of the instrument, the output and the reference signals are each heterodyned in the signal processor to translate the signals to 212 kHz. The two signals are then more easily correlated to determine phase difference. Both the in-phase and the quadrature signals of the phase comparator are recorded. This results in an easily discernible unambiguous differential height measuring ability of 3.05 m. A 10-kHz filter in the signal processor allows this instrument to follow waves with heights of 2.5 cm.

Glitter Patterns. If the sea surface were smooth like a mirror, an observer would be able to see the reflected image of the sun under the proper conditions of sun angle, sea surface tilt, and observer position. The sea surface can be thought of as composed of many small mirror-like areas, each capable of reflecting sunlight. A photograph of such a surface shows which segments of the surface at that moment are at the proper angle relative to the camera position and sun angle. Such a pattern is called a glitter pattern, and is a method whereby a statistical distribution of sea surface slopes can be measured.

The mathematics of glitter pattern analysis has been developed [7, 8, 9, 22]. Slope-distribution statistics from glitter pictures show that the slope of the sea surface is generally in an upwind direction, probably due to the effects of wind stress.

Stereo Photogrammetry. Stereo photogrammetry techniques are well developed [6, 17]. Stereo photography has been used for many years to produce contour maps of land masses. High-quality cameras are available using 12.5 and 23-cm films and a number of highly sophisticated aerial films that combine high speed with high resolution. Land masses remain stationary with time so that an aircraft can take a series of overlapping pictures, with the overlapping portions of each pair of contiguous pictures forming a stereo image of that area. For mapping the ocean surface, the two pictures have to be made at the same instant since the surface is constantly in motion, and two cameras are required.

Measurement of capillary waves using stereo photogrammetry techniques has been accomplished using a pair of modified aerial cameras on a special platform extending from the bow of a ship. The cameras were placed 3.7 m apart, 6 m in front of the ship, with a mean height above the water surface of 6 m, to produce a 60% overlap of the areas covered by the individual cameras. Vertical resolution under the best lighting conditions and minimum tilt of the ship was a maximum of 0.8 mm.

It is important that there be a minimum of glitter in stereo pictures because such sections of the photographs are unusable in this technique. Similarly, white-caps and other such phenomena are bad data points.

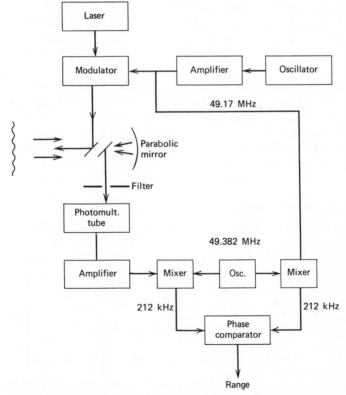

Figure 4.31-8 Simplified block diagram of laser profilometer.

Fourier Transforms. This technique uses an optical transform to determine directional spectra of the sea surface [23]. High-quality photographs of the sea surface are taken with an aerial camera from an airborne platform. That part of the sky acting as the light source for the section of ocean viewed by the camera must be either clear or uniformly overcast. As in the stereo technique, glitter is an unwanted phenomenon.

Optical density of the image on photographic film is directly related to the incident light intensity. Light intensity variations from points on the surface of the water in a direction toward the camera therefore produce corresponding optical density variations on the film. Because the desired light source produces a monotonically decreasing luminance with zenith angle, there results a one-to-one mapping of the surface normal into the photographic film.

The optical analysis consists of determining and measuring the Fourier transform of the optical density of the photographic transparency. The transparency is placed in a focal plane of an optical bench and illuminated with monochromatic, spatially coherent light as shown in Figure 4.31-9. This light is passed through a thin lens, and the Fourier transform appears in the focal plane of the lens. The pattern produced has a light amplitude proportional to the Fourier transform of the light amplitude in the data transparency because the basic transfer function for a lens has the form of a Fourier transform.

Photographic recording is responsive to energy density, and the energy spectrum is recorded as optical density on a photographic film placed at the focal plane. Thus, the distance from the focal point in the transform plane is proportional to the wave number of the wave in the original transparency. The azimuth angle corresponds to the wave direction, and the optical density is functionally related to the spectral amplitude. A picture of waves and the corresponding Fourier transform is shown in Figure 4.31-10.

4.31.6 Acoustic Sensors

Upward-Looking Sonar. High-frequency sound waves sent out in short pulses from an electroacoustic transducer are reflected back from a large

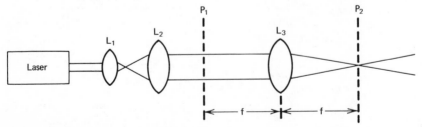

Figure 4.31-9 Optical system for processing Fourier transforms. P_1 is the plane of the wave data transparency; P_2 is the Fourier transform plane.

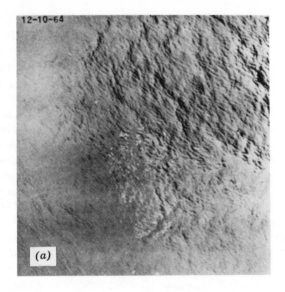

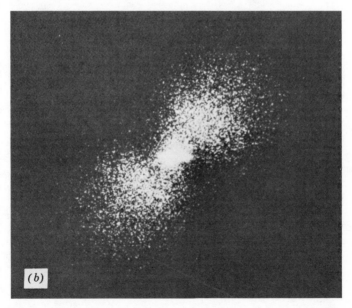

Figure 4.31-10 (*a*) A typical ocean wave data photograph: (*b*) the corresponding Fourier transform picture.

change in refractive index such as exists at the air-sea interface. The time required for the pulse to travel to the reflecting interface and back, together with the knowledge of the speed of propagation in the medium, allows the computation of the distance of the interface from the instrument. A sonar system that is vertically stabilized in the ocean can measure a profile of the waves passing overhead.

One instrument operates at a frequency of 200 kHz, with 10 W peak power driving a barium titanate transducer [12]. The high frequency helps keep the acoustic beamwidth to 2.5° at the half-power points. Pulses are transmitted with a duration of 375 μs and a repetition rate of 40 pulses per second.

Acoustic echoes can be caused by anomalies in the water, bubbles, fish, and flotsam. At this high frequency, nearby reflecting objects can produce false readings in the instrument, so received signals are blocked during the first 3.5 ms corresponding to a range of 2.5 m. False returns from distant objects are usually too low in amplitude to trigger the receiver detector.

The output of the instrument is a dc pulse whose amplitude is fixed and whose duration is a function of travel time or height of the sea-air interface above the transducer. Successive pulses are integrated to produce a more usable output.

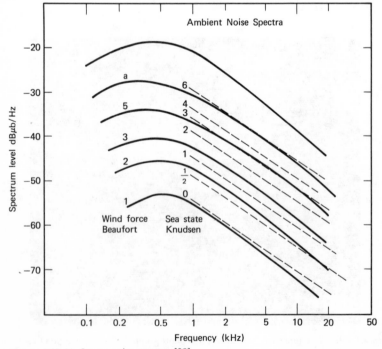

Figure 4.32-11 Ocean noise spectra [29].

The design of this instrument illustrates some of the critical design parameters in an upward looking sonar. Physical placement of the system (depth below the surface) and measurement resolution determine the frequency, the pulse length, and the power necessary for meaningful measurements. The acoustic spreading of the beam and the resulting area of sea surface illuminated is one of the most critical parameters.

In use, the instrument is placed 6 to 10 m below the trough of the waves. The linear dc output voltage proportional to wave height is accurate to $\pm 5\%$ of full scale over a range of 0- to 6-m waves.

Ambient Ocean Noise. Ambient noise in the ocean is generated predominantly by the action of waves at the surface in the frequency range of 100 to 10,000 Hz. These are, in turn, a function of the winds, duration, and fetch. Thus ambient ocean noise is a function of sea state. Noise measurements, as seen in Figure 4.31-11 [28, 29], are fairly good measures of the Beaufort Sea State.

Ultrasonic Altimeter. An ultrasonic altimeter has been used to obtain a sea surface profile from aboard ship [10]. The instrument uses ultrasonic pulses to determine the height of the unit above the sea surface. It was mounted on a boom extending over the bow of the ship on a gyro stabilized platform. The heaving motions of the ship were compensated by subtracting the vertical displacement of ship movement.

References

1. Ayres, R. A., and D. J. Cretzler, "A Resistance Wire Water Level Measurement System," *Marine Sciences Instrumentation*, Vol. 2, 93, Plenum Press, New York, 1963.
2. Barber, N. F., *Measurement of Sea Conditions by the Motion of a Floating Buoy*, Admiralty Research Laboratory Report 103.40/N.2/W, Teddington (bound as one of "Four Theoretical Notes on the Estimation of Sea Conditions," by M. S. Longuet-Higgins and N. F. Barber, 1946).
3. Barber, N. F., "The Directional Resolving Power of an Array of Wave Detectors," *Ocean Wave Spectra*, pp. 137–150, Prentice-Hall, Englewood Cliffs, New Jersy, 1963.
4. Bascom, W., *Waves and Beaches and the Dynamics of the Ocean Surface*, Doubleday, New York, 1964.
5. Campbell, W. S., "Liquid Level Measuring Apparatus," United States Patent Office Patent No. 2,817,234, December 24, 1957.
6. Chase, J., et al., "The Directional Spectrum of a Wind-Generated Sea as Determined from Data Obtained by the Stereo Wave Observation Project," New York University College of Engineering Report, July, 1957.
7. Cox, C., and W. Munk, "Measurement of the Roughness of the Sea Surface from Photographs of the Sun's Glitter," *J. Opt. Soc. Am.* **44**, 838–850, 1954.
8. Cox, C., and W. Munk, "Slopes of the Sea Surface Deduced from Photographs

of Sun Glitter," *Bulletin of the Scripps Institution of Oceanography*, Vol. 6, No. 9, pp. 401–488, Univ. of Cal. Press, Los Angeles, 1956.

9. Cox, C., and W. Munk, "Statistics of the Sea Surface Derived from Sun Glitter," *J. Marine Res.* **13**, No. 2, 198–227, 1954.

10. DeLeonibus, P. S., and A. Moskios, "Performance of a Shipboard Wave Height Sensor," U. S. Naval Oceanographic Office Informal Manuscript Report No. 0-4-65, 1965.

11. Kinsman, B., *Wind Waves—Their Generation and Propagation on the Ocean Surface*, Prentice-Hall, Englewood Cliffs, New Jersey, 1965.

12. Kronengold, M., J. M. Lowenstein, and G. A. Berman, "Sensors for the Observation of Wave Height and Wind Direction," *Marine Sciences Instrumentation*, Vol. 3, Plenum Press, New York, 1965.

13. Knight, A. M., *Modern Seamanship*, 13th ed., pp. 423–448, D. Van Nostrand, Princeton, New Jersey, 1960.

14. Knudsen, V. O., R. S. Alford, and J. W. Emling, "Underwater Ambient Noise," *J. Marine Res.* **7**, 410, 1948.

15. Lefcort, M. D., "Vibrating Wire Pressure Transducer Technology," *J. Ocean Tech.* **2**, No. 2, 37–44, 1968.

16. Longuet-Higgins, M. S., D. E. Cartwright, and N. D. Smith, "Observations of the Directional Spectrum of Sea Waves Using the Motions of a Floating Buoy," *Ocean Wave Spectra*, pp. 111–136, Prentice-Hall, Englewood Cliffs, New Jersey, 1963.

17. *Manual of Photogrammetry*, 3rd ed., American Society of Photogrammetry, Falls Church, Virginia, 1966 (two volumes).

18. Morrow, C. M., "Ocean Wave Profiling Radar Systems," Naval Research Laboratory Report No. 6052, April, 1964.

19. Olsen, W. S., and R. M. Adams, "A Laser Profilometer," *J. Geophys. Res.* **75**, No. 12, 2185–2187, April 20, 1970.

20. Pederson, A. M., "A Digital Depth Reference," *J. Ocean Tech.* **2**, No. 2, 28–36, 1968.

21. Rolfe, R. C., "Vibrating Wire Pressure Transducer Electronics," *J. Ocean Tech.* **2**, No. 2, 45–48, 1968.

22. Schooley, A. H., "A Simple Optical Method for Measuring the Statistical Distributions of Water Surface Slopes," *J. Opt. Soc. Am.* **44**, 37–40, 1954.

23. Stilwell, Jr., D., "Directional Energy Spectra of the Sea from Photographs," *J. Geophys. Res.* **74**, No. 8, 1974–1980, April, 1969.

24. Stogryn, A., "The Apparent Temperature of the Sea at Microwave Frequencies," *IEEE Trans. Antennas and Propagation* **AP-15**, 278–286, March, 1967.

25. *Wind Waves and Swell*, U. S. Navy Hydrographic Office Publication No. 604, Washington, D. C., 1851.

26. Tucker, M. J. and H. Charnok, "A Capacitance Wire Recorder for Small Waves," in *Proceedings of the 5th Conference on Coastal Engineering*.

27. Tucker, M. J., "The Accuracy of Wave Measurements Made with Vertical Accelerometers," *Deep-Sea Res.* **5**, 185–192, March, 1959.

28. Urick, R. J., *Principles of Underwater Sound for Engineers*, pp. 160–186, McGraw-Hill Book Company, New York, 1967.

29. Wenz, G. M., "Acoustic Ambient Noise in the Ocean; Spectra and Sources," *J. Acoust. Soc. Am.* **34**, 1936–1952, 1962.

30. Wilkerson, J., and J. J. Schule, "An Oceanographic Aircraft," Informal Manuscript No. 66–26, U. S. Naval Oceanographic Office, 1967.
31. Yerman, A. J., "An Ocean Depth Sensor Utilizing a Tunnel Diode as the Pressure Sensing Element," in *Marine Sciences Instrumentation*, Vol. 4, pp. 541–554, Plenum Press, New York, 1968.

H. Dean Parry

4.32 WIND SENSORS

Wind is atmospheric motion, a vector quantity. Meteorological convention usually defines wind as the horizontal component of air motion. "Surface" winds are measured by ground (or sea surface) based sensors. "Upper winds" are measured by tracking objects floating with the moving air or by remote sensing techniques. Surface winds are measured in the boundary layer near the earth's surface. The World Meteorological Organization recommends that surface winds be measured at a height of 10 m above the earth's surface. Usual practice in the United States is to mount surface wind measuring equipment 6 m above the ground.

4.32.1 Surface Wind Sensors

Transduction of the air motion to some other form of signal is much more difficult than the further processing of this signal. Most wind transducers consist of a physical device which is intended to follow atmospheric motions. The inertia and aerodynamic properties of the device keep it from exactly following the atmospheric motions so that the output depends on the atmospheric motion and the physical characteristics of the device. The device itself can be read out magnetically, optically, or electrically with virtually no error. Consequently most of the concern is with that portion of the wind sensor which interfaces directly with the wind.

The speed direction technique is most commonly used for surface wind measurements. The device which measures speed is called an anemometer while the direction measuring device is called a wind vane.

Wind Vane. The wind vane (also called a weather vane or weather cock) is the standard device for measuring wind direction. Any bar which is free to rotate in the horizontal plane and which has a vertical plate affixed to one end of it, such a device [1, 9, 18]. If the wind vane were a massless, frictionless device which responded linearly to aerodynamic effects, it would exactly follow the atmospheric motions and be a perfect transducer. The physical vane has none of these properties and therefore does not follow the true air motion exactly.

The response of the vane is represented to a good approximation by the equation of motion for a second-order system as

$$\ddot{\theta} + 2\zeta\omega_n\dot{\theta} + \omega_n^2\theta = f(t), \tag{4.32-1}$$

where ζ is the damping ratio, ω_n is the natural frequency of the system, θ is the response, and $f(t)$ is the forcing function.

For a step change and a normally designed vane the solution to this equation is

$$\theta = \gamma \left[1 - e^{nt}\cos\omega_1 t - \frac{n}{\omega_1}\sin\omega_1 t \right], \tag{4.32-2}$$

where t is time, γ is the angular difference between wind direction and the vane's tail, $n = -\zeta\omega_n$, and $\omega_1 = \omega_n(1 - \zeta^2)^{1/2}$,

Only the periodic solution is considered. Real vanes have $\zeta < 1$ hence their motion is damped oscillatory. The step change used in Eq. 4.32-2 is a simple forcing function. In the real atmosphere the forcing function is atmospheric turbulence. To be precise the analytic expression for the forcing function of the turbulence should be used, but this is not feasible since no single and unique expression represents atmospheric turbulence.

The response of a second order system to a step change is shown in Figure 4.32-1. Note that the greater the damping ratio the longer the time required to reach the new value of the step change.

When a subcritically damped wind vane oscillates around the new value of the step change it is said to overshoot. In doing this it indicates directions which do not exist physically in the atmosphere. The degree of damping is selected on the basis of the use for the data and the manner in which the data is processed. The optimum damping ratio for virtually all applications probably lies between 0.4 and 0.6. It can be shown that [18]

$$\zeta \approx 0.395 \left(\frac{a_v R^3 S}{J} \right), \tag{4.32-3}$$

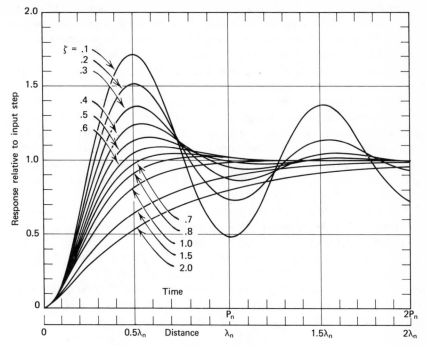

Figure 4.32-1 Second-order system response to a step input.

where, a_v is a factor accounting for the effect of the shape of the tail plate on torque, R is the distance from vane pivot to the aerodynamic center of pressure of the tail, S is the area of the tail, and J is the moment of inertia of the vane about the pivot point.

It is difficult to build a wind vane which can withstand strong winds (50 m/s) and has a 0.5 damping ratio. To achieve this degree of damping the moment of inertia, and hence the weight of all components (especially those at some distance from the axis of rotation), must be minimized. Decreasing the weight for a given material also decreases the strength. The damping is also increased if the aerodynamic torque (tail area and length) is increased. Again there is a conflict because increasing these increases the moment of inertia. Several wind vane designs have achieved a damping ratio near 0.5 with moderately good ruggedness. These vanes can withstand wind-tunnel winds of 60 m/s, which is equivalent to a natural wind of 50 m/s. The natural wind is more destructive because it is turbulent and generates larger stresses than the laminar flow in a wind tunnel.

Except for negligible second-order effects, the monovane described above measures only the direction of the horizontal flow. For some applications measurement of three-dimensional flow is needed. To obtain a three-dimensional wind direction the shaft of the wind vane is made to rotate in the vertical plane while retaining its ability to rotate in the horizontal. To this shaft two plates are attached. One is mounted vertically in the conventional manner. The other is mounted horizontally. The second vane aggravates the moment of inertia problem and decreases the responsiveness for the same degree of ruggedness.

The output signal of the vane is the position of the vane or of the shaft on which the vane turns. Techniques used to read out the shaft position include contacting switches, magnetic reed switches, photocells shaded by shutters attached to the shaft, a potentiometer giving an analog change in resistance with change in direction, stationary plate and one attached to the shaft giving an analog change in capacitance with change in direction (this requires a 2:1 gearing to remove ambiguities), synchro systems, and a shift of phase with change in direction.

Pressure-Plate Anemometer. The pressure plate is probably the oldest anemometer (Hooke 1667). A plate is held normal to the wind direction by the action of a large vane. In one version the plate is mounted so that the force of the wind compresses a spring [10, p. 151–2]. The amount of compression is a measure of the wind force on the plate and the wind speed. The pressure-plate anemometer has many problems and is not in general use. The wind force on a plate normal to the wind is

$$F = \tfrac{1}{2} C_D A \rho V^2 , \qquad (4.32\text{–}4)$$

where C_D is the drag coefficient, A is the plate area, ρ is the air density, and V is the wind speed.

Rotating-Cup Anemometer. The most common anemometer is the rotating cup. The first device employed four hemispherical cups supported by four horizontal cross-arms. Modern instruments have three truncated cone cups. Beading around the edges of the cups makes the rate of rotation more linear with wind speed.

These cups turn on a shaft which is as near frictionless as can be achieved. The cups are symmetrical in the horizontal plane so the rate of rotation is not affected by changes of direction in the horizontal plane. The output signal is rate of shaft rotation. Cup design has achieved good linearity ($< \pm 0.25$ m/s deviation) between wind speed and shaft rotation for wind speeds between 0 and 50 m/s. As in the case of the wind vane, many techniques have been used to change shaft rotation (a mechanical signal) to an electrical signal. The most common methods are driving an ac or dc generator with the shaft, attaching one or more magnets to the shaft which activate reed switches, and attaching a shade to the shaft that passes between a photocell and a light source.

The response time of nearly all kinds* of wind measuring instruments is a function of the magnitude of the measurement. The time constant, lag coefficient, or response time depends on wind speed. For cup anemometers and vanes the response time varies inversely as the mean wind speed. It is convenient to define a *distance constant* (in lieu of a time constant and independent of wind speed) for rotating anemometers *as the length of wind which must pass the anemometer to cause it to indicate 63 % of a step* change [16]. An analogous characteristic for the wind vane is the delay distance usually defined as the length of wind which must pass the vane to cause it to indicate 50 % of a step change. The anemometer can be matched to the vane by making the distance constant equal to the delay distance [9].

Errors inherent in the cup anemometer can be grouped into three classes. The first is the error due to friction and loading (if any) imposed by the telemetering technique. The frictional error is most troublesome when the anemometer is stopped in a calm and has to be started by a light breeze. Many modern anemometers have starting speeds of less than 0.5 m/s. The second class of error is due to the fact that the air motion may not be exclusively in the horizontal plane. Winds which are nonhorizontal or which vary around the horizontal can affect the horizontal wind measurement by 10% [9]. The third class of error is dynamic error. This error does not exist in steady-state nonturbulent conditions. Under non-steady-state conditions the average rate of rotation departs considerably from the average wind. The magnitude of this error depends principally on the distance constant of the anemometer and the character of the turbulence, and it ranges from a few percent to as much as 20 percent. The error is due to the fact that the distance "constant" is not constant but is smaller for an accelerating wind than for a decelerating wind [5, 15, 16].

The average distance constant can be computed from the expression

$$L = \frac{2J}{\rho N_c A R^2 C \gamma},$$ (4.32-5)

where J is the moment of inertia of the cup assembly around its axis of rotation; ρ is the density of ambient air; A is the frontal area of the cup; R is the cup moment arm; N_c is the number of cups; C is the ratio of air speed, V, to cup center speed, U, under equilibrium conditions; and γ, the coefficient in the equation,

$$J \frac{dU}{dt} = \tfrac{1}{2} \rho N_c U^2 A R \gamma,$$ (4.32-6)

relates to the aerodynamic property of the cup. It can be measured in the laboratory.

* The sonic anemometer is an example of an exception.

Equation 4.32–5 shows that an increase in the moment of inertia results in an increase in L. The only other variables on the right side of this equation which the designer can change are N_c, A, and R. Increasing any of these decreases L but unfortunately also increases J. Cup anemometers which withstand 50 m/s and have an L value of 2 m or less have been built.

Propeller Anemometer. A propeller with two or more blades can be designed to have a rotation rate nearly proportional to wind speed. The motion of the propeller is closely approximated by a first-order differential equation, and expressions for dynamic characteristics closely parallel those of the cup anemometer. The distance constant of the propeller is

$$L \approx 426 \, \frac{t(\text{SG})}{A_b},$$ (4.32–7)

where t is the blade thickness, A_b is the approximate blade lift curve slope, and SG is the specific gravity of blade material [5].

While the cup anemometer is omnidirectional in the horizontal plane, the propeller responds correctly only if its axis of rotation is pointed into the wind. Consequently the propeller is mounted on a wind vane which continually points it into the wind. This complicates response problems by adding to the moment of inertia of the vane. Even though the wind is not horizontal, the bivane keeps the propeller pointing into the wind to produce acceptable windspeed measurements of the three orthogonal wind components.

Fluidics Anemometer. These devices measure the difference in pressure between receiver orifices in the path of one or more air jets exposed to the wind. The wind changes the jet speed and direction which change the receiver pressures [11].

Hot-Wire Anemometer. The amount of heat carried away from a heated wire is a function of the speed of the air flowing across the wire [3, 13]. Normally the power required to maintain the wire at a constant temperature is measured and the measurement is converted into air speed. Hot-wire anemometers can be designed to have extremely rapid response. Variations of wind speed having frequencies greater than 5 kHz can be measured with hot-wire anemometers. The equation relating current, i, to wind speed, V, for a constant temperature anemometer is known as King's law. It is

$$i^2 = i_0^2 + K \sqrt{V},$$ (4.32–8)

where i_0 and K are experimentally determined constants.

Pitot-Tube Anemometer. The pitot-tube anemometer measures wind by using Bernoulli's principle. A vane turns one orifice into the wind and a second orifice at 90° to the first has wind blowing parallel to its face [6, pp. 196–199; 10, pp. 152–162]. The difference in pressure between these two orifices is a measure of the wind speed. The analog of this device is the airspeed sensor on an aircraft. This instrument has a fast response and is frequently used to measure gustiness. The pressure output is difficult to automate and difficult to average. Consequently the application of this instrument is limited. The basic equation for this instrument is

$$\Delta P = \tfrac{1}{2}\rho V^2(1 - C),\cdot \qquad (4.32\text{--}9)$$

where ρ is the density, ΔP the differential pressure, C a constant, and V the speed. Since ΔP is proportional to V^2, the device is not effective at low speeds.

Sonic Anemometer. The velocity of sound through a medium is equal to the vector sum of sound velocity in an undisturbed medium and the velocity of the medium. The sonic anemometer measures the difference in time required for sound to pass in opposite directions between two fixed receiver-transmitters. This time difference must be corrected for temperature since the velocity of sound in air is proportional to the square root of the absolute temperature. After correction the time difference can be used to compute the component of the velocity of sound along the line connecting the receiver-transmitters. For information along more than one line, additional pairs of receiver-transmitters are used. Up to three pairs are often used. Sonic anemometers have no inertia and therefore measure fluctuating winds much more precisely than mechanical wind measuring devices [8].

Aerodynamic Drag Anemometer. A number of anemometers which measure the aerodynamic drag of the wind on a sphere or cylinder have been built. Typically this drag is read out by two or three mutually perpendicular strain gages. Since the drag is proportional to the square of the wind speed, it is difficult to obtain a large dynamic range with this instrument. Since this instrument typically has been difficult to keep in calibration and has several inherent errors, it has not enjoyed wide acceptance for operational use.

Wind Component Meter. Components of wind speed are read from a standard wind vane and cup anemometer by a photoelectric readout technique. Light from a source passes through two cylindrical shutters onto four photoelectric cells spaced 90° apart. The vane, the anemometer, and two cylindrical shutters all have a common vertical axis. One cylinder has a row of holes in the form of a sine wave and is turned by the vane. The other has horizontal rows of holes and rotates with the anemometer. The two shutters are designed so that pulses of light strike one photocell in accordance with $V \sin \theta$ and the other cell in accordance with $V \cos \theta$, where V is speed and θ is the direction angle. The number of pulses impinging on one pair of cells is proportional to the component of wind in the east-west direction. The number of pulses impinging on the other pair of cells is proportional to the component of wind in the north-south direction. The principal advantage of the system is that the pulses can be counted over an interval of time to obtain a true average component of speed along each of the two orthogonal directions. These two average components can be combined by simple vector addition to obtain a true time average wind speed and direction. Since wind is a vector the simple average of speed and direction does not give a true time average wind. A true time average of wind velocity is important for many applications, particularly climatology and air pollution studies.

Two directional anemometers mounted orthogonally can be used to measure wind components only if their response is proportional to the product of wind speed and the cosine of the angle between anemometer and wind directions. Pressure plate and pitot tube anemometers do not obey this cosine law. It is possible to design propellers and fluidic anemometers that closely approximate a cosine response.

4.32.2 Upper Wind Measurements

Theodolite Techniques. In wind finding by tracking an airborne object (balloon or parachute), two successive positions of the tracked object are projected onto a sphere which is concentric with the earth and has a diameter equal to the average sea level diameter of the earth. In Figure 4.32-2 the balloon's projection moves from P to Q, during the time interval t, and the wind is determined by evaluating the vector PQ. This vector is evaluated by making a set of measurements identified by the subscript 1 at time t_1 while the balloon is above P and a second set of measurements identified by subscript 2 at time t_2 while the balloon is above Q. Several sets of measurements can be made and used to evaluate the vector PQ. The measurement set most commonly in use is the azimuth angle, A, the elevation angle, E, and the height, H, shown in Figure 4.32-2. The height is calculated using the hydrostatic equation and measurements of pressure and temperature made by the ascending radiosonde. Angle measurements are made by a radiotheodolite. This technique is called the AEH technique. The convention that $t_2 - t_1 = t$, $A_2 - A_1 = A$ is adopted. The wind speed, V, from the AEH technique is given by [12]

$$
V = \frac{R}{t} \cos^{-1} \left\{ \left[\cos \left(\cos^{-1} \frac{R \cos E_1}{R + H_1} \right) - E_1 \right] \right.
$$

$$
\left[\cos \left(\cos^{-1} \frac{R \cos E_2}{R + H_2} \right) - E_2 \right] + \left[\sin \left(\cos^{-1} \frac{R \cos E_1}{R + H_1} \right) - E_1 \right]
$$

$$
\left. \left[\sin \left(\cos^{-1} \frac{R \cos E_2}{R + H_2} \right) - E_2 \right] \cos A \right\}, \tag{4.32–10}
$$

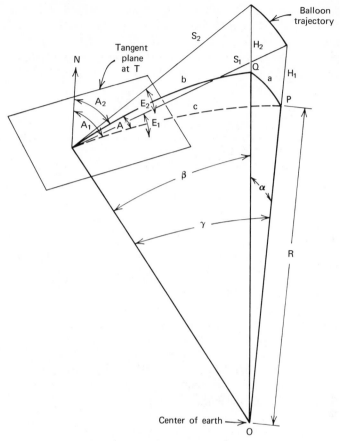

Figure 4.32-2 Balloon tracking geometry using theodolite.

where R is the mean radius of the earth. The direction, D, is given to a sufficient approximation by

$$D = 180 - \tan^{-1} \frac{(-\beta \sin A_2 + \gamma \sin A_1)}{(\beta \cos A_2 - \gamma \cos A_1)}, \qquad (4.32\text{--}11)$$

where β and γ are defined in Figure 4.32-2.

It can be shown that [12]

$$\beta = \left[\cos^{-1}\left(\frac{R}{R + H_2} \cos E_2 \right) \right] - E_2, \qquad (4.32\text{--}12)$$

and

$$\gamma = \left[\cos^{-1}\left(\frac{R}{R + H_1} \cos E_1 \right) \right] - E_1. \qquad (4.32\text{--}13)$$

The slant distance, S, from the radiotheodolite position, T, to the balloon can be measured by a radar technique or by a transponder type radiosonde. This measurement makes possible two additional techniques for fixing the vector **PQ**. One is the azimuth-elevation-slant range (AES) technique. For these measurements it can be shown that β and γ are given by

$$\beta = \sin^{-1} \frac{S_2 \cos E_2}{G_2}, \qquad (4.32\text{--}14)$$

where

$$G_n = (R^2 - 2S_n R \sin E_n + S_n^2)^{1/2}, \qquad (4.32\text{--}15)$$

$$\gamma = \sin^{-1} \frac{S_1 \cos E_1}{G_1}, \qquad (4.32\text{--}16)$$

and

$$V = \frac{R}{t} \cos^{-1} \left[\cos \left\{ \sin^{-1} \left(\frac{S_1 \cos E_1}{G_1} \right) \right\} \cos \left\{ \sin^{-1} \left(\frac{S_2 \cos E_2}{G_2} \right) \right\} + \left(\frac{S_1 \cos E_1}{G_1} \right) \left(\frac{S_2 \cos E_2}{G_2} \right) \cos A \right].$$

$$(4.32\text{--}17)$$

As before D is given by Eq. 4.32-11 using the new values of β and γ.

Another technique using the slant range is the slant range-height-azimuth (SHA) technique. With these measurements

$$\beta = \cos^{-1} \frac{R^2 + (R + H_2)^2 - S_2^2}{2R(R + H_2)}, \qquad (4.32\text{--}18)$$

$$\gamma = \cos^{-1} \frac{R^2 + (R + H_1)^2 - S_1^2}{2R(R + H_1)}, \qquad (4.32\text{--}19)$$

and

$$V = \frac{R}{t} \cos^{-1} \left[\frac{(R^2 + (R + H_1)^2 - S_1^2)\,(R^2 + (R + H_2)^2 - S^2)}{4R^2(R + H_1)(R + H_2)} \right.$$
$$+ \sin \cos^{-1} \left(\frac{R^2 + (R + H_1)^2 - S_1^2}{2R(R + H_1)} \right)$$
$$\left. \sin \cos^{-1} \left(\frac{R^2 + (R + H_2)^2 - S_2^2}{2R(R + H_2)} \right) \cos A \right]. \qquad (4.32\text{--}20)$$

The direction is given by Eq. 4.32–11 using the appropriate values of β and γ. Computer programs exist for reducing winds from AEH and SHA measurement sets [12].

Errors in upper wind measurements depend on the system used, the cotangent of the elevation angle (whether measured or not), the time interval between measurements, the error of angular measurement, and the error of height and slant range measurement (if made). Table 4.32-1 [17] shows rms wind errors for various techniques for an angular error of $0.1°$ for various values of Q, the cotangent of the elevation angle, for a slant range error of 20 m, a height error equivalent to 1 millibar, and a time interval of 1 min. The SHA and AES systems have an accuracy much greater than the AEH system. Improved accuracy is attained by using the significantly more costly transponder type radiosonde.

Table 4.32-1. Vector of Wind Error, in m/s, as a Function of Height and cot e

cot e	Height							
	5 km		10 km		15 km		20 km	
TECHNIQUE[a]	1	2	1	2	1	2	1	2
1	0.5	0.5	0.6	1	1.2	1.5	1.2	3
2	0.5	1	1	2.5	1.5	4	2	6.5
3	0.5	2	1.5	4.5	2	7	2.5	11
4	1	3.5	1.5	7	2.5	11	3	16
5	1	5.5	2	11	3	16	4	24

[a]Technique #1 is AES and SHA; technique #2 is AEH.

An error common to all balloon tracking systems is caused by failure of the balloon to exactly follow atmospheric motions. It is known that a sphere rising vertically through the atmosphere sheds vortices from one side and then another. This imparts a horizontal component of motion with respect to the atmosphere. A constant pressure balloon (ROBIN) used to measure wind above 30 km has dimples impressed in its skin to prevent formation of mature vortices.

Navigational Aid Techniques. Navigational fixes can be computed from the time/phase difference in signals received from the Loran or Omega navigational aid systems [4]. Such position fixes can be used to measure winds aloft [2, 7] by computing the vector difference, $\Delta\mathbf{R}$, between positions at time t_1 and at some later time t_2. The direction of wind during the interval t_1 to t_2 is the direction of $\Delta\mathbf{R}$, and the speed is $|\Delta\mathbf{R}|/(t_2 - t_1)$. In practice speed and direction can not be computed from a single vector. It is necessary to compute a sequence of vectors and then apply an appropriate smoothing algorithm to obtain satisfactory accuracy.

The *difference* between two successive positions is used in computing each vector. The absolute error in positioning a navigational fix by Loran is about 100 m. Wind speeds and directions computed from two such successive fixes would not have acceptable accuracy, but by computing the difference in position it is possible to obtain an accuracy of better than 0.25 m/s. Accuracy obtainable from Omega is less, being limited by the stronger interference from atmospheric electricity at Omega frequencies and the greater baseline differences used by Omega. Sophisticated signal processing techniques are required to obtain suitable winds from the Omega network.

Figure 4.32-3 shows the geometery. Points P_1 and P_2 are projections on a plane, tangent to the geocentric sphere through point P_1, of successive positions of the balloon floating with the wind and carrying the Loran C/Omega receiver. Thus P_1 and P_2 are the opposite ends of an individual wind vector, ΔR, x and y are its rectilinear components, S_a is the projection on this plane of a Loran/Omega station location, a_1 and a_2 are the distances from S_a to P_1 and P_2 respectively, and α_1 and α_2 are the azimuth of these two lines. Now, from plane geometry [14]

$$a_1 = a_2 \cos(\Delta\alpha) + x \sin\alpha_1 + y \cos\alpha_1. \qquad (4.32\text{-}21)$$

$\Delta\alpha$ is $\alpha_1 - \alpha_2$ and is a very small angle. Hence $\cos \Delta\alpha \approx 1$.

$$a_2 = a_1 - x \sin\alpha_1 - y \cos\alpha_1. \qquad (4.32\text{-}22)$$

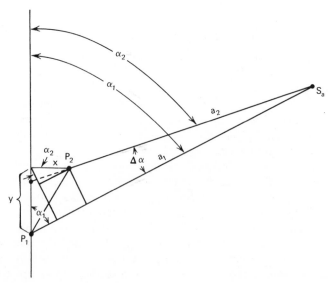

Figure 4.23-3 Balloon tracking geometry using navigation aids.

Similarly for other nav-aid stations S_b and S_c,

$$b_2 = b_1 - x \sin \beta_1 - y \cos \beta_1, \qquad (4.32\text{-}23)$$

and

$$c_2 = c_1 - x \sin \gamma_1 - y \cos \gamma_1. \qquad (4.32\text{-}24)$$

The last three equations are called the range up-date equations.

Also

$$a_2 \sin (\Delta \alpha) = y \sin \alpha_1 - x \cos \alpha_1. \qquad (4.32\text{-}25)$$

Since $\sin (\Delta \alpha) \approx \Delta \alpha$ and $\Delta \alpha = \alpha_1 - \alpha_2$,

$$\alpha_2 = \alpha_1 - (y \sin \alpha_1 - x \cos \alpha_1) \frac{1}{a_2}. \qquad (4.32\text{-}26)$$

Similarly for stations S_b and S_c

$$\beta_2 = \beta_1 - (y \sin \beta_1 - x \cos \beta_1) \frac{1}{b_2}, \qquad (4.32\text{-}27)$$

$$\gamma_2 = \gamma_1 - (y \sin \gamma_1 - x \cos \gamma_1) \frac{1}{c_2}. \qquad (4.32\text{-}28)$$

These are the azimuth up-date equations.

Now

$$c[\Delta T_{ab_2} + B_1] = a_2 - b_2, \qquad (4.32\text{-}29)$$

where c is the velocity of radio propagation, ΔT_{ab_2} is the measured difference in propagation time between the balloon at P_2, and station S_a and S_b, and B_1 is the total bias or fixed time differential between the transmission starts from stations S_a and S_b. Subtracting Eq. 4.32-23 from Eq. 4.32-22 and substituting Eq 4.32-29 in the result gives

$$c[\Delta T_{ab_2} + B_1] = a_1 - b_1 - x[\sin \alpha_1 - \sin \beta_1] - y[\cos \alpha_1 - \cos \beta_1],$$

$$(4.32\text{-}30)$$

$$c[\Delta T_{cb_2} + B_2] = c_1 - b_1 - x[\sin \gamma_1 - \sin \beta_1] - y[\cos \gamma_1 - \cos \beta_1].$$

$$(4.32\text{-}31)$$

ΔT_{ab_2} and ΔT_{cb_2} are measured. The system constants B_1 and B_2 are known. For the first iteration $a_1, b_1, c_1, \alpha_1, \beta_1,$ and γ_1 are calculated from the known geographic relationship between the radiosonde ground station and the navigational aid stations, S_a, S_b and S_c. Equations 4.32-30 and 4.32-31 are simultaneous linear equations in x and y and are solved for these variables.

$\Delta \mathbf{R}$ is obtained from x and y. Knowing x, y, a_2, b_2 and c_2, α_2, β_2, and γ_2 are obtained from the respective up-date equations. Then ΔT_{ab_3} and ΔT_{cb_3} are measured for the balloon at P_3 and x and y for the vector between P_2 and P_3 is computed. This iterative process is continued through the entire flight and wind data are derived from properly smoothed sequences of vectors obtained by the iteration.

The navigational aid wind finding technique has a significant advantage over radiotheodolite techniques. As is seen from Table 4.32-1 these latter techniques, particularly those techniques not using the much more expensive transponder sonde, have very poor accuracy at intermediate and great heights during high winds for which the cotangent of the elevation angle is large. In contrast the accuracy of the navigational aid wind finding technique is almost independent of the height and the distance from the radiosonde ground station.

References

1. Arakawa, H., "On the Windvane of the RAE Pattern"; *Geophys. Mag.* **4**, 53–59, 1931.
2. Beukers, John M., "Windfinding Using the Loran-C and Omega Long Range Navigation Systems" *IEEE Transactions, Geosci. Electron.* August 1968.
3. "Bulletin TN-5," Thermosystems Corp., St. Paul, Minnesota.
4. Campbell, A. C., *Loran to Geographic Conversion and Geographic to Loran Conversion IR No. N-3-64*, Naval Oceanographic Office, U. S. Navy, 1964.
5. Deacon, E. L., "The Overestimation Error of Cup Anemometers in Fluctuating Winds," *J. Sci. Instr.* **28**, 231–234, Aug. 1951.
6. *Handbook of Meteorological Instruments*, Part I, "Instruments for Surface Observations," Her Majesty's Stationery Office, London, 1956.
7. Harmantas, Christos, "Winds Aloft Measurement through Loran-C Navigation Aid," ESSA Technical Memorandum WBTM. EDL 6, U. S. Department of Commerce, Environmental Science Services Administration, 1968.
8. Kaimal, J. C. "A Continuous Wave Sonic Anemometer—Thermometer," *J. Appl. Meteorology* **2**, 156–164, 1963 (see also pp. 180–186).
9. MacCready, Paul B., Jr., and Henry R. Jex, "Response Characteristics and Meteorological Utilization of Propeller and Wind Vane Sensors," *J. Appl. Meteorology* **3**, 182–193, 1964.
10. Middleton, W. E., and A. F. Spilhaus, *Meteorological Instruments*, 3rd ed., University of Toronto Press, Toronto, 1953.
11. Neradka, V. F., "Fluidic Wind Sensor Research Leading to a Flight Test Model," Final Report on NASA CR-111808, NASA/Langley Research Center, Hampton, Virginia, December 1970.
12. Parry, H. D., "Semi-Automatic Computation of Rawinsondes," ESSA Technical Memorandum, WBTM EDL 10, U. S. Department of Commerce/ Environmental Science Services Administration, October 1969.
13. *Proceedings of the University of Maryland International Symposium on Hot Wire*

Anemometers, Ed. W. L. Melnik and J. R. Weske, USAF OSR Final Report #68-1492, 1968.

14. Sanders, M. J., Jr., Private Communication, Oct. 1973.
15. Schrenk, O., "Ueber die Tragheitsfehler des Schalenkreuz-Anemometers bei schwankender Windstaerke," *Zeits. f. Technische Physik,* **10**, No. 2, 57–66, 1929.
16. Schubauer, G. B., and G. H. Adams, "Lag of Anemometers," NBS Report 3245, 1954.
17. World Meteorological Organization, *Guide to Meteorological Instrument and Observing Practices,* 3rd ed., WMO-No. 8, TP. 3, Geneva, Switzerland.
18. Wreringa, V., "Evaluation and Design of Wind Vanes," *J. Appl. Meteorology* **6**, 1114–1122, 1967.

Chapter 5

Remote Sensors

Edward A. Wolff

5.1 SENSING TECHNIQUES

Remote sensors are used to measure parameters located at some distance from the sensing instrument. These sensors make use of the fact that the measured parameter has an effect on the fields or waves detected by the remote sensor. Astronomy is an example of an early science which made use of a remote sensor, a telescope on earth, to observe celestial bodies.

Remote instruments can be either active or passive. An active instrument such as a radar generates a field and measures the effects on this field. The radar (Section 5.6) measures the characteristics of the energy reflected from the object under study to determine the characteristics of that object. A passive instrument such as a radiometer measures the effects on a field that is generated elsewhere. The radiometer (Section 5.7) measures the temperature of various portions of the sky to imply information on the characteristics of the sources or the intervening media.

The remote sensors in this chapter have been classified according to the techniques used for the sensing. In addition to the radar and radiometer, other remote sensors measure audio frequency acoustic (Section 5.2) and electromagnetic (Section 5.3) waves, magnetic fields (Section 5.3), gamma rays (Section 5.4), and optical waves (Section 5.5).

Remote sensors are usually not as useful as in situ sensors for those applications requiring high accuracy at a specific, well-defined, accessible point. Remote sensors do have advantages for several situations. They are often the only way of obtaining information from inaccessible places such as distant stars. They are useful for surveying a broad field to obtain an overview unobtainable from an in situ sensor. Photographs of the cloud cover of the earth from a satellite are an example of the utility of such an overview. Remote sensors can provide a method of studying phenomena that would be affected by the presence of an in situ sensor. Remote sensors also offer speed and cost advantages when it is necessary to collect data over a large area.

The problem of determining the many characteristics of a remote sensor such as resolution, accuracy, and response time is the same as for in situ sensors as discussed in Section 4.1.

Some of the available remote sensing techniques are discussed in this chapter. These include acoustic echo sounders to measure atmospheric characteristics, electromagnetic and magnetic sensors used for mineral exploration, gamma ray sensors used for measuring radioactivity, optical sensors for studying the surface of the earth and atmospheric particles, radar sensors for studying the earth and precipitation, and radiometers for studying the earth and the atmosphere.

C. Gordon Little

5.2 ACOUSTIC ECHO-SOUNDERS

Sound propagation is strongly affected by atmospheric conditions. Acoustic echoes from the internal structure of the atmosphere can be used to derive information on the internal structure and processes of the lower atmosphere (thermal plumes, temperature inversions, internal gravity waves, breaking waves) and to measure important atmospheric parameters such as wind and turbulence profiles [2,4].

5.2.1 Theory of Scatter of Sound Waves by Atmospheric Inhomogeneities and Turbulence

The turbulence and temperature fluctuations in the lower atmosphere result in an inhomogeneous medium capable of weakly scattering some of the acoustic energy incident upon it [4]. The sound is scattered by the fluctuations in local phase velocity arising from the nonuniform velocity and temperature fields.

The acoustical scattering cross section per unit volume of the atmosphere due to the velocity fluctuations (i.e., to the small-scale turbulence) is given by

$$\eta_v = 8\pi^2 k^4 \cos^2\theta \, \frac{1}{C^2} \, E(\mathbf{K}) \cos^2 \frac{\theta}{2}, \qquad (5.2\text{-}1)$$

where $k = 2\pi/\lambda$ is the wave number of the acoustic wave, θ is the angle through which the acoustic wave is scattered, C is the mean phase velocity of sound in the volume, and $E(\mathbf{K})$ is the spectral intensity of the velocity fluctuations at the effective wave number $\mathbf{K}$ at which acoustic waves of wave number k interrogate the medium when scattered through an angle θ.

The corresponding scattering cross section per unit volume due to the temperature fluctuations is given by

$$\eta_T = 8\pi^2 k^4 \cos^2\theta \frac{1}{4T^2} \Phi(\mathbf{K}), \qquad (5.2\text{--}2)$$

where $\Phi(\mathbf{K})$ is the spectral intensity of the temperature fluctuations at wave number $\mathbf{K}$ and T is the mean temperature.

For the inertial subrange of a Kolmogorov spectrum of turbulence, and assuming that the velocity and temperature fluctuations of wave number $\mathbf{K}$ are uncorrelated, the scattering cross section per unit volume from an atmosphere containing both velocity and temperature fluctuations is given by

$$\eta = 0.20 k^{1/3} \cos^2\theta \left[1.85 \frac{C_v^2}{C^2} \cos^2\frac{\theta}{2} + \frac{C_T^2}{4T^2} \right] \left(\sin\frac{\theta}{2} \right)^{-4/3}, \qquad (5.2\text{--}3)$$

where C_v^2 is a measure of the turbulent energy (the mean square difference in velocity at two points unit distance apart) and C_T^2 is a similar measure of the intensity of the temperature fluctuations.

Equation 5.2–3 shows that the scattered acoustic power resulting from illumination of a Kolmogorov spectrum of turbulence varies relatively weakly with wavelength ($\eta \propto \lambda^{-1/3}$). This scattered acoustic power is the sum of two terms, one due to the wind fluctuations (normalized by the mean velocity of sound in the medium) and one due to the temperature fluctuations (normalized by the mean temperature of the medium). Both wind and temperature scattering terms are multiplied by $\cos^2\theta$, which means that no power is scattered at an angle of 90°. The wind term includes a $\cos^2(\theta/2)$ multiplying term, which means that the wind fluctuations produce no scatter in the backward direction ($\theta = 180°$). Both the wind and temperature components of the scatter are multiplied by a $(\sin\theta/2)^{-4/3}$ factor so that most of the power scattered is scattered into the forward hemisphere.

This equation indicates that a full measurement of the scattered power as a function of wave number and scatter angle permits the following measurements:

1. $\Phi(\mathbf{K})$, the spectral intensity of temperature fluctuations at the three-dimensional wave number $\mathbf{K}$, can be measured as a function of direction, wave number, height, and time. This parameter is of considerable communication and atmospheric importance, being directly propor-

tional to the intensity of the refractive index fluctuations which are responsible for the amplitude and phase scintillations of optical signals.

2. $E(\mathbf{K})$, the spectral intensity of velocity fluctuations at wave number $\mathbf{K}$, can be measured as a function of direction, wave number, height, and time. This three-dimensional spectrum is of immediate concern to the meteorologist and those interested in atmospheric turbulence, diffusion, and pollution.

3. The mean wind speed and direction can be measured as a function of height, using Doppler or other techniques [2].

5.2.2 Engineering Problems in Acoustic Echo-Sounding

Radar Equation. The acoustic radar equation, applied to the monostatic case (colocated transmitter and receiver) gives the received power as

$$P_r = P_t \frac{\eta}{4\pi} \frac{C\tau}{2} \frac{A_r}{R^2} \alpha_t \alpha_r \exp\left[-2 \int \alpha(R)\, dR \right], \qquad (5.2\text{--}4)$$

where P_r is the received electrical power, P_t is the input electrical power to the transmitting element, α_t is the acoustic transducer transmission efficiency for the conversion of electrical power to acoustic power, and α_r the transducer reception efficiency for conversion of acoustic power to electrical power, τ is the pulse length, C is the phase velocity of sound, R is the range, A_r is the effective collecting area of the antenna, and $\alpha(R)$ is the acoustic attenuation constant of the air as a function of range.

Maximizing Received Echo Strength. Equation 5.2–4 shows that, since η is only weakly dependent upon wavelength, the received echo strength from a given range is maximized by maximizing the product

$$P_t \tau A_r \alpha_t \alpha_r \exp\left[-2 \int \alpha(R)\, dR \right]. \qquad (5.2\text{--}5)$$

The first three terms are usually independent of operating frequency, while the last three terms are more frequency sensitive. The high-power loudspeakers commonly used as the acoustical transducer (for both transmission and reception) typically have peak values of α_t and α_r of about 0.1 in the range 1000 to 5000 Hz, but are usually less efficient above and below those frequencies. Under normal atmospheric conditions, the atmospheric absorption increases monotonically with acoustic frequency and generally becomes limiting at frequencies of about 1000 Hz if the range is 1000 m, or about 5000 Hz if the range is 100 m.

Minimizing System Noise. Five sources of noise potentially limit the sensitivity of an acoustic echo sounder. These are the statistical pressure fluctuations at the microphone due to the random thermal motion of the atmospheric molecules, the Johnson noise due to the random motion of the electrons in the resistors of the preamplifier, the acoustic waves reaching the microphone which have been generated at a distance by atmospheric turbulence, the pressure fluctuations affecting the microphone due to local wind noise on the microphone and antenna structure, and ambient acoustic noise level at the antenna due to sources such as vehicles and insects.

Even under ideal circumstances, the first two sources are present. The minimum possible noise level in the receiving system is therefore given by

$$P_{min} = kTB, \qquad (5.2-6)$$

where k is Boltzmann's constant, T is the temperature of air and B is the receiving bandwidth. For normal atmospheric temperatures, this corresponds to a noise level of about 4.2×10^{-21} W/Hz.

Very little is known about the noise from naturally occurring acoustic waves of atmospheric origin at the frequencies used by acoustic echo sounders, except that it is apparently low compared with typical ambient levels. Wind noise can be reduced to low levels, at least under moderate wind conditions, by placing the acoustic antenna close to or below the surface of the earth and by screening the microphone.

The remaining (and usually dominant) source of system noise is the response of the antenna to the ambient noise level. Since most of the noise originates very close to the surface of the earth, careful control of the antenna sidelobes in these directions can greatly minimize its effect. The ambient noise level per Hz decreases by about 8 dB per octave increase in frequency. For this reason, it is usually desirable to use as high an operating frequency as atmospheric attenuation permits.

By following these techniques, system noise levels of from 10 to 20 dB above the theoretical limit are attainable much of the time, and under favorable conditions noise levels as low as 3 dB have been obtained.

Minimizing Noise Pollution. An important consideration in acoustic echo sounding is the noise pollution created by the sounder. By careful design of the antenna sidelobe structure, it is possible to reduce the sound level across the ground in the neighborhood of the antenna to acceptable levels. Distances of as little as 100 m reduce the transmitted signal leaking across the ground from a well-designed antenna to below detectable levels in many urban environments [5].

Echo-Sounder Description. Figure 5.2-1 is a block diagram of an echo sounder. The acoustic antenna is a 120-cm-diameter horn reflector antenna originally designed for microwave work.

The main parameters of the acoustic echo sounder are shown in Table 5.2-1. The echo-sounder sensitivity for a range of 300 m is shown in Table 5.2-2. Under the assumed conditions, unity or higher signal-to-noise ratios are achieved at a range of 300 m whenever the rms difference in temperature at two points 1 cm apart exceeds 4×10^{-4}°K. Such conditions are experienced essentially all the time. Similarly, in the bistatic case, scattering from the turbulence is detectable whenever the root mean square difference in velocity at two points 1 cm apart exceeds 1.5×10^{-2} cm/s.

Table 5.2-1.   Echo Sounder Characteristics

Frequency	1000 to 5000 Hz
Pulse power (input)	40 W (electrical)
Pulse power (output)	~ 5 W (acoustical)
Pulse length	10 ms to 1 s
Pulse repetition frequency	0.1, 0.2, 0.5, 1.0 per s
Range resolution	1.7 to 170 m
Maximum range without ambiguity	170 to 1700 m
Receiver bandwidth	10 or 100 Hz
Antenna: horn reflector	120-cm-diam aperture
Antenna gain[a]	26 dB
Antenna beamwidth[a]	$\pm 4°$ to -3 dB points
Transducer: PA loudspeaker	Type ID-40
Antenna recovery time	150 ms
Minimum range	< 30 m

[a] At 2000 Hz.

Table 5.2-2.   Echo-Sounder Sensitivity at 300 m Range, Assuming $f = 2000$ Hz, $\tau = 20$ ms, $B = 100$ Hz, antenna noise level $= 15$ dB above kTB, (i.e., $P_{min} = 1.3 \times 10^{-17}$ W, and no attenuation

Minimum detectable scattering cross section		3.4×10^{-4} cm^2
Minimum detectable reflectivity		$7 \ \times 10^{-14}$ cm^2
Minimum detectable C_n^2		4.6×10^{-13} cm$^{-2/3}$
Minimum detectable C_T		$4 \ \times 10^{-4}$ cm$^{-1/3}$
Minimum detectable C_v	$\theta = 178°$	1.3 cm/s cm$^{-1/3}$
	$\theta = 120°$	$7 \ \times 10^{-2}$ cm/s cm$^{-1/3}$
	$\theta = \ \ 60°$	1.5×10^{-2} cm/s cm$^{-1/3}$

The operation of the echo sounder is analogous to that of any simple pulse-modulated electromagnetic radar, and can be followed from the block diagram of Figure 5.2-1. The tone burst and timing generator acts as the master control unit of the sounder. The audio oscillator is a cw audio frequency oscillator whose output is gated by the tone burst and timing generator to provide well-controlled audio-frequency pulses of specified frequency, duration, and pulse repetition frequency to the power amplifier. The amplified pulses are used to drive the loudspeaker transducer at the throat of the horn reflector antenna. The back-to-back diodes between the power amplifier and the loudspeaker transducer and between the transducer and the preamplifier act as transmit-receive switches to isolate the preamplifier

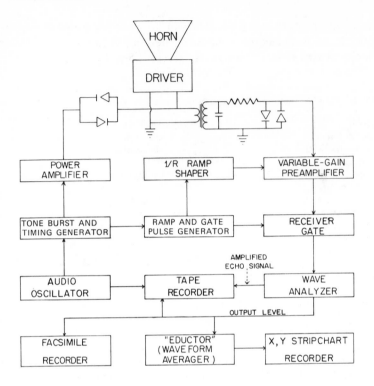

Figure 5.2-1 Acoustic echo sounder block diagram.

from the antenna during the transmit phase, and the transmitter from the antenna during the receiving phase. The broad-band variable-gain preamplifier is preceded by a high step-up ratio transformer to enhance the microphone signal relative to the internal noise of the preamplifier. The voltage gain of the preamplifier is swept by a suitable waveform, initiated by the timing generator, to provide automatic compensation for the $1/R^2$ divergence of the acoustic beam.

The output of the preamplifier is fed to a narrow-band wave analyzer through a receiver gate which serves to isolate the main part of the receiver (the wave analyzer) from any leak-through during the transmitted pulse or the succeeding period (~ 150 ms) of ringing of the antenna. This also serves to prevent overloading the receiver with any short-range echoes due to nearby structures. The operation of the gate is controlled by the timing generator.

The wave analyzer acts as a narrow-band, tunable receiver to further amplify the acoustic echoes received by the acoustic antenna. The video signal is recorded directly on one channel of a magnetic tape recorder, for later Doppler analysis. The detected receiver output is used to operate the facsimile recorder (see Figures 5.2-2 through 5.2-4) and is also tape recorded for later digitization and quantitative analysis. An alternative receiver output display which provides intergration over an adjustable number of pulses makes use of the eductor and X, Y recorder.

The sounder has proven to be a flexible and reliable research tool. Further work has been done to reduce the sidelobes of the antenna [5].

5.2.3 System Performance

Measurement of Profiles of C_T. Equation 5.2–4 indicates that, after correcting for attenuation and the range-squared divergence of the antenna beam, a zenithally directed monostatic acoustic echo sounder can record the time variations of the height profile of C_T. Examples of C_T profiles taken in this way are depicted in the form of a time section in Figure 5.2-5. The isolines represent 6–s averages of C_T over depths of about 14 m, and clearly show the development of a moderately strong thermal plume just prior to 13:02:00 followed by a brief quiet period and a second weaker plume at about 13:02:10. This plume is followed by a second region of low C_T and finally at 13:03:00 another moderately strong plume appears. Less smoothing of the original data and equipment modification would make it possible to gain greater resolution and also to observe the structure below 30 m. The relative values of C_T are thought to be accurate to about 20%; the absolute value to within a factor of 2.

The Doppler effect can be used to measure the radial component of the wind velocity, since the small temperature eddies responsible for the backscattered echo are transported by the mean wind. Figure 5.2-6 shows contours of vertical velocity observed on a fixed zenithally directed sounder during thermal plume conditions [1]. The velocities are believed to be accurate to about 0.2 m/s.

Measurement of Atmospheric Structure. Experience with acoustic echo sounders has shown three main types of acoustic echo patterns. During thermally unstable, clear-sky daytime conditions, thermal plumes are clearly seen (Figure 5.2-2). During nighttime ground-based radiation inversion conditions (thermally stable conditions) echoes are again obtained, often showing marked oscillations in height of the inversion layer (Figure 5.2-3). Under either condition, discrete elevated layers are often seen. These layers may be multiple and often show strong coherent oscillations in height. On occasion these waves are seen to take a form attributed to breaking Kelvin-Helmholtz waves (Figure 5.2-4).

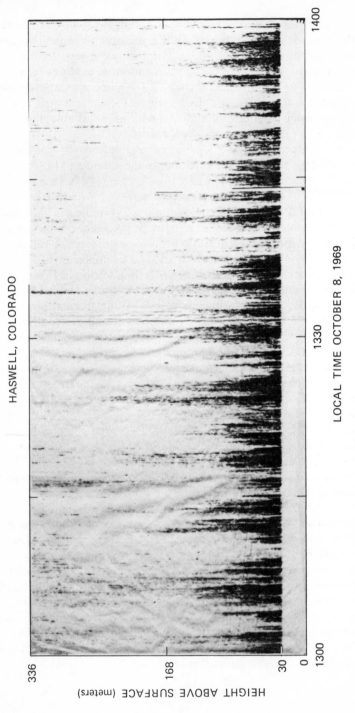

Figure 5.2-2 Acoustic echoes obtained from thermal plumes during unstable convective conditions.

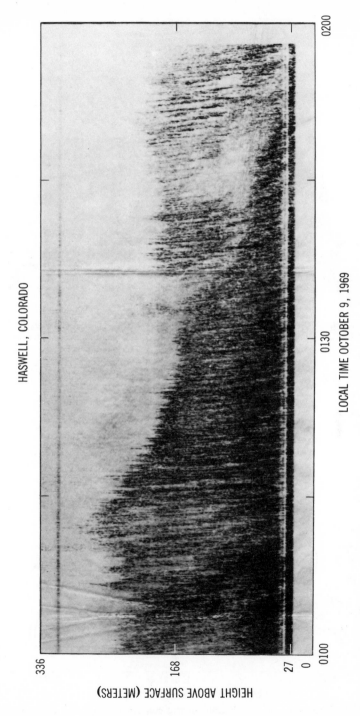

Figure 5.2-3 Acoustic echoes obtained during stable, ground-based radiation inversion conditions.

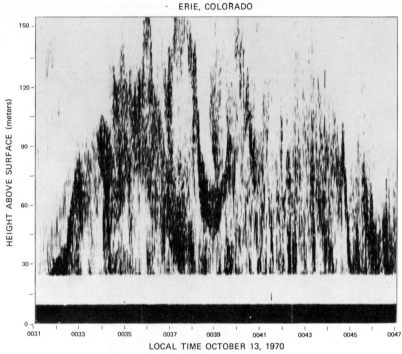

Figure 5.2-4 Breaking Kelvin–Helmholtz waves as recorded by acoustic echo sounder.

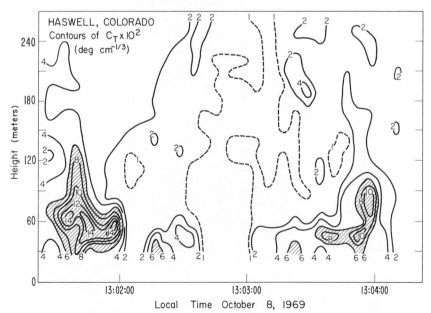

Figure 5.2-5 Time section of C_T obtained during convective conditions. Contours of C_T $\times$ 10^2 (deg cm$^{-1/3}$) are shown as solid lines. Region of $C_T > 0.06$ deg cm$^{-1/3}$ are shaded.

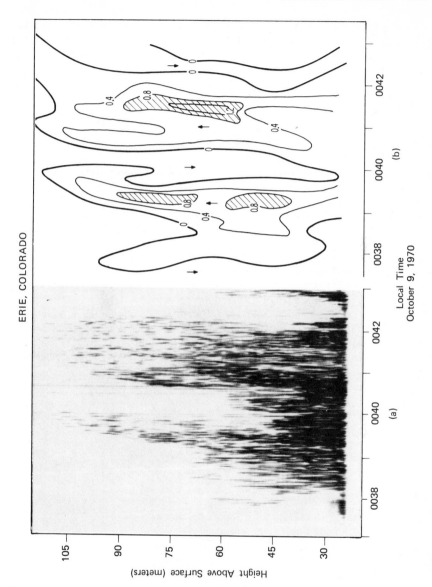

Figure 5.2-6 *a* Facsimile record of thermal plume. *b* Contours of vertical velocity of acoustic scatterers [taken from the same record as (a)] in the height time plane measured using the Doppler effect. Values given on contours are in m/s, arrows indicate the sense of velocity.

5.2.4 Potential Studies with Acoustic Echo-Sounders

The Doppler effect can be used to measure the radial component of velocity of the acoustic scatterers. If these measurements can be made with sufficient resolution, it should be possible to use the technique to measure the intensity of the turbulence as a function of height and time, and to derive considerable information on the three-dimensional spectrum of turbulence. The vertical flux of momentum associated with the turbulence in the boundary layer would also be measurable using acoustic analogs of microwave radars [3, 6].

References

1. Beran, D. W., C. G. Little, and B. C. Willmarth, "Acoustic Doppler Measurements of Vertical Velocities in the Atmosphere," *Nature* **230**, 160–162 March 19, 1971.
2. Derr, V. E., and C. G. Little, "A Comparison of Remote Sensing of the Clear Atmosphere by Optical, Radio, and Acoustic Radar Techniques," *Appl. Optics* **9**, 1976–1992, 1970.
3. Lhermitte, R. M., "Turbulent Air Motion As Observed by Doppler Radar," *Proc. of 13th Radar Meteorology Conference*, 1968.
4. Little, C. G., "Acoustic Methods for the Remote Probing of the Lower Atmosphere," *Proc. IEEE* **57**, 4, 571–578, 1969.
5. Simmons, W. R., and J. W. Wescott, "Acoustic Echo Sounding as Related to Air Pollution in Urban Environment," NOAA Technical Report, 1971.
6. Wilson, D. A., "The Doppler Radar Studies of Boundary Layer Wind Profile and Turbulence in Snow Condition," *Proc. of 14th Radar Meteorology Conference*, 1970.

A. R. Barringer

5.3 RADIO FREQUENCY, AUDIO FREQUENCY, AND MAGNETIC SENSORS

5.3.1 Radio Frequency Instrumentation

The utilization of radio frequencies (rf) has been confined mainly to waves propagated from distant VLF transmitters. However, some work has been done with underground rf transmitters operated in conjunction with receivers in drillholes [13].

Electromagnetic radiation at radio frequencies is characterized by relatively shallow penetration into the surface of the earth as compared with the audio frequencies which are more commonly used for geophysical exploration. Figure 5.3-1 shows skin depths of penetration at frequencies in the broadcast band, in the VLF region and at 500 Hz.

Although the penetration at radio frequencies is relatively small compared with that at audio frequencies as used in geophysical exploration, the penetration is substantial compared with most other remote sensing techniques.

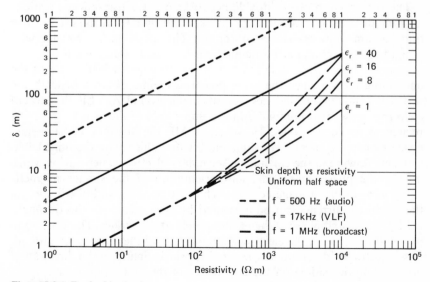

Figure 5.3-1 Earth skin depth.

In general these techniques measure surface or quasi-surface phenomena and can provide subsurface information only by inference.

Geological variations, both in the near-surface overburden and in underlying bedrock, are accompanied by marked changes in electrical resistivity which can be mapped using propagating electromagnetic waves. The measurement of the apparent impedance of the ground at two or more frequencies, such as 1 MHz in the broadcast band (MF) and 20 Hz in the VLF band, can provide information on the variation of resistivity with depth. Such information is directly related to many economic geological problems such as detection of gravel deposits in clay, location of aquifers, and exploration for manganese ores.

Instrumentation designed to operate in conjunction with distant transmitters can be grouped into four categories: field strength measuring equipment employing magnetic dipole detectors such as loops on ferrite rods or electric field detectors, such as short monopole or dipole antennas [9,10]; tilt angle measuring equipment for measuring the tilt of the (normally zero) vertical magnetic field vector when the field propagates across conductive bodies or discontinuities [11, 12]; devices for measuring the field strength of the horizontal magnetic or electric component and splitting this component into two time components, one in-phase and one out-of-phase with reference to the phase of the field detected by a vertical electric dipole [1, 2]; and systems for measuring field strength of the electric field component and the tilt of the electric vector as the field propagates over ground of varying conductivity [3].

The construction of world-wide networks of VLF transmitting stations for long-distance communications has spurred interest in the use of these transmissions for geoscience purposes. Tilt angle measurements with ground equipment and airborne methods have had extensive use.

Measurements of the components of the propagated field relate to the underlying terrain since the wave penetrates some distance beneath the earth-air boundary and induces currents in the ground. At VLF, the effect of dielectric constant is minimal and the impedance of the ground is closely related to its dc resistivity. However, at MF the dielectric constant and hence moisture content have some modifying influence on the impedance. Currents flowing in the earth have secondary electromagnetic fields which modify the phase, amplitude, and orientation of the surface magnetic component of the propagated wave. Thus the presence and distribution of these ground currents can be measured by their effects on the tilt angle, field strength, and phase of the resultant field at the surface. There are complementary effects in the electric components, and the propagated ground-wave electric field component is tilted forward slightly in the direction of propagation depending on the impedance of the underlying terrain.

Theoretical and laboratory modeling can provide interpretational data

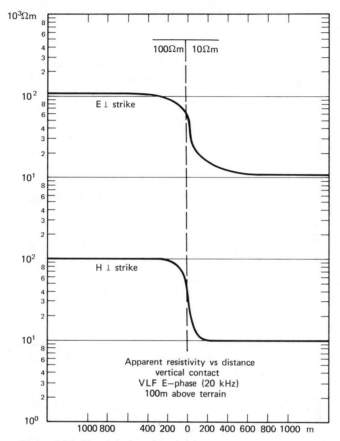

Figure 5.3-2 Theoretical earth resistivity.

on the response of simple structures with contrasting resistivity. Figure 5.3-2 illustrates theoretical calculations of resistivity as derived from measurements of the horizontal electric field of a plane wave propagating both perpendicular and parallel to the strike of a vertical contact. The contact separates regions of resistivity 10 and 100 Ωm in an infinite half-space, and the calculations are performed at an altitude of 100 m above the surface to simulate the response of an airborne system. It is seen that the apparent resistivity reflects quite accurately the true resistivity. Figure 5.3-3 shows the results of tank modeling experiments in which a vertical slab of a conducting material (to simulate an ore body or conductive geological structure) was immersed in a tank of salt water (to simulate a moderately conductive host rock) and measurements made of the in-phase (I) and quadrature-phase (Q) components of the horizontal magnetic field. Such

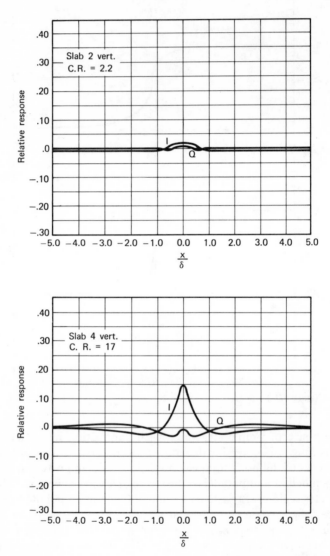

Figure 5.3-3 Plane-wave tank modeling results. Model: vertical slab, just submerged. Varying parameter: slab conductivity.

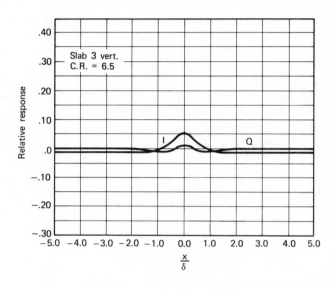

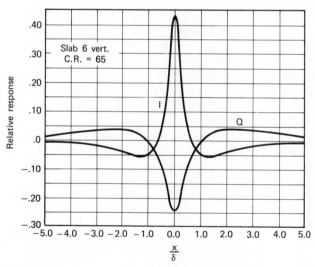

Figure 5.3-3 (continued).

experiments illustrate that measurement of these and other field parameters are diagnostic of the physical and electrical properties of the conductive anomalies. Actual ground situations are obviously much more complex, particularly with regard to the magnetic field components. However, the contouring of airborne surveys of rf magnetic component parameters often generates patterns which clearly relate to major geological structures such as faults, shear zones, and contacts between rock types of contrasting resistivity. These data are complementary to aeromagnetic maps and provide a different expression of geology than that obtained with magnetics.

Measurements made of the electric field components are particularly useful for mapping near-surface resistivities in alluvium and glacial terrain. Airborne measurements of tilt angle of the electric vector are virtually impossible to obtain due to the very small angles involved and the instability of the platform. However, it is possible to measure the horizontal electric field component which is in quadrature phase with respect to the vertical electric field component. The ratio of the horizontal to vertical components (Figure 5.3–4) is given by

$$\frac{E_x}{E_z} = (j\omega\varepsilon_0\rho)^{1/2} \qquad (5.3-1)$$

where $\omega = 2\pi f$, f is the frequency in Hertz, ε_0 is the permittivity of free space (8.854×10^{-12} F/m), ρ is the ground resistivity in ohm meters, and $j = (-1)^{1/2}$.

Over a homogeneous earth, the phase shift between the horizontal and vertical components is 45° through a wide range of ground resistivities and consequently the horizontal quadrature component can be used to derive a fairly good approximation of the ground resistivity. Such a measurement is feasible in an aircraft even under relatively turbulent conditions and maps prepared by this technique correlate well with ground resistivity maps. Figure 5.3-5 shows two examples of an area surveyed at MF and VLF. The factor of 50 between the two frequencies causes a difference of a factor of 7 in the average penetration depth which greatly assists in determining the depth of potential deposits. The targets in this case were gravel and sand deposits in an area in which there was widespread silt and clay. High-

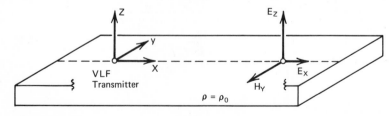

Figure 5.3-4 Field vectors, lossy earth.

resistivity zones represent target areas for exploration. Interpretation of electric field measurements of the above type is generally less complex than for magnetic fields, and maps can be produced which are contoured in resistivity units Ωm.

In a typical helicopter installation for airborne electric field measurements, a nose boom carries vertical and horizontal electric dipole antennas. The vertical antenna provides the phase reference for measurements on the horizontal dipole. In addition, a ferrite cored coil can be installed in the boom to sense the magnetic field components.

5.3.2 Audio Frequency Electromagnetic Methods

Audio-frequency electromagnetic methods are used both for detecting locally conductive bodies such as mineral deposits beneath the surface and for determining the geometry and electrical properties of horizontally stratified layers in the earth [14, 15]. They operate in the inductive field and utilize transmitters generally consisting of multiturn loops carrying alternating currents at audio frequencies. Both ground and airborne systems are used. The ground systems vary in size from small portable units that can be easily carried by one man to much larger systems that require the use of a truck. Similarly, airborne systems range in size from those which can be slung beneath a small helicopter to those that require relatively large aircraft. Frequencies are normally between 100 and 3000 Hz, the lower frequencies being used in regions where high surface conductivities are encountered and the higher frequencies in regions where ground resistivities are high.

Current practice with ground electromagnetic geophysical exploration equipment is to employ separate transmitters and receivers separated by 25 to 500 m. In one of the simplest of the ground methods shown in Figure 5.3-6 an audio-frequency field generated by a transmitting coil held in predetermined orientation is detected in a receiving coil spaced apart from the transmitter by a fixed distance. Traverses are made during which the tilt angle of the null position in the receiving coil is monitored by the receiver operator. When a local conductive body such as a mineralized zone containing metallic sulphides lies within the vicinity of the transmitter-receiver coil pair, eddy currents are induced in this conductive body and a secondary field is generated. This secondary field combines with the primary field from the transmitter to form a resultant field which is distorted with respect to the primary field orientation. Thus, if systematic measurements are taken with pairs of coils over the region to be surveyed, areas where distortions occur can be plotted to provide target zones for more detailed prospecting and perhaps drilling.

Measurements of phase distortions as well as orientation distortions of the primary field can be made if a reference link is provided between the

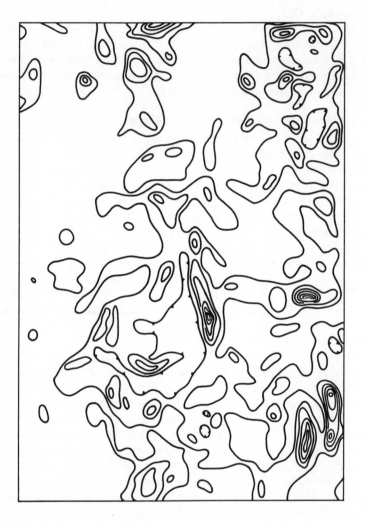

Figure 5.3-5   (*a*) Broadcast-band E-phase survey, apparent resistivity contours. Contour interval: 250 Ωm.

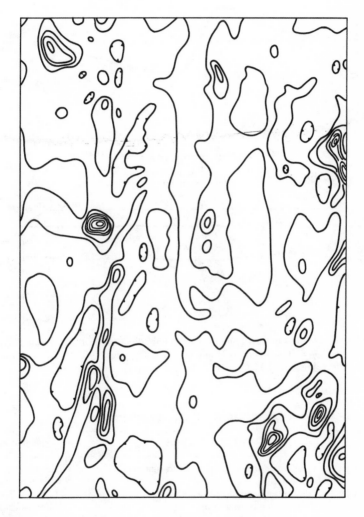

Figure 5.3-5 (*b*) VLF E-phase survey, apparent resistivity contours. Contour interval 250 Ωm.

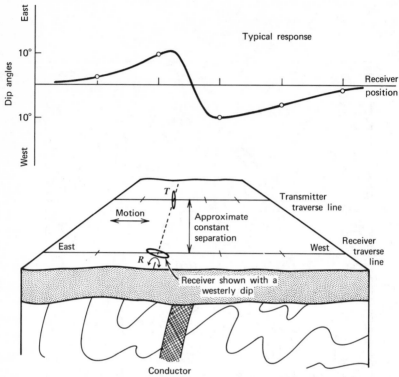

Figure 5.3-6 Illustrating the vertical-loop, parallel-line, dip-angle method. (From *Interpretation Theory in Applied Geophysics* by Grant and West. Copyright 1965, McGraw-Hill Book Co. Used with permission of McGraw-Hill Book Company.)

Table 5.3-1. Airborne Electromagnetic Prospecting Systems

Aircraft Type	Coil Characteristics	Signal Type
1. Fixed wing	Transmitter and receiver coil with rigid mutual orientation	1 or more audio frequencies
2. Fixed wing	Transmitting and receiving coils in nonrigid mutual orientation (towed bird)	1 or more audio frequencies
3. Fixed wing	Transmitting and receiving coils in nonrigid mutual orientation (towed bird)	Pulse type frequencies
4. Helicopter	Transmitting and receiving coils in rigid mutual orientation in towed boom	1 or more audio frequencies
5. Helicopter	Transmitting and receiving coils in rigid mutual orientation mounted on boom fixed to helicopter	1 or more audio frequencies
6. Helicopter	Transmitting and receiving coils in nonrigid mutual orientation (towed bird)	1 or more audio frequencies

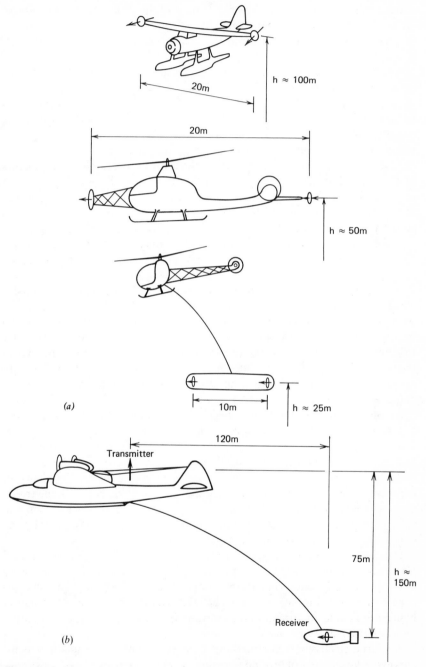

Figure 5.3-7 (*a*) Airborne electromagnetic systems of the double-dipole type; (*b*) disposition of the receiver and transmitter in a quadrature measuring airborne electromagnetic system. (From *Interpretation Theory in Applied Geophysics* by Grant and West. Copyright 1965, McGraw-Hill Book Co. Used with permission of McGraw-Hill Book Company.)

transmitter and the receiver. The most common method is to employ a cable which is dragged between the two coils during traversing, but equipment has been constructed using radio-frequency telemetry to provide the reference link.

Airborne electromagnetic prospecting systems fall under a number of classes as shown in Table 5.3-1 and Figure 5.3-7.

Referring to Type 1 installations in Table 5.3-1, the usual approach is to mount the transmitting coil on one wingtip and the receiving coil on the opposite wingtip. Measurements are normally made at one frequency typically in the vicinity of 400 Hz. The primary field picked up in the receiving coil is cancelled out and the in-phase and out-of-phase components of the secondary field detected when flying at low altitude are recorded continuously in flight. Traverses are flown on a systematic basis typically at an altitude of 60 m and line spacing of 300 m. The sensitivity of such systems is a function of the amount of flexing that occurs in the wings (thus changing the inter coil spacing) during aircraft maneuvers and in turbulence. Noise levels vary between 10 and 35 ppm of the primary field detected by the receiving coil.

A second class of system employs a towed bomb-like container called a "bird" which contains the receiving coil. Since this changes orientation with the transmitting coil, it is usual to detect only that component received in the coil which is out-of-phase with the primary excitation field. In order to provide two parameters of information and to make up for the loss of the in-phase component of information, which is recorded in the rigid systems, two frequencies are generally used and conductivities of bodies detected are expressed in terms of ratios of responses at the two frequencies.

In another system, the same basic type of configuration of transmitting and receiving coils is employed but a pulse waveform is used and the characteristics of shape and amplitude of the transient response of conductive bodies in the ground are measured. The transient responses are detected after termination of the primary pulse, which minimizes problems in compensating for changes in orientation of the receiving coil with the transmitting coil in the plane. The system is in effect a multifrequency system and six parameters of information are recorded, representing different time delays following termination of the primary pulse.

Helicopter-mounted systems operate at one or more frequencies and generally employ rigid booms to hold transmitting and receiving coils spaced apart in fixed orientation. These booms are fixed directly to the helicopter or towed 3 m below. The helicopter systems have the advantage of being applicable in mountainous terrain but are generally considerably more expensive to operate.

Airborne electromagnetic methods have seen wide usage as a means for surveying large areas of ground in order to detect the presence of subsurface conductive bodies. The ground electromagnetic methods are frequently applied in conjunction with airborne surveys and are used to further resolve the geometrical and electrical properties of bodies located in such airborne surveys and to locate airborne anomalies on the ground precisely. Numerous commercial ore deposits have been found using these techniques.

Audio-frequency methods are differentiated from radio-frequency methods in that they operate with locally generated inductive fields as opposed to plane-wave propagated fields from distant transmitters. The results obtained are entirely different since they tend to accentuate localized conductive bodies, whereas the radio-frequency plane-wave methods emphasize large regional structures.

5.3.3 Magnetic Instrumentation

A variety of types of magnetometers have seen service for geoscience purposes both for ground and airborne applications [5, 7, 8, 16], and some are described in Section 4.15. These types include the following:

1. Magnetic balance systems of the Schmidt type, now largely replaced by electronic magnetometers.

2. Fluxgate magnetometers employing permeable coils and bridge networks. Current systems have drift and noise of less than one gamma. These systems have been instrumented for highly portable operation and have also been extensively used in airborne applications.

3. Proton magnetometers based on the measurement of the phenomenon known as nuclear free precession. These magnetometers are entirely devoid of instrumental drift and are fundamental magnetometers in the sense that the quantity measured is the frequency of precession of atomic nuclei in the magnetic field. Counting of this frequency provides a digital readout which is immutable. Proton magnetometers are now dominant in oceanographic applications and have also become increasingly favored over fluxgate magnetometers for airborne use.

4. Optically pumped magnetometers. These are based on electron spin phenomena in alkali metal vapors or in helium and are characterized by extremely high sensitivity. They lack the basic simplicity of the proton magnetometer, but are used for those requirements calling for sensitivities of the order of 0.01 gamma. This includes vertical gradient measurements of magnetic fields and high resolution magnetic mapping over sedimentary areas for oil exploration.

5. Miscellaneous magnetometers. Other methods have been used from time to time for measuring magnetic fields including instruments for measuring torsional forces on suspended magnets, deflection of electron beams, and recently the Josephson junction effect. Hall-effect magnetometers are generally too insensitive for geoscience purposes. Spinning coil magnetometers have been largely restricted to the laboratory and magnetic observa-

tory for the measurements of magnetic field dip angles and the magnetic properties of rock samples.

Large areas of the earth have been covered by aeromagnetic surveys and interpretation of the results has been invaluable in the elucidation of deep and near-surface geological structures and other features [6]. Computer techniques [4] are being applied increasingly to airborne magnetic data to solve interpretational problems. These include the generation of power spectral-density curves to assist in determining depth to basement rocks and two-dimensional filtering in order to separate near-surface from deep magnetic structures or to accentuate geological structures having certain specific directional characteristics.

References

1. Barringer, A. R., and J. D. McNeill, "RADIOPHASE™—A New System of Conductivity Mapping," in *Proceedings of the 5th Symposium on Remote Sensing of Environment*, pp. 157–167, 1968.
2. Barringer, A. R., and J. D. McNeill, "The Airborne RADIOPHASE™ System—A Review of Experience," Paper presented at Canadian Institute of Mining and Metallurgy, Toronto, 1970.
3. Barringer, A. R., and J. D. McNeill, "E-PHASE—A New Remote Sensing Technique for Resistivity Mapping," *Proceedings of the 7th Symposium on Remote Sensing of Environment,* 1971.
4. Bhattacharyya, B. K., "Semi-Automatic Methods of Interpretation of Magnetic Data, Mining and Groundwater Geophysics," Geological Survey of Canada Economic Report No. 26, p. 32–45, 1967.
5. Bhattacharyya, B. K., and B. Raychaudhuri, "Aeromagnetic and Geological Interpretation of a Section of the Appalachian Belt in Canada," *Can. J. Earth Sci.* **4**, 1015–37, 1967.
6. Boyd, D., "The Contribution of Airborne Magnetic Surveys to Geological Mapping, Mining and Groundwater Geophysics," Geological Survey of Canada Economic Report No. 26, p. 213–27, 1967.
7. Hood, P., "Magnetic Surveying Instrumentation—A Review of Recent Advances, Mining and Groundwater Geophysics," Geological Survey of Canada Economic Report No. 26, pp. 3–31, 1967.
8. Jensen, H., "Geophysical Surveying with the Magnetic Airborne Detector," AN/ASQ-3A, U. S. Naval Lab. Report, pp. 937–93, 1945.
9. Keller, G. V., and F. C. Frischknecht, *Electrical Methods in Geophysical Prospecting,* Chapter 7, Pergamon Press, New York, 1966.
10. Maley, S. W., "Radio Wave Methods of Measuring the Electrical Parameters of the Earth," in *Electromagnetic Probing in Geophysics*, Ed. J. R. Wait, Geolem Press, Boulder Colo., 1971.
11. Paal, G., "Ore Prospecting Based on VLF Radio Signals," *Geoexploration* **3**, 139–147, 1965.

12. Patterson, N. R., and V. Ronka, "Five Years of Surveying with the VLF-EM Method," *Geoexploration* **9**, 726, 1971.
13. Petrovsky, A. D., "Underground Radio Wave Exploration," Lecture read at the International Seminar on Geophysical Methods of Prospecting, Moscow, July 1967. Auspices UNO.
14. Ward, S. H., "Electromagnetic Theory for Geophysical Applications," in *Mining Geophysics,* Vol. II, pp. 10–196, Society of Exploration Geophysicists, 1967.
15. Ward, S. H., "The Electromagnetic Method," in *Mining Geophysics,* Vol. II, pp. 224–372, Society of Exploration Geophysicists, 1967.
16. Wyckoff, R. D., "The Gulf Airborne Magnetometer," *Geophysics* **13**, 182–208, 1948.

Quentin Bristow

5.4 GAMMA-RAY SENSORS

5.4.1 Radioactivity Fundamentals

Composition of the Atom. Radioactivity is a result of processes occurring in the atomic nucleus. An atom consists of a nucleus (radius approximately 10^{-13} cm) surrounded by one or more "shells" of orbital electrons. The number and configuration of these orbital electrons makes the atoms of one element chemically distinguishable from those of another. The electron is the lightest of the subatomic particles and carries unit negative charge (1.6×10^{-19} C).

The atomic nucleus consists of protons and neutrons. The proton has a mass approximately 2000 times that of an electron and carries unit positive charge. The neutron has a mass close to that of a proton and no charge. There are as many protons in the nucleus as there are orbital electrons, thus ensuring an electrically neutral atom. Atomic radii are approximately 10^{-8} cm or about 10^{5} times that of the nucleus.

The atomic number, Z, of an element is the number of protons in the nucleus of that element, and the arrangement of elements in the periodic table is in order of increasing Z. The mass number, A (directly related to the atomic weight), is the combined number of protons and neutrons in the atomic nucleus of that element.

Isotopes of an element are in effect subelements, the atoms of which are identical in all respects except that their atomic nucleii contain different numbers of neutrons; they are chemically indistinguishable. Isotopes of an element have the same atomic number but different mass numbers.

Radioactive Decay Processes. When the nucleus of an atom is unstable it decays to a stable state by the emission of particles and electromagnetic radiation, the most important of which are characterized as follows:

Alpha (α) particles: Doubly positively charged helium nucleii consisting of two protons and two neutrons.

Beta (β) particles: Electrons carrying unit negative charge.

Positrons: Particles having the same mass as electrons but carrying unit positive charge.

Gamma (γ) radiation: Electromagnetic energy emitted as quanta, a series of discontinuous emissions of specific energies.

The energies of these various emissions are measured in terms of electron volts.

The unit of radioactivity is the curie, Ci, the quantity of any radioactive nuclide in which the number of disintegrations is 3.7×10^{10} per second. As an example, the activity in a luminous watch is less than 1 μCi.

Radioactive decay decreases exponentially with time and follows the law

$$\text{final activity} = (\text{initial activity}) \exp - \lambda t, \qquad (5.4\text{--}1)$$

where λ is the decay constant. The half-life is the time required for an amount of activity to decrease to one-half of its initial value; this time is a constant independent of the amount of activity and is equal to $0.693/\lambda$.

5.4.2 Characteristics of Gamma Radiation

Gamma radiation is electromagnetic in nature and, as predicted by quantum theory, is emitted as discreet amounts or quanta. This behavior makes it similar in many respects to corpuscular radiation and the term photon is used for a quantum in this context.

Gamma-ray photons interact with matter in three ways:

Photoelectric effect: An orbital electron in an atom of material is ejected and receives all of the energy originally contained by the photon, less the electron binding energy.

Compton scatter: An electron receives only part of the photon energy, the remainder is scattered at an angle to the original line of incidence as a photon of reduced energy. The greater the angle the less energy the scattered photon has. The scattered photon is still a gamma ray and can still continue to interact with matter.

Pair production: The incident gamma-ray photon generates an electron-positron pair. When the positron combines with an electron they are both converted to radiant energy, and two gamma photons are generated, each having an energy equivalent to the rest mass of the electron (0.511 MeV). These photons are often referred to as annihilation radiation and travel along the same line in opposite directions. Pair production can only occur at energies greater than 1.02 MeV.

The relative probability of these three processes as a function of the Z number of the material is shown in Figure 5.4-1.

Gamma radiation intensity in vacuum due to a point source varies inversely as the square of the distance from the source. For a given energy of radiation and Z number of absorbing material, the intensity I_0 of a collimated beam of radiation is related to the intensity I at a distance x into the material by

$$I = I_0 e^{-\mu x}, \tag{5.4-2}$$

where μ is the absorption coefficient. A half-value layer of absorber material can be defined as that thickness which causes attenuation of the original parallel beam intensity by a factor of 2. The value depends on the beam energy as well as the Z number of material. It is important to note that the value of I in the above equation applies only to those photons which reach point x having escaped interaction by any of the three processes described earlier.

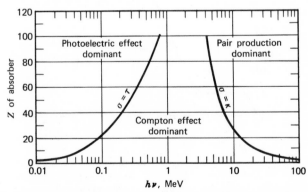

Figure 5.4-1 Relative importance of the three major types of gamma-ray interaction. τ, σ, κ, are the partial attenuation coefficients for the photoelectric effect, compton scatter, and pair production, respectively [7]. (From R. D. Evans, *The Atomic Nucleus.* Copyright 1955 by McGraw-Hill, Inc.; reproduced courtesy of the McGraw-Hill Book Company.)

5.4.3 Gas Filled Radiation Detectors

Detector Types. Gas-filled detectors are based on the principle of charge collection from electrodes in a gas which is ionized by radiation (alpha, beta, or gamma). Radiation events ionize gas molecules into ion pairs (free electrons and positive ions). The energy required for this process is approximately 35 eV per ion pair. When a voltage is applied across the electrodes the charge carriers are collected and used either to actuate counting circuitry corresponding to individual radiation events (proportional counters and Geiger Muller tubes), or can be integrated to provide a mean dc level indication (ionization chambers).

At very low polarizing voltages (< 50 V) the ion pairs tend to recombine before collection. Beyond this level there is a plateau over which virtually all the charge is collected (ionization chamber region). At higher voltages (> 200 V) liberated electrons acquire sufficient kinetic energy to cause secondary ionization, resulting in a final collected charge much greater than, but still proportional to, the original number of ion pairs liberated and to the energy of incident radiation. Detectors employing this phenomenon (gas multiplication) are called proportional counters and are useful

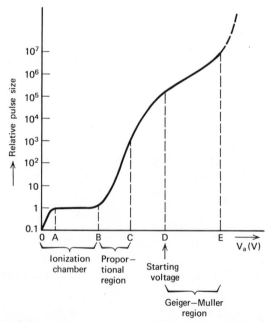

Figure 5.4-2 The relative pulse size as a function of the anode voltage V_a of a radiation counter tube. (From "Tubes for Radiation Detection and Measurements," Philips Electron Tube Division; reproduced courtesy of Philips Electron Tube Division.)

at low energies. They have been largely superseded by silicon semiconductor detectors. Beyond a certain critical polarizing field, secondary ionization, and hence the quantity of charge collected, reaches a saturation level for every radiation event regardless of its energy. This is known as the Geiger Muller region of operation. Figure 5.4-2 illustrates these various regions.

    Geiger Muller Tubes as Gamma-Ray Sensors. Of the three types of gas-filled detector, ionization chamber, proportional counter and Geiger Muller (GM) counter, the last is the only one used to any extent as a gamma-ray sensor, although miniature ionization chambers continue to be used for the monitoring of extremely high gamma fields. A cutaway view of a typical GM tube is show in Figure 5.4-3. The center electrode is positive and the metal can forms the cathode. A typical gas filling is a neon-argon mixture with about 1% of a halogen quenching gas added to prevent after-pulsing

Figure 5.4-3 Cut-away view of a modern radiation counter tube. The metal strip around the cathode serves as a cathode connection. (Photo courtesy of Philips Electron Tube Division.)

caused by positive ions liberating sufficient energy when neutralized at the cathode to reinitiate the avalanche cycle. Organic quenching gases (e.g., ethyl formate) are sometimes used; however, they dissociate and are not reformed as are halogen quench gases, and this limits the life of the tube to about 10^8 counts. Halogen quenched tubes have a virtually indefinite life ($>10^{10}$ counts). The deionization time following a radiation event is typically 80 μs, known as the paralysis time during which the counter is insensitive to radiation.

For a constant radiation intensity there is a specified range of polarizing voltage over which the count rate is virtually constant known as the GM counter plateau. The lower limit is the starting voltage. The actual voltage limits are a function of the external circuit parameters (particularly shunt capacitance) among other factors. Data for a typical tube are shown in Table 5.4-1.

Table 5.4-1. Data for a typical commercial Geiger Muller Tube, The Philips Model 18503

Wall thickness	250 mg/cm^2
Effective Length	40 mm
Outside diameter	15 mm
Material	Cr Fe
Gas filling	Neon/argon (halogen quenched)
Maximum starting voltage	325 V
Plateau	375–600 V
Plateau slope	2 % per 100 V
Maximum paralysis time	100 μs
Ambient temperature range	− 55 to 75°C
Weight	7 g
Life expectancy	$> 10^{10}$ counts

Recommended circuit:

Source. Reproduced by courtesy of Philips Electron Tube Division.

The intrinsic efficiency of GM tubes for gamma-ray sensing (counts recorded per number of incident photons) is of the order of 1 %, so that the absolute detection efficiency for a point source of gamma radiation at a distance of 20 cm from the tube would be about 0.01 % in terms of counts recorded per number of photons emitted by the source. Almost all detected events are due to electrons ejected from the tube wall by photoelectric and Compton scatter processes ionizing the gas, rather than direct ionization by gamma photons. Surrounding the tube with compensating foils to enhance this effect can give substantially constant efficiencies over wide gamma-ray energy ranges. Table 5.4-2 provides a brief summary of GM counter tubes as gamma ray sensors.

Table 5.4-2. Geiger Muller Tubes as Gamma-Ray Sensors—Summary

Applications
   Radiation alarms
   Area and personnel monitoring
   Level gages
   Radioisotope tracer studies

Advantages
   Inexpensive
   Wide operating temperature range
   Simple and inexpensive external circuitry
   Rugged
   Lightweight

Disadvantages
   Low gamma ray detection efficiency
   No energy resolution
   Low count rate capability

5.4.4 Scintillation Detectors

These devices are covered in some detail in the Section 4.16 on particle detection.

Certain solids and liquids when doped with traces of activating impurities scintillate (emit visible light) when subjected to ionizing radiation, the intensity of individual scintillations being proportional to the energies of the gamma photons causing them. The scintillation intensities can be converted to electrical charge pulses, the amplitudes of which are also proportional to photon energies, by means of a photomultiplier tube.

Scintillation detectors, comprising a thallium activated sodium iodide scintillation crystal [Na I(Tl)] and a photomultiplier tube, are in very wide use where gamma-ray energy analysis is required as opposed to recording total count.

Typical applications are airborne and ground gamma radiation surveys, radioactive tracer studies (industrial and medical), gamma-ray coincidence studies in conjunction with Ge(Li) detectors, radiochemistry, radioactive mineral assay, bore hole logging, personnel monitoring, and certain types of radioisotope gages.

Complete gamma-ray energy spectra can be acquired using a scintillation detector with the aid of a multichannel pulse-height analyzer. A good pulse-height analysis system can maintain acceptable spectral energy resolution for input count rates in excess of 5×10^4 counts/s. Scintillation crystals for gamma-ray sensing are usually in the form of a right circular cylinder and sizes range from 3.8 cm diameter by 2.5 cm deep to 28 cm diameter by 10 cm deep. A summary of the characteristics of a typical 7.5 × 7.5 cm detector complete with the photomultiplier tube and its mu metal shield is given in Table 5.4-3 and a picture of the detector is shown in Figure 5.4-4.

Table 5.4-3.   7.5 × 7.5 cm. NaI Scintillation Detector General Characteristics

Efficiency: High, approximately 75 % of photons normally incident on front face are detected at 1.0 MeV.

External circuitry for detector operation: Minimum requirement is well-regulated (0.01 %) high-voltage supply in the 700 to 1200 V range at approximately 200 μA for the photomultiplier tube, plus high input impedance preamplifier.

Energy resolution: Typically 8 % full width at half-maximum (FWHM.) at 0.662 MeV; i.e., Gaussian-shaped 0.662 energy photopeak in recorded gamma-energy spectrum of radioactive isotope ^{137}Cs has width of 52.96 keV at half maximum peak height.

Sodium iodide crystal: Hermetically sealed in aluminum or stainless steel skin with polished transparent window on one face. Hygroscopic, deteriorates rapidly if ingress of moisture occurs. Fragile, easily cracked or fractured by mechanical or thermal shock. Usable over wide temperature range subject to 10°C/hr rate of change. Life virtually indefinite if seal maintained.

Photomultiplier tube: Gain affected by presence of magnetic field; gain and noise are temperature dependent. Relatively fragile, can be damaged by vibration and mechanical shock but ruggedized versions available. Life normally many thousands of hours. Operational temperature range 0–75°C. Hence complete detector is limited to operation in this range.

Weight: Complete detector approximately 4.5 kg.

5.4.5   Lithium Drifted Germanium (Ge(Li)) Detector

The Ge(Li) detector provides the ultimate in energy resolution, far exceeding that of scintillation detectors. It is widely used in nuclear research, and has advanced the technique of neutron activation analysis to the point where it is being used industrially. The main drawbacks are price, the the requirement for liquid-nitrogen cooling, and the low efficiency, typically one-twentieth of a 7.5 × 7.5 cm scintillation detector.

Fundamentals. This type of detector can be considered as the solid-state analog of a gas ionization chamber. Radiation events liberate ion pairs in a

Figure 5.4-4 A typical 7.5 × 7.5 cm. Na I (Tl) scintillation counter. (Photo courtesy of Harshaw Chemical Company.)

region of intrinsically pure germanium across which a high field strength (typically 100 V/mm) is maintained by electrodes which allow charge collection. Electrons ejected from atoms bound in the crystal lattice are raised to the conduction energy band and move relatively freely toward the positive electrode. The vacancies they leave, known as holes, progress stepwise toward the negative electrode as the vacancies are filled and new ones created.

The high field strength necessary for charge collection is maintained without an intolerably large leakage current by making the detector in the form of a PIN junction diode and operating it under reverse bias of 1000 V or more. The leakage current at liquid-nitrogen temperatures (necessary for operation and storage) can be held to the nanoampere level.

A planar detector starts as a right circular cylinder of P type germanium which contains impurity atoms lacking a necessary valence electron and resulting in a net excess of holes. Lithium, an N-type donor material having an excess of electrons is diffused into one end face. It is then allowed to drift through the P-type germanium under the action of a reverse bias voltage until a region is formed where the P-type germanium is compensated by the N-type lithium. This is the intrinsic region with no excess of either holes or electrons. The resulting sandwich consists of heavily doped N material at the surface where the lithium was diffused in, I (intrinsic) material in the compensated region and P-type material at the other end of the cylinder. Since the lithium atoms have significant mobility at room temperatures, the detector must be maintained permanently at a low temperature after the lithium drifting process. Liquid nitrogen at 77°K more than meets this requirement and is universally used as the coolant for Ge(Li) detectors. In practice the PIN diode detector is housed in an evacuated cryostat (with molecular sieve and possibly an ion pump to ensure the hard vacuum is maintained) and attached to a metal cold finger which is in contact with the liquid nitrogen. Electrical contacts are made via vacuum seals to the diode. A thin window of low-Z material (e.g., aluminum) is built into the cryostat close to the detector to permit passage of radiation with minimum attenuation.

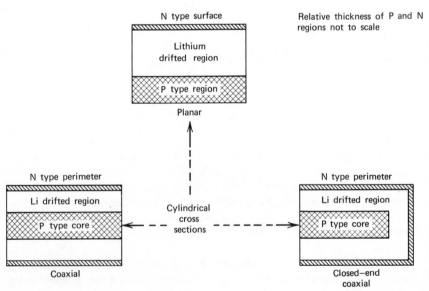

Figure 5.4-5 Ge(Li) detector configurations.

Figure 5.4-5 shows the commom configurations of Ge(Li) detector available. In the case of true coaxial detectors, lithium drifting is done from the perimeter only, while for the closed-end or wrap-around version it is done from the perimeter and one end. The radiation sensitive or active volume is the intrinsic (depletion) region and this varies from about 3 cm^3 for a small planar to 60 cm^3 and more for a closed coaxial type.

Efficiency. A Ge(Li) detector has a useful detection efficiency from about 0.1 to 10 MeV. The generally accepted method of specification is Relative Photopeak Efficiency: Counts per s measured in the 1.33-MeV photopeak of the ^{60}Co spectrum as a percentage of the theoretical count obtained in the same photopeak using a 7.5 × 7.5 cm sodium iodide scintillation detector (44.4 counts/s/μCi) with the source at a distance of 25 cm in both cases. Large coaxial detectors have efficiencies of 12 % and more based on this definition.

Ge(Li) Detector Resolution. A multichannel analyzer energy spectrum recorded for a truly monoenergetic beam using a mythical ideal detector would show all detected events in the channel corresponding to the beam energy. In real detectors the charge generated by successive events of the same energy has a statistical fluctuation which is a Poisson distribution modified by the Fano factor (approximately 0.13). As a result the recorded energy peak has a finite width and the figure of merit used to classify detector performance is the full width of this peak at half-maximum height (FWHM) measured in keV. The theoretical limit of resolution as a function of photopeak energy due to the statistical consideration alone is given by

$$\text{FWHM (keV)} = 1.44[E(\text{MeV})]^{1/2}$$
$$= 1.66 \text{ keV for 1.33-MeV } {}^{60}\text{Co photopeak.} \quad (5.4\text{–}3)$$

Other factors affecting the resolution are total detector plus preamplifier input capacitance, preamplifier and main amplifier noise, incomplete charge collection, and detector leakage current noise.

Coaxial detectors are now being made with resolution as high as 2.2 keV at 1.33 MeV.

Instrumental Aspects of Gamma Spectrometry with Ge(Li) Detectors. The block diagram of a typical pulse-height analysis system is shown in Figure 5.4-6. The pulses from the preamplifier have rise times of the order of 100 nanoseconds and an exponential tail corresponding approximately to a 50-μs time constant. For optimum resolution these are shaped in the main amplifier to an almost Gaussian form having a time to peak of between 2 and 3 μs. The preamp pulses are differentiated by a short-time-constant RC network and undershoot due to the long tail is eliminated by a technique known as pole zero cancellation. Essentially a resistor is added in parallel with the differentiating capacitor such that their time constant is equal to that of the original pulse tail. Successive integrations (or an effective simulation of this process in an active network) are then employed to produce unipolar near Gaussian pulses. A baseline restorer is frequently used to avoid the problem of baseline shift due to count rate variation and the effects of overload pulses which would otherwise cause errors in the pulse-height analysis and consequent degradation of the energy resolution.

The pulse-height analysis is performed in a high-speed A/D converter of the linear ramp or successive approximation type. Clock speeds of

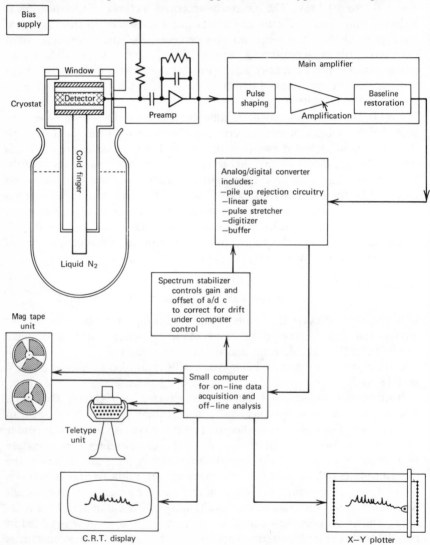

Figure 5.4-6 Typical Ge(Li) detector high-resolution gamma-ray spectrometry system.

200 MHz are available in the linear ramp type, giving an encoding time for a 4192 channel unit of 21 μs per pulse. Successive approximation converters are considerably faster (5 μs and less per pulse), however, good linearity is harder to achieve than with the linear ramp converter.

The A/D converter can be part of a hard-wired multichannel analyzer, complete with core memory for spectrum storage and facilities for punched tape, magnetic tape and type output plus CRT display of the spectrum; or it can be a module interfacing to an on-line computer.

References

1. Bertolini, G., and A. Coche, *Semiconductor Detectors,* North-Holland, Amsterdam, 1968.
2. Birks, J. B., *Scintillation Counters,* McGraw-Hill Book Company, 1953.
3. Birks, J. B., *The Theory and Practice of Scintillation Counting*, Pergamon Press, London, 1967.
4. Brown, W. L., W. A. Higinbotham, G. L. Miller, and R. L. Chase, "Semiconductor Nuclear Particle Detectors and Circuits," Publication 1593, National Academy of Sciences, Washington, D. C., 1969.
5. Camp, D. C., "Application and Optimization of the Lithium Drifted Germanium Detector System," Lawrence Radiation Lab., Univ. of Calif. Livermore, publication UCRL-50156, 1967.
6. Ewan, G. T., "Semiconductor Spectrometers", *Progress in Nuclear Techniques and Instrumentation*, Vol. 3, John Wiley and Sons, New York, 1968.
7. Evans, R. D., *The Atomic Nucleus*, McGraw-Hill Book Company, New York, 1955.
8. Fairstein, E., and J. Hahn, "Nuclear Pulse Amplifiers-Fundamentals and Design Practice," Part I *Nucleonics* **23**, No. 7, July, 1965; Part 2 *Nucleonics* **23**, No. 9 Sept., 1965; Part 3 *Nucleonics* **23**, No. 11 Nov., 1965; Part 4 *Nucleonics* **24**, No. 1 June, 1966; Part 5 *Nucleonics* **24**, No. 3 Mar., 1966.
9. Ferrari, A. M. R., and E. Fairstein, "Nuclear Preamplifiers and the Pulse Pile-up Problem," *Nucl. Inst. Methods* **63**, 218–220, 1968.
10. Friedlander, G., J. W. Kennedy, and J. M. Miller, *Nuclear and Radiochemistry,* John Wiley and Sons, New York, 1964.
11. Glasstone, S., and D. Lewis, *Elements of Physical Chemistry*, Chapters 6 and 18, D. Van Nostrand, Princeton, 1960.
12. Goulding, F. S., J. Walton, and D. F. Malone, "Opto-Electronic Feedback Preamplifiers for High Resolution Nuclear Spectroscopy", Lawrence Radiation Lab. publication UCRL 18698, January, 1969.
13. Heath, R. L., "Gamma Ray Spectroscopy and Automated Data Systems for Activation Analysis", in *Modern Trends in Activation Analysis*, Vol. II, p. 959, U. S. Dept. of Commerce, Nat. Bureau of Standards special Publication 312, Washington, D. C., 1969.
14. Hofker, W. K., "Semiconductors for Ionising Radiation", *Philips Tech. Rev.* **27**, No. 12, 323–336, 1966.

15. "Lithium Drifted Germanium Detectors," International Atomic Energy Agency STI/Pub. 132, Vienna, 1966.
16. Knowlin, C. A., and J. L. Blekinship, "Elimination of Undesirable Undershoot in the Operation and Testing of Nuclear Pulse Amplifiers," *Rev. Sci. Instr.* **36**, No. 12, 1830, 1965.
17. Kowalski, E., *Nuclear Electronics*, Springer-Verlag, New York, 1970.
18. Kurz, R., "Fast A. D. C. for Pulse Height Analysis," presented at Nuclear Electronics Symposium held in Ispra, Italy, 1969.
19. Lyons, W. S., E. Ricci and H. H. Ross, "Nucleonics Annual Review," *Anal. Chem.* **42**, No. 5, 123R, 1970.
20. McKenzie, J. M., *Index to the Literature of Semiconductor Detectors,* National Academy of Sciences, Washington, D. C., 1969 (1227 references including computer permuted title index).
21. Paradellis, T., and S. Hantzeas, "A Semi-empirical Efficiency Equation for Ge(Li) Detectors," *Nucl. Inst. Methods* **73**, 210–214, 1969.
22. *Tubes for Radiation Detection and Measurements,* Philips Electron Tube Division Bulletin 20/415/D/E-12-59.
23. Price, W. J., *Nuclear Radiation Detection*, McGraw-Hill Book Company, New York, 1963.
24. *Ge(Li) Handbook*, 3rd ed., 1970, available from Princeton Gamma Tech., Inc., Princeton, New Jersey.
25. Robinson, L. B., and F. S. Goulding, "An Inexpensive Gain Stabiliser Controlled by a Time Shared Computer," *Nucl. Inst. Methods* **75**, 117–120, 1969.
26. Santhanem, S., and S. Monaro, "A Well-type Ge(Li) Detector for Sum-Coincidence Studies," *Nucl. Inst. Methods* **75**, 322–327, 1969.
27. Strauss, M. G., F. R. Lenkszus, and J. J. Eicholz, "Simple and Accurate Calibration Technique for Measuring Gamma-Ray Energies and Ge(Li) Detector Linearity," *Nucl. Inst. Methods* **76**, 285–294, 1969.
28. Sutterfield, M. M., G. R. Dyer, and W. J. McClain, "An Overload Cancellation Circuit for a Charge Sensitive Preamplifier," *Nucl. Inst. Methods* **75**, 312–316, 1969.
29. Wallace, G., and G. E. Coote, "Efficiency Calibration of Ge(Li) Detectors using a Radium Source," *Nucl. Inst. Methods* **74**, 353–354, 1969.

A. R. Barringer

5.5 OPTICAL SENSORS

Optical instrumentation for geoscience measurements covers the spectrum from the ultraviolet to the infrared and includes airborne line scanning equipment producing imagery of the underlying terrain in one or more optical bands which can extend from 3000 Å in the ultraviolet to 13 μm in the infrared; infrared radiometers for portable or airborne operation in a nonscanning mode; airborne and portable correlation spectrometers and interferometers operating from the ultraviolet to the infrared in a non-scanning mode and carrying out spectral correlations with stored references in real time; infrared spectral scanning radiometers operating in the vicinity of 10 μm and measuring characteristic spectral emission curves of terrain surfaces in truck mounted and airborne applications; multi-band photographic methods employing conventional photographic techniques at a number of filtered optical bands; and Fraunhofer line luminescence systems which detect luminescence of materials at the surface of the earth or ocean in daylight when such luminescence is invisible to the eye. They are based upon the detection of changes in the apparent line depth of Fraunhofer lines in the solar spectrum on reflection of sunlight from the earth's surface.

5.5.1 Line Scanning Instruments

The line-scanning systems are capable of producing imagery of near photographic quality and some of the more sophisticated equipment can produce imagery in up to 24 spectral bands simultaneously [20]. One such system has a spatial resolution of 2 mrad and an active scan angle of 80° [7, 33]. It is used in conjunction with a ground data station which provides processing, displaying, and recording facilities for the multi-spectral data. Computerized pictures of specific locations can be seen on a standard 525-line studio-type color television monitor. Additional channels are provided in the near infrared to estimate the precipitable water vapor between the scanner and target in order to provide corrections for measurements made of reflectance and emissivity in the infrared. Data can be transferred from digital tapes made in flight to computer tape in order to allow various types of processing to be carried out on the imagery.

Applications include determining the vigor of crops, identifying oil films on water, predicting and assessing flood damage, monitoring volcanic eruptions and landslides, and identifying and enhancing geological structures. Line scan imagery in the thermal infrared in particular provides types of data which are not available from normal photographic imagery. This includes identifying fault zones by changes in moisture content and their associated heat budget modifications, differentiating between rock types by virtue of differences in emissivity or thermal capacity, and detecting hot springs and other thermal emissions in geothermal areas [26].

A variety of detecting elements are used in line scanning equipment, depending upon the wavelength band in which measurements are being made. Photomultipliers are normally used between 3,000 and 5,000 Å with silicon photodiodes sometimes used in place of photomultipliers between 0.5 and 1 μm. Germanium or lead sulphide detectors are used between 1 and 2 μm and indium antimonide is used between 2 and 5 μm. Mercury-doped germanium is one of the sensitive detectors used between 5 and 13 μm. The infrared detectors are operated at liquid-nitrogen and liquid-helium temperatures.

The majority of airborne line scanning optical instruments use a rotating mirror to scan successive strips across the ground with the aircraft motion providing the orthogonal axis. Systems have also been constructed which scan lines electronically using television type camera tubes such as the vidicon. Television camera tubes can be constructed to cover the wavelength range between 3000 Å and 2.8 μm, and when used in conjunction with a grating it is possible to provide a line scan at a number of wavelengths simultaneously and with good resolution in a system having no moving parts. Such an approach has a number of advantages when operating in the near infrared, visible, and ultraviolet such as providing a lighter and simpler system than a comparable mechanical line scanning equipment [4]. The main disadvantage of this approach is its inability to provide thermal infrared information. A system of this type is shown in Figure 5.5-1.

The choice of spectral bands in which to make line scan measurements is a function of the location of atmospheric windows which limit the portions of the spectrum available in the infrared and short ultraviolet. The atmospheric windows cover the following bands:

(i)	0.29 to 0.93 μm		(v)	2.10 to 2.40 μm
(ii)	1.00 to 1.10 μm		(vi)	3.00 to 4.20 μm
(iii)	1.20 to 1.38 μm		(vii)	4.40 to 5.40 μm
(iv)	1.60 to 1.80 μm		(viii)	8.00 to 13.0 μm
			(ix)	16.00 to 19.0 μm

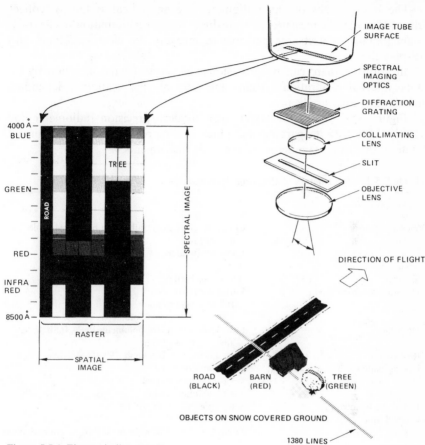

Figure 5.5-1 Electronic line scanner.

5.5.2 Infrared Radiometers

Radiometers are, in general, devices for measurement of radiation but the term is usually reserved for devices which operate in the infrared region with wide spectral bandwidths.

Radiometers have been used to measure temperature and emissivity of terrain with high sensitivity and in many cases serve in a complementary calibration role to thermal line scan imagery. In offshore applications the radiometer has frequently been used to measure small changes in water temperature as an aid to oceanographic and fishery studies. Some attempts have been made to employ radiometers to sense possible thermal changes over oxidizing mineral deposits but results reported have been negative.

The basic blocks of any radiometer are an optical system to collect the radiation to be measured, a transducer to convert radiation to electrical signals, and an electronic package to measure the electrical signal and display the result.

In addition there can be provisions for spectral band selection by filters, internal calibration systems and field of view scanning, depending on the intended application.

Figure 5.5-2 shows the layout of a modern precision radiometer and Table 5.5-1 lists its parameters, which are representative of the current state of the art in commercially available instruments.

Table 5.5-1. Parameters of Precision Radiometers

Optical system	f/2.5 Cassegrain
Focus range	1.5 m to infinity
Viewing system	Chopped reflex, no parallax (can be viewed while measuring)
Field of view	1.5 mrad, standard
Accuracy	$\pm 1\%$
Stability	$\pm 0.3\%$ or better
Dynamic range (energy)	Varies with detector:
	10^5 with lead sulfide
	350 with thermistor bolometer
Detector	Lead sulfide, standard bolometer PbSe, InSb, InAs and others available
Reference	Fast-slewing true blackbody, 99.9% emissive
Optical bandwidth	Varies with detector:
	2 to 2.6 μm standard
	1 to 20 μm available
Electronic bandwidth	Selectable 0.01, 0.1, 1, 10, 100 Hz
Control unit	
Size	18 cm H $\times$ 48cm W $\times$ 40 cm D (Std. 48 cm rack panel)
Weight	13.5 kg
Power requirement	105–125 Vac, 60 Hz, 100 W

A radiometer can be an absolute or a differential instrument. Absolute radiometers require a transducer which has an accurately known and stable responsivity (usually measured as volts out/watt of radiation). However, since the detector itself emits as well as absorbs, the response is to the net radiation flux (incident minus emitted) and thus the accuracy of absolute calibration depends on the accuracy with which the detector temperature can be measured. Further, the responsivity of most presently available transducers is not stable with time. As a result, there is no radiation transducer which is accepted as a standard in the same way as, for example, a blackbody is accepted as a standard radiation source. Consequently almost all radiometers are operated as differential instruments in which the incoming radiation is compared to some internal

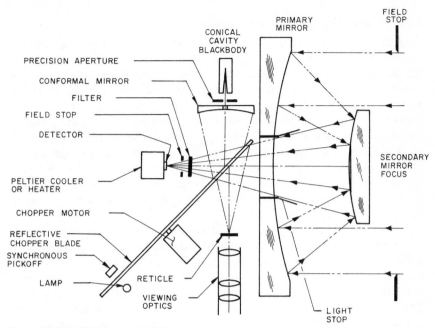

Figure 5.5-2 Radiometer optical diagram.

standard. This mode also has the advantage of permitting ac electrical measurements which are much more stable than dc, especially where very low signal levels are involved.

In radiometers not requiring high accuracy or for measurements of bodies well above ambient temperature, the reference source can be the back of the chopper blade which periodically interrupts the incoming radiation. Alternatively, if a refrigerated transducer is used, the chopper can be made highly reflective so that the detector sees itself as reference and if it is at, say, 77° K this is a good approximation to zero radiation. However, even if the reference level does not vary, stability of calibration of these devices still depends on the stability of transducer responsivity.

For the most accurate applications radiometers are operated as nulling instruments used to balance the output of some adjustable source against the received radiation. Figure 5.5-2 shows an instrument of this type in which the incoming radiation is nulled against the output of a servo-controlled blackbody. In practice it may not be feasible to obtain an exact null and then the small difference signal is measured by assuming a responsivity for the transducer. If the difference signal is only a small correction to the total measurement, then the effect of an error in transducer calibration is proportionally reduced in the final answer.

5.5.3 Correlation Spectrometers and Interferometers

Correlation spectrometers and interferometers are a group of optical sensors which depend upon the identification of spectral signatures [10, 23, 24, 25]. Solar radiation reflected from the surface of the earth or infrared radiation emitted from the earth passes upwards through the overlying atmosphere and undergoes selective absorption at various wavelengths depending upon the presence of various atmospheric gases. Thus, a pollutant gas such as sulphur dioxide impresses its banded absorption spectra in the vicinity of 3100 Å on sunlight reflected from the earth's surface, and this characteristic absorption spectrum can be identified looking downwards from aircraft or spacecraft. A list of gases occurring in the atmosphere which can be measured by these means is given in Table 5.5-2.

The basic correlation spectrometer concept involves the dispersion of incoming light reflected or emitted from a distant object through a spectrometer of conventional design onto an exit plane in which one or more masks can be inserted. In a typical system mask patterns can be introduced sequentially into the exit plane by means of a spinning correlation disc. This arrangement is illustrated in Figure 5.5-3 and typical sensor parameters are listed in Table 5.5-3. The correlation function between the dispersed spectrum of the incoming light and the mask is simply the radiant power passing through the mask from the dispersed power spectrum to a photodetector situated behind the mask.

The spectrum is sampled with at least two masks. In the presence of the absorption spectrum of the chosen gas, the difference in level of radiant power sampled by the masks increases as the concentration of the gas increases. The output can be expressed as

$$O = I_1 R \left\{ 1 - \frac{P_2}{P_1} \right\}, \qquad (5.5\text{--}1)$$

and in a more general fashion as

$$O = I_1 R \{ 1 - \phi(\xi) \} + I_1 R \phi(\xi) \{ 1 - e^{-(a_2 - a_1)L_C} \}, \qquad (5.5\text{--}2)$$

where $I_1 R$ is an instrument constant (volts), $\phi(\xi)$ is the ratio of the radiant power P_1 and P_2 passing through masks 1 and 2 when the chosen gas is not present at the relative position (ξ) between the masks of the spectrum, a_1 is the average value of the absorption coefficients of the gas as seen by mask 1, a_2 is the same for the mask 2, and L_C is the concentration pathlength of the gas. $\phi(\xi)$, by appropriate mask design, can be made to have a value of unity or a constant value. When measuring with solar illumination $\phi(\xi)$ varies slowly with the variation of spectral distribution.

Table 5.5-2.   Table of Ambient Levels of Atmospheric Gases

Gas	Urban $\mu g/m^3$	Nonurban $\mu g/m^3$	Spectral Absorption Range or Locations	Refs.
SO_2	370	0.6	290–320 μm 3.9–4.5 μm	[2] Arithmetic Average for Chicago, 1965 [12, 18, 19]
NO_2	106	10.6	400–510 nm 2.9, 3.2, 3.4, 3.8, 5.7 6.8, 7.9 μm	[1][a]
NO	23.4	2.34	2.6, 5.3 μm	[1]
CO	10×10^6 $- 75 \times 10^6$	125	2.3 μm 4.6 μm	[28] [15][b]
CO_2		595×10^3 -629×10^3	2.7 μm 4.3 μm	[28][c]
I_2		0.13–1.95 15.5	510–610 nm	[16][d] [16][e]
Br_2	No data available		530–590 nm	
O_3	4.3×10^5 $- 14 \times 10^5$	38	310–325 nm 4.8, 9.5 μm	[1, 15][b]
CH_4	3570	1100	2.3 μm 3.1–4.0 μm 7.1–8.5 μm	[8] Averages Dec. 1969, equivalent methane [15]
NH_3 (ammonia)		4.6 15.2	2.2, 2.9, 6.1, 9–13 μm	[13, 14, 18]
Terpenes		3–6	Infrared	[18][f]
Olefins (Ethylene, C_2H_4)	0.4–0.8		2.7, 3.2, 10.5 μm	[9] Pt. Barrow, Alaska.

[a]2-hr discrete values 1.2 m above ground.
[b]Typical of smoggy day in Los Angeles.
[c]At points far removed from vegetation and urban activity.
[d]Nonmarine.
[e]Coastal.
[f]As aerosol.

The first term in Eq. 5.5–2 is a zero offset and the second term is the instrument output. The necessity of knowing $\phi(\xi)$ is eliminated by the use of calibration cells of known concentrations.

This type of correlation spectrometer is well suited to operation in the ultraviolet, visible, and near infrared. In order to achieve extremely large light throughputs in the thermal or far infrared, however, a correlation interferometer has some advantages over a spectrometer. A conventional Michelson interferometer modified so that the pathlength in one arm of the interferometer is modulated by rotating the compensating plate pro-

Table 5.5-3. Correlation Spectrometer Typical Sensor Parameters

Target gases	SO_2 and NO_2
Spectral band	SO_2, 300–315 μm
	NO_2, 425–450 μm
Field of view	Two interchangable telescopes; 1° × 1° or
	1 mrad × 10 mrad
Dynamic range[a]	SO_2, low concentration, 1–1000 ppm m
	SO_2, high concentration, 100–10,000 ppm m
Sensitivity	SO_2, 5 ppm m threshold NO_2, } for 1 s
	1 ppm m threshold }
	Integration time (S/N–1:1)
Detectors	Photomultipliers; XP11118 for SO_2, XP1110
	for NO_2
Mechanical	Dimensions: 71 × 43 × 30 cm
	Weight: 17 kg
Power	10 W, 12 or 24 Vdc

[a] High and low concentration SO_2 masks are interchangable.

vides a simple correlation interferometer [11]. An optical or magnetic recording plate is fixed to the rotating compensation plate and correlation patterns can be recorded on this plate corresponding to the Fourier transforms of gas spectra being detected. A system of this type can employ optics up to 10 cm in diameter in order to achieve large light throughputs and can be constructed of ultra-low-expansion glass having an expansion coefficient only a fraction of that of quartz, thereby providing the required thermal stability.

Table 5.5-4. Typical Correlation Spectrometer Parameters

Aperture	
Interferometer	6.6 cm dia.
Telescope	22.0 cm dia.
Field of View	
Interferometer	0.12 radian dia.
Telescope	0.34 radian dia.
Spectral band	4240–4340 cm^{-1}
Delay scan range	2.5–4.0 mm
No. of sample points	.0–32
Sample length	1–15 fringes
Scan rate	1 Hz
No. of scans accumulated	1–100
Noise equivalent power	1.6 × 10^{-11} W/Hz
Noise equivalent CO amount	0.004 atm cm
(2% albedo, $\zeta = 1$ s)	
Detector	LN_2 cooled PbS immersed on Sr Ti O_3
Weight	
Interferometer	4.5 kg
Telescope	9kg
Power	100 W

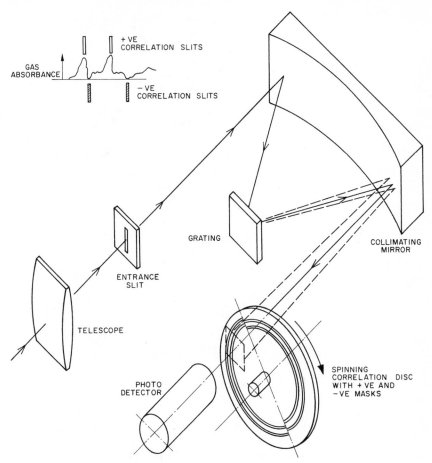

Figure 5.5-3 Correlation spectrometer with spinning disc correlator.

Figure 5.5-4 shows the layout of a correlation interferometer and Table 5.5-4 lists typical design and performance parameters.

Another technique for correlating against gas spectra is the use of sample reference cells of the gas being detected. Sequential introduction of gas cells in the light beam can be used to detect certain gases at relatively high sensitivity, and this method can be satisfactory providing that the gas has sufficiently high absorptivity and stability to be useful when contained in small-volume reference cells.

The applications of correlation spectroscopy in remotely sensing vapors include the detection of volcanic emissions, gaseous emissions from oxidizing ore deposits, iodine vapor over oil fields and off-shore accumulations of plankton, and atmospheric pollution. Experimental traverses

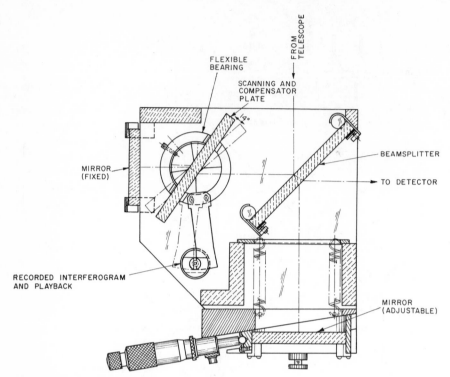

Figure 5.5-4 Correlation interferometer layout.

have been made over Chicago at a height of 35 km in the detection of sulfur dioxide and nitrogen dioxide in the ultraviolet and visible spectrum, respectively, thereby establishing that meaningful semiquantitative measurements of low-altitude pollutants can be made through the greater part of the earth's atmosphere and ozonosphere [5, 6, 22].

5.5.4 Infrared Spectral Scanning Radiometers

Another type of spectral signature that can be identified by remote sensing techniques is the thermal emission of silicate rocks [21, 31]. In the general vicinity of 10 μm, the emission spectra of rocks vary significantly according to the silica content and useful classification of rock types can be carried out by this technique under suitable conditions. Special types of spectral scanning radiometers have been employed for this purpose in which relatively low spectral resolution is adequate to resolve the structure. One simple type employs a spinning circular variable passband interference filter. The output of the scanning radiometer is recorded through successive scans on magnetic tape for subsequent computer analysis and classification. Since the output is sensitive to the silica content of rocks and this is a key parameter by which geologists classify these rocks, the method has definite potential application in providing an aid to geological mapping by remote sensing.

5.5.5 Multiband Photography

Multiband photographic techniques are employed extensively for geoscience applications. The photographic spectrum in solar illumination at the earth's surface covers a range from 3,000 to 9,000 Å, and through this range the reflectivity of different terrain materials and vegetation varies substantially. For this reason the contrast ratios between different materials is a function of the film and filter combination used for the photography. Quartz lenses are required for photography in the ultraviolet, but apart from this it is possible to use conventional cameras for multiband aerial photography. Banks of high quality 35-mm and 70-mm cameras have frequently been used for this purpose and for those types of camera fitted with electric winding mechanisms, it is not difficult to assemble an acceptable system. Special cameras are also manufactured with multiple lenses which can carry at least two film types simultaneously. Some skills are required in achieving appropriate adjustment of exposures for differing film and filter combinations [27].

In order to facilitate interpretation of multiband cameras, special projectors have been constructed which allow superposition of several transparencies in projection and the manipulation of combinations by the interpreter [32]. Color television techniques have also been employed for analyzing and superimposing different combinations of pictures. The computer analysis of multiband photographs is being developed to achieve automatic classification of many types of crops and facilitate the recognition of crop disease and nutritional deficiency problems [3].

5.5.6 Fraunhofer Line Luminescence

The Fraunhofer line depth technique for detecting luminescence under natural daylight conditions is another optical method having geoscience applications. The solar spectrum is characterized by a series of absorption lines known as Fraunhofer lines. These lines, which are related to the absorption by an atomic species in the chromosphere of the sun, place a spectral fingerprint on solar illumination which is retained when daylight is reflected from nonluminescent surfaces. However, if there is a contribution of luminescence from a surface, this modifies the contrast between the energy at the center of a Fraunhofer line and at the edges. Precise measurements made of the spectral distribution of energy at the center and edges of Fraunhofer lines therefore provide a means of detecting luminescence using natural daylight as the illumination source. Actual systems have been constructed [17, 29, 30] utilizing this phenomenon and have been employed to establish the feasibility of tracing rhodomine dyes in water. This dye, which is used as a marker for tracing movements of water in hydrogeological problems, can be detected in aqueous concentrations as low as 3 ppb in sunlight.

The full potential of the Fraunhofer line depth luminescence detection system has not been evaluated in terms of its response over a variety of terrain conditions. Several ore minerals and most crude oils exhibit luminescence. It is possible that the system could be quite sensitive in detecting oil seeps when used in airborne surveys. Chlorophyll also exhibits luminescence and there are possibilities that a method of imaging luminescence in vegetation could be of value as a complementary tool to infrared photography in recognizing plant stress.

Instrumentation employed for Fraunhofer line depth measurements calls for spectral resolution of considerably better than 1 Å. Measurements are required to be made at at least two wavelengths, representing the center and edge of the Fraunhofer lines, and at the same time it is necessary to monitor the shape of the Fraunhofer line by direct monitoring of solar illumination before reflection from the target area. This is because the dynamics of the sun's chromosphere are such that there are minor fluctuations of the Fraunhofer line profiles taking place continuously. Unless the instrumentation provides normalization against these fluctuations, sensitivity is limited by high noise levels. These instruments employ special types of very narrow-band interference filters and a chopping cycle which allows for alternate viewing of the sun and underlying terrain.

References

1. *Air Poll. Control Assn. J.,* 1970, pp. 589–592.
2. "Air Quality—Data from the National Air Sampling Networks and Contributing State and Local Networks," U.S. Department of Health, Education and Welfare, 1966.
3. Anuta, P. E., S. J. Dristof, D. W. Levendowski, R. B. Macdonald, and T. L. Phillips, "Crop, Soil and Geological Mapping from Digitized Multispectral Satellite Photography," in 7th International Symposium on Remote Sensing of the Environment, Ann Arbor, Michigan.
4. Bailey, J. S., and P. G. White, "Remote Sensing of Ocean Color," *Trans. Instr. Soc. Am.* 1969.
5. Barringer, A. R., and A. J. Moffat, "The Remote Sensing of Vapours of Marine Organic Origin," Presented at the Symposium on Remote Sensing in Marine Biology and Fishing Resources, Texas A&M University, College Station, Texas, January 25, 26 1971.
6. Barringer Research Final Report, "Absorption Spectrometer Balloon Flight and Iodine Investigations," NASA Contract NAS9-9492, Barringer Research Report No. TR70-148, August 1970.

7. Bendix Tech. J. "Earth Resources Exploration," **3**, No. 2, 20–32, Summer/Autumn, 1970.

8. California (1970)—"California Air Quality Data," Vol. II, No. 2, State of California Air Resources Board, June, 1970.

9. Cavanagh, L. A., Stanford Research Institute unpublished data, 1967.

10. Davies, J. H., "Correlation Spectroscopy," Presented at the 11th Annual Eastern Analytical Symposium, New York, November 20, 1969.

11. Dick, R., G. Levy, "Correlation Interferometry," Aspen International Conference on Fourier Spectroscopy, 1970, Air Force Cambridge Research Laboratories, AFCRL-71-0019, Special Report No. 114, pp. 353–360. January 5, 1971.

12. Erickson, E., Part I, *Tellus* **11**, 375–403, 1959; Part II *Tellus* **12**, 63–109, 1960; *Tellus* **68**, 4001, 1963.

13. Erickson, E., *Tellus* **4**, 215, 1952.

14. Georgii, H. W. *J. Geophys. Res.* **68**, 3963, 1963.

15. Goldberg, L., and E. A. Mueller, *J. Opt. Soc. Am.* **43**, 1033, 1953.

16. Goldschmidt, V. M., *Geochemistry*, Oxford Clarendon Press, Oxford, 1958.

17. Hemphill, W. R., G. E. Stoertz, and D. A. Markle, "Remote Sensing of Luminescent Materials," *Proceedings of the 6th International Symposium on Remote Sensing of Environment*, pp. 565–585, Vol. 1, October 1969.

18. Junge, C. E., *Air Chemistry and Radioactivity*, p. 123, Academic Press, New York, 1963.

19. Lodge, J. P., and J. B. Pate, *Science* **153**, 408, 1966.

20. Lowe, D. S., "Line Scan Devices and Why We Use Them," *Proceedings of the 5th Symposium on Remote Sensing of Environment*, pp. 77–101, April 1968.

21. Lyon, R. J. P., and J. W. Paterson, "Infrared Spectral Signature—Field Geological Tool," *Proceedings of the 6th Symposium on Remote Sensing of Environment*, pp. 215–230, 1966.

22. Millan, M. M., "Feasibility of Air Pollutant Detection from a Balloon Platform Using the Barringer Correlation Spectrometer," University of Toronto Institute of Aerospace Studies Report 156, July, 1970.

23. Millan, M. M., S. J. Townsend, and J. H. Davies, "Study of the Barringer Refractor Plate Spectrometer as a Remote Sensing Instrument," University of Toronto Institute of Aerospace Studies Report 146, August 1, 1970.

24. Moffat, A. J., and A. R. Barringer, "Recent Progress in the Remote Detection of Vapours and Gaseous Pollutants," *Proceedings of the 6th Symposium on Remote Sensing of Environment*, pp. 379–413, Vol. 1, October, 1969.

25. Newcomb, J. S., and M. M. Millan, "Theory, Applications and Results of the Long-Line Correlation Spectrometer," *IEEE Trans. Geosci. Electron.* **GE-8**, No. 3, 149–157, July 1970.

26. Sabins, F. F., "Infrared Imagery and Geologic Aspects," *Photogrammetric Eng.* **33**, 743–750, 1967.

27. Sorem, A. L., "Principles of Aerial Color Photography," *Photogrammetric Eng.* **33**, 1008–18, 1967.

28. Stern, A. C., *Air Pollution*, Vol. 1, pp. 28–33, Academic Press, New York, 1968.

29. Stoertz, G. E., W. R. Hemphill, and D. A. Markle, "Airborne Fluorometer

Applicable to Marine and Estuarine Studies," *Marine Tech. Soc. J.* **3**, No. 6, 11–26, November-December 1969.

30. Stoertz, G. E., W. R. Hemphill, and D. A. Markle, "Remote Analysis of Fluorescence by a Fraunhofer Line Discriminator," Presented at the Marine Technology Society Annual Meeting, Washington, D.C., June 25–July 1, 1970.

31. Vickers, R. S., and R. J. P. Lyon, "Infrared Sensing from Spacecraft—A Geological Interpretation," Am. Inst. Aeronautics and Astronautics Paper, No. 67–284, 1967.

32. Yost, E. F., and S. Wendereth, "Multi Spectral Color Aerial Photography," *Photogrammetric Eng.* **33**, 1020–1023, 1967.

33. Zaitaeff, E. M., C. L. Korb and C. L. Wilson, "MSDS and Experimental 24-Channel Multispectral Scanner System," *IEEE, Trans. on Geosci. Elect,* **GE-9**, No. 3, 114–120, July, 1971.

Richard K. Moore

5.6 RADAR SENSORS

Radar is an electromagnetic sensor that provides its own source of waves that return to the receiver after changes by the environment. The term radar stands for *ra*dio *d*etection *a*nd *r*anging.

5.6.1 Basic Principles of Radar

Radar Equation. The performance of a radar is described by the radar equation relating the received power to the properties of the radar, the target, and the path between them. When path losses can be neglected, the received power is

$$W_R = \left(\frac{W_T G_T}{4\pi R_T^2}\right)\sigma\left(\frac{1}{4\pi R_R^2}\right)A_R = \frac{W_T G_T A_R \sigma}{(4\pi R^2)^2}. \qquad (5.6-1)$$

The first parentheses enclose the terms determining the power density at the target, including the transmitter power W_T, the transmitting antenna gain G_T, and the distance from radar to targer R_T. The radar cross-section, σ, determines the effective power radiated per unit solid angle toward the receiver. It combines the receiving cross-section of the target, its losses, and its directivity. The next parentheses enclose the spherical divergence factor associated with power reaching the receiver a distance R_R from the target. A_R is the effective area of the receiving antenna. For most radars the transmitter and receiver are at the same place, so that only one distance need be considered, as indicated in the right-hand expression. One antenna is normally used both for transmitting and receiving, so the received power, W_R, can be expressed either in terms of antenna gain or antenna effective area:

$$W_R = \frac{W_T G^2 \lambda^2}{(4\pi)^3 \, R^4} \sigma = \frac{W_T A^2 \sigma}{4\pi \lambda^2 R^4}, \qquad (5.6\text{--}2)$$

where λ is the radar wavelength.

Most geoscience radars are used to observe the earth. The radar cross-section of the surface is normalized to cross-section per unit area. A radar looking at the earth receives signals back from many different facets. Usually, the phases of signals from different facets are statistically independent and the total received power is the sum of the power returned from each of the facets. For this case the return from area-extensive targets is

$$W_R = \sum_i \frac{W_T G^2 \lambda^2}{(4\pi)^3 \, R_i^4} \sigma_i = \sum_i \frac{W_T G^2 \lambda^2}{(4\pi)^3 \, R_i^4} \sigma_i \frac{\Delta A_i}{\Delta A_i}, \qquad (5.6\text{--}3)$$

where ΔA_i is the incremental area of the surface associated with each facet. The differential scattering cross-section is defined by an average as

$$\sigma^0 = \left\langle \frac{\sigma_i}{\Delta A_i} \right\rangle, \qquad (5.6\text{--}4)$$

and the total return power from the surface is

$$W_R = \frac{\lambda^2}{(4\pi)^3} \int \frac{W_T(t - 2R/c) \, G^2 (\mathbf{R}/R) \, \sigma^0 \, dA}{R^4}. \qquad (5.6\text{--}5)$$

The transmitted power is usually a function of time, and the power received at a time t is due to the power that was transmitted at a time $2R/c$ earlier, where c is the velocity of light.

For targets distributed over a volume there is a differential scattering cross-section per unit volume, σ_v, and Eq. 5.6–5 becomes

$$W_R = \frac{\lambda^2}{(4\pi)^3} \int \frac{W_T(t - 2R/c)\, G^2(\mathbf{R}/R)\, \sigma_v\, dv}{R^4}. \tag{5.6-6}$$

Factors Governing σ and σ^0. Radar scatter is governed by the geometry and the dielectric (or conducting) properties of the target. The significant size of geometric variations and the dielectric properties depend on wavelength.

Small targets scatter nearly uniformly in all directions and backscatter from a sphere is totally independent of the direction of the illumination; cloud and raindrops are small dielectric spheres. If the circumference of the sphere is significantly less than a wavelength, the Rayleigh scattering expression applies [3]:

$$\sigma = 64\pi^5 |K|^2 \frac{a^6}{\lambda^4}, \tag{5.6-7}$$

where a is the diameter, and K is given by

$$K = \frac{m^2 - 1}{m^2 + 2}, \qquad m^2 = \eta - j\kappa, \tag{5.6-8}$$

where η is the refractive index, and κ is the loss term. Equation 5.6–7 shows the cross-section is sensitive to small changes in diameter or wavelength.

Scattering is more directional for larger targets. The size of the scattering cross-section σ depends on the directivity of the target, and the angular orientation of the scattered beam with respect to the receiver. Flat surfaces radiate in the specular direction determined by Snell's law, so backscatter is only large near normal incidence. For incident angles well away from normal, rough surfaces beam larger signals back toward the radar, but send a weaker signal in the specular direction.

The lobe structure for scatter is like that for an antenna. The minimum lobe width is given by

$$\text{Minimum lobe width} = \lambda/2L \text{ rad}, \tag{5.6-9}$$

where L is the effective length of the target.

Signal fluctuation (*fading*) occurs when position or frequency are varied. The minimum width for a "lobe" on the frequency axis is given by

$$\Delta f = c/2L_r, \tag{5.6-10}$$

where L_r is the critical dimension in the range direction.

The geometry of many radar targets can be approximated by combinations of cylinders and corner reflectors. Figure 5.6-1 shows the radar cross section of a cylinder [10] with dimensions large compared with wavelength. The lobes indicate the large signal fluctuation as this cylinder rotates relative to the radar. Figures 5.6-2 and 5.6-3 show that a corner reflector has much less cross-section variation than a cylinder. Corner reflectors are common in man-made ground targets wherever a vertical wall and a horizontal surface intersect.

Dielectric properties affect the scattering coefficient primarily through the effect on the reflection coefficient. For good conductors and for dielectric materials having large permittivity, the reflection coefficient is about unity. Most dry natural materials have relative permittivities less than 8, and wet materials have considerably larger permittivities determined principally by the amount of moisture present as shown in Figure 5.6-4 [12].

Facets and surface roughness can be used as descriptors for large-area targets. Theories describe roughness by the probability density function of height and the two-dimensional spectrum, or correlation function of height. However, a mathematical description of most *real* land surfaces is difficult.

Since vegetation scatters some of the signal and permits other parts to

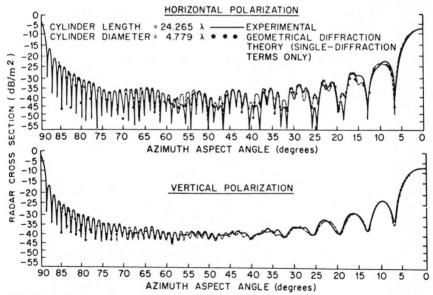

Figure 5.6-1 Radar cross section of a right-circular cylinder for vertical and horizontal polarizations. Ordinate is the cross section in decibels relative to one square meter [14]. (Courtesy *Proc. IEE*.)

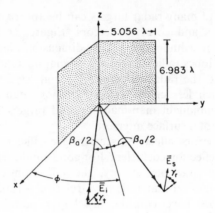

Figure 5.6-2 Dihedral corner reflector whose radar cross section is shown in Figure 5.6-3 [15]. (Courtesy *IEEE*.)

pass through, it is a volume scatterer. Furthermore, only part of the energy striking the soil is scattered from the surface; some of it penetrates and is scattered from rocks, voids, and rough boundaries beneath the visible surface. This is difficult to describe mathematically and to measure in the field.

Resolution. Resolution is the ability to discriminate between two objects separated by the resolution distance. This ability depends on the relative amplitudes of the two returns and on their contrast with the background. In radar a resolution cell is defined as the region on the ground contributing to the signal observed at an instant, and its boundaries are separated by the "resolution."

Passive sensors resolve only on the basis of angle. Radar has additional resolution options because it can measure range and speed as shown in

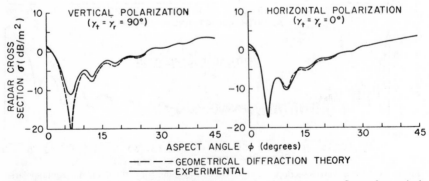

Figure 5.6-3 Monostatic ($\beta_a = 0°$) radar cross section of a dihedral corner reflector for vertical and horizontal polarizations. $\lambda = 0.861$ cm [16]. (Courtesy, *IEEE*.)

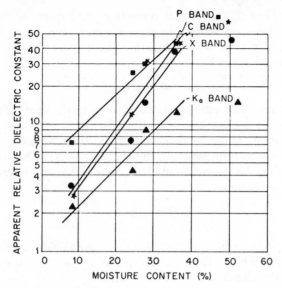

Figure 5.6-4 Apparent relative dielectric constant versus moisture content (Richfield silt loam) [12].

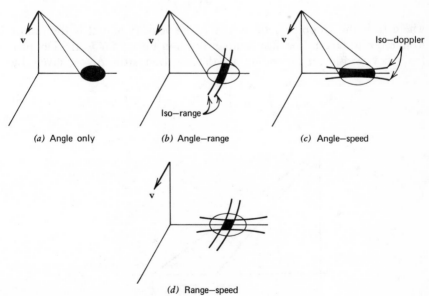

(a) Angle only   (b) Angle–range   (c) Angle–speed

(d) Range–speed

Figure 5.6-5 Resolution techniques for radar.

Figure 5.6-5. Four different radar resolution techniques are shown in the figure. The angle-only technique depends upon the beamwidth of the antenna, which in turn depends on the antenna size. The beamwidth of a linear antenna is given by

$$\beta \approx \lambda/l, \qquad (5.6-11)$$

where β is the antenna beamwidth in radians and l is the length of the antenna aperture.

Radar measures range by determining the time between reception of an echo and the transmission causing the echo. For a short transmitted pulse of length τ, the resolution in slant range is ΔR, and in ground range is Δr_r, which are given by

$$\Delta R = c\tau/2, \qquad \Delta r_r = \frac{c\tau}{2\sin\theta}, \qquad (5.6-12)$$

where the various quantities are defined in Figure 5.6-6 and the situation is that of Figure 5.6-5b. The area resolved is a ring segment that is much smaller than the elliptical area of the beam.

Radar establishes speed by measuring the Doppler frequency shift of the return echo. This frequency shift is due to radial motion between the radar and the target and is given by

$$f_d = \left(\frac{2\mathbf{v}}{\lambda}\right)\left(\frac{\mathbf{R}}{R}\right), \qquad (5.6-13)$$

where f_d is the Doppler frequency, $\mathbf{v}$ is the velocity and $\mathbf{R}$ is the radius from radar to target. For the example shown in Figure 5.6-5c, the Doppler frequency width corresponding to the resolved strip Δr_s is given by

$$\Delta f_d = \left(\frac{2v}{\lambda}\right)\left(\frac{\Delta r_s}{R}\right), \qquad \Delta r_s = \left(\frac{R\lambda}{2v}\right)\Delta f_d. \qquad (5.6-14)$$

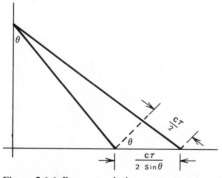

Figure 5.6-6 Range resolution geometry.

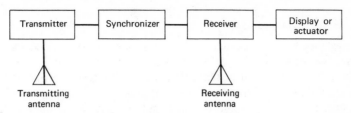

Figure 5.6-7 Elements of a radar system.

Since both the range and speed measurements can narrow the resolved area (in orthogonal directions for the geometry shown in Figure 5.6-5), they can be combined to greatly reduce the size of the resolved area as shown in Figure 5.6-5*d*.

Basic Radar System Elements. Any radar has the elements shown in Figure 5.6-7. The transmitter is the source of the illuminating electromagnetic wave, which is coupled to space through the transmitting antenna. The signal travels to the target and returns through the receiving antenna to a receiver. The synchronizer permits establishing the relation between transmitted and received waveforms. The result is displayed or caused to actuate a device. The type of signal transmitted depends on the radar function.

Most radars use pulse transmission, although some use frequency modulation, binary phase code modulation, or even noise-amplitude or frequency modulation. A typical pulse radar is shown in Figure 5.6-8. The transmitter is pulsed on for short intervals by the modulator. The synchronizer provides triggers to the modulator and oscilloscope at the proper times. Most radars use superheterodyne receivers.

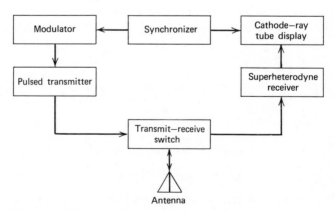

Figure 5.6-8 Elements of a pulsed radar system.

5.6.2 Radar Meteorology

Attenuation in the Atmosphere. Radio waves are attenuated in passing through the atmosphere. The attenuation is larger for higher frequencies and there are absorption peaks due to molecular resonances. Absorption and scatter by water droplets in clouds, fog, or rain, increase rapidly with increasing frequency. Meteorological radars ordinarily operate at frequencies of 3 GHz or higher to obtain sufficient scatter. Meteorological radars seldom operate above 10 GHz because the absorption from nearby clouds is too great.

The lowest-frequency molecular absorption line at about 22 GHz is due to water vapor. The next absorption line at about 60 GHz is due to oxygen. These are shown in Figure 5.6-9 [4]. The oxygen absorption is almost independent of location and time, whereas the water vapor absorption, even at a particular spot, is highly time variable.

Attenuation in clouds is due to droplets that are smaller than the wavelength. This attenuation is approximately proportional to the mass density of liquid water in clouds and is given by

$$\alpha = KM, \tag{5.6-15}$$

where M is the density, and K is a constant that depends on wavelength, temperature, and the water state. Figure 5.6-10 shows the variation in the attenuation coefficient due to water state, frequency and temperature [5].

Attenuation in rain usually is considerably greater than attenuation in clouds. As raindrops increase in size or the wavelength decreases to the size of the larger drops, Eq. 5.6-7 no longer applies. Figure 5.6-11 shows a set of curves calculated for different precipitation rates, assuming a standard drop-size distribution [5]. The attenuations can be large for short distances when the precipitation rate is high. Fortunately, high precipitation rates do not persist over long distances.

Precipitation as a Radar Target. The radar equation for volume scatter is given in Eq. 5.6-6. The volume scattering coefficient, using Eq. 5.6-7 for each drop, is given by

$$\sigma_v = \sum_i \sigma_i \approx \sum 64 \frac{\pi^5}{\lambda^4} |K|^2 a_i^6 = 64 \frac{\pi^5}{\lambda^4} |K|^2 \sum a_i^6, \tag{5.6-16}$$

where a_i is the radius of the ith drop, and K is the constant, related to dielectric properties, defined by Eq. 5.6-8. For wavelengths larger than 1.24 cm and temperatures above freezing [3, p. 28]

$$0.906 < |K|^2 < 0.934. \tag{5.6-17}$$

Thus, the principal variable in determining scatter for a particular wavelength is the mean value of the sum of the *sixth* powers of the drop radii

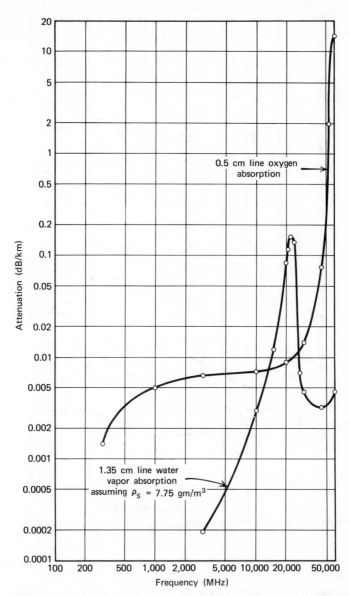

Figure 5.6-9 Atmospheric absorption by the 1.35-cm line of water vapor and the 0.5-cm line of oxygen. [4]

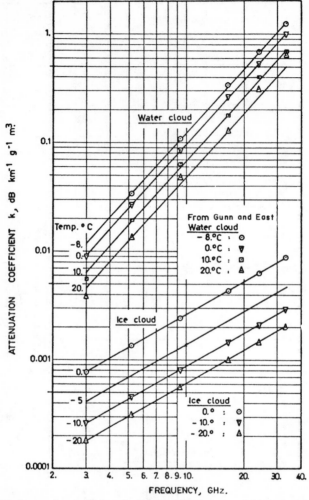

Figure 5.6-10 Attenuation coefficient for clouds [5]. (Courtesy of *Microwave Journal*.)

within a volume. This factor is strongly dependent on the drop size distribution.

For a pencil beam meteorological radar transmitting rectangular pulses, Eq. 5.6–6 reduces to

$$W_R = \frac{\lambda^2 W_T G^2 \sigma_v}{(4\pi)^3 R^4} \left(\frac{R^2 \theta_0 \phi_0 c\tau}{2} \right), \qquad (5.6-18)$$

where θ_0 and ϕ_0 are beamwidths in orthogonal directions, and τ is the

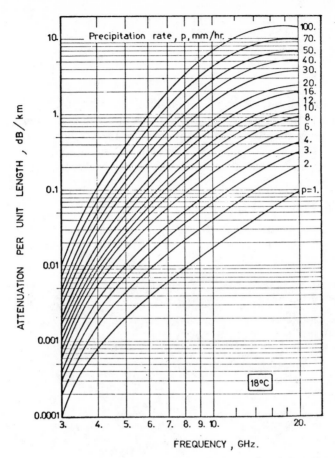

Figure 5.6-11 Attenuation (dB/km) due to rain [5]. (Courtesy of *Microwave Journal*.)

pulse length. Combining Eq. 5.6–18 with 5.6–16 gives the power returned from the volume illuminated at an instant as

$$W_R = \frac{\pi^2 |K|^2 \theta_0 \phi_0 c\tau}{2\lambda^2 R^2} W_T G^2 \sum a_i^6. \qquad (5.6\text{–}19)$$

Frequently the summation is expressed in terms of a quantity Z:

$$Z = \sum (2a_i)^6 \ \text{mm}^6/\text{m}^3. \qquad (5.6\text{–}20)$$

Equation 5.6–19 ignores the effect of attenuation of the wave traveling through the cloud or rain which must be taken into account.

The value of Z depends on the rainfall rate and the type of rainfall. Representative values are [3, p. 55]

$$\text{rain} \qquad Z = 200R^{1.60} \text{ mm}^6/\text{m}^3, \qquad (5.6–21)$$

$$\text{snow} \qquad Z = 2000R^{2.0} \text{ mm}^6/\text{m}^3, \qquad (5.6–22)$$

where the rainfall rate, R, is given in millimeters per hour.

The backscatter cross section for snow is much greater than that for rain with the same precipitation rate. Because the crystalline structure of snowflakes is often surrounded by a thin layer of water, these water-coated particles are much larger than raindrops, so the backscatter cross-section is also much greater for a given total amount of water.

5.6.3 Side-Looking Airborne Radar

Side-looking airborne radars (SLAR) produce high-quality images used like aerial photographs.

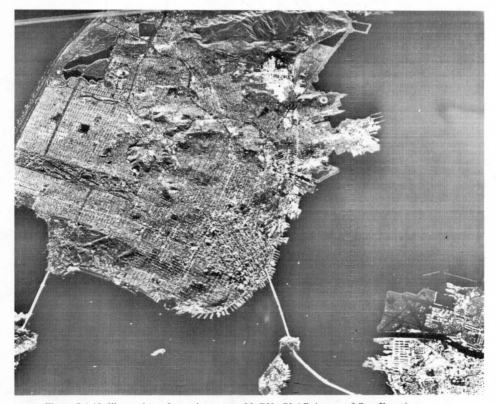

Figure 5.6-12 Illustration of a real-aperture 35 GHz SLAR image of San Francisco.

Real-Aperture SLAR. Figure 5.6-12 illustrates a radar image produced by a real-aperture SLAR operating at a wavelength just under 1 cm. The basic radar block diagram of Figure 5.6-8 is appropriate for the SLAR. The signal strength intensity-modulates the beam in the cathode ray tube as the beam is deflected in one direction to produce a single line on the tube. The image is produced by translating a film past this line at a rate synchronized to the speed of the vehicle as illustrated in Figure 5.6-13.

The over-all view of a farming area (Figure 5.6-14) appears to have the quality of an aerial photograph. At the expanded scale the rather grainy texture and poorer resolution are evident. The grainy characteristic is due to fading to the signal, as discussed in Section 5.6.3, p. 666. Since the resolution in the along-track direction is determined by the antenna beam-width, a very large antenna must be used to achieve fine resolution at aircraft distances. To overcome this, the synthetic-aperture technique is used.

Synthetic-Aperture SLAR. The synthetic aperture (SA) uses the coherence of the transmitted signal to permit processing the returns received as the radar flies past the target for finer resolution [9]. Since the effective antenna is much larger than the real antenna, it is said to be a "synthetic aperture." This system can be viewed as a broadside array with element

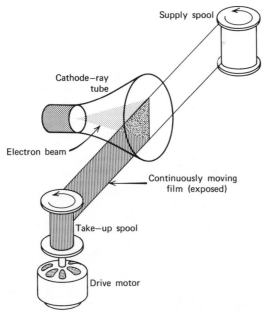

Figure 5.6-13 Diagram of recording technique for side-looking airborne radar. Electron beam scans vertically with position proportional to distance from flight track.

summing in a computer, as a correlation device, and as a filtering technique applied to the received Doppler spectrum.

Consider a radar flying past targets T_1 and T_2 and transmitting pulses at points *a* through *f* in Figure 5.6-15. For a real antenna array a distant target located on line cT_1 perpendicular to *a-e* returns echoes to each of the antenna elements with the same phase, and the phasor sum equals the algebraic sum. A target located along a different line returns signals to points *a-e* along paths that are longer for some elements than for others so the phasor sum is less than the algebraic sum of the returns. The maximum of the antenna pattern is in the direction of T_1, and weaker signals are returned from T_2. The synthetic-aperture radar achieves the same result without actually transmitting simultaneously from all elements *a-e*. It transmits from them sequentially and stores the returns with phase information. After the radar has passed point *e*, the returns are combined in phase as they would be in a real antenna.

The technique described is appropriate for an antenna focused at infinity, which is the case for most real-aperture antennas. For the problem illustrated in Figure 5.6-15 the antenna is focused at the point T_1, and a phase correction is inserted prior to adding the returns. A synthetic-aperture radar making this correction is said to be "focused," whereas one that does not make the correction is said to be "unfocused," and is, in fact, focused at infinity.

Figure 5.6-14 SLAR image of Garden City, Kansas farming area at 35 GHz. (*a*) Normal presentation scale; area shown in (*b*) is outlined.

Since the returns from both T_1 and T_2 appear at almost the same time at points *b-e*, the signals from the two targets are separated largely because of the different phase corrections. For the problem illustrated, the signals from five different targets are observed simultaneously at each location of the radar, and each stored signal is used five times to produce five different

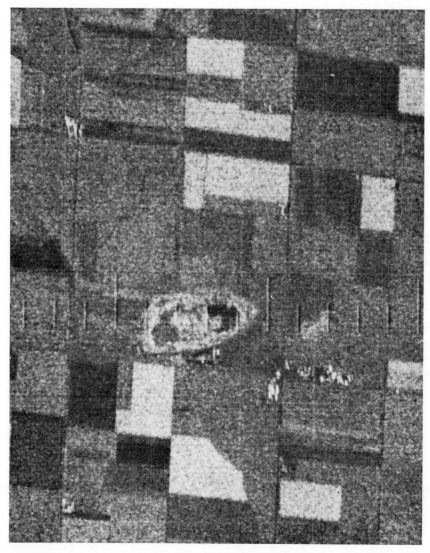

Figure 5.6-14 (*b*) Expanded scale by photographic enlargement.

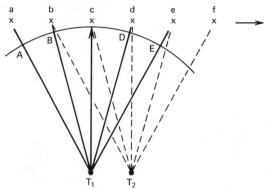

Figure 5.6-15 Principle of synthetic aperture viewed as array antenna.

synthetic apertures. In reality, the number of synthetic apertures produced by a SLAR and stored for simultaneous processing is much greater.

The synthetic aperture can be considered a correlation process. The Doppler frequency shift is zero at right angles to the line of flight. For small angles the Doppler frequency varies linearly with off-normal displacement. The returned frequency is increased for points ahead of the normal and is decreased for points behind the normal as illustrated in Figure 5.6-16a for the two targets of the previous example.

The output frequency of a mixer combining the received signal with a local oscillator signal varying in the same linear fashion is zero if the times for the variations are synchronized. This is illustrated in Figure 5.6-16. The local oscillator is synchronized with the return from T_1 (solid line on Figure 5.6-16a) but it is therefore *not* synchronized with return from T_2. Thus the output of the mixer (or correlator) is at zero frequency for T_1 and at f_{T_2} for T_2.

A synthetic-aperture radar requires a coherent detector that preserves phase so subsequent processing can add phasors from returns received at

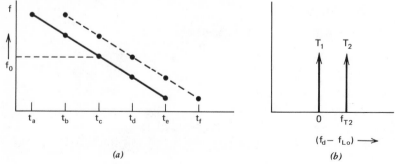

Figure 5.6-16 Principle of synthetic aperture viewed as Doppler correlation.

different times. The coherent detector usually is a mixer whose output frequencies are centered on zero or a low frequency. The local oscillator is phase coherent with the transmitted signal.

The coherently detected signals can be recorded with any medium having sufficient bandwidth. A common technique uses film recording like that for the ordinary real-aperture SLAR as shown in Figure 5.6-13.

The signal recorded on film for the SA radar is similar to a hologram in one dimension, and an image can be made from it by optical techniques similar to those for hologram reconstruction. Figure 5.6-17 shows one arrangement [7, pp. 18–23].

The resolution for a synthetic-aperture radar can be made far superior to that for a real-aperture radar. The real-aperture ground resolution in the azimuth direction (along the flight track) is given approximately by

$$r_a = \frac{\lambda R}{D}, \qquad (5.6\text{–}23)$$

where D is the length of the antenna along the flight track and R is the slant range from radar to target. The arc length along the ground is the product of the beamwidth and slant range. For the synthetic aperture operated in a fully focused mode, the resolution is [7, pp. 23–4]

$$r_a = \frac{D}{2}, \qquad (5.6\text{–}24)$$

which is just half the length of the real antenna.

The degree with which the ideal resolution of Eq. 5.6–24 can be approached depends on the precision with which the moving vehicle position and the return signal phase are known.

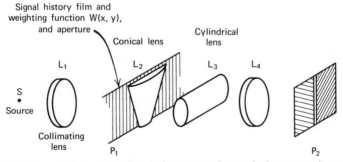

Figure 5.6-17 Layout of one type of optical processor for synthetic-aperture imagery. The conical lens changes the depth of focus to accomodate the different ranges. The cylindrical lens is necessary because the Fourier transform is obtained only in the along-track direction. [10] (Courtesy of McGraw-Hill Book Co.)

Scintillation and Averaging. Radar images are speckled in appearance because of scintillation resulting from phase interference. Each resolution cell is a collection of independent scatterers. The phasor sum of the fields due to the different scatterers depends upon their relative distances and the look angle.

Each scatterer can be considered an oscillator whose frequency is set by Doppler shift of the carrier frequency. The set of Doppler frequencies forms a continuum and the statistics of this signal are like that for noise.

The noise and the radar signal can be considered as the sum of sine and cosine components whose amplitudes are independent and statistically determined by a Gaussian distribution. The voltage envelope of this signal is Rayleigh distributed and its power is exponentially distributed (see Section 4.1). Each pulse return is a *single sample* from this distribution.

For the exponential distribution the standard deviation σ and mean μ are equal:

$$\sigma = \mu. \qquad (5.6-25)$$

This implies that individual samples can be widely spaced about the mean. The result is a speckled appearance with stronger signals causing brighter spots. The speckling can be reduced by averaging several independent samples of this exponential distribution. The standard deviation for an average of N samples is

$$\sigma_N = \frac{\mu}{\sqrt{N}}. \qquad (5.6-26)$$

For the along-track direction and a real-aperture radar, the number of independent samples averaged is approximately equal to the independent sample rate (the Doppler bandwidth) multiplied by the time to pass the target and is given by

$$N = \frac{2r_a^2}{\lambda R} = \frac{2\lambda R}{D^2}. \qquad (5.6-27)$$

For averaging in the cross-track or range direction, the range resolution is given by

$$r_R = \frac{1.5 \times 10^8 \tau}{\sin \theta} \ \text{m}. \qquad (5.6-28)$$

If τ is made short enough, the range resolution is less than that required and several resolution cells can be averaged to reduce the standard deviation. Using the bandwidth associated with the shorter pulse but retaining a pulse duration associated with the final resolution desired permits the same amount of averaging. In this case, the final range resolution is given by

$$r_R = \frac{1.5 \times 10^8 k}{B \sin \theta} \ \text{m}, \qquad (5.6-29)$$

where k is the ratio of the actual bandwidth B to that required to achieve this resolution.

For many geoscience applications, averaging is not as important as resolution, since only the shape and context of the items imaged are important. On the other hand, automatic spectral recognition and many other applications necessarily depend on good gray scale and require significant averaging.

Multiple Polarization and Multiple Frequency. The radar signal scattered from a relatively smooth surface has essentially the same polarization as that transmitted. Differences exist between polarizations because of differences in the reflection coefficients. The reflection coefficient for horizontal polarization is almost independent of angle, whereas the Brewster-angle effect for vertical polarization causes large variations in reflection coefficient with angle. Most natural surfaces are rough enough to depolarize the scattered wave significantly. Since different surfaces depolarize by different amounts, comparing the size of the received signal having the transmitted polarization to the cross-polarized signal can be useful. Figure 5.6-18 illustrates significant differences in the depolarization

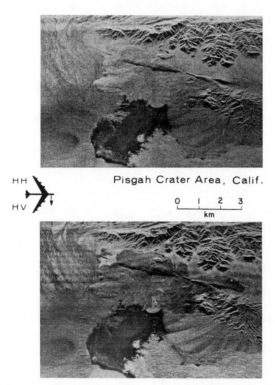

Figure 5.6-18 Like- and cross-polarized radar images (35 GHz) of Pisgah Crater area, California. Note the differences in response of some lava areas adjacent to the dry lake bed. HH means Horizontal Transmit-Horizontal Receive; HV means Horizontal Transmit-Vertical Receive.

by two different lava flows. The difference in cross polarization allows distinction between the two different kinds of lava.

Wavelength variations cause signal variations from physical resonances due to objects of wavelength size. Part of the effect is also caused by slow variations in the dielectric properties. Several wavelengths can be used together to distinguish terrain elements [8].

Some Recent Usage of Imaging Radar. The use of radar for mapping was demonstrated where cloudy weather prevented photography [16]. The results include contour maps with an interval of 100 m, a map of land form regions, and a geologic map.

Radar has been used for mapping vegetation [14], mapping drainage nets [11], locating old snow [17], assessing engineering properties of soils [2], and mapping polar sea ice [1].

5.6.4 Radar Scatterometers

A scatterometer is a calibrated radar system that measures scattering coefficient. The scattering coefficient of the ocean at 2 cm is a measure of wind speed. Measuring the scattering coefficient of a point at different angles indicates more about the terrain. Scatterometers provide information for design of imaging radars and other specialized systems, such as altimeters and Doppler navigators.

Measurement Problems. A terrain image can be useful with little averaging and a large uncertainty of the scattering coefficient. Since the scatterometer is designed to measure the scattering coefficient, it requires averaging. The scatterometer sacrifices resolution to achieve more samples and adequate averaging.

Calibration of the scatterometer is usually limited by the accuracy with which the antenna gain and pattern can be established.

Ground-Based Systems. Numerous scatterometer measurements have been made using ground-based systems [6] transmitting short pulses or cw to obtain spectral responses of targets.

Ground-based scatterometers are restricted to very small resolution cells. This limits the type of target that can be observed to one containing many scatterers in a small patch such as grass, pavements, small grains, and certain types of rock surfaces.

Airborne Scatterometers. Two principal types of airborne scatterometer systems have been used: the pencil beam (narrow conical pattern) system and the along-track fan beam system. With a pencil beam, the antenna is pointed at some incident angle, the aircraft flies far enough to achieve sufficient averaging, and the flight path is repeated for different incident angles.

For the along-track fan beam system illustrated in Figure 5.6-19, the returns can be collected from different parts of the terrain at different

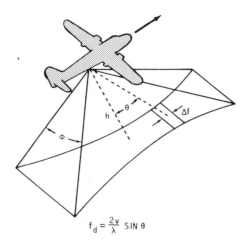

$$f_d = \frac{2v}{\lambda} \text{ SIN } \theta$$

Where: f_d = Doppler frequency Φ = Lateral beam width angle
     v = True ground speed h = Altitude above terrain
     λ = Wavelength Δf = Doppler resolution,
     θ = Incidence angle,   determined by adjustment
       measured from nadir   of spectrum-analyzer filter
                               bandwidth during data reduction

Figure 5.6-19 Arrangement of antenna beam for the fan-beam scatterometer.

incident angles simultaneously and can be correlated to obtain scattering-coefficient-versus-angle curves. The system illustrated uses Doppler to distinguish between returns from ahead of the aircraft and those coming from behind.

The Doppler frequency along the flight track is given by

$$f_d = \frac{2v}{\lambda} \sin\theta \quad \text{Hz}, \tag{5.6–30}$$

where v is the ground speed and θ is the angle of incidence. For a cw system transmitting with an antenna pattern like that of the figure, returns from a patch of the ground lying between two angles of incidence, θ_1 and θ_2, can be isolated by filtering the return. The Doppler bandwidth is

$$\Delta f_d = \frac{2v}{\lambda} (\sin\theta_2 - \sin\theta_1). \tag{5.6–31}$$

For a Gaussian process independent samples are spaced by a time interval equal to the reciprocal of the Doppler bandwidth. The trade-off between accuracy and resolution in the along-track system is given by

$$\Delta\rho \left(\frac{\sigma}{\mu} \right) = \left(\frac{\lambda h}{2 \cos^3\theta} \right)^{1/2}, \tag{5.6–32}$$

where $\Delta\rho$ is the length of the cell on the ground between θ_1 and θ_2.

References

1. Anderson, V. H., "High-Altitude Side Looking Radar Images of Sea Ice in the Arctic," *Proceedings 4th Symposium on Remote Sensing of Environment*, pp. 845–857, University of Michigan, Ann Arbor, 1966.
2. Barr, D. J., and R. D. Miles, "SLAR Imagery and Site Selection," *Photogrammetric Eng.* **36**, 1155–1170, Nov. 1970.
3. Battan, L. J., *Radar Meteorology*, Chapter 4, "Radar Detection of Spherical Particles," pp. 24–31, and Chapter 7, "Use of Radar for Quantitative Perception Measurements," pp. 51–63, University of Chicago Press, Chicago, 1959.
4. Bean, B. R., and E. J. Dutton, *Radio Meteorology*, National Bureau of Standards Monograph 92, p. 271, U.S. Government Printing Office, Washington, D.C., 1966.
5. Benoit, A., "Signal Attenuation Due to Neutral Oxygen and Water Vapor, Rain and Clouds," *Microwave J.* **11**, No. 11, 73–80, Nov. 1968.
6. Cosgriff, R. L., W. H. Peake, and R. C. Taylor, "Terrain Scattering Properties for the Sensor System Design," *Terrain Handbook II*, Engineering Experimental Station, Ohio State University, Columbus, 1959.
7. Cutrona, L. J., "Synthetic Aperture Radar," Chapter 23, pp. 23-1 to 23-25, *Radar Handbook*, Ed. M. I. Skolnik, McGraw-Hill Book Company, New York, 1970.
8. Dellwig, L. F., "An Evaluation of Multifrequency Radar Imagery of the Pisgah Crater Area, California," *Mod. Geology* **1**, 65–73, Nov. 1969.
9. Harger, R. O., *Synthetic Aperture Radar Systems*, Academic Press, New York, 1970.
10. Kell, R. E., and R. A. Ross, "Radar Cross Section of Targets," *Radar Handbook*, pp. 27-27 and 27-36, Ed. M. I. Skolnik, McGraw-Hill Book Company, New York, 1970.
11. McCoy, R. M., "Drainage Network Analysis with K-Band Radar Imagery," *Geographical Rev.* **59**, 493–512, Oct. 1959.
12. Moore, R. K., "Ground Echo," *Radar Handbook*, pp. 25–26, Ed. M. I. Skolnik, McGraw-Hill Book Company, New York, 1970.
13. Morain, S. A., and D. S. Simonett, "K-Band Radar in Vegetation Mapping," *Photogrammetric Eng.* **33**, 730–740, July 1967.
14. Ross, R. A., "Scattering by a Finite Cylinder," *Proc. IEE* (London) **114**, 864–868, 1967.
15. Ross, R. A., "Application of Geometrical Diffraction Theory to Reflex Scattering Centers," *1968 Inter. Ant. and Prop. Symp. Digest*, pp. 94–99.
16. Viksne, A., T. C. Liston, and C. D. Sapp, "SLR Reconnaissance of Panama," *Photogrammetric Eng.* **36**, 253–259, March 1970.
17. Waite, W. P., and H. C. MacDonald, "Snowfield Mapping with K-Band Radar," *Remote Sensing of Environment* **1**, 143–150, March 1970.
18. Waite, W. P., "Broad-Spectrum Electromagnetic Backscatter," Ph.D. Thesis, University of Kansas, Lawrence, 1970.

Harold I. Ewen

5.7 MICROWAVE RADIOMETRIC SENSORS

The development of passive microwave remote sensing techniques and capabilities was spawned by the science of radio astronomy. Microwave radiometers became commercially available in the early 1950s, and the first passive microwave images from an airborne platform were reported in the mid-1950s. During the 1960s the microwave radiometric sensor became well established as the passive counterpart of the active microwave radar.

The wavelength range of interest in geoscience applications extends from approximately 40 cm (0.75 GHz) to 1 mm (300 GHz). In the 40 to 3 cm portion of this wavelength region, the all-weather penetration capability of microwaves allows remote sensing of terrestrial surface materials from air and space platforms under complete overcast conditions. In the wavelength range from 3 cm to 1 mm the atmosphere is semitransparent with the exception of isolated opaque regions associated with strong oxygen and water-vapor gas resonances. The degree of transparency in the "windows" between these resonant lines is dependent on the conditions of cloud cover and precipitation at the time of observation. Meteorological applications represent the prime use of microwave radiometric sensors in this upper portion of the spectrum.

Although the size and complexity of a microwave radiometric sensor are determined by the wavelength of observation and the specific application, the functional parts include an antenna, receiver, and output indicator. The antenna selects electromagnetic waves from specific directions with a particular polarization and presents at its output terminals, for delivery to the receiver, power extracted from these waves. The receiver provides the amplification of the low-level noise power signal received from the antenna needed to drive the output indicator system. The output indicator provides a visual or permanent record of the noise power delivered by the antenna to the receiver, as a function of the time of observation. The output indicator is normally calibrated in temperature units of degrees Kelvin as a consequence of the "temperature-sensing" characteristics of the antenna.

5.7.1 The Antenna as a Temperature-Sensing Device

The antenna in a microwave radiometric sensor is analogous to the lens or mirror of an optical telescope. It is the region of transition between a free-space wave and a guided wave. The response of an antenna as a function of direction can be expressed in terms of the field intensity (field pattern) The power pattern can be visualized in polar coordinates as consisting of several lobes, the largest of which is known as the main lobe.

The function and performance characteristics of the antenna in a radiometric sensor system can be visualized by considering an antenna placed in a hohlraum (blackbody) in which the walls of the hohlraum are sufficiently removed from the antenna structure such that an increase in the size of the hohlraum introduces no change in the angular characteristics of the antenna pattern. If the output of the antenna is fed via a lossless transmission line to a matched resistive termination, then under conditions of thermal equilibrium, the power extracted from the hohlraum and delivered to the resistive termination must be equivalent to the power generated by the resistive termination and transmitted to the antenna (Second Law of thermodynamics).

The noise power per unit bandwidth available at the terminals of the resistor is given by kT, where k is Boltzmann's constant and T is the absolute temperature of the resistor expressed in degrees Kelvin. Since an equivalent amount of power must be extracted from the hohlraum by the antenna and delivered to the resistor, the power per unit bandwidth extracted from the hohlraum by the antenna must also be equivalent to kT. Since the power received by the antenna from the hohlraum is proportional to the temperature of the hohlraum and is independent of the frequency of observation, it is convenient to describe the noise power received by an antenna in equivalent temperature units.

A useful parameter descriptive of the performance of an antenna in a radiometric sensor system is solid-angle efficiency, defined as the percentage of total power received by an antenna placed in a hohlraum, as a function of the solid angle measured about the antenna boresight. The main beam angle efficiency is the value of the solid-angle efficiency measured at the first nulls in the power pattern. This definition assumes circular symmetry of the power pattern about the boresight. Circular symmetry of the power pattern is an important requirement in geoscience applications, since the observed source radiation is frequently polarization sensitive. It is important therefore that the observed signal in either polarization be obtained from the same footprint, that is, projection of the antenna power pattern on the source. The polarization sensitivity of terrestrial material radiation, when viewed at other than normal incidence, further requires that the antenna have a minimum cross-polarization response.

The cross-polarization response and pattern symmetry can be obtained directly from power pattern measurements performed on an antenna test range. The definition of symmetry and solid angle efficiency normally requires that power patterns be measured in the half-diagonal (22.5°)

and diagonal (45°) planes, as well as the orthogonal principal planes about the boresight axis of the antenna. To obtain a useful measure of solid angle efficiency, the response in the back hemisphere of the antenna must be measured to a level at least 50 dB and preferably 60 dB below the main lobe response. This requirement places severe restraints on the multiple reflection properties of the antenna test range since most radar and communication antenna test ranges normally provide a 40- to 50-dB dynamic range capability.

5.7.2 Function of the Receiver

The prime function of the receiver in a microwave radiometric sensor is to amplify the noise power at the output terminal of the antenna, raising it to a level sufficient to drive the output indicator. The antenna can be considered to be replaced by an equivalent resistive load at the receiver input. An antenna with a temperature T_A gives the same noise power input to the receiver as the resistive load in a thermal bath at a temperature T_A. The need for signal amplification is apparent from the fact that the average noise power per unit bandwidth produced by a resistor at an ambient temperature of 290°K is of the order of 10^{-20} W. Most radiometric receivers provide sufficient amplification to record temperature changes of the order of 0.1°K (power per unit bandwidth change of 10^{-24} W).

All receivers, regardless of design, add noise to the input signal in the process of amplification. The receiver output, therefore, consists of the amplified input noise signal received from the antenna combined with an additional noise component generated internally in the receiver. Consequently, the receiver output signal-to-noise ratio (S/N) is always lower than its input signal-to-noise ratio. This degradation in signal-to-noise ratio is described by the receiver noise figure, which is the ratio of the input S/N to the output S/N. A useful parameter derivable directly from the measured value of receiver noise figure is receiver noise temperature. The relationship between noise figure F and noise temperature T_{RT} is

$$T_{RT} = (F - 1) T_0, \qquad (5.7\text{--}1)$$

where $T_0 = 290°K$.

The total system noise temperature T_{sys} referred to the input terminals of the receiver (including input transmission line) is therefore the sum of the matched input resistive load temperature T_A and the internal receiver noise temperature T_{RT} or

$$T_{sys} = T_A + T_{RT}. \qquad (5.7\text{--}2)$$

The ability to detect a change ΔT_A in the input signal noise temperature

T_A implies a signal-to-noise detection capability given by

$$\frac{S}{N} = \frac{\Delta T_A}{T_A + T_{RT}}. \tag{5.7-3}$$

For example, for a signal detection capability of 0.1°K, a signal noise temperature T_A of 290°K, and a receiver with a measured noise figure F of 10 dB, the receiver must be capable of detecting a change in the input noise signal which is 45 dB below the system noise level. This example clearly demonstrates the significance of receiver noise temperature and gain stability. If the receiver is noiseless, $T_{RT} = 0$, then the S/N detection capability is -35 dB, rather than -45 dB. Even in this case, however, the required receiver gain stability is 0.03% during the observing period. This is beyond the capability of most receivers, even for observing periods of a few hours. A means for reducing the effect of gain instabilities is described in Section 5.7.3 [2].

For a perfectly gain-stable receiver, the limit on minimum detectable signal capability is determined by the statistical nature of noise. The output noise power of a resistive input termination is associated with the thermal agitation of electrons within the resistive conductor which produces electronic collisions. As the thermal temperature T_A of the resistor is increased, the thermal agitation increases and the number of collisions per unit time increases. The resultant noise power output per hertz is directly proportional to the absolute temperature of the resistor. The proportionality factor is Boltzmann's constant, k. In this sense, a radio measurement of the thermal temperature of the input resistor is a measurement of the electron collision frequency within the resistor. Since the collisions are random, the number per second varies. However, the mean of an infinite number of one-second samples leads to an exact value for the collision frequency. From statistical theory, the probable error in the measurement of a quantity of this type is inversely proportional to the square root of the number of measurements. The electronic collisions within a resistor are measured using an amplifier of finite bandwidth Δv. The number of independent collisions per second which can be counted is equivalent to the receiver bandwidth. Hence the error in determining the mean value of the noise temperature (which is proportional to the collision frequency) is inversely proportional to the square root of the receiver bandwidth. If the averaging process is extended over τ seconds rather than one second, there will be on the average $\tau \, \Delta v$ independent collisions in each interval of τ seconds and

$$\frac{\overline{\Delta T_A}}{T_A} \propto \frac{1}{(\tau \Delta v)^{1/2}}. \tag{5.7-4}$$

In most radiometric applications the magnitude of the signal temperature T_A is small compared with the receiver noise temperature, T_{RT}, which describes the noise power added to the received signal by the various circuits within the receiver. Hence the minimum detectable change in the input signal temperature is given by

$$\overline{\Delta T_A} \propto \frac{T_A + T_{RT}}{(\tau \Delta v)^{1/2}}. \tag{5.7-5}$$

The constant of proportionality is determined by predetection and post detection signal processing parameters of the receiver and is usually in the range 1.0 to 2.5.

5.7.3 Dicke Mode

In the analysis of minimum detectable signal capability in the preceding section, it was assumed that the receiver gain was constant. In practice, gain variations are unavoidable. By carefully stabilizing all supply voltages and operating equipment temperatures, gain stabilities of the order of 0.1 to 1% can be achieved over a period of a few hours.

Gain stability is critically important since an increase in signal power cannot be distinguished from an increase in the receiver predetection gain. If, for example, the total system noise temperature is 3,000°K, a 0.1% change in receiver gain results in a 3° change in the indicated output.

A marked reduction in receiver gain variations can be accomplished if the receiver input is continuously switched between the antenna output terminal and a comparison noise source at a switching frequency high enough to assure that the receiver gain does not change during one cycle [1]. This technique is shown in Figure 5.7-1. One of the input ports of the switch is connected to the output antenna terminal and the other to a resistive load held at a constant temperature T_C. The switch is driven sequentially in a square-wave fashion at a frequency considerably higher than that at which a substantial receiver gain variation occurs (typically 30 to 1,000 Hz). With the switch in operation, a signal at the switching or modulation frequency is presented at the input terminals of the receiver with an amplitude proportional to the temperature difference $T_A - T_C$. Because of the rapid switching rate, any receiver gain variation operates equally on $T_A + T_{RT}$ during one-half of the switching cycle and on $T_C + T_{RT}$ during the other half, with the result that it operates only on the difference $T_A - T_C$. If, for example, the difference $T_A - T_C$ is 1°K, the effect of a 1% receiver gain variation referred to the output indicator is 0.01°K.

The circuitry following the rf amplification portion of the receiver usually consists of a square-law detector, followed by a narrow-band video amplifier tuned to the modulation frequency. The fundamental component

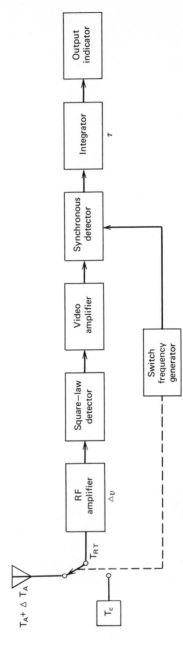

Figure 5.7-1 Switched or Dicke receiver.

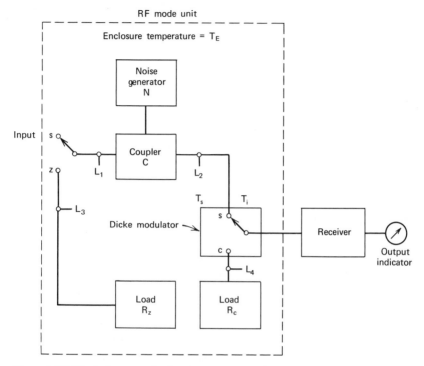

Figure 5.7-2 Block diagram, absolute temperature mode.

of the modulation signal is then synchronously detected to provide a dc voltage to the output indicator which is proportional to the input temperature difference $T_A - T_C$.

5.7.4 Absolute Temperature Mode

The radiometric mode of operation which provides an absolute temperature calibration of the output indicator without the use of cold loads or sky horns during operation is frequently referred to as the "absolute temperature mode." In this mode of operation, the output indicator is calibrated on an absolute temperature scale in degrees Kelvin referenced to the input signal port.

A functional block diagram of the absolute temperature mode is shown in Figure 5.7-2. The losses associated with the individual sections of transmission line within the RF mode unit are indicated as L_1, L_2, L_3, and L_4. The important feature is that all components contained within the RF mode unit are maintained at the same thermometric temperature T_E.

With the input switch connected to position z and with the noise generator N in the off condition, the temperature at the input port s of the modulator is derived from the resistive load R_z and the interconnecting sections of transmission line. The value of the noise temperature T_s at port s is

$$T_s = \frac{T_E}{L_1 L_2 L_3} + \left(1 - \frac{1}{L_1 L_2 L_3}\right) T_E = T_E, \tag{5.7-6}$$

where the insertion loss of coupler, C, is contained in either L_1 or L_2.

In a similar manner, the signal at port c of the modulator is given by

$$T_c = \frac{T_E}{L_4} + \left(1 - \frac{1}{L_4}\right) T_E = T_E. \tag{5.7-7}$$

The circuitry following the input ports of the modulator is a Dicke receiver. Hence the indicated output for the condition described above is zero.

The output indicator scale is calibrated in degrees Kelvin per volt by the introduction of a cold load at a known temperature T_K at the input signal port. With the input switch connected to the input terminal, the effective temperature at port s of the modulator is given by

$$\frac{T_K}{L_1 L_2} + \left(1 - \frac{1}{L_1 L_2}\right) T_E. \tag{5.7-8}$$

The temperature at the comparison port c of the modulator is given by Eq. 5.7-7. Hence, the indicated output signal (input difference) for this case is

$$\text{output} = \frac{1}{L_1 L_2} (T_K - T_E), \tag{5.7-9}$$

which provides the output indicator scale value in degrees Kelvin per volt.

An important feature to note in reference to Eq. 5.7-9 is that the indicated output is proportional to the temperature difference between the source T_K at the input signal terminal and the temperature of the environmental enclosure T_E. The proportionality constant is not important and need not be measured. However, it is of interest to note that it is determined by the attenuating effects of losses L_1 and L_2 in the signal path.

Calibration of the output indicator on an absolute scale is accomplished by turning on the noise generator and adjusting the noise injection level T_i via the coupler to obtain an indicated output equivalent to T_K measured in the scale units determined in the prior step. The value of T_i required to achieve this result is given by the condition

$$\frac{T_K}{L_1 L_2} + \left(1 - \frac{1}{L_1 L_2}\right) T_E + T_i - T_E = \frac{T_K}{L_1 L_2} \tag{5.7-10}$$

$$T_i = \frac{T_E}{L_1 L_2}. \tag{5.7-11}$$

The value of T_i at port s of the modulator is proportional to T_E and the proportionality coefficient is the same as that obtained in the prior step.

It is not required that the actual value of this coefficient be measured. The only parameter values which require careful measurement are the cold-load temperature T_K and the environmental enclosure temperature T_E.

The significance of maintaining all components in the RF mode unit at the same thermometric temperature can be seen by introducing slight temperature differences for each of the losses L_1 through L_4 in Eqs. 5.7–8, to 5.7–10. The magnitude of these individual losses and their distribution as well as their temperature differences relative to T_E determine the magnitude of the associated errors.

In practical applications, the output terminal of the antenna system cannot be directly connected to the input port of the radiometric receiver without the introduction of passive components such as transmission lines and polarizers. The resistive losses of these passive components attenuate the received signal and introduce a radiated noise signal. The combination of these two effects on the received signal is observed by the radiometric receiver, with the result that the input signal temperature indicated by the receiver is not the antenna temperature.

For several geoscience applications, the required output indication is the antenna temperature on an absolute temperature scale. This can be accomplished with a high degree of accuracy through the application of absolute temperature mode concepts which allow the translation of an absolute temperature measurement capability from the input of the radiometric receiver to the output terminal of the antenna. The efficacy of these techniques is predicated on judicious temperature control of critical portions of the passive circuitry between the antenna output terminal and the radiometric receiver input.

5.7.5 Calibration

In most geoscience applications, the terrestrial material under surveillance represents an extended source covering a solid angle significantly larger than the main beam angle of the sensor antenna. Angular resolution of selected areas is therefore determined by the power pattern characteristics of the antenna. A prime parameter of interest is the brightness temperature of the source. The observed antenna temperature represents a summation of the spatially distributed brightness temperatures of all sources viewed by the antenna, appropriately weighted by the power pattern response. Since it is antenna temperature which is measured by the receiver and recorded by the output indicator, it is desirable that antenna temperature be recorded on an absolute temperature scale in degrees Kelvin. The antenna temperature of interest is that component of the total antenna temperature delivered at the output terminals of the antenna which has been extracted from the source radiation incident on the antenna aperture.

The receiver is incapable of discriminating between this component of antenna temperature and the total value of antenna temperature even if the receiver is configured in the absolute temperature mode.

The distinction between total antenna temperature and that portion extracted from the radiation incident on the antenna aperture is best understood by considering the measurement and calibration procedures which are normally applied to the antenna and radiometric receiver separately before they are connected as a complete system.

The important characteristics of the antenna are solid angle efficiency, symmetry, and cross polarization. These parameters are derived from power pattern measurements performed on an antenna range. If the transmission line from the antenna feed element to the output terminal of the antenna contains a resistive loss of 0.1 dB, it introduces a negligible effect on the measured power patterns and cross-polarization response even under the most optimum antenna range conditions. This resistive loss, however, contributes nearly 6°K to the total antenna temperature when measured by the receiver, which would be indistinguishable from the contribution to antenna temperature provided by external radiation incident on the antenna aperture.

Small antenna mismatches which are inevitable also contribute a noise temperature term since the input circuitry, even if equipped with a ferrite isolator, radiates noise toward the antenna, and a small fraction of this noise is reflected back into the receiver and appears as a component of the total antenna temperature.

There is no simple well-defined procedure for the identification and measurement of these small noise bias terms which enter into the system when the antenna is connected to the receiver. A rigorous solution would be provided by calibration of the sensor system with the antenna placed first in a cold hohlraum followed by a hot hohlraum similar to the method used in calibration of the radiometric receiver alone. This is impractical for antennas of large aperture or significant far field requirements. Approximations to hohlraum-type measurements have been used [4, 6].

The development of the absolute temperature mode represents a significant advancement in microwave radiometric technology since it provides a mode of operation for recording the total antenna temperature on an absolute scale without the use of cryogenic equipments associated with cold-load standards during normal operation. It does not, however, solve the remaining problem of distinguishing between that portion of the antenna temperature extracted from the incident wave on the antenna aperture and the bias temperature component contributed by the antenna.

References

1. Dicke, R. H., "The Measurement of Thermal Radiation at Microwave Frequencies," *Rev. Sci. Instr.* **17**, 268–275, July 1946.
2. Jansky, K. G., "Directional Studies of Atmospherics at High Frequencies," *Proc. IRE* **20**, 1920–1932, December 1932.
3. Kraus, J. D., *Radio Astronomy*, McGraw-Hill Book Company, New York, 1966.
4. Kuz'min, A. D., and A. E. Salomonovich, *Radioastronomical Methods of Antenna Measurements*, Academic Press, New York, 1966.
5. Pawsey, J. L., and R. N. Bracewell, *Radio Astronomy*, Clarendon Press, Oxford, 1955.
6. Penzias, A. A., and R. W. Wilson, "A Measurement of Excess Antenna Temperature at 4080 Mc/s," *Astrophys. J.* **142**, 419–421, 1965.
7. Young, Leo, *Advances in Microwaves*, Vol. 5, Academic Press, New York, 1970.

Chapter 6

Signal Processing

Robert W. Rochelle

6.1 SYSTEM CONSIDERATIONS

Most sensors react to the environment and produce an electrical signal as an output. It would be quite wasteful of the radio-frequency spectrum — a most valuable resource—if these sensor outputs were transmitted directly. Instead, the signals at the sensor output are processed in the instrument platform to conserve bandwidth. An example is an energetic particle scintillation detector that produces 50 million pulses per second that are processed in a counter which is read out once per second. Thirty-two bits could describe the total number of counts received in 1 s. This information is transmitted over a radio link in a bandwidth of approximately 64 Hz. This represents a tremendous saving in bandwidth over the case in which the pulses are directly transmitted to the receiving station. The output of the counter can be compressed further to save bandwidth if only the logarithm of the count each second is sent back. The 32 data bits could be compressed to 16 bits with little loss of information. Thus, signal processing is very instrumental in producing systems which can accomplish difficult measurements with high efficiency. Because of power and spectrum limitations, signal processing must be employed to convert the sensor outputs to a signal format which is compatible with the transmission channel from the instrument platform to the receiving site.

There are two types of signal processing used on an instrument platform. The processing of the sensor output to put the signal in a standard format is known as signal conditioning or source encoding. The output of the source encoder can be analog or digital. A typical analog signal is in the range of 0 to + 5 V; a digital signal is one of two possible voltage levels. These source encoded signals pass into the second phase of the signal processor, the channel encoder. Here the signals are encoded to match the characteristics of the channel to obtain a higher efficiency and lower probability of data errors.

A block diagram of a typical instrument system is shown in Figure 6.1-1. Since all the blocks are in series, any improvement in the performance of a block improves the system performance by that same amount. The source encoder is the unit with the greatest potential for bandwidth reduction and system efficiency improvement. Improvements in channel encoding have produced gains of up to 10 dB over uncoded transmissions, and some encoders are within less than 5 dB of the theoretical limit set by Shannon's channel-capacity theorem.

6.2 SIGNAL CONDITIONING AND SOURCE ENCODING

The signal conditioner or source encoder often has the most influence on the instrument system design. Unfortunately this point is not always recognized early in a system design, resulting in deficiencies such as poor output signal-to-noise ratio, aliasing, higher probability of error and lower information rate. This unit offers the greatest potential of reducing the data rate without affecting the desired information rate. The sensor output often contains redundancy which adds little improvement to the output. However, redundancy can be judiciously added in the channel encoder to improve the communications link. Thus, redundancy is often removed in the source encoding process and added in the channel encoding process.

6.2.1 Analog Signals

Many sensors produce a voltage or a current as their output signal. When a complex instrument platform has many sensors, it is desirable to standardize on the maximum dynamic range of the output voltage so that the sensors with their signal conditioners are compatible with other signal

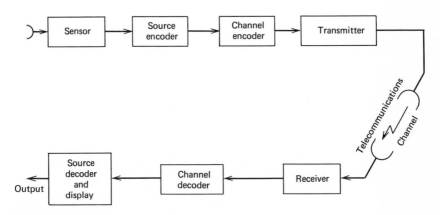

Figure 6.1-1 System block diagram.

processing units. An output of from 0 to + 5 V is common. If the output is below this range, a linear amplifier is used to raise the voltage.

Signal conditioning for analog sensors with slowly varying outputs is relatively easy. A thermistor temperature sensor requires a stable voltage and a resistive network as its signal conditioner. Figure 6.2-1a shows the circuit diagram for a temperature sensor and Figure 6.2-1 b gives the relation between temperature and output voltage.

For an analog sensor output with a poor signal-to-noise power ratio or a wide frequency band, on-board signal processing can be impractical. For these cases a special-purpose transmitter is used to relay the signal to the receiving site for processing.

6.2.2   Digital Signals

The early geoscience instrument sensors yielded an analog output. With the development of digital logic and the digital computer there was an impetus

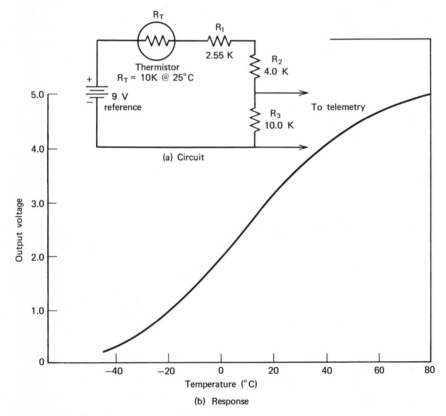

(a) Circuit

(b) Response

Figure 6.2-1   Temperature sensor circuit and response.

to design sensors which produced digital signals directly without having to convert from an analog to a digital signal. The all-digital sensor and signal processor has advantages in cost, efficiency, and reliability.

An example of the all-digital system is the optical aspect encoding system

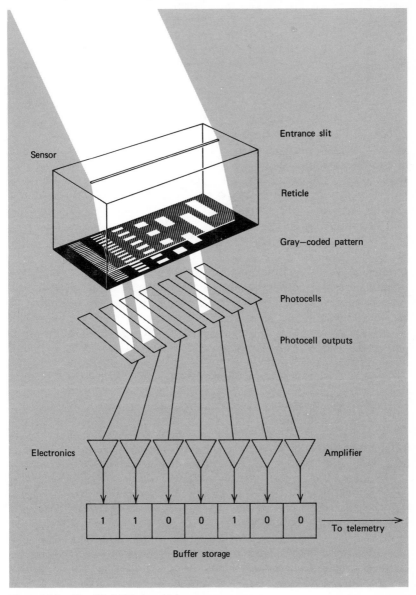

Figure 6.2-2 Simplified digital optical aspect sensor.

for the Small Scientific Spacecraft (Explorer 45) [15]. A simplified diagram of the optical aspect sensor is shown in Figure 6.2-2. As the spacecraft rotates about its spin axis, the sun's rays pass through a light mask to a set of photo diodes. Whether a diode sees the sun or not is a function of the light mask and the angle between the spin axis and a line to the sun. When a diode sees the sun, a logical *one* is set into a shift register. The register is read out as a seven-bit representation of the sun angle. Such a sensor has an accuracy of $\pm 2°$ for the least significant bit and can read over a 180° field of view. Connected to this sensor is a device to generate timing and gating functions in synchronization with the spinning spacecraft [16]. The time between pulses from a photo diode (one revolution of the spacecraft) is divided into 512 parts. The output of a sensor during any one of the 512 segements is stored in a register. If this register is used to store data from that sensor only during that particular segment of the revolution, an averaging or accumulation process occurs. The output of the register is the average value of the sensor output integrated over a number of revolutions of the spacecraft. This saves considerable ground computer data processing time, because the telemetry sampling rate is not normally in synchronization with the spacecraft spin rate.

6.2.3 Data Compression

In many cases the communications channel does not have the capacity to support the sensor output data rate. An examination of the output statistics of the sensor will suggest ways to reduce the data flow with little or no loss of information content. Some forms of data compression are information preserving, others are information destroying. This latter case of irreversible operations (called entropy-reducing data compression) on a data source reduces the original signal to a code that enables the necessary information to be encoded using fewer bits.

Entropy-reducing transformations can be employed to achieve substantial gains in the reduction of the data rate. Although the original signal cannot be reconstructed having only the output signal, the statistics of the input signal are retained, and it can be reconstructed to within a certain specified tolerance or peak error.

The quantizer is one of the most useful entropy-reducing devices. The dynamic range of the input analog signal is divided into N steps. When the input is sampled, a digital word is developed which indicates the step containing the signal amplitude. An n-bit word then represents the amplitude of the signal where $2^n = N$. Information is lost both by increasing the time between samples and by decreasing the number of steps. Increasing the time between samples to effect data compression can lead to serious problems such as aliasing [5].

The digital counter is another very useful entropy-reducing device. The ripple counter is the simplest form, and further compression can be achieved by converting it to a floating-point counter or a logarithmic counter [19]. Other entropy-reducing devices are logarithmic amplifiers, limiters, threshold monitors, and moment estimators.

Polynomial data compressors are entropy-reducing devices but are considered for some applications to be information preserving within a specified tolerance [28]. The two types of polynomial data compressors are the predictor and the interpolator [10]. The zero- and first-order predictors and interpolators are used in data compression schemes. Higher-order systems are more subject to noise and are inherently more complex.

The algorithm for the zero-order predictor sets a tolerance around the first transmitted data point. The succeeding data points are not transmitted unless they fall outside the tolerance value. When a data point falls outside the tolerance value, it is transmitted and the tolerance value is applied to this transmitted value.

A typical sampling sequence is shown in Figure 6.2-3*a*. Since the time of the sample is not known, the time must also be transmitted. For rapidly varying data (high-frequency components) the bit rate can approach twice the uncompressed data rate. To avoid this, the first-order predictor is sometimes used.

In the first-order predictor the first data point is transmitted. The algorithm constructs an extension to the line drawn through the first and second data points. The prediction is made that the third data point will fall within a specified tolerance of the extended line. If it does, the point is not transmitted. Each successive point is checked to see if it falls within the specified tolerance on the extended line. When a data point falls outside the tolerance, that point is transmitted along with the time, a new line is projected from that data point through the following one, and the process is repeated. Figure 6.2-3*b* shows the sampling for the first-order predictor.

In the zero-order interpolator a tolerance is set around both the first and second data points. If the two tolerance windows overlap, an upper bound is formed by the lesser of the two upper bounds, and a new lower bound is formed by the greater of the two lower bounds. Figure 6.2-3*c* illustrates the selection of the sampled points. The tolerance is set up around the third point and the check is made to see if the new window overlaps the generated window. If it does not, the average value of the upper and the lower bound of the generated window is transmitted, and a new run is started with the next data point.

In the first-order interpolator the first data point is transmitted, a tolerance is set about the second point, and lines are drawn from the first data point through the tolerance limits on the second data point forming a fan. If the third data point is within the fan, a new fan is formed by lines from the first data point through the tolerance values on the third data point. This process

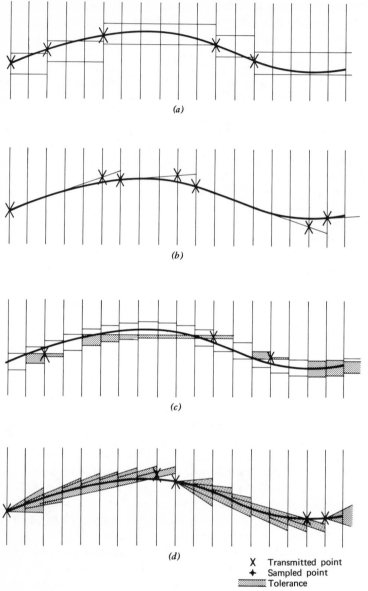

Figure 6.2-3 Data-compression sampling sequences. (*a*) zero-order predictor; (*b*) first-order predictor; (*c*) zero-order interpolator; (*d*) first-order interpolator.

is continued until the next data point falls outside the fan. The mean value of the last fan is transmitted, and the process is repeated starting with the last actual data point. Figure 6.2-3*d* illustrates the selection of the sampled values for the first-order interpolator.

The choice of the proper algorithm depends upon the type of data to be compressed. The zero-order predictor works well for compressing step functions while the first-order algorithm is better for a saw-tooth waveform.

6.3 CHANNEL ENCODING

The channel provides the communication link between the platform and the data acquisition facility. If the distance is short and the platform is fixed in location, a simple wire transmission line can serve as the channel. The designer has some control of the characteristics of the channel. When the platform is moving and the distance is great, radio propagation is used to form the channel. The designer has little control over the characteristics of this type of channel and therefore must use the proper modulation and coding techniques to overcome unfavorable channel characteristics.

A number of modulation techniques are available to the designer. Frequency modulation dominated in early telemetry systems since it offered noise immunity, simplicity, and acceptable linearity. As the required number of channels increased, pulse-amplitude modulation was used. With a continued capacity increase and the availability of microelectronic circuitry, pulse-code modulation has become widely used in modern telemetry systems.

6.3.1 Multiplexing

Sensor signals (analog or digital) at the output of the source encoder are put in a standard form for acceptance by the channel encoder. When several signals are present at one time, multiplexing is used to accomplish the simultaneous transmission of the signals over the single channel.

Frequency-division multiplexing and time-division multiplexing are commonly used in telemetry systems. In general, any orthogonal set of waveforms can be used in the multiplexing process, but frequency-division and time-division multiplexing are by far the most common.

Frequency-Division Multiplexing. In frequency-division multiplexing the allocated frequency spectrum is divided into several frequency bands, and a subcarrier is assigned to each band. The subcarrier converts the input waveform into a frequency spectrum which is contained entirely within the assigned band. In the demodulation process a filter encompasses each band so that signals in one band do not interfere with signals in the other bands. This orthogonality is the basis of any multiplexing scheme. To increase the orthogonality and prevent cross-talk, guard bands are inserted between the subcarrier bands. A description of the assignment of these bands is contained in Section 6.3.2.

Time-Division multiplexing. In time-division multiplexing each sensor is allocated a predetermined length of time for transmission of a sample of its encoded data. A switch selects a sample from each sensor in sequence. The complete switch sequence is called a frame. The sensor sample can be either a pulse amplitude or a set of binary pulses which represents the amplitude of the sensor output at the sampling time. Since the encoded sensor outputs are commutated sequentially, each output is transmitted with the full power of the transmitter. Systems of this type are easier to implement since intermodulation distortion is negligible with only a single signal connected to the output at one time.

The rate of sampling for a particular sensor can be increased by multiple sampling per frame. This is referred to as supercommutation or cross-sampling. Conversely, the rate of sampling can be decreased to accommodate more sensors by sampling a sensor less than once per telemetry frame using subcommutation.

6.3.2 Base-Band Frequency Modulation

A block diagram of a typical base-band FM telemetry system is shown in Figure 6.3-1. Each sensor is connected to a subcarrier oscillator through a source encoder. The source encoder converts the sensor output to match the input voltage range of the subcarrier oscillator. The source encoder is not needed if the sensor output voltage range is correct. The oscillator produces a frequency proportional to the input voltage.

Each sensor is connected to a separate subcarrier oscillator, and the oscillator outputs are summed to produce a composite signal for modulation of the radio-frequency carrier. The subcarrier oscillators are assigned specific frequency bands which do not overlap to prevent distortion. The standard bands designated by the Inter-Range Instrumentation Group (IRIG) are listed in Table 6.3-1 [6]. The bandwidth of each band is proportional to the center frequency. The frequency deviation of the subcarriers is limited to ± 7.5 percent for bands 1 through 18 and ± 15 percent for bands A through H.

The proportional subcarrrier channel assignments are inadequate when telemetering a number of channels each with the same frequency response. A different standard set of bands, the constant-bandwidth subcarrier channels, developed to cover this case is given in Table 6.3-2. The A channels have a nominal frequency response of 400 Hz with a deviation of ± 2 kHz. The B channels have twice the frequency response and occupy the bandwidth of two A channels. The C channels accommodate four times the frequency response of the A channels and occupy four A channels each. The constant-bandwidth channel assignments are used for many identical sensors

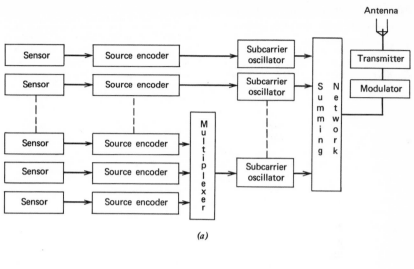

(a)

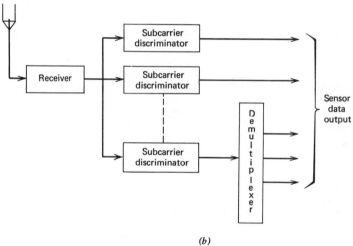

(b)

Figure 6.3-1 FM telemetry system.

gathering the same type of data. In most data gathering tasks different types of measurements are made with different data rates. In these applications the proportional subcarrier channel assignments are used with the lower-frequency channels assigned to the low data rates.

Sensor outputs can be commutated or multiplexed to time share the subcarrier oscillator and expand the number of available channels as shown in the lower half of Figure 6-5a. The subcommutation is accomplished in a

Table 6.3-1. Proportional Subcarrier Channels

Channel	Center Frequency (Hz)	Lower Deviation Limit[a] (Hz)	Upper Deviation Limit[a] (Hz)	Nominal Frequency Response (Hz)	Nominal Rise Time (ms)	Maximum Frequency Response[b] (Hz)[a]	Minimum Rise Time[b] (ms)
			± 7.5% *Channels*				
1	400	370	430	6	58	30	11.7
2	560	518	602	8	42	42	8.33
3	730	675	785	11	32	55	6.40
4	960	888	1,032	14	24	72	4.86
5	1,300	1,202	1,398	20	18	98	3.60
6	1,700	1,572	1,828	25	14	128	2.74
7	2,300	2,127	2,473	35	10	173	2.03
8	3,000	2,775	3,225	45	7.8	225	1.56
9	3,900	3,607	4,193	59	6.0	293	1.20
10	5,400	4,995	5,805	81	4.3	405	0.864
11	7,350	6,799	7,901	110	3.2	551	0.635
12	10,500	9,712	11,288	160	2.2	788	0.444
13	14,500	13,412	15,588	220	1.6	1,088	0.322
See paragraph 5.2.2.1 of Ref. 6							
14	22,000	20,350	23,650	330	1.1	1,650	0.212
15	30,000	27,750	32,250	450	0.78	2,250	0.156
16	40,000	37,000	43,000	600	0.58	3,000	0.117
17	52,500	48,562	56,438	790	0.44	3,938	0.089
18	70,000	64,750	75,250	1050	0.33	5,250	0.067
19	93,000	86,025	99,975	1395	0.25	6,975	0.050
See paragraph 5.2.3 of Ref. 6							
20[c]	124,000	114,700	133,300	1860	0.19	9,300	0.038
21[c]	165,000	152,625	177,375	2475	0.14	12,375	0.029
			± 15% *Channels*[d]				
A	22,000	18,700	25,300	660	0.53	3,300	0.106
B	30,000	25,500	34,500	900	0.39	4,500	0.078
C	40,000	34,000	46,000	1200	0.29	6,000	0.058
D	52,500	44,625	60,375	1575	0.22	7,875	0.044
E	70,000	59,500	80,500	2100	0.17	10,500	0.033
F	93,000	79,050	106,950	2790	0.13	13,950	0.025
G[c]	124,000	105,400	142,600	3720	0.09	18,600	0.018
H[c]	165,000	140,250	189,750	4950	0.07	24,750	0.014

[a] Rounded off to nearest Hz.

[b] The indicated maximum data frequency response and minimum rise time are based upon the maximum theoretical response that can be obtained in a bandwidth between the upper and lower frequency limits specified for the channels.

[c] Recommended for use in UHF transmission systems only.

[d] Channels A through H can be used by omitting adjacent lettered and numbered channels. Channels 13 and A can be used together with some increase in adjacent channel interference.

Table 6.3-2. Constant-Bandwidth Subcarrier Channels

A Channels		B Channels		C Channels	
Deviation Limits = ± 2kHz Nominal Frequency Response = 0.4kHz Maximum Frequency Response = 2kHz[a]		Deviation Limits = ± 4kHz Nominal Frequency Response = 0.8kHz Maximum Frequency Response = 4kHz[a]		Deviation Limits = ± 8kHz Nominal Frequency Response = 1.6kHz Maximum Frequency Response = 8KHz[a]	
Channel	Center Frequency (kHz)	Channel	Center Frequency (kHz)	Channel	Center Frequency (kHz)
1A	16				
2A	24				
3A	32	3B	32		
4A	40				
5A	48	5B	48		
6A	56				
7A	64	7B	64	7C	64
8A	72				
9A	80	9B	80		
10A	88				
11A	96	11B	96	11C	96
12A	104				
13A	112	13B	112		
14A	120				
15A	128	15B	128	15C	128
16A[b]	136				
17A[b]	144	17B[b]	144		
18A[b]	152				
19A[b]	160	19B[b]	160	19C[b]	160
20A[b]	168				
21A[b]	176	21B[b]	176		

[a] The indicated maximum frequency response is based upon the maximum theoretical response that can be obtained in a bandwidth between deviation limits specified for the channel.

[b] Recommended for use in UHF transmission systems only.

standard manner to avoid special-purpose decommutation equipment by using the pulse-amplitude-modulation standard developed by the IRIG.

6.3.3 Pulse-Amplitude Modulation

When a group of sensors is subcommutated onto a single subcarrier, pulse-amplitude modulation (PAM) results. The IRIG standard allows two duty cycles, 50 percent and 100 percent. Figure 6.3-2 shows the formats for

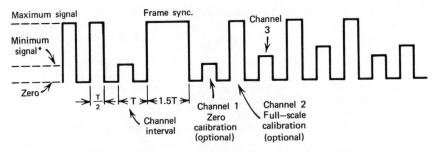

* 20 to 25 % deviation reserved, for pulse synchronization is recommended

(a)

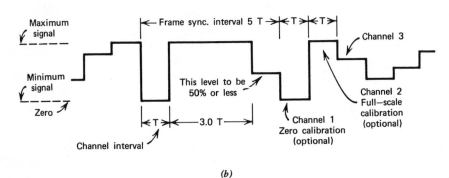

(b)

Figure 6.3-2 Pulse-amplitude-modulation format, (a) 50% duty cycle; (b) 100% duty cycle.

these duty cycles. In the 50-percent duty cycle the sensor is connected to the subcarrier oscillator for one-half the sample period and the oscillator input is returned to a zero baseline for the second half of the sample period. There is no return to zero for the 100-percent duty cycle. Frame synchronization is accomplished by producing a pulse either 1.5 or 3.0 times the sample period length. Both zero and full-scale calibrations are available in channels 1 and 2 of both duty cycles. The standard stipulates that no more than 128 pulses be used per frame since synchronization would be difficult to achieve with more pulses per frame. PAM permits an existing FM/FM system to be updated and increased in channel capacity by adding a PAM channel in one of the FM/FM bands.

6.3.4 Pulse-Code Modulation

With the integration of the small general-purpose digital computer into instrument data processing systems, the use of pulse-code modulation (PCM) for telemetry has increased over the use of analog forms of modulation because the digital PCM form is readily acceptable by the digital computer.

PCM has its greatest advantage where data are gathered from a large number of sensors at relatively low sampling rates for each. It is excellent for a noisy environment since only the presence or absence of the pulse has to be detected, not its height. However, it occupies a wider bandwidth than a corresponding pulse-amplitude modulation signal, but this increase in bandwidth is compensated for by the extra performance in the noisy channel. When accuracies better than 1 percent are needed, PCM is usually the preferred choice of modulation since maintaining calibration of an FM system to better than 1 percent is very difficult.

General Description. In a pulse-code-modulation system the transducer waveform is sampled periodically [26]. Each sample is converted into a set of binary symbols (binary word) which represents the quantized sample amplitude and is transmitted serially over the channel. The received data can be left in digital form to facilitate data processing or it can be converted back to its original analog form for display. The analog-to-digital (A/D) converter (quantizer), described in Section 6.2.3, p. 686, generates the binary symbols. The output of the A/D converter is stored in a shift register and read out at predetermined times. The input to the A/D converter can be switched to connect to different sensors at different times. The control logic which sequentially switches the input of the A/D converter and outputs the shift register into the telemetry bit stream comprises the majority of the PCM encoder. Figure 6.3-3 is a diagram of a typical PCM telemetry system.

The binary symbol represents two states and is called a "bit" which is the contraction of "binary digit." The set of bits which is the binary representation of the sample amplitude is the PCM word. The bit period is a fixed length of time, and its reciprocal, the bit frequency, determines the bandwidth needed to send the PCM waveform.

A typical PCM word consisting of eight bits can be further subdivided into syllables. Several sensors can contribute to one word with each sensor's data being a syllable of the word. A group of words or telemetry channels comprises a frame with the frame synchronization word included at the beginning of each frame. These are called prime or major frames with typical frame lengths of 1024 bits. Generally several of the channels are subcommutated to increase the number of measurement points to be telemetered. The group of major frames which include the full cycle of the subcommutator is called the minor frame or subframe. Also, channels can be cross-strapped to increase the sampling rate for a particular sensor. This is termed supercommutation.

PCM Code Formats. The output of the A/D converter is a non-return-to-zero (NRZ) waveform with a "one" represented by one voltage and a "zero" by another voltage. This waveform is often used to directly modulate the

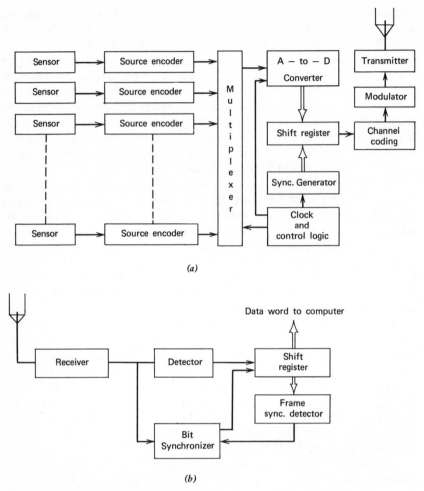

(a)

(b)

Figure 6.3-3 PCM telemetry system.

transmitter. It has the advantage that the spectrum occupancy is relatively small allowing the transfer of more data within a given bandwidth. The spectrum of the NRZ waveform with random data is shown in Figure 6.3-4. Modulation of a carrier with this waveform produces a spectrum which has the data energy at and close to the carrier. Figure 6.3-4 shows this spectrum if the carrier is inserted at the zero-frequency mark. Having the data energy close to the carrier can substantially interfere with most demodulators. If a phase-locked loop is used to demodulate and track the carrier frequency, the data sidebands within the loop bandwidth cause cycle slippage and loss of lock under noisy conditions.

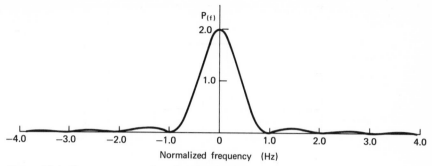

Figure 6.3-4 Power spectrum of random PCM-NRZ data.

To prevent this degradation in performance, the sidebands are moved away from the carrier by using a subcarrier. A square-wave subcarrier with a frequency equal to the reciprocal of the bit period (one cycle per bit) moves the sideband energy sufficiently far from the carrier to remove the carrier interference. This results in a split-phase or Manchester code format. Figure 6.3-5 illustrates the power spectrum of the split-phase code format.

The IRIG telemetry standards define a set of seven PCM waveforms illustrated in Figure 6.3-6 [6]. The NRZ-level and the bi-phase-level are the two most commonly used waveforms. The NRZ-mark and NRZ-space were originally used for recording PCM data on magnetic-tape recorders, but a more common use is evolving as differential-phase-shift-keying (DPSK) [23]. Demodulation is simple in that the level of the previous bit is compared to the level of the present bit to determine if a change in level has occurred. The return-to-zero (RZ) format is seldom used.

The bi-phase mark and bi-phase space formats each have a transition at the beginning of a bit period and the information is carried by the presence or absence of a transition in the middle of the bit period. The spectrum is identical to that for the bi-phase-level waveform for random data. Like the NRZ-mark and NRZ-space waveforms, the demodulation can be accom-

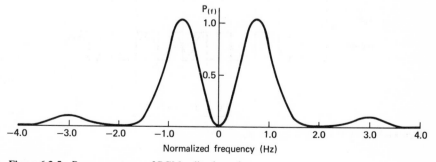

Figure 6.3-5 Power spectrum of PCM split-phase data.

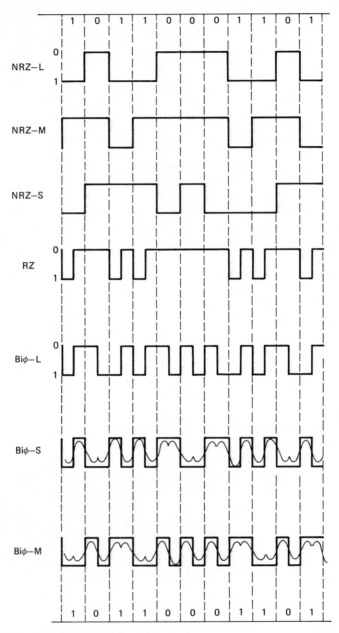

Figure 6.3-6 PCM standard waveforms.

NRZ—level (or NRZ Change)
"One" is represented by one level.
"Zero " is represented by the other level.

NRZ—mark
"One" is represented by a change in level.
"Zero" is represented by no change in level.

NRZ—space
"One" is represented by no change in level.
"Zero" is represented by a change in level.

RZ
"One" is represented by a half—bit wide pulse.
"Zero" is represented by no pulse condition.

Bi—phase—level (or split phase, Manchester 11 + 180°)
"One" is represented by a 10.
"Zero" is represented by a 01.

Bi—phase space
A transition occurs at the beginning of every bit period.
"One" is represented by no phase shift (i.e., a zero crossing)
at the middle of the bit period.
"Zero" is represented by a 180° phase shift (i.e., no transition
or zero crossing) at the middle of the bit period.

Bi—phase mark
A transition occurs at the beginning of every bit period.
"One" is represented by a 180° phase shift (i.e., no transition
or zero crossing) at the middle of the bit period.
"Zero" is represented by no phase shift (i.e., a zero crossing)
at the middle of the bit period.

plished with a very simple detector that detects the presence or absence of a transition in the middle of a bit period.

Another waveform similar to the bi-phase mark and bi-phase space and not presently included in the IRIG standard is delay modulation (Miller encoding) [9]. The bi-phase mark becomes delay modulation if the transition at the beginning of a bit period is deleted only when the preceding bit has a transition in the middle of the bit period. Figure 6.3-7a shows the waveform for delay modulation and Figure 6.3-7b illustrates the very favorable power spectrum occupancy. This type of encoding has found a very worthwhile application in magnetic tape recorders, since the spectrum is confined to a relatively narrow band. While it is excellent for tape recorders, its performance is not good in a noisy environment.

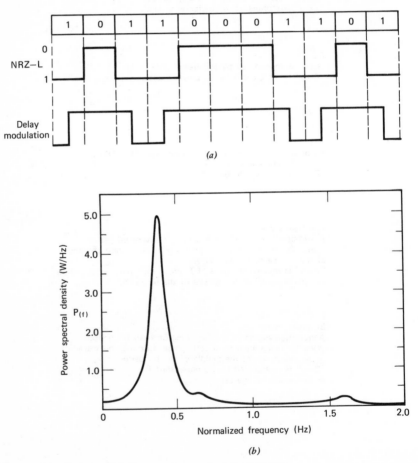

Figure 6.3-7   Delay modulation waveform and power spectrum.

PCM Performance Measurement. The performance of a PCM system is evaluated in terms of the number of errors which occur as a function of the signal-to-noise power ratio [22]. Since the noise in most systems is a statistical and random quantity, the number of errors in the data is also statistical and random. The error rate is indicated by the probability of bit error, the average number of errors which occur in a block of data divided by the total number of bits in that block. As an example, for a group of 500 blocks of data with 1024 bits per block, and an average number of errors per block of 2.05, the error probability is 2.05/1024 which is 2.0×10^{-3} or an average of two errors in 1000 bits.

Since sensor data is organized into groups of words, the probability of error in a word becomes more meaningful than the bit-error probability. At very low error rates the word-error probability is equal to the bit error probability multiplied by the number of bits in the word. At higher error rates two or more errors can occur within a single word so that the statistics are slightly modified.

A common parameter used to make performance comparisons between systems is the ratio of the quantity of energy (E_b) that arrives at the demodulator for each information bit divided by the single-sided noise power density (N_0), the noise power measured in a 1-Hz bandwith. This parameter, E_b/N_0, is completely independent of the demodulator characteristics so that comparisons can be made between various demodulation configurations. Figure 6.3-8 illustrates the bit-error probability for a phase-coherent demodulator and a non-phase-coherent demodulator. The phase-coherent curve is generally used as the basis of comparison with any other demodulator performance curve.

Coded PCM. Noise in the transmission link causes errors in the sensor data. One way to combat this noise and thus reduce the number of errors in the data is to increase the transmitter power. Remote unattended sensor systems are usually battery powered and cannot afford an increase in dc power to support a higher transmitter power. Another way to lower the error rate with a given transmitter power at the same information rate is to code the PCM data. Redundancy is added to the bit stream in the form of extra bits in a predetermined manner. While this increases the data rate for the same information rate, the number of errors in the information message is decreased. Various schemes have been devised which prescribe rules for encoding the information bit stream. In the block diagram of Figure 6.1-1 the channel encoder inserts the necessary redundancy to improve the error rate. The decoding is accomplished in the channel decoder, and its output is the best estimate of the output of the source encoder.

Codes have been developed for both the noiseless channel and the noisy channel. Coding for the noiseless channel minimizes the average number of binary digits in a message. Examples are the Shannon-Fano code [20] and

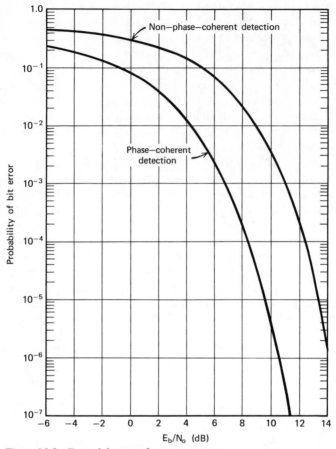

Figure 6.3-8 Demodulator performance curves.

the Huffman code [17]. If a sensor has an output for which the levels are not equi-probable, these codes could be advantageous for encoding the sensor output for storage directly on a tape recorder. The recorder can be considered an element of a noiseless channel since the recorded signals are not power limited. Coding for the noisy channel reduces the probability of error in the information being transmitted by the addition of extra bits into the bit stream, increasing the redundancy of the message.

Hamming Codes. The simplest error check is the addition of a single bit to the end of each PCM word at the time of transmission. The value of this parity bit depends upon the value of the preceding information bits. When the value assigned to the parity bit makes the total number of "ones" in the word even, the parity is said to be even. Likewise, if the parity bit value is assigned to make the number of "ones" an odd number, it is odd parity. Table 6.3-3(a) lists the values of a three-bit code which is capable of describing eight amplitude levels. Column number four is the parity bit added to make the four-digit word contain an even number of "ones." Table 6.3-3(b) lists the same three-bit code with the additional column for odd parity.

The parity code can be implemented as shown in Figure 6-13a. The block labeled D is a single-stage shift register. The circle with the + sign denotes modulo-2 addition ($0 \oplus 0 = 0, 1 \oplus 0 = 1, 0 \oplus 1 = 1, 1 \oplus 1 = 0$) and is a simple binary half-adder. To develop the code of words of Table 6.3-3(a), each of the information bits of any one of the eight words is shifted into the encoder one bit at a time with the switch in the I position and appears in the output. After the three information bits have been shifted into the encoder, the switch is moved to the P position. The fourth or parity bit is the value stored in the shift register and appearing at terminal P. The shift register is reset to zero after each word.

To decode the received message, a device very similar to the encoder is used as shown in Figure 6.3-9b. The four data bits are shifted into the decoder one at a time. The first three bits are transferred to storage if a binary zero

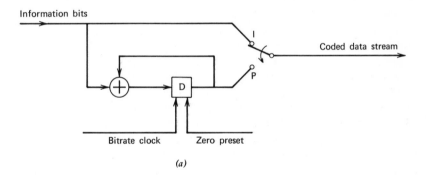

(a)

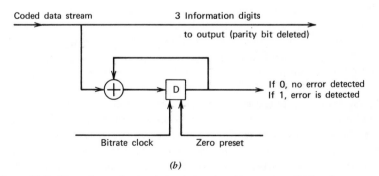

(b)

Figure 6.3-9 Hamming code encoder and decoder. (*a*) encoder; (*b*) decoder.

Table 6.3-3. Parity

Amplitude	Data 1	Data 2	Data 3	Parity 4	Amplitude	Data 1	Data 2	Data 3	Parity 4
1	0	0	0	0	1	0	0	0	1
2	0	0	1	1	2	0	0	1	0
3	0	1	0	1	3	0	1	0	0
4	0	1	1	0	4	0	1	1	1
5	1	0	0	1	5	1	0	0	0
6	1	0	1	0	6	1	0	1	1
7	1	1	0	0	7	1	1	0	1
8	1	1	1	1	8	1	1	1	0

(a) Even (b) Odd

is detected at the output of the shift register. If a binary one is detected, the three information bits are deleted.

Since this parity check code has a four-digit codeword length with three digits of information, the conventional notation is given as (4,3). The general code notation is (n, k) where n is the codeword length and k is the number of information bits.

A convenient way of evaluating the error performance of a code is the use of a parameter called the Hamming distance [7]. The Hamming distance is the number of disagreements obtained when two codewords of equal length are compared digit for digit. In Table 6.3-4 (a) codeword 1 and codeword 2 disagree in two of the four digits, yielding a Hamming distance of two. Without a parity bit the minimum Hamming distance from any codeword to any other in the set of eight is one. Adding the parity bit, the minimum Hamming distance between any two codewords is two. It would take two errors in a single word to convert the correct codeword to one of the other seven and cause an *undetected error*. A single error in one of the four bits still leaves the word a distance of at least one away from any of the other words in the set. Unless the received word matches exactly with one of the eight codewords, it is generally deleted from the data set. Thus, a simple parity check detects the presence of a single error but allows a double error to pass undetected. In most data systems this is acceptable since single errors occur with low probability (i.e., $P_{se} = 10^{-4}$), and the probability of double errors is much less ($P_{de} = 10^{-8}$).

In the Hamming code the parity checks are not placed at the end of the word but are placed in positions corresponding to powers of two (i.e., 1, 2, 4, 8, 16, ...) with data in the remaining positions. The code is most efficient when the number of bits in a word equals a power of two minus one or $2^n - 1$. Table 6.3-4 illustrates a seven-digit (7, 4) Hamming code. Columns 3, 5, 6, and 7 contain the uncoded data while columns 1, 2, and 4 contain

Table 6.3-4. **(7, 4) Hamming Code**

Amplitude	Parity 1	Parity 2	Data 3	Parity 4	Data 5	Data 6	Data 7
0	0	0	0	0	0	0	0
1	1	1	0	1	0	0	1
2	0	1	0	1	0	1	0
3	1	0	0	0	0	1	1
4	1	0	0	1	1	0	0
5	0	1	0	0	1	0	1
6	1	1	0	0	1	1	0
7	0	0	0	1	1	1	1
8	1	1	1	0	0	0	0
9	0	0	1	1	0	0	1
10	1	0	1	1	0	1	0
11	0	1	1	0	0	1	1
12	0	1	1	1	1	0	0
13	1	0	1	0	1	0	1
14	0	0	1	0	1	1	0
15	1	1	1	1	1	1	1

parity bits. Column 1 is even parity for columns 1, 3, 5, and 7. Column 2 is even parity for columns 2, 3, 6, and 7. Column 4 is even parity for columns 4, 5, 6, and 7.

The actual position of an error in a word can be determined by determining which of the three parity checks indicate an error. A three-bit word can be formed to indicate exactly which bit is in error. The least significant bit of the word is a zero if the column 1 parity check is satisfactory and a one if it is not. The column 2 parity check determines the value of the next to the least significant bit, and the column 4 parity check determines the most significant bit. As an example, if the columns 4 and 1 parity checks are not correct and column 2 is correct, the three-bit location word is 101 or decimal 5 indicating that the error is in the fifth bit. Knowing error location permits error correction.

BCH Codes. The Bose-Chaudhuri-Hocquenghem (BCH) codes [14] are used to correct more than a single error. These codes are easy to implement with shift registers and simple logic circuitry. The number of digits in a codeword is $2^m - 1$ where m is any integer. The codes can correct t errors by using only mt check digits and can detect $2t$ errors.

An example is the BCH (15, 7) code which has $mt = 8$ check digits. Since $m = 4$, $(2^m - 1 = 15)$, $t = 2$, and the code is capable of correcting two errors. Figure 6.3-10a illustrates an encoder for generating the coded data. Switch 1 is kept in the I position for the seven information digits while the information bits enter the encoder through the half-adder and through

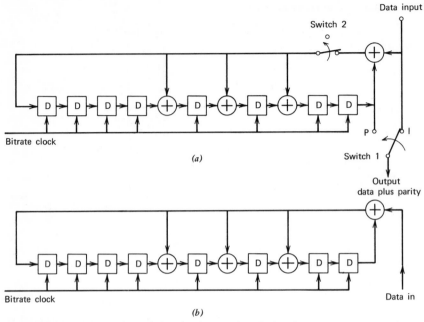

Figure 6.3-10 BCH encoder and decoder. (*a*) encoder; (*b*) decoder.

switch 2. After the last information bit has entered the encoder, switch 1 is thrown to the parity position and switch 2 is opened. The bitrate clock signal shifts the contents of the shift register through switch 1 to the output.

The errors in the data can be located by feeding the data into the decoder of Figure 6.3-10*b*. After all 15 data bits have been loaded into the encoder, the states of the shift registers are examined. If the states are all zero, no errors are indicated. If the states are not all zero, then the shift register values can be used to locate and correct up to two errors.

Pseudo-Noise Codes. Sensor systems on moving platforms have to contend with two basic problems, the need for ranging for platform position location and the need for reliable communications. The pseudo-noise (PN) codes provide an excellent tool to solve both of these problems [25].

The PN code is very easy to implement. In Figure 6.3-11 a four-stage shift register is loaded with the four information bits to be encoded. The clocking pulses are started, and the shift register stages are shifted 15 times. The codeword that appears on the output is the original four bits of information plus 11 bits of parity. As the shift register is shifted to the left, the content of the right-hand shift register is replaced by the modulo-2 sum of the first two stages. On the fifteenth shift the original four-bit word appears back in the shift register. Table 6.3-5(a) lists the contents of the shift register after each

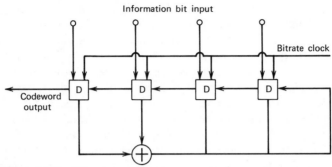

Figure 6.3-11 PN code generator.

shift. Column 1 is the code word which would be forwarded to the transmitter if 0001 had been the four information bits loaded in the shift register. The second column is the code word resulting when 0010 is the information word loaded in the register. Since the code is cyclic, it is easy to determine the codeword for any other four-bit word loaded in the shift register. The all-zero word is not allowed. If it is set into the register, only zeros will be shifted out. Table 6.3-5 (b) lists the fifteen 15-bit codewords which follow when any four information bits of Table 6.3-5 (a) are loaded in the shift register. For an n-bit shift register, the number of codewords that can be formed is $2^n - 1$, and because of the cyclic nature of the code, this is also the number of bits in a codeword.

This code has a very interesting property that permits ranging for location. The Hamming distance between any two codewords is equal to one-half of

Table 6.3-5. Pseudo-Noise Code

(a) Information word				(b) Code word														
0	0	0	1	0	0	0	1	0	0	1	1	0	1	0	1	1	1	1
0	0	1	0	0	0	1	0	0	1	1	0	1	0	1	1	1	1	0
0	1	0	0	0	1	0	0	1	1	0	1	0	1	1	1	1	0	0
1	0	0	1	1	0	0	1	1	0	1	0	1	1	1	1	0	0	0
0	0	1	1	0	0	1	1	0	1	0	1	1	1	1	0	0	0	1
0	1	1	0	0	1	1	0	1	0	1	1	1	1	0	0	0	1	0
1	1	0	1	1	1	0	1	0	1	1	1	1	0	0	0	1	0	0
1	0	1	0	1	0	1	0	1	1	1	1	0	0	0	1	0	0	1
0	1	0	1	0	1	0	1	1	1	1	0	0	0	1	0	0	1	1
1	0	1	1	1	0	1	1	1	1	0	0	0	1	0	0	1	1	0
0	1	1	1	0	1	1	1	1	0	0	0	1	0	0	1	1	0	1
1	1	1	1	1	1	1	1	0	0	0	1	0	0	1	1	0	1	0
1	1	1	0	1	1	1	0	0	0	1	0	0	1	1	0	1	0	1
1	1	0	0	1	1	0	0	0	1	0	0	1	1	0	1	0	1	1
1	0	0	0	1	0	0	0	1	0	0	1	1	0	1	0	1	1	1
0	0	0	1	0	0	0	1	0	0	1	1	0	1	0	1	1	1	1

the quantity one plus the number of bits in the codeword. In this case (Table 6.3-5) the Hamming distance is eight, which ensures that the correct codeword is selected with high probability at the receiving station for transmission to the platform where it is turned around and retransmitted to the receiving station. Since the position of the platform is not known, the time of arrival at the receiving station of the leading bit in the repeating codeword is not known. To find this time of arrival, each bit and the 14 following bits are compared with the transmitted codes and the Hamming distance noted. When the received codeword has a Hamming distance of zero when compared with a replica of the transmitted codeword, the codeword is identified in time. Any other comparison yields a Hamming distance of eight. In the presence of noise the codeword is considered identified when the Hamming distance appears to be less than four. An ambiguity in the distance calculation can result if the codeword is not sufficiently long in time.

Convolutional Codes. Convolutional encoding [12] combined with sequential detection [11] is an extremely powerful technique that allows a reduction in the transmitter power of up to 8 dB with the other system parameters remaining the same. It is used in almost every deep space probe.

In convolutional encoding the information bit stream flows continuously through the encoder. The output data stream is a logical computation on the information bits in the encoder at that particular time. If the information bits appear unaltered and interspersed with parity bits, the code is called systematic. If the information bits do not appear in the output data stream, the code is nonsystematic. The rate r of the code is the ratio of the input information rate to the output data rate. High rate codes preserve bandwidth but are less efficient than low rate codes. A common rate is one-half. The constraint length L of the code is the number of stages in the shift register over which the parity computation is made. In general, the longer the constraint length, the more efficient is the code.

Figure 6.3-12 shows a block diagram of a short-constraint-length convolutional encoder. As each information bit enters the shift register, two parity bits are computed by means of the series of half-adders connected to the shift register stages. The switch moves from P_1 to P_2 and back to P_1 for each information bit shifted into the encoder.

Decoding is done either algebraically [2, 13] using a decoder similar to the encoder or probabilistically by sequential detection [18]. For short-constraint-length codes, Viterbi decoding is very effective [27].

6.4 CARRIER MODULATION

Modulation of the sensor platform radio-frequency carrier provides the means for conveying the sensor information from the platform to the data

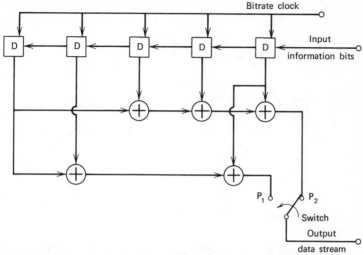

Figure 6.3-12 Convolutional encoder.

acquisition station. There are three basic forms of carrier modulation: amplitude, phase, and frequency (abbreviated AM, PM, and FM). The waveform which modulates the carrier can itself be a subcarrier which has been modulated. The term FM/PM denotes a carrier phase modulated with an FM subcarrier.

The general equation for the modulated carrier voltage is

$$e(t) = A(t) \cos \left[\omega_0 t + \phi_0 + \psi(t) \right], \qquad (6.4\text{--}1)$$

where $A(t)$ is the amplitude of the modulated carrier, ω_0 is the angular frequency of the modulated carrier, ϕ_0 is the phase of unmodulated carrier, and $\psi(t)$ is the phase contribution due to modulation.

Figure 6.4-1 is a plot of a typical baseband analog sensor signal which exemplifies a typical modulation waveform.

6.4.1 Amplitude Modulation

For amplitude modulation [8] the general expression of Eq. 6.4-1 reduces to

$$e_{\text{AM}}(t) = A_0 \left[1 + M_a g(t) \right] \cos \omega_0 t, \qquad (6.4\text{--}2)$$

where A_0 is the amplitude of the unmodulated carrier, M_a is the modulation index $(0 < M_a \leq 1)$, and $g(t)$ is the modulating waveform.

Figure 6.4-2 illustrates the waveform that results from modulating an rf carrier of angular frequency ω_0 with the modulating signal of Figure 6.4-1. If the baseband signal is a cosine waveform given by

$$G(t) = \cos \omega_a t, \qquad (6.4\text{--}3)$$

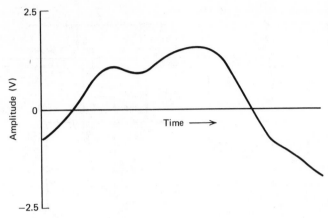

Figure 6.4-1 Analog signal.

where ω_a is the modulation angular frequency, then the AM voltage waveform becomes

$$e_{\text{AM}}(t) = A_0(1 + M_a \cos \omega_a t) \cos \omega_0 t$$

$$= A_0 \cos \omega_0 t + \tfrac{1}{2} A_0 M_a \cos(\omega_0 - \omega_a) t \qquad (6.4\text{--}4)$$

$$+ \tfrac{1}{2} A_0 M_a \cos(\omega_0 + \omega_a) t.$$

The modulated waveform has three frequency components, the carrier at angular frequency ω_0, a lower-sideband at $\omega_0 - \omega_a$ and an upper sideband at $\omega_0 + \omega_a$.

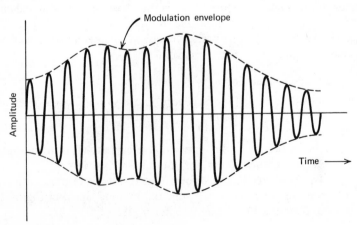

Figure 6.4-2 Radio-frequency waveform.

The distribution of the power in the carrier and the sidebands (assuming a 1-Ω load) is

$$P_{AM} = \overline{[e_{AM}(t)]^2} = \frac{A_0^2}{2} + \frac{A_0^2 M_a^2}{4} = P_0 \frac{1 + M_a^2}{2}, \qquad (6.4\text{-}5)$$

where P_0 is the power in the unmodulated carrier. When the modulation index is unity, the carrier power is exactly twice the sideband power.

Double-Sideband Suppressed-Carrier Modulation. Because of the relatively large amount of carrier power with conventional amplitude modulation, the transmitter power is not effectively utilized. The carrier component can be suppressed to put more power into the sidebands and effectively double the sideband power. The demodulator must insert a carrier so that the modulation can be detected without distortion. The bandwidth is the same as in conventional AM. A Costas loop [4] can be used for demodulation.

Single-Sideband Suppressed-Carrier Modulation. If one of the sidebands can be removed from the double-sideband suppressed-carrier transmission, the bandwidth can be reduced by a factor of 2. This is important for effective utilization of the spectrum. The sideband is often removed with crystal filters. The power in the remaining sideband can then be boosted by a factor of 2 to maintain the same average power.

In the demodulator the carrier reconstruction is not as critical as it is in the case of double-sideband suppressed-carrier modulation. For voice communications small errors in the frequency location of the reconstructed carrier cause a small change in the pitch of the voice. The single-sideband signal [24] is harder to generate than the double-sideband suppressed-carrier signal, but it is easier to demodulate for some applications.

6.4.2 Phase Modulation

The phase-modulated signal can be expressed in the time domain by the equation

$$e_{PM}(t) = A_0 \cos[\omega_0 t + \psi(t)], \qquad (6.4\text{-}6)$$

where A_0 is the amplitude of the rf carrier, $\psi(t) = Kg(t)$, and K is the maximum phase deviation in radians. If the modulating signal, $g(t)$, is a sinusoid such as $-\cos \omega_a t$, then

$$e_{PM}(t) = A_0 \cos(\omega_0 t - \Delta\phi \cos \omega_a t), \qquad (6.4\text{-}7)$$

where $\Delta\phi$ is the maximum phase deviation, ω_0 is the carrier frequency, and ω_a is the modulation frequency. This can also be expressed as

$$e_{PM}(t) = \text{Re}\,[A_0 \exp(+j\omega_0 t)\exp(-j\Delta\phi \cos \omega_a t], \qquad (6.4\text{-}8)$$

but from Bessel's equation

$$\exp(-j\Delta\phi\cos\omega_a t) = \sum_{-\infty}^{\infty} J_n(\Delta\phi)\, j^n \exp(jn\omega_a t). \qquad (6.4\text{–}9)$$

Also,

$$J_n(-\Delta\phi) = -1^n J_n(\Delta\phi). \qquad (6.4\text{–}10)$$

so that Eq. 6.4–8 can be reduced to

$$\begin{aligned}
e_{PM}(t) = A_0 \{ & J_0(\Delta\phi)\cos\omega_0 t \qquad\qquad\qquad\qquad (6.4\text{–}11) \\
& + J_1(\Delta\phi)\left[\sin(\omega_0 + \omega_a)t + \sin(\omega_0 - \omega_a)t\right] \\
& - J_2(\Delta\phi)\left[\cos(\omega_0 + 2\omega_a)t + \cos(\omega_0 - 2\omega_a)t\right] \\
& - J_3(\Delta\phi)\left[\sin(\omega_0 + 3\omega_a)t + \sin(\omega_0 - 3\omega_a)t\right] \\
& + \cdots \}.
\end{aligned}$$

The graph of the Bessel function is shown in Figure 6.4-3. The spectrum of the waveform, unlike that of amplitude modulation, contains an infinite number of harmonics in the upper and lower sidebands. The magnitude of

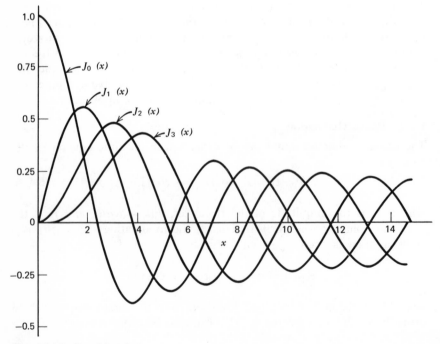

Figure 6.4-3 Bessel functions.

these harmonics is insignificant for frequencies which are further away from the carrier than the product of the modulation index and the highest modulation frequency. That is, all Bessel functions in which the order is higher than the argument can be neglected.

6.4.3 Frequency Modulation

If the modulating signal is

$$g(t) = \sin \omega_0 t,$$

(6.4–12)

then,

$$\psi(t) = M\Delta\omega \int^t \sin \omega_a \tau \, d\tau,$$

(6.4–13)

$$= -\frac{M\Delta\omega}{\omega_a} \cos \omega_a t.$$

Substituting into Eq. 6.4–1 gives

$$e_{FM}(t) = A_0 \cos\left(\omega_0 t - \frac{M\Delta\omega}{\omega_a} \cos \omega_a t\right).$$

(6.4–14)

This is the same equation as that for phase modulation except that the peak phase deviation $\Delta\phi$ is replaced by $M\Delta\omega/\omega_a$. The modulation index is $\Delta\omega/\omega_a$ with M considered to be unity [21].

6.5 CALIBRATION

One of the most important aspects of a signal processing system is its calibration. This is especially true for applications involving remote platforms unattended for long periods of time. Seasonal trends can be masked by system aging unless a means for calibration is provided. Additional information on calibration is given in Section 8.4.

The resolution of a measurement is the ability to detect a small change between two consecutive measurements. The *accuracy* is the difference between a particular measurement anywhere in the scale and an absolute errorless measurement. These can be expressed as a percent of the full-scale value.

Ideally it is desirable to calibrate during the testing phase, during the operational phase, and after recovery of the instrument. Frequent calibration during the testing phase uncovers trends, establishes the range of accuracies available during the operational phase, and establishes a master calibration curve.

Calibration during the operational phase involves transmitting, at periodic intervals, known reference voltages near the maximum and the minimum of the scale. In most systems these two measurements suffice since the points between the maximum and minimum reference points can be obtained from the test calibration curves by interpolation. In the data reduction process the raw data are converted to engineering units by scaling the test calibration curve using the in-flight reference points. The raw data are then converted to engineering units through a table look-up using values from the scaled calibration curve.

In time-multiplexed systems where one A/D converter is used on a number of sensors, an excellent way of calibrating is to provide four calibration points. The first two are at zero and full scale. Two calibrations at mid-scale connected through two impedances such as 10 and 100 kΩ monitor the impedance changes in the multiplexing switches as a function of aging.

For A/D converters which saturate at both ends of the scale (zero and full scale) the calibration points should be several quantized steps away from zero and full scale. This allows all drifts in calibration to be detected without being masked by the saturation effect.

Care should be exercised in applying calibration measurements to the data. As an example, if one of the calibration voltages is close to the edge of a quantized step, a very small drift can move it into the next step. The new calibration curve gives a step function to the data.

6.6 COMMAND AND CONTROL

After a geoscience instrument platform has been placed in operation, it may be desirable to remotely control its operation from a central ground station. The amount of control to be implemented is a cost-related factor. Instruments with no control are termed "random" systems while those that can be turned on only by command are "ordered" systems.

6.6.1 Random Systems

In a random system [1] the instrument platform is commanded on by a timer aboard the platform. The timer is set, for example, to turn the transmitter on for one second each minute. With this low-duty cycle the probability of two platforms being on at the same time is low so that the multi-access problem is partially solved. The probability that two or more are on at one time is

$$P_I = 1 - \exp(- N\tau/T), \qquad (6.5-1)$$

where P_I is the probability of interference, N is the number of platforms in view, τ is the length of time of the transmission, and T is the time between transmissions.

The interference problem for systems with large numbers of platforms in view at once can be minimized by separating the transmissions in frequency. For platforms communicating with low orbiting earth satellites, the Doppler effect introduces a frequency dispersion which, with proper filtering, can be used to separate transmissions which occur simultaneously.

At 400 MHz the Doppler effect for a 1000-km orbit altitude causes a frequency dispersion of $\pm$ 7.5 kHz. For a data rate of 100 bits per second, each platform occupies a frequency band of approximately 500 Hz so there are 30 frequency bands to prevent interference when transmissions occur simultaneously. A comb filter at the receiver can differentiate between signals in adjacent frequency bands.

Another means of obtaining orthogonality is by signal strength. The capture effect of a large signal against a weak signal is used to retain the strongest signal.

The crystals which derive the stable frequency source for the platform transmission provide another means to prevent interference. The probability of mutual interference can be reduced by having a random spread in the center frequencies of the transmissions. A reasonable specification which reduces the cost of the crystals is to allow a deviation of as much as $\pm$ 3 kHz in the 400-MHz transmitter center frequency. If the distribution is uniform, this spread is equivalent to adding twelve 500-Hz frequency channels for simultaneous use.

6.6.2 Ordered Systems

When the number of platforms in use is large, an ordered system is desirable. With the ordered scheme the mutual interference is eliminated since both the time and frequency of transmission can be controlled. The random system is a nonefficient way of utilizing the frequency spectrum since only a small percentage of the transmissions are actually received. This amounts to a pollution of the radio spectrum when large numbers of platforms are involved.

The fundamental command to an ordered system is the turn-on command. One system accomplishes this in an ingenious manner with a controller that transmits a PN code of 511 bits [3]. Each bit in the code is the first bit of a particular platform's address. The next eight bits are the remaining bits in the address for that platform. Each platform is synchronized with the 511-bit PN code so that when its address of nine bits appears, it returns its data. Each platform utilizes 18 bits of the PN code even though only 9 are necessary for identification. The remaining 9 bits are needed for error detection. A command of six zeros and six ones follows the 18 address bits to energize the transmitter.

References

1. Arndt, A. E., et al., "System Study for the Random Access Measurement System (RAMS)," NASA—Goddard Space Flight Center, Greenbelt, Md., October, 1970.

2. Berlekamp, E. R., *Algebraic Coding Theory*, pp. 390–391, McGraw-Hill Book Company, New York, 1968.

3. Bourdeau, J. P., et al., "Communications Between Satellite and Balloons for the Eole Mission," International Telemetering Conference, pp. 614–630, Los Angeles, Calif., October 1968.

4. Costas, J. P., "Synchronous Communications," *Proc. IRE* **44**, 1713–1718, December 1956.

5. Downing, J. J., *Modulation Systems and Noise,* pp. 140–143, Prentice-Hall, Englewood Cliffs, N. J., 1964.

6. Document 106–71, "Telemetry Standards," Inter-Range Instrumentation Group, White Sands Missile Range, New Mexico, January 1971.

7. Hamming, R. W., "Error Detecting and Error Correcting Codes," *Bell Sys. Tech. J.* **29**, 147–160, April 1950.

8. Hancock, J. C., *An Introduction to the Principles of Communication Theory,* pp 34–47, McGraw-Hill Book Company, New York, 1961.

9. Hecht, M., and A. Guida, "Delay Modulation," *Proc. IEEE* **57**, 1314–1316, July 1969.

10. Hogg, G., "A Data Compression Primer," NASA—Goddard Space Flight Center, Greenbelt, Md., X-521-65-320, August 1965.

11. Jacobs, I. M., "Sequential Decoding for Efficient Communication from Deep Space," *IEEE Trans. Commun. Tech.* **COM-15**, 492–501, 1967.

12. Lin, S., and H. Lyne, "Some Results on Binary Convolutional Code Generators," *IEE Trans. Inform. Theory* **IT-11**, 90–100, 1965.

13. Massey, J. L., *Threshold Decoding,* pp. 54–59, M.I.T. Press, Cambridge, Mass., 1963.

14. Peterson, W. W., *Error Correcting Codes,* pp. 162–167, M.I.T. Press, Cambridge, Mass., 1961.

15. Pyle, E. J., and Stewart, J. R., "Optical Aspect System for the Small Scientific Satellite (S^3-A)," NASA—Goddard Space Flight Center, Greenbelt, Md., X-711-70-174, April 1970.

16. Pyle, E. J., *A Spin-Synchronous Clock for Spin-Stabilized Vehicles,* NASA—Goddard Space Flight Center, Greenbelt, Md., G-988, May 1970.

17. Reza, F. M., *An Introduction to Information Theory,* pp. 155–158, McGraw-Hill Book Company, New York, 1961.

18. Saliga, T. V., "A Comparison of Sequential Decoding Metrics by Computer Simulation," National Aeronautics and Space Administration, Washington, D.C., Technical Report R-294, January 1969.

19. Schaefer, D. H., and Snively, J. W., Jr., "On-Board Plasma Data Processor," *Rev. Scientific Instruments,*" **39**, No. 4, pp. 455–456, April 1968.

20. Schwartz, L. S., *Principles of Coding, Filtering, and Information Theory,* pp. 17–19, Spartan Books, Baltimore, Md., 1963.

21. Schwartz, M., *Information Transmission, Modulation and Noise,* pp. 111–132, McGraw-Hill Book Company, New York, 1959.

22. Strong, J. P., III, and T. V. Saliga, "Comparison of Phase-Coherent and Non-Phase-Coherent Coded Communications," NASA–Goddard Space Flight Center, Greenbelt, Md., G-671.

23. Taub, H., and D. L. Schilling, *Principles of Communication Systems,* pp. 224–227, McGraw-Hill Book Company, New York, 1971.
24. Taub, H., and D. L. Schilling, *ibid.* pp. 98–104.
25. Titsworth, R. C., "The Role of Pseudorandom Codes in Communications," Jet Propulsion Lab. Tech. Rept. 33185, Pasadena, Calif., August 3, 1964.
26. Tyler, J., "Basics of Pulse Code Modulation Telemetry," *Telemetry J.* **4,** No. 5, August/September, 1969.
27. Jacobs, I. M., and J. A. Heller, "Performance Study of Viterbi Decoding as Related to Space Communications," AD-738 213, Final Technical Report in Contract No. DAAB07-71-C-0148, United States Army Satellite Communications Agency, Fort Monmouth, N. J., August 31, 1971.
28. Weber, D., "A Synopsis on Data Compression," National Telemetry Conference, 1965.

Chapter 7

Data Processing

A. A. J. Hoffman

The measurement of geoscience parameters often involves recording data in large quantities, at high rates, over long periods of time, with high precision or combinations of these. Furthermore, it may be necessary to analyze data quickly so that adjustments can be made to obtain optimal response to current conditions.

Data processing involves the introduction of "original" data into a device capable of performing a prearranged sequence of operations on the data. This results in a new set of data which is more useful and/or more easily interpreted than the original data. The availability, flexibility, accuracy, and relatively low cost of the high-speed electronic digital computer often precludes the use of other (analog or hybrid) data-processing equipment.

7.1 CHARACTERISTICS OF DATA

Geoscience data processing can be classified into two main types depending on whether the data is presented to the computer from a storage device upon which the data has been collected or whether the data is to be processed continually while the sensor or data-acquisition device is in operation. The former case is referred to as "batch" processing, since a batch of data is collected before processing begins. The latter case is referred to as "on-line" or "real-time" processing, since data are processed before data collection is completed. Once the computer begins processing the data, real-time and batch input are essentially the same relative to the computer. There are exceptions to this when the computer is connected in real-time to several different data-gathering devices or when the computer is programmed to control signals back to the data-gathering devices. In these cases, according to prearranged logic, the computer input is monitored in such a way as to interrupt itself, change the sequence of operations, and respond to the particular input device. Systems arranged in this manner are referred to as on-line real-time data acquisition and control systems and are most useful in either a controlled laboratory situation or in process control.

Geoscience instrumentation sensor signals are either analog or digital. An analog signal is a continuous function of time. A digital signal is a series of discrete data values each associated with a particular instant of time.

Analog signals must be converted to digital signals before presentation to the digital computer as input. Furthermore, both converted analog signals and digital signals must be in a digital code (format) which is acceptable to the computer. If the data originates in analog form, the data must be changed via an analog to digital (A/D) converter. The A/D conversion process must accommodate several characteristics including volume (amount), rate of acquisition, dynamic range, duration, noise level, and whether or not the data is continuous or intermittent [1]. Such characteristics are relative and interrelated, and an understanding of their effect can be obtained from information theory [7]. Sensing and recording involve energy conversion which is information transmission. Data contains information if it changes with time. Data processing requires knowledge of the information content and information rate of the expected signal. The important parameters are signal voltage range, frequency range (bandwidth), required precision, signal-to-noise ratio, and duration of recording [13]. Selection of transducers, signal conditioning and transmission methods, and recording equipment depends on these parameters.

Regardless of the sophistication of the A/D device, the digital signal cannot contain all of the information present in the original analog signal [1, 15]. Nevertheless, it is often possible for the digital signal to contain all of the pertinent information. One of the most important decisions to be made when designing an A/D process is the selection of a sampling rate. In converting from a continuous signal to one derived from sampling the continuous signal at, for example, equi-spaced intervals of time, a selection of the time interval (Δt) must be made such that the pertinent information content of the continuous function is transferred to the sample data.

7.1.1 Sample Data Considerations

A continuous function of time is completely determined by its values at equally spaced intervals provided that the continuous function contains no frequencies higher than f Hz and the ordinates are given at points spaced $\frac{1}{2}f$ seconds apart, the series of values extending for all time [15]. Usually the analysis is based on sampled values obtained from continuous records that are not infinite in extent and are not band limited. Analysis based on finite amounts of data is common to statistical work. The fact that the original functions are not band limited is discussed in connection with the problem of aliasing. Power spectral density computations (Section 7.3) are meaningless without an analysis of the data relative to aliasing. Figure 7.1-1 shows portions of two sine waves of different frequencies. The equally spaced sample data values are the same for each sine wave. Thus, given only the sample values, a sine wave of a given frequency can be confused with a sine wave of higher frequency. If a harmonic time function $x(t)$ is sampled at equally spaced time intervals Δt, then a frequency

$$f_N = \frac{1}{2\Delta t}, \tag{7.1-1}$$

called the Nyquist or folding frequency [2] exists such that the functions with frequencies

$$f + nf_N \quad \text{for } n = 0, 2, 4, \cdots, \tag{7.1-2}$$

are not distinguishable. For example, consider the function

$$x(f, t) = \sin 2\pi f t, \tag{7.1-3}$$

sampled at $k\Delta t$ $(k = 0, 1, 2, \cdots)$ so that the sampled values of x are

$$x(f, k\Delta t) = \sin 2\pi f k \, \Delta t. \tag{7.1-4}$$

Now consider the function

$$x(f \pm nf_N, t) = \sin 2\pi (f + nf_N)t, \tag{7.1-5}$$

where

$$f_N = \frac{1}{2\Delta t} \tag{7.1-6}$$

and

$$n = 0, 2, 4, 6, \cdots, \tag{7.1-7}$$

sampled at equi-spaced intervals Δt so that the sampled values are

$$x(f \pm nf_N, k\Delta t) = \sin 2\pi (f \pm nf_N)k\Delta t. \tag{7.1-8}$$

Use of simple trigonometric identities reveals that

$$x(f, k\Delta t) = x(f \pm nf_N, k\Delta t). \tag{7.1-9}$$

Thus, power contributed to a power spectrum at a given frequency f cannot be distinguished from power contributed by frequencies $f \pm nf_N$. This translation of frequencies is known as aliasing [4]. If the data actually contain power at frequencies greater than f_N, this power will be "folded

Figure 7.1-1 Sine waves of different frequencies with the same set of equally spaced sample values.

back" into the principal band which extends from 0 to f_N. Power that is folded back results in a distortion of the true power spectrum in the principal band.

To make the effect of aliasing negligible it is necessary to select a sampling interval small enough to place the Nyquist frequency beyond all significant power contributions.

On the other hand, selection of a smaller than necessary sampling interval results in more data values which in turn increase the possibility that round-off error will result in inaccurate results. There are techniques involving digital filtering [10] which allow taking samples small enough to protect against aliasing yet allow a subsequent smoothing of the data and reduction of the number of sample values in computation.

7.1.2 Other Characteristics of Data

Noise and spurious signals often contaminate geoscience instrument data. These can originate in the sensor environment, in the sensor, or elsewhere in the instrument system. Sometimes these can be eliminated at the source, otherwise they must be eliminated by signal processing (see Chapter 6) or removed during the computer processing of the data. Because of its random nature, noise is the easiest to remove [3]. A spurious signal has to be found by establishing that such a signal is not part of the physical phenomena under consideration.

Another type of data contamination is caused by temporary malfunction in the instrument system. This is often referred to as signal drop out. Usually this is handled by data editing during the data processing.

7.2 CHARACTERISTICS OF DATA PROCESSING SYSTEM

7.2.1 Computer

The computer consists of input and output devices, a central processor (for logic, control, and arithmetic functions), and memory as shown in Figure 7.2-1. Equally important is the software which is the program of step-by-step instructions which cause the computer to perform useful work. The computer accepts information-carrying data through its input devices, combines this information according to the sequence of operations dictated by the program stored in memory (also combines with information stored in memory), and sends the resultant information back to the user through its output devices. The reason for storing instructions in the memory of the machine is to provide these instructions at a rate commensurate with the speed at which the computer can execute the instructions.

7.2.2 Recording and Storage of Data for Computer Processing

Recording and processing large amounts of geoscience data reliably can be difficult. Many options are often available and compromises must be made

[14, 16]. For example, if the sensor produces an analog signal, a choice must be made between recording on an analog tape or performing real time A/D conversion and recording in digital format on magnetic tape [1]. Whereas the dynamic range of analog tape is only 20 dB (about four decimal digits plus sign), the digital accuracy could be several times greater. On the other hand, there is at least a five-to-one ratio for the amount of tape required to store in digital mode versus analog mode. Certain techniques have been developed to improve this ratio [13].

Geoscience data can be recorded and stored for batch processing or the data can be processed in real-time, in which case the data moves from the sensor (perhaps through a communications link) through the input device directly into the memory of the computer. In the latter case, the storage media must either match or be converted to a format and media form acceptable to the available input device. Typical input devices read punched cards, punched paper tape or magnetic tape [5].

Data stored on mass storage units are readily accessible to the central processing unit. The principal types of mass storage are magnetic tape and magnetic disk. These differ in cost, type of access (serial or random), rate of transfer, and volume of data stored. Typically, disk units are more expensive, have random access (it takes as long to access any data item in a set of data as any other data item in the set), have a higher transfer rate, and store less than magnetic tapes. Since most geoscience data is recorded and processed serially, magnetic tapes are used more than disks. Tape has a higher storage density, is more portable, and takes considerably less space to store.

Standard digital tape reels are commercially available containing tape from 60 to 730 m in length. One 730-m reel stores 20,000,000 characters of information. The reels are physically interchangeable, but tapes prepared

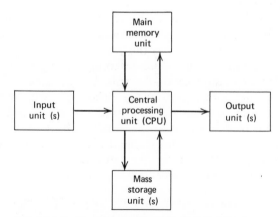

Figure 7.2-1 Functional units of the digital computer.

on one computer cannot be expected to run on another. The parameters in tape recording are many and varied. For example, the tape can be recorded with either 7 or 9 tracks, at recording densities of 100, 200, 315, or 630 bits per cm, and can be recorded in binary code, BCD code, or EBCDIC code [5].

7.2.3 Programming Languages

The computer sequentially executes instructions found in the instruction register—a storage register in the central processing unit. These instructions typically consist of an operation code (command), one or more memory addresses, and/or instruction modifiers. The command is recognized (decoded) by the logic circuitry and the addresses contain the places from or to which information flows. For example, if the contents of two memory locations are to be added, typically the contents of the first are copied into an accumulator, the contents of the second address are copied and added to the accumulator, then the contents of the accumulator (the sum) are copied into a third location in memory for further use. This particular example requires at least three instructions. Typically a computer is designed with 20 to 200 different instructions. The class of programs written in the code directly acceptable to the machine is called machine language. Generally, each machine (even by the same manufacturer) has a different language. Because of the time required to learn the many instructions, the complexity of writing programs consisting of thousands of instructions, and the tremendous bookkeeping problem associated with keeping track of the location (memory address) of each data item, few programs are written in machine language.

For convenience, a higher-level language class called assembly language was developed. In this case, the programmer writes instructions in assembly language and uses a program (written in machine language and generally provided by the manufacturer) to translate the application program from assembly language into machine language. The process of translating assembly language (source language) into machine language is called assembling the program. A second step is required to execute the machine language program. Assembly language is characterized by having symbolic command codes (such as L for load the accumulator instead of a numonic code like 15 in a decimal machine or 1111 on a binary machine) and symbolic addresses instead of actual addresses for memory. For example, in a serial list of data, the first address could be X1 and the second X2 or relative addressing could be used such as $X, X + 1, X + 2, \cdots, X + I$. As with machine language coding, the programmer must know what each instruction does. However, the programmer uses familiar instruction code names (instead of code numbers), and is relieved of a tremendous amount of bookkeeping (i.e., knowing the exact address of each data value). Assembly language is widely used in programming minicomputers (where memory is relatively limited) or in certain special cases on standard computers. Minicomputers are used mainly for data acquisition and control rather than for data processing.

General processing of geoscience data as described in Section 7.3 usually involves a high-level mathematical problem oriented computer language such as FORTRAN. FORTRAN is an algebraic language relatively easy to learn and efficient to use. The instructions are written in an algebra-like form which are translated into machine language using a manufacturer-supplied program referred to as a compiler. The user does not have to know the instruction codes for the computer nor the memory location of any data. A statement (the name used to refer to a FORTRAN instruction) written in FORTRAN often results in several machine language instructions. With slight modifications FORTRAN programs can be transferred from one computer to another. It is useful to learn a high-level language such as FORTRAN to utilize the power of the computer.

7.2.4 System Selection

Minicomputers for data acquisition and control vary widely in price and performance. Some of the most important factors to consider are data rate, word size (accuracy), cost, size, environmental requirements, power requirements, weight, heat dissipation, ease of programming, available peripherals (input, output, storage), memory size, interrupt capability, storage protect, power failure problems, communications capability, ease of operation, indexing, indirect addressing, synchronous or asynchronous input/output channels, internal speed, multiply-divide hardware, floating point hardware, available software, and reliability [14].

7.3 MATHEMATICAL METHODS AND COMPUTER PROCESSING

Once data is acquired and/or stored in a form accessible to the computer, computer programs are needed to extract the desired information. This involves selection of appropriate mathematical techniques and preparation of a computer program.

7.3.1 Selection of Mathematical Method

Measurements or observed data from a physical system are often classified as either deterministic or nondeterministic (random). The choice of analysis method depends on which classification of data is at hand [3]. If it is expected that the data can be described by an explicit mathematical relationship, then the data is said to be deterministic. There are many phenomena of geoscientific interest which produce data which can be represented with reasonable accuracy by an explicit mathematical relationship. A good example of this is the flight path of a satellite in orbit around the earth.

Other data such as wave height as a function of time at any particular location in the ocean, cannot be described by an explicit mathematical relation. Such data are random (i.e., nondeterministic) in nature and must be described in statistical terms.

There are data which do not clearly fall into one or the other of these classifications. However, due to the differences in the mathematical methods associated with such classification of data, a decision must be made regarding classification into either deterministic or nondeterministic.

Figure 7.3-1 shows classifications of both deterministic and random (nondeterministic) data.

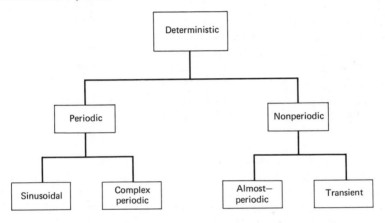

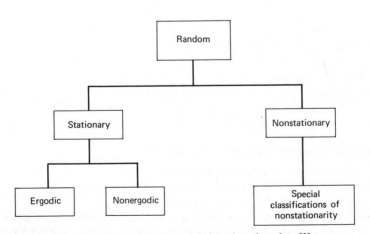

Figure 7.3-1 Classifications for deterministic and random data [3].

7.3.2 Mathematical Methods for Deterministic Data

As indicated schematically in Figure 7.3-1, deterministic data can be either periodic or nonperiodic. Periodic data are analyzed by use of Fourier-series techniques [9, 11]. The explicit function involved is a sine function or a linear combination of sine functions. The mathematical expression by itself is useful and of interest, but a graph of the frequency domain representation (amplitude versus frequency) is most useful for interpretation of the physical phenomena [11] because the occurrence of periodic data is coupled with a frequency dependent physical situation.

There is an almost endless list of mathematical methods for finding an explicit mathematical relation to describe nonperiodic data with reasonable accuracy. To select a procedure it is necessary to either guess or have some idea of the form of the expected relation. Deciding which relation fits the data "best" requires consideration. The most common practice is to select the relation which gives the best fit in the least-squares sense [9].

For example, suppose the data is assumed to fit a linear relationship. The form of the explicit mathematical relation is

$$x = at + b, \tag{7.3-1}$$

where the parameters a and b need to be selected such that the relation is a best fit to the data. In other words, the relation $x = at + b$ represents a family of lines and the best fit is the one selected according to some criteria.

Generally data is not error free and thus does not fall on a single straight line. Also, the data may be from an almost linear portion of an otherwise nonlinear relation. In either case, there is no line which passes through all of the data points. Points on the least-squares graph are often regarded as the smoothed values or expected values of the original data. The least-squares principle states that the best curve is the one for which the sum of the squares of the deviations of the data from their expected values is a minimum.

Therefore, the usefulness of the least-squares method is in the selection of the best curve from a family of curves. The important task is to select the family. Polynomial curve fitting is useful in many situations, but often does not provide physical insight into the phenomena. It should be noted that the Fourier coefficients give the best least-squares fit if the data is fit with an orthonormal set of functions [9].

7.3.3 Mathematical Methods for Random Data

Random (nondeterministic) data cannot be described by an explicit mathematical relationship since each observation is unique and represents only one of many possible results. A single time history of a random phenomenon is referred to as a sample record and a collection of such records is referred to as a random process. The term stochastic process is used interchangeably with random process. The description and interpretation of such processes require the use of statistical methods.

It is convenient to categorize random processes as either stationary or nonstationary (Figure 7.3-1). If the statistical description of a sample record is independent of the starting time, then the process is said to be stationary; otherwise it is nonstationary. Functions which represent stationary random processes have probability density functions which do not change with time. There are many methods for determining the properties of a nonstationary random process, but there is controversy concerning the validity of such methods and no proof that they work in practical cases. Often, geoscientific data varies so little from being stationary that it is assumed stationary for all practical purposes.

The properties of interest in a stationary random process are mean square value, autocorrelation function, and power spectral density. The mean square value gives information about the intensity of the data. The autocorrelation function gives time domain properties of the data. In general, the properties of most interest result from examination of the power spectral density which is the frequency domain representation.

7.3.4 Power Spectral Analysis

Computational procedures are available which yield the statistical variance spectrum of a time series. Other names for the resultant computation are power density spectrum, second-degree spectrum, and quadratic spectrum. All of these names refer to the distribution of variance as a function of frequency. Formal derivations of the formulas used in power spectral computation are available [4, 6, 8, 12]. The engineering term, power, is used in connection with statistical variance because the variance for a time series with zero mean is equal to the ac power dissipated in a 1-Ω resistance [2].

For deterministic functions such as periodic and aperiodic functions, a harmonic analysis is usually carried out by Fourier-series analysis and by Fourier integral analysis, respectively. The discrete line spectrum for a periodic function and the continuous spectrum for the aperiodic function can be determined analytically because these deterministic functions are, by definition, known for all values of time.

Random series [12] are a class of functions which are not deterministic and do not lend themselves to harmonic analysis techniques. Functions which represent stationary random processes are characterized by probability distributions that do not change with time, and the time average over a sample function of an ensemble of functions is equivalent to an average over the ensemble of functions. Records of the earth's magnetic and electric fields can be selected which are assumed to approximate closely the above conditions. It should be noted that proof of the ergodic hypothesis in a practical case appears to be impossible [6].

The autocorrelation function, $\phi(\tau)$, defined by [17]

$$\phi(\tau) = \lim_{T \to \infty} \frac{1}{T} \int_{-T/2}^{T/2} x(t)\, x(t + \tau)\, dt, \quad -\infty < \tau < +\infty, \quad (7.3\text{-}2)$$

where $x(t)$ is a stationary time series, has a Fourier transform $\Phi(f)$, given by

$$\Phi(f) = \int_{-\infty}^{\infty} \phi(\tau) e^{-j2\pi f\tau} d\tau, \quad (7.3\text{-}3)$$

which is the power density spectrum of $x(t)$. Stated another way, the power density spectrum of $x(t)$, defined by

$$\Phi(f) = \lim_{T \to \infty} \frac{1}{T} \left| \int_{-T/2}^{T/2} x(t)\, e^{-j2\pi f t}\, dt \right|^2, \quad (7.3\text{-}4)$$

has a Fourier transform

$$\phi(\tau) = \int_{-\infty}^{\infty} \Phi(f) e^{j2\pi f t}\, df, \quad (7.3\text{-}5)$$

which is the autocorrelation function for $x(t)$.

Practical power spectral analysis is based on special computational methods [4] that consist of estimating the spectrum of a discrete time series by computing numerical approximations to Eqs. 7.3–2 and 7.3–3. Since Eq. 7.3–2 is an even function of τ, Eq. 7.3–3 can be simplified to

$$\Phi(f) = 2 \int_{0}^{\infty} \phi(\tau) \cos 2\pi f\tau\, d\tau. \quad (7.3\text{-}6)$$

An important feature of the method is that it provides a means of estimating the power spectral density from finite sets of sampled data with a knowledge of the accuracy of the estimate [4].

An experimental data record of finite length is sampled at equally spaced intervals (Δt) yielding a discrete time series

$$X_1, X_2, X_3, \cdots, X_N, \quad (7.3\text{-}7)$$

which will be referred to as the raw data. As has already been shown (Section 7.1), the selection of the length of the sampling interval for a particular record has important implications. To avoid trouble from aliasing, a minimal requirement on the choice of Δt would be such that no more than one relative maximum or minimum occur in any interval Δt.

The average value for the series,

$$\langle X(t) \rangle = \frac{1}{N} \sum_{n=1}^{N} X_n, \quad (7.3\text{-}8)$$

where

$$X_n = X(n\Delta t), \quad n = 1, 2, \cdots, N, \qquad (7.3\text{–}9)$$

is subtracted from each X_n using the relation

$$\overline{X}_n = X_n - \langle X(t) \rangle \qquad (7.3\text{–}10)$$

to remove the dc or mean value from the series.

The estimates of the unnormalized autocorrelation at lag τ ($= n\Delta t$) are computed by

$$C_\tau = \frac{1}{N - \tau} \sum_{P=0}^{N-\tau} X_p \cdot X_p + \tau, \quad n = 0, 1, 2, \cdots, m, \qquad (7.3\text{–}11)$$

where C_τ is the computed autocorrelation at lag τ ($= n\Delta t$), N is the number of raw data values, m is the number of lags, and X_p are the data values after the mean value is removed. The C_τ coefficients are often referred to as the sample serial products, mean lagged products, or autocorrelation coefficients.

The raw spectral density estimates are computed by applying a discrete finite cosine series transform to the mean lagged products according to the formula

$$Q_r = \left[C_0 + 2 \sum_{q=1}^{m-1} C_q \cos \frac{qr\pi}{m} + C_m \cos\pi r \right] \Delta t, \quad r = 0, 1, 2, \cdots, m. \qquad (7.3\text{–}12)$$

It is necessary to compute Q_r only for r in the above range since Q_r is symmetric in r. The frequency corresponding to r is

$$f = \frac{r}{2m\Delta t}. \qquad (7.3\text{–}13)$$

The systematic statistical errors resulting from use of a finite length of data appear in the values of Q_r. The technique to obtain improved spectral estimates involves a smoothing or refining operation performed on the raw estimates. The specific smoothing operation used here is referred to as hanning (Eqs. 7.3–14 to 7.3–16). The refined spectral estimates, P'_r, are computed using the relations

$$P'_r(0) = 0.5[Q_r(0) + Q_r(1)], \qquad (7.3\text{–}14)$$

$$P'_r(r) = 0.5Q_r(r) + 0.25[Q_r(r-1) + Q_r(r+1)], \qquad (7.3\text{–}15)$$

and

$$P'_r(m) = 0.5[Q_r(m) + Q_r(m-1)], \qquad (7.3\text{–}16)$$

where $r = 1, 2, \cdots, m - 1$.

The computational formulas given by Eqs. 7.3–8 through 7.3–16 produce a frequency analysis of the data. For large volumes of data such computations are virtually impossible to do by manual methods. On the other hand, the formulas lend themselves very well to computerization [10].

References

1. "Analog-Digital Conversion Handbook (E-5100)," Digital Equipment Corporation, Maynard, Massachusetts, 1964.
2. Bendat, Julius S., *Principles and Applications of Random Noise Theory*, John Wiley and Sons, New York, 1958.
3. Bendat, J. S., and A. G. Persol, *Measurement and Analysis of Random Data*, John Wiley and Sons, New York, 1966.
4. Blackman, R. B., and J. W. Tukey, *The Measurement of Power Spectra*, Dover Publications, New York, 1959.
5. Chapin, Ned, *Computers—A Systems Approach*, Van Nostrand Reinhold, New York, 1971.
6. Davenport, Jr., W. B., R. A. Johnson, and D. Middleton, "Statistical Errors in Measurements on Random Time Functions," *J. Appl. Phys.* **23**, 377–388, 1952.
7. Goldman, Stanford, *Information Theory*, Prentice-Hall, Englewood Cliffs, New Jersey, 1953.
8. Goodman, N. R., "On the Joint Estimation of the Spectra, Cospectrum, and Quadrature Spectrum of a Two-Dimensional Stationary Gaussian Process," Ph. D. thesis submitted to Princeton University, March, 1957.
9. Hamming, R. W., *Numerical Methods for Scientists and Engineers*, McGraw-Hill Book Company, New York, 1962.
10. Hoffman, A. A. J., "Geophysical Interpretations of Power Spectral Analysis of Low Frequency Fluctuations in the Magnetotelluric Field," Ph. D. Dissertation submitted to the University of Texas, University Micro-films, Inc., Ann Arbor, Michigan, 1962.
11. Lanczos, C., *Applied Analysis,* Prentice-Hall, Englewood Cliffs, New Jersey, 1956.
12. Laning, Jr., J. Halcombe, and Richard H. Battin, *Random Processes in Automatic Control,* McGraw-Hill Book Company, 1956.
13. Melton, Ben S., "Digital Recording of Extended Range Signals," *IEEE Trans. Geosci. Electron.* **GE-6,** No. 2, 110–123, May 1968.
14. "Principles of Data Acquisition Systems," E20–0090, IBM Corporation, Data Processing Division, 112 East Post Road, White Plains, New York.
15. Shannon, Claude E., "Communication in the Presence of Noise," *Proc. IRE* **37**, 10–21, January 1949.
16. Weber, Paul J., "The Tape Recorder as an Instrumentation Device," Ampex Corporation, 1963.
17. Weiner, N., *Extrapolation, Interpolation, and Smoothing of Stationary Time Series*, John Wiley and Sons, New York, 1949.

Chapter 8

Design Problems

Edward A. Wolff

8.1 SYSTEM PROBLEMS

The preceding chapters contain discussions of many of the design problems applicable to specific portions of a geoscience instrument system. This chapter is devoted to some of the overall design problems applicable to the entire system. Many of these problems have been alluded to in the previous chapters. The problems of primary importance described below are electromagnetic compatibility, reliability and maintainability, calibration and test, human factors, and economic factors.

The electromagnetic compatability problem is the problem of designing an instrument to minimize its susceptibility to interfering signals and to minimize its generation of interfering signals. These interference signals can consist of radiated fields as well as interfering signals conducted into and out of the instrument through connecting wires.

Reliability and maintainability involve considerations in the design of long-life equipment that can be easily repaired or calibrated when necessary. These considerations become increasingly more important as the cost of constructing and operating the instruments increases and as the cost of parts and labor for maintenance increases. As instrumentation becomes more complex it requires more highly trained people for maintenance with an attendant increased maintenance cost. Often, however, instruments are dispatched to the depth of the sea or the remote regions of outer space where the normal concept of maintainability loses its meaning.

Human factors are important considerations in the interaction of the user, operator, and analyst with the geoscience instrument system. Human factors considerations become increasingly important as the value of the instruments increases and the cost of the humans involved increases.

The design problems described above all involve engineering compromises between performance and cost. The instrument designer therefore is involved in the consideration of the economic factors involved in all of his engineering compromises. A discussion of the considerations involved in the economic factors of geoscience instrumentation is given in Section 8.5.

Donald R. J. White

8.2 ELECTROMAGNETIC COMPATIBILITY

Electromagnetic compatibility (EMC) is the gainful operation of electric, electromechanical, and/or electronic devices, equipments, and/or systems in a common environment so that no degradation of performance exists due to either internally or externally generated emission noise or electromagnetic interference (EMI).

Examples of EMI include electric shavers that jam nearby radios, automobile ignition noise that causes interference to TV pictures, appliances that interfere with heart pacer operation, and telephone conversations jammed by high electrical or interfering noise level backgrounds.

It is usually more economical to design the instrument to be interference-free than to try to fix an existing instrument. Design considerations involve frequency management, shielding, cabling and wiring, bonding and grounding, and filtering.

8.2.1 Emission and Susceptibility

EMC is the discipline of containing the otherwise damaging effects of EMI. This is best done in the conception and design stages of instruments. The implementation depends on whether or not EMI develops from within a system or equipment (intra-system interference) or between two or more remote systems (inter-system interference). Figure 8.2-1 is a block diagram involving both forms of interference. The transmitter on the left is attempting to communicate with the receiver in the center of the figure. The receiver is also subject to a number of interfering sources including other emitters shown on the right and different coupling paths.

In Figure 8.2-1 the sensed instrumentation information originates at the information source. The information is converted by a source transducer to the proper electronic format for direct transmission. EMI can develop as radiation from the higher power levels of the transmitter E or conduction through ground current loops or a common power supply G. These routes are within the system and are called intra-system interference. Other examples of intra-system interference are shown as E and G within the receiver system.

After signal processing the modified source information is transmitted from the antenna. The receiver antenna which picks up the desired trans-

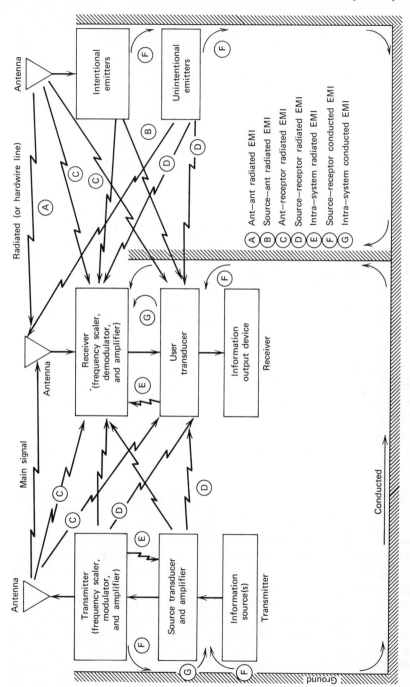

Figure 8.2-1 Transmitter and reciever exhibiting both intra-system and inter-system EMI.

mitter radiations also picks up undesired antenna-radiated interference from other intentional emitters *A* such as communications, navigation and radar. The receiver antenna can also intercept emissions from unintentional radiators *B* such as dielectric heaters and motor brushes. Interference resulting from *A* and *B* is called inter-system interference because it involves two or more systems.

EMI manifests itself in disturbances or lost information at the receiver output. Figure 8.2-1 shows other paths for EMI including radiated routes *C* and *D* and conducted routes *F* which can couple by a common co-site grounding, power supply, or cable distribution system.

8.2.2 Terms and Definitions

Ambient electromagnetic environment. Atmospheric and manmade electromagnetic noise and signals covering a wide spectrum. The ambient is almost always scintillating or changing with time and the profile is sometimes better described statistically.

Broadband interference. A spectrum of energy covering a wide frequency range. The spectrum occupancy of a signal or interference source is broadband relative to a receiver bandwidth when tuning the receiver by an increment equal to its own 3-dB bandwidth results in a signal change of less than 3 dB.

Conducted. The coupling mode by which a voltage, current, or power is directly coupled with attenuation from one network to another.

Degradation. An unwanted change in the operational performance of an EMI victim. Degradation does not necessarily mean malfunction or catastrophic failure.

Intermodulation. A nonlinear distortion; the undesired generation of energy at frequencies equal to the sums and differences of integer multiples of the individual input frequencies.

Narrowband interference. A term used to describe the modulation bandwidth of an interference signal relative to the predetector bandwidth of a receiver. Narrowband corresponds to a signal-modulation 3-dB bandwidth smaller than the receiver 3-dB bandwidth. For a pulse type signal, $B > 2/\tau$, where B is the receiver bandwidth and τ is the source pulsewidth.

Radiated. The coupling mode by which either an electric or magnetic field induces or an electromagnetic field couples between two or more wires, antennas, or equipments.

RI-FI receiver. The radio interference and field intensity (RI-FI) receiver is a precision instrument used to measure the amplitude of an electromagnetic signal or interference, either broadband or narrowband, conducted or radiated.

Shielded enclosure. A screened or metallic housing for isolating the internal and external electromagnetic environments to prevent outside ambient electromagnetic fields from causing false test readings and to prevent susceptibility tests from causing interference to outside activities.

Spurious emissions. Emissions at frequencies which are outside the necessary band, and which can be reduced without affecting the corresponding transmission of intelligence.

Spurious response. The undesired response of a receiver, due to its circuitry and construction, to off-frequency signals.

Transient interference. Either single-shot impulses characterized by coherent amplitude, frequencies, and phases which are related by a Fourier integral, or a signal of low repetition rate.

8.2.3   Instrument EMC Performance Assessment

There exist a number of specifications and standards on EMI and EMC for evaluating radiated or conducted emissions and susceptibility.

Emission Measurement. Conducted emissions from an instrument involve undesired potentially interfering signals or noise on the power, control, signal, and antenna cables. The measurement of these conducted emission transferring sources is performed inside a shielded enclosure and typically employs either a current probe or a line impedance stabilization network to couple the energy existing on the instrument leads. The output from this emission sensor is coupled to a calibrated receiver for conducted emission level measurements.

Instruments sometimes radiate directly from their housings. This can be due to substantial magnetic fields from the use of internal transformers, motors, generators, relays and solenoids. Electric or electromagnetic fields can be emitted from joints where sheet metal panels or members are mated, cover plates, glass front panel meters, digital displays, at shaft openings for control knobs, and ventilation apertures. Magnetic field testing, using a loop or ferrite sensor, is generally performed inside a shielded enclosure over the frequency range from 30 Hz to 50 kHz. Electric field and electromagnetic field testing, using a rod, biconical or conical log spiral antenna, is performed over the applicable frequency range.

Susceptibility Measurement. The conducted susceptibility tests parallel the conducted emission tests with the process reversed. Specified signal levels are injected into the instrument external cables over the applicable frequency spectrum.

Radiated susceptibility measurements parallel radiated emission tests. The instrument is irradiated with a magnetic, electric, or electromagnetic field, as applicable. Test compliance is evidenced by no degradation in output performance when the field intensities are established at the prescribed limits. Magnetic field radiated susceptibility tests are performed over the applicable spectrum. Electric and electromagnetic field irradiation tests are also performed at the required levels.

8.2.4 Frequency Management

Frequency management consists of minimizing emission spectrum and receiver bandwidths, controlling oscillator frequencies and levels, pulse rise times, and harmonic content sidebands. Frequency management attempts to assure that the least signal power, out-of-band emissions, and bandwidth required to transfer intelligence are established. For example, it is often desired to produce pulse shapes with sharp rise times. From an interference point-of-view, this implies reciprocally related bandwidths and frequency response requirements of amplifiers to accommodate these emissions. If the bandwidths of critical low-level circuits can be made the least possible to permit system performance, then the probability of interference is concomitantly reduced. This necessitates, wherever practical without degrading equipment performance, that pulse widths and especially rise times be made as long as possible.

8.2.5 Shielding and Gasketing

Electromagnetic waves from a source are partially reflected from the surface of a metallic barrier, and the remainder of the field is attenuated in passing through the barrier. Shielding efficiency, S, measured in decibels, describes the fraction of impinging energy which penetrates the enclosure, and is given by

$$S = 20 \log\left(\frac{E_1}{E_2}\right), \tag{8.2-1}$$

where E_1 is the impinging field intensity in volts per meter, and E_2 is the exiting field intensity in volts per meter.

The effectiveness of a conducting shield in reducing the energy of an electromagnetic field is the result of the absorption loss A incurred in passing through the conducting medium, the reflection losses R occurring at each surface of discontinuity, and internal reflection losses B. Thus,

$$S = A + R + B \text{ dB}. \tag{8.2-2}$$

The factor B can usually be neglected for an electrically thick barrier that provides more than 5 dB penetration attenuation.

The shielding factors are functions of frequency and certain intrinsic properties of the metal being used for shielding, such as the metal's conductivity or effective resistance, its permeability, and its thickness. The absorption loss is

$$A = 1.315t(fG\mu)^{1/2} \text{ dB}, \tag{8.2-3}$$

where t is the thickness of the barrier in cm, f is the frequency in Hz, G is the conductivity relative to copper, and μ is the magnetic permeability relative to vacuum.

Equation 8.2–3 is applicable independent of the type of wave impinging on the shield, but the computation of reflection loss is more complex since it depends on the type of wave encountered.

As an electromagnetic wave propagates from air into a metallic barrier, the wave impedance changes suddenly and reflection occurs at the boundary of the two media. For a plane electromagnetic wave of 377 Ω characteristic impedance, the reflection loss, R, is:

$$R = 108 + 10 \log \frac{G \times 10^6}{\mu f} \text{ dB.} \qquad (8.2\text{--}4)$$

Where there is significant difference in the impedance of the incident wave and the shielding barrier, reflection at the boundary is substantial and good shielding is obtained. This occurs relative to the barrier impedance with high- or low-impedance waves. The former is an electric field and its reflection loss is

$$R = 362 + 10 \log \frac{G}{f^3 \mu r^2} \text{ dB,} \qquad (8.2\text{--}5)$$

where r is the distance from source to barrier in cm. For the case of low-impedance or magnetic waves, the reflection loss is

$$R = 20 \log [0.0354 + (1.175/r)(\mu/fG)^{1/2} + (0.346/r)(Gf/\mu)^{1/2}].$$
$$(8.2\text{--}6)$$

Good shielding efficiency for electric (high-impedance) fields is obtained by use of shields of high conductivity, such as copper and aluminum. As Eq. 8.2–5 shows, the shielding efficiency for electric fields is infinite at zero frequency and decreases with an increase of frequency. However, magnetic fields, Eq. 8.2–6, are more difficult to shield since the reflection loss can approach zero at certain combinations of material and frequency. With decreasing frequency, the losses for reflection and absorption for nonmagnetic materials such as aluminum steadily decrease. At high frequencies, the shielding efficiency is good due to the reflections at the media discontinuity and the rapid dissipation of the field by absorption.

In general, for plane waves, magnetic materials provide better penetration loss, whereas good conductors provide better reflection loss. Figure 8.2-2 shows comparative data for shielding effects of copper and iron, neglecting the effects of the internal reflection factor, B.

Perfectly clean and matched mating surfaces seemingly would produce perfect radio frequency (rf) conductivity and hence high shielding effectiveness. However, bowing or waviness between actual contact points can produce a radial waveguide. Such gaps are most commonly produced by spot-weld or screw fastenings. Conductive gaskets or spring contact fingers are used to provide good contact between removable panels.

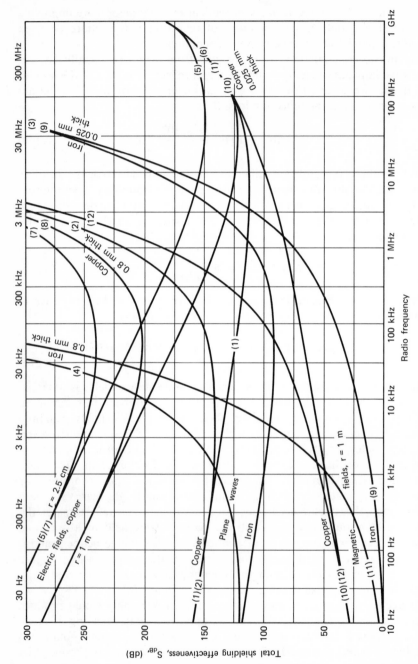

Figure 8.2-2 Total shielding effectiveness versus frequency for electric and magnetic fields and plane waves.

It is necessary to keep ventilation holes small in effective electrical area since such holes reduce shielding efficiency. The leakage intensity is proportional to the cube of the hole dimensions. A small hole is one which is less than one-twentieth of the operating wavelength in dimension. Larger holes should be covered by fine mesh copper screen. Where an aperture is used to view panel meters, dial faces or digital displays, conductive glass should be avoided for use above 100 MHz where its attenuation is reduced.

Waveguide attenuators can be used to shield large holes. Waveguides pass all frequencies above the cutoff frequency and substantially attenuate below that frequency determined by the dimensions of the guide. For 100 dB of low-impedance wave (TM_{11} or TM_{01}) attenuation, the length of the waveguide must be three times the diameter of the hole.

8.2.6 Cabling, Wiring, and Grounding

Cable shielding and grounding must take into consideration the impedances of both the radiating and the susceptibility instrument or circuit, the frequency range of the emissions relative to the desired signals, the levels of the emissions relative to the desired signals and the length of the cables.

Inductive Emission Sources. High-impedance emitters are conductors which have a large series impedance (usually in excess of 1000 Ω). Voltages can develop between the conductor and ground with comparatively little current. A field is produced which has a high wave impedance with a large electric and a small magnetic field component. This field can induce large rf voltages in adjacent high-impedance circuits. The field induces little current in adjacent low impedance circuits. Shielding is relatively effective in containment of this field.

Low-impedance emitters are conductors in a closed loop which permit the flow of large rf current at low voltage. Such emitters have a large magnetic and a small electric field component. The field can induce large currents in low-impedance surfaces or circuits. This field induces little voltage in high-impedance circuits. Shielding to contain such fields is difficult to effect.

Susceptible Circuits. The susceptibility of a circuit to inductive emissions depends on the termination impedance of the circuit and the impedance of the interfering field. A non-static magnetic field is always accompanied by an electric field, and the impedance is the ratio of the two field components. This is 377 Ω for plane waves in free space but is higher for electric and lower for magnetic fields in the near field.

For high-impedance circuits, the coupling is mainly capacitive in nature. A single shield grounded at both ends can give from 30 to 40 dB of attenuation to high-impedance fields. Conventional copper braid shielding is effective for prevention of capacitive coupling. For low-

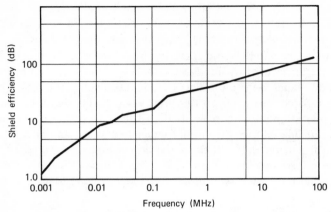

Figure 8.2-3 Typical measured shield insertion loss for electromagnetic coupling copper braid shield.

impedance (less than 10 Ω), the coupling is primarily magnetic in nature. The limit of conventional shielding is about 10 dB for low-frequency, low-impedance fields. Conventional copper braid shielding is ineffectual for prevention of magnetic coupling although a permeable braid can be used.

Magnetic Coupling. Allowing currents to flow in a grounding device or ground plane can introduce a voltage in adjacent conductors. The induced voltage is produced by a change in flux linkages within a circuit loop of finite area. Increasing the separation of the circuits or reducing the pick-up loop area reduces the induced voltage. Circuit configurations can be employed in which equal and opposite self-cancelling voltages are induced

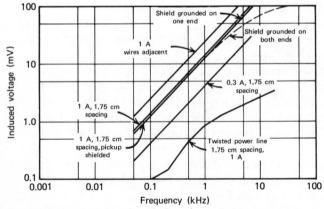

Figure 8.2-4 Electromagnetic coupling between two No. 18 wires, 3 m length. Load resistance: 1000 Ω.

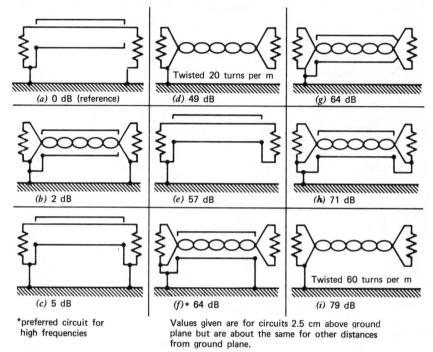

Figure 8.2-5 Relative susceptibility of circuits to magnetic interference at 100 kHz.

in the circuit. This is the principle by which magnetic coupling is reduced using twisted-pair wiring.

Figure 8.2-3 illustrates the magnetic shielding effectiveness of a copper braid shield. The coupling between parallel wires is shown in Figure 8.2-4 and the effectiveness of twisted-pair wires is shown in Figure 8.2-5.

Electric Coupling. No appreciable electric field effects appear on closed-loop secondary lines, but great changes can occur on high-impedance lines or under open-circuit conditions. Consider two parallel conductors carrying signals referenced to ground with the voltage of one high with respect to the other. There is a capacity between the two conductors which is a function of the conductor size, the distance between the conductors, and the dielectric constant of the intervening material. If the first circuit is of low impedance with respect to the second, the voltage appearing on the second conductor is a result of the voltage division between the source and the impedance of the first circuit, the capacity existing between the conductors, and the impedance of the second circuit. This coupling capacity forms a voltage divider as shown in Figure 8.2-6. The coupling between the two circuits increases with frequency, since the reactance of the coupling capacity decreases with frequency.

Low frequency control and instrumentation leads which do not normally carry radio-frequency currents can conduct interference because of pickup from adjacent hot cables. It is good practice, therefore, to separate pulse cables from low-level cables and shield them separately. The cables should also be terminated in separate low-impedance connectors, well bonded to the shields. For shielded conductors carrying rf currents, the shield should be terminated at both ends in a good ground connection.

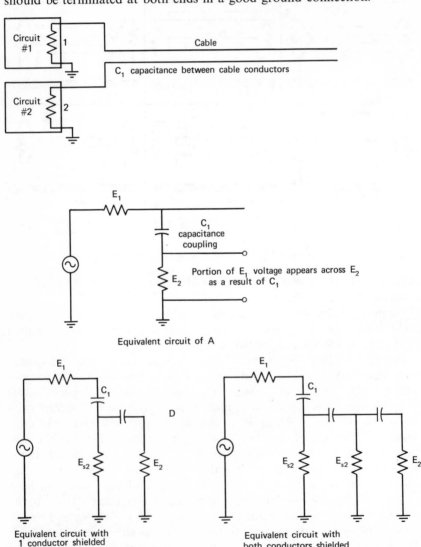

Figure 8.2-6 Electric field coupling.

Ground Currents and Shield Grounding. Electromagnetic compatibility in a complex instrument often depends on the shielding and the grounding of the shields of interconnecting leads.

Grounding of the shields can be accomplished as single-point grounding or multi-point grounding depending on the interference signal frequencies involved, the length of the line, the relative sensitivity of the circuit to either high or low-impedance fields, and the existence of ground currents. Ground currents can cause a voltage drop between two points in a ground plane. If a shield is grounded at two points of different potential a current can flow through the shield, enabling coupling to take place between the shield and the center conductor. Grounding the shield at each end can make the system susceptible to magnetic fields, since currents can be induced in the shield. Grounding the shield at only one point does not offer satisfactory electrostatic field protection to the wire at high frequencies.

Cable Grouping and Harnessing. In order to ensure that EMI does not become a problem in instruments having interconnecting cables and harnesses, the dc and ac power, signal, and control cables are separated.

Units designed to supply voltages to electronic circuits should have a low output impedance. This impedance at the output terminals of the power supply is not the impedance at the far end of the interconnecting leads. The dc resistance is increased by the ohmic value of the leads. The ac impedance is a function of the lead inductance as well as the dc resistance and increases with frequency. With increasing frequency, therefore, the leads become more susceptible to electric field pickup. Some means must be used to lower the impedance of the dc supply line at the instrument circuits to protect these circuits against the introduction of higher frequency components.

Most ac power lines carry large currents and are potential sources of magnetic field interference coupling to adjacent lines. Twisted lines should be used for all ac power circuits. The routing of these lines should be away from susceptible lines. Power circuits in which switching transients are expected can beneficially use shielding to contain the higher frequency components of the transient. Such shielding should be over the entire power wiring bundle and should be grounded at both ends.

For short distances the use of a twisted-pair may prove adequate. For long runs, the use of twisted-pair shielded wire is mandatory for unbalanced as well as balanced output devices. Where signals from high-impedance output devices must be transmitted over an appreciable distance, an impedance transforming device is used to lower the transmission circuit impedance to 50 Ω or less. Signals of high level are generally not bothered by susceptibility of circuits but can be a source of interference to lower level signal lines so that shielding should be used.

Control signals are usually switched on or off in a step manner and are potential interference generators requiring separation and shielding.

Signals within a unit are routed so that coupling between input and output circuits is minimized. Outside the unit, shielding is used if necessary to prevent excessive radiation from the wire or excessive pickup on the wire. Grounding of such shields is multi-point for rf signals.

8.2.7 Electrical Bonding and Corrosion

Electrical bonding is the process in which the elements of an assembly are electrically connected by means of a low-impedance conductor. The purpose is to make the structure homogenous with respect to the flow of rf currents to avoid electric potentials between metallic parts which can produce EMI.

Bonding Theory. Any two points on a metallic structure, whether electrically connected or not, can develop a potential difference at some frequency. When the structural dimensions are on the order of a quarter wavelength, potential differences exist in the presence of an electric or magnetic field. At lower frequencies the potential difference between two points in the structure is proportional to the impedance between the points.

A high potential difference between different portions of a structure can cause spark discharges in a strong electric field. When the two points are bonded by a low-impedance path, a conduction current exists with a comparatively weak electric and stronger magnetic field. The magnitude of this conduction current is usually negligible, and the magnetic field is preferable to the possible generation of spark discharges. Poor bonds can produce EMI due to varying impedance under shock or vibration. They can also produce EMI coupling paths between circuits having the common bond impedance. There is little correlation between the direct-current resistance of a bond and its rf impedance.

Types of Bonds. The most desirable bond consists of a permanent, direct, metal-to-metal contact, such as is provided by welding, brazing, sweating, or swaging. The best soldered joints have appreciable contact resistance. Semipermanent joints, such as provided by bolts or rivets, can provide effective bonding, but care must be taken that relative motion of the joined members does not reduce the bonding effectiveness by introducing a varying impedance with its attendant EMI.

Joints that are press-fitted or joined by screws of the self-tapping or sheet metal type cannot be relied on to provide low-impedance rf paths.

Corrosion. When two metals are placed in contact (bonded) in the presence of moisture, corrosion can take place through two chemical processes. The first process is termed galvanic corrosion, and is due to the formation of a voltaic cell with the moisture acting as electrolyte. The degree of resultant corrosion depends on the relative positions of the metals in the electrochemical series. If the metals differ appreciably in this series, such as aluminum and copper, the resulting electromotive force causes a continuous ion stream with an accompanying decomposition of the more active metal (high in the series) as it gradually goes into solution.

The second chemical process is termed electrolytic corrosion, and while it also requires two metals in contact through an electrolyte, the metals need not be of different electrochemical activity. In this case, the decomposition is due to the presence of local electrical currents.

Since the mating of bare metal to bare metal is essential for a satisfactory bond, metallic conductive coatings such as cadmium, tin, gold, or silver are used.

8.2.8   Electrical Filtering

Filters are useful when the interfering voltage is separated in frequency from the desired voltage [14]. To specify an EMI filter for instrumentation use, both the source and termination impedances must be examined at the lowest frequency to be protected. Consider the noise source as either a voltage generator (low impedance with respect to the filter input impedance) or a current generator (high impedance). Consider the load as either a voltage-sensitive device (high impedance with respect to the filter output impedance) or a current-sensitive device (low impedance). The filter input and output elements should be the inverse match of the impedances of the system.

A single capacitor in shunt with the dc or ac power line provides low-pass filtering action when its reactance approaches that of either the source or load impedance, whichever is less. Thereafter, above cutoff, it offers attenuation to conducted emissions at the rate of 6 dB/octave.

A three-stage, pi-type, feed-through filter is often preferred for isolating leads which must penetrate a bulkhead. Not only does it offer 18 dB/octave above the cutoff frequency, but the series inductor is important for low-impedance sources such as regulated dc power mains.

A different approach to filtering is to make the conductor itself a filter in which its dielectric insulation material has high loss tangent or the equivalent series inductance is significantly increased. The former absorbs rf energy and dissipates it in the form of heat while the latter reflects energy.

Beads and rods are small ferrite devices which can readily be slipped over the end of a conductor and spread along its length. Since they concentrate the magnetic flux surrounding the wire into the ferrite bead or rod, they act as small series inductors on the conductor.

Where it is desired to provide easy retrofit to existing conductors having known EMI problems, flexible lossy tubing can be slipped over one or more wires. It is more effective than the beads and rods. The tubing attenuates undesired rf energy by the absorptive-dissipation process above a few MHz.

References

1. Air Force Systems Command, "Design Handbook AFSC DH 1–4, Electromagnetic Compatibility," USAF, 1973.
2. Electronic Industries Association, "Designer's Guide on Electromagnetic Compatibility," Washington, D.C., 1965–1967.
3. MIL-E-6051D, "Electromagnetic Compatibility Requirements, Systems," 7 September 1967, Department of Defense issue.
4. MIL-STD-461A, "Electromagnetic Interference Characteristics Requirements for Equipment," 1 August 1968, Department of Defense.
5. MIL-STD-462, "Electromagnetic Interference Characteristics, Measurement of," 31 July 1967, Department of Defense issue.
6. NASA, "Electromagnetic Compatibility Principles and Practices," Apollo Design Reliability Series, 1965.
7. NAVAIR/SAE, course notes, "Electromagnetic Compatibility Program Management," Washington, D.C., 1970.
8. Pearlston, C. B., "Case and Cable Shielding, Bonding, and Grounding Considerations in Electromagnetic Interference," *IRE Trans. Radio-Frequency Interference,* pp. 1–16, October 1962.
9. White, D. R. J., "Vol. 1., Electrical Noise and Electromagnetic Interference Specifications," *A Handbook Series on Electromagnetic Interference and Compatibility*, Don White Consultants, Germantown, Maryland, 1971.
10. White, D. R. J., "Vol. 2., EMI Test Methods and Procedures," loc cit.
11. White, D. R. J., "Vol. 3., EMI Control Methods and Techniques," loc cit.
12. White, D. R. J., "Vol. 4., EMI Test Instruments and Systems," loc cit.
13. White, D. R. J., "Vol. 5., EMC Prediction and Analysis," loc cit.
14. White, D. R. J., *Electrical Filters—Theory and Practice*, Don White Consultants, Germantown, Maryland, 1963 and 1970.

George R. Grainger

8.3 RELIABILITY AND MAINTAINABILITY

8.3.1 Basic Concepts

The problem of developing a specific geoscience instrumentation system that works is often overshadowed by the companion design problem of making it work with a desired level of reliability [7]. Reliability is the probability that the system will perform its intended function for a specific period of time under a given set of operating conditions. Alternatively, reliability is stated in terms of the system's failure rate λ or its mean time between failures (MTBF), Θ. If maintenance is not feasible, then MTBF means the mean time before the first failure of the system; otherwise it means the mean time between failures.

For instruments for which maintenance is possible, another performance characteristic of the system is its maintainability. Maintainability is the probability that the failed system is restored to operable condition within a specified time when maintenance is performed in accordance with prescribed procedures and resources. Alternatively, maintainability is expressed as the system's repair rate μ or its mean time to repair (MTTR).

System reliability and maintainability are often combined into a single performance characteristic, availability, which measures the probability that a maintainable system is operable at any specified instant of time. Availability is important to earth-based instruments and to satellite-borne systems that require an operable ground station during the normally brief periods that the satellite-borne sensors are in view.

Both reliability and maintainability (and hence availability) are performance parameters inherent in the design and operation of the system. Their achievement becomes a challenging design problem whenever the system is required to operate in a hostile environment or for a long period of time, its function requires that the system design be very complex, it is provided limited maintenance, or any combination of these factors.

8.3.2 Reliability Problems and Solutions

The effectiveness of early electronic systems was severely limited because of the high unreliability of piece parts. A reliability of the order of 20 percent (i.e., probability of success equal to 0.2) was common and one user found it necessary to supply a million replacement parts each year to keep 160,000 pieces of equipment in operation.

Parts improvement programs have resulted in remarkable improvement in both the performance capability and reliability of piece parts. During this period of parts miniaturization and subminiaturization, the electron tube gave way to the transistor and diode which, in turn, are being replaced by microelectronic devices. Nevertheless, achieving an acceptable level of over-all system reliability continues to be one of the primary design problems. The reason for this is best understood from the basic mathematics of reliability.

If an equipment consists of n piece parts, the ith part having failure rate λ_i, the reliability $r(t)$ of the equipment after t hours of operation is given* by

$$r(t) = \exp\left[-\left(\sum_{i=1}^{n} \lambda_i\right)t\right]. \qquad (8.3\text{--}1)$$

Denoting the summation in Eq. 8.3–1 by λ and its reciprocal by Θ, the

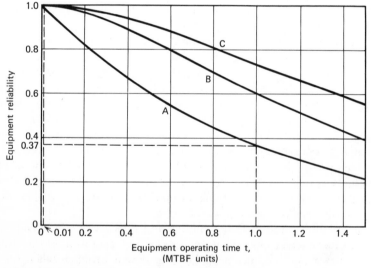

Figure 8.3-1 Universal graphs of equipment reliability. Curve A: single equipment, $r(t)$; Curve B: active redundancy, $r_1(t)$; Curve C: standby redundancy, $r_2(t)$.

---

*This formulation assumes that the parts' failure rates are all constant during the time period of interest. If not, other basic reliability formulas have been developed. However, the above formulation applies reasonably well to most equipment. The failure rate λ is the reciprocal of the MTBF, Θ, only in the constant failure rate case.

basic equation for equipment reliability can be written simply as

$$r(t) = e^{-\lambda t} = e^{-t/\Theta}. \tag{8.3-2}$$

Curve A of Figure 8.3–1 is a universal graph of $r(t)$ wherein equipment operating time t is expressed in MTBF units. From Curve A, it is seen that if an equipment's planned mission time T is set equal to the MTBF, Θ, the equipment's mission reliability is only 0.37. Since such a low value of reliability is usually unsatisfactory, it is clear that the equipment's mission time must be a small fraction of its MTBF to achieve a high level of reliability. For example, if 0.99 reliability is desired, then Curve A shows that the mission time T must equal 0.01 Θ. Or conversely, to ensure a 0.99 chance of surviving a mission of T hours, the equipment's MTBF must be at least $100T$ hr.

Figure 8.3-2 illustrates the impact of these reliability-mission time-MTBF relationships on the typical instrument system shown in Figure 1-2. The lines between the blocks for the major system elements indicate that the elements are "in series" in the sense that failure of any one causes system failure. For this series arrangement, the reliability $R_s(t)$ of the instrument system is given by the well-known reliability product rule

$$R_s(t) = r_1(t)\, r_2(t)\, r_3(t)\, r_4(t)\, r_5(t)\, r_6(t)\, r_7(t), \tag{8.3-3}$$

where the subscripts on the element reliabilities $r_i(t)$ correspond to the block numbers shown in Figure 8.3-2 (e.g., $r_2(t)$ is signal processor reliability).

If each element's reliability is 0.99 for a given mission time, then the system's reliability is $(0.99)^7$ or 0.932. If each element's reliability is as low as 0.9, then system reliability is only 0.48. Thus, the product rule of Eq. 8.3–3 is a very harsh rule wherein system reliability is *never* larger than the lowest elemental reliability in the product. It follows that designers of reliable instrument systems must ensure that the reliability of the weakest element is acceptable and that the reliabilities of the other elements do not collectively degrade this lowest reliability to an unacceptable level of system reliability.

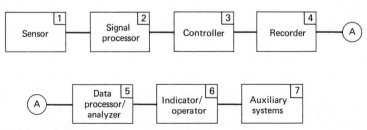

Figure 8.3-2 Reliability block diagram for typical instrument system.

To obtain near-unity reliability for each instrument system element, the designer should let the basic reliability equations guide early design decisions. First, from Eq. 8.3–1 it is desirable to select parts with the lowest possible failure rate λ_i to keep the total failure rate, λ, small. It is desirable to minimize the number n of parts, *simplifying* the design as much as possible to keep n and hence the total failure rate small. Simplification can stem from either a more *efficient* design or less *exacting* performance requirements (e.g., does the instrument really require the accuracy sought or is such accuracy only desirable?). Reduced performance requirements usually mean a less complex design. As noted above, the steepness of the initial portion of Curve A of Figure 8.3-1, makes it possible to increase reliability simply by *shortening* the desired mission duration, *T*.

After reducing the failure rate, design complexity, and mission time, the instrument can still have an unacceptable level of reliability. Further improvement is often possible by incorporating redundancy into the design. Two common types of redundancy are active redundancy (all units energized) and standby redundancy (the backup units not energized). The basic reliability equations for two-unit (single backup unit) redundancies are

$$\text{active redundancy:} \quad r_1(t) = 1 - (1 - e^{-t/\Theta})^2 = 2e^{-t/\Theta} - e^{-2t/\Theta},$$

$$\text{standby redundancy:} \quad r_2(t) = e^{-t/\Theta}(1 + t/\Theta). \tag{8.3–5}$$

These equations are graphed in Figure 8.3-1 as Curves B and C, respectively. Clearly, the real value of either type of redundancy is that it drastically raises and flattens out the initial steep portion of the nonredundant Curve A. For example, from Figure 8.3-1, it is seen that if the mission time T were 0.1Θ then the equipment's reliability is approximately 0.905, but if it is backed up with an identical actively redundant unit, the reliability of the two units increases to about 0.991, a ten-fold decrease in the unreliability, $1 - r(t)$, of the single equipment. Introduction, instead, of a standby redundant unit increases the reliability to 0.995, approximately a 20-fold decrease in the unreliability of the single equipment or about half the unreliability of the actively redundant pair. Equations for more complex redundancies are available [2, 4, 9, 20, 24, 29, 37, 39].

Redundancy does not help when failures are not caused by catastrophic part failures. Some sensors have a relatively short effective life; hence active redundancy is of no avail and standby redundancy is often of limited value. In other redundant situations, failure can be caused by external agents (foreign objects, outgassing, or electromagnetic interference) that adversely affect all redundant units. The feasibility of redundancy also rests on such nonreliability oriented factors as increased cost, weight, power requirements, and the possible need for switching equipment.

Another key to high geoscience instrument system reliability lies in the selection and application of component parts. Over the years, component part failure rate curves have been developed for a wide range of part types and applications. These curves show significant decreases in part failure rates if the designer "derates" their application. Derating entails applying each part so that its operating stresses are a fraction of the rated stresses specified by the manufacturer. In addition, the designer should provide environmental protection to the piece parts by eliminating local hot spots, potting assemblies and subassemblies, providing shielding and hermetic sealing, isolating shock and vibration, guarding against radiation and outgassing effects, and avoiding magnetic materials. Information is available on how to formulate appropriate system and subsystem reliability models and how to assign numerical values to the piece part failure rates appearing in these models to quantitatively predict the reliability of an instrument system [32, 38, 40]. Design guidelines for identifying adverse environmental effects and for designing appropriate environmental protection measures are also available [3, 10, 20, 30, 34, 38, 39, 46, 49].

In designing an instrument system, it is desirable to know which environments are most likely to affect the system adversely and contribute the greatest percentage of failures. Compilation of data on over 3000 development and preflight certification tests on major assemblies yielded the following table of percentages of failures occurring in various environments [20, p. 9–5]:

Environments	Percent Failures
Shock and vibration	28.7
Low temperature	24.1
High temperature	21.3
Humidity	13.9
Altitude	4.2
Acceleration	3.2
Salt spray	1.9
Others	2.7
Total	100.0

At significant milestones in the design of the instrument, formal design reviews should be conducted [17, 21, 48]. Depending on the complexity and importance of the instrument, representation on the design review team should include design engineers not previously associated with the design, reliability engineers, quality control and manufacturing representatives, test and field engineers, materials and packaging engineers, and other specialists (e.g., human factors and value engineers), as required. A powerful design evaluation tool that should be applied prior to the design reviews is a thorough failure mode effects and critical analysis (FMECA). The FMECA is a systematic procedure that examines all potential failure modes and mechanisms and determines their effect on over-all system operation. This analysis procedure has yielded significant increases in reliability and is well documented [18, 36, 42, 48]. Analysis of hardware of all kinds has identified literally thousands of failure modes. Examples of some of the most common failure modes for piece parts (transistors, diodes, relays, resistors, and potentiometers, capacitors, connectors, and mechanical devices) are available [5, 20, 23, 50].

The value of standard developmental and environmental tests is widely recognized. Reliability testing, on the other hand, is often a small part of the over-all test program or omitted completely because of the added cost and difficulty of "proving" that the specified system reliability has been achieved. For example, a widely used failure rate document [40] cites a nominal failure rate of 0.10 failures per million hours for transistors. To demonstrate this failure rate at a 90 percent level of confidence requires a total test time of 2.3×10^7 hr with no failures. If failures occur, the total test time required is considerably longer. To accumulate 2.3×10^7 test hr, it is possible to trade test specimens versus test time by testing approximately 2600 transistors for 1 yr, 5200 for 6 months or 31,200 for 1 month. For instrument systems with high reliability objectives, such statistical reliability demonstrations are normally not feasible because very few systems are fabricated and most are utilized for their intended purpose.

To reduce the risk of building a low reliability instrument the design reliability guidelines described above should be followed and supplemented with a failure-detection-oriented reliability test program. During the design phase, the orientation is that of developing an instrument that performs a specified function. During the design reviews, two questions are raised, "How can it fail?" and "What is the effect of each anticipated failure?" During subsequent reliability testing, additional questions arise: "Does the system still have unanticipated failure modes?", "How likely are they?", and "How does the system perform over time (e.g., do certain performance parameters drift?) and with changes in its operating environment?" Many of these questions can be addressed during normal engineering development and evaluation testing which can also be reliability testing. An effective over-all test program should include provisions for extending certain of the normal engineering tests so that they are also reliability tests.

Typical reliability tests are overstress tests, life tests, and reliability acceptance tests. Overstress tests elevate the environments or input conditions above the normal use conditions to determine unanticipated failure modes and to establish margins of safety. Failure modes identified by overstress tests are eliminated by appropriate redesign when necessary.

Analytical methods have been developed to quantitatively predict the reliability of systems and equipments based on overstress test data gathered on a limited sample of test items [13, 14, 19, 46]. One type of overstress test known as test-to-failure attempts to establish a sample distribution of strengths via test and then statistically predict reliability by estimating the probability that a randomly selected equipment's strength exceeds the environmental stresses to which it is likely to be subjected. Alternatively, the mean values of the distribution of strengths and the distribution of stresses (environments) are calculated and a suitable measure of the margin of safety calculated. The magnitude of this margin is then used as a rough indicator of the level of reliability achieved. For explosive and other one-shot items, up-and-down sensitivity tests have been developed [11, 22, 24].

Some form of life testing is usually required to assure, at preestablished risk levels, the producer that a satisfactory product is not rejected and to assure the user that the product has achieved a minimum allowable level of reliability (usually measured in terms of MTBF). Such life testing is generally performed in the laboratory using specially designed equipment that operates continuously and automatically controls the test environment. Test costs and time can be markedly reduced by employing sequential life tests. For equipment known or assumed to follow the exponential failure law (i.e., Eq. 8.3–2 applies), sequential test plans having preselected producer's and user's risks are available [41]. For nonexponential distributions, fewer life plans are available [28, 43].

Reliability acceptance testing takes many forms. If formal life tests are possible, these become the reliability acceptance tests. In other cases, overstress tests with a limited sample size and agreed-on accept/reject criteria are used. For development programs for which very few end items are produced, reliability acceptance testing is normally extended to the piece part level in the form of a wide variety of parts screening and burn-in tests [1, 16, 31, 33]. These tests do not quantitatively measure design reliability, but do greatly increase confidence that the system's inherent design reliability will be achieved because all parts exhibiting manufacturing defects or a tendency to drift are discarded during the screening process. Modern piece parts are so sophisticated and their fabrication process is so difficult to control that an effective parts screening program is an indispensable prerequisite to fabrication of instrument systems that have in service the inherent reliability designed into them.

An important aspect of all reliability testing is to conduct a complete analysis of each failed item [25]. This should include careful and thorough disassembly of the failed item to determine the true failure causes. This important concept has been extended to the pre-test period, from which the notion and theory of the physics of failure has developed [8, 12, 44, 45].

8.3.3 Maintainability Problems and Solutions

One of the most frequent maintainability problems is the semantic one that uses "maintainability" synonymously with "maintenance." Like reliability, maintainability is an inherent feature of the design and is concerned with those design aspects that facilitate maintenance of the instrument system; the time, ease, cost, and man hours necessary to keep the system operable. Not all instrument systems (e.g., satellite-borne sensors) are maintainable in use. However, the desirability of incorporating maintainability into the design to facilitate required maintenance during the sometimes extensive pre-operational use period should be considered. The instrument's interfaces with other systems should be studied to determine their effect on its maintainability.

The two primary reasons for maintainability are the unreliability of component parts and the presence of limited-life items in the design. Good maintainability design ranks the failure rates of subsystems and components, then provides for ease of maintenance for those having the highest failure rates. Organic materials such as rubber and other elastomers or materials having early wear-out characteristics should be avoided in the design. If this is not possible, maintainability provisions for them should be incorporated into the design.

Preventive as well as corrective maintenance should be included in the maintenance philosophy, and the maintainability provisions should consider the probable frequency of maintenance and its impact on the system's operability. Preventive maintenance not only provides for optimum replacement of limited-life items but also provides for regular and systematic inspection, and, as required, correction of other, usually high failure rate, items. Corrective maintenance occurs on a nonscheduled basis to repair failures that have caused the instrument to fail catastrophically or have caused it to degrade unacceptably.

The formal definition of maintainability given in Section 8.3.1 is concerned with the probability of restoration within a specified period of time. Not all instruments have an urgent restoration time requirement; however, for those that do, one of the predominant maintainability design problems is that of accurately predicting the restoration time and restoration probability. A normal pitfall is to not include certain necessary maintenance actions in the predictions or to underestimate the time required for those that are included. Typical maintenance actions include fault diagnosis and isolation, removal and repair or replacement, cleaning and lubrication, securing parts and materials, and reassembly and checkout. These actions involve many steps, each of which contributes a finite time increment to the over-all restoration time. Typical maintainability design guidelines for electrical designers, mechanical designers, and those concerned with the human-engineering aspects of system operation and maintenance are available [20].

The effort required to determine the distribution of down (restoration) times and the above-defined maintainability probability is often prohibitive. In lieu of this, the analytical approach taken usually assumes the form of the down time distribution (experience supports a log-normal distribution for most systems) and directs attention towards estimating certain of its statistical parameters (e.g., its mean, standard deviation, and 95th percentile) [6, 27, 35, 39].

Designers of spacecraft sensors must have confidence that each maintainable ground station is operable during designated passes of the spacecraft. Similarly, maintainable ground-based sensors must be operable during specified time periods. The measure of their operability during these times is their availability, A, usually defined as

$$A = \frac{\text{MTBF}}{\text{MTBF} + \text{MTTR}}, \qquad (8.3-6)$$

where MTBF is the mean time between failures and MTTR is the mean time to repair. This mean-value definition of availability provides a useful estimate of the steady-state probability that the system will be available at any random point of time. Alternative formulations of specific types of availability (inherent, operational) as well as interval availability $A(\tau)$ are available [15, 26, 47, 48]. $A(\tau)$ is the probability that the instrument or platform is operable at a random point in time *and* remains operable for τ units of time.

References

1. "A Comparison of Burn-In and Bake as Semiconductor Screening Techniques for the Nimbus Spacecraft Program," NASA-GSFC-X-650-650105, Goddard Space Flight Center, Greenbelt, Maryland, March, 1965.

2. Barlow, R. E., and F. Proshan, *Mathematical Theory of Reliability*, John Wiley and Sons, New York, 1963.

3. Barta, F. A., "Designing for Reliability," *Proceedings, 1967 Annual Symposium on Reliability*, Washington, D. C., January 1967.

4. Bazovsky, Igor, *Reliability Theory and Practice*, Prentice-Hall, Englewood Cliffs, New Jersey, 1961.

5. Bean, E. E., and C. E. Bloomquist, "Reliability Data from Inflight Spacecraft," January 1968 and updated in November 1971, Planning Research Corporation, Los Angeles, California.

6. Blanchard, B. S., and E. E. Lowery, *Maintainability*, McGraw-Hill Book Company, New York, 1969.

7. Bonner, W. J., "Reliability and Maintainability Experience with Air Pollution

Monitoring Systems," *Ann. Reliability and Maintainability*, **10**, July 1971, Tenth Reliability and Maintainability Conference, Los Angeles, California.

8. Bretts, G., *et al.* "Failure Physics and Accelerated Testing," *Proceedings of Physics of Failure in Electronics Symposium*, Volume 2, September 1963.

9. Calabro, S. R., *Reliability Principles and Practice*, McGraw-Hill Book Company New York, 1962.

10. Corliss, W. R., *Scientific Satellites*, NASA SP-133, U. S. Government Printing Office, Washington, D. C., 1967.

11. Dixon, W. J., and A. M. Mood, "A Method for Obtaining and Analyzing Sensitivity Data," *J. Am. Statistical Assn.* **43,** 1948.

12. Eisenberg, P. H., and C. W. Scott, "Reliability Physics Investigation of Integrated Circuit Failures," *Proceedings, 1970 Annual Symposium of Reliability*, Los Angeles, California, Febuary 1970.

13. Endicott, H. S., and T. M. Walsh, "Accelerated Testing of Component Parts," *Proceedings, 1966 Annual Symposium on Reliability*, San Francisco, California, January 1966.

14. Evans, R. A., "Accelerated Testing of Electronic Parts," NASA-CR-97207, Research Triangle Institute, Durham, North Carolina.

15. Faragher, W. E., and H. S. Watson, "Availability Analysis—A Realistic Methodology," *Proceedings, Tenth National Symposium on Reliability and Quality Control*, January 1964.

16. Fink, R. W., "Screening for Reliability Growth," *Proceedings 1971 Annual Symposium on Reliability*, Washington, D.C., January 1971.

17. Franciscovich, P. J., "Technique for Reliability Circuit Design Review in Space Electronics," *Annals of Reliability and Maintainability, Sixth Reliability and Maintainability Conference*, Cocoa Beach, Florida, July 1967.

18. Greene, K. and T. J. Cunningham, "Failure Mode, Effects, and Criticality Analyses," *Proceedings, 1968 Annual Symposium on Reliability*, Boston, Massachusetts, January 1968.

19. Guzski, D. P., and Albert Fox, "Reliability Technology in Accelerated Testing," *Proceedings, 1968 Annual Symposium on Reliability*, Boston, Massachusetts, January 1968.

20. Ireson, W. G., *et al., Reliability Handbook*, McGraw-Hill Book Company, New York, 1966.

21. Jacobs, R. M., "Implementing Formal Design Review," *Proceedings, 1967 Annual Symposium on Reliability*, Washington, D. C., January 1967.

22. Langlie, H. J., "A Reliability Test Method for 'One-Shot' Items," Report U-1972, Aeronautic, A Division of Ford Motor Company, Newport Beach, California, 1962.

23. Lauffenburger, H. A., "LSI Reliability Assessment and Prediction," *Proceedings, 1970 Annual Symposium on Reliability*, Los Angeles, California, February 1970.

24. Lloyd, D. K., and M. Lipow, *Reliability Management, Methods and Mathematics*, Prentice-Hall, Englewood Cliffs, New Jersey, 1962.

25. Lythe, W. J., and O. J. McAtlee, "Characteristic Traits of Semiconductor Failures," *Proceedings, 1970 Annual Symposium on Reliability*, Los Angeles, California, February 1970.

26. "Maintainability Engineering," RADC TDR-63-85, February 1963, RCA Service Company, a division of Radio Corporation of America.
27. "Maintainability Prediction," MIL-STD-472, 24 May 1966.
28. Martin, C. A., "Rationale and Use of Military Sampling Handbooks," *Proceedings, 1968 Annual Symposium on Reliability*, January 1968, Boston, Massachusetts.
29. Meyers, R. H., K. L. Wong, and H. M. Gordy, Eds., *Reliability Engineering for Electronic Systems*, John Wiley and Sons, New York, 1964.
30. "NEL Reliability Design Handbook," PB 121839, U. S. Naval Electronics Laboratory, San Diego, California.
31. Neuner, G. E., "Parts Selection and Screening for Long Life Spacecraft," Volume 2, *Proceedings of the Symposium on Long Life Hardware for Space*, Marshall Space Flight Center, Huntsville, Alabama, March 1969.
32. "Nonelectronic Reliability Notebook," RADC TR-69-458 (AD 686372) Rome Air Development Center, Griffiss Air Force Base, New York, 1970.
33. Nowak, T. J., "Reliability of Integrated Circuits by Screening," *Proceedings, 1967 Annual Symposium on Reliability*, Washington, D. C., January 1967.
34. Nowak, T. J., "Reliability Physics for Microelectronics," *Proceedings, 1968 Annual Symposium on Reliability*, Boston, Mass., January 1968.
35. Peterson, E. L., and H. B. Loo, "Maintainability Risk Analysis Using The Analytical Maintainance Mode," *Annals of Reliability and Maintainability*, Sixth Reliability and Maintainability Conference, Cocoa Beach, Florida, July 1967.
36. "Procedure for Failure Mode, Effects, and Criticality Analysis (FMECA)," RA-006-013-1A, National Aeronautics and Space Administration, Washington, D. C., August 1966.
37. "Quality Assurance Reliability Handbook," AMC Pamphlet No. 702-3, Headquarters, U.S. Army Materiel Command, 1968.
38. "RADC Reliability Notebook," Volume 1, RADC TR-58, 11 (AD-148868), 1959, Volume II, RADC TR-67-108, 1967, Rome Air Development Center, Griffiss Air Force Base, New York.
39. *Reliability Engineering*, ARINC Research Corporation, Prentice-Hall, Englewood Cliffs, New Jersey, 1964.
40. "Reliability Stress and Failure Rate Data for Electronic Equipment," MIL-HDBK-217A, 1 December 1965.
41. "Reliability Tests: Exponential Distribution," MIL-STD-781B, November 1967.
42. Russell, B. H., "Practical Benefits of Reliability Analysis for Long Life Designs," Volume 1, *Proceedings of the Symposium on Long Life Hardware for Space*, Marshall Space Flight Center, Huntsville, Alabama, March 1969.
43. "Sampling Procedures and Tables for Life and Reliability Testing, Based on the Werbull Distribution," Quality Control and Reliability, Technical Reports TR3 (Mean Life Criterion, 1961) and TR 4 (Hazard Rate Criterion, 1962), Office of the Assistant Secretary of Defense (Installations and Logistics).
44. Vacarro, J., and J. S. Smith, "Methods of Reliability Physics," *Proceedings, 1966 Annual Symposium on Reliability*, San Francisco, California, January, 1966.

45. Vacarro, J., "Reliability Physics—An Assessment," *Proceedings, 1970 Annual Symposium on Reliability*, Los Angeles, California, February 1970.
46. Vaccaw, J., and H. C. Gorton, Eds., "Reliability Physics Notebook," RADC TR-65-330, Rome Air Development Center, Griffiss Air Force Base, New York, 1965.
47. Wilson, M. A., "Effects of Maintenance and Support Factors on Availability of Systems," *Annals of Reliability and Maintainability* Sixth Reliability and Maintainability Conference, Cocoa Beach, Florida, July 1967.
48. Winlund, E. S., and C. S. Thomas, "Reliability and Maintainability Training Handbook," NAVSHIPS 0900-002-3000, prepared for the U. S. Department of the Navy, Bureau of Ships, by General Dynamics/Astronautics, San Diego, California, 1965.
49. Woodson, W. E., and D. W. Conver, *Human Engineering Guide for Equipment Designers,* 2nd ed., University of California Press, Berkeley, California, 1964.
50. Workman, W., *Failure Modes of Integrated Circuits and Their Relationship to Reliability*, Microelectronics and Reliability, Vol. 7, p. 257–264, Pergamon Press, New York, 1968.

Lawrence Chase, Barbara Pijanowski, and Knute Berstis

8.4 CALIBRATION AND TEST

Testing is the investigation of the entire performance of an instrument including environmental effects, accuracy, and variations with time. Calibration, a crucial part of the test, is the determination of the difference between the indicated output of the instrument and the true value of the parameter being measured. The calibration procedure produces data used to correct the instrument output to values within a stated calibration accuracy.

Complete testing is usually performed only once or twice on an instrument to determine its characteristics or the expected behavior of all similar models of the same instrument. Calibration is performed routinely on most instruments. Common laboratory equipment is calibrated on a routine schedule. Field equipment is adjusted and calibrated prior to use and recalibrated afterwards.

A list of terms used most frequently in testing is presented in Table 8.4-1.

8.4.1 Test Information

The tests used to evaluate any instrument are chosen according to the design and function of the individual instrument components. The information considered basic for most instrumentation is overall accuracy, calibration, effect of environment (storage and operational), hysteresis, operating life, storage life, linearity, maximum allowable overload, repeatability, resolution, response time, sensitivity, stability (short term and long term), threshold, time constant of sensor and of electronics, vibration effect, and

Table 8.4-1. Instrumentation Testing Terminology (ISA S37.1)

ACCURACY—The ratio of the ERROR to the FULL SCALE OUTPUT or the ratio of the ERROR to the OUTPUT, as specified, expressed in percent. Note: Accuracy may be expressed in terms of units of MEASURAND, or as within $\pm$ — percent of FULL SCALE OUTPUT.

AMBIENT CONDITIONS—The conditions (pressure, temperature, etc.) of the medium surrounding the TRANSDUCER.

CALIBRATION—A test during which known values of MEASURAND are applied to the TRANSDUCER and corresponding OUTPUT readings are recorded under specified conditions.

CALIBRATION CURVE—The graphical representation of the CALIBRATION RECORD.

CALIBRATION RECORD—A record (e.g., table or graph) of the measured relationship of the TRANSDUCER OUTPUT to the applied MEASURAND over the TRANSDUCER RANGE. Note: CALIBRATION RECORDS may contain additional calculated points so identified.

COMPENSATION—Provision of a supplemental device, circuit or special materials to counteract known sources of ERROR.

CONTINUOUS RATING—The rating applicable to specified operation for a specified uninterrupted length of time.

DIGITAL OUTPUT—TRANSDUCER OUTPUT that represents the magnitude of the MEASURAND in the form of a series of discrete quantities coded in a system of notation.

DRIFT—An undesired change in OUTPUT over a period of time, which is not a function of the MEASURAND.

END POINTS—The outputs at the specified upper and lower limits of the RANGE.

ENVIRONMENTAL CONDITIONS—Specified external conditions (shock, vibration, temperature, etc.) to which a TRANSDUCER may be exposed during shipping, storage, handling, and operation.

Table 8.4-1. (continued)

ERROR—The algebraic difference between the indicated value and the true value of the MEASURAND. Note: It is usually expressed in percent of the FULL SCALE OUTPUT, sometimes expressed in percent of the OUTPUT reading of the TRANSDUCER.

ERROR BAND—The band of maximum deviations of OUTPUT values from a specified reference line or curve due to those causes attributable to the TRANSDUCER. Note: The band of allowable deviations is usually expressed as " $\pm$ — percent of FULL SCALE OUTPUT," whereas in test and calibration reports the band of actual maximum deviations is expressed as " $+$ — percent, $-$ — percent of FULL SCALE OUTPUT."

ERROR CURVE—A graphical representation of ERRORS obtained from a specified number of calibration cycles.

EXCITATION—The external electrical voltage and/or current applied to a TRANSDUCER for its proper operation.

FREQUENCY-MODULATED OUTPUT—An OUTPUT in the form of a frequency deviation from a center frequency, where the deviation is a function of the applied MEASURAND.

FULL-SCALE OUTPUT—The algebraic difference between the END POINTS. Note: Sometimes expressed as " $\pm$ (half the algebraic difference)" e.g., " $\pm$ 2.5 V."

HYSTERESIS—The maximum difference in OUTPUT, at any MEASURAND value within the specified RANGE, when the value is approached first with increasing and then with decreasing MEASURAND. Note: HYSTERESIS is expressed in percent of FULL-SCALE OUTPUT, during any one calibration cycle. Friction error is included with HYSTERESIS unless dithering is specified.

INTERMITTENT RATING—The rating applicable to specified operation over a specified number of time intervals of specified duration: the length of time between these intervals must also be specified.

LIFE OPERATING—The specified minimum length of time over which the specified CONTINUOUS and INTERMITTENT RATING of a TRANSDUCER applies without change in TRANSDUCER performance beyond specified tolerances.

LINEARITY—The closeness of a CALIBRATION CURVE to a specified straight line. Note: LINEARITY is expressed as the maximum deviation of any CALIBRATION point on a specified straight line, during any one calibration cycle. It is expressed as "within $\pm$ – percent of FULL SCALE OUTPUT."

MEASURAND—A physical quantity, property or condition which is measured.

OUTPUT—The electrical quantity, produced by a transducer, which is a function of the applied MEASURAND.

OVERLOAD—The maximum magnitude of MEASURAND that can be applied to a TRANSDUCER without causing a change in performance beyond specified tolerance.

PRECISION—(See REPEATABILITY and STABILITY.)

RANGE—The MEASURAND values, over which a TRANSDUCER is intended to measure, specified by their upper and lower limits.

REPEATABILITY—The ability of a TRANSDUCER to reproduce OUTPUT readings when the same MEASURAND value is applied to it consecutively under the same conditions, and in the same direction. Note: REPEATABILITY is expressed as the maximum difference between OUTPUT readings; it is expressed as "within

Table 8.4-1.   (continued)

—percent of FULL SCALE OUTPUT." Two (2) calibration cycles are used to determine REPEATABILITY unless otherwise specified.

RESOLUTION—The magnitude of OUTPUT step changes as the MEASURAND is continuously varied over the RANGE. Note: It is usually expressed as percent of FULL SCALE OUTPUT.

RESPONSE TIME—The length of time required for the OUTPUT of a TRANSDUCER to rise to a specified percentage of its final value as a result of a step change of MEASURAND. Note: See also TIME CONSTANT.

ROOM CONDITIONS—Ambient ENVIRONMENTAL CONDITIONS for conducting operational tests, which have been established as follows:
  a.   Temperature: $25 \pm 10°C$
  b.   Relative Humidity: 90% or less
  c.   Barometric Pressure: 660 to 810 mm Hg.

SENSING ELEMENT—That part of the TRANSDUCER which responds directly to the MEASURAND. Note: This term is preferred to "Primary Element," "Primary Detector," or "Primary Detecting Element."

SENSITIVITY—The ratio of the change in TRANSDUCER OUTPUT to a change in the value of the MEASURAND.

SENSITIVITY SHIFT—A change in the slope of the CALIBRATION CURVE due to a change in SENSITIVITY.

SPAN—The algebraic difference between the limits of the RANGE.

STABILITY—The ability of the TRANSDUCER to retain its performance characteristics for a relatively long period of time. Note: Unless otherwise stated, STABILITY is the ability of a TRANSDUCER to reproduce OUTPUT readings obtained during its initial CALIBRATION, at ROOM CONDITIONS, for a specified period of time: it is then typically expressed as "within — percent of FULL SCALE OUTPUT for a period of — months."

STATIC CALIBRATION—A CALIBRATION performed under ROOM CONDITIONS and in the absence of any vibration, shock or acceleration (unless one of these is the MEASURAND).

TEMPERATURE ERROR—The maximum change in OUTPUT, at any MEASURAND value within the specified RANGE, when the TRANSDUCER temperature is changed from ROOM TEMPERATURE to specified temperature extremes.

THRESHOLD—The smallest change in the MEASURAND that will result in a measurable change in TRANSDUCER OUTPUT. Note: When the THRESHOLD is influenced by the MEASURAND values, these values must be specified.

TIME CONSTANT—The length of time required for the OUTPUT of a TRANSDUCER to rise to 63% of its final value as a result of a step change of MEASURAND.

TRANSDUCER—A device which provides a usable OUTPUT in response to a specified MEASURAND.

ZERO SHIFT—A change in the zero-MEASURAND OUTPUT over a specified period of time and at ROOM CONDITIONS.

All upper case terms are defined in this table.

compensation error (if instrument is internally compensated for temperature, pressure, salinity).

Accuracy tests determine errors resulting from hysteresis, nonlinearity, nonrepeatability, temperature, pressure, motion and other environmental effects. The results indicate the precision of the instrument output before corrections other than calibration are applied. By applying the calibration information, the user corrects the instrument output to the accuracy of the calibration data.

Environmental tests evaluate the performance of the instrument under the conditions of temperature, pressure and humidity that would normally be encountered by the instrument in operation and during storage. For example, oceanographic in situ instruments are generally tested through an operational temperature range -2 to $50°C$ and humidities from 0% to 95% RH, the extreme conditions that can be expected. Equipment to be used on the deck of a ship is subjected to salt spray testing.

Tests for storage conditions vary according to the storage environment expected. Many instrument failures can be traced to abnormally high or low temperatures encountered during storage.

Operating and storage life tests are often impossible to perform on a short term basis for complete instruments and systems. However, this type of information is obtainable for some components such as batteries, fuel cells, and chemical solutions.

Most instruments are designed with a margin of safety with respect to overload. Inadvertent overloads can result in subtle changes to the output which are not detected by the user.

Repeatability measurements are easily obtainable in the laboratory, but duplication of laboratory results during field test is often impossible because of the changing test medium.

The response of the system is an important parameter. Ideally for moving instruments, the rate of travel should be determined by the response time of the instrument. It is also important for measuring time-varying parameters. The execution of any testing program is dependent on the reference standards against which the tested equipment is compared. Very few field and in situ reference standards are clearly defined or accepted by the entire geoscience community. For this reason, it is possible for test results to vary when reported by different laboratories.

8.4.2 Test Program

A geoscience instrument which operates according to prescribed specifications on location provides mute testimony to the adequacy of the test program which governed its development and the validity of the tests to which the unit, its predecessor models, and component parts were subjected. Proper operation implies that the instrument performs satisfactorily in its environment, comprised of the platform and surrounding medium (which may be land, sea, air, or space). The instrument must also survive the rigors of transport from its point of assembly or final test to its working site, the degradations of storage between fabrication and use, and prior environmental and operational exposures.

Testing is important to ensure that the instrument is adequate for its intended purpose and to prevent the unwarranted rejection of an acceptable unit. A thoroughly executed test program results in elimination of faulty components, confidence in the life expectancy of the assembly, and interface compatibility with the platform and the environments to which the instrument will be exposed. The importance of testing depends on the accessibility of the unit in the event repairs are necessary, its unattended life span, and the importance of the instrument to science or human life. Testing is important for instruments expected to last for protracted periods of time on spacecraft.

While it is not possible to perform tests which duplicate all environmental, time, and interface parameters, it is possible to plan a test program to minimize the risk of overlooking an exposure which could decrease the probability of success. Well-designed tests reveal and forestall potential failure modes.

8.4.3 Units Subjected to Tests

Tests are conducted on the geoscience instrument, on the prototype of the instrument (a prototype is an exact duplicate of the operational unit that is not primarily intended for operational use), and on preliminary units, frequently referred to as engineering, breadboard, design, or development models. The preliminary unit is usually not configured like the final unit and does not necessarily contain the high quality components used in prototype and operational models. Component parts such as diodes, transistors, printed curcuit boards, subassemblies, and sensors are also given a variety of tests. The test program includes integrating the instrument into the platform and subjecting the total package to final systems tests. Spare instruments, because they are intended to replace the operational instrument in the event of loss or irreparable failure, receive the same tests as operational models.

8.4.4 Types of Tests

The type of tests conducted vary with the model: preliminary, prototype,or operational. Further variations are introduced due to differences in the complexity of the instrument, quantities of the instrument to be produced, and relative state-of-the-art of components, designs, and fabrication techniques.

Component Tests. Component and sub-assembly level tests are performed to disclose imperfections due to materials, manufacturing processes, or faulty workmanship. Components and sub-assemblies are also subjected to tests which simulate the environmental exposure range of the final instrument. Such tests result in early selection of suitable component parts and minimize the chance of failures in assembled instruments. Components destined for space flight are completely tested while the same components earmarked for noncritical applications often receive only random sample tests. The thoroughness of testing influences component price and delivery time. Geoscience instrument designers often specify additional component tests to supplement the manufacturer's test program.

Design/Development Tests. An operational geoscience instrument is usually preceded by preliminary breadboard, engineering, development, or design models. These are tested to determine the feasibility of a design, confirm design analyses, determine instrument response, reveal failure modes which may warrant use of redundant sub-assemblies, and compare different configurations or designs. The need for such tests and the diversity of tests required are governed by the degree of confidence in the design, the configuration of the assembly, the components used, and the fabrication techniques. Often, final quality sensors are used in the preliminary model for calibration and interface compliance tests. Preliminary units are used for the early detection of electrical, magnetic, and other interference characteristics. Because the preliminary model is usually not in final configuration and does not contain high quality components, it is seldom subjected to a full range of environmental tests. Tests of the preliminary unit are frequently referred to as design/development tests.

Qualification Tests. The prototype model of the geoscience instrument is exactly the same in component quality, workmanship, and configuration as the model which will be used in operation. Prototypes have been used as operational spares. Prototypes are subject to the same types of tests as the operational model of the instrument, but the exposure levels and duration times are considerably higher than expected in normal service. Frequently, special additional tests are conducted on the prototype. The more severe prototype tests generate confidence that the operational models can survive the less severe operational environment. For space instruments vibration levels fifty percent higher than expected from the launch vehicle and the spacecraft are imposed on the prototype. Similarly, the prototype is expected to operate satisfactorily at temperatures about 10°C below and above expected minimum and maximum temperatures respectively. If the environment is unknown, the prototype may be tested to failure to determine its limiting capability. Prototype tests are frequently referred to as qualification tests.

Acceptance Tests. The operational model of the geoscience instrument is tested to less severe levels than the prototype to determine that materials, workmanship, and tolerances meet design specifications and that the unit is suitable for operational use. These tests designed to provide evidence of acceptability are called acceptance tests.

Systems Tests. After the geoscience instrument has passed acceptance tests, it is integrated into the platform on which it will operate. The entire system then receives a complete series of environmental and functional tests to determine that materials, workmanship, and tolerances meet design specifications and that the acceptable subsystems together constitute an acceptable system. In cases involving long life or critical applications, prototype instruments and subsystems are mounted on a prototype platform and the resultant system is subjected to qualification tests. Space vehicles, missile systems, aircraft, and underwater manned platforms are examples where complete prototype systems are qualification tested.

8.4.5 Instrument Monitoring During Test

During all tests it is desirable to simultaneously supply a signal input to the sensor, conduct the test, and monitor the output to simulate actual operational conditions. Space instruments, for example, are often designed to permit immersion of the detector in a vacuum to simulate space during ground tests. In cases where environmental simulation is not feasible, a signal equivalent to that expected from the sensor is injected into the instrument.

If the performance of the instrument cannot be monitored during the test, it is necessary to perform a test of the instrument before and after each exposure to determine whether the test has changed the characteristics of the instrument. These before-and-after checks are frequently referred to as functional tests.

The increased use of computers in geoscience instrumentation has improved monitoring capabilities during tests. Many instruments have meaningful outputs only when measurements are statistically averaged or otherwise processed. The output of some instruments is so sensitively coupled to precise input signals that accurate monitoring depends on simultaneous readings of input and output signals. As instrumentation and telemetry systems become more sophisticated, data rates increase. Satellites sometimes transmit data at rates in excess of 64,000 bits per second. Computers have proved invaluable in keeping abreast of high data rates, in quickly performing data analyses, and in providing convenient coincidence curves between input and output signals. Because of the variety of tests performed, individual computer programs are often written to accommodate the test program. The large number of systems level test exposures create a need for separate computer programs to accommodate specific tests such as vibration, shock, and radiation and to provide meaningful outputs during tests of varying time duration.

8.4.6 Test Sequence

The test program is designed to duplicate environmental and platform exposures and to apply these exposures in a sequence consistent with that which will actually be experienced. An equatorial oceanographic instrument must first survive air or surface passage to the test site and is tested accordingly. A space instrument is exposed to launch vehicle accelerations and vibrations before it encounters a vacuum environment.

8.4.7 Specific Tests

Geoscience instrumentation is used in so many environments and disciplines that no single list of specific tests is universally applicable. Time, cost, quantity, and other factors influence the test program. Table 8.4-2 summarizes the purpose and exposure levels of tests as a function of the unit under test. Table 8.4-3 is a guide for generating a test program. Neither table indicates the sequence of tests, which is often of vital importance. The references provide specific test requirements, exposure limits, and sequences applicable to a variety of geoscience instruments and platforms. Table 8.4-4 [2] lists the environments requiring tests for a manned space vehicle during its development.

Table 8.4-2. Summary of Tests Performed on Geoscience Instrument Models

Hardware	Test	Exposure Levels	Purpose (to verify)	Use
Components and subassemblies	Component	Operational to destruction	Fabrication Workmanship Specification adherence Life expectancy	Preliminary unit Design development unit Prototype model Operational unit
Preliminary unit (Breadboard, engineering, design, or development models)	Design-development	Operational to destruction	Design Design comparisons Sensor compatibility Sensor/instrument interface compatibility	Nonoperational
Prototype	Qualification	Beyond operational to destruction	Design Life duration Workmanship Components Specification adherence Life expectancy	Sometimes spare operational
Operational	Acceptance	Operational	Workmanship Components Specification adherence	Operational
System (Instrument and Platform)	Qualification acceptance	Same as prototype and operational	Same as prototype and operational. Interference (mechanical, electrical, other)	Prototype purposes Operational

Table 8.4-3. Specific Tests Conducted on Geoscience Instrumentation Hardware

Test	Environments Simulated	Limitations
1. Static load	In situ, transport to site, fabrication, assembly	Lack of advanced knowledge for exact duplication of environment
2. Dynamic load	In situ, transport to site, fabrication, assembly	Same as 1
3. Pressure	In situ, transport to site	Same as 1
4. Vibration	In situ, transport to site	Same as 1 plus transmissibility coefficients
5. Acoustic	In situ	Same as 1
6. Shock/impact	In situ, transport to site, assembly, fabrication	Same as 1
7. Thermal	In situ, transport to site, storage, assembly, fabrication	Same as 1 plus solar radiation duplication, radiative heat transfer duplication
8. Temperature	In situ, transport to site, storage, assembly, fabrication	Same as 1
9. Vacuum	In situ	Exact duplication depending on size of instrument of system, size of vacuum chamber, vacuum limitation of vacuum chamber
10. Humidity	In situ, transport to site, storage, assembly, fabrication	None
11. Magnetic	In situ, influence of vibration	Elimination of earth's magnetic field
12. Electrical interference	In situ platform environment	Exact duplication of environment
13. Life	In situ	Time to perform
14. Corrosion	In situ, transport to site, storage	Time to perform
16. Radiation	Solar radiation Radioactive sources, electromagnetic, other	Exact duplication of environment

Table 8.4-4. Structural Environments [68]

Life Phase	Environment															
	Atmospheric Properties	Winds and Gusts	Rain	Hail	Blowing Land and Dust	Salt Air	Humidity	Fungus	Atmospheric Contaminants	Atmospheric Electricity	Solar Thermal Radiation	Albedo	Electron Radiation	Meteoroids	Noise	Runway Roughness
Manufacturing	X				X		X		X							
Storage	X	X	X	X	X	X	X	X	X	X						
Transportation and ground handling	X	X	X	X	X	X	X		X	X						X
Prelaunch	X	X	X		X	X	X		X	X	X					
Launch	X	X	X				X			X	X				X	
Ascent	X	X	X		X					X					X	
Space											X	X	X	X	X	
Entry	X	X	X							X					X	
Atmospheric flight	X	X	X	X	X	X	X		X	X					X	
Landing and horizontal takeoff	X	X	X	X		X	X		X	X					X	X

References

1. "Apollo Test Requirements," NHB 8080.1, March 1967.
2. "Apollo Applications Test Requirements," NHB 8080.3, Oct. 13, 1967.
3. Himelblau, H., C. M. Fuller, and T. Scharton, "Assessment of Space Vehicle Aeroacoustic-Vibration Prediction, Design, and Testing," NASA Report CR-1596, July 1970.
4. "Environmental Test Methods," MIL-STD-810B, Sept. 29, 1969.
5. "Quality Program Provisions for Space System Contractors," NASA Report NPC 200-2, April 20, 1962.
6. "Recommended Practice for Combined, Simulated Space-Environment Testing of Thermal Control Materials," *1969 Book of ASTM Standards*, pp. 1122–1129, Pt. 30, ASTM E 332-67, Sept. 8, 1967.
7. "Reliability Program Provisions for Aeronautical and Space System Contractors," NASA Handbook NHB 5300.4(1B), April 1970.
8. Trout, Otto F., Jr., "Vacuum Leakage Tests of a Simulated Lightweight Spacecraft Air Lock," NASA Report TN D-5864, July 1970.

Joseph G. Wohl

8.5 HUMAN FACTORS

Human factors must be considered in the context of a systems approach to instrumentation [8, 11, 22, 24] that involves development of performance requirements; identification of technical and economic constraints; analysis of alternative instrumentation techniques for sensing, processing, and actuating; identification of required human roles and functions; allocation of functions and tasks between human and machine; and design, fabrication, installation, and evaluation. Table 8.5-1 summarizes guidelines for integrating human factors into instrument system development. The progressive narrowing and limiting of design alternatives open to the instrumentation

Table 8.5-1. Guidelines for Integration of Human Factors into Instrument System Development [24]

Phases in System Development	End Product	System Engineering Activities	Relevant Human Factors Activities
1. Definition of system operational requirements	Specific operational requirement	Mission requirements determination and analysis; identification of operating, weight, space, and other constraints	Establishment of personnel implications and constraints (e.g., number, type, training capabilities, operating environment)
2. Determination of functional requirements	Development plan	Performance of gross trade-off studies; definition of system and subsystem types, hardware techniques and performance, reliability, and maintainability goals; determination of information transfer requirements throughout the system; establishment of operational, maintenance, and support concepts; development of system evaluation requirements	Contribution to operational and maintenance concepts and maintainability goals Participation in gross trade-off studies Gross assignment of subsystem functions to personnel and equipment Gross division of duties among personnel Determination of implications for manning and training Determination of implications for system evaluation
3. Preliminary design of system and subsystems	System and subsystem specifications	Evaluation of alternative design concepts with respect to preceding step (e.g., analog vs. digital, mechanical vs. electrical); determination of installation requirements; performance of detailed trade-off studies; preparation of system and subsystem specifications	Analysis of information transfer requirements throughout the system Detailed assignment of subsystem functions to personnel and equipment for each alternative design concept Translation of information transfer requirements into display, control, and processing requirements for each alternative design concept Evaluation of each with respect to human factors (e.g., implications for number and types of personnel, training, job aids, human-machine interface design, data processing requirements) General location of operator and maintenance stations General design of functional procedures Preparation of human-machine portions of subsystem specifications
4. Detailed design of system and	Equipment and	Reliability-maintainability analysis and prediction;	Translation of control and display requirements into hardware

Table 8.5-1. (continued)

Phases in System Development	End Product	System Engineering Activities	Relevant Human Factors Activities
subsystem	component design specifications	determination of system and subsystem test and checkout requirements; logistic support analysis; generation of system evaluation plans, and of performance, reliability and maintainability reporting requirements and procedures; establishment of quality assurance and equipment modification programs; definition of personnel and training requirements; establishment of documentation (i.e., technical manual) requirements	requirements Specification of location and environment for operator stations Selection and design, where required, of control and display components Design of operator panels, consoles, and workspaces Selection and location of communication equipment Specification of operating and maintenance procedures for normal and non-normal modes of operation Participation in system evaluation planning Personnel manning and training requirements determination
5. Fabrication and testing of system	System prototype (or first model) and technical manuals	Continuing review of quality assurance program results; early performance, reliability, and maintainability reports; definition of spare parts procurement, inventory, and transportation requirements; initiation and/or approval of minor modifications	Evaluation of prototype system from human factors standpoint Recommendations of modifications to designs or procedures for operation and maintenance Establishment of training courses Familiarization of personnel with system
6. Installation and initial operation of system	Completed system, ready for trials	Analysis of installation problems and reports; initiation and/or approval of minor modifications	Review of installation reports from human factors standpoint; recommendations of minor design changes Training of personnel
7. Operational evaluation of system	Test and evaluation reports	Analysis of individual test and evaluation reports; initiation and/or approval of minor modifications; continuing review of system performance, reliability, and maintainability	Review of test and evaluation reports from human factors standpoint Recommendations of design or procedural modifications Determination of training adequacy and recommendation of changes
8. Deployment and operational utilization	Operational deficiency reports	Analysis of operational deficiencies; initiation and/or approval of minor modifications; continuing review of system performance, reliability, and maintainability	Review of operational deficiency reports from human factors standpoint Interviews with operating and maintenance personnel to obtain additional data Recommendation of design or procedural modifications

engineer underscores the importance of early consideration of human factors in each development stage.

8.5.1 Human-Machine Tasks and Task Elements

The various types of functions which can be performed by humans and/or machines can be grouped into decision, operating, checkout and maintenance, and administrative tasks. Table 8.5-2 identifies basic task types within each class and the human skills required.

The three basic elements common to every task, whether performed by human, machine, or both, are sensing, processing, and actuating. These elements are summarized in Figures 8.5-1a and b.

Table 8.5-2. Human-Machine Tasks

Task Category	Skill Requirements if Done by Humans
Decision Tasks	
Operational decisions	Semi- or highly skilled
Tactical decisions	Highly skilled
Strategic decisions	Highly skilled
Policy decisions	Highly skilled
Operating Tasks	
Materials handling and processing	Low or semiskilled
Continuous control	Highly skilled
Periodic or discrete control	Semi- or highly skilled
Monitoring	Semi- or highly skilled
Data processing and computation	Semi- or highly skilled
Communication	Semiskilled
Checkout and Maintenance Tasks	
Preventive maintenance	Semiskilled
Test and checkout	Highly skilled
Diagnosis	Highly skilled
Fault isolation	Highly skilled
Restore, replace or repair	Semi- or highly skilled
Administrative Tasks	
Planning	Highly skilled
Scheduling	Highly skilled
Reporting	Semiskilled

8.5.2 Designing the Human-Machine Interface

The interactions possible between people and instruments occur via control and display interfaces as indicated in Figure 8.5-2. In designing these interfaces, the designer must consider the capabilities and limitations of humans and machines relative to the tasks to be performed. Table 8.5-3 indicates some of the more important advantages of humans and machines with respect to the basic building blocks.

Sensing. Sensing is the detection of physical energy (information or signals) within the environment of a system. These signals can originate external to the system as in radar or sonar returns or the human senses of vision, hearing, and smell. They can also be developed within the system as in the

Table 8.5-3. Summary of Significant Human-Machine Factors [24, 25]

Advantages of Humans	Task Element	Advantages of Machines
Detect low levels of energy	S	Sensitivity to stimuli outside of human ability
	E	
Sensitivity to a wide variety of stimuli	N	
	S	Insensitivity to extraneous factors
Perceive patterns and generalize from them	I	
	N	Monitoring of other machines or people
Detect signals in a high noise environment	G	
Store and recall large amounts of information	P	Respond quickly to control signals
	R	
	O	Store and recall large amounts of data for short periods
Exercise judgment	C	
	E	
Improvise and adopt flexible procedures	S	Computing ability
	S	
Handle low-probability events	I	Handling of highly complex operations
	N	
Arrive at new and different solutions to problems	G	Deductive logical ability
Profit from experience		
Track under a wide variety of situations		
Perform when overloaded		
Reason inductively		
Perform fine manipulations	A	Perform routine, repetitive, precise tasks
	C	
	T	
	U	Exert large amounts of force smoothly and precisely
	A	
	T	
	I	
	N	
	G	

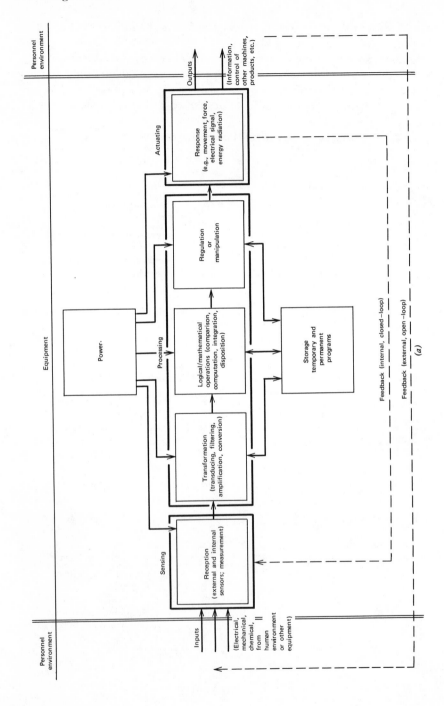

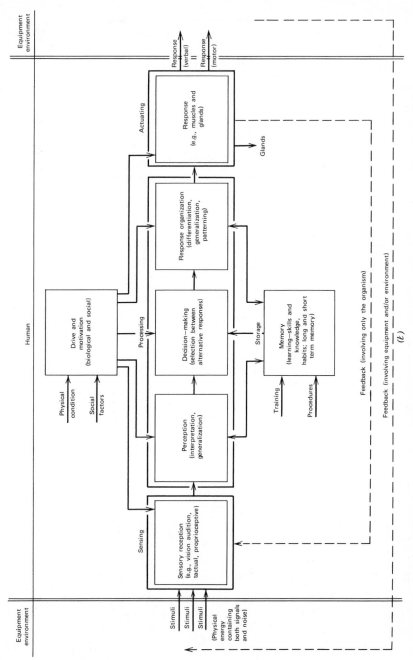

Figure 8.5-1 (*a*) Generalized functions of equipment [24]. (*b*) Generalized behavioral functions of personnel (single person) [24].

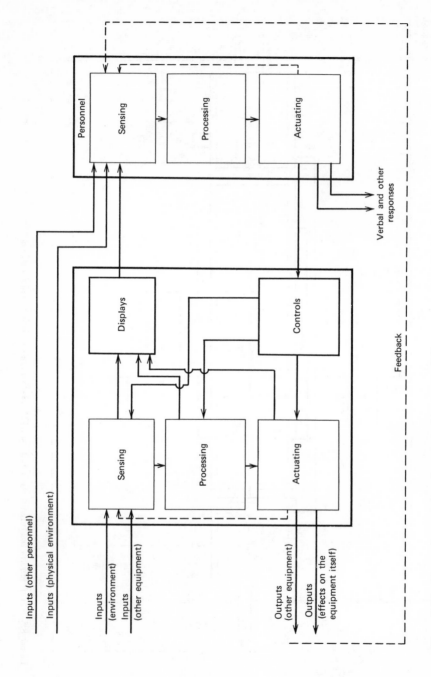

Figure 8.5-2 Human-instrument or human-machine system [24].

result of switch closures in equipment of the kinesthetic sense in humans. In either case, sensors are required to receive these signals as inputs to the system. The sensing function is often coupled with transducing, converting, filtering, or amplifying functions.

Table 8.5-4 summarizes capabilities and limitations of human sensory functions. The dynamic range of human senses is very great while the frequency range is relatively small and resolution is moderate.

An important characteristic associated with human senses is the phenomenon of attention. Although humans constantly receive sensations from many sources, they are able to select and concentrate on only those which are of importance to them, much as equipment sensors are able to filter out unwanted signals or noise or to select desired signals. Attention and the reception of stimuli are influenced by a person's physical condition and the range of sensitivity of the senses.

Processing. The information processing function in an instrument system can involve measuring or comparing external signals, integrating signals, and storing information. Transducing, amplifying, converting, and filtering can also be involved.

Humans process information based on their perception of signals received and on stored information. This integration of information for selection of appropriate courses of action is often called decision-making.*

The processes which support human decision-making capabilities include abilities for qualitative estimations, comparisons, judgment, transformation, coding and decoding, and inductive and deductive reasoning. Because of these capabilities people are more flexible as data processors than machines. With training, people are enabled to deal with changing situations and unforeseen problems in the absence of a specific program. Unlike a computer, humans can continuously develop and modify their own programming. In other words, they can learn. Closely linked with this function is the human memory or storage capability. Memory is the ability to retain and retrieve what is learned, and, conversely, forgetting is the failure to do so. People's ability to remember and to modify their behavior through learning accounts for much of their flexibility as programmers of computers. What is retained is in the form of words, numbers, or images which represent abstractions or symbolizations of what is learned. This capability for "chunking" [18], i.e., for abstraction and conversion to symbols of large

---

*The term "decision" is used in this section to describe an activity which can be performed by person or machine which can involve sensing and actuating and which always involves processing. The term describes the process of selection from among alternative courses of action; alternatively, it describes the process of "mapping" input sets upon output sets. Thus while it is correct to talk about "having made a decision" (past tense), when a person "is deciding" (present tense), they are *processing*.

Table 8.5-4.   Summary of the Capabilities and Limitations of the Human Senses

Parameter	Vision	Audition
Sufficient stimulus	Light: radiated electromagnetic energy in the visible spectrum	Sound: vibratory energy, usually airborne
Spectral (or (frequency) range	Wavelengths from 300 to 1500 nm absolute maximum range (ultraviolet to infrared)	20 Hz to 20,000 Hz
Spectral (or frequency) resolution	120 to 160 steps in wavelength (hue) varying from 1 to 20 nm	$\sim$ 3 Hz (20 to 1000 Hz) 0.3% (above 1000 Hz)
Dynamic range	$\sim$ 90 dB (useful range) for rods = 0.00001 mL to 0.004 mL; cones = 0.004 mL to 10,000 mL	$\sim$ 140 dB (0 dB = 0.0002 dyne/cm^2)
Amplitude resolution $\Delta I/I$	$\Delta I/I \simeq 0.015$ (for cone vision)	0.5 dB (1000 Hz at 20 dB or above)
Spatial acuity	1 minute of arc visual angle	Unknown
Temporal acuity	0.05 s normal, 0.25 s feeble illumination	Clicks $\simeq$ 0.001 s, tone bursts $\simeq$ 0.01 s
Reaction time for simple muscular movement to stimulus	$\sim$ 0.22 s	$\sim$ 0.19 s
Optimal operating range	400 to 700 nm (visible spectrum for adequate color discrimination) (green-yellow) 10 to 200 footcandles	300 to 6000 Hz 40 to 80 dB
Indications for use	Spatial orientation required Spatial scanning or search required Simultaneous comparisons required Multidimensional material presented High ambient noise levels	Warning or emergency signals Interruption of attention required Small temporal relations important Poor ambient lighting High vibration or G forces present Confirming or warning feedback on system operation required via alternate (nonvisual) mode

Mechanical Vibration	Kinesthetic
Tissue displacement by physical means	Joint movements
1 Hz at 0.5 cm to 100 Hz at 0.01 cm	Unknown
Δpps/pps $\simeq 0.10$	Unknown

Mechanical Vibration	Kinesthetic
$\sim$ 40 dB 0.01 mm to 10 mm Δp/p $\sim$ 0.15 at a pressure of 2 g/mm Two point acuity = 0.1 mm (tongue) to 50 mm (back)	Lower limit 0.2° to 0.7° at a rage of 10°/min Upper limit unknown Larger joints are most sensitive
Touches sensed as discrete down to 0.05 s	Unknown
$\sim$ 0.15 s (for finger motion, if finger is the one stimulated)	$\sim$ 0.15 s
For fingertip area 250 Hz	Unknown
Conditions unfavorable for both vision and and audition Visual and auditory senses disabled Confirming or warning feedback on system operation required via alternate mode	No method yet exists for directly utilizing this sensory mode Confirming or warning feedback on system operation required via alternate mode

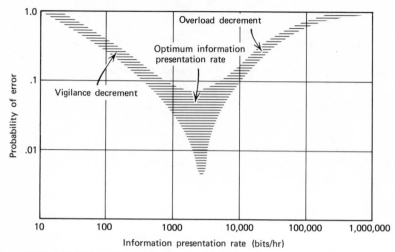

Figure 8.5-3 Idealized graph showing error probability of a human-machine channel as a function of information presentation rate, based on the results of a number of varied experiments [9, 16, 18, 21]. Vigilance decrement is characteristic of human only.

amounts of information, accounts for much of the human superiority over machines in decision making.

Figure 8.5-3 summarizes human-machine information-handling capability. It indicates a high error probability at both very high and very low information presentation rates as compared to a perfect or noiseless channel (in the Shannon sense). There appears to be a broad optimum presentation rate for minimizing errors in human-machine channels, in the vicinity of $\frac{1}{3}$ to 3 bits per second. The reader is invited to sort playing cards without error by color (1 bit per card) or suit (2 bits per card), being careful to subtract card handling time, and measure his processing rate. He will find, as is usual in self-paced tasks of this nature, that his processing rate is within this range. In designing instrumentation systems to be monitored by human operators, every effort should be made to achieve an average information presentation rate within the optimum range shown in Figure 8.5-3.

Actuating. Once a desired action has been selected it is necessary to implement. it. This can involve the function of regulation, such as controlling the rate and time at which the action is performed, or other forms of control. Regulation in humans involves the organization or patterning of their responses so that they occur at the proper time, in the proper sequence, or in the proper combinations. Eventually these responses become sufficiently learned that the procedure is performed rapidly and accurately as a perfected and organized *skill*. As skills are mastered, they are performed more automatically and involve less conscious effort and thought. This is particularly evident in the learning of sequential responses, as in keyset operation (typing), or continuous responses, as in steering or tracking. For the learning of complex *knowledge* required for complex tasks, a similar organization of responses can take place if similar situations occur often enough; otherwise, people can exhibit considerable variability and disorganization in their behavior.

The concept of *feedback* is also related to the regulation function. Instrument systems can have one or more sensing or monitoring circuits which feed back information on the operation of the system to provide a basis for regulation and further action. Such feedback loops are the distinguishing feature of *closed-loop* systems. In many instrument systems a human operator or monitor is depended on to close the loop.

Finally, all systems have one or more actuation functions that require a supply of energy, In humans, this final phase of the behavior process is the evocation of a muscle response such as a verbal command, a movement of an extremity to activate a control, or a movement of the eyes to view a display.

Drive and motivation are other functions peculiar to humans. Drives include hunger and thirst and are related to the physiological requirements of the human organism. Motives arise out of the individual's learning experiences, primarily those involving interaction with other people. Drives and motives are energizers of human behavior analogous to the power required by hardware systems for activation (see Figure 8.5-1).

Special human factors problems associated with the task of actuating in geoscience instrument design arise in special environments such as space and underwater. Much more *time* must be allotted to even the simplest actuating tasks performed in these environments than performed normally on the earth's surface.

8.5.3 Applicable Tools and Techniques

Consideration of human factors in geoscience instrument system design requires the application of a number of tools and techniques for analysis and synthesis. These range from the classical time-and-motion-study techniques of industrial engineering [3, 20] through the flow-charting techniques of human factors engineering called operational sequence diagramming [6, 13]; and from the mechanistic mathematical modelling of the human as a servomechanism [4, 9, 12] to human-performance-oriented, computer-based modelling of complex human-machine systems [23].

The intelligent application of human factors considerations requires a knowledge of human capabilities and limitations with respect to physical, physiological, psychological, and psychophysiological requirements imposed by mission, system- and use-determined factors.

Example: Modelling the Sonar Operator's Detection Process. An analysis of the visual aspects of the sonar operator's performance in search and detection [23] resulted in the development of a mathematical model of the interaction among the sonar observer, the ship, the equipment, the ocean environment, and the target. This model was used to predict the effects of human performance changes on sonar system performance and to demonstrate the feasibility of using a computer model as a diagnostic and descriptive task-analytic tool.

The core of the model involved quantification of the interaction between the sonar operator and his PPI scope, and included a model of the differential brightness sensitivity of the retina [5, 14, 15]. The model was programmed for simulation on a computer and was exercised on a Monte Carlo basis to explore the effects of a large number of variables. These included selected electrical characteristics of the sonar; human operating techniques and visual scanning protocols; search geometry; ship, target and ocean environment characteristics; and CRT energy conversion efficiency.

Several measures of system performance were taken, including range and probability of detection. The model was subjected to a sensitivity analysis to determine the relative significance of trainable aspects of human search and detection performance. Among those found to be of greatest significance were receiver gain and CRT intensity control settings, human visual scanning pattern and fixation time, and false contact reporting rate. Where possible, optimal values (standards) were established, and new operating procedures and doctrine were devised for field evaluation.

Optimizing the receiver gain setting and the reporting doctrine for sonar contacts indicated a potential improvement in range and probability of detection upwards of 50%.

Example: Use of Phase-Plane Displays for Incipient Failure Detection in Instrument System Testing. The methods used to analyze servo response data do not provide adequate information on the existence of marginal conditions, incipient failures, or anomalies in the servo response.

Determining the status of a complex servo system by assessment of its dynamic response is formidable. Response data are recorded in amplitude-versus-time formats as shown in Figure 8.5-4. Time-domain servo specifications (delay time, rise time, time of occurrence of maximum overshoot, settling time) are checked, and data are examined for anomalies in the waveform which warn of incipient, non-normal conditions.

The amplitude-versus-time format is not suitable for distinguishing between normal and marginal or non-normal conditions because the visual appearance of the data does not change *noticeably* with the introduction of small amplitude variations in the response data.

Table 8.5-5 shows the results of research into a human's ability to make

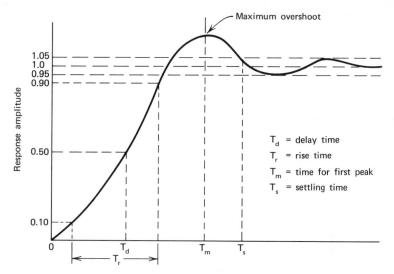

Figure 8.5-4   Time domain step response and performance criteria [1].

absolute judgments about various aspects of a visual display [18]. Table 8.5-5 shows that people can distinguish 15 positions along a linear interval but only 10 angles of inclination. Thus, there is a marked reduction in a human's ability to obtain derivative (slope) data as opposed to ordinate data from a graphic display. Data in Table 8.5-5 indicate a considerable further reduction in a human's ability to discriminate curvature.

Thus, a human's ability to percieve amplitude, rate, and higher-order derivative information in graphic displays is a decreasing function of the derivative order. For derivative data, it is advantageous to generate the derivatives *explicitly* from the source data and encode these along the *position* stimulus dimension, rather than require the instrument observer to sense the derivative information from a display of amplitude versus time.

A computer simulation study [1] of the "step" response of an underdamped nonlinear second-order servo system was undertaken. A "normal" step response was generated and displayed in both the conventional amplitude-time format (Figure 8.5-5) and the acceleration-plane format (Figure 8.5-6), $\ddot{X}$ versus X where $\ddot{X} = d^2x/dt^2$. The step response was repeated with known degradations in performance, and comparisons were made to determine the extent to which the appearance of the response changed for each format.

The amplitude response of a system with erratic position feedback is given in Figure 8.5-7 and has the same appearance as the response of the normal system.

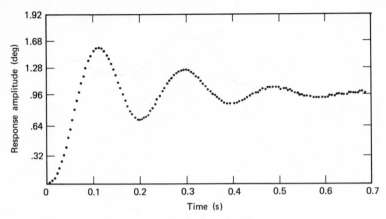

Figure 8.5-5   Normal system: amplitude-time plot [1].

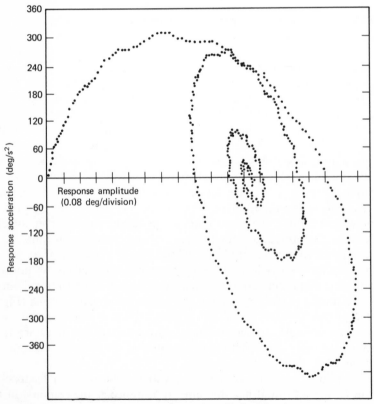

Figure 8.5-6   Normal system: acceleration-plane display [1].

Table 8.5-5. **The Capacity for Absolute Judgment of Several Aspects of Visual Displays [18]**

Visual Display Aspect	N	B
Position along a linear interval	15 (9)	3.9 (3.2)
Angle of inclination	10 (7)	3.3 (2.8)
Curvature (arc length constant)	— (4.6)	— (2.2)
Curvature (chord length constant)	— (3)	— (1.6)

N = Number of distinct absolute judgments possible.
B = Number of bits of information (channel capacity of the observer's stimulus dimension, viz., $B = \log_2 N$).
Values in parentheses for 1/40-s exposure, others for 5 s.

The appearance of the acceleration-plane plot of Figure 8.5-8 is noticeably different than the plot of Figure 8.5-6.

Data in Table 8.5-5 imply that a human is capable of resolving amplitude to approximately 7 percent, slope to approximately 10 percent, and curvature to approximately 15 percent. The changes in position and curvature for the time plots are 1 and 9 percent of their respective maxima, well below respective human resolution thresholds of 7 and 15 percent. The acceleration-plane format, however, yields changes which exceed these threshold values and are therefore readily detectable.

The phase-plane display technique utilizes the human capability for complex pattern recognition and offers several advantages. The appearance of response data in the phase-plane display format is sensitive to the presence of small amplitude anomalies. The response can be displayed compactly on a CRT in real time. Responses from systems of any kind or complexity,

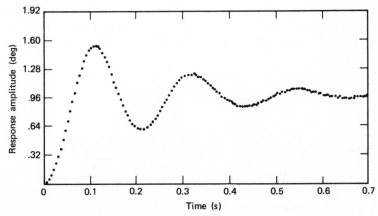

Figure 8.5-7 Erratic position feedback: amplitude-time plot

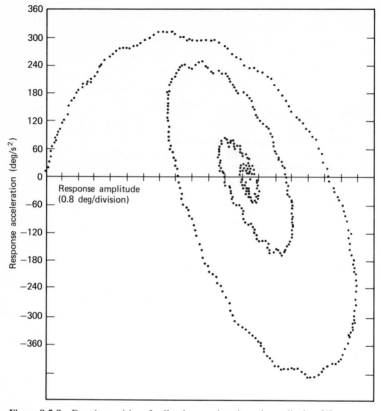

Figure 8.5-8 Erratic position feedback: acceleration-plane display [1].

whether linear or nonlinear, can be analyzed with equal facility. Generation of the phase-plane displays requires neither large-scale computing facilities nor extensive computer time. Third- and higher-order phase-plane displays can be generated for a given instrument system as their diagnostic utility would indicate.

References

1. Affinito, F. J., and J. G. Wohl, "Data Conditioning and Display for APOLLO Prelaunch Checkout: Servo Response Display Technique," Dunlap and Associates, Inc., Darien, Conn., 675-TM-7, SSD-66-353, January 1967.
2. Alden, D. G., R. W. Daniels, and A. F. Kanarick, "Human Factors Principles for Keyboard Design and Operation—A Summary Review," prepared for U. S. Post Office Department, Bureau of Research and Engineering, 26 March 1970.
3. Barnes, R. M., *Motion and Time Study*, John Wiley and Sons, New York, 1949.

4. Birmingham, H. P., and F. V. Taylor, "A Design Philosophy for Man-Machine Control Systems," *Proc. IRE* **42**, 12, 1954.
5. Blackwell, H. R., "Psychophysical Thresholds—Experimental Studies of Methods of Measurement," Bull. No. 36, Engineering Research Institute, University of Michigan, 1953.
6. Brooks, F. A., Jr., "Operational Sequence Diagrams," *IRE Trans. Human Factors in Electronics*, **1**, 33, March 1960.
7. Channell, R. C., and M. A. Tolcott, "Arrangement of Equipment," in *Supplement to Human Factors in Undersea Warfare*, National Academy of Science–National Research Council, Washington, D. C., 1954.
8. Fitts, P. M., Ed., *Human Engineering for an Effective Air Navigation and Traffic Control System*, National Research Council, Washington, D. C., 1951.
9. Fogel, L. J., *Biotechnology: Concepts and Applications*, Prentice-Hall, Englewood Cliffs, New Jersey, 1963.
10. Garvey, W. D., "The Effects of 'Task-Induced Stress' on Man-Machine System Performance," Naval Research Lab., Washington, D. C., NRL Report 5015, September 1957.
11. Goode, H. H., and R. E. Machol, *System Engineering*, McGraw-Hill Book Company, New York, 1957.
12. Kelley, C. R., *Manual and Automatic Control*, John Wiley and Sons, New York, 1968.
13. Kurke, M. I., "Operational Sequence Diagrams in System Design," *Human Factors*, 66–73, March 1961.
14. Lamar, E. S., S. Hecht, S. Shlaer, and C. S. Hendley, "Size, Shape and Contrast in Detection of Targets by Daylight Vision. I. Data and Analytical Description," *J. Opt. Soc. Am.* **37**, No. 7, 531–545, 1947.
15. Lamar, E. S., S. Hecht, C. D. Hendley, and S. Shlaer, "Size, Shape and Contrast in Detection of Targets by Daylight Vision. II. Frequency of Seeing and the Quantum Theory of Cone Vision," *J. Opt. Soc. Am.* **38**, No. 9, 741–755, 1948.
16. Mackworth, N. H., "Researches on the Measurement of Human Performance," Medical Research Council Report Series, H. M. Stationery Office, No. 268, 1950.
17. McGrath, J. J., A. Harabedian, and D. N. Buckner, *Review and Critique of the Literature on Vigilance Performance*. Human Factors Research, Inc., Los Angeles, California, "Human Factors Problems in Antisubmarine Warfare," Technical Report No. 1, December 1959.
18. Miller, G. A., "The Magic Number Seven, Plus or Minus Two: Some Limits on Our Capacity for Processing Information," *Psychological Rev.* **63**, No. 2, March 1956.
19. Miller, R. B., "A Method for Man-Machine Task Analysis," USAF, Wright Air Dev. Cent., Technical Report No. 53-137, 1953.
20. Mundell, M. E., *Motion and Time Study: Principles and Practice*, Prentice-Hall, Englewood Cliffs, New Jersey, 1955.
21. Sinaiko, H. W., Ed., *Selected Papers on Human Factors in the Design and Use of Control Systems*, Dover Publications, New York, 1961.
22. Teeple, J. B., "System Design and Man-Computer Function Allocation,"

Presented at ORSA-TIMS meeting, Seattle, Washington, 19–21 April 1961.

23. Wohl, J. G., "Modelling the Sonar Operator's Detection Process," in *Technical Digest, 8th Annual IEEE Symposium on Human Factors in Electronics*, Palo Alto, California, May 1967.

24. Wohl, J. G., Ed., *Human Factors Design Standards for the Fleet Ballistic Missile Weapon System, Volumes 1 and 2*, Bureau of Naval Weapons Publication NAVWEPS OD 18413 A&B, May 1963.

25. Wohl, J. G., and A. D. Swain, "Factors Affecting Degree of Automation in Test and Checkout Equipment," Dunlap and Associates, Inc. Report TR-60-36F, Stamford, Connecticut, 1 March 1961.

26. Woodson, W. E., and D. W. Conover, *Human Engineering Guide for Equipment Designers*, 2nd. ed., University of California Press, Berkeley, California, 1966.

Joseph W. Noah

8.6 ECONOMIC CONSIDERATIONS

Systems analysis encompasses a definition of the problem, postulation of alternative solutions, identification of relevant parameters, development of a comprehensive analytical procedure, and performance of economic analyses leading to a clear understanding of the alternatives. The criteria for making a choice must be explicit, and intangibles and uncertainties must be considered.

8.6.1 Major Considerations in Systems Analysis

Systems analysis compares alternatives to discover which is preferred on quantitative and qualitative grounds. The quantitative analysis is referred to as cost-benefit or cost-effectiveness analysis.

The procedure for assessing alternatives can take either of two fundamental forms. If the desired level of effectiveness is specified, the criterion for selection is *minimum cost*. If a level of expenditure is specified, the criterion for selection is *maximum effectiveness*. In most cases there are several meaningful effectiveness elements, and the specified-cost procedure is employed. This procedure does not provide a means of evaluating the relative desirability of the several elements of achievement; it merely permits translation from the measurement of resources to the measurement of effectiveness elements for each system examined.

Economic analysis is frequently unsuccessful because too little thought is given at the outset to a clear understanding of the problem. Economic analysis usually deals with problems of choice. A choice must be made from among a number of different and competing alternative courses of action. The problem is best understood when the alternatives are clearly and explicitly stated so that their implications can be examined over an appropriate period of time.

The criterion problem is one of devising definitive tests of preferredness. Usually proximate rather than ultimate tests have to be used. It is desirable to choose that course of action which maximizes what is obtained from available resources. To measure what can be expected to be obtained from the available resources, it is usually necessary to suboptimize and establish proximate criteria.

There are no clear rules for determining alternatives, just as there are none for devising appropriate criteria. However, there are a few general precautions. The system in which the possible actions take place must be adaptable to those actions. Ingenuity in the design and redesign of alternatives is very important. Different project scales and different combinations of measures should be treated; the addition or deletion of extra features and size changes create relevant alternatives. The interrelationships among alternatives and the limitations that attach to piecemeal analysis must be considered. The timing of events associated with each alternative should be carefully scheduled. It is usually not practical to list all feasible alternatives.

Some costs and benefits cannot be quantified and others cannot be valued in any market sense. For example, it is impossible to value a reduction in lives lost, but the alternative costs of averting such a loss can be compared.

Quantitative information about intangibles can be presented in terms other than money. Even if the intangibles are not commensurable with other costs and gains, clues to their impact can often be given. If both gains and costs are computed, it may be possible to show the minimum dollar value that must be attached to the intangibles if one project is preferred to the other alternatives considered. Suppose that transportation system A costs $1 billion (all amounts in present values), yields $2 billion over its lifetime, but degrades the scenery along the right-of-way; while transportation system B costs $1 billion, yields $1.5 billion over its lifetime, and does not degrade the scenery. The choice of system B attaches the value of at least $0.5 billion to the scenic degradation. This procedure ties no price tags to the intangible effects but brings out explicitly the minimum valuation that would be implicit in choosing system **B**.

Accurate cost estimates early in the planning stage are important because they help the evaluation of alternatives. Absolute costs are less important in those cases where the relevant issue is how to fulfill an agreed objective.

The preference for spending money later and receiving benefits sooner is common. The use of a positive discount rate implies such a time preference. Economists differ in their interpretation of what economic analysis implies about applying a discount rate in comparing nonmarketable programs. Systems analysis concentrates on improving estimates of costs and benefits and testing the sensitivity of alternative streams of costs and benefits to a range of interest rates.

8.6.2 An Example in Geoscience Instrumentation

The example summarized here is a partial study; it does not include an evaluation of all interesting and feasible alternatives, a constraint necessitated by the time and resources available for the study [7]. It deals with the costs and benefits of applying geoscience instrumentation to the solution of two problems: improving the management of specific water resources and agricultural activities.

Purpose. The purpose of the study was to explore the feasibility and practicality of using space systems for meeting existing and projected needs on earth. Economic tradeoffs between space systems and aircraft systems were examined for obtaining information for improvements in managing water resource and agricultural activities. Improvements arising from better information were compared with noninformation improvements such as the construction of new dams. These alternatives were not compared with non-space, non-aircraft information systems. Thus, benefits identified are not incremental to those alternatives but represent estimated total benefits. The results indicate economical directions and priorities for research and development activities in space applications.

The study constitutes an analysis of a possible future operational concept for a satellite-assisted information system. The objectives of the concept were to identify management and participant information requirements in water management, wheat production, and wheat rust prevention; to conceptualize a satellite-supported information system capable of supplying appropriate effective information; to estimate total system benefits and costs, including research and development costs; and to examine alternative cost and benefit relationships to provide guidance for the formulation of resource allocation policy.

Major Limitations and Assumptions. In addition to those limitations mentioned above, there were two others of major importance: point design and cost allocation to users.

In developing a concept for a satellite-assisted information system, a point design approach was used. That is, the missions—water management and agriculture—were selected and information requirements were derived from analysis of the decisions made or actions taken by the users. The space segment was selected by judgment and limited analysis of feasible components available to support the information requirements. The basis for the conceptual design was determined by study of selected geographical regions and potential information improvements. The sensor package assumed to be placed on the spacecraft was not necessarily a definitive design. However, a point design was needed for the estimation of costs, and that design was selected to satisfy specific information requirements.

Farmers concerned with wheat production control and rust prevention and managers of river basins and dams concerned with hydropower production, flood control, and navigation were assumed to be the users. No attempt was made to allocate portions of the costs to individual users or groups of users, since the space segment of the conceptual systems was designed to support all the users.

Study Plan. The schematic shown in Figure 8.6-1 depicts the concept followed by this study effort. Specific conditions are observed on the earth by remote sensors placed on satellites. Data from the sensors are transmitted to the ground stations for analysis and interpretation. Those results are combined with results from complementary information systems. The processed data are used in earth resources models to provide information and forecasts for water and agricultural resource managers. Application of information and forecasts in user-decision models results in resource

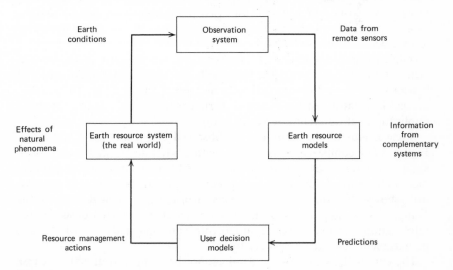

Figure 8.6-1 Model of study concept.

management actions. These actions and changing natural phenomena provide new sets of earth conditions for subsequent observation; thus the process is a continuing or closed-loop one as indicated in Figure 8.6-1.

The study was oriented toward the information needs of the user to assure that the resulting system concept is responsive to its fundamental purpose. There can be alternative solutions that are even more suited to that purpose. The specific cases selected as a basis for developing the satellite-assisted information system were water management in the Pacific Northwest, wheat crop management in the United States, and wheat rust prevention in the United States.

In each case the system was compared to an aircraft system that provided similar information. Since benefits can be achieved by means other than improved information to users, the satellite-assisted information system benefits were also compared to non-information alternatives. These were dam construction and nuclear-powered plants, a wheat storage program, and research on rust-resistant wheat varieties.

The general concept discussed above was used to develop the study plan shown in Figure 8.6-2.

Costs and Benefits. The costs for the conceptual system developed in this study were calculated by estimating a dollar value (in present dollars) for each year for the next 20 years. This stream of total system costs was then discounted to present value. Benefits were separately estimated for each of the three cases selected for study. To make benefits commensurable with total system costs, the benefit streams were also discounted to present value.

Costs incurred prior to a decision to acquire and operate the system must be considered sunk; that is, they are irrelevant to a consideration of prospective costs and benefits—the two parameters that permit making valid comparisons of alternative courses of action. Hence, it was necessary to fix a decision year to begin the cost and benefit stream and to estimate a schedule for implementation. The present year was selected as the decision year, making the third year the year the benefits would begin. All cost and benefit streams were discounted to their present values at $7\frac{1}{2}$, 10, and $12\frac{1}{2}$ percent to identify sensitivity and to permit planners latitude in comparing implications of different courses of action.

Some results of the study can be observed in Figures 8.6-3 and 8.6-4. Figure 8.6-3 compares costs for the entire satellite-assisted information system, discounted at 10 percent, to expected benefits from water management in the United States and to expected worldwide benefits. The comparison shows that the benefits expected from water management in the United States alone are substantially greater than expected costs for the information system capable of supporting water management, wheat crop management, and wheat rust control.

The curves shown in Figure 8.6-4 cannot be compared directly because

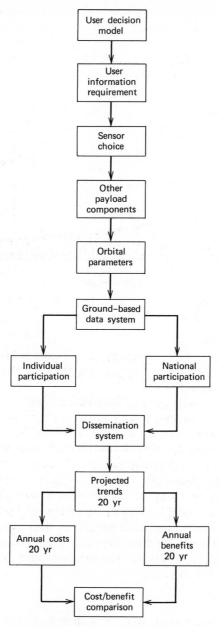

Figure 8.6-2 Study plan.

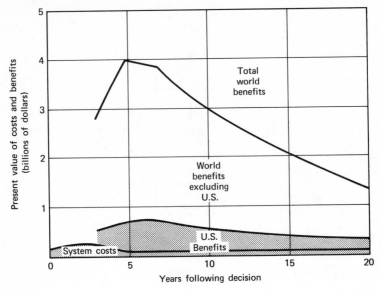

Figure 8.6-3 Annual U.S. and world water management costs and benefits discounted to present value at 10 percent (present dollars).

the aircraft-assisted system is single-purpose—designed for the performance required. The satellite-assisted system, on the other hand, is designed to serve all three sets of users.

Conclusions. A satellite-assisted information system for water management, wheat crop management, and wheat rust control, employing remote sensors in unmanned spacecraft is technically feasible.

The system conceptualized in this study is multipurpose and could provide substantial benefits to different groups of users. Although alternative systems can be less expensive for some specific applications, the total costs for a multipurpose satellite-assisted system are expected to be considerably less than the sum of benefits to the various user groups. In this study the anticipated water management benefits in the Pacific Northwest alone would be sufficient to pay for the entire satellite-assisted information system. Benefits from other applications studied would then be obtained without significant additional costs.

8.6.3 Summary

The example application demonstrates that it is most difficult and time consuming to apply economic analysis principles comprehensively and in a detailed way to most problems of choice. Nevertheless, an effort to analyze the economic implications of decisions that influence prospective receipts

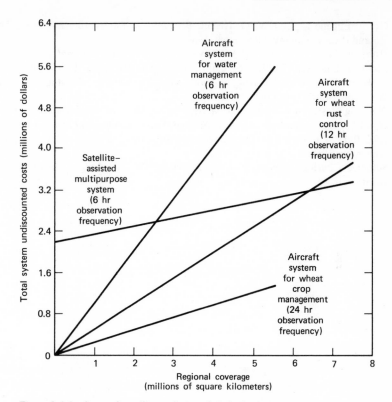

Figure 8.6-4 Costs of satellite and aircraft information systems.

and disbursements will usually have a large influence on the successful management of resources. Mitigating against this sort of activity is the fact that it is impossible to predict the future with accuracy. However, it is important to recognize that relative accuracy among alternative courses of action provides useful insight into the problem of choosing from among myriad alternatives, and the mere act of defining those alternatives tends to require thought in terms of relevant comparisons.

References

1. Breckner, Norman V., and J. W. Noah, "Costing of Systems," in *Defense Management*, Ed. Stephen Enke, Chapter 3, Prentice-Hall, Englewood Cliffs, New Jersey, 1967.
2. Dean, Joel, *Capital Budgeting*, Columbia University Press, New York, 1951.
3. Fisher, Gene H., *Cost Considerations in Systems Analysis*, American Elsevier Publishing Company, New York, 1971.

4. Grant, Eugene L., and W. G. Ireson, *Principles of Engineering Economy*, The Roland Press Company, New York, 1964.

5. Hinrichs, Harley H., and G. M. Taylor, *Program-Budgeting and Benefit-Cost Analysis*, Goodyear Publishing Company, Pacific Palisades, California, 1969.

6. McKean, Roland N., *Efficiency in Government Through Systems Analysis*, John Wiley and Sons, New York, 1958.

7. Planning Research Corporation, *A Systems Analysis of Applications of Earth Orbital Space Technology to Select Cases in Water Management and Agriculture*, PRC R-1224, prepared for the National Aeronautics and Space Administration, November 1969.

8. Schultze, Charles L., *Setting National Priorities, The 1971 Budget*, The Brookings Institution, 1970.

Author Index

Subject Index